Prefixes and Numerical Values for SI Units

Prefix	Symbol	Numerical value	Power of 10 equivalent
exa	E	1,000,000,000,000,000,000	10^{18}
peta	P	1,000,000,000,000,000	10^{15}
tera	T	1,000,000,000,000	10^{12}
giga	G	1,000,000,000	10^{9}
mega	M	1,000,000	10^{6}
kilo	k	1,000	10^{3}
hecto	h	100	10^{2}
deka	da	10	10^{1}
—	—	1	10^{0}
deci	d	0.1	10^{-1}
centi	c	0.01	10^{-2}
milli	m	0.001	10^{-3}
micro	μ	0.000001	10^{-6}
nano	n	0.000000001	10^{-9}
pico	p	0.000000000001	10^{-12}
femto	f	0.000000000000001	10^{-15}
atto	a	0.000000000000000001	10^{-18}

SI Units and Conversion Factors

Length

SI unit: meter (m)

1 meter	=	1000 millimeters
	=	1.0936 yards
1 centimeter	=	0.3937 inch
1 inch	=	2.54 centimeters (exactly)
1 kilometer	=	0.62137 mile
1 mile	=	5280 feet
	=	1.609 kilometers
1 angstrom	=	10^{-10} meter

Mass

SI unit: kilogram (kg)

1 kilogram	=	1000 grams
	=	2.20 pounds
1 gram	=	1000 milligrams
1 pound	=	453.59 grams
	=	0.45359 kilogram
	=	16 ounces
1 ton	=	2000 pounds
	=	907.185 kilograms
1 ounce	=	28.3 grams
1 atomic mass unit	=	1.6606×10^{-27} kilograms

Volume

SI unit: cubic meter (m^3)

1 liter	=	1000 milliliters
	=	$10^{-3}\ m^3$
	=	1 dm^3
	=	1.0567 quarts
1 gallon	=	4 quarts
	=	8 pints
	=	3.785 liters
1 quart	=	32 fluid ounces
	=	0.946 liter
	=	4 cups
1 fluid ounce	=	29.6 mL

Temperature

SI unit: kelvin (K)

$$0\ K = -273.15°C$$
$$= -459.67°F$$
$$K = °C + 273.15$$
$$°C = \frac{(°F - 32)}{1.8}$$
$$°C = \frac{5}{9}(°F - 32)$$
$$°F = 1.8(°C) + 32$$

Energy

SI unit: joule (J)

1 joule	=	1 kg m^2/s^2
	=	0.23901 calorie
1 calorie	=	4.184 joules

Pressure

SI unit: pascal (Pa)

1 pascal	=	1 kg/(ms^2)
1 atmosphere	=	101.325 kilopascals
	=	760 torr
	=	760 mm Hg
	=	14.70 pounds per square inch (psi)

FOUNDATIONS OF
COLLEGE CHEMISTRY

FOUNDATIONS OF
COLLEGE CHEMISTRY

FOURTEENTH EDITION

Morris Hein
Mount San Antonio College

Susan Arena
University of Illinois, Urbana-Champaign

WILEY

VP & Publisher: Kaye Pace
Associate Publisher: Petra Recter
Acquisition Editor: Nicholas Ferrari
Senior Product Designer: Geraldine Osnato
Marketing Manager: Kristine Ruff
Senior Production Editor: Elizabeth Swain
Design Director: Harry Nolan
Senior Designer: Maureen Eide
Content Editor: Alyson Rentrop
Editorial Assistant: Ashley Gayle
Media Specialist: Evelyn Brigandi
Senior Photo Editor: Mary Ann Price
Production Management Services: cMPreparé, CSR Assunta Petrone
Cover: Frank Krahmer/Radius Images/Getty Images, Inc.

This book was typeset in 10/12 Times New Roman at cMPreparé and printed and bound by
R. R. Donnelley/Jefferson City. The cover was printed by R. R. Donnelley/Jefferson City.

The paper in this book was manufactured by a mill whose forest management programs include sustained yield—harvesting of its timberlands. Sustained yield harvesting principles ensure that the number of trees cut each year does not exceed the amount of new growth.

This book is printed on acid free paper. ∞

Main text   ISBN 978-1-118-13355-2
Binder-ready version ISBN 978-1-118-14018-5

Printed in the United States of America
10 9 8 7 6 5 4 3 2 1

ABOUT THE AUTHORS

Morris Hein earned a BS and MS in Chemistry at the University of Denver and his PhD at the University of Colorado, Boulder. He is Professor Emeritus of Chemistry at Mt. San Antonio College, where he regularly taught the preparatory chemistry course and organic chemistry. He is the original author of *Foundations of College Chemistry*, and his name has become synonymous with clarity, meticulous accuracy, and a step-by-step approach that students can follow. Over the years, more than three million students have learned chemistry using a text by Morris Hein. In addition to *Foundations of College Chemistry*, Fourteenth Edition, he is co-author of *Introduction to General, Organic, and Biochemistry*, Tenth Edition, and *Introduction to Organic and Biochemistry*. He is also co-author of *Foundations of Chemistry in the Laboratory*, Fourteenth Edition, and *Introduction to General, Organic and Biochemistry in the Laboratory*, Tenth Edition.

Susan Arena earned a BS and MA in Chemistry at California State University-Fullerton. She has taught science and mathematics at all levels, including middle school, high school, community college, and university. At the University of Illinois she developed a program for increasing the retention of minorities and women in science and engineering. This program focused on using active learning and peer teaching to encourage students to excel in the sciences. She has coordinated and led workshops and progrms for science teachers from elementary through college levels that encourage and support active learning and creative science teaching techniques. For several years she was director of an Institute for Chemical Education (ICE) field center in Southern California. In addition to *Foundations of College Chemistry*, Fourteenth edition, she is co-author of *Introduction to General, Organic and Biochemistry*, Tenth edition. Susan enjoys reading, knitting, traveling, classic cars, and gardening in her spare time when she is not playing with her grandchildren.

BRIEF CONTENTS

1 An Introduction to Chemistry   1

2 Standards for Measurement   13

3 Elements and Compounds   44

4 Properties of Matter   62

5 Early Atomic Theory and Structure   82

6 Nomenclature of Inorganic Compounds   98

7 Quantitative Composition of Compounds   121

8 Chemical Equations   143

9 Calculations from Chemical Equations   167

10 Modern Atomic Theory and the Periodic Table   191

11 Chemical Bonds: The Formation of Compounds from Atoms   212

12 The Gaseous State of Matter   248

13 Liquids   282

14 Solutions   305

15 Acids, Bases, and Salts   337

16 Chemical Equilibrium   363

17 Oxidation–Reduction   390

18 Nuclear Chemistry   417

19 Introduction to Organic Chemistry   441

20 Introduction to Biochemistry   483

    Appendices   A-1

    Glossary   G-1

    Index   I-1

CONTENTS

1 An Introduction to Chemistry   1

1.1 The Nature of Chemistry   2
     Thinking Like a Chemist   2
1.2 A Scientific Approach to Problem Solving   3

CHEMISTRY IN ACTION
Egyptians, the First Medicinal Chemists   4

     The Scientific Method   4
1.3 The Particulate Nature of Matter   5
     Physical States of Matter   6
1.4 Classifying Matter   7
     Distinguishing Mixtures from Pure Substances   8
     Review   9
     Review Questions   10
     Paired Exercises, Additional Exercises   11
     Answers to Practice Exercises   12

2 Standards for Measurement   13

2.1 Scientific Notation   14
2.2 Measurement and Uncertainty   15
2.3 Significant Figures   16
     Rounding Off Numbers   17
2.4 Significant Figures in Calculations   18
     Multiplication or Division   18
     Addition or Subtraction   19
2.5 The Metric System   21
     Measurement of Length   22
     Unit Conversions   23
     Measurement of Mass   24

CHEMISTRY IN ACTION
Keeping Track of Units   25

     Measurement of Volume   26
2.6 Dimensional Analysis: A Problem-
     Solving Method   27
2.7 Measurement of Temperature   30

CHEMISTRY IN ACTION
Setting Standards   32

CHEMISTRY IN ACTION
Taking the Temperature of Old Faithful   33

2.8 Density   34
     Review   37
     Review Questions   38
     Paired Exercises   39
     Additional Exercises   41
     Challenge Exercises, Answers to Practice
     Exercises   43

3 Elements and Compounds   44

3.1 Elements   45
     Natural States of the Elements   45
     Distribution of Elements   46
     Names of the Elements   47
     Symbols of the Elements   47

CHEMISTRY IN ACTION
Naming Elements   48

3.2 Introduction to the Periodic Table   49
     Metals, Nonmetals, and Metalloids   50
     Diatomic Elements   51

CHEMISTRY IN ACTION
Atomic Oxygen, Friend or Foe?   52

3.3 Compounds and Formulas   52
     Molecular and Ionic Compounds   52
     Writing Formulas of Compounds   54
     Composition of Compounds   55
     Review   57
     Review Questions   58
     Paired Exercises   59
     Additional Exercises   60
     Challenge Exercises, Answers to
     Practice Exercises   61

4 Properties of Matter   62

4.1 Properties of Substances   63

CHEMISTRY IN ACTION
Making Money   64

4.2 Physical and Chemical Changes   65
4.3 Learning to Solve Problems   68

4.4 Energy 68
Energy in Chemical Changes 69
Conservation of Energy 70
4.5 Heat: Quantitative Measurement 70
4.6 Energy in the Real World 72

CHEMISTRY IN ACTION
Popping Popcorn 73
Review 74
Review Questions 75
Paired Exercises 76
Additional Exercises 77
Challenge Exercises, Answers to
Practice Exercises 78

PUTTING IT TOGETHER
CHAPTERS **1–4 REVIEW** 79

5 Early Atomic Theory and Structure 82

5.1 Dalton's Model of the Atom 83
5.2 Electric Charge 84
Discovery of Ions 84
5.3 Subatomic Parts of the Atom 85
5.4 The Nuclear Atom 87
General Arrangement of Subatomic Particles 88
Atomic Numbers of the Elements 89
5.5 Isotopes of the Elements 89

CHEMISTRY IN ACTION
Isotope Detectives 91

5.6 Atomic Mass 92
Review 93
Review Questions 94
Paired Exercises 95
Additional Exercises 96
Challenge Exercise, Answers to
Practice Exercises 97

6 Nomenclature of Inorganic Compounds 98

6.1 Common and Systematic Names 99
6.2 Elements and Ions 100

CHEMISTRY IN ACTION
What's in a Name? 101

**6.3 Writing Formulas from Names of
Ionic Compounds** 103
6.4 Naming Binary Compounds 105
Binary Ionic Compounds Containing a Metal
Forming Only One Type of Cation 105

Binary Ionic Compounds Containing a Metal
That Can Form Two or More Types of Cations 106
Binary Compounds Containing Two Nonmetals 108
**6.5 Naming Compounds Containing
Polyatomic Ions** 109
6.6 Acids 111
Binary Acids 111
Naming Oxy-Acids 112
Review 114
Review Questions 115
Paired Exercises 116
Additional Exercises 117
Challenge Exercise, Answers to
Practice Exercises 118

PUTTING IT TOGETHER
CHAPTERS **5–6 REVIEW** 119

**7 Quantitative Composition of
Compounds** 121

7.1 The Mole 122
7.2 Molar Mass of Compounds 126
7.3 Percent Composition of Compounds 129
Percent Composition from Formula 130

CHEMISTRY IN ACTION
Vanishing Coins? 132

7.4 Calculating Empirical Formulas 133
**7.5 Calculating the Molecular Formula from
the Empirical Formula** 135
Review 138
Review Questions, Paired Exercises 139
Additional Exercises 141
Challenge Exercises, Answers to
Practice Exercises 142

8 Chemical Equations 143

8.1 The Chemical Equation 144
Conservation of Mass 145
8.2 Writing and Balancing Chemical Equations 145
Information in a Chemical Equation 149
8.3 Types of Chemical Equations 150
Combination Reaction 150

CHEMISTRY IN ACTION
CO Poisoning—A Silent Killer 151

Decomposition Reaction 151
Single-Displacement Reaction 152
Double-Displacement Reaction 153

8.4 Heat in Chemical Reactions   156

8.5 Global Warming: The Greenhouse Effect   159

CHEMISTRY IN ACTION
Decreasing Carbon Footprints   160

Review   161

Review Questions, Paired Exercises   163

Additional Exercises   165

Challenge Exercise, Answers to
Practice Exercises   166

9 Calculations from Chemical Equations   167

9.1 Introduction to Stoichiometry   168
A Short Review   168

9.2 Mole–Mole Calculations   170

9.3 Mole–Mass Calculations   173

9.4 Mass–Mass Calculations   174

9.5 Limiting Reactant and Yield Calculations   176

CHEMISTRY IN ACTION
A Shrinking Technology   177

Review   182

Review Questions, Paired Exercises   183

Additional Exercises   185

Challenge Exercises, Answers to
Practice Exercises   187

PUTTING IT TOGETHER
CHAPTERS **7–9 REVIEW**   **188**

10 Modern Atomic Theory and the Periodic Table   191

10.1 Electromagnetic Radiation   192
Electromagnetic Radiation   192

CHEMISTRY IN ACTION
You Light Up My Life   193

10.2 The Bohr Atom   193

10.3 Energy Levels of Electrons   195

CHEMISTRY IN ACTION
Atomic Clocks   197

10.4 Atomic Structures of the
First 18 Elements   198

10.5 Electron Structures and the
Periodic Table   201

CHEMISTRY IN ACTION
Collecting the Elements   202

Review   206

Review Questions   207

Paired Exercises   208

Additional Exercises   210

Challenge Exercises, Answers to
Practice Exercises   211

11 Chemical Bonds: The Formation of Compounds from Atoms   212

11.1 Periodic Trends in Atomic Properties   213
Metals and Nonmetals   213
Atomic Radius   214
Ionization Energy   214

11.2 Lewis Structures of Atoms   216

11.3 The Ionic Bond: Transfer of Electrons
from One Atom to Another   217

11.4 Predicting Formulas of Ionic Compounds   222

11.5 The Covalent Bond: Sharing Electrons   224

11.6 Electronegativity   226

CHEMISTRY IN ACTION
Trans-forming Fats   228

11.7 Lewis Structures of Compounds   229

CHEMISTRY IN ACTION
Strong Enough to Stop a Bullet?   232

11.8 Complex Lewis Structures   232

11.9 Compounds Containing Polyatomic Ions   234

CHEMISTRY IN ACTION
Chemistry or Art?   235

11.10 Molecular Shape   235
The Valence Shell Electron Pair Repulsion
(VSEPR) Model   235

Review   239

Review Questions   240

Paired Exercises   241

Additional Exercises   243

Challenge Exercises   244

Answers to Practice Exercises   245

PUTTING IT TOGETHER
CHAPTERS **10–11 REVIEW**   **246**

12 The Gaseous State of Matter   248

12.1 Properties of Gases   249
Measuring the Pressure of a Gas   249
Pressure Dependence on the Number
of Molecules and the Temperature   251

CHEMISTRY IN ACTION
What the Nose Knows   252

12.2 Boyle's Law   252
12.3 Charles' Law   256
12.4 Avogadro's Law   259
12.5 Combined Gas Laws   260
      Mole–Mass–Volume Relationships
      of Gases 262
12.6 Ideal Gas Law   264
      The Kinetic-Molecular Theory 266
      Real Gases 266

CHEMISTRY IN ACTION
Air Quality   267

12.7 Dalton's Law of Partial Pressures   267

CHEMISTRY IN ACTION
Getting High to Lose Weight?   268

12.8 Density of Gases   270
12.9 Gas Stoichiometry   270
      Mole–Volume and Mass–Volume
      Calculations 270
      Volume–Volume Calculations 272
      Review   274
      Review Questions   276
      Paired Exercises   277
      Additional Exercises   279
      Challenge Exercises, Answers to
      Practice Exercises   281

13 Liquids   282

13.1 States of Matter: A Review   283
13.2 Properties of Liquids   283
      Surface Tension 283
      Evaporation 284
      Vapor Pressure 285
13.3 Boiling Point and Melting Point   286

CHEMISTRY IN ACTION
Chemical Eye Candy   288

13.4 Changes of State   288
13.5 Intermolecular Forces   290
      Dipole–Dipole Attractions 290
      The Hydrogen Bond 291

CHEMISTRY IN ACTION
How Sweet It Is!   293

      London Dispersion Forces 294
13.6 Hydrates   295

13.7 Water, a Unique Liquid   297
      Physical Properties of Water 297

CHEMISTRY IN ACTION
Reverse Osmosis?   298

      Structure of the Water Molecule 298
      Sources of Water for a Thirsty World 299
      Review   300
      Review Questions   301
      Paired Exercises   302
      Additional Exercises   303
      Challenge Exercises, Answers to
      Practice Exercises   304

14 Solutions   305

14.1 General Properties of Solutions   306
14.2 Solubility   307
      The Nature of the Solute and Solvent 308
      The Effect of Temperature on Solubility 309
      The Effect of Pressure on Solubility 310
      Saturated, Unsaturated, and Supersaturated
      Solutions 310
14.3 Rate of Dissolving Solids   311
14.4 Concentration of Solutions   312
      Dilute and Concentrated Solutions 313
      Mass Percent Solution 313
      Mass/Volume Percent (m/v) 315
      Volume Percent 315
      Molarity 315
      Dilution Problems 319
14.5 Colligative Properties of Solutions   320

CHEMISTRY IN ACTION
The Scoop on Ice Cream   324

14.6 Osmosis and Osmotic Pressure   325
      Review   327
      Review Questions   328
      Paired Exercises   329
      Additional Exercises   332
      Challenge Exercises, Answers to
      Practice Exercises   333

PUTTING IT TOGETHER
CHAPTERS 12–14 REVIEW   334

15 Acids, Bases, and Salts   337

15.1 Acids and Bases   338

CHEMISTRY IN ACTION
Drug Delivery: An Acid–Base Problem   341

15.2 Reactions of Acids and Bases 342

Acid Reactions 342

Base Reactions 343

15.3 Salts 343

CHEMISTRY IN ACTION
A Cool Fizz 344

15.4 Electrolytes and Nonelectrolytes 344

Dissociation and Ionization of
Electrolytes 345

Strong and Weak Electrolytes 346

Colligative Properties of Electrolyte
Solutions 348

Ionization of Water 348

15.5 Introduction to pH 349

CHEMISTRY IN ACTION
Ocean Corals Threatened by Increasing
Atmospheric CO_2 Levels 351

15.6 Neutralization 352

15.7 Writing Net Ionic Equations 354

15.8 Acid Rain 356

Review 357

Review Questions, Paired Exercises 359

Additional Exercises 361

Challenge Exercises, Answers to
Practice Exercises 362

16 Chemical Equilibrium 363

16.1 Rates of Reaction 364

16.2 Chemical Equilibrium 365

Reversible Reactions 365

16.3 Le Châtelier's Principle 366

CHEMISTRY IN ACTION
New Ways in Fighting Cavities and
Avoiding the Drill 367

Effect of Concentration on
Equilibrium 368

Effect of Volume on Equilibrium 370

Effect of Temperature on Equilibrium 371

Effect of Catalysts on Equilibrium 372

16.4 Equilibrium Constants 373

16.5 Ion Product Constant for Water 374

16.6 Ionization Constants 376

16.7 Solubility Product Constant 378

16.8 Buffer Solutions: The Control of pH 381

CHEMISTRY IN ACTION
Exchange of Oxygen and Carbon
Dioxide in the Blood 382

Review 383

Review Questions 384

Paired Exercises 385

Additional Exercises 387

Challenge Exercises, Answers to
Practice Exercises 389

17 Oxidation–Reduction 390

17.1 Oxidation Number 391

Oxidation–Reduction 393

17.2 Balancing Oxidation–Reduction Equations 395

17.3 Balancing Ionic Redox Equations 398

CHEMISTRY IN ACTION
Sensitive Sunglasses 400

17.4 Activity Series of Metals 401

17.5 Electrolytic and Voltaic Cells 403

CHEMISTRY IN ACTION
Superbattery Uses Hungry Iron Ions 407

Review 407

Review Questions 409

Paired Exercises 410

Additional Exercises 412

Challenge Exercises, Answers to
Practice Exercises 413

PUTTING IT TOGETHER
CHAPTERS **15–17 REVIEW** 414

18 Nuclear Chemistry 417

18.1 Discovery of Radioactivity 418

Natural Radioactivity 419

18.2 Alpha Particles, Beta Particles, and
Gamma Rays 421

Alpha Particles 421

Beta Particles 422

Gamma Rays 422

18.3 Radioactive Disintegration Series 424

Transmutation of Elements 425

Artificial Radioactivity 425

Transuranium Elements 426

18.4 Measurement of Radioactivity 426

18.5 Nuclear energy 427

Nuclear Fission 427

CHEMISTRY IN ACTION
Does Your Food Glow in the Dark? 428

Nuclear Power   430
The Atomic Bomb   431
Nuclear Fusion   432

18.6 Mass–Energy Relationship in
Nuclear Reactions   433
18.7 Biological Effects of Radiation   434

Acute Radiation Damage   434
Long-Term Radiation Damage   434
Genetic Effects   434

CHEMISTRY IN ACTION
A Window into Living Organisms   435

Review   436
Review Questions   437
Paired Exercises   438
Additional Exercises   439
Challenge Exercises, Answers to
Practice Exercises   440

19 Introduction to Organic
Chemistry   **441**

19.1 The Beginnings of Organic Chemistry   442
19.2 Why Carbon?   442

CHEMISTRY IN ACTION
Biodiesel: Today's Alternative Fuel   444

Hydrocarbons   444
19.3 Alkanes   445

Structural Formulas and Isomerism   446
Naming Alkanes   448
19.4 Alkenes and Alkynes   452

Naming Alkenes and Alkynes   453
Reactions of Alkenes   455
Addition   455
19.5 Aromatic Hydrocarbons   456

Naming Aromatic Compounds   457
Monosubstituted Benzenes   457
Disubstituted Benzenes   457
Tri- and Polysubstituted Benzenes   458
19.6 Hydrocarbon Derivatives   459

Alkyl Halides   460
19.7 Alcohols   461

Methanol   462
Ethanol   463
Naming Alcohols   463
19.8 Ethers   465

Naming Ethers   466
19.9 Aldehydes and Ketones   467

Naming Aldehydes   467
Naming Ketones   468

19.10 Carboxylic Acids   469
19.11 Esters   471

CHEMISTRY IN ACTION
Getting Clothes CO_2 Clean!   472

19.12 Polymers—Macromolecules   473
Review   474
Review Questions, Paired Exercises   477
Additional Exercises   481
Answers to Practice Exercises   482

20 Introduction to Biochemistry   **483**

20.1 Chemistry in Living Organisms   484
20.2 Carbohydrates   484

Monosaccharides   484
Disaccharides   486
Polysaccharides   488
20.3 Lipids   488
20.4 Amino Acids and Proteins   492

CHEMISTRY IN ACTION
The Taste of Umami   496

20.5 Enzymes   497
20.6 Nucleic Acids, DNA, and Genetics   499

DNA and Genetics   502
Review   504
Review Questions and Exercises   505
Answers to Practice Exercises   507

PUTTING IT TOGETHER
CHAPTERS **18–20 REVIEW**   508

APPENDICES

I.   Mathematical and Review   A-1
II.   Using a Scientific Calculator   A-10
III. Units of Measurement   A-14
IV.   Vapor Pressure of Water at Various
Temperatures   A-15
V.   Solubility Table   A-16
VI.   Answers to Selected Exercises   A-17
VII. Answers to Putting It Together Review
Exercises   A-33

GLOSSARY   G-1

INDEX   I-1

• PREFACE

This new Fourteenth Edition of *Foundations of College Chemistry* presents chemistry as a modern, vital subject and is designed to make introductory chemistry accessible to all beginning students. The central focus is the same as it has been from the first edition: to make chemistry interesting and understandable to students and teach them the problem-solving skills they will need. In preparing this new edition, we considered the comments and suggestions of students and instructors to design a revision that builds on the strengths of previous editions. We have especially tried to relate chemistry to the real lives of our students as we develop the principles that form the foundation for the further study of chemistry, and to provide them with problem-solving skills and practice needed in their future studies.

Foundations of College Chemistry, 14th Edition, is intended for students who have never taken a chemistry course or those who have had a significant interruption in their studies but plan to continue with the general chemistry sequence. Since its inception, this book has helped define the preparatory chemistry course and has developed a much wider audience. In addition to preparatory chemistry, our text is used extensively in one-semester general purpose courses (such as those for applied health fields) and in courses for nonscience majors.

Customization and Flexible Options to Meet Your Needs

Wiley Custom Select allows you to create a textbook with precisely the content you want, in a simple, three-step online process that brings your students a cost-efficient alternative to a traditional textbook. Select from an extensive collection of content at **http://customselect.wiley.com**, upload your own materials as well, and select from multiple delivery formats—full color or black and white print with a variety of binding options, or eBook. Preview the full text online, get an instant price quote, and submit your order; we'll take it from there. *WileyFlex* offers content in flexible and cost-saving options to students. Our goal is to deliver our learning materials to our customers in the formats that work best for them, whether it's traditional text, eTextbook, *WileyPLUS*, loose-leaf binder editions, or customized content through Wiley Custom Select.

Development of Problem-Solving Skills

We all want our students to develop real skills in solving problems. We believe that a key to the success of this text is the fact that our problem-solving approach works for students. It is a step-by-step process that teaches the use of units and shows the change from one unit to the next. We have then used this problem-solving approach in our examples throughout the text to encourage students to think their way through each problem. In this edition we continue to use examples to incorporate fundamental mathematical skills, scientific notation, and significant figures. We have added Problem-Solving Strategy boxes in the text to highlight the steps needed to solve chemistry problems. Painstaking care has been taken to show each step in the problem-solving process and to use these steps in solving example problems. We continue to use four significant figures for atomic and molar masses for consistency and for rounding off answers appropriately. We have been meticulous in providing answers, correctly rounded, for students who have difficulty with mathematics.

PROBLEM-SOLVING STRATEGY: For Writing and Balancing a Chemical Equation

1. **Identify the reaction.** Write a description or word equation for the reaction. For example, let's consider mercury(II) oxide decomposing into mercury and oxygen.

$$\text{mercury(II) oxide} \xrightarrow{\Delta} \text{mercury} + \text{oxygen}$$

2. **Write the unbalanced (skeleton) equation.** Make sure that the formula for each substance is correct and that reactants are written to the left and products to the right of the arrow. For our example,

$$\text{HgO} \xrightarrow{\Delta} \text{Hg} + \text{O}_2$$

Problem-Solving Strategy

PRACTICE 4.3

Calculate the quantity of energy needed to heat 8.0 g of water from 42.0°C to 45.0°C.

PRACTICE 4.4

A 110.0-g sample of metal at 55.5°C raises the temperature of 150.0 g of water from 23.0°C to 25.5°C. Determine the specific heat of the metal in J/g°C.

Practice Problems

FOSTERING STUDENT SKILLS *Attitude* plays a critical role in problem solving. We encourage students to learn that a systematic approach to solving problems is better than simple memorization. Throughout the book we emphasize the use of our approach to problem solving to encourage students to think through each problem. Once we have laid the foundations of concepts, we highlight the steps so students can locate them easily. Important rules and equations are highlighted for emphasis and ready reference.

STUDENT PRACTICE Practice problems follow the examples in the text, with answers provided at the end of the chapter. The end of each chapter begins with a *Chapter Review* and *Review Questions* section, which help students review key terms and concepts, as well as material presented in tables and figures. This is followed by *Paired Exercises*, covering concepts and numerical exercises, where two similar exercises are presented side by side. The final section, *Additional Exercises*, includes further practice problems presented in a more random order. In our new edition we have changed a significant number of exercises per chapter, added many new practice problems and Examples throughout the text and added many new exercises which emphasize practical and real-life applications of the chemical principles discussed in the text. In addition, for many of these examples enhanced versions have been developed for use with either the online book or *WileyPlus*.

Organization

We continue to emphasize the less theoretical aspects of chemistry early in the book, leaving the more abstract theory for later. This sequence seems especially appropriate in a course where students are encountering chemistry for the very first time. Atoms, molecules, and reactions are all an integral part of the chemical nature of matter. A sound understanding of these topics allows the student to develop a basic understanding of chemical properties and vocabulary.

Chapters 1 through 3 present the basic mathematics and the language of chemistry, including an explanation of the metric system and significant figures. In Chapter 4 we present chemical properties—the ability of a substance to form new substances. Then, in Chapter 5, students encounter the history and language of basic atomic theory.

We continue to present new material at a level appropriate for the beginning student by emphasizing nomenclature, composition of compounds, and reactions in Chapters 6 through 9 before moving into the details of modern atomic theory. Some applications of the Periodic Table are shown in early chapters and discussed in detail in Chapters 10 and 11. Students gain confidence in their own ability to identify and work with chemicals in the laboratory before tackling the molecular models of matter. As practicing chemists we have little difficulty connecting molecular models and chemical properties. Students, especially those with no prior chemistry background, may not share this ability to connect the molecular models and the macroscopic properties of matter. Those instructors who feel it is essential to teach atomic theory and bonding early in the course can cover Chapters 10 and 11 immediately following Chapter 5.

New to This Edition

In Fourteenth Edition we have tried to build on the strengths of the previous editions. We have added a new contributor to our author team, **Cary Willard** from Grossmont College in California. Cary has added a new perspective on problem-solving and technology which has substantially added new content to our instructional package. Cary revised the end-of-chapter materials and added new real-world exercises, including many applications in fields of interest for our students. She designed and authored the new Enhanced Examples and worked to correlate the end-of-chapter material to the concept modules and learning objectives as well as updating the solutions for this material. We are delighted to have her join our team.

We continually strive to keep the material at the same level so that students can easily read and use the text and supplemental material to learn chemistry. With a focus on problem solving, student engagement, and clarity, some of the specific changes are highlighted below:

- **CONCEPT MODULES.** Each chapter was reviewed and reorganized into **concept modules** to facilitate student learning. Each module is structured to ensure that students are focused on learning the key concepts and skills presented within the module. While the chapters may look different in their section titles, the content remains essentially the same; it has simply been repackaged to improve the flow and place conceptual material together.

- Each module begins with a **Learning Objective** which states the goal for the section and prompts students to recognize the important concepts, providing the scaffolding for understanding.

- New **Examples** model problem solving for students, and **Practice Problems** have been added to provide students with opportunities to apply their knowledge. **Key Terms** in each section are highlighted in bold and listed at the beginning of each section as well as in the chapter Review. Each Key Term is also defined in the **Glossary**.

- **NEW ENHANCED EXAMPLES** within *WileyPLUS* extend the examples presented within the Concept Modules and are designed to provide more detailed, interactive examples and practice for students.

  - *Interactive concept questions* help the student understand the underlying concepts presented in the text example. The student then has the opportunity to use the concepts to solve interactive problems closely related to the text examples.

  - *Question assistance* If the student gets a problem incorrect, immediate feedback is given to assist the student in working through the problem.

  - *GOTutorials* In more complex examples the students are also given a GOTutorial to aid them by providing step-by-step guided questions to work toward the correct solution to the problem.

EXAMPLE 12.1

WileyPLUS
ENHANCED EXAMPLE

The average atmospheric pressure at Walnut, California, is 740. mm Hg. Calculate this pressure in (a) torr and (b) atmospheres.

SOLUTION

Let's use conversion factors that relate one unit of pressure to another.

(a) To convert mm Hg to torr, use the conversion factor

760 torr/760 mm Hg (1 torr/1 mm Hg):

$$(740. \text{ mm Hg})\left(\frac{1 \text{ torr}}{1 \text{ mm Hg}}\right) = 740. \text{ torr}$$

(b) To convert mm Hg to atm, use the conversion factor

1 atm/760. mm Hg:

$$(740. \text{ mm Hg})\left(\frac{1 \text{ atm}}{760. \text{ mm Hg}}\right) = 0.974 \text{ atm}$$

PRACTICE 12.1

A barometer reads 1.12 atm. Calculate the corresponding pressure in (a) torr, (b) mm Hg, and (c) kilopascals.

Enhanced Examples have been clustered together to reinforce the connection between similar problems and to help students recognize how a single idea or relationship can be utilized to solve a variety of problems. The problems within the Enhanced Examples are randomized to provide multiple opportunities for a student to practice solving various types of problems. Enhanced Examples are designated by a margin note and icon.

- **Chemistry In Action** boxes have been updated and new boxes have been added to include different applications of the concepts in the text.

- A new concept module has been added to **Chapter 13** on **Intermolecular Forces** which includes dipole-dipole forces, hydrogen bonding, and London dispersion forces.

- Some of the older industrial chemistry applications have been removed and newer applications added as appropriate throughout the text.

- **New, modern design.** The entire text has been redesigned to foster greater accessibility and increase student engagement.

- The **Illustration Program** has been improved to include new photos and updated art, with an emphasis on molecular art to illustrate for the student what happens at the microscopic level.

Learning Aids

To help the beginning student gain the confidence necessary to master technical material, we have refined and enhanced a series of learning aids:

- **Learning Objectives** highlight the concept being taught in each section. These objectives are tied to **Example**, **Practice Problems**, **Review Exercises**, and **Exercises** to assist the student in mastering each concept module and objective.

- Important **terms** are set off in bold type where they are defined, and are listed in grey at the beginning of each section. All **Key Terms** listed in the **Chapter Review** are also defined in the **Glossary**.

- Worked **examples** show students the how of problem solving using **Problem-Solving Strategies** and **Solution Maps** before they are asked to tackle problems on their own.

- **Practice problems** permit immediate reinforcement of a skill shown in the example problems. Answers are provided at the end of the chapter to encourage students to check their problem solving immediately.

- **Marginal notations** help students understand basic concepts and problem-solving techniques. These are printed in blue to clearly distinguish them from text and vocabulary terms.

LEARNING AIDS: MATH SKILLS For students who may need help with the mathematical aspects of chemistry, the following learning aids are available:

- A **Review of Mathematics**, covering the basic functions, is provided in Appendix I.
- **Math Survival Guide: Tips and Tricks for Science Students**, 2nd Edition, by Jeffrey R. Appling and Jean C. Richardson, a brief paperback summary of basic skills that can be packaged with the text, provides an excellent resource for students who need help with the mathematical aspects of chemistry.

Supplements Package

FOR THE STUDENT Study Guide by Rachael Henriques Porter is a self-study guide for students. For each chapter, the **Study Guide** includes a self-evaluation section with student exercises, a summary of chapter concepts, one or more "challenge problems," and answers and solutions to all **Study Guide** exercises.

Solutions Manual by Morris Hein, Cary Willard, Susan Arena, and Kathy Mitchell includes answers and solutions to all end-of-chapter questions and exercises.

Math Survival Guide: Tips and Tricks for Science Students, 2nd Edition, by Jeffrey Appling and Jean Richardson, is a paperback summary of basic skills with practice exercises in every chapter.

Foundations of Chemistry in the Laboratory, 14th Edition, by Morris Hein, Judith N. Peisen, and Robert L. Miner includes 28 experiments for a laboratory program that may accompany the lecture course. Featuring updated information on waste disposal and emphasizing safe laboratory procedures, the lab manual also includes study aids and exercises.

FOR THE INSTRUCTOR Test Bank, by Raymond Sadeghi, includes chapter tests with additional test questions and answers to all test questions.

Computerized Test Bank. The test bank contains true-false, multiple-choice, and open-ended questions, and is available in two formats.

ALTERNATE EDITIONS For the convenience of instructors and to accommodate the various lengths of academic terms, two versions of this book are available. *Foundations of College Chemistry*, 14th Edition, includes 20 chapters and is our main text. *Foundations of College Chemistry*, Alternate 14th Edition, provides a shorter, 17-chapter text in paperback with the same material, but without the nuclear, organic, and biochemistry chapters.

WileyPLUS

WileyPLUS is an innovative, research-based online environment for effective teaching and learning. *WileyPLUS* builds students' confidence because it takes the guesswork out of studying by providing students with a clear roadmap: **what to do, how to do it, if they did it right**. This interactive approach focuses on:

CONFIDENCE: Research shows that students experience a great deal of anxiety over studying. That's why we provide a structured learning environment that helps students focus on **what to do**, along with the support of immediate resources.

MOTIVATION: To increase and sustain motivation throughout the semester, *WileyPLUS* helps students learn **how to do it** at a pace that's right for them. Our integrated resources—available 24/7—function like a personal tutor, directly addressing each student's demonstrated needs with specific problem-solving techniques.

SUCCESS: *WileyPLUS* helps to assure that each study session has a positive outcome by putting students in control. Through instant feedback and study objective reports, students know **if they did it right**, and where to focus next, so they achieve the strongest results.

With *WileyPLUS*, our efficacy research shows that students improve their outcomes by as much as one letter grade. *WileyPLUS* helps students take more initiative, so you'll have greater impact on their achievement in the classroom and beyond.

What Do Students Receive with *WileyPLUS*?

- The complete digital textbook, saving students up to 60% off the cost of a printed text.
- Question assistance, including links to relevant sections in the online digital textbook.
- Immediate feedback and proof of progress, 24/7.

WileyPLUS addresses different learning styles, different levels of proficiency, and different levels of preparation—each of your students is unique. *WileyPLUS* empowers them to take advantage of their individual strengths:

- Students receive timely access to resources that address their demonstrated needs, and get immediate feedback and remediation when needed.
- Integrated, multimedia resources—including audio and visual exhibits, demonstration problems, and much more—provide multiple study paths to fit each student's learning preferences and encourage more active learning.
- *WileyPLUS* includes many opportunities for self-assessment linked to the relevant portions of the text. Students can take control of their own learning and practice until they master the material.
- Complete online version of the textbook, including relevant student study tools and learning resources to ensure positive learning outcomes, including:
  - Enhanced Examples—these interactive examples are designed to promote conceptual understanding and practice quantitative problem-solving skills. They are designated with this icon **WileyPLUS** in the text.
    ENHANCED EXAMPLE
  - MathSkills module provides remedial review.
  - Interactive Periodic Table

What Do Instructors Receive with *WileyPLUS*?

- Reliable resources that reinforce course goals inside and outside of the classroom.
- The ability to easily identify those students who are falling behind.
  - WileyPLUS Quickstart assignments and presentations come preloaded for every chapter. Use as is, edit, or start from scratch.
  - Lecture Notes PowerPoint™ Slides to summarize chapter content.
  - Image Gallery includes all line art, tables, and photos in jpeg format. Leader lines and labels have been enhanced for projection display.
  - Test Bank questions for use during in-class exams or as online quizzes—Student Area
  - Classroom Response System (Clickers) questions
  - Instructor's Solutions Manual
  - All content is tagged to Learning Objectives
  - New visualizations of key concepts

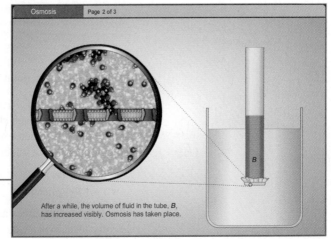

After a while, the volume of fluid in the tube, *B*, has increased visibly. Osmosis has taken place.

- WileyPLUS Includes a robust assessment package with options created specifically to support undergraduate education in Chemistry.
- All end-of-chapter questions are available for assignment and automatic grading in WileyPLUS.
- Every end-of-chapter question has several forms of assistance that are released to students at the instructor's discretion. This assistance can include:
  - Hints
  - Step-by-step tutorials
  - Wrong answer feedback
  - Links to the textbook or other media
- Contextual help as they are working
- Test Bank
- Select end-of-chapter questions which may include review questions, problems, and additional exercises
- CATALYST: true concept mastery assignments for over 150 topics in the course
- GO (Guided Online) Tutorial problems

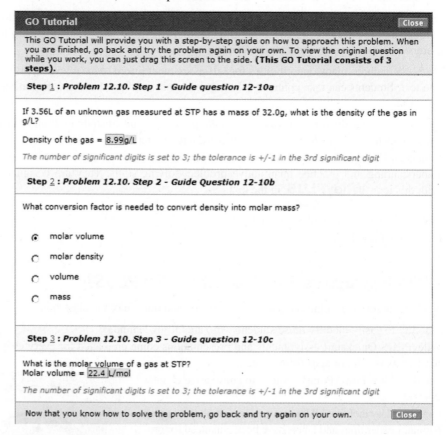

- Prelecture Checkpoint questions

STUDY OBJECTIVE REPORTING delivers an individualized report on how the student has been doing against each study objective that's been covered within a specific assignment. It enables students to determine where they should focus their time.

Instructors can see students' performance on assignments and identify problem areas at a glance.

Learn More.
www.wileyplus.com

All Instructor Resources are available within *WileyPLUS*, or they can be accessed by contacting your local Wiley Sales Representative. Many of the assets are located on the book companion site, www.wiley.com/college/hein

FOR THE LABORATORY **Foundations of Chemistry in the Laboratory, 14th Edition, by Morris Hein, Judith Peisen, and Robert Miner**, has been completely updated and revised to reflect the most current terminology and environmental standards. Instructors can customize their own lab manual to meet the distinct needs of their laboratory by selecting from any of the 28 experiments, adding their own experiments or exercises.

INSTRUCTOR'S COMPANION WEB SITE Instructors have access to all:

- **Digital Image Library**: Images from the text are available online in JPEG format. Instructors may use these to customize their presentations and to provide additional visual support for quizzes and exams.
- **Test Bank**: true-false, multiple-choice, and free-response questions.
- **Power Point Lecture Slides**: Created by Chris Bradley, these slides contain lecture outlines and key topics from each chapter of the text, along with supporting artwork and figures from the text.

Acknowledgments

Books are the result of a collaborative effort of many talented and dedicated people. We particularly want to thank our editor, Nick Ferrari, who guided the project through a complex revision. We are grateful to Mary Ann Price for finding new and interesting photos which add so much to the pages of our text. We also want to thank Elizabeth Swain, our production editor who kept us on track and kept all of the pieces of production running smoothly. Assunta Petrone, of cMPreparé in Italy was amazing as our compositor and turned our revisions into the beautiful text in our 14th edition. We also especially appreciate the work of Geraldine Osnato, Senior Product Designer who helped us figure out how to bring our text into the world of WileyPLUS and provided so much support in the process of adding enhanced examples and improving our end-of-chapter online materials. Thanks to others who provided valuable assistance on this revision including Marketing Manager Kristine Ruff, Senior Designer Maureen Eide, Media Editor Evelyn Brigandi, Content Editor Alyson Rentrop, and Editorial Assistant Ashley Gayle. We are grateful for the many helpful comments from colleagues and students who, over the years, have made this book possible. We hope they will continue to share their ideas for change with us, either directly or through our publisher.

We are especially thankful for the help and support of Tom Martin, our developmental editor. His positive attitude, attention to detail, efficiency, good humor, and willingness to help in any way were indispensable in this revision. Special thanks as well to Cary for all her hard work and quick mind. She has added to our team immeasurably.

Our sincere appreciation goes to the following reviewers who were kind enough to read and give their professional comments.

Reviewers

For the 13th Edition:

Madeline Adamczeski
San Jose City College

Edward L. Barnes, Jr.
Fayetteville Technical Community College

Sean Birke
Jefferson College

Jing-Yi Chin
Suffolk Community College

Joe David Escobar
Jr., Oxnard College

Theodore E. Fickel
Los Angeles Valley College

Melodie Graber
Oakton Community College

Dawn Richardson
Collin College

Lydia Martinez Rivera
The University of Texas—San Antonio

Karen Sanchez
Florida Community College—Jacksonville

Ali O. Sezer
California University of Pennsylvania

David B. Shaw
Madison Area Technical College

Joy Walker
Truman College

For the 14th Edition:

Jeffrey Allison
Austin Community College

Jeanne Arquette
Phoenix College

Rebecca Broyer
University of Southern California

Michael Byler
Community College of Philadelphia

Kevin Cannon
Penn State Abington

Rong Cao
Community College of Allegheny City

Ken Capps
College of Central Florida

Charles Carraher
Florida Atlantic University

Loretta Dorn
Fort Hays State University

Robert Eves
Southern Utah University

Mitchel Fedak
Duquesne University

Paul Fox
Danville Community College

Erick Fuoco
Richard J. Daley College

Amy Grant
El Camino College

Tamara Hanna
Texas Tech Lubbock

Chris Hamaker
Illinois State University

Claudia Hein
Diablo Valley College

Donna Iannotti
Brevard Community College

Crystal Jenkins
Santa Ana College

Jodi Kreiling
University of Nebraska at Omaha

Julie Larson
Bemidji State University

Anne Lerner
Santa Fe College

Lauren McMills
Ohio University—Main Campus

Mitchel Millan
Casper College

Timothy Minger
Mesa Community College

Franklin Ow
East LA City College

Ethel (April) Owusu
Santa Fe College

Fumin Pan
Mohawk Valley Community College

David Peitz
Wayne State College

Sharadha Sambasivan
Suffolk Community College

Hussein Samha
Southern Utah University

Mary Shoemaker
Pennsylvania State University—University Park

Lee Silverberg
Penn State—Schuylkill

Gabriela Smeureanu
Hunter College

Sunanda Sukumar
Albany College of Pharmacy

Paris Svoronos
QCC of Cuny

Susan Thomas
University of Texas—San Antonio

Sergey Trusov
Oxnard College

Elaine Vickers
Southern Utah University

Loretta Vogel
Ocean County College

Liwen Yu
Inver Hills Community College

Karl Wallace
The University of Southern Mississippi

Morris Hein and Susan Arena

FOUNDATIONS OF
COLLEGE CHEMISTRY

The spectacular colors of the aurora borealis are the result of chemistry in our atmosphere.

© Stocktrek Images/SuperStock

CHAPTER **1**

AN INTRODUCTION TO CHEMISTRY

Do you know how the beautiful, intricate fireworks displays are created? Have you ever wondered how a tiny seedling can grow into a cornstalk taller than you in just one season? Perhaps you have been mesmerized by the flames in your fireplace on a romantic evening as they change color and form. And think of your relief when you dropped a container and found that it was plastic, not glass. These phenomena are the result of chemistry that occurs all around us, all the time. Chemical changes bring us beautiful colors, warmth, light, and products to make our lives function more smoothly. Understanding, explaining, and using the diversity of materials we find around us is what chemistry is all about.

CHAPTER OUTLINE

1.1 The Nature of Chemistry

1.2 A Scientific Approach to Problem Solving

1.3 The Particulate Nature of Matter

1.4 Classifying Matter

1.1 THE NATURE OF CHEMISTRY

KEY TERM
chemistry

Key terms are highlighted in bold to alert you to new terms defined in the text.

A knowledge of chemistry is useful to virtually everyone—we see chemistry occurring around us every day. An understanding of chemistry is useful to engineers, teachers, health care professionals, attorneys, homemakers, business people, firefighters, and environmentalists, just to name a few. Even if you're not planning to work in any of these fields, chemistry is important and is used by us every day. Learning about the benefits and risks associated with chemicals will help you to be an informed citizen, able to make intelligent choices concerning the world around you. Studying chemistry teaches you to solve problems and communicate with others in an organized and logical manner. These skills will be helpful in college and throughout your career.

What is chemistry? One dictionary gives this definition: "**Chemistry** is the science of the composition, structure, properties, and reactions of matter, especially of atomic and molecular systems." Another, somewhat simpler definition is "Chemistry is the science dealing with the composition of *matter* and the changes in composition that matter undergoes." Neither of these definitions is entirely adequate. Chemistry and physics form a fundamental branch of knowledge. Chemistry is also closely related to biology, not only because living organisms are made of material substances but also because life itself is essentially a complicated system of interrelated chemical processes.

The scope of chemistry is extremely broad. It includes the whole universe and everything, animate and inanimate, in it. Chemistry is concerned with the composition and changes in the composition of matter and also with the energy and energy changes associated with matter. Through chemistry we seek to learn and to understand the general principles that govern the behavior of all matter.

The chemist, like other scientists, observes nature and attempts to understand its secrets: What makes a tulip red? Why is sugar sweet? What is occurring when iron rusts? Why is carbon monoxide poisonous? Problems such as these—some of which have been solved, some of which are still to be solved—are all part of what we call chemistry.

A chemist may interpret natural phenomena, devise experiments that reveal the composition and structure of complex substances, study methods for improving natural processes, or synthesize substances. Ultimately, the efforts of successful chemists advance the frontiers of knowledge and at the same time contribute to the well-being of humanity.

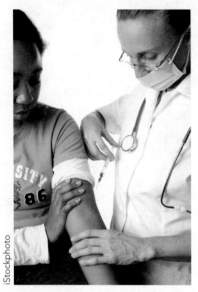

iStockphoto

A health care professional needs to understand chemistry in order to administer the correct dose of medication.

Thinking Like a Chemist

Chemists take a special view of things in order to understand the nature of the chemical changes taking place. Chemists "look inside" everyday objects to see how the basic components are behaving. To understand this approach, let's consider a lake. When we view the lake from a distance, we get an overall picture of the water and shoreline. This overall view is called the *macroscopic* picture.

As we approach the lake we begin to see more details—rocks, sandy beach, plants submerged in the water, and aquatic life. We get more and more curious. What makes the rocks and sand? What kind of organisms live in the water? How do plants survive underwater? What lies hidden in the water? We can use a microscope to learn the answers to some of these questions. Within the water and the plants, we can see single cells and inside them organelles working to keep the organisms alive. For answers to other questions, we need to go even further inside the lake. A drop of lake water can itself become a mysterious and fascinating *microscopic* picture full of molecules and motion. (See **Figure 1.1**.) A chemist looks into the world of atoms and molecules and their motions. Chemistry makes

Image Source/SuperStock

Figure 1.1
Inside a drop of lake water we find water molecules, some dissolved substances.

the connection between the *microscopic* world of molecules and the *macroscopic* world of everyday objects.

Think about the water in the lake. On the surface it has beauty and colors, and it gently laps the shore of the lake. What is the microscopic nature of water? It is composed of tiny molecules represented as

In this case H represents a hydrogen atom and O an oxygen atom. The water molecule is represented by H_2O since it is made up of two hydrogen atoms and one oxygen atom.

EXAMPLE 1.1

You are given eight oxygen atoms and fifteen hydrogen atoms. How many water molecules can you make from them?

SOLUTION

From the model shown above for a water molecule you can see that one molecule of water contains one atom of oxygen and two atoms of hydrogen. Using this model as reference, you can make eight water molecules from eight oxygen atoms. But you can make only seven water molecules from fifteen hydrogen atoms with one H atom and one O atom left over. The answer is seven water molecules.

PRACTICE 1.1

You are given ten hydrogen atoms and eight oxygen atoms. How many water molecules can you make from them?

1.2 A SCIENTIFIC APPROACH TO PROBLEM SOLVING

Describe the steps involved in the scientific method.

● LEARNING OBJECTIVE

KEY TERMS
scientific method
hypothesis
theory
scientific laws

One of the most common and important things we do every day is to solve problems. For example,

- You have two exams and a laboratory report due on Monday. How should you divide your time?
- You leave for school and learn from the radio that there is a big accident on the freeway. What is your fastest alternate route to avoid the traffic problem?
- You need to buy groceries, mail some packages, attend your child's soccer game, and pick up the dry cleaning. What is the most efficient sequence of events?

We all face these kinds of problems and decisions. A logical approach can be useful for solving daily problems:

1. Define the problem. We first need to recognize we have a problem and state it clearly, including all the known information. When we do this in science, we call it *making an observation.*
2. Propose possible solutions to the problem. In science this is called making a *hypothesis.*
3. Decide which is the best way to proceed or solve the problem. In daily life we use our memory of past experiences to help us. In the world of science we *perform an experiment.*

Using a scientific approach to problem solving is worthwhile. It helps in all parts of your life whether you plan to be a scientist, doctor, business person, or writer.

>CHEMISTRY *IN ACTION*

Egyptians, the First Medicinal Chemists

Look at any images of the ancient Egyptians and notice the black eyeliner commonly worn at that time. As chemists analyzed the composition of a sample of this eyeliner in the antiquities collection at the Louvre Museum in Paris, they were appalled to discover the high concentration of lead in the samples. Today lead is routinely removed from most consumer products because it is very toxic even in low concentrations. It is toxic to many organs and can cause symptoms such as abdominal pain, dementia, anemia, seizures, and even death. It turns out the lead compounds found in the Egyptian eyeliner are not found in nature but must be synthesized. The synthesis of these lead salts is complicated and the products are not lustrous. This led chemists to question why the Egyptians would add these compounds to their eyeliner. The answer was revealed after reading some of the ancient manuscripts from

age fotostock/SuperStock

Terra cotta sculpture of Nefertiti

that time. Lead salts were synthesized for use in treating eye ailments, scars, and discolorations. So even if the lead salts were not the best ingredients for beauty, they were added for the perceived health benefits.

Since we now know that lead compounds are very toxic, Christian Amatore, an analytical chemist at the Ecole Normale Supérieure in Paris, wondered if the lead compounds in Egyptian eyeliner could have actually conferred any health benefits. He introduced lead salts into samples of human tissue growing in the laboratory and observed that the cells began forming compounds that trigger an immune response. Perhaps the ancient Egyptians did know something about medicinal chemistry after all. So should we follow the Egyptians example and add lead to our cosmetics? This is probably not a good idea because the risks associated with prolonged lead exposure outweigh the benefits.

Scott Bauer/Courtesy USDA

Scientists employ the scientific method every day in their laboratory work.

The Scientific Method

Chemists work together and also with other scientists to solve problems. As scientists conduct studies they ask many questions, and their questions often lead in directions that are not part of the original problem. The amazing developments from chemistry and technology usually involve what we call the **scientific method**, which can generally be described as follows:

1. **Collect the facts or data** that are relevant to the problem or question at hand. This is usually done by planned experimentation. The data are then analyzed to find trends or regularities that are pertinent to the problem.

2. **Formulate a hypothesis** that will account for the data and that can be tested by further experimentation.

3. **Plan and do additional experiments to test the hypothesis.**

4. **Modify the hypothesis** as necessary so that it is compatible with all the pertinent data.

Confusion sometimes arises regarding the exact meanings of the words *hypothesis*, *theory*, and *law*. A **hypothesis** is a tentative explanation of certain facts that provides a basis for further experimentation. A well-established hypothesis is often called a **theory** or model. Thus, a theory is an explanation of the general principles of certain phenomena with considerable evidence or facts to support it. Hypotheses and theories explain natural phenomena, whereas **scientific laws** are simple statements of natural phenomena to which no exceptions are known under the given conditions.

These four steps are a broad outline of the general procedure that is followed in most scientific work, but they are not a "recipe" for doing chemistry or any other science (**Figure 1.2**). Chemistry is an experimental science, however, and much of its progress has been due to application of the scientific method through systematic research.

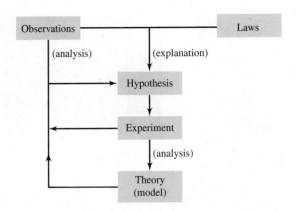

Figure 1.2
The scientific method.

We study many theories and laws in chemistry; this makes our task as students easier because theories and laws summarize important aspects of the sciences. Certain theories and models advanced by great scientists in the past have since been substantially altered and modified. Such changes do not mean that the discoveries of the past are any less significant. Modification of existing theories and models in the light of new experimental evidence is essential to the growth and evolution of scientific knowledge. Science is dynamic.

1.3 THE PARTICULATE NATURE OF MATTER

Describe the characteristics of matter, including the states of matter. ● **LEARNING OBJECTIVE**

KEY TERMS

matter
solid
amorphous
liquid
gas

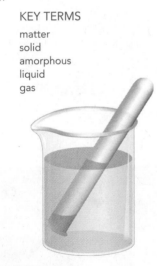

The entire universe consists of matter and energy. Every day we come into contact with countless kinds of matter. Air, food, water, rocks, soil, glass, and this book are all different types of matter. Broadly defined, **matter** is *anything* that has mass and occupies space.

Matter may be quite invisible. For example, if an apparently empty test tube is submerged mouth downward in a beaker of water, the water rises only slightly into the tube. The water cannot rise further because the tube is filled with invisible matter: air (see **Figure 1.3**).

To the macroscopic eye, matter appears to be continuous and unbroken. We are impressed by the great diversity of matter. Given its many forms, it is difficult to believe that on a microscopic level all of matter is composed of discrete, tiny, fundamental particles called *atoms* (**Figure 1.4**). It is truly amazing to understand that the fundamental particles in ice cream are very similar to the particles in air that we breathe. Matter is actually discontinuous and is composed of discrete, tiny particles called *atoms*.

Figure 1.3

An apparently empty test tube is submerged, mouth downward, in water. Only a small volume of water rises into the tube, which is actually filled with invisible matter–air.

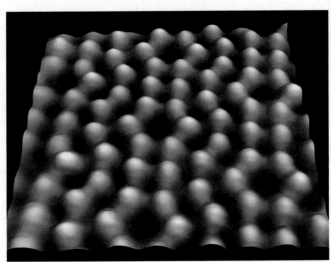

© Andrew Dunn/Alamy

Figure 1.4

Silicon atoms on a silicon chip produced this image using a scanning tunneling microscope.

Figure 1.5

The three states of matter.
(a) Solid—water molecules are held together rigidly and are very close to each other.
(b) Liquid—water molecules are close together but are free to move around and slide over each other. (c) Gas—water molecules are far apart and move freely and randomly.

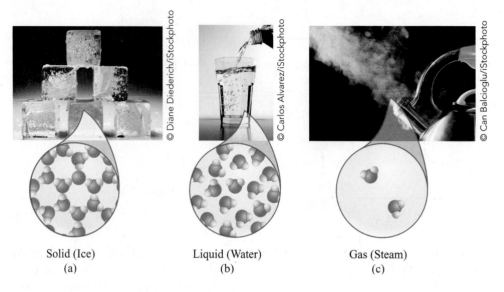

Solid (Ice)   Liquid (Water)   Gas (Steam)
(a)   (b)   (c)

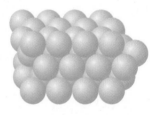

Crystalline solid

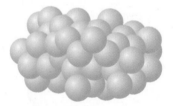

Amorphous solid

Physical States of Matter

Matter exists in three physical states: solid, liquid, and gas (see **Figure 1.5**). A **solid** has a definite shape and volume, with particles that cling rigidly to one another. The shape of a solid can be independent of its container. In Figure 1.5a we see water in its solid form. Another example, a crystal of sulfur, has the same shape and volume whether it is placed in a beaker or simply laid on a glass plate.

Most commonly occurring solids, such as salt, sugar, quartz, and metals, are *crystalline*. The particles that form crystalline materials exist in regular, repeating, three-dimensional, geometric patterns (see **Figure 1.6**). Some solids such as plastics, glass, and gels do not have any regular, internal geometric pattern. Such solids are called **amorphous** solids. (*Amorphous* means "without shape or form.")

A **liquid** has a definite volume but not a definite shape, with particles that stick firmly but not rigidly. Although the particles are held together by strong attractive forces and are in close contact with one another, they are able to move freely. Particle mobility gives a liquid fluidity and causes it to take the shape of the container in which it is stored. Note how water looks as a liquid in Figure 1.5b.

A **gas** has indefinite volume and no fixed shape, with particles that move independently of one another. Particles in the gaseous state have gained enough energy to overcome the attractive forces that held them together as liquids or solids. A gas presses continuously in all directions on the walls of any container. Because of this quality, a gas completely fills a container. The particles of a gas are relatively far apart compared with those of solids and liquids. The actual volume of the gas particles is very small compared with the volume of the space occupied by the gas. Observe the large space between the water molecules in Figure 1.5c compared to ice and liquid water. A gas therefore may be compressed into a very small volume or expanded almost indefinitely. Liquids cannot be compressed to any great extent, and solids are even less compressible than liquids.

If a bottle of ammonia solution is opened in one corner of the laboratory, we can soon smell its familiar odor in all parts of the room. The ammonia gas escaping from the solution demonstrates that gaseous particles move freely and rapidly and tend to permeate the entire area into which they are released.

Figure 1.6

A large crystal of table salt. A salt crystal is composed of a three-dimensional array of particles.

Na⁺

Cl⁻

Although matter is discontinuous, attractive forces exist that hold the particles together and give matter its appearance of continuity. These attractive forces are strongest in solids, giving them rigidity; they are weaker in liquids but still strong enough to hold liquids to definite volumes. In gases, the attractive forces are so weak that the particles of a gas are practically independent of one another. Table 1.1 lists common materials that exist as solids, liquids, and gases. Table 1.2 compares the properties of solids, liquids, and gases.

TABLE 1.1 Common Materials in the Solid, Liquid, and Gaseous States of Matter

Solids	Liquids	Gases
Aluminum	Alcohol	Acetylene
Copper	Blood	Air
Gold	Gasoline	Butane
Polyethylene	Honey	Carbon dioxide
Salt	Mercury	Chlorine
Sand	Oil	Helium
Steel	Vinegar	Methane
Sulfur	Water	Oxygen

TABLE 1.2 Physical Properties of Solids, Liquids, and Gases

State	Shape	Volume	Particles	Compressibility
Solid	Definite	Definite	Rigidly clinging; tightly packed	Very slight
Liquid	Indefinite	Definite	Mobile; adhering	Slight
Gas	Indefinite	Indefinite	Independent of each other and relatively far apart	High

1.4 CLASSIFYING MATTER

Distinguish among a pure substance, a homogeneous mixture, and a heterogeneous mixture.

● **LEARNING OBJECTIVE**

KEY TERMS
substance
homogeneous
heterogeneous
phase
system
mixture

The term *matter* refers to all materials that make up the universe. Many thousands of distinct kinds of matter exist. A **substance** is a particular kind of matter with a definite, fixed composition. Sometimes known as *pure substances*, substances are either elements or compounds. Familiar examples of elements are copper, gold, and oxygen. Familiar compounds are salt, sugar, and water. We'll discuss elements and compounds in more detail in Chapter 3.

We classify a sample of matter as either *homogeneous or heterogeneous* by examining it. **Homogeneous** matter is uniform in appearance and has the same properties throughout. Matter consisting of two or more physically distinct phases is **heterogeneous**. A **phase** is a homogeneous part of a system separated from other parts by physical boundaries. A **system** is simply the body of matter under consideration. Whenever we have a system in which visible boundaries exist between the parts or components, that system has more than one phase and is heterogeneous. It does not matter whether these components are in the solid, liquid, or gaseous states.

A pure substance may exist as different phases in a heterogeneous system. Ice floating in water, for example, is a two-phase system made up of solid water and liquid water. The water in each phase is homogeneous in composition, but because two phases are present, the system is heterogeneous.

A **mixture** is a material containing two or more substances and can be either heterogeneous or homogeneous. Mixtures are variable in composition. If we add a spoonful of sugar to a glass of water, a heterogeneous mixture is formed immediately. The two phases are a solid (sugar) and a liquid (water). But upon stirring, the sugar dissolves to form a homogeneous mixture or solution.

(a) Water is the liquid in the beaker, and the white solid in the spoon is sugar. (b) Sugar can be dissolved in the water to produce a solution.

(a)

(b)

Flowcharts can help you to visualize the connections between concepts.

Both substances are still present: All parts of the solution are sweet and wet. The proportions of sugar and water can be varied simply by adding more sugar and stirring to dissolve. Solutions do not have to be liquid. For example, air is a homogeneous mixture of gases. Solid solutions also exist. Brass is a homogeneous solution of copper and zinc.

Many substances do not form homogeneous mixtures. If we mix sugar and fine white sand, a heterogeneous mixture is formed. Careful examination may be needed to decide that the mixture is heterogeneous because the two phases (sugar and sand) are both white solids. Ordinary matter exists mostly as mixtures. If we examine soil, granite, iron ore, or other naturally occurring mineral deposits, we find them to be heterogeneous mixtures. **Figure 1.7** illustrates the relationships of substances and mixtures.

Figure 1.7

Classification of matter. A pure substance is always homogeneous in composition, whereas a mixture always contains two or more substances and may be either homogeneous or heterogeneous.

Distinguishing Mixtures from Pure Substances

Single substances—elements or compounds—seldom occur naturally in a pure state. Air is a mixture of gases; seawater is a mixture of a variety of dissolved minerals; ordinary soil is a complex mixture of minerals and various organic materials.

Mixture

1. A mixture always contains two or more substances that can be present in varying amounts.

2. The components of a mixture do not lose their identities and may be separated by physical means.

Pure Substance

1. A pure substance (element or compound) always has a definite composition by mass.

2. The elements in a compound lose their identities and may be separated only by chemical means.

PRACTICE 1.2

Which of the following is a mixture and which is a pure substance? Explain your answer.

(a) vinegar (4% acetic acid and 96% water)
(b) sodium chloride (salt) solution
(c) gold
(d) milk

How is a mixture distinguished from a pure substance? A mixture always contains two or more substances that can be present in varying concentrations. Let's consider two examples.

Homogeneous Mixture Homogeneous mixtures (solutions) containing either 5% or 10% salt in water can be prepared simply by mixing the correct amounts of salt and water. These mixtures can be separated by boiling away the water, leaving the salt as a residue.

Heterogeneous Mixture The composition of a heterogeneous mixture of sulfur crystals and iron filings can be varied by merely blending in either more sulfur or more iron filings. This mixture can be separated physically by using a magnet to attract the iron.

(a)   (b)

(a) When iron and sulfur exist as pure substances, only the iron is attracted to a magnet. (b) A mixture of iron and sulfur can be separated by using the difference in magnetic attraction.

CHAPTER **1 REVIEW**

1.1 THE NATURE OF CHEMISTRY

- Chemistry is important to everyone because chemistry occurs all around us in our daily lives.
- Chemistry is the science dealing with matter and the changes in composition that matter undergoes.
- Chemists seek to understand the general principles governing the behavior of all matter.
- Chemistry "looks inside" ordinary objects to study how their components behave.
- Chemistry connects the macroscopic and microscopic worlds.

KEY TERM

chemistry

1.2 A SCIENTIFIC APPROACH TO PROBLEM SOLVING

- Scientific thinking helps us solve problems in our daily lives.
- General steps for solving problems include:
  - Defining the problem
  - Proposing possible solutions
  - Solving the problem
- The scientific method is a procedure for processing information in which we:
  - Collect the facts
  - Formulate a hypothesis
  - Plan and do experiments
  - Modify the hypothesis if necessary

KEY TERMS

scientific method
hypothesis
theory
scientific laws

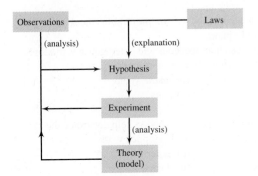

1.3 THE PARTICULATE NATURE OF MATTER

KEY TERMS

matter
solid
amorphous
liquid
gas

- Matter is anything with the following two characteristics:
  - Has mass
  - Occupies space
- On the macroscopic level matter appears continuous.
- On the microscopic level matter is discontinuous and composed of atoms.
- Solid—rigid substance with a definite shape
- Liquid—fluid substance with a definite volume that takes the shape of its container
- Gas—takes the shape and volume of its container

1.4 CLASSIFYING MATTER

KEY TERMS

substance
homogeneous
heterogeneous
phase
system
mixture

- Matter can be classified as a pure substance or a mixture.
- A mixture has variable composition:
  - Homogeneous mixtures have the same properties throughout.
  - Heterogeneous mixtures have different properties in different parts of the system.
- A pure substance always has the same composition. There are two types of pure substances:
  - Elements
  - Compounds

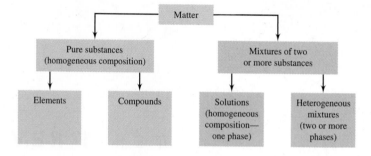

REVIEW QUESTIONS

1. Explain the difference between
   (a) a hypothesis and a theory
   (b) a theory and a scientific law
2. Consider each of the following statements and determine whether it represents an observation, a hypothesis, a theory, or a scientific law:
   (a) The battery in my watch must be dead since it is no longer keeping time.
   (b) My computer must have a virus since it is not working properly.
   (c) The air feels cool.
   (d) The candle burns more brightly in pure oxygen than in air because oxygen supports combustion.
   (e) My sister wears red quite often.
   (f) A pure substance has a definite, fixed composition.
3. Determine whether each of the following statements refers to a solid, a liquid, or a gas:
   (a) It has a definite volume but not a definite shape.
   (b) It has an indefinite volume and high compressibility.
   (c) It has a definite shape.
   (d) It has an indefinite shape and slight compressibility.
4. Some solids have a crystalline structure, while others have an amorphous structure. For each of the following 5 descriptions, determine whether it refers to a crystalline solid or an amorphous solid:
   (a) has a regular repeating pattern
   (b) plastic
   (c) has no regular repeating pattern
   (d) glass
   (e) gold

5. Define a phase.
6. How many phases are present in the graduated cylinder?
7. What is another name for a homogeneous mixture?
8. Which liquids listed in Table 1.1 are not mixtures?
9. Which of the gases listed in Table 1.1 are not pure substances?
10. When the stopper is removed from a partly filled bottle containing solid and liquid acetic acid at 16.7°C, a strong vinegar-like odor is noticeable immediately. How many acetic acid phases must be present in the bottle? Explain.
11. Is the system enclosed in the bottle in Question 10 homogeneous or heterogeneous? Explain.
12. Is a system that contains only one substance necessarily homogeneous? Explain.
13. Is a system that contains two or more substances necessarily heterogeneous? Explain.
14. Distinguish between homogeneous and heterogeneous mixtures.
15. Which of the following are pure substances?
    (a) sugar   (d) maple syrup
    (b) sand   (e) eggs
    (c) gold
16. Use the steps of the scientific method to help determine the reason that your cell phone has suddenly stopped working:
    (a) observation   (c) experiment
    (b) hypothesis   (d) theory

Richard Megna/
Fundamental Photographs

Most of the exercises in this chapter are available for assignment via the online homework management program, WileyPLUS (www.wileyplus.com)
All exercises with blue numbers have answers in Appendix VI.

PAIRED EXERCISES

1. Refer to the illustration and determine which state(s) of matter are present.

2. Refer to the illustration and determine which states(s) of matter are present.

3. Look at the photo and determine whether it represents a homogeneous or heterogeneous mixture.

Richard Megna/Fundamental Photographs

4. Look at the maple leaf below and determine whether it represents a homogeneous or heterogeneous mixture.

© Justin Horrocks/iStockphoto

5. For each of the following mixtures, state whether it is homogeneous or heterogeneous:
(a) tap water
(b) carbonated beverage
(c) oil and vinegar salad dressing
(d) people in a football stadium

6. For each of the following mixtures, state whether it is homogeneous or heterogeneous:
(a) stainless steel
(b) motor oil
(c) soil
(d) a tree

ADDITIONAL EXERCISES

7. At home, check your kitchen and bathroom cabinets for five different substances; then read the labels and list the first ingredient of each.

8. During the first week of a new semester, consider that you have enrolled in five different classes, each of which meets for 3 hours per week. For every 1 hour that is spent in class, a minimum of 1 hour is required outside of class to complete assignments and study for exams. You also work 20 hours per week, and it takes you 1 hour to drive to the job site and back home. On Friday nights, you socialize with your friends. You are fairly certain that you will be able to successfully complete the semester with good grades. Show how the steps in the scientific method can help you predict the outcome of the semester.

Use the following food label to answer Exercises 9–10.

Nutrition Facts	Amount/serving	%DV*	Amount/serving	%DV*
Serv. Size 1 cup (249g)	Total Fat 12g	18%	Sodium 940mg	39%
	Sat. Fat 6g	30%	Total Carb. 24g	8%
Servings About 3	Polyunsat. Fat 1.5g		Dietary Fiber 1g	4%
Calories 250 Fat Cal. 110	Monounsat. Fat 2.5g		Sugars 1g	
*Percent Daily Values (DV) are based on a 2,000 calorie diet.	Cholest. 60mg	20%	Protein 10g	20%
	Vitamin A 0% • Vitamin C 0% • Calcium 6% • Iron 8%			

INGREDIENTS: WATER, CHICKEN STOCK, ENRICHED PASTA (SEMOLINA WHEAT FLOUR, EGG WHITE SOLIDS, NIACIN, IRON, THIAMINE MONONITRATE [VITAMIN B1], RIBOFLAVIN [VITAMIN B2] AND FOLIC ACID), CREAM (DERIVED FROM MILK), CHICKEN, CONTAINS LESS THAN 2% OF: CHEESES (GRANULAR, PARMESAN AND ROMANO PASTE [PASTEURIZED COW'S MILK, CULTURES, SALT, ENZYMES], WATER, SALT, LACTIC ACID, CITRIC ACID AND DISODIUM PHOSPHATE), BUTTER (PASTEURIZED SWEET CREAM [DERIVED FROM MILK] AND SALT), MODIFIED CORN STARCH, SALT, WHOLE EGG SOLIDS, SUGAR, RICE STARCH, GARLIC, SPICE, XANTHAM GUM, MUSTARD FLOUR, ISOLATED SOY PROTEIN AND SODIUM PHOSPHATE.

9. Identify the following ingredients as either a pure substance or a mixture.
   (a) water
   (b) chicken stock (the liquid that remains in the pot after cooking the chicken)
   (c) salt
   (d) mustard flour

10. Using the food label, make a hypothesis regarding the nutritional quality of this food. Propose a way to determine whether your hypothesis is valid.

11. Read the following passage (from *Science News*) and identify the observation and hypothesis.

   Could an invisibility cloak, like the one used by Harry Potter, really exist? For scientists in Cambridge, Massachusetts, it may be a reality. Researchers at MIT have developed an invisibility cloak for small objects such as an ant or a grain of sand. Calcite crystals have the ability to reflect light around an object, rendering it invisible, or at least "unseeable." If a larger crystal were to be used, it should be able to hide larger objects. The invisibility cloak works only with laser light aimed directly at the crystal. Future work will improve the effectiveness of these invisibility cloaks.

12. Various chemical elements are pictured. For each one a microscopic view is also shown. Determine the number of phases for each substance below and identify them.
   (a) iodine

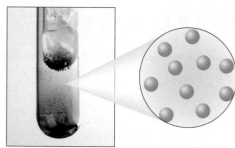

© 2001 Richard Megna/ Fundamental Photographs

   (b) bromine

© 1995 Richard Megna/ Fundamental Photographs

   (c) sulfur

Stocktrek Images/Richard Roscoe/Getty Images, Inc.

ANSWERS TO PRACTICE EXERCISES

1.1   five water molecules: H_2O, H_2O, H_2O, H_2O, H_2O

1.2   (a) mixture; concentration can be changed by adding more acetic acid or more water. (b) mixture; concentration can be changed by adding more or less salt. (c) pure substance; gold is 100% gold. (d) mixture; milk contains several substances.

Careful and accurate measurements for each ingredient are essential when baking or cooking as well as in the chemistry laboratory.

CHAPTER **2**

STANDARDS FOR MEASUREMENT

Doing an experiment in chemistry is very much like cooking a meal in the kitchen. It's important to know the ingredients and the amounts of each in order to have a tasty product. Working on your car requires specific tools in exact sizes. Buying new carpeting or draperies is an exercise in precise and accurate measurement for a good fit. A small difference in the concentration or amount of medication a pharmacist gives you may have significant effects on your well-being. As we saw in Chapter 1, observation is an important part of the scientific process. Observations can be *qualitative* (the substance is a blue solid) or *quantitative* (the mass of the substance is 4.7 grams). A quantitative observation is called a **measurement**. Both a number and a unit are required for a measurement. For example, at home in your kitchen measuring 3 flour is not possible. We need to know the unit on the 3. Is it cups, grams, or tablespoons? A measurement of 3 cups tells us both the amount and the size of the measurement. In this chapter we will discuss measurements and the rules for calculations using these measurements.

CHAPTER OUTLINE

2.1 Scientific Notation
2.2 Measurement and Uncertainty
2.3 Significant Figures
2.4 Significant Figures in Calculations
2.5 The Metric System
2.6 Dimensional Analysis: A Problem-Solving Method
2.7 Measurement of Temperature
2.8 Density

2.1 SCIENTIFIC NOTATION

LEARNING OBJECTIVE ● Write decimal numbers in scientific notation.

KEY TERMS

measurement
scientific notation

Scientists often use numbers that are very large or very small in **measurements**. For example, the Earth's age is estimated to be about 4,500,000 (4.5 billion) years. Numbers like these are bulky to write, so to make them more compact scientists use powers of 10. Writing a number as the product of a number between 1 and 10 multiplied by 10 raised to some power is called **scientific notation**.

To learn how to write a number in scientific notation, let's consider the number 2468. To write this number in scientific notation:

1. Move the decimal point in the original number so that it is located after the first nonzero digit.

   $$2468 \rightarrow 2.468 \qquad \text{(decimal moves three places to the left)}$$

2. Multiply this new number by 10 raised to the proper exponent (power). The proper exponent is equal to the number of places that the decimal point was moved.

   $$2.468 \times 10^3$$

3. The sign on the exponent indicates the direction the decimal was moved.

   $$\text{moved right} \rightarrow \text{negative exponent}$$

   $$\text{moved left} \rightarrow \text{positive exponent}$$

Examples show you problem-solving techniques in a step-by-step form. Study each one and then try the Practice Exercises.

WileyPLUS

ENHANCED EXAMPLE

EXAMPLE 2.1

Write 5283 in scientific notation.

SOLUTION   5283.   Place the decimal between the 5 and the 2. Since the decimal
             ‿‿‿   was moved three places to the left, the power of 10 will be 3
              3   and the number 5.283 is multiplied by 10^3.

$$5.283 \times 10^3$$

EXAMPLE 2.2

Write 4,500,000,000 in scientific notation (two digits).

SOLUTION   4 500 000 000.   Place the decimal between the 4 and the 5. Since the
             ‿‿‿‿‿‿‿‿‿   decimal was moved nine places to the left, the power of
                 9   10 will be 9 and the number 4.5 is multiplied by 10^9.

$$4.5 \times 10^9$$

Scientific notation is a useful way to write very large numbers, such as the distance between the Earth and the moon, or very small numbers, such as the length of these *E. coli* bacteria (shown here as a colored scanning electron micrograph × 14,000).

NASA

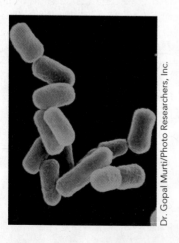
Dr. Gopal Murti/Photo Researchers, Inc.

EXAMPLE 2.3

Write 0.000123 in scientific notation.

SOLUTION   0.000123   Place the decimal between the 1 and the 2. Since the deci-
                            mal was moved four places to the right, the power of 10
                            will be -4 and the number 1.23 is multiplied by 10^{-4}.

1.23×10^{-4}

Answers to Practice Exercises are found at the end of each chapter.

PRACTICE 2.1

Write the following numbers in scientific notation:

(a) 1200 (four digits)   (c) 0.0468
(b) 6,600,000 (two digits)   (d) 0.00003

2.2 MEASUREMENT AND UNCERTAINTY

Explain the significance of uncertainty in measurement in chemistry and how significant figures are used to indicate a measurement's certainty.

● **LEARNING OBJECTIVE**

KEY TERM
significant figures

To understand chemistry, it is necessary to set up and solve problems. Problem solving requires an understanding of the mathematical operations used to manipulate numbers. Measurements are made in an experiment, and chemists use these data to calculate the extent of the physical and chemical changes occurring in the substances that are being studied. By appropriate calculations, an experiment's results can be compared with those of other experiments and summarized in ways that are meaningful.

A measurement is expressed by a numerical value together with a unit of that measurement. For example,

A measurement always requires a unit.

$$\overbrace{70.0 \text{ kilograms}}^{\text{numerical value}} = \underbrace{154 \text{ pounds}}_{\text{unit}}$$

Whenever a measurement is made with an instrument such as a thermometer or ruler, an estimate is required. We can illustrate this by measuring temperature. Suppose we measure temperature on a thermometer calibrated in degrees and observe that the mercury stops between 21 and 22 (see **Figure 2.1a**). We then know that the temperature is at least 21 degrees and is less than 22 degrees. To express the temperature with one more digit, we estimate that the mercury is about two-tenths the distance between 21 and 22. The temperature is therefore 21.2 degrees. The last digit (2) has some uncertainty because it is an estimated value. Because the last digit has some uncertainty (we made a visual estimate), it may be different when another person makes the same measurement. If three more people make this same reading, the results might be

Person	Measurement
1	21.2
2	21.3
3	21.1
4	21.2

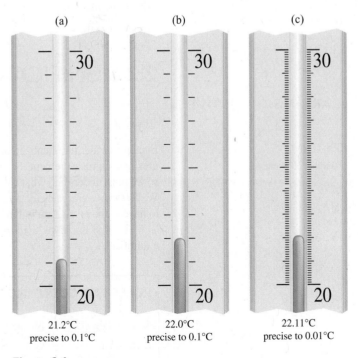

(a)   (b)   (c)

21.2°C   22.0°C   22.11°C
precise to 0.1°C   precise to 0.1°C   precise to 0.01°C

Figure 2.1
Measuring temperature (°C) with various degrees of precision.

Notice that the first two digits of the measurements did not change (they are *certain*). The last digit in these measurements is *uncertain*. The custom in science is to record all of the certain digits and the first uncertain digit.

Numbers obtained from a measurement are never exact values. They always have some degree of uncertainty due to the limitations of the measuring instrument and the skill of the individual making the measurement. It is customary when recording a measurement to include all the digits that are known plus one digit that is estimated. This last estimated digit introduces some uncertainty. Because of this uncertainty, every number that expresses a measurement can have only a limited number of digits. These digits, used to express a measured quantity, are known as **significant figures**.

Now let's return to our temperature measurements. In Figure 2.1a the temperature is recorded as 21.2 degrees and is said to have three significant figures. If the mercury stopped exactly on the 22 (Figure 2.1b), the temperature would be recorded as 22.0 degrees. The zero is used to indicate that the temperature was estimated to a precision of one-tenth degree. Finally, look at Figure 2.1c. On this thermometer, the temperature is recorded as 22.11°C (four significant figures). Since the thermometer is calibrated to tenths of a degree, the first estimated digit is the hundredths.

EXAMPLE 2.4

The length of a ballpoint pen was recorded by three students as 14.30 cm, 14.33 cm, and 14.34 cm.

(a) What are the estimated digits in these measurements?
(b) How many figures are certain?

SOLUTION

(a) The estimated digits are 0, 3, and 4, respectively.
(b) Three figures are certain in each measurement

PRACTICE 2.2

Three measurements for the boiling point of water are 100.4°C, 100.1°C, and 100.0°C. The accepted boiling point is 100.0°C.

(a) What are the estimated digits?
(b) How many figures are certain?

2.3 SIGNIFICANT FIGURES

LEARNING OBJECTIVE ● Determine the number of significant figures in a given measurement and round measurements to a specific number of significant figures.

KEY TERM
rounding off numbers

Because all measurements involve uncertainty, we must use the proper number of significant figures in each measurement. In chemistry we frequently do calculations involving measurements, so we must understand what happens when we do arithmetic on numbers containing uncertainties. We'll learn several rules for doing these calculations and figuring out how many significant figures to have in the result. You will need to follow these rules throughout the calculations in this text.

The first thing we need to learn is how to determine how many significant figures are in a number.

RULES FOR COUNTING SIGNIFICANT FIGURES

1. *Nonzero digits.* All nonzero digits are significant.
2. *Exact numbers.* Some numbers are exact and have an infinite number of significant figures. Exact numbers occur in simple counting operations; when you count 25 dollars, you have exactly 25 dollars. Defined numbers, such as

12 inches in 1 foot, 60 minutes in 1 hour, and 100 centimeters in 1 meter, are also considered to be exact numbers. Exact numbers have no uncertainty.

3. *Zeros.* A zero is *significant* when it is

- between nonzero digits:

  205 has three significant figures (2, 0, 5)

  2.05 has three significant figures (2, 0, 5)

  61.09 has four significant figures (6, 1, 0, 9)

- at the end of a number that includes a decimal point:

  0.500 has three significant figures (5, 0, 0)

  25.160 has five significant figures (2, 5, 1, 6, 0)

  3.00 has three significant figures (3, 0, 0)

  20. has two significant figures (2, 0)

A zero is *not significant* when it is

- before the first nonzero digit. These zeros are used to locate a decimal point:

  0.0025 has two significant figures (2, 5)

  0.0108 has three significant figures (1, 0, 8)

- at the end of a number without a decimal point:

  1000 has one significant figure (1)

  590 has two significant figures (5, 9)

Rules for significant figures should be memorized for use throughout the text.

One way of indicating that these zeros are significant is to write the number using scientific notation. Thus if the value 1000 has been determined to four significant figures, it is written as 1.000×10^3. If 590 has only two significant figures, it is written as 5.9×10^2.

WileyPLUS

ENHANCED EXAMPLE

EXAMPLE 2.5

How many significant figures are in each of the following measurements?

(a) 45 apples
(b) 0.02050 cm
(c) 3500 ft

SOLUTION

(a) The number is exact and has no uncertainty associated with it (Rule 2).
(b) Four significant figures. The zeroes before the 2 are not significant, the 2 is significant, the zero between the 2 and the 5 is significant, the 5 is significant, and the final zero is significant.
(c) Two significant figures. The zeroes at the end of a number without a decimal point are not significant.

PRACTICE 2.3

How many significant figures are in each of these measurements?

(a) 4.5 inches
(b) 3.025 feet
(c) 125.0 meters
(d) 0.001 mile
(e) 25.0 grams
(f) 12.20 liters
(g) 100,000 people
(h) 205 birds

Rounding Off Numbers

When we do calculations on a calculator, we often obtain answers that have more digits than are justified. It is therefore necessary to drop the excess digits in order to express the answer with the proper number of significant figures. When digits are dropped from a number, the value of the last digit retained is determined by a process known as **rounding off numbers**. Two rules will be used in this book for rounding off numbers.

Not all schools use the same rules for rounding. Check with your instructor for variations in these rules.

RULES FOR ROUNDING OFF

1. When the first digit after those you want to retain is 4 or less, that digit and all others to its right are dropped. The last digit retained is not changed. The following examples are rounded off to four digits:

74.693 = 74.69   1.00629 = 1.006
    This digit is dropped.   These two digits are dropped.

2. When the first digit after those you want to retain is 5 or greater, that digit and all others to the right are dropped and the last digit retained is increased by one. These examples are rounded off to four digits:

1.026868 = 1.027   18.02500 = 18.03
    These three digits are dropped.   These three digits are dropped.
    This digit is changed to 7.   This digit is changed to 3.

12.899 = 12.90
    This digit is dropped.
    These two digits are changed to 90.

PRACTICE 2.4

Round off these numbers to the number of significant figures indicated:

(a) 42.246 (four)   (d) 0.08965 (two)
(b) 88.015 (three)   (e) 225.3 (three)
(c) 0.08965 (three)   (f) 14.150 (three)

2.4 SIGNIFICANT FIGURES IN CALCULATIONS

LEARNING OBJECTIVE ● Apply the rules for significant figures, in calculations involving addition, subtraction, multiplication, and division.

The results of a calculation based on measurements cannot be more precise than the least precise measurement.

Multiplication or Division

In calculations involving multiplication or division, the answer must contain the same number of significant figures as in the measurement that has the least number of significant figures. Consider the following examples:

Use your calculator to check your work in the examples. Compare your results to be sure you understand the mathematics.

WileyPLUS
ENHANCED EXAMPLE

EXAMPLE 2.6

$(190.6)(2.3) = 438.38$

SOLUTION

The value 438.38 was obtained with a calculator. The answer should have two significant figures because 2.3, the number with the fewest significant figures, has only two significant figures.

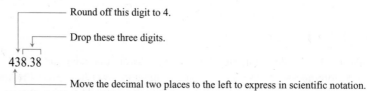

Round off this digit to 4.

Drop these three digits.

438.38

Move the decimal two places to the left to express in scientific notation.

The correct answer is 440 or 4.4×10^2.

EXAMPLE 2.7

$$\frac{(13.59)(6.3)}{12} = 7.13475$$

SOLUTION

The value 7.13475 was obtained with a calculator. The answer should contain two significant figures because 6.3 and 12 each have only two significant figures.

Drop these four digits.

7.13475

This digit remains the same.

The correct answer is 7.1.

PRACTICE 2.5

(a) $(134 \text{ in.})(25 \text{ in.}) = ?$

(b) $\dfrac{213 \text{ miles}}{4.20 \text{ hours}} = ?$

(c) $\dfrac{(2.2)(273)}{760} = ?$

(d) $0.0321 \times 42 = ?$

(e) $\dfrac{0.0450}{0.00220} = ?$

(f) $\dfrac{1.280}{0.345} = ?$

Addition or Subtraction

The results of an addition or a subtraction must be expressed to the same precision as the least precise measurement. This means the result must be rounded to the same number of decimal places as the value with the fewest decimal places (blue line in examples).

EXAMPLE 2.8

Add 125.17, 129, and 52.2.

SOLUTION

$$\begin{array}{r} 125.|17 \\ 129 \\ 52.|2 \\ \hline 306.|37 \end{array}$$

The number with the least precision is 129. Therefore the answer is rounded off to the nearest unit: 306.

EXAMPLE 2.9

Subtract 14.1 from 132.56.

SOLUTION

$$\begin{array}{r} 132.5|6 \\ -\ 14.1 \\ \hline 118.4|6 \end{array}$$

The number with the least precision is 14.1. Therefore the answer is rounded off to the nearest tenth: 118.5.

EXAMPLE 2.10

Subtract 120 from 1587.

SOLUTION

$$\begin{array}{r} 158|7 \\ - \ 12|0 \\ \hline 146|7 \end{array}$$

The number with the least precision is 120. The zero is not considered significant; therefore the answer must be rounded to the nearest ten: 1470 or 1.47×10^3.

EXAMPLE 2.11

Add 5672 and 0.00063.

SOLUTION

$$\begin{array}{r} 5672 \ | \\ + \ \ \ \ 0.|00063 \\ \hline 5672.|00063 \end{array}$$

The number with the least precision is 5672. So the answer is rounded off to the nearest unit: 5672.

Note: When a very small number is added to a large number, the result is simply the original number.

EXAMPLE 2.12

$$\frac{1.039 - 1.020}{1.039} = ?$$

SOLUTION

$$\frac{1.039 - 1.020}{1.039} = 0.018286814$$

The value 0.018286814 was obtained with a calculator. When the subtraction in the numerator is done,

$$1.039 - 1.020 = 0.019$$

the number of significant figures changes from four to two. Therefore the answer should contain two significant figures after the division is carried out:

Drop these six digits.

0.018286814

This digit remains the same.

The correct answer is 0.018, or 1.8×10^{-2}.

PRACTICE 2.6

If you need to brush up on your math skills, refer to the "Mathematical Review" in Appendix I.

How many significant figures should the answer in each of these calculations contain?

(a) (14.0)(5.2)

(b) (0.1682)(8.2)

(c) $\dfrac{(160)(33)}{4}$

(d) 8.2 + 0.125

(e) 119.1 − 3.44

(f) $\dfrac{94.5}{1.2}$

(g) 1200 + 6.34

(h) 1.6 + 23 − 0.005

> ### RULES FOR SIGNIFICANT FIGURES IN CALCULATIONS
>
> **Multiplication or Division**
> The answer contains the same number of significant figures as the measurement with the *least number* of significant figures.
>
> **Addition or Subtraction**
> The answer contains the same number of significant figures as the *least precise* measurement.

Additional material on mathematical operations is given in Appendix I, "Mathematical Review." Study any portions that are not familiar to you. You may need to do this at various times during the course when additional knowledge of mathematical operations is required.

2.5 THE METRIC SYSTEM

Name the units for mass, length, and volume in the metric system and convert from one unit to another.

● **LEARNING OBJECTIVE**

The **metric system**, or **International System** (**SI**, from *Système International*), is a decimal system of units for measurements of mass, length, time, and other physical quantities. Built around a set of standard units, the metric system uses factors of 10 to express larger or smaller numbers of these units. To express quantities that are larger or smaller than the standard units, prefixes are added to the names of the units. These prefixes represent multiples of 10, making the metric system a decimal system of measurements. Table 2.1 shows the names, symbols, and numerical values of the common prefixes. Some examples of the more commonly used prefixes are

$$1 \; kilo\text{meter} = 1000 \text{ meters}$$
$$1 \; kilo\text{gram} = 1000 \text{ grams}$$
$$1 \; milli\text{meter} = 0.001 \text{ meter}$$
$$1 \; micro\text{second} = 0.000001 \text{ second}$$

The common standard units in the International System, their abbreviations, and the quantities they measure are given in Table 2.2. Other units are derived from these units. The metric system, or International System, is currently used by most of the countries in the world, not only in scientific and technical work but also in commerce and industry.

KEY TERMS

metric system or International System (SI)
meter (m)
conversion factor
solution map
mass
weight
kilogram (kg)
volume
liter (L)

The prefixes most commonly used in chemistry are shown in bold in Table 2.1.

TABLE 2.1 Common Prefixes and Numerical Values for SI Units

Prefix	Symbol	Numerical value	Power of 10 equivalent
giga	G	1,000,000,000	10^9
mega	M	1,000,000	10^6
kilo	k	1,000	10^3
hecto	h	100	10^2
deka	da	10	10^1
—	—	1	10^0
deci	d	0.1	10^{-1}
centi	c	0.01	10^{-2}
milli	m	0.001	10^{-3}
micro	μ	0.000001	10^{-6}
nano	n	0.000000001	10^{-9}
pico	p	0.000000000001	10^{-12}
femto	f	0.000000000000001	10^{-15}

Most products today list both systems of measurement on their labels.

TABLE 2.2 International System's Standard Units of Measurement

Quantity	Name of unit	Abbreviation
Length	meter	m
Mass	kilogram	kg
Temperature	kelvin	K
Time	second	s
Amount of substance	mole	mol
Electric current	ampere	A
Luminous intensity	candela	cd

Measurement of Length

The standard unit of length in the metric system is the **meter (m)**. When the metric system was first introduced in the 1790s, the meter was defined as one ten-millionth of the distance from the equator to the North Pole, measured along the meridian passing through Dunkirk, France. The latest definition describes a meter as the distance that light travels in a vacuum during 1/299,792,458 of a second.

A meter is 39.37 inches, a little longer than 1 yard. One meter equals 10 decimeters, 100 centimeters, or 1000 millimeters (see **Figure 2.2**). A kilometer contains 1000 meters. **Table 2.3** shows the relationships of these units.

The nanometer (10^{-9} m) is used extensively in expressing the wavelength of light, as well as in atomic dimensions. See the inside back cover for a complete table of common conversions.

Common length relationships:

$1\text{ m} = 10^6\,\mu\text{m} = 10^{10}\,\text{Å}$
$\qquad = 100\text{ cm} = 1000\text{ mm}$
$1\text{ cm} = 10\text{ mm} = 0.01\text{ m}$
$1\text{ in.} = 2.54\text{ cm}$
$1\text{ mile} = 1.609\text{ km}$

See inside back cover for a table of conversions.

1 in.   **Inches**

2.54 cm
25.4 mm   **Centimeters**

Figure 2.2

Comparison of the metric and American systems of length measurement: 2.54 cm = 1 in. A playing card measures 2 ½ in. or 6.3 cm.

TABLE 2.3 Metric Units of Length

Unit	Abbreviation	Meter equivalent	Exponential equivalent
kilometer	km	1000 m	10^3 m
meter	m	1 m	10^0 m
decimeter	dm	0.1 m	10^{-1} m
centimeter	cm	0.01 m	10^{-2} m
millimeter	mm	0.001 m	10^{-3} m
micrometer	μm	0.000001 m	10^{-6} m
nanometer	nm	0.000000001 m	10^{-9} m
angstrom	Å	0.0000000001 m	10^{-10} m

Unit Conversions

One of the benefits of using the metric system is the ease with which we can convert from one unit to another. We do this by using the relationships between units—math equations called equivalent units. Table 2.4 shows common equivalent units for length in the metric system.

To convert from one unit to another we must use a **conversion factor**. A conversion factor is a ratio of equivalent quantities. Where do we find these conversion factors? They are formed from equivalent units. For example, since 1 m = 100 cm (Table 2.4) we could write two ratios: $\dfrac{1 \text{ m}}{100 \text{ cm}}$ or $\dfrac{100 \text{ cm}}{1 \text{ m}}$. In both cases the ratio is 1 since the units are equal.

We can convert one metric unit to another by using conversion factors.

$$\textbf{unit}_1 \times \textbf{conversion factor} = \textbf{unit}_2$$

If you want to know how many millimeters are in 2.5 meters, you need to convert meters (m) to millimeters (mm). Start by writing

$$\text{m} \times \text{conversion factor} = \text{mm}$$

This conversion factor must accomplish two things: It must cancel (or eliminate) meters, *and* it must introduce millimeters—the unit wanted in the answer. Such a conversion factor will be in fractional form and have meters in the denominator and millimeters in the numerator:

$$\cancel{\text{m}} \times \frac{\text{mm}}{\cancel{\text{m}}} = \text{mm}$$

We know that 1 m = 1000 mm. From this relationship we can write two conversion factors.

$$\frac{1 \text{ m}}{1000 \text{ mm}} \quad \text{and} \quad \frac{1000 \text{ mm}}{1 \text{ m}}$$

Choosing the conversion factor $\dfrac{1000 \text{ mm}}{1 \text{ m}}$, we can set up the calculation for the conversion of 2.5 m to millimeters:

$$(2.5 \ \cancel{\text{m}}) \left(\frac{1000 \text{ mm}}{1 \ \cancel{\text{m}}} \right) = 2500 \text{ mm} \quad \text{or} \quad 2.5 \times 10^3 \text{ mm} \quad \text{(two significant figures)}$$

Note that, in making this calculation, units are treated as numbers; meters in the numerator are canceled by meters in the denominator.

TABLE 2.4
Length Equivalent Units
1 m = 10 dm
1 m = 100 cm
1 m = 1000 mm
1 kg = 1000 g
1 cm = 10 mm

Important equations and statements are boxed or highlighted in color.

We know that multiplying a measurement by 1 does not change its value. Since our conversion factors both equal 1, we can multiply the measurement by the appropriate one to convert units.

EXAMPLE 2.13

Change 215 centimeters to meters.

SOLUTION

We will use a **solution map** to outline our path for the conversion for this example:

Solution map: cm → m

We start with

$$\text{cm} \times \text{conversion factor} = \text{m}$$

The conversion factor must have centimeters in the denominator and meters in the numerator:

$$\cancel{\text{cm}} \times \frac{\text{m}}{\cancel{\text{cm}}} = \text{m}$$

From the relationship 100 cm = 1 m, we can write a factor that will accomplish this conversion:

$$\frac{1 \text{ m}}{100 \text{ cm}}$$

WileyPLUS

ENHANCED EXAMPLE

Now set up the calculation using all the data given:

$$(215 \text{ cm})\left(\frac{1 \text{ m}}{100 \text{ cm}}\right) = 2.15 \text{ m}$$

PRACTICE 2.7

Convert the following measurements:

(a) 567 mm to m
(b) 68 cm to mm
(c) 125 m to km

Measurement of Mass

Although we often use mass and weight interchangeably in our everyday lives, they have quite different meanings in chemistry. In science we define the **mass** of an object as the amount of matter in the object. Mass is measured on an instrument called a balance. The **weight** of an object is a measure of the effect of gravity on the object. Weight is determined by using an instrument called a scale, which measures force against a spring. This means mass is independent of the location of an object, but weight is not. In this text we will use the term *mass* for all of our metric mass measurements. The gram is a unit of mass measurement, but it is a tiny amount of mass; for instance, a U.S. nickel has a mass of about 5 grams. Therefore the *standard unit* of mass in the SI system is the **kilogram (kg)** (equal to 1000 g). The amount of mass in a kilogram is defined by international agreement as exactly equal to the mass of a platinum-iridium cylinder (international prototype kilogram) kept in a vault at Sèvres, France. Comparing this unit of mass to 1 lb (16 oz), we find that 1 kg is equal to 2.205 lb. A pound is equal to 453.6 g (0.4536 kg). The same prefixes used in length measurement are used to indicate larger and smaller gram units (see Table 2.5). A balance is used to measure mass. Two examples of balances are shown in Figure 2.3.

TABLE 2.5	Metric Units of Mass		
Unit	**Abbreviation**	**Gram equivalent**	**Exponential equivalent**
kilogram	kg	1000 g	10^3 g
gram	g	1 g	10^0 g
decigram	dg	0.1 g	10^{-1} g
centigram	cg	0.01 g	10^{-2} g
milligram	mg	0.001 g	10^{-3} g
microgram	μg	0.000001 g	10^{-6} g

Figure 2.3

(a) Digital electronic, top-loading balance with a precision of 0.001 g (1 mg); and (b) a digital electronic analytical balance with a precision of 0.0001 g (0.1 mg).

(a)   (b)

© Sciencephotos/Alamy

© 2005 Richard Megna/Fundamental Photographs

>CHEMISTRY *IN ACTION*

Keeping Track of Units

Why do scientists worry about units? The National Aeronautics and Space Administration (NASA) recently was reminded of just why keeping track of units is so important. In 1999 a $125 million satellite was lost in the atmosphere of Mars because scientists made some improper assumptions about units. NASA's scientists at the Jet Propulsion Lab (JPL) in Pasadena, California, received thrust data from the satellite's manufacturer, Lockheed Martin Aeronautics in Denver, Colorado. Unfortunately, the Denver scientists used American units in their measurements and the JPL scientists assumed the units were metric. This mistake caused the satellite to fall

Mars climate orbiter

100 km lower into the Mars atmosphere than planned. The spacecraft burned up from the friction with the atmosphere.

Measuring and using units correctly is very important. In fact, it can be critical, as we have just seen. For example, a Canadian jet almost crashed when the tanks were filled with 22,300 pounds (instead of kilograms) of fuel. Calculations for distance were based on kilograms, and the jet almost ran out of fuel before landing at the destination. Correct units are also important to ensure perfect fits for household purchases such as drapes, carpet, or appliances. Be sure to pay attention to units in both your chemistry problems and in everyday life!

EXAMPLE 2.14

Change 25 grams to milligrams.

SOLUTION

We use the conversion factor 1000 mg/g.

Solution map: g → mg

$$(25 \ \cancel{g})\left(\frac{1000 \text{ mg}}{1 \ \cancel{g}}\right) = 25{,}000 \text{ mg} \qquad (2.5 \times 10^4 \text{ mg})$$

Note that multiplying a number by 1000 is the same as multiplying the number by 10^3 and can be done simply by moving the decimal point three places to the right:

$$(6.428)(1000) = 6428 \qquad (6.428) \atop 3$$

PRACTICE 2.8

Convert the following measurements:

(a) 560 mg to g
(b) 525 g to kg
(c) 175 g to mg

Common mass relationships:
1 g = 1000 mg
1 kg = 1000 g
1 kg = 2.205 lb
1 lb = 453.6 g

To change milligrams to grams, we use the conversion factor 1 g/1000 mg. For example, we convert 155 mg to grams as follows:

$$(155 \ \cancel{\text{mg}})\left(\frac{1 \text{ g}}{1000 \ \cancel{\text{mg}}}\right) = 0.155 \text{ g}$$

Measurement of Volume

Volume, as used here, is the amount of space occupied by matter. The SI unit of volume is the *cubic meter* (m^3). However, the liter (pronounced *lee-ter* and abbreviated L) and the milliliter (abbreviated mL) are the standard units of volume used in most chemical laboratories. A **liter (L)** is usually defined as 1 cubic decimeter (1 kg) of water at 4°C.

The most common instruments or equipment for measuring liquids are the graduated cylinder, volumetric flask, buret, pipet, and syringe, which are illustrated in **Figure 2.4**. These pieces are usually made of glass and are available in various sizes. Let's try some examples.

Common volume relationships:
$$1\ L = 1000\ mL = 1000\ cm^3$$
$$1\ mL = 1\ cm^3$$
$$1\ L = 1.057\ qt$$
$$946.1\ mL = 1\ qt$$

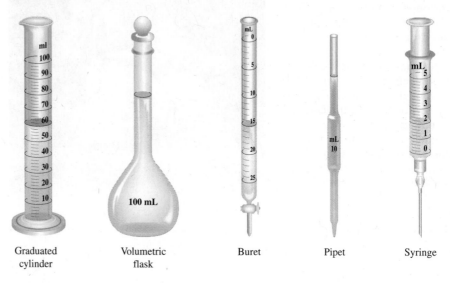

| Graduated cylinder | Volumetric flask | Buret | Pipet | Syringe |

Figure 2.4

Calibrated glassware for measuring the volume of liquids.

EXAMPLE 2.15

How many milliliters are contained in 3.5 liters?

SOLUTION

The conversion factor to change liters to milliliters is 1000 mL/L:

Solution map: $L \rightarrow mL$

$$(3.5\ \cancel{L})\left(\frac{1000\ mL}{\cancel{L}}\right) = 3500\ mL \qquad (3.5 \times 10^3\ mL)$$

Liters may be changed to milliliters by moving the decimal point three places to the right and changing the units to milliliters:

$$1.500\ L = 1500.\ mL$$

PRACTICE 2.9

Convert the following measurements:

(a) 25 mL to L
(b) 1.55 L to mL
(c) 4.5 L to dL

2.6 DIMENSIONAL ANALYSIS: A PROBLEM-SOLVING METHOD

Use dimensional analysis to solve problems involving unit conversions.

● **LEARNING OBJECTIVE**

See Appendix II for help in using a scientific calculator.

Many chemical principles can be illustrated mathematically. Learning how to set up and solve numerical problems in a systematic fashion is *essential* in the study of chemistry.

Usually a problem can be solved by several methods. But in all methods it is best to use a systematic, orderly approach. The *dimensional analysis method* is emphasized in this book because it

- provides a systematic, straightforward way to set up problems.
- gives a clear understanding of the principles involved.
- trains you to organize and evaluate data.
- helps to identify errors, since unwanted units are not eliminated if the setup of the problem is incorrect.

Dimensional analysis converts one unit to another unit by using conversion factors.

Label all factors with the proper units.

The following examples show the conversion between American units and metric units. Mass conversions from American to metric units are shown in Examples 2.16 and 2.21.

WileyPLUS

ENHANCED EXAMPLE

EXAMPLE 2.16

A 1.50-lb package contains how many grams of baking soda?

SOLUTION

We are solving for the number of grams equivalent to 1.50 lb. Since 1 lb = 453.6 g, the conversion factor is 453.6 g/lb:

Solution map: lb → g

$$(1.50 \text{ lb})\left(\frac{453.6 \text{ g}}{1 \text{ lb}}\right) = 680. \text{ g}$$

PRACTICE 2.10

A tennis ball has a mass of 65 g. Determine the American equivalent in pounds.

The volume of a cubic or rectangular container can be determined by multiplying its length × width × height. Thus a box 10 cm on each side has a volume of (10 cm)(10 cm)(10 cm) = 1000 cm³.

EXAMPLE 2.17

How many cubic centimeters are in a cube that is 11.1 inches on a side?

SOLUTION

First we change inches to centimeters; our conversion factor is 2.54 cm/in.:

Solution map: in. → cm

$$(11.1 \text{ in.})\left(\frac{2.54 \text{ cm}}{1 \text{ in.}}\right) = 28.2 \text{ cm on a side}$$

Then we determine volume (length × width × height):

$$(28.2 \text{ cm})(28.2 \text{ cm})(28.2 \text{ cm}) = 22,426 \text{ cm}^3 \qquad (2.24 \times 10^4 \text{ cm}^3)$$

When doing problems with multiple steps, you should round only at the end of the problem. We are rounding at the end of each step in example problems to illustrate the proper significant figures.

EXAMPLE 2.18

How many cubic centimeters (cm^3) are in a box that measures 2.20 in. by 4.00 in. by 6.00 in.?

SOLUTION

First we need to determine the volume of the box in cubic inches ($in.^3$) by multiplying the length × width × height:

$$(2.20 \text{ in.})(4.00 \text{ in.})(6.00 \text{ in.}) = 52.8 \text{ in.}^3$$

Now we convert $in.^3$ to cm^3 by using the inches and centimeters relationship (1 in. = 2.54 cm) three times:

Solution map: $(in. \rightarrow cm)^3$

$$in.^3 \times \frac{cm}{in.} \times \frac{cm}{in.} \times \frac{cm}{in.} = cm^3$$

$$(52.8 \text{ in.}^3)\left(\frac{2.54 \text{ cm}}{1 \text{ in.}}\right)\left(\frac{2.54 \text{ cm}}{1 \text{ in.}}\right)\left(\frac{2.54 \text{ cm}}{1 \text{ in.}}\right) = 865 \text{ cm}^3$$

Some problems require a series of conversions to reach the correct units in the answer. For example, suppose we want to know the number of seconds in 1 day. We need to convert from the unit of days to seconds using this solution map:

Solution map: day → hours → minutes → seconds

Each arrow in the solution map represents a conversion factor. This sequence requires three conversion factors, one for each step. We convert days to hours (hr), hours to minutes (min), and minutes to seconds (s). The conversions can be done individually or in a continuous sequence:

$$day \times \frac{hr}{day} \longrightarrow hr \times \frac{min}{hr} \longrightarrow min \times \frac{s}{min} = s$$

$$day \times \frac{hr}{day} \times \frac{min}{hr} \times \frac{s}{min} = s$$

Inserting the proper factors, we calculate the number of seconds in 1 day to be

$$(1 \text{ day})\left(\frac{24 \text{ hr}}{1 \text{ day}}\right)\left(\frac{60 \text{ min}}{1 \text{ hr}}\right)\left(\frac{60 \text{ s}}{1 \text{ min}}\right) = 86,400. \text{ s}$$

All five digits in 86,400. are significant, since all the factors in the calculation are exact numbers.

EXAMPLE 2.19

How many centimeters are in 2.00 ft?

SOLUTION

The stepwise conversion of units from feet to centimeters may be done in this sequence. Convert feet to inches; then convert inches to centimeters:

Solution map: ft → in. → cm

The conversion factors needed are

$$\frac{12 \text{ in.}}{1 \text{ ft}} \quad \text{and} \quad \frac{2.54 \text{ cm}}{1 \text{ in.}}$$

$$(2.00 \text{ ft})\left(\frac{12 \text{ in.}}{1 \text{ ft}}\right) = 24.0 \text{ in.}$$

$$(24.0 \text{ in.})\left(\frac{2.54 \text{ cm}}{1 \text{ in.}}\right) = 61.0 \text{ cm}$$

NOTE   Since 1 ft and 12 in. are exact numbers, the number of significant figures allowed in the answer is three, based on the number 2.00.

EXAMPLE 2.20

How many meters are in a 100.-yd football field?

SOLUTION

The stepwise conversion of units from yards to meters may be done by this sequence, using the proper conversion factors:

Solution map: yd → ft → in. → cm → m

$$(100. \text{ yd})\left(\frac{3 \text{ ft}}{1 \text{ yd}}\right) = 300. \text{ ft}$$

$$(300. \text{ ft})\left(\frac{12 \text{ in.}}{1 \text{ ft}}\right) = 3600 \text{ in.}$$

$$(3600 \text{ in.})\left(\frac{2.54 \text{ cm}}{1 \text{ in.}}\right) = 9144 \text{ cm}$$

$$(9144 \text{ cm})\left(\frac{1 \text{ m}}{100 \text{ cm}}\right) = 91.4 \text{ m} \quad \text{(three significant figures)}$$

PRACTICE 2.11

(a) How many meters are in 10.5 miles?
(b) What is the area of a 6.0-in. × 9.0-in. rectangle in square meters?

Examples 2.19 and 2.20 may be solved using a linear expression and writing down conversion factors in succession. This method often saves one or two calculation steps and allows numerical values to be reduced to simpler terms, leading to simpler calculations. The single linear expressions for Examples 2.19 and 2.20 are

$$(2.00 \text{ ft})\left(\frac{12 \text{ in.}}{1 \text{ ft}}\right)\left(\frac{2.54 \text{ cm}}{1 \text{ in.}}\right) = 61.0 \text{ cm}$$

$$(100. \text{ yd})\left(\frac{3 \text{ ft}}{1 \text{ yd}}\right)\left(\frac{12 \text{ in.}}{1 \text{ ft}}\right)\left(\frac{2.54 \text{ cm}}{1 \text{ in.}}\right)\left(\frac{1 \text{ m}}{100 \text{ cm}}\right) = 91.4 \text{ m}$$

Using the units alone (Example 2.20), we see that the stepwise cancellation proceeds in succession until the desired unit is reached:

$$\text{yd} \times \frac{\text{ft}}{\text{yd}} \times \frac{\text{in.}}{\text{ft}} \times \frac{\text{cm}}{\text{in.}} \times \frac{\text{m}}{\text{cm}} = \text{m}$$

EXAMPLE 2.21

A driver of a car is obeying the speed limit of 55 miles per hour. How fast is the car traveling in kilometers per second?

SOLUTION

Several conversions are needed to solve this problem:

Solution map: mi. → km

hr → min → s

To convert mi → km,

$$\left(\frac{55 \text{ mi}}{\text{hr}}\right)\left(\frac{1.609 \text{ km}}{1 \text{ mi}}\right) = 88\frac{\text{km}}{\text{hr}}$$

Next we must convert hr → min → s. Notice that hours is in the denominator of our quantity, so the conversion factor must have hours in the numerator:

$$\left(\frac{88 \text{ km}}{\text{hr}}\right)\left(\frac{1 \text{ hr}}{60 \text{ min}}\right)\left(\frac{1 \text{ min}}{60 \text{ s}}\right) = 0.024\frac{\text{km}}{\text{s}}$$

PRACTICE 2.12

How many cubic meters of air are in a room measuring 8 ft × 10 ft × 12 ft?

EXAMPLE 2.22

Suppose four ostrich feathers weigh 1.00 lb. Assuming that each feather is equal in mass, how many milligrams does a single feather weigh?

SOLUTION

The unit conversion in this problem is from 1.00 lb/4 feathers to milligrams per feather. Since the unit *feathers* occurs in the denominator of both the starting unit and the desired unit, the unit conversions are

Solution map: lb → g → mg

$$\left(\frac{1.00 \text{ lb}}{4 \text{ feathers}}\right)\left(\frac{453.6 \text{ g}}{1 \text{ lb}}\right)\left(\frac{1000 \text{ mg}}{1 \text{ g}}\right) = \frac{113,400 \text{ mg}}{\text{feather}}(1.13 \times 10^5 \text{ mg/feather})$$

PRACTICE 2.13

You are traveling in Europe and wake up one morning to find your mass is 75.0 kg. Determine the American equivalent (in pounds) to see whether you need to go on a diet before you return home.

PRACTICE 2.14

A bottle of excellent Chianti holds 750 mL. What is its volume in quarts?

PRACTICE 2.15

Milk is often purchased by the half gallon. Determine the number of liters equal to this amount.

2.7 MEASUREMENT OF TEMPERATURE

LEARNING OBJECTIVE • Convert measurements among the Fahrenheit, Celsius, and Kelvin temperature scales.

KEY TERMS

thermal energy
temperature
heat

Thermal energy is a form of energy associated with the motion of small particles of matter. Depending on the amount of thermal energy present, a given system is said to be hot or cold. **Temperature** is a measure of the intensity of thermal energy, or how hot a system is, regardless of its size. The term **heat** refers to the flow of energy due to a temperature difference. Heat always flows from a region of higher temperature to one of lower temperature. The SI unit of temperature is the kelvin. The common laboratory instrument for measuring temperature is a thermometer (see **Figure 2.5**).

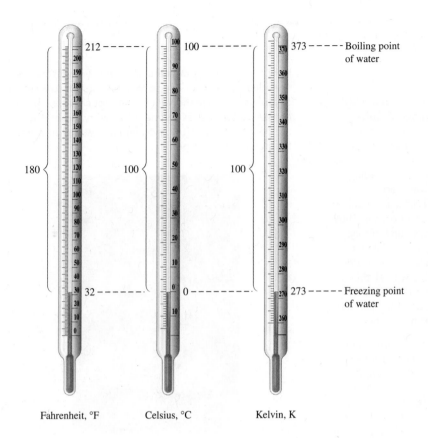

Figure 2.5

Comparison of Celsius, Kelvin, and Fahrenheit temperature scales.

The temperature of a system can be expressed by several different scales. Three commonly used temperature scales are Celsius (pronounced *sell-see-us*), Kelvin (also called absolute), and Fahrenheit. The unit of temperature on the Celsius and Fahrenheit scales is called a *degree*, but the size of the Celsius and the Fahrenheit degree is not the same. The symbol for the Celsius and Fahrenheit degrees is °, and it is placed as a superscript after the number and before the symbol for the scale. Thus, 10.0°C means 10.0 *degrees Celsius*. The degree sign is not used with Kelvin temperatures:

$$\text{degrees Celsius} = °C$$
$$\text{Kelvin (absolute)} = K$$
$$\text{degrees Fahrenheit} = °F$$

On the Celsius scale the interval between the freezing and boiling temperatures of water is divided into 100 equal parts, or degrees. The freezing point of water is assigned a temperature of 0°C and the boiling point of water a temperature of 100°C. The Kelvin temperature scale is known as the absolute temperature scale because 0 K is the lowest temperature theoretically attainable. The Kelvin zero is 273.15 kelvins below the Celsius zero. A kelvin is equal in size to a Celsius degree. The freezing point of water on the Kelvin scale is 273.15 K. The Fahrenheit scale has 180 degrees between the freezing and boiling temperatures of water. On this scale the freezing point of water is 32°F and the boiling point is 212°F:

$$0°C \cong 273 \, K \cong 32°F \qquad 100°C \cong 373 \, K \cong 212°F$$

The three scales are compared in Figure 2.5. Although absolute zero (0 K) is the lower limit of temperature on these scales, temperature has no upper limit. Temperatures of several million degrees are known to exist in the sun and in other stars.

By examining Figure 2.5, we can see that there are 100 Celsius degrees and 100 kelvins between the freezing and boiling points of water, but there are 180 Fahrenheit degrees between these two temperatures. Hence, the size of the Celsius degree and the kelvin are the same, but 1 Celsius degree is equal to 1.8 Fahrenheit degrees.

$$\frac{180}{100} = 1.8$$

>CHEMISTRY *IN ACTION*

Setting Standards

The kilogram is the standard base unit for mass in the SI system. But who decides just what a kilogram is? Before 1880, a kilogram was defined as the mass of a cubic decimeter of water. But this is difficult to reproduce very accurately since impurities and air are dissolved in water. The density of water also changes with temperature, leading to inaccuracy. So in 1885 a Pt-Ir cylinder was made with a mass of exactly 1 kilogram. This cylinder is kept in a vault at Sèvres, France, outside Paris, along with six copies. In 1889 the kilogram was determined to be the base unit of mass for the metric system and has been weighed against its copies three times in the past 100 years (1890, 1948, 1992) to make sure that it is accurate. The trouble with all this is that the cylinder itself can vary. Experts in the science of measurement want to change the definition of the kilogram to link it to a property of matter that does not change.

The kilogram is the only one of seven SI base units defined in terms of a physical object instead of a property of matter. The meter (redefined in 1983) is defined in terms of the speed of light (1 m = distance light travels in $\frac{1}{299,792,458}$ s). The second is defined in terms of the natural vibration of the cesium atom. Unfortunately, three of the SI base units (mole, ampere, and candela) depend on the definition of the kilogram. If the mass of the kilogram is uncertain, then all of these units are uncertain.

How else could we define a kilogram? Scientists have several ideas. They propose to fix the kilogram to a set value of a physical constant (which does not change). These constants connect experiment to theory. Right now they are continuing to struggle to find a new way to determine the mass of the kilogram.

Kilogram artifact in Sèvres, France.

From these data, mathematical formulas have been derived to convert a temperature on one scale to the corresponding temperature on another scale:

$$K = {}^\circ C + 273.15$$

$$\,^\circ F = (1.8 \times {}^\circ C) + 32$$

$$\,^\circ C = \frac{{}^\circ F - 32}{1.8}$$

WileyPLUS

ENHANCED EXAMPLE

EXAMPLE 2.23

The temperature at which table salt (sodium chloride) melts is 800.°C. What is this temperature on the Kelvin and Fahrenheit scales?

SOLUTION

To calculate K from °C, we use the formula

$$K = {}^\circ C + 273.15$$

$$K = 800.{}^\circ C + 273.15 = 1073 \text{ K}$$

To calculate °F from °C, we use the formula

$$\,^\circ F = (1.8 \times {}^\circ C) + 32$$

$$\,^\circ F = (1.8)(800.{}^\circ C) + 32$$

$$\,^\circ F = 1440 + 32 = 1472\,^\circ F$$

Summarizing our calculations, we see that

$$800.{}^\circ C = 1073 \text{ K} = 1472\,^\circ F$$

Remember that the original measurement of 800.°C was to the units place, so the converted temperature is also to the units place.

>CHEMISTRY *IN ACTION*

Taking the Temperature of Old Faithful

If you have ever struggled to take the temperature of a sick child, imagine the difficulty in taking the temperature of a geyser. Such are the tasks scientists set for themselves! In 1984, James A. Westphal and Susan W. Keiffer, geologists from the California Institute of Technology, measured the temperature and pressure inside Old Faithful during several eruptions in order to learn more about how a geyser functions. The measurements, taken at eight depths along the upper part of the geyser were so varied and complicated that the researchers returned to Yellowstone in 1992 to further investigate Old Faithful's structure and functioning. To see what happened between eruptions, Westphal and Keiffer lowered an insulated 2-in. video camera into the geyser. Keiffer had assumed that the vent (opening in the ground) was a uniform vertical tube, but this is not the case. Instead, the geyser appears to be an east–west crack in the Earth that extends downward at least 14 m. In some places it is over 1.8 m wide, and in other places it narrows to less than 15 cm. The walls of the vent contain many cracks, allowing water to enter at several depths. The complicated nature of the temperature data is explained by these cracks. Cool water enters the vent at depths of 5.5 m and 7.5 m. Superheated water and steam blast into the vent 14 m underground. According to Westphal, temperature increases of up to 130°C at the beginning of an eruption suggest that water and steam also surge into the vent from deeper geothermal sources.

During the first 20–30 seconds of an eruption, steam and boiling water shoot through the narrowest part of the vent at near the speed of sound. The narrow tube limits the rate at which the water can shoot from the geyser. When the pressure falls below a critical value, the process slows and Old Faithful begins to quiet down again.

The frequency of Old Faithful's eruptions is not on a precise schedule, but varies from 45 to 105 minutes—the average is about 79 minutes. The variations in time between eruptions depend on the amount of boiling water left in the fissure. Westphal says, "There's no real pattern except that a short eruption is always followed by a long one." Measurements of temperatures inside Old Faithful have given scientists a better understanding of what causes a geyser to erupt.

Old Faithful eruptions shoot into the air an average of 130 feet.

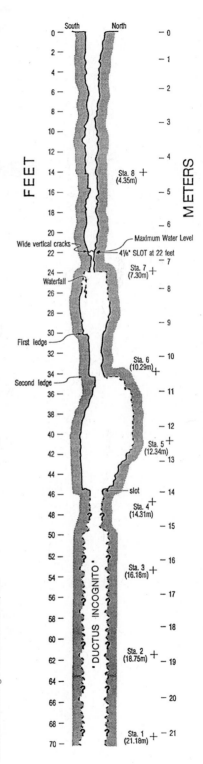

EXAMPLE 2.24

The temperature on December 1 in Honolulu, Hawaii, was 110.°F, a new record. Convert this temperature to °C.

SOLUTION

We use the formula

$$°C = \frac{°F - 32}{1.8}$$

$$°C = \frac{110. - 32}{1.8} = \frac{78}{1.8} = 43°C$$

EXAMPLE 2.25

What temperature on the Fahrenheit scale corresponds to −8.0°C? (Notice the negative sign in this problem.)

SOLUTION

$$°F = (1.8 \times °C) + 32$$
$$°F = (1.8)(-8.0) + 32 = -14 + 32$$
$$°F = 18°F$$

Temperatures used throughout this book are in degrees Celsius (°C) unless specified otherwise. The temperature after conversion should be expressed to the same precision as the original measurement.

PRACTICE 2.16

Helium boils at 4 K. Convert this temperature to °C and then to °F.

PRACTICE 2.17

"Normal" human body temperature is 98.6°F. Convert this to °C and K.

2.8 DENSITY

LEARNING OBJECTIVE • Solve problems involving density

KEY TERMS

density
specific gravity

Density (d) is the ratio of the mass of a substance to the volume occupied by that mass; it is the mass per unit of volume and is given by the equation

$$d = \frac{mass}{volume}$$

Density is a physical characteristic of a substance and may be used as an aid to its identification. When the density of a solid or a liquid is given, the mass is usually expressed in grams and the volume in milliliters or cubic centimeters.

$$d = \frac{mass}{volume} = \frac{g}{mL} \quad or \quad d = \frac{g}{cm^3}$$

Since the volume of a substance (especially liquids and gases) varies with temperature, it is important to state the temperature along with the density. For example, the volume of 1.0000 g of water at 4°C is 1.0000 mL; at 20°C, it is 1.0018 mL; and at 80°C, it is 1.0290 mL. Density therefore also varies with temperature.

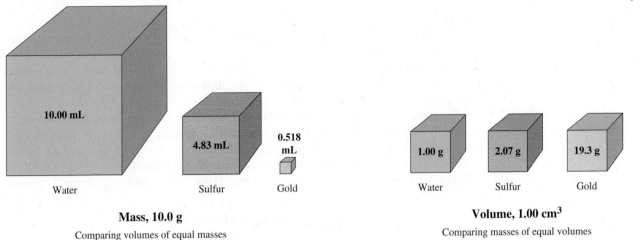

Figure 2.6
Comparison of mass and volume.

The density of water at 4°C is 1.0000 g/mL, but at 80°C the density of water is 0.9718 g/mL:

$$d^{4°C} = \frac{1.0000 \text{ g}}{1.0000 \text{ mL}} = 1.0000 \text{ g/mL}$$

$$d^{80°C} = \frac{1.0000 \text{ g}}{1.0290 \text{ mL}} = 0.97182 \text{ g/mL}$$

The density of iron at 20°C is 7.86 g/mL.

$$d^{20°C} = \frac{7.86 \text{ g}}{1.00 \text{ mL}} = 7.86 \text{ g/mL}$$

The densities of a variety of materials are compared in **Figure 2.6**.

Densities for liquids and solids are usually represented in terms of grams per milliliter (g/mL) or grams per cubic centimeter (g/cm³). The density of gases, however, is expressed in terms of grams per liter (g/L). Unless otherwise stated, gas densities are given for 0°C and 1 atmosphere pressure (discussed further in Chapter 13). **Table 2.6** lists the densities of some common materials.

TABLE 2.6   Densities of Some Selected Materials

Liquids and solids		Gases	
Substance	**Density (g/mL at 20°C)**	**Substance**	**Density (g/L at 0°C)**
Wood (Douglas fir)	0.512	Hydrogen	0.090
Ethyl alcohol	0.789	Helium	0.178
Vegetable oil	0.91	Methane	0.714
Water (4°C)	1.000*	Ammonia	0.771
Sugar	1.59	Neon	0.90
Glycerin	1.26	Carbon monoxide	1.25
Karo syrup	1.37	Nitrogen	1.251
Magnesium	1.74	Air	1.293*
Sulfuric acid	1.84	Oxygen	1.429
Sulfur	2.07	Hydrogen chloride	1.63
Salt	2.16	Argon	1.78
Aluminum	2.70	Carbon dioxide	1.963
Silver	10.5	Chlorine	3.17
Lead	11.34		
Mercury	13.55		
Gold	19.3		

*For comparing densities, the density of water is the reference for solids and liquids; air is the reference for gases.

When an insoluble solid object is dropped into water, it will sink or float, depending on its density. If the object is less dense than water, it will float, displacing a *mass* of water equal to the mass of the object. If the object is more dense than water, it will sink, displacing a *volume* of water equal to the volume of the object. This information can be used to determine the volume (and density) of irregularly shaped objects.

The **specific gravity** (sp gr) of a substance is the ratio of the density of that substance to the density of another substance, usually water at 4°C. Specific gravity has no units because the density units cancel. The specific gravity tells us how many times as heavy a liquid, a solid, or a gas is as compared to the reference material. Since the density of water at 4°C is 1.00 g/mL, the specific gravity of a solid or liquid is the same as its density in g/mL without the units.

$$\text{sp gr} = \frac{\text{density of a liquid or solid}}{\text{density of water}}$$

Sample calculations of density problems follow.

EXAMPLE 2.26

What is the density of a mineral if 427 g of the mineral occupy a volume of 35.0 mL?

SOLUTION

We need to solve for density, so we start by writing the formula for calculating density:

$$d = \frac{\text{mass}}{\text{volume}}$$

Then we substitute the data given in the problem into the equation and solve:

$$\text{mass} = 427 \text{ g} \qquad \text{volume} = 35.0 \text{ mL}$$

$$d = \frac{\text{mass}}{\text{volume}} = \frac{427 \text{ g}}{35.0 \text{ mL}} = 12.2 \text{ g/mL}$$

EXAMPLE 2.27

The density of gold is 19.3 g/mL. What is the mass of 25.0 mL of gold? Use density as a conversion factor, converting

Solution map: $\text{mL} \rightarrow \text{g}$

The conversion of units is

$$\text{mL} \times \frac{\text{g}}{\text{mL}} = \text{g}$$

$$(25.0 \text{ mL})\left(\frac{19.3 \text{ g}}{\text{mL}}\right) = 483 \text{ g}$$

EXAMPLE 2.28

Calculate the volume (in mL) of 100. g of ethyl alcohol.

SOLUTION

From Table 2.6 we see that the density of ethyl alcohol is 0.789 g/mL. This density also means that 1 mL of the alcohol has a mass of 0.789 g (1 mL/0.789 g).

For a conversion factor, we can use either

$$\frac{\text{g}}{\text{mL}} \quad \text{or} \quad \frac{\text{mL}}{\text{g}}$$

Solution map: g → mL,

$$(100. \,\cancel{g})\left(\frac{1 \text{ mL}}{0.789 \,\cancel{g}}\right) = 127 \text{ mL of ethyl alcohol}$$

EXAMPLE 2.29

The water level in a graduated cylinder stands at 20.0 mL before and at 26.2 mL after a 16.74-g metal bolt is submerged in the water. (a) What is the volume of the bolt? (b) What is the density of the bolt?

SOLUTION

(a) The bolt will displace a volume of water equal to the volume of the bolt. Thus, the increase in volume is the volume of the bolt:

$$
\begin{aligned}
26.2 \text{ mL} &= \text{volume of water plus bolt} \\
-20.0 \text{ mL} &= \text{volume of water} \\
\hline
6.2 \text{ mL} &= \text{volume of bolt}
\end{aligned}
$$

(b) $d = \dfrac{\text{mass of bolt}}{\text{volume of bolt}} = \dfrac{16.74 \text{ g}}{6.2 \text{ mL}} = 2.7 \text{ g/mL}$

PRACTICE 2.18

Pure silver has a density of 10.5 g/mL. A ring sold as pure silver has a mass of 18.7 g. When it is placed in a graduated cylinder, the water level rises 2.0 mL. Determine whether the ring is actually pure silver or whether the customer should contact the Better Business Bureau.

PRACTICE 2.19

The water level in a metric measuring cup is 0.75 L before the addition of 150. g of shortening. The water level after submerging the shortening is 0.92 L. Determine the density of the shortening.

CHAPTER 2 REVIEW

2.1 SCIENTIFIC NOTATION

- Quantitative observations consist of a number and a unit and are called measurements.
- Very large and very small numbers can be represented compactly by using scientific notation:
  - The number is represented as a decimal between 1 and 10 and is multiplied by 10 raised to the appropriate exponent.
  - The sign on the exponent is determined by the direction the decimal point is moved in the original number.

KEY TERMS
measurement
scientific notation

2.2 MEASUREMENT AND UNCERTAINTY

- All measurements reflect some amount of uncertainty, which is indicated by the number of significant figures in the measurement.
- The significant figures include all those known with certainty plus one estimated digit.

KEY TERM
significant figures

2.3 SIGNIFICANT FIGURES

KEY TERM

rounding off numbers

- Rules exist for counting significant figures in a measurement:
  - Nonzero numbers are always significant.
  - Exact numbers have an infinite number of significant figures.
  - The significance of a zero in a measurement is determined by its position within the number.
- Rules exist for rounding off the result of a calculation to the correct number of significant figures.
  - If the first number after the one you want to retain is 4 or less, that digit and all those after it are dropped.
  - If the first number after the one you want to retain is 5 or greater, that digit and all those after it are dropped and the last digit retained is increased by one.

2.4 SIGNIFICANT FIGURES IN CALCULATIONS

- The results of a calculation cannot be more precise than the least precise measurement.
- Rules exist for determining the correct number of significant figures in the result of a calculation.

2.5 THE METRIC SYSTEM

KEY TERMS

metric system or
    International System (SI)
meter (m)
conversion factor
solution map
mass
weight
kilogram (kg)
volume
liter (L)

- The metric system uses factors of 10 and a set of standard units for measurements.
- Length in the metric system is measured by the standard unit of the meter.
- The standard unit for mass in the metric system is the kilogram. In chemistry we often use the gram instead, as we tend to work in smaller quantities.
- Volume is the amount of space occupied by matter.
- The standard unit for volume is the cubic meter. In chemistry we usually use the volume unit of the liter or the milliliter.

2.6 DIMENSIONAL ANALYSIS: A PROBLEM-SOLVING METHOD

- Dimensional analysis is a common method used to convert one unit to another.
  - Conversion factors are used to convert one unit into another:

$$\text{unit}_1 \times \text{conversion factor} = \text{unit}_2$$

  - A solution map is used to outline the steps in the unit conversion.

2.7 MEASUREMENT OF TEMPERATURE

KEY TERMS

thermal energy
temperature
heat

- There are three commonly used temperature scales: Fahrenheit, Celsius, and Kelvin.
- We can convert among the temperature scales by using mathematical formulas:
  - $K = {}^\circ C + 273.15$
  - ${}^\circ F = (1.8 \times {}^\circ C) + 32$
  - ${}^\circ C = \dfrac{{}^\circ F - 32}{1.8}$

2.8 DENSITY

KEY TERMS

density
specific gravity

- The density of a substance is the amount of matter (mass) in a given volume of the substance:

$$d = \frac{\text{mass}}{\text{volume}}$$

- Specific gravity is the ratio of the density of a substance to the density of another reference substance (usually water).

REVIEW QUESTIONS

1. When a very large number is written in scientific notation, should the exponent be a positive or a negative number?
2. To write 635.2×10^{-8} in the proper scientific notation form, should the exponent increase or decrease?
3. Explain why the last digit in a measurement is uncertain.
4. How can the number 642,000 g be written to indicate that there are four significant figures?
5. How are significant zeroes identified?

6. State the rules used in this text for rounding off numbers.
7. Is it possible for the calculated answer from a multiplication or division problem to contain more significant figures than the measurements used in the calculation?
8. Is it possible for the calculated answer from an addition or subtraction problem to contain more significant figures than the measurements used in the calculation?
9. How many nanometers are in 1 cm? (Table 2.3)
10. How many milligrams are in 1 kg?
11. Why does an astronaut weigh more on Earth than in space when his or her mass remains the same in both places?
12. What is the relationship between milliliters and cubic centimeters?
13. What is the metric equivalent of 3.5 in.?
14. Distinguish between heat and temperature.
15. Compare the number of degrees between the freezing point of water and its boiling point on the Fahrenheit, Kelvin, and Celsius temperature scales. (Figure 2.5)

16. Describe the order of the following substances (top to bottom) if these three substances were placed in a 100-mL graduated cylinder: 25 mL glycerin, 25 mL mercury, and a cube of magnesium 2.0 cm on an edge. (Table 2.6)
17. Arrange these materials in order of increasing density: salt, vegetable oil, lead, and ethyl alcohol. (Table 2.6)
18. Ice floats in vegetable oil and sinks in ethyl alcohol. The density of ice must lie between what numerical values? (Table 2.6)
19. Distinguish between density and specific gravity.
20. Ice floats in water, yet ice is simply frozen water. If the density of water is 1.0 g/mL, how is this possible?
21. If you collect a container of oxygen gas, should you store it with the mouth up or down? Explain your answer. (Table 2.6)
22. Which substance has the greater volume; 25 g gold or 25 g silver? (Table 2.6)

Most of the exercises in this chapter are available for assignment via the online homework management program, WileyPLUS (www.wileyplus.com). All exercises with blue numbers have answers in Appendix VI.

PAIRED EXERCISES

1. What is the numerical meaning of each of the following?
   (a) kilogram
   (b) centimeter
   (c) microliter
   (d) millimeter
   (e) deciliter

2. What is the correct unit of measurement for each of the following?
   (a) 1000 m
   (b) 0.1 g
   (c) 0.000001 L
   (d) 0.01 m
   (e) 0.001 L

3. State the abbreviation for each of the following units:
   (a) gram
   (b) microgram
   (c) centimeter
   (d) micrometer
   (e) milliliter
   (f) deciliter

4. State the abbreviation for each of the following units:
   (a) milligram
   (b) kilogram
   (c) meter
   (d) nanometer
   (e) angstrom
   (f) microliter

5. Determine whether the zeros in each number are significant:
   (a) 2050
   (b) 9.00×10^2
   (c) 0.0530
   (d) 0.075
   (e) 300.
   (f) 285.00

6. Determine whether the zeros in each number are significant:
   (a) 0.005
   (b) 1500
   (c) 250.
   (d) 10.000
   (e) 6.070×10^4
   (f) 0.2300

7. How many significant figures are in each of the following numbers?
   (a) 0.025
   (b) 22.4
   (c) 0.0404
   (d) 5.50×10^3

8. State the number of significant figures in each of the following numbers:
   (a) 40.0
   (b) 0.081
   (c) 129,042
   (d) 4.090×10^{-3}

9. Round each of the following numbers to three significant figures:
   (a) 93.246
   (b) 0.02857
   (c) 4.644
   (d) 34.250

10. Round each of the following numbers to three significant figures:
   (a) 8.8726
   (b) 21.25
   (c) 129.509
   (d) 1.995×10^6

11. Express each of the following numbers in exponential notation:
   (a) 2,900,000
   (b) 0.587
   (c) 0.00840
   (d) 0.0000055

12. Write each of the following numbers in exponential notation:
   (a) 0.0456
   (b) 4082.2
   (c) 40.30
   (d) 12,000,000

13. Solve the following problems, stating answers to the proper number of significant figures:
   (a) $12.62 + 1.5 + 0.25 = ?$
   (b) $(2.25 \times 10^3)(4.80 \times 10^4) = ?$
   (c) $\dfrac{(452)(6.2)}{14.3} = ?$
   (d) $(0.0394)(12.8) = ?$
   (e) $\dfrac{0.4278}{59.6} = ?$
   (f) $10.4 + 3.75(1.5 \times 10^4) = ?$

14. Evaluate each of the following expressions. State the answer to the proper number of significant figures:
   (a) $15.2 - 2.75 + 15.67$
   (b) $(4.68)(12.5)$
   (c) $\dfrac{182.6}{4.6}$
   (d) $1986 + 23.84 + 0.012$
   (e) $\dfrac{29.3}{(284)(415)}$
   (f) $(2.92 \times 10^{-3})(6.14 \times 10^5)$

15. Change these fractions into decimals. Express each answer to three significant figures:

(a) $\dfrac{5}{6}$ (b) $\dfrac{3}{7}$ (c) $\dfrac{12}{16}$ (d) $\dfrac{9}{18}$

16. Change each of the following decimals to fractions in lowest terms:
(a) 0.25   (b) 0.625   (c) 1.67   (d) 0.8888

17. Solve each of these equations for x:

(a) $3.42x = 6.5$ (c) $\dfrac{0.525}{x} = 0.25$

(b) $\dfrac{x}{12.3} = 7.05$

18. Solve each equation for the variable:

(a) $x = \dfrac{212 - 32}{1.8}$ (c) $72°F = 1.8x + 32$

(b) $8.9\dfrac{g}{mL} = \dfrac{40.90\ g}{x}$

Unit Conversions

19. Complete the following metric conversions using the correct number of significant figures:
(a) 28.0 cm to m
(b) 1000. m to km
(c) 9.28 cm to mm
(d) 10.68 g to mg
(e) 6.8×10^4 mg to kg
(f) 8.54 g to kg
(g) 25.0 mL to L
(h) 22.4 L to μL

20. Complete the following metric conversions using the correct number of significant figures:
(a) 4.5 cm to Å
(b) 12 nm to cm
(c) 8.0 km to mm
(d) 164 mg to g
(e) 0.65 kg to mg
(f) 5.5 kg to g
(g) 0.468 L to mL
(h) 9.0 μL to mL

21. Complete the following American/metric conversions using the correct number of significant figures:
(a) 42.2 in. to cm
(b) 0.64 m to in.
(c) 2.00 in.^{2} to cm^2
(d) 42.8 kg to lb
(e) 3.5 qt to mL
(f) 20.0 L to gal

22. Make the following conversions using the correct number of significant figures:
(a) 35.6 m to ft
(b) 16.5 km to mi
(c) 4.5 in.^{3} to mm^3
(d) 95 lb to g
(e) 20.0 gal to L
(f) 4.5×10^4 ft^3 to m^3

23. After you have worked out at the gym on a stationary bike for 45 min, the distance gauge indicates that you have traveled 15.2 mi. What was your rate in km/hr?

24. A competitive college runner ran a 5-K (5.0-km) race in 15 min, 23 s. What was her pace in miles per hour?

25. A pharmacy technician is asked to prepare an antibiotic IV solution that will contain 500. mg of cephalosporin for every 100. mL of normal saline solution. The total volume of saline solution will be 1 L. How many grams of the cephalosporin will be needed for this IV solution?

26. An extra-strength aspirin tablet contains 0.500 g of the active ingredient, acetylsalicylic acid. Aspirin strength used to be measured in grains. If 1 grain = 60 mg, how many grains of the active ingredient are in 1 tablet? (Report your answer to three significant figures.)

27. The maximum speed recorded for a giant tortoise is 0.11 m/sec. How many miles could a gaint tortoise travel in 5.0 hr?

28. How many days would it take for a crepe myrtle tree to grow 1 cm in height if it grows 3.38 feet per year? (Report your answer to three significant figures.)

29. A personal trainer uses calipers on a client to determine his percent body fat. After taking the necessary measurements, the personal trainer determines that the client's body contains 11.2% fat by mass (11.2 lb of fat per 100 lb of body mass). If the client weighs 225 lb, how many kg of fat does he have?

30. The weight of a diamond is measured in carats. How many pounds does a 5.75-carat diamond weigh? (1 carat = 200. mg)

31. A competitive high school swimmer takes 52 s to swim 100. yards. What is his rate in m/min?

32. In 2005, Jarno Trulli was the pole winner of the U.S. Grand Prix Race with a speed of 133 mi per hr. What was his speed in cm/s?

33. In 2006, Christian Stengl climbed to the top of Mount Everest, elevation 29,035 ft, from a starting point of 21,002 ft in a record time of 16 hr, 42 min. Determine his average rate of climb in
(a) miles per minute   (b) meters per second

34. The *Alvin*, a submersible research vessel, can descend into the ocean to a depth of approximately 4500 m in just over 5 hr. Determine its average rate of submersion in
(a) feet per minute   (b) kilometers per second

35. The world's record for the largest cup of coffee was broken on October 15, 2010, with a 2010-gal cup of coffee in Las Vegas, Nevada. If a cup of coffee contains 473 mL of coffee, how many cups of coffee would be required to fill this coffee cup?

PRNewsFoto/GourmetGift Baskets.com/AP/Wide World Photos

36. Tilapia is rapidly becoming an important source of fish around the world. This is because it can be farmed easily and sustainably. In 2011, 475 million lb of tilapia were consumed by Americans. If the average tilapia has a mass of 535 g, how many tilapia were consumed in 2011?

37. Assuming that there are 20. drops in 1.0 mL, how many drops are in 1.0 gallon?

38. How many liters of oil are in a 42-gal barrel of oil?

39. Calculate the number of milliliters of water in a cubic foot of water.

40. Oil spreads in a thin layer on water called an "oil slick." How much area in m^2 will 200 cm^3 of oil cover if it forms a layer 0.5 nm thick?

41. A textbook is 27 cm long, 21 cm wide and 4.4 cm thick. What is the volume in:
(a) cubic centimeters?   (c) cubic inches?
(b) liters?

42. An aquarium measures 16 in. $\times$ 8 in. $\times$ 10 in. How many liters of water does it hold? How many gallons?

43. A toddler in Italy visits the family doctor. The nurse takes the child's temperature, which reads 38.8°C.
(a) Convert this temperature to °F.
(b) If 98.6°F is considered normal, does the child have a fever?

44. Driving to the grocery store, you notice the temperature is 45°C. Determine what this temperature is on the Fahrenheit scale and what season of the year it might be.

45. Make the following conversions and include an equation for each one:
(a) 162°F to °C   (c) −18°C to °F
(b) 0.0°F to K   (d) 212 K to °C

46. Make the following conversions and include an equation for each one:
(a) 32°C to °F   (c) 273°C to K
(b) −8.6°F to °C   (d) 100 K to °F

47. At what temperature are the Fahrenheit and Celsius temperatures exactly equal?

48. At what temperature are Fahrenheit and Celsius temperatures the same in value but opposite in sign?

49. The average temperature on Venus is 460°C. What is this temperature in °F?

50. The average temperature at the top of Jupiter's clouds is −244°F. What is this temperature in °C?

51. What is the density of a sample of 65.0 mL of automobile oil having a mass of 59.82 g?

52. A 25.2-mL sample of kerosene was determined to have a mass of 20.41 g. What is the density of kerosene?

53. A student weighed an empty graduated cylinder and found that it had a mass of 25.23 g. When filled with 25.0 mL of an unknown liquid, the total mass was 50.92 g. What is the density of the liquid?

54. A total of 32.95 g of mossy zinc was placed into a graduated cylinder containing 50.0 mL of water. The water level rose to 54.6 mL. Determine the density of the zinc.

55. Linseed oil has a density of 0.929 g/mL. How many mL are in 15 g of the oil?

56. Glycerol has a density of 1.20 g/mL. How many mL are in 75 g of glycerol?

ADDITIONAL EXERCISES

57. You calculate that you need 10.0123576 g of NaCl for an experiment. What amount should you measure out if the precision of the balance is
(a) + or − 0.01 g?   (c) + or − 0.0001 g?
(b) + or − 0.001 g?

58. Often small objects are measured by mass in order to count them.
(a) If the mass of 1 Skittle is 1.134, what mass of Skittles should be packaged in a bag containing 175 Skittles?
(b) The volume of exactly 6 Skittles is 5.3 mL as measured in a 10.0-mL graduated cylinder. What volume (in liters) of Skittles should be packaged in a bag containing 175 Skittles?
(c) Determine the number of Skittles expected in 325.0-g bag of Skittles.
(d) Determine the number of Skittles expected in a beaker containing 0.550 L of Skittles.
(e) Five bags were filled with 350.0 g of Skittles and 5 more bags were filled with 0.325 L of Skittles. The Skittles in each bag were then counted and the data are tabulated below:

Mass Skittles	Number of Skittles	Volume Skittles	Number of Skittles
350.0 g	310	0.325 L	392
350.0 g	313	0.325 L	378
350.0 g	308	0.325 L	401
350.0 g	309	0.325 L	369
350.0 g	312	0.325 L	382

Which measurement method is more accurate? More precise? Explain your reasoning.

59. Suppose you want to add 100 mL of solvent to a reaction flask. Which piece of glassware shown in Figure 2.3 would be the **best** choice for accomplishing this task and why?

60. A reaction requires 21.5 g of $CHCl_3$. No balance is available, so it will have to be measured by volume. How many mL of $CHCl_3$ need to be taken? (Density of $CHCl_3$ is 1.484 g/mL.)

61. A 25.27-g sample of pure sodium was prepared for an experiment. How many mL of sodium is this? (Density of sodium is 0.97 g/mL.)

62. In the United States, coffee consumption averages 4.00×10^8 cups per day. If each cup of coffee contains 160 mg caffeine, how many pounds of caffeine are being consumed each day?

63. In a role-playing video game (RPG) your character is a Human Paladin that can carry 115 lb of gear. Your character is carrying 92 lb of gear and a vial of strength potion (which allows you to carry an additional 50.0 lb of gear). If you find a cave filled with mass potions (used for resisting strong winds), after using the strength potion, how many vials can you collect if the vials each contain 50.0 mL of mass potion with a density of 193 g/mL? The vials have negligible mass.

64. A cape is designed for Lady Gaga's concert with 4560 sequins. If a sequin has a volume of 0.0241 cm^3 and the sequins have a density of 41.6 g/cm^3, what is the mass in lb and kg of sequins on Lady Gaga's cape?

65. Will a hollow cube with sides of length 0.50 m hold 8.5 L of solution? Depending on your answer, how much additional solution would be required to fill the container or how many times would the container need to be filled to measure the 8.5 L?

66. The accepted toxic dose of mercury is 300 µg/day. Dental offices sometimes contain as much as 180 µg of mercury per cubic meter of air. If a nurse working in the office ingests 2×10^4 L of air per day, is he or she at risk for mercury poisoning?

67. Hydrogen becomes a liquid at 20.27 K. What is this temperature in (a) °C?　　　　　(b) °F?

68. Scientists led by Rob Eagle at CalTech have examined 150-million-year-old fossilized teeth of sauropods, huge four-legged dinosaurs, to determine their average body temperature. They analyzed the type and number of carbon-oxygen bonds, leading them to conclude that the sauropods' internal temperature was between 36 and 38°C. Determine this temperatue range in °F and compare the body temperature to other modern animals. Which of the following animals do these dinosaurs most closely resemble?

Animal	Temperature
dogs	100.5–102.5°F
tropical fish	78°F
bottlenose dolphin	97–99°F
tortoises	78–82°F
humans	99°F
birds	105°F
cows	102°F

69. According to the National Heart, Lung, and Blood Institute, LDL-cholesterol levels of less than 130 mg of LDL-cholesterol per deciliter of blood are desirable for heart health in humans. On the average, a human has 4.7 L of whole blood. What is the maximum number if grams of LDL-cholesterol that a human should have?

70. You have been sent to buy gold from prospectors in the West. You have no balance but you do have a flask full of mercury. You have been told that many prospectors will bring you fool's gold or iron pyrite in the hopes that you cannot tell the difference. Given the following densities for gold, mercury, and iron pyrite, how could you determine whether the samples brought to you are really gold?

gold	18.3 g/mL
mercury	13.6 g/mL
iron pyrite	5.00 g/mL

71. The height of a horse is measured in hands (1 hand = exactly 4 in.). How many meters is a horse that measures 14.2 hands?

72. Camels have been reported to drink as much as 22.5 gal of water in 12 hr. How many liters can they drink in 30. days?

73. You are given three cubes, A, B, and C; one is magnesium, one is aluminum, and the third is silver. All three cubes have the same mass, but cube A has a volume of 25.9 mL, cube B has a volume of 16.7 mL, and cube C has a volume of 4.29 mL. Identify cubes A, B, and C.

74. When a chunk of wood burns, much more than just smoke is produced. In addition nanotubes made of pure carbon are formed that they have a structure resembling a roll of chicken wire. These nanotubes are stronger than steel, resistant to fire, and make good heat conductors. If a typical carbon nanotube has a diameter of 1.3 nm, how many nanotubes would need to be laid side by side to construct a bridge 40.0 feet wide?

75. In the United States, land is measured in acres. There are 43,560 ft^2 in each acre. How many km^2 are in 125 acres?

76. A cube of aluminum has a mass of 500. g. What will be the mass of a cube of gold of the same dimensions?

77. A 25.0-mL sample of water at 90°C has a mass of 24.12 g. Calculate the density of water at this temperature.

78. The mass of an empty container is 88.25 g. The mass of the container when filled with a liquid ($d = 1.25$ g/mL) is 150.50 g. What is the volume of the container?

79. Which liquid will occupy the greater volume, 50 g of water or 50 g of ethyl alcohol? Explain.

80. The Sacagawea gold-colored dollar coin has a mass of 8.1 g and contains 3.5% manganese. What is its mass in ounces (1 lb = 16 oz), and how many ounces of Mn are in this coin?

81. The density of sulfuric acid is 1.84 g/mL. What volume of this acid will weigh 100. g?

82. Dark-roasted coffee contains increased amonts of N-methylpyridinium, or NMP, a ringed compound that is not present in green unroasted coffee beans. This compound seems to decrease the stomach acid production normally associated with drinking coffee. In coffee made from dark-roasted coffee, the concentration of NMP is 31.4 mg/L. How many milligrams of NMP would you consume if you drank 2.00 large cups of dark-roasted coffee? (One coffee cup contains 10.0 fluid ounces of liquid.)

83. The density of palladium at 20°C is 12.0 g/mL, and at 1550°C the density is 11.0 g/mL. What is the change in volume (in mL) of 1.00 kg Pd in going from 20°C to 1550°C?

84. As a solid substance is heated, its volume increases, but its mass remains the same. Sketch a graph of density versus temperature showing the trend you expect. Briefly explain.

85. The first Apple computer had 5.0 Mbytes of storage space on its hard drive and the cost of this computer was $9995. An Apple iPad II has 64 Gbytes of storage for a cost of $699. Calculate the cost per byte for each of these two Apple products. Which is a better buy?

(a) Apple II computer

(b) iPad

Rama & Musée Bolo/Wikimedia/http://en.wikipedia.org/wiki/File: Apple-II.jpg

© Christine Glade/iStockphoto

86. A gold bullion dealer advertised a bar of pure gold for sale. The gold bar had a mass of 3300 g and measured 2.00 cm × 15.0 cm × 6.00 cm. Was the bar pure gold? Show evidence for your answer.

87. A 35.0-mL sample of ethyl alcohol (density 0.789 g/mL) is added to a graduated cylinder that has a mass of 49.28 g. What will be the mass of the cylinder plus the alcohol?

88. Several years ago, pharmacists used the Apothecary System of Measurement. In this system, 1 scruple is equal to 20 grains (gr). There are 480 gr in 1 oz, and in this system, there are 373 g in 12 oz. How many scruples would be in 695 g?

CHALLENGE EXERCISES

89. You have just purchased a 500-mL bottle of a decongestant medication. The doctor prescribed 2 teaspoons 4 times a day for 10 days. Have you purchased enough medication? (1 teaspoon (tsp) = 5 mL)

90. Your boss found a piece of metal in the lab and wants you to determine what the metal is. She is pretty sure that the metal is either lead, aluminum, or silver. The lab bench has a balance and a 100-mL graduated cylinder with 50 mL of water in it. You decide to weigh the metal and find that it has a mass of 20.25 g. After dropping the piece of metal into the graduated cylinder containing water, you observe that the volume increased to 57.5 mL. Identify the metal.

91. Forgetful Freddie placed 25.0 mL of a liquid in a graduated cylinder with a mass of 89.450 g when empty. When Freddie placed a metal slug with a mass of 15.454 g into the cylinder, the volume rose to 30.7 mL. Freddie was asked to calculate the density of the liquid and of the metal slug from his data, but he forgot to obtain the mass of the liquid. He was told that if he found the mass of the cylinder containing the liquid and the slug, he would have enough data for the calculations. He did so and found its mass to be 125.934 g. Calculate the density of the liquid and of the metal slug.

92. Neutrinos are subatomic particles with a very low mass. Recent work at CERN, Europe's particle-physics lab near Geneva, Switzerland, suggests that neutrinos may have the ability to travel faster than the speed of light. If the speed of light is 1.86×10^8 mi/hr, how many nanoseconds should it take for light to travel from CERN to the Gran Sasso National Lab in Italy, a 730.0-km journey? If a neutrino can travel the same distance 60 nsec faster, how many significant figures would you need to detect the difference in speed?

ANSWERS TO PRACTICE EXERCISES

2.1 (a) $1200 = 1.200 \times 10^3$
(left means positive exponent)

(b) $6,600,000 = 6.6 \times 10^6$
(left means positive exponent)

(c) $0.0468 = 4.68 \times 10^{-2}$
(right means negative exponent)

(d) $0.00003 = 3 \times 10^{-5}$
(right means negative exponent)

2.2 (a) the estimated digits are 4, 1, 0; (b) 3

2.3 (a) 2; (b) 4; (c) 4; (d) 1; (e) 3; (f) 4; (g) 1; (h) 3

2.4 (a) 42.25 (Rule 2); (b) 88.0 (Rule 1); (c) 0.0897 (Rule 2); (d) 0.090 (Rule 2); (e) 225 (Rule 1); (f) 14.2 (Rule 2)

2.5 (a) 3350 in.^{2} = 3.4×10^3 in.^{2}; (b) 50.7 mi/hr; (c) 0.79; (d) 1.3; (e) 20.5; (f) 3.71

2.6 (a) 2; (b) 2; (c) 1; (d) 2; (e) 4; (f) 2; (g) 2; (h) 2

2.7 (a) 0.567 m; (b) 680; (c) 0.125

2.8 (a) 0.560 g; (b) 0.525 kg; (c) 1.75×10^5 mg

2.9 (a) 0.025 L; (b) 1550 mL; (c) 45 dL

2.10 0.14 lb

2.11 (a) 1.69×10^4 m; (b) 3.5×10^{-2} m^2

2.12 30 m^3 or 3×10^1 m^3

2.13 165 lb

2.14 0.793 qt

2.15 1.89 L (the number of significant figures is arbitrary)

2.16 $-269°C$, $-452°F$

2.17 37.0°C, 310. K

2.18 The density is 9.35 g/mL; therefore the ring is not pure silver. The density of silver is 10.5 g/mL.

2.19 0.88 g/mL

This reclining Buddha in the Grand Palace in Bangkok, Thailand, is made of gold.

CHAPTER **3**

ELEMENTS AND COMPOUNDS

In Chapter 1 we learned that matter can be divided into the broad categories of pure substances and mixtures. We further learned that pure substances are either elements or compounds. The chemical elements are very important to us in our daily lives. In tiny amounts they play a large role in our health and metabolism. Metallic elements are used for the skin of airplanes, buildings, and sculpture. In this chapter we explore the nature of the chemical elements and begin to learn how chemists classify them.

CHAPTER OUTLINE

3.1 Elements
3.2 Introduction to the Periodic Table
3.3 Compounds and Formulas

3.1 ELEMENTS

Define an element and write the chemical symbol for an element when given its name.

LEARNING OBJECTIVE

KEY TERMS

element
atom
symbols

See the periodic table on the inside front cover.

All words in English are formed from an alphabet consisting of only 26 letters. All known substances on Earth—and most probably in the universe, too—are formed from a sort of "chemical alphabet" consisting of over 100 known elements. An **element** is a fundamental or elementary substance that cannot be broken down by chemical means to simpler substances. Elements are the building blocks of all substances. The elements are numbered in order of increasing complexity beginning with hydrogen, number 1. Of the first 92 elements, 88 are known to occur in nature. The other four—technetium (43), promethium (61), astatine (85), and francium (87)—either do not occur in nature or have only transitory existences during radioactive decay. With the exception of number 94, plutonium, elements above number 92 are not known to occur naturally but have been synthesized, usually in very small quantities, in laboratories. The discovery of trace amounts of element 94 (plutonium) in nature has been reported. No elements other than those on Earth have been detected on other bodies in the universe.

Most substances can be decomposed into two or more simpler substances. Water can be decomposed into hydrogen and oxygen. Sugar can be decomposed into carbon, hydrogen, and oxygen. Table salt is easily decomposed into sodium and chlorine. An element, however, cannot be decomposed into simpler substances by ordinary chemical changes.

If we could take a small piece of an element, say copper, and divide it and subdivide it into smaller and smaller particles, we would finally come to a single unit of copper that we could no longer divide and still have copper (see **Figure 3.1**). This smallest particle of an element that can exist is called an **atom**, which is also the smallest unit of an element that can enter into a chemical reaction. Atoms are made up of still smaller subatomic particles. However, these subatomic particles (described in Chapter 5) do not have the properties of elements.

Photodisc/Getty Images, Inc.

Figure 3.1
The surface of a penny is made up of tiny identical copper atoms packed tightly together.

Natural States of the Elements

Most substances around us are mixtures or compounds. Elements tend to be reactive, and they combine with other elements to form compounds. It is rare to find elements in nature in pure form. There are some exceptions, however. Gold, for example, can be found as nuggets. Silver and platinum can also be found in nature in pure form. In fact, these metals are sometimes called the *noble metals* since they have a low reactivity. The noble gases are also not reactive and can be found in nature in uncombined form. Helium gas, for example, consists of tiny helium atoms moving independently.

 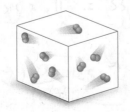

Nitrogen and oxygen gases are composed of molecules (N_2, ⬤) and (O_2, ⬤).

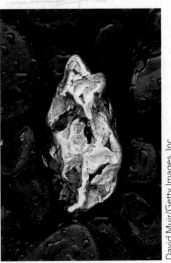

David Muir/Getty Images, Inc.

Gold is one of the few elements found in nature in an uncombined state.

Helium causes these balloons to float in air.

Air can also be divided into its component gases. It is mainly composed of nitrogen and oxygen gases. But when we "look inside" these gases, we find tiny molecules (N_2 and O_2) instead of independent atoms like we see in the noble gases.

Distribution of Elements

At the present time, 118 elements are known and only 88 of these occur naturally. At normal room temperature only two of the elements, bromine and mercury, are liquids (see Figure 3.2a). Eleven elements—hydrogen, nitrogen, oxygen, fluorine, chlorine, helium, neon, argon, krypton, xenon, and radon—are gases (see Figure 3.2b). The elements are distributed unequally in nature.

Ten elements make up about 99% of the mass of the Earth's crust, seawater, and atmosphere. The distribution of the elements is listed in order of their abundance in Table 3.1.

Oxygen, the most abundant of these elements, accounts for about 20% of the atmosphere and is found in virtually all rocks, sand, and soil. In these places, oxygen is not present as O_2 molecules but as part of compounds usually containing silicon and aluminum atoms. The mass percents given in Table 3.1 include the Earth's crust to a depth of 10 miles, the oceans, fresh

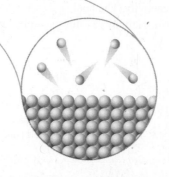

TABLE 3.1 Mass Percent of the Most Abundant Elements in the Earth's Crust, Oceans, and Atmosphere

Element	Mass percent	Element	Mass percent
Oxygen	49.2	Titanium	0.6
Silicon	25.7	Chlorine	0.19
Aluminum	7.5	Phosphorus	0.11
Iron	4.7	Manganese	0.09
Calcium	3.4	Carbon	0.08
Sodium	2.6	Sulfur	0.06
Potassium	2.4	Barium	0.04
Magnesium	1.9	Nitrogen	0.03
Hydrogen	0.9	Fluorine	0.03
		All others	0.49

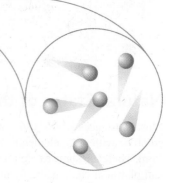

(b) Neon creates the light in these signs.

(a) Mercury is sometimes used in thermometers.

Figure 3.2

Examples of a liquid and a gaseous element.

water, and the atmosphere. It does not include the mantle and core of the Earth, which consist primarily of nickel and iron.

The list of elements found in living matter is very different from those of the Earth's crust, oceans, and atmosphere. The major elements in living organisms are shown in Table 3.2.

The major biologically important molecules are formed primarily from oxygen, carbon, hydrogen, and nitrogen. Some elements found in the body are crucial for life, although they are present in very tiny amounts (trace elements). Some of these trace elements include chromium, copper, fluorine, iodine, and selenium.

Names of the Elements

The names of the elements come to us from various sources. Many are derived from early Greek, Latin, or German words that describe some property of the element. For example, iodine is taken from the Greek word *iodes*, meaning "violetlike," and iodine is certainly violet in the vapor state. The name of the metal bismuth originates from the German words *weisse masse*, which means "white mass." Miners called it *wismat*; it was later changed to *bismat*, and finally to bismuth. Some elements are named for the location of their discovery—for example, germanium, discovered in 1886 by a German chemist. Others are named in commemoration of famous scientists, such as einsteinium and curium, named for Albert Einstein and Marie Curie, respectively.

Symbols of the Elements

We all recognize Mr., N.Y., and Ave. as abbreviations for mister, New York, and avenue, respectively. In a like manner, each element also has an abbreviation; these are called **symbols** of the elements. Fourteen elements have a single letter as their symbol, and the rest have two letters. A symbol stands for the element itself, for one atom of the element, and (as we shall see later) for a particular quantity of the element.

RULES FOR SYMBOLS OF ELEMENTS

1. Symbols have either one or two letters.

2. If one letter is used, it is capitalized.

3. If two letters are used, only the first is capitalized.

Examples: Iodine I   Barium Ba

The symbols and names of all the elements are given in the table on the inside front cover of this book. Table 3.3 lists the more commonly used elements and their symbols. Examine

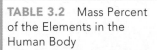

TABLE 3.2 Mass Percent of the Elements in the Human Body

Element	Mass percent
Oxygen	65
Carbon	18
Hydrogen	10
Nitrogen	3
Calcium	1.4
Phosphorus	1
Magnesium	0.5
Potassium	0.34
Sulfur	0.26
Sodium	0.14

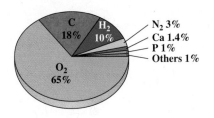

TABLE 3.3 Symbols of the Most Common Elements

Element	Symbol	Element	Symbol	Element	Symbol
Aluminum	Al	Gold	Au	Platinum	Pt
Antimony	Sb	Helium	He	Plutonium	Pu
Argon	Ar	Hydrogen	H	Potassium	K
Arsenic	As	Iodine	I	Radium	Ra
Barium	Ba	Iron	Fe	Silicon	Si
Bismuth	Bi	Lead	Pb	Silver	Ag
Boron	B	Lithium	Li	Sodium	Na
Bromine	Br	Magnesium	Mg	Strontium	Sr
Cadmium	Cd	Manganese	Mn	Sulfur	S
Calcium	Ca	Mercury	Hg	Tin	Sn
Carbon	C	Neon	Ne	Titanium	Ti
Chlorine	Cl	Nickel	Ni	Tungsten	W
Chromium	Cr	Nitrogen	N	Uranium	U
Cobalt	Co	Oxygen	O	Xenon	Xe
Copper	Cu	Palladium	Pd	Zinc	Zn
Fluorine	F	Phosphorus	P		

Aluminum is used for recyclable cans.

© Photononstop/SuperStock

Chunks of sulfur are transported in baskets from a mine in East Java.

TABLE 3.4   Symbols of the Elements Derived from Early Names*

Present name	Symbol	Former name
Antimony	Sb	Stibium
Copper	Cu	Cuprum
Gold	Au	Aurum
Iron	Fe	Ferrum
Lead	Pb	Plumbum
Mercury	Hg	Hydrargyrum
Potassium	K	Kalium
Silver	Ag	Argentum
Sodium	Na	Natrium
Tin	Sn	Stannum
Tungsten	W	Wolfram

*These symbols are in use today even though they do not correspond to the current name of the element.

Table 3.3 carefully and you will note that most of the symbols start with the same letter as the name of the element that is represented. A number of symbols, however, appear to have no connection with the names of the elements they represent (see Table 3.4). These symbols have been carried over from earlier names (usually in Latin) of the elements and are so firmly implanted in the literature that their use is continued today.

Special care must be taken in writing symbols. Capitalize only the first letter, and use a lowercase second letter if needed. This is important. For example, consider Co, the symbol for the element cobalt. If you write CO (capital C and capital O), you will have written the two elements carbon and oxygen (the *formula* for carbon monoxide), *not* the single element cobalt. Also, make sure that you write the letters distinctly; otherwise, Co (for cobalt) may be misread as Ca (for calcium).

Knowledge of symbols is essential for writing chemical formulas and equations and will be needed in the remainder of this book and in any future chemistry courses you may take. One

>CHEMISTRY *IN ACTION*

Naming Elements

Have you ever wondered where the names of the elements came from? Some of the elements were named for famous scientists or where they were first discovered and others for their characteristics. Some of the elemental symbols refer to old and rarely used names for the elements.

Elements named for places
- Americium (Am)—first created in America.
- Berkelium (Bk)—first created in Berkeley, Califorina.
- Francium (Fr)—discovered at Curie Institute in France.
- Scandium (Sc)—discovered and mined in Scandinavia.

Elements named for planets and the sun
- Helium (He)—named for the sun (helios) because this was the first place helium was detected.
- Uranium (U)—named for Uranus, which was discovered just before uranium.
- Plutonium (Pu)—named for the former planet Pluto.

Elements named for mythological characters
- Thorium (Th)—named after Thor, the Scandinavian god of war. Thorium is used as a fuel for nuclear weapons.

- Titanium (Ti)—named after Titans, supermen of Greek mythology. Titanium is a super-element because it is very resistant to acid.
- Vanadium (V)—named after the Scandinavian goddess of beauty, Vandis, because vanadium compounds form such beautiful colors.
- Tantalum (Ta) and Niobium (Nb)—named after the mythological Greek king, Tantalus, and his daughter, Princess Niobe. Like the king and his daughter, tantalum and niobium are often found together.

Elements named for their properties
- Chlorine (Cl)—from Greek *chloros* meaning "green." Chlorine is a greenish yellow gas.
- Iodine (I)—from Greek *iodos* meaning "violet."
- Argon (Ar)—from Greek *argos* meaning "lazy." Argon is an extremely unreactive gas, so it was thought too lazy to react.
- Cobalt (Co)—from Greek *kobold* meaning "goblin" or "evil spirit." Cobalt miners often died suddenly and unexpectedly, which was probably due to ingesting arsenic found with the cobalt ore in the German mines.

way to learn the symbols is to practice a few minutes a day by making flash cards of names and symbols and then practicing daily. Initially, it is a good plan to learn the symbols of the most common elements shown in Table 3.3.

EXAMPLE 3.1

Write the names and symbols for the elements that have only one letter as their symbol. (Use the periodic table on the inside front cover of your text.)

SOLUTION

Boron, B; Carbon, C; Fluorine, F; Hydrogen, H; Iodine, I; Nitrogen, N; Oxygen, O; Phosphorus, P; Potassium, K; Sulfur, S; Uranium, U; Vanadium, V; Tungsten, W; Yttrium, Y.

PRACTICE 3.1

Write the name, symbol, and vertical column location for the elements whose symbols begin with C. [Use the periodic table on the inside front cover of your text. (1A, 2A, etc.)]

WileyPLUS

ENHANCED EXAMPLE

3.2 INTRODUCTION TO THE PERIODIC TABLE

Explain the arrangement of the elements on the periodic table and classify elements as metal, nonmetal, or metalloid.

LEARNING OBJECTIVE

KEY TERMS

groups
noble gases
alkali metals
alkaline earth metals
halogens
representative elements
transition elements
metals
nonmetals
metalloids
diatomic molecules

Almost all chemistry classrooms have a chart called the *periodic table* hanging on the wall. It shows all the chemical elements and contains a great deal of useful information about them. As we continue our study of chemistry, we will learn much more about the periodic table. For now let's begin with the basics.

A simple version of the periodic table is shown in Table 3.5. Notice that in each box there is the symbol for the element and, above it, a number called the *atomic number*. For example nitrogen is $\begin{array}{c}7\\N\end{array}$ and gold is $\begin{array}{c}79\\Au\end{array}$.

TABLE 3.5   The Periodic Table

Noble gases

1 H																	2 He
3 Li	4 Be	Metals			Metalloids		Nonmetals					5 B	6 C	7 N	8 O	9 F	10 Ne
11 Na	12 Mg											13 Al	14 Si	15 P	16 S	17 Cl	18 Ar
19 K	20 Ca	21 Sc	22 Ti	23 V	24 Cr	25 Mn	26 Fe	27 Co	28 Ni	29 Cu	30 Zn	31 Ga	32 Ge	33 As	34 Se	35 Br	36 Kr
37 Rb	38 Sr	39 Y	40 Zr	41 Nb	42 Mo	43 Tc	44 Ru	45 Rh	46 Pd	47 Ag	48 Cd	49 In	50 Sn	51 Sb	52 Te	53 I	54 Xe
55 Cs	56 Ba	57 La*	72 Hf	73 Ta	74 W	75 Re	76 Os	77 Ir	78 Pt	79 Au	80 Hg	81 Tl	82 Pb	83 Bi	84 Po	85 At	86 Rn
87 Fr	88 Ra	89 Ac†	104 Rf	105 Db	106 Sg	107 Bh	108 Hs	109 Mt	110 Ds	111 Rg	112 Cn	113 Uut	114 Fl	115 Uup	116 Lv	117 Uus	118 Uuo

*	58 Ce	59 Pr	60 Nd	61 Pm	62 Sm	63 Eu	64 Gd	65 Tb	66 Dy	67 Ho	68 Er	69 Tm	70 Yb	71 Lu
†	90 Th	91 Pa	92 U	93 Np	94 Pu	95 Am	96 Cm	97 Bk	98 Cf	99 Es	100 Fm	101 Md	102 No	103 Lr

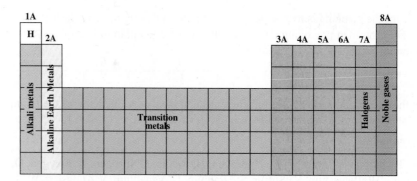

Figure 3.3

Some groups of elements have special names.

Noble Gases

2 He
10 Ne
18 Ar
36 Kr
54 Xe
86 Rn

The elements are placed in the table in order of increasing atomic number in a particular arrangement designed by Dimitri Mendeleev in 1869. His arrangement organizes the elements with similar chemical properties in columns called families or **groups**. An example of this is the column in the margin.

These elements are all gases and nonreactive. The group is called the **noble gases**. Other groups with special names are the **alkali metals** (under 1A on the table), **alkaline earth metals** (Group 2A), and **halogens** (Group 7A).

The tall columns of the periodic table (1A–7A and the noble gases) are known as the **representative elements**. Those elements in the center section of the periodic table are called **transition elements**. **Figure 3.3** shows these groups on a periodic table.

Metals, Nonmetals, and Metalloids

The elements can be classified as metals, nonmetals, and metalloids. Most of the elements are metals. We are familiar with them because of their widespread use in tools, construction materials, automobiles, and so on. But nonmetals are equally useful in our everyday life as major components of clothing, food, fuel, glass, plastics, and wood. Metalloids are often used in the electronics industry.

The **metals** are solids at room temperature (mercury is an exception). They have high luster, are good conductors of heat and electricity, are *malleable* (can be rolled or hammered into sheets), and are *ductile* (can be drawn into wires). Most metals have a high melting point and a high density. Familiar metals are aluminum, chromium, copper, gold, iron, lead, magnesium, mercury, nickel, platinum, silver, tin, and zinc. Less familiar but still important metals are calcium, cobalt, potassium, sodium, uranium, and titanium.

Metals have little tendency to combine with each other to form compounds. But many metals readily combine with nonmetals such as chlorine, oxygen, and sulfur to form compounds such as metallic chlorides, oxides, and sulfides. In nature, minerals are composed of the more reactive metals combined with other elements. A few of the less reactive metals such as copper, gold, and silver are sometimes found in a native, or free, state. Metals are often mixed with one another to form homogeneous mixtures of solids called alloys. Some examples are brass, bronze, steel, and coinage metals.

Nonmetals, unlike metals, are not lustrous, have relatively low melting points and densities, and are generally poor conductors of heat and electricity. Carbon, phosphorus, sulfur, selenium, and iodine are solids; bromine is a liquid; and the rest of the nonmetals are gases. Common nonmetals found uncombined in nature are carbon (graphite and diamond), nitrogen, oxygen, sulfur, and the noble gases (helium, neon, argon, krypton, xenon, and radon).

Nonmetals combine with one another to form molecular compounds such as carbon dioxide (CO_2), methane (CH_4), butane (C_4H_{10}), and sulfur dioxide (SO_2). Fluorine, the most reactive nonmetal, combines readily with almost all other elements.

Several elements (boron, silicon, germanium, arsenic, antimony, tellurium, and polonium) are classified as **metalloids** and have properties that are intermediate between those of metals and those of nonmetals. The intermediate position of these elements is shown in Table 3.5. Certain metalloids, such as boron, silicon, and germanium, are the raw materials for the semiconductor devices that make the electronics industry possible.

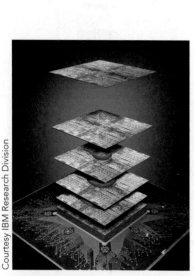

These tiny chips containing silicon, a metalloid, are being glued together to form a microprocessor many times more powerful than current chips.

EXAMPLE 3.2

Which of the following elements are not metals?

Na, Mo, Cl, S, Mg, Pt, Kr, I, C, Cu

SOLUTION

Nonmetals are generally found on the right-hand side of the periodic table: Cl, S, Kr, I, and C.

PRACTICE 3.2

Write the chemical symbols for
(a) five elements that are metals.
(b) two elements that are liquids at normal room temperature.
(c) five elements that are gases (not noble gases) at normal room temperature.

WileyPLUS

ENHANCED EXAMPLE

Diatomic Elements

Diatomic molecules each contain exactly two atoms (alike or different). Seven elements in their uncombined state are **diatomic molecules**. Their symbols, formulas, and brief descriptions are listed in Table 3.6. Whether found free in nature or prepared in the laboratory, the molecules of these elements always contain two atoms. The formulas of the free elements are therefore always written to show this molecular composition: H_2, N_2, O_2, F_2, Cl_2, Br_2, and I_2.

Container of hydrogen molecules.

TABLE 3.6   Elements That Exist as Diatomic Molecules

Element	Symbol	Molecular formula	Normal state
Hydrogen	H	H_2	Colorless gas
Nitrogen	N	N_2	Colorless gas
Oxygen	O	O_2	Colorless gas
Fluorine	F	F_2	Pale yellow gas
Chlorine	Cl	Cl_2	Greenish-yellow gas
Bromine	Br	Br_2	Reddish-brown liquid
Iodine	I	I_2	Bluish-black solid

It is important to see that symbols can designate either an atom or a molecule of an element. Consider hydrogen and oxygen. Hydrogen gas is present in volcanic gases and can be prepared by many chemical reactions. Regardless of their source, all samples of free hydrogen gas consist of diatomic molecules.

Free hydrogen is designated by the formula H_2, which also expresses its composition. Oxygen makes up about 20% by volume of the air that we breathe. This free oxygen is constantly being replenished by photosynthesis; it can also be prepared in the laboratory by several reactions. The majority of free oxygen is diatomic and is designated by the formula O_2. Now consider water, a compound designated by the formula H_2O (sometimes HOH). Water contains neither free hydrogen (H_2) nor free oxygen (O_2). The H_2 part of the formula H_2O simply indicates that two atoms of hydrogen are combined with one atom of oxygen to form water.

H_2 (gray) and O_2 (red) molecules.

Symbols are used to designate elements, show the composition of molecules of elements, and give the elemental composition of compounds.

Figure 3.4 summarizes the classification of elements.

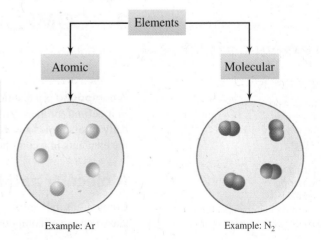

Figure 3.4

Elements fall into two broad categories.

Example: Ar

Example: N_2

>CHEMISTRY *IN ACTION*

Atomic Oxygen, Friend or Foe?

When scientists at NASA first began sending spacecraft into orbit, they discovered that the spacecraft were being corroded by the environment of the upper atmosphere. The culprit was the atomic oxygen formed by the interaction of ultraviolet radiation with oxygen gas, or O_2. Researchers using atomic oxygen to test the durability of satellite coatings found that atomic oxygen could remove carbon-based materials from the surface of objects without damaging them. Furthermore, NASA engineers discovered that atomic oxygen would not react with and damage glass composed of silicon dioxide because this silicon had already reacted with oxygen and would not react further. Armed with this knowledge, they began to coat the space station and shuttle parts with a thin layer of silicon dioxide to protect them from the corrosive effects of atomic oxygen.

Back in the lab, NASA researchers looked for new ways to utilize the power of atomic oxygen. They discovered several useful applications.

SURGICAL IMPLANTS When implanting foreign objects in a patient, a surgeon must be sure the parts are clean and sterile. It is easy to sterilize these parts and kill all bacteria, but occasionally there is organic debris left behind from the dead bacteria. This organic material can cause postoperative inflammation. Treatment of the sterile implants with atomic oxygen dissolves away the organic matter, leading to more successful and less painful recovery times.

CLEANING OF DAMAGED ART Artwork that has been damaged by soot can be cleaned using atomic oxygen. The oxygen atoms react with the soot and cause it to vaporize away. The chief conservator at the Cleveland Museum of Art tried the atomic oxygen treatment on two paintings damaged in a church fire. Although the paintings were not extremely valuable (and so were good subjects for an experiment), all other attempts to restore them had failed. Atomic oxygen proved to work wonders, and the soot and char were removed to reveal the original image. Because the treatment is a gas, the underlying layers were not harmed. The treatment doesn't work on everything and won't replace other techniques altogether, but the conservator was impressed enough to continue to work with NASA on the process.

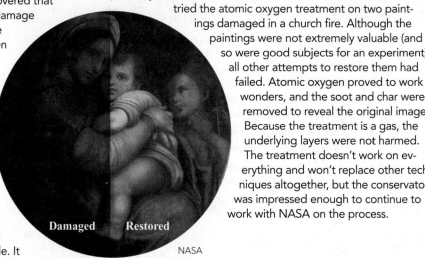

Damaged   Restored

NASA

PRACTICE 3.3

Identify the physical state of each of the following elements at room temperature (20°C):

H, Na, Ca, N, S, Fe, Cl, Br, Ne, Hg

Hint: You may need to use a resource (such as the Internet or a chemical handbook) to assist you.

PRACTICE 3.4

Identify each of the following elements as a nonmetal, metal, or metalloid:

Na, F, Cr, Mo, Kr, Si, Cu, Sb, I, S

3.3 COMPOUNDS AND FORMULAS

LEARNING OBJECTIVE ● Distinguish between molecular and ionic compounds and write chemical formulas for compounds.

KEY TERMS

compound
molecule
ion
cation
anion
chemical formula
subscripts
natural law
law of definite composition
law of multiple proportions

A **compound** is a distinct substance that contains two or more elements chemically combined in a definite proportion by mass. Compounds, unlike elements, can be decomposed chemically into simpler substances—that is, into simpler compounds and/or elements. Atoms of the elements in a compound are combined in whole-number ratios, never as fractional parts.

Molecular and Ionic Compounds

Compounds fall into two general types, *molecular* and *ionic*. **Figure 3.5** illustrates the classification of compounds.

A **molecule** is the smallest uncharged individual unit of a compound formed by the union of two or more atoms. Water is a typical molecular compound. If we divide a drop of water into smaller and smaller particles, we finally obtain a single molecule of water consisting of two hydrogen atoms bonded to one oxygen atom, as shown in **Figure 3.6**. This molecule is the ultimate particle of water; it cannot be further subdivided without destroying the water molecule and forming hydrogen and oxygen.

An **ion** is a positively or negatively charged atom or group of atoms. An ionic compound is held together by attractive forces that exist between positively and negatively charged ions. A positively charged ion is called a **cation** (pronounced *cat-eye-on*); a negatively charged ion is called an **anion** (pronounced *an-eye-on*).

Sodium chloride is a typical ionic compound. The ultimate particles of sodium chloride are positively charged sodium ions and negatively charged chloride ions, shown in Figure 3.6. Sodium chloride is held together in a crystalline structure by the attractive forces existing between these oppositely charged ions. Although ionic compounds consist of large aggregates of cations and anions, their formulas are normally represented by the simplest possible ratio of the atoms in the compound. For example, in sodium chloride the ratio is one sodium ion to one chloride ion, so the formula is $NaCl$.

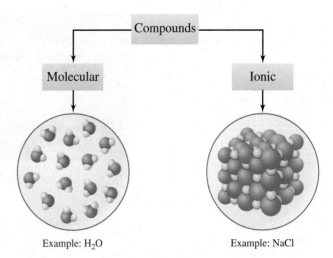

Figure 3.5

Compounds can be classified as molecular and ionic. Ionic compounds are held together by attractive forces between their positive and negative charges. Molecular compounds are held together by covalent bonds.

Two hydrogen atoms combined with an oxygen atom to form a molecule of water (H_2O).

A positively charged sodium ion and a negatively charged chloride ion form the compound sodium chloride ($NaCl$).

Figure 3.6

Representation of molecular and ionic (nonmolecular) compounds.

There are more than 50 million known registered compounds, with no end in sight as to the number that will be prepared in the future. Each compound is unique and has characteristic properties. Let's consider two compounds, water and sodium chloride, in some detail. Water is a colorless, odorless, tasteless liquid that can be changed to a solid (ice) at 0°C and to a gas (steam) at 100°C. Composed of two atoms of hydrogen and one atom of oxygen per molecule, water is 11.2% hydrogen and 88.8% oxygen by mass. Water reacts chemically with sodium to produce hydrogen gas and sodium hydroxide, with lime to produce calcium hydroxide, and with sulfur trioxide to produce sulfuric acid. When water is decomposed, it forms hydrogen and oxygen molecules (see **Figure 3.7**). No other compound has all these exact physical and chemical properties; they are characteristic of water alone.

Sodium chloride is a colorless crystalline substance with a ratio of one atom of sodium to one atom of chlorine. Its composition by mass is 39.3% sodium and 60.7% chlorine. It does not conduct electricity in its solid state; it dissolves in water to produce a solution that conducts electricity. When a current is passed through molten sodium chloride, solid sodium and

Figure 3.7

A representation of the decomposition of water into oxygen and hydrogen molecules.

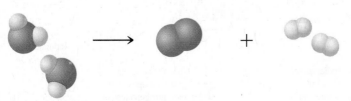

Water molecules $\longrightarrow$ Oxygen molecule   +   Hydrogen molecules

Sodium chloride $\longrightarrow$ Sodium metal   +   Chlorine gas

Figure 3.8

When sodium chloride is decomposed, it forms sodium metal and chlorine gas.

gaseous chlorine are produced (see **Figure 3.8**). These specific properties belong to sodium chloride and to no other substance. Thus, a compound may be identified and distinguished from all other compounds by its characteristic properties. We consider these chemical properties further in Chapter 4.

Writing Formulas of Compounds

Chemical formulas are used as abbreviations for compounds. A **chemical formula** shows the symbols and the ratio of the atoms of the elements in a compound. Sodium chloride contains one atom of sodium per atom of chlorine; its formula is NaCl. The formula for water is H_2O; it shows that a molecule of water contains two atoms of hydrogen and one atom of oxygen.

The formula of a compound tells us which elements it is composed of and how many atoms of each element are present in a formula unit. For example, a unit of sulfuric acid is composed of two atoms of hydrogen, one atom of sulfur, and four atoms of oxygen. We could express this compound as HHSOOOO, but this is cumbersome, so we write H_2SO_4 instead. The formula may be expressed verbally as "H-two-S-O-four." Numbers that appear partially below the line and to the right of a symbol of an element are called **subscripts**. Thus, the 2

Figure 3.9

Explanation of the formulas NaCl, H_2SO_4, and $Ca(NO_3)_2$.

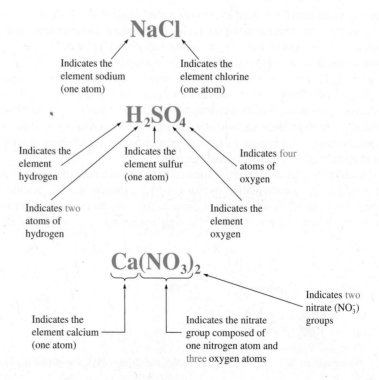

and the 4 in H_2SO_4 are subscripts (see **Figure 3.9**). Characteristics of chemical formulas are as follows:

1. The formula of a compound contains the symbols of all the elements in the compound.

2. When the formula contains one atom of an element, the symbol of that element represents that one atom. The number 1 is not used as a subscript to indicate one atom of an element.

3. When the formula contains more than one atom of an element, the number of atoms is indicated by a subscript written to the right of the symbol of that atom. For example, the 2 in H_2O indicates two atoms of H in the formula.

4. When the formula contains more than one of a group of atoms that occurs as a unit, parentheses are placed around the group, and the number of units of the group is indicated by a subscript placed to the right of the parentheses. Consider the nitrate group, NO_3^-. The formula for sodium nitrate, $NaNO_3$, has only one nitrate group, so no parentheses are needed. Calcium nitrate, $Ca(NO_3)_2$, has two nitrate groups, as indicated by the use of parentheses and the subscript 2. $Ca(NO_3)_2$ has a total of nine atoms: one Ca, two N, and six O atoms. The formula $Ca(NO_3)_2$ is read as "C-A [pause] N-O-three taken twice."

5. Formulas written as H_2O, H_2SO_4, $Ca(NO_3)_2$, and $C_{12}H_{22}O_{11}$ show only the number and kind of each atom contained in the compound; they do not show the arrangement of the atoms in the compound or how they are chemically bonded to one another.

PRACTICE 3.5

How would you read these formulas aloud?
(a) KBr   (b) $PbCl_2$ (c) $CaCO_3$ (d) $Mg(OH)_2$

EXAMPLE 3.3

Write formulas for the following compounds; the atomic composition is given.

(a) hydrogen chloride: 1 atom hydrogen + 1 atom chlorine;
(b) methane: 1 atom carbon + 4 atoms hydrogen;
(c) glucose: 6 atoms carbon + 12 atoms hydrogen + 6 atoms oxygen.

SOLUTION

(a) First write the symbols of the atoms in the formula: H Cl.
    Since the ratio of atoms is one to one, we bring the symbols together to give the formula for hydrogen chloride as HCl.
(b) Write the symbols of the atoms: C H.
    Now bring the symbols together and place a subscript 4 after the hydrogen atom. The formula is CH_4.
(c) Write the symbols of the atoms: C H O.
    Now write the formula, bringing together the symbols followed by the correct subscripts according to the data given (six C, twelve H, six O). The formula is $C_6H_{12}O_6$.

PRACTICE 3.6

Write formulas for the following compounds from the compositions given.

(a) potassium bromide: 1 atom potassium + 1 atom bromine
(b) sodium chlorate: 1 atom sodium + 1 atom chlorine + 3 atoms oxygen
(c) sulfurous acid: 2 atoms hydrogen + 1 atom sulfur + 3 atoms oxygen
(d) aluminum carbonate: 2 aluminum atoms + 3 carbonate ions

Composition of Compounds

A large number of experiments extending over a long period have established the fact that a particular compound always contains the same elements in the same proportions by mass. For example, water always contains 11.2% hydrogen and 88.8% oxygen by mass

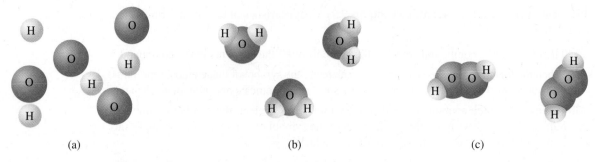

Figure 3.10

(a) Dalton's atoms were individual particles, the atoms of each element being alike in mass and size but different in mass and size from other elements. (b) and (c) Dalton's atoms combine in specific ratios to form compounds.

(see **Figure 3.10**). The fact that water contains hydrogen and oxygen in this particular ratio does not mean that hydrogen and oxygen cannot combine in some other ratio but rather that a compound with a different ratio would not be water. In fact, hydrogen peroxide is made up of two atoms of hydrogen and two atoms of oxygen per molecule and contains 5.9% hydrogen and 94.1% oxygen by mass; its properties are markedly different from those of water (see Figure 3.10).

	Water	**Hydrogen peroxide**
	11.2% H	5.9% H
	88.8% O	94.1% O
Atomic composition	2 H + 1 O	2 H + 2 O

We often summarize our general observations regarding nature into a statement called a **natural law**. In the case of the composition of a compound, we use the **law of definite composition**, which states that a compound always contains two or more elements chemically combined in a definite proportion by mass.

Let's consider two elements, oxygen and hydrogen, that form more than one compound. In water, 8.0 g of oxygen are present for each gram of hydrogen. In hydrogen peroxide, 16.0 g of oxygen are present for each gram of hydrogen. The masses of oxygen are in the ratio of small whole numbers, 16 : 8 or 2 : 1. Hydrogen peroxide has twice as much oxygen (by mass) as does water. Using Dalton's atomic model, we deduce that hydrogen peroxide has twice as many oxygen atoms per hydrogen atom as water. In fact, we now write the formulas for water as H_2O and for hydrogen peroxide as H_2O_2. See Figure 3.10b and c.

The **law of multiple proportions** states that atoms of two or more elements may combine in different ratios to produce more than one compound.

Some examples of the law of multiple proportions are given in Table 3.7. The reliability of this law and the law of definite composition is the cornerstone of the science of chemistry. In essence, these laws state that (1) the composition of a particular substance will always be the same no matter what its origin or how it is formed, and (2) the composition of different compounds formed from the same elements will always be unique.

TABLE 3.7   Selected Compounds Showing Elements That Combine to Give More Than One Compound

Compound	Formula	Percent composition
Copper(I) chloride	CuCl	64.2% Cu, 35.8% Cl
Copper(II) chloride	$CuCl_2$	47.3% Cu, 52.7% Cl
Methane	CH_4	74.9% C, 25.1% H
Octane	C_8H_{18}	85.6% C, 14.4% H
Methyl alcohol	CH_4O	37.5% C, 12.6% H, 49.9% O
Ethyl alcohol	C_2H_6O	52.1% C, 13.1% H, 34.7% O
Glucose	$C_6H_{12}O_6$	40.0% C, 6.7% H, 53.3% O

You need to recognize the difference between a *law* and a *model* (*theory*). A law is a summary of observed behavior. A model (theory) is an attempt to explain the observed behavior. This means that laws remain constant—that is, they do not undergo modification—while theories (models) sometimes fail and are modified or discarded over time.

CHAPTER **3 REVIEW**

3.1 ELEMENTS

- All matter consists of about 100 elements.
- An element is a fundamental chemical substance.
- The smallest particle of an element is an atom.
- Elements cannot be broken down by chemical means to a simpler substance.
- Most elements are found in nature combined with other elements.
- Elements that are found in uncombined form in nature include gold, silver, copper, and platinum as well as the noble gases (He, Ne, Ar, Kr, Xe, Rn).
- Chemical elements are not distributed equally in nature.
- Hydrogen is the most abundant element in the universe.
- Oxygen is the most abundant element on the Earth and in the human body.
- Names for the chemical elements come from a variety of sources, including Latin, location of discovery, and famous scientists.
- Rules for writing symbols for the elements are:
  - One or two letters
  - If one letter, use a capital
  - If two letters, only the first is a capital

KEY TERMS

element
atom
symbols

3.2 INTRODUCTION TO THE PERIODIC TABLE

- The periodic table was designed by Dimitri Mendeleev and arranges the elements according to their atomic numbers and in groups by their chemical properties.
- Elements can be classified as representative or as transition elements.
- Elements can also be classified by special groups with similar chemical properties. Such groups include the noble gases, alkali metals, alkaline earth metals, and halogens.
- Elements can be classified as metals, nonmetals, or metalloids.
  - Most elements are metals.
  - Metals have the following properties:
    - High luster
    - Good conductors of heat and electricity
    - Malleable
  - Nonmetals have the following properties:
    - Not lustrous
    - Poor conductors of heat and electricity
- Diatomic molecules contain exactly two atoms (alike or different).
- Seven elements exist as diatomic molecules—H_2, N_2, O_2, F_2, Cl_2, Br_2, and I_2.

KEY TERMS

groups
noble gases
alkali metals
alkaline earth metals
halogens
representative elements
transition elements
metals
nonmetals
metalloids
diatomic molecules

```
              ┌──── Elements ────┐
              │                  │
          ┌───────┐          ┌──────────┐
          │ Atomic │          │ Molecular │
          └───────┘          └──────────┘
              │                  │
```

Example: Ar   Example: N_2

3.3 COMPOUNDS AND FORMULAS

KEY TERMS

compound
molecule
ion
cation
anion
chemical formula
subscripts
natural law
law of definite composition
law of multiple proportions

- A compound is a substance that contains two or more elements chemically combined in a definite proportion by mass.
- There are two general types of compounds:
  - Molecular—formed of individual molecules composed of atoms
  - Ionic—formed from ions that are either positive or negative
    - Cation—positively charged ion
    - Anion—negatively charged ion
- A chemical formula shows the symbols and the ratios of atoms for the elements in a chemical compound.
- Characteristics of a chemical formula include:
  - It contains symbols of all elements in the compound.
  - The symbol represents one atom of the element.
  - If more than one atom of an element is present, the number of atoms is indicated by a subscript.
  - Parentheses are used to show multiple groups of atoms occurring as a unit in the compound.
  - A formula does not show the arrangement of the atoms in the compound.
- The law of definite composition states that a compound always contains two or more elements combined in a definite proportion by mass.
- The law of multiple proportions states that atoms of two or more elements may combine in different ratios to form more than one compound.

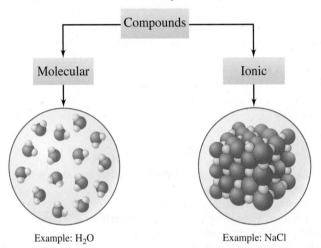

Example: H_2O Example: NaCl

REVIEW QUESTIONS

1. Are there more atoms of silicon or hydrogen in the Earth's crust, seawater, and atmosphere? Use Table 3.1 and the fact that the mass of a silicon atom is about 28 times that of a hydrogen atom.
2. Give the symbols for each of the following elements:
   (a) manganese   (e) chlorine
   (b) fluorine   (f) vanadium
   (c) sodium   (g) zinc
   (d) helium   (h) nitrogen
3. Give the names for each of the following elements:
   (a) Fe   (e) Be
   (b) Mg   (f) Co
   (c) C   (g) Ar
   (d) P   (h) Hg
4. What does the symbol of an element stand for?
5. List six elements and their symbols in which the first letter of the symbol is different from that of the name. (Table 3.4)
6. Write the names and symbols for the 14 elements that have only one letter as their symbol. (See periodic table on inside front cover.)
7. Of the six most abundant elements in the human body, how many are metals? nonmetals? metalloids? (Table 3.2)

8. Write the names and formulas for the elements that exist as diatomic molecules. (Table 3.6)
9. Interpret the difference in meaning for each pair:
   (a) CO and Co   (c) S_8 and 8 S
   (b) H_2 and 2 H   (d) CS and Cs
10. Distinguish between an element and a compound.
11. How many metals are there? nonmetals? metalloids? (Table 3.5)
12. Of the ten most abundant elements in the Earth's crust, seawater, and atmosphere, how many are metals? nonmetals? metalloids? (Table 3.1)
13. Give the names of (a) the solid diatomic nonmetal and (b) the liquid diatomic nonmetal. (Table 3.6)
14. Distinguish between a compound and a mixture.
15. What are the two general types of compounds? How do they differ from each other?
16. What is the basis for distinguishing one compound from another?
17. What is the major difference between a cation and an anion?

Most of the exercises in this chapter are available for assignment via the online homework management program, WileyPLUS (www.wileyplus.com). All exercises with blue numbers have answers in Appendix VI.

PAIRED EXERCISES

1. Which of the following are diatomic molecules?
   (a) HCl
   (b) O_2
   (c) N_2O
   (d) P_4
   (e) SiO_2
   (f) H_2O_2
   (g) CH_4
   (h) ClF

2. Which of the following are diatomic molecules?
   (a) O_3
   (b) H_2O
   (c) CO_2
   (d) HI
   (e) S_8
   (f) Cl_2
   (g) CO
   (h) NH_3

3. What elements are present in each compound?
   (a) potassium iodide KI
   (b) sodium carbonate Na_2CO_3
   (c) aluminum oxide Al_2O_3
   (d) calcium bromide $CaBr_2$
   (e) acetic acid $HC_2H_3O_2$

4. What elements are present in each compound?
   (a) magnesium bromide $MgBr_2$
   (b) carbon tetrachloride CCl_4
   (c) nitric acid HNO_3
   (d) barium sulfate $BaSO_4$
   (e) aluminum phosphate $AlPO_4$

5. Write the formula for each compound (the composition is given after each name):
   (a) zinc oxide   1 atom Zn, 1 atom O
   (b) potassium chlorate   1 atom K, 1 atom Cl, 3 atoms O
   (c) sodium hydroxide   1 atom Na, 1 atom O, 1 atom H
   (d) ethyl alcohol   2 atoms C, 6 atoms H, 1 atom O

6. Write the formula for each compound (the composition is given after each name):
   (a) aluminum bromide   1 atom Al, 3 atoms Br
   (b) calcium fluoride   1 atom Ca, 2 atoms F
   (c) lead(II) chromate   1 atom Pb, 1 atom Cr, 4 atoms O
   (d) benzene   6 atoms C, 6 atoms H

7. Many foods contain interesting compounds that give them their flavors and smells. Some of these compounds are listed below with the chemical composition given after each name. Write the formulas for these compounds.
   (a) Allicin gives garlic its flavor (6 carbons, 10 hydrogens, 1 oxygen, and 2 sulfurs).
   (b) Capsaicin gives peppers their heat (18 carbons, 27 hydrogens, 1 nitrogen, and 3 oxygens).
   (c) Limonene gives oranges their fragrance (10 carbons and 16 hydrogens).

8. Many plants contain interesting compounds that sometimes have medicinal properties. Some of these compounds are listed below with the chemical composition given after each name. Write the formulas for these compounds.
   (a) Aescin from horse chestnuts has anti-inflammatory properties (55 carbons, 86 hydrogens, and 24 oxygens).
   (b) Proanthocyanidins found in cranberries help to prevent urinary tract infections. Proanthocyanidins are polymers composed of epicatechin units (15 carbons, 14 hydrogens, and 6 oxygens).
   (c) Betulinic acid found in the common birch tree is an antimalarial drug (30 carbons, 48 hydrogens, and 3 oxygens).

9. Write the name and number of atoms of each element in each of the following compounds:
   (a) Fe_2O_3
   (b) $Ca(NO_3)_2$
   (c) $Co(C_2H_3O_2)_2$
   (d) CH_3COCH_3
   (e) K_2CO_3
   (f) $Cu_3(PO_4)_2$
   (g) C_2H_5OH
   (h) $Na_2Cr_2O_7$

10. Write the name and number of atoms of each element in each of the following compounds:
    (a) $HC_2H_3O_2$
    (b) $(NH_4)_3PO_4$
    (c) $Mg(HSO_3)_2$
    (d) $ZnCl_2$
    (e) $NiCO_3$
    (f) $KMnO_4$
    (g) $CH_3CH_2CH_2CH_3$
    (h) $PbCrO_4$

11. How many total atoms are represented in each formula?
    (a) $Co(ClO_3)_2$
    (b) $(NH_4)_2SO_3$
    (c) CH_3CH_2COOH
    (d) $C_{12}H_{22}O_{11}$

12. How many total atoms are represented in each formula?
    (a) CH_3CH_2OH
    (b) $KAl(SO_4)_2$
    (c) $NH_4C_2H_3O_2$
    (d) $C_6H_4Cl_2$

13. How many hydrogen atoms are represented in each formula?
    (a) $Al(C_2H_3O_2)_3$
    (b) $CH_3CH_2CH_2OH$
    (c) NH_4OH
    (d) $C_6H_5CH_2CH_3$

14. How many oxygen atoms are represented in each formula?
    (a) $Fe(C_2H_3O_2)_3$
    (b) H_3PO_4
    (c) $Ba(ClO_3)_2$
    (d) $Fe_2(Cr_2O_7)_3$

15. Determine whether each of the following is a pure substance or a mixture:
    (a) hot tea
    (b) beach sand
    (c) carbon dioxide
    (d) cement
    (e) zinc
    (f) vinegar

16. Determine whether each of the following is a pure substance or a mixture:
    (a) dirt
    (b) salad dressing
    (c) tungsten
    (d) dinitrogen monoxide
    (e) brass
    (f) egg

17. For Question 15, state whether each pure substance is an element or a compound.

18. For Question 16, state whether each pure substance is an element or a compound.

19. Classify each of the following as an element, a compound, or a mixture:

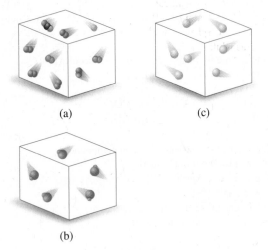

(a)

(c)

(b)

20. Classify each of the following as an element, a compound, or a mixture:

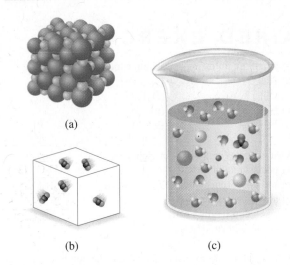

(a)

(b)

(c)

21. Is there a pattern to the location of the gaseous elements on the periodic table? If so, describe it.

22. Is there a pattern to the location of the liquid elements on the periodic table? If so, describe it.

23. What percent of the first 36 elements on the periodic table are metals?

24. What percent of the first 36 elements on the periodic table are solids at room temperature?

25. In Section 3.3, there is a statement about the composition of water. It says that water (H_2O) contains 8 grams of oxygen for every 1 gram of hydrogen. Show why this statement is true.

26. In Section 3.3, there is a statement about the composition of hydrogen peroxide. It says that hydrogen peroxide (H_2O_2) contains 16 grams of oxygen for every 1 gram of hydrogen. Show why this statement is true.

ADDITIONAL EXERCISES

27. You accidentally poured salt into your large grind pepper shaker. This is the only pepper that you have left and you need it to cook a meal, but you don't want the salt to be mixed in. How can you successfully separate these two components?

28. On the periodic table at the front of this book, do you notice anything about the atoms that make up the following ionic compounds: NaCl, KI, and $MgBr_2$? (*Hint*: Look at the position of the atoms in the given compounds on the periodic table.)

29. How many total atoms are present in each of the following compounds?
    (a) CO
    (b) BF_3
    (c) HNO_3
    (d) $KMnO_4$
    (e) $Ca(NO_3)_2$
    (f) $Fe_3(PO_4)_2$

30. The formula for vitamin B_{12} is $C_{63}H_{88}CoN_{14}O_{14}P$.
    (a) How many atoms make up one molecule of vitamin B_{12}?
    (b) What percentage of the total atoms are carbon?
    (c) What fraction of the total atoms are metallic?

31. It has been estimated that there is 4×10^{-4} mg of gold per liter of seawater. At a price of $19.40/g, what would be the value of the gold in 1 km^3(1×10^{15} cm^3) of the ocean?

32. Calcium dihydrogen phosphate is an important fertilizer. How many atoms of hydrogen are there in ten formula units of $Ca(H_2PO_4)_2$?

33. How many total atoms are there in one molecule of $C_{145}H_{293}O_{168}$?

34. Name the following:
    (a) three elements, all metals, beginning with the letter M
    (b) four elements, all solid nonmetals
    (c) five elements, all solids in the first five rows of the periodic table, whose symbols start with letters different from the element name

35. How would you separate a mixture of sugar and sand and isolate each in its pure form?

36. How many total atoms are there in seven dozen formulas of nitric acid, HNO_3?

37. Make a graph using the following data. Plot the density of air in grams per liter along the *x*-axis and temperature along the *y*-axis.

Temperature (°C)	Density (g/L)
0	1.29
10	1.25
20	1.20
40	1.14
80	1.07

(a) What is the relationship between density and temperature according to your graph?
(b) From your plot, find the density of air at these temperatures:
    5°C   25°C   70°C

38. These formulas look similar but represent different things.

8 S S_8

Compare and contrast them. How are they alike? How are they different?

39. Write formulas for the following compounds that a colleague read to you:
    (a) NA–CL
    (b) H2–S–O4
    (c) K2–O
    (d) Fe2–S3
    (e) K3–P–O4
    (f) CA (pause) CN taken twice
    (g) C6–H12–O6
    (h) C2–H5 (pause) OH
    (i) CR (pause) NO3 taken three times

40. The abundance of iodine in seawater is 5.0×10^{-8} percent by mass. How many kilograms of seawater must be treated to obtain 1.0 g of iodine?

41. When the tongue detects something sweet, it begins a chain of reactions to tell the body to expect calorie-containing foods. If we consume noncaloric foods containing artificial sweeteners, our bodies expect food and will tell our brains to find more if the sweet food does not contain anything to digest. Some common sweeteners are listed below. Write the name and number of atoms of each element in the sweeteners.

Compound	Common name	Chemical formula
(a) Sucrose	Table sugar	$C_{12}H_{22}O_{11}$
(b) Saccharin	Sweet 'N Low	$C_7H_5O_3NS$
(c) Aspartame	NutraSweet, Equal	$C_{14}H_{18}O_5N_2$
(d) Acesulfame-K	Sunett, Sweet One	$C_4H_4O_3NS\text{-}K$
(e) Sucralose	Splenda	$C_{12}H_{19}O_8Cl_3$

42. Should the elemental symbol for cobalt be written as Co or CO? What confusion could arise if the wrong notation were used?

43. Caffeine has the chemical structure illustrated in the image given below. Carbon atoms are represented as grey spheres, nitrogen atoms as blue spheres, oxygen atoms as red spheres, and hydrogen atoms as white spheres. Write a chemical formula for caffeine.

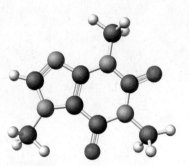

CHALLENGE EXERCISES

44. Write the chemical formulas of the neutral compounds that would result if the following ions were combined. Use the Charges of Common Ions table on the inside back cover of the book to help you.
(a) ammonium ion and chloride ion
(b) hydrogen ion and hydrogen sulfate ion
(c) magnesium ion and iodide ion
(d) iron(II) ion and fluoride ion
(e) lead(II) ion and phosphate ion
(f) aluminum ion and oxide ion

45. Write formulas of all the compounds that will form between the first five of the Group 1A and 2A metals and the oxide ion.

46. Unsaturated fats and oils are often used in cooking. For each pair of fatty acids listed below, tell the number of carbon and hydrogen atoms in each molecule.
(a) arachidic acid (saturated) $CH_3(CH_2)_{18}COOH$ and arachidonic acid (unsaturated) $CH_3(CH_2)_4(CH\!=\!CHCH_2)_4(CH_2)_2COOH$
(b) stearic acid (saturated) $CH_3(CH_2)_{16}COOH$ and linoleic acid (unsaturated) $CH_3(CH_2)_4(CH\!=\!CHCH_2)_2(CH_2)_6COOH$
(c) What is the ratio of hydrogen to carbon atoms for each pair?
(d) Propose a possible explanation for describing linoleic and stearic acids as being saturated.

ANSWERS TO PRACTICE EXERCISES

3.1 Cesium, Cs, IA; Calcium, Ca, 2A; Chromium, Cr, 6B; Cobalt, Co, 8B; Copper, Cu, 1B; Cadmium, Cd, 2B; Carbon, C, 4A; Chlorine, Cl, 7A; Cerium, Ce, 4B; Copernicium, Cn, 2B; Curium, Cm, 8B; Californium, Cf, 1B.

3.2 (a) any of the transition metals
(b) Br and Hg
(c) H, N, O, F, Cl

3.3 gases   H, N, Cl, Ne
liquids   Br, Hg
solids   Na, Ca, S, Fe

3.4 nonmetal   F, Kr, I, S
metal   Na, Cr, Mo, Cu
metalloid   Si, Sb

3.5 (a) K–BR
(b) PB–CL2
(c) CA (pause) CO3
(d) MG (pause) OH taken twice

3.6 (a) KBr
(b) $NaClO_3$
(c) H_2SO_3
(d) $Al_2(CO_3)_3$

A burning log undergoes chemical change resulting in the release of energy in the form of heat and light. The physical properties of the log change during the chemical reaction.

© Jenny Swanson/iStockphoto

CHAPTER **4**

PROPERTIES OF MATTER

The world we live in is a kaleidoscope of sights, sounds, smells, and tastes. Our senses help us to describe these objects in our lives. For example, the smell of freshly baked cinnamon rolls creates a mouthwatering desire to gobble down a sample. Just as sights, sounds, smells, and tastes form the properties of the objects around us, each substance in chemistry has its own unique properties that allow us to identify it and predict its interactions.

These interactions produce both physical and chemical changes. When you eat an apple, the ultimate metabolic result is carbon dioxide and water. These same products are achieved by burning logs. Not only does a chemical change occur in these cases, but an energy change occurs as well. Some reactions release energy (as does the apple or the log) whereas others require energy, such as the production of steel or the melting of ice. Over 90% of our current energy comes from chemical reactions.

CHAPTER OUTLINE

4.1 Properties of Substances
4.2 Physical and Chemical Changes
4.3 Learning to Solve Problems
4.4 Energy
4.5 Heat: Quantitative Measurement
4.6 Energy in the Real World

4.1 PROPERTIES OF SUBSTANCES

Compare the physical and chemical properties of a substance.

How do we recognize substances? Each substance has a set of **properties** that is characteristic of that substance and gives it a unique identity. Properties—the "personality traits" of substances—are classified as either physical or chemical. **Physical properties** are the inherent characteristics of a substance that can be determined without altering its composition; they are associated with its physical existence. Common physical properties include color, taste, odor, state of matter (solid, liquid, or gas), density, melting point, and boiling point (see **Figure 4.1**). **Chemical properties** describe the ability of a substance to form new substances, either by reaction with other substances or by decomposition.

Let's consider a few of the physical and chemical properties of chlorine. Physically, chlorine is a gas at room temperature about 2.4 times heavier than air. It is greenish yellow in color and has a disagreeable odor. Chemically, chlorine will not burn but will support the combustion of certain other substances. It can be used as a bleaching agent, as a disinfectant for water, and in many chlorinated substances such as refrigerants and insecticides. When chlorine combines with the metal sodium, it forms a salt called sodium chloride (see **Figure 4.2**). These properties, among many others, help us characterize and identify chlorine.

Figure 4.1

Physical property. The boiling point of water is a physical property. At its boiling point water changes from a liquid to a gas, but the molecules remain water molecules. They are still water but are farther apart.

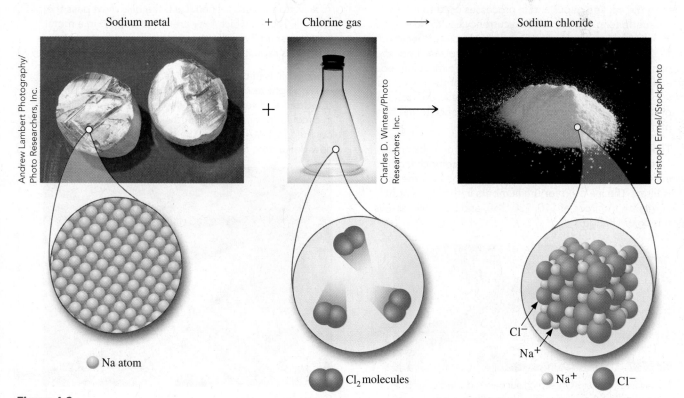

Sodium metal   +   Chlorine gas   ⟶   Sodium chloride

Na atom

Cl₂ molecules

Cl⁻
Na⁺

Na⁺   Cl⁻

Figure 4.2

Chemical property. When sodium metal reacts with chlorine gas, a new substance called sodium chloride forms.

TABLE 4.1   Physical Properties of Selected Substances

Substance	Color	Odor	Physical state	Melting point (°C)	Boiling point (°C)
Chlorine	Greenish yellow	Sharp, suffocating	Gas (20°C)	−101.6	−34.6
Water	Colorless	Odorless	Liquid	0.0	100.0
Sugar	White	Odorless	Solid	—	Decomposes 170–186
Acetic acid	Colorless	Like vinegar	Liquid	16.7	118.0
Nitrogen dioxide	Reddish brown	Sharp, suffocating	Gas	−11.2	21.2
Oxygen	Colorless	Odorless	Gas	−218.4	−183

Many chemists have reference books such as the *Handbook of Chemistry and Physics* to use as a resource.

Substances, then, are recognized and differentiated by their properties. **Table 4.1** lists six substances and several of their common physical properties. Information about physical properties, such as that given in Table 4.1, is available in handbooks of chemistry and physics. Scientists don't pretend to know all the answers or to remember voluminous amounts of data, but it is important for them to know where to look for data in the literature and on the Internet.

No two substances have identical physical and chemical properties.

>CHEMISTRY *IN ACTION*

Making Money

Chemists are heavily involved in the manufacture of our currency. In fact, in a very real way, the money industry depends on chemistry and finding substances with the correct properties. The most common paper currency in the United States is the dollar bill. Chemistry is used to form the ink and paper and in processes used to defeat counterfeiters. The ink used on currency has to do a variety of things. It must be just the right consistency to fill the fine lines of the printing plate and release onto the paper without smearing. The ink must dry almost immediately, since the sheets of currency fly out of the press and into stacks 10,000 sheets tall. The pressure at the bottom of the stack is large, and the ink must not stick to the back of the sheet above it.

The security of our currency also depends on substances in the ink, which is optically variable. This color-changing ink shifts from green to black depending on how the bill is tilted. The ink used for the numbers in the lower right corner on the front of $10, $20, $50, and $100 bills shows this color change.

Once the currency is printed, it has to undergo durability tests. The bills are soaked in different solvents for 24 hours to make sure that they can stand dry cleaning and chemical exposure to household items such as gasoline. Then the bills must pass a washing-machine test to be sure that, for example, your $20 bill is still intact in your pocket after the washer is through with it. Paper currency is a blend of 75% cotton and 25% linen. Last, the bills must pass the crumple test, in which they are rolled up, put in a metal tube, crushed by a plunger, removed, and flattened. This process is repeated up to 50 times.

Once a bill is in circulation, Federal Reserve banks screen it using light to measure wear. If the bill gets too dirty, it is shredded and sent to a landfill. The average $1 bill stays in circulation for only 18 months. No wonder Congress decided in 1997 to revive the dollar coin by creating a new dollar coin to succeed the Susan B. Anthony dollar.

Sacagawea dollar coin, which succeeded the Susan B. Anthony dollar, is still not in wide circulation today.

Ryan McVay/Getty Images, Inc.

Currency in stacks.

Corbis Digital Stock

WileyPLUS

ENHANCED EXAMPLE

EXAMPLE 4.1

You have three flasks that contain chlorine gas, nitrogen dioxide gas, and oxygen gas. Describe how you would use physical properties to identify the contents in each flask.

SOLUTION

Use Table 4.1 to compare the compounds and note their colors. You will find that chlorine is a greenish yellow color, nitrogen dioxide is reddish brown, and oxygen is colorless. Now you have the information to identify the contents in each flask.

PRACTICE 4.1

You are given three samples of silver compounds: AgCl, AgBr, and AgI. Arrange these silver compounds in the order of their percent silver by mass (highest to lowest). Describe how you arrived at your arrangement. *Hint:* Use the periodic table inside the front cover of your text.

4.2 PHYSICAL AND CHEMICAL CHANGES

Compare the physical and chemical changes in a substance.

● **LEARNING OBJECTIVE**

KEY TERMS

physical change
chemical change
chemical equation
reactants
products

Matter can undergo two types of changes, physical and chemical. **Physical changes** are changes in physical properties (such as size, shape, and density) or changes in the state of matter without an accompanying change in composition. The changing of ice into water and water into steam are physical changes from one state of matter into another. No new substances are formed in these physical changes.

When a clean platinum wire is heated in a burner flame, the appearance of the platinum changes from silvery metallic to glowing red. This change is physical because the platinum can be restored to its original metallic appearance by cooling and, more importantly, because the composition of the platinum is not changed by heating and cooling.

In a **chemical change**, new substances are formed that have different properties and composition from the original material. The new substances need not resemble the original material in any way.

When a clean copper wire is heated in a burner flame, the appearance of the copper changes from coppery metallic to glowing red. Unlike the platinum wire, the copper wire is not restored to its original appearance by cooling but instead becomes a black material. This black material is a new substance called copper(II) oxide. It was formed by a chemical change when copper combined with oxygen in the air during the heating process. The unheated wire

Sawing wood produces physical change.

© Josh McCulloch/All Canada Photos/SuperStock

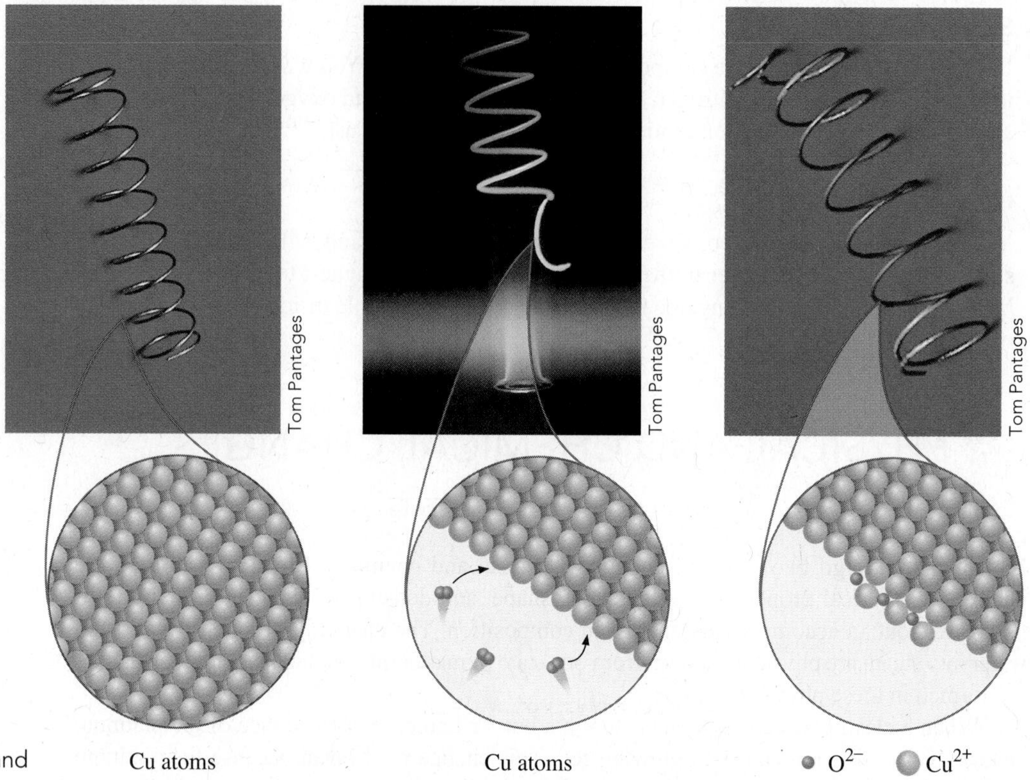

Before heating, the wire is 100% copper (1.000 g).

Copper and oxygen from the air combine chemically when the wire is heated.

After heating, the wire is black copper(II) oxide (79.9% copper, 20.1% oxygen) (1.251 g).

Tom Pantages

Tom Pantages

Tom Pantages

Figure 4.3

Chemical change: Forming of copper(II) oxide from copper and oxygen.

Cu atoms

Cu atoms
O₂ molecules

• O²⁻ ● Cu²⁺

was essentially 100% copper, but the copper(II) oxide is 79.9% copper and 20.1% oxygen. One gram of copper will yield 1.251 g of copper(II) oxide (see **Figure 4.3**). The platinum is changed only physically when heated, but the copper is changed both physically and chemically when heated.

Water can be decomposed chemically into hydrogen and oxygen. This is usually accomplished by passing electricity through the water in a process called *electrolysis*. Hydrogen collects at one electrode while oxygen collects at the other (see **Figure 4.4**). The composition and the physical appearance of the hydrogen and the oxygen are quite different from those of water. They are both colorless gases, but each behaves differently when a burning wooden splint is placed into the sample: The hydrogen explodes with a pop while the flame brightens considerably in the oxygen (oxygen supports and intensifies

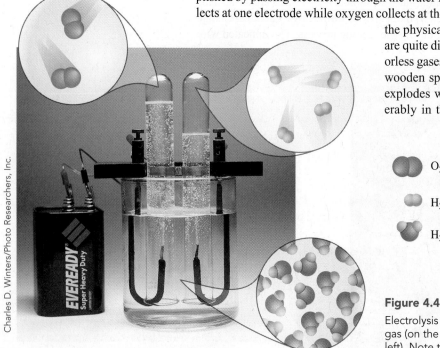

Charles D. Winters/Photo Researchers, Inc.

O₂ molecules

H₂ molecules

H₂O molecules

Figure 4.4

Electrolysis of water produces hydrogen gas (on the right) and oxygen gas (on the left). Note the ratio of the gases is 2:1.

the combustion of the wood). From these observations, we conclude that a chemical change has taken place.

Chemists have devised a shorthand method for expressing chemical changes in the form of **chemical equations**. The two previous examples of chemical changes can be represented by the following molecular representations, word and symbol equations. In the electrolysis reaction, water decomposes into hydrogen and oxygen when electrolyzed. In the formation reaction, copper plus oxygen when heated produce copper(II) oxide. The arrow means "produces," and it points to the products. The Greek letter delta (Δ) represents heat. The starting substances (water, copper, and oxygen) are called the **reactants**, and the substances produced [hydrogen, oxygen, and copper(II) oxide] are called the **products**. We will learn more about writing chemical equations in later chapters.

Formation of Copper(II) Oxide

Type of equation	Reactants			Products
Word	copper	+	oxygen $\xrightarrow{\Delta}$	copper(II) oxide
Molecular   		+	$\xrightarrow{\Delta}$	
Symbol (formula)	2 Cu	+	O$_2$ $\xrightarrow{\Delta}$	2 CuO

Electrolysis of Water

Type of equation	Reactants		Products		
Word	water $\xrightarrow{\text{electrical energy}}$		hydrogen	+	oxygen
Molecular	 $\xrightarrow{\text{electrical energy}}$			+	
Symbol (formula)	2 H$_2$O $\xrightarrow{\text{electrical energy}}$		2 H$_2$	+	O$_2$

Physical change usually accompanies a chemical change. Table 4.2 lists some common physical and chemical changes; note that wherever a chemical change occurs, a physical change also occurs. However, wherever a physical change is listed, only a physical change occurs.

TABLE 4.2 Physical or Chemical Changes of Some Common Processes

Process taking place	Type of change	Accompanying observations
Rusting of iron	Chemical	Shiny, bright metal changes to reddish brown rust.
Boiling of water	Physical	Liquid changes to vapor.
Burning of sulfur in air	Chemical	Yellow, solid sulfur changes to gaseous, choking sulfur dioxide.
Boiling of an egg	Chemical	Liquid white and yolk change to solids.
Combustion of gasoline	Chemical	Liquid gasoline burns to gaseous carbon monoxide, carbon dioxide, and water.
Digestion of food	Chemical	Food changes to liquid nutrients and partially solid wastes.
Sawing of wood	Physical	Smaller pieces of wood and sawdust are made from a larger piece of wood.
Burning of wood	Chemical	Wood burns to ashes, gaseous carbon dioxide, and water.
Heating of glass	Physical	Solid becomes pliable during heating, and the glass may change its shape.

EXAMPLE 4.2

In a distillation, a chemist is boiling pure ethyl alcohol and observes that the boiling point is 78.5°C. Using an apparatus to condense and recover the alcohol vapors back to a liquid, she finds that the boiling point of the condensed liquid also is 78.5°C. Describe whether this experiment is a physical or chemical change.

SOLUTION

The process of changing a liquid to a vapor and back to a liquid is a physical change verified by the fact that both liquids have the same boiling points.

PRACTICE 4.2

When ice melts, is this a physical or chemical change?

4.3 LEARNING TO SOLVE PROBLEMS

LEARNING OBJECTIVE ● List the basic steps in solving chemistry problems.

Now that we have learned some of the basics of chemistry including names and symbols for the elements, dimensional analysis, and a good amount of chemical terminology, we are ready to begin learning to solve chemistry problems.

One of the great joys of studying chemistry is learning to be a good problem solver. The ability to solve complicated problems is a skill that will help you greatly throughout your life. It is our goal to help you learn to solve problems by thinking through the problems on your own. To this end the basic steps in solving problems are:

Read Read the problem carefully. Determine what is known and what is to be solved for and write them down. It is important to label all factors and measurements with the proper units.

Plan Determine which principles are involved and which unit relationships are needed to solve the problem. You may refer to tables for needed data. Set up the problem in a neat, organized, and logical fashion, making sure all unwanted units cancel. Use the examples as guides for setting up the problem.

Calculate Proceed with the necessary mathematical operations. Make certain that your answer contains the proper number of significant figures.

Check Check the answer to see if it is reasonable.

At first we will help you through this process in each example. Then later in the text as you learn more about solving problems and have lots of practice, we'll help you less. By the end of your course you'll be solving problems on your own.

A few more words about problem solving. Don't allow any formal method of problem solving to limit your use of common sense and intuition. If a problem is clear to you and its solution seems simpler by another method, by all means use it.

If you approach problem solving using this general method, you will be able to learn to solve more and more difficult problems on your own. As you gain confidence you will become an independent, creative problem solver and be able to use this skill wherever you go in "real life."

4.4 ENERGY

LEARNING OBJECTIVE ● List the various forms of energy, explain the role of energy in chemical changes, and state the law of conservation of energy.

KEY TERMS
energy
potential energy
kinetic energy
law of conservation of energy

From the early discovery that fire can warm us and cook our food to our discovery that nuclear reactors can be used to produce vast amounts of controlled energy, our technical progress has been directed by our ability to produce, harness, and utilize energy. **Energy** is the capacity of matter to do work. Energy exists in many forms; some of the more familiar forms are mechanical, chemical, electrical, heat, nuclear, and radiant or light energy. Matter can have both potential and kinetic energy.

Potential energy (PE) is stored energy, or energy that an object possesses due to its relative position. For example, a ball located 20 ft above the ground has more potential energy than when located 10 ft above the ground and will bounce higher when allowed to fall. Water backed up behind a dam represents potential energy that can be converted into useful work in the form of electrical or mechanical energy. Gasoline is a source of chemical potential energy. When gasoline burns (combines with oxygen), the heat released is associated with a decrease in potential energy. The new substances formed by burning have less chemical potential energy than the gasoline and oxygen.

Kinetic energy (KE) is energy that matter possesses due to its motion. When the water behind the dam is released and allowed to flow, its potential energy is changed into kinetic energy, which can be used to drive generators and produce electricity. Moving bodies possess kinetic energy. We all know the results when two moving vehicles collide: Their kinetic energy is expended in the crash that occurs. The pressure exerted by a confined gas is due to the kinetic energy of rapidly moving gas particles.

Energy can be converted from one form to another form. Some kinds of energy can be converted to other forms easily and efficiently. For example, mechanical energy can be converted to electrical energy with an electric generator at better than 90% efficiency. On the other hand, solar energy has thus far been directly converted to electrical energy at an efficiency approaching 30%. In chemistry, energy is most frequently released as heat.

The mechanical energy of falling water is converted to electrical energy at the hydroelectric plant at Niagara Falls. As the water falls, potential energy is converted to kinetic energy and turns a turbine to produce electrical energy.

Energy in Chemical Changes

In all chemical changes, matter either absorbs or releases energy. Chemical changes can produce different forms of energy. For example, electrical energy to start automobiles is produced by chemical changes in the lead storage battery. Light energy is produced by the chemical change that occurs in a light stick. Heat and light are released from the combustion of fuels. For example, the space shuttle uses the reaction between hydrogen and oxygen-producing water vapor to help power it (see **Figure 4.5**). All the energy needed for our life processes—breathing, muscle contraction, blood circulation, and so on—is produced by chemical changes occurring within the cells of our bodies.

Conversely, energy is used to cause chemical changes. For example, a chemical change occurs in the electroplating of metals when electrical energy is passed through a salt solution in which the metal is submerged. A chemical change occurs when radiant energy from the sun is used by green plants in the process of photosynthesis. And as we saw, a chemical change occurs when electricity is used to decompose water into hydrogen and oxygen. Chemical changes are often used primarily to produce energy rather than to produce new substances. The heat or thrust generated by the combustion of fuels is more important than the new substances formed.

Figure 4.5

The external fuel tank (orange) contained liquid hydrogen and liquid oxygen which fueled the shuttle's main engine. The exhaust gas from the shuttle was water.

EXAMPLE 4.3

Methane (CH_4) is the major compound in natural gas and is used in our homes for heat and cooking. The combustion of methane in oxygen (air) produces considerable heat energy. The chemical reaction is

$$CH_4 \ + \ 2\,O_2 \ \rightarrow \ CO_2 \ + \ 2\,H_2O \ + \ 802.5\ kJ$$

methane   oxygen   carbon dioxide   water   heat

If it were possible to make methane from carbon dioxide and water, how much heat energy would it take to make methane? Write the equation.

SOLUTION

From the equation it takes 802.5 kJ to make methane. The chemical equation for the reaction is

$$CO_2 + 2\,H_2O + 802.5\ kJ \rightarrow CH_4 + 2\,O_2$$

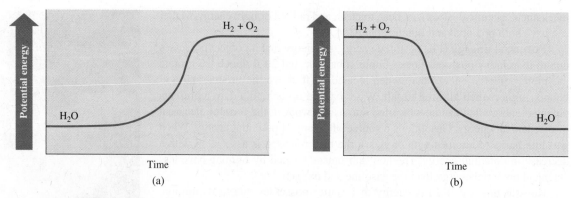

Figure 4.6

(a) In electrolysis of water, energy is absorbed by the system, so the products H_2 and O_2 have a higher potential energy. (b) When hydrogen is used as a fuel, energy is released and the product (H_2O) has lower potential energy.

Conservation of Energy

An energy transformation occurs whenever a chemical change occurs (see **Figure 4.6**). If energy is absorbed during the change, the products will have more chemical potential energy than the reactants. Conversely, if energy is given off in a chemical change, the products will have less chemical potential energy than the reactants. Water can be decomposed in an electrolytic cell by absorbing electrical energy. The products, hydrogen and oxygen, have a greater chemical potential energy level than that of water (see Figure 4.6a). This potential energy is released in the form of heat and light when the hydrogen and oxygen are burned to form water again (see Figure 4.6b). Thus, energy can be changed from one form to another or from one substance to another and, therefore, is not lost.

The energy changes occurring in many systems have been thoroughly studied. No system has ever been found to acquire energy except at the expense of energy possessed by another system. This is the **law of conservation of energy**: Energy can be neither created nor destroyed, though it can be transformed from one form to another.

4.5 HEAT: QUANTITATIVE MEASUREMENT

LEARNING OBJECTIVE ● Calculate the amount of heat lost or gained in a given system.

KEY TERMS
joule
calorie
specific heat

The SI-derived unit for energy is the joule (pronounced *jool*, rhyming with *tool*, and abbreviated J). Another unit for heat energy, which has been used for many years, is the calorie (abbreviated cal). The relationship between joules and calories is

$$4.184\ J = 1\ cal \quad (exactly)$$

To give you some idea of the magnitude of these heat units, 4.184 **joules**, or 1 **calorie**, is the quantity of heat energy required to change the temperature of 1 gram of water by 1°C, usually measured from 14.5°C to 15.5°C.

Since joules and calories are rather small units, kilojoules (kJ) and kilocalories (kcal) are used to express heat energy in many chemical processes. The kilocalorie is also known as the nutritional Calorie (spelled with a capital C and abbreviated Cal). In this book, heat energy will be expressed in joules.

1 kJ = 1000 J
1 kcal = 1000 cal = 1 Cal

The difference in the meanings of the terms *heat* and *temperature* can be seen by this example: Visualize two beakers, A and B. Beaker A contains 100 g of water at 20°C, and beaker B contains 200 g of water also at 20°C. The beakers are heated until the temperature of the water in each reaches 30°C. The temperature of the water in the beakers was raised by exactly the same amount, 10°C. But twice as much heat (8368 J) was required to raise the temperature of the water in beaker B as was required in beaker A (4184 J).

TABLE 4.3	Specific Heat of Selected Substances	
Substance	**Specific heat (J/g°C)**	**Specific heat (cal/g°C)**
Water	4.184	1.000
Ethyl alcohol	2.138	0.511
Ice	2.059	0.492
Aluminum	0.900	0.215
Iron	0.473	0.113
Copper	0.385	0.0921
Gold	0.131	0.0312
Lead	0.128	0.0305

In the middle of the eighteenth century, Joseph Black (1728–1799), a Scottish chemist, was experimenting with the heating of elements. He heated and cooled equal masses of iron and lead through the same temperature range. Black noted that much more heat was needed for the iron than for the lead. He had discovered a fundamental property of matter—namely, that every substance has a characteristic heat capacity. Heat capacities may be compared in terms of specific heats. The **specific heat** of a substance is the quantity of heat (lost or gained) required to change the temperature of 1 g of that substance by 1°C. It follows then that the specific heat of liquid water is 4.184 J/g°C (or 1.000 cal/g°C). The specific heat of water is high compared with that of most substances. Aluminum and copper, for example, have specific heats of 0.900 J/g°C and 0.385 J/g°C, respectively (see Table 4.3). The relation of mass, specific heat, temperature change (Δt), and quantity of heat lost or gained by a system is expressed by this general equation:

$$\left(\begin{array}{c}\text{mass of}\\\text{substance}\end{array}\right)\left(\begin{array}{c}\text{specific heat}\\\text{of substance}\end{array}\right)(\Delta t) = \text{heat}$$

Thus the amount of heat needed to raise the temperature of 200. g of water by 10.0°C can be calculated as follows:

$$(200.\ \cancel{g})\left(\frac{4.184\ \text{J}}{\cancel{g}\ \cancel{°C}}\right)(10.0\ \cancel{°C}) = 8.37 \times 10^3\ \text{J}$$

> The greek letter Δ is used to mean "change in."
>
> Lowercase t is used for Celsius and upper case T for Kelvin.
>
> Mass is in grams, specific heat is in cal/g°C or J/g°C and Δt is in °C.

EXAMPLE 4.4

Calculate the specific heat of a solid in J/g°C and cal/g°C if 1638 J raises the temperature of 125 g of the solid from 25.0°C to 52.6°C.

SOLUTION

Read • **Knowns:** 125 g of the solid
$\Delta t = 52.6 - 25.0 = 27.6$°C
Heat = 1638 J

Solving for: specific heat of the solid

Plan • Use the equation

(mass)(specific heat)(Δt) = heat

solving for specific heat

$$\text{specific heat} = \frac{\text{heat}}{\text{g} \times \Delta t}$$

Calculate • $\text{specific heat} = \dfrac{1638\ \text{J}}{125\ \text{g} \times 27.6\text{°C}} = 0.475\ \text{J/g°C}$

Convert joules to calories using 1.00 cal/4.184 J

$$\text{specific heat} = \left(\frac{0.475\ \cancel{J}}{\text{g°C}}\right)\left(\frac{1.000\ \text{cal}}{4.184\ \cancel{J}}\right) = 0.114\ \text{cal/g°C}$$

Check • Note that the units in the answer agree with the units for specific heat.

WileyPLUS

ENHANCED EXAMPLE

EXAMPLE 4.5

A sample of a metal with a mass of 212 g is heated to 125.0°C and then dropped into 375 g water at 24.0°C. If the final temperature of the water is 34.2°C, what is the specific heat of the metal? (Assume no heat losses to the surroundings.)

SOLUTION

Read •   **Knowns:**   mass (metal) = 212 g

mass (water) = 375 g

Δt (water) = 34.2 − 24.0 = 10.2°C

Δt (metal) = 125.0 − 34.2 = 90.8°C

Solving for: specific heat of the metal

Plan •   When the metal enters the water, it begins to cool, losing heat to the water. At the same time, the temperature of the water rises. This continues until the temperature of the metal and the water is equal (34.2°C), after which no net flow of heat occurs.

Heat lost or gained by a system is given by

(mass)(specific heat)(Δt) = energy change (heat)

heat gained by the water = heat lost by the metal

heat gained by the water = (mass water)(specific heat)(Δt)

$$= (375\ g)\left(\frac{4.184\ J}{g\ °C}\right)(10.2\ °C) = 1.60 \times 10^4\ J$$

Calculate • heat gained by the water = heat lost by the metal = 1.60×10^4 J

(mass metal)(specific heat metal)(Δt) = 1.60×10^4 J

(212 g)(specific heat metal)(90.8°C) = 1.60×10^4 J

To determine the specific heat of the metal, we rearrange the equation

(mass)(specific heat)(Δt) = heat

solving for the specific heat of the metal

$$\text{specific heat metal} = \frac{1.60 \times 10^4\ J}{(212\ g)(90.8\ °C)} = 0.831\ J/g°C$$

Check •   Note the units in the answer agree with units for specific heat.

PRACTICE 4.3

Calculate the quantity of energy needed to heat 8.0 g of water from 42.0°C to 45.0°C.

PRACTICE 4.4

A 110.0-g sample of metal at 55.5°C raises the temperature of 150.0 g of water from 23.0°C to 25.5°C. Determine the specific heat of the metal in J/g°C.

4.6 ENERGY IN THE REAL WORLD

LEARNING OBJECTIVE • Define a hydrocarbon compound and explain its role in the world's energy supply.

Coal, petroleum, natural gas, and woody plants provide us with a vast resource of energy, all of which is derived from the sun. Plants use the process of photosynthesis to store the sun's energy, and we harvest that energy by burning the plants or using the decay products of the plants. These decay products have been converted over millions of years to fossil fuels. As the plants died and decayed, natural processes changed them into petroleum deposits that we now use in the forms of gasoline and natural gas.

>CHEMISTRY *IN ACTION*

Popping Popcorn

What's the secret to getting all those kernels in a bag of microwave popcorn to pop? Food scientists have looked at all sorts of things to improve the poppability of popcorn. In every kernel of popcorn there is starch and a little moisture. As the kernel heats up, the water vaporizes and the pressure builds up in the kernel. Finally, the kernel bursts and the starch expands into the white foam of popped corn.

Researchers in Indiana tested 14 types of popcorn to determine the most poppable. They found a range of 4–47% of unpopped kernels and determined that the kernels that held moisture the best produced the fewest unpopped kernels. Further analysis led them to the conclusion that the determining factor was the pericarp.

The pericarp is the hard casing on the popcorn kernel. It is made of long polymers of cellulose. When the kernel is heated up, the long chains of cellulose line up and make

a strong crystalline structure. This strong structure keeps the water vapor inside the kernel, allowing the pressure to rise high enough to pop the kernel.

Getting the most pops comes from the molecular changes resulting from heating the kernel. Energy and chemistry combine to produce a tasty snack!

Popcorn

Although we do not really understand completely how petroleum deposits were formed, they most likely came from the remains of marine organisms living over 500 million years ago. Petroleum is composed of hydrocarbon compounds (which contain only carbon and hydrogen). Carbon is a unique element in that it can produce chains of atoms in a molecule of different lengths. Table 4.4 gives the names and formulas of some simple hydrocarbons. Natural gas is associated with petroleum deposits and consists mainly of methane with some ethane, propane, and butane mixed in. Large resources of natural gas are now being made available in the United States through a process called fracking, (hydraulic fracturing).

Coal was formed from the remains of plants that were buried and subjected to high pressure and heat over many years. Chemical changes lowered the oxygen and hydrogen content of these plant tissues and produced the black solid we recognize as coal. The energy available from a mass of coal increases as the carbon content increases. Coal is an important energy source in the United States.

TABLE 4.4 Names and Formulas for Simple Hydrocarbons

Names	Formula
Methane	CH_4
Ethane	C_2H_6
Propane	C_3H_8
Butane	C_4H_{10}
Pentane	C_5H_{12}
Hexane	C_6H_{14}
Heptane	C_7H_{16}
Octane	C_8H_{18}
Nonane	C_9H_{20}
Decane	$C_{10}H_{22}$

Alternative energy sources include wind, solar, and biofuels.

We are currently facing an energy crisis that has caused us to search for new renewable energy sources for our future. We will need to consider economic, supply, and climatic factors to find fuels to serve us as we continue to grow our human population. There are a variety of potential resources: solar, nuclear, biomass, wind, and synthetic fuels. We now are beginning to focus on developing these resources and using them in the most economic way to lower our energy dependence globally.

EXAMPLE 4.6

Much of the energy used in the world is dependent on coal. Large deposits of coal are found all over the world. Coal supplies about 20% of the energy used in the United States. The combustion of coal, which is mainly carbon and reacts with oxygen in the air, produces carbon dioxide and much heat energy. Write the chemical equation for this reaction.

SOLUTION

$$C + O_2 \rightarrow CO_2 + \text{heat (393 kJ)}$$

PRACTICE 4.5

The major compounds in gasoline are hydrocarbons. Without petroleum (from which hydrocarbons are distilled to make gasoline), industry would come to a halt and millions of automobiles would be idled. Devise a general formula (C_xH_y) that represents the hydrocarbons in Table 4.4.

CHAPTER 4 REVIEW

4.1 PROPERTIES OF SUBSTANCES

KEY TERMS

properties
physical properties
chemical properties

- Each substance has a set of properties that is characteristic of the substance and gives it a unique identity.
- Physical properties are inherent in the substance and can be determined without altering the composition of the substance:
  - Color
  - Taste
  - Odor
  - State of matter
  - Density
  - Melting point
  - Boiling point
- Chemical properties describe the ability of a substance to interact with other substances to form different substances.

4.2 PHYSICAL AND CHEMICAL CHANGES

KEY TERMS

physical change
chemical change
chemical equations
reactants
products

- A physical change is a change in the physical properties or a change in the state of matter for a substance without altering the composition of the substance.
- In a chemical change different substances are formed that have different properties and composition from the original material.
- Chemical changes can be represented by chemical equations:
  - Word equation
  - Molecular equation
  - Symbol (formula) equation

4.3 LEARNING TO SOLVE PROBLEMS

Read	Read the problem carefully, determining what is known and what is to be solved for.
Plan	Determine what is required to solve the problem and set up the problem in a neat, organized manner using the examples as guides.
Calculate	Complete the necessary calculations making sure your answer contains the proper units and significant figures.
Check	Check to see if your answer is reasonable.

4.4 ENERGY

- Energy is the capacity to do work.
- Potential energy is energy that results from position or is stored within a substance.
- Kinetic energy is the energy matter possesses as a result of its motion.
- Energy can be converted from one form to another.
- Common forms of energy:
  - Mechanical
  - Chemical
  - Electrical
  - Heat
  - Nuclear
  - Light
- In chemistry, energy is most frequently expressed as heat.
- In all chemical changes matter either absorbs or releases energy.
- Energy can be used to cause chemical change.
- Chemical changes are often used to produce energy rather than new substances.
- Energy can neither be created nor destroyed. It can be transformed from one form to another.
- The energy released or absorbed in a chemical reaction can be summarized in graphical form.

KEY TERMS

energy
potential energy
kinetic energy
law of conservation of energy

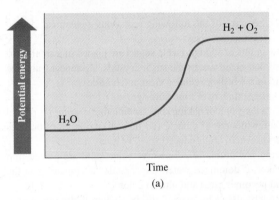

(a)

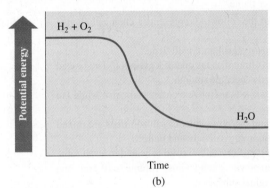

(b)

4.5 HEAT: QUANTITATIVE MEASUREMENT

- The SI unit for heat is the joule
  $$4.184 \text{ J} = 1 \text{ cal}$$
- The calorie is defined as the amount of heat required to change the temperature of 1 gram of water 1°C.
- Every substance has a characteristic heat capacity:
  - The specific heat of a substance is a measure of its heat capacity.
  - The specific heat is the quantity of heat required to change the temperature of 1 gram of the substance by 1°C.
- Heat lost or gained by a system can be calculated by
  heat = (mass of substance)(specific heat of substance)(Δt).

KEY TERMS

joule
calorie
specific heat

REVIEW QUESTIONS

1. In what physical state does acetic acid exist at 393 K? (Table 4.1)
2. In what physical state does chlorine exist at −65°C? (Table 4.1)
3. Calculate the boiling point of acetic acid in
   (a) kelvins
   (b) degrees Fahrenheit (Table 4.1)
4. Calculate the melting point of acetic acid in (Table 4.1)
   (a) degrees Fahrenheit
   (b) kelvins
5. What evidence of chemical change is visible when electricity is run through water? (Figure 4.4)
6. What is the fundamental difference between a chemical change and a physical change?
7. What is the critical first step in solving any chemistry problem?
8. Is the calculation step the last step in solving a chemistry problem? Why or why not?

9. Distinguish between potential and kinetic energy.
10. How is chemistry calorie (cal) different from a food calorie (Cal)?
11. Which requires more energy to heat from 20°C to 30°C, iron or gold? (See Table 4.3.)
12. Predict which molecule would produce more carbon dioxide when burned, a molecule of ethane or a molecule of octane.
13. What is the common characteristic of all hydrocarbons?
14. Which is the most common element used to produce energy for human use?
15. Why might the combustion of fossil fuels not be the best solution to the planet's energy needs? What are some other energy sources that may become important in the future?

Most of the exercises in this chapter are available for assignment via the online homework management program, WileyPLUS (www.wileyplus.com). All exercises with blue numbers have answers in Appendix VI.

PAIRED EXERCISES

1. Determine whether each of the following represents a physical property or a chemical property:
   (a) Vinegar has a pungent odor.
   (b) Carbon cannot be decomposed.
   (c) Sulfur is a bright yellow solid.
   (d) Sodium chloride is a crystalline solid.
   (e) Water does not burn.
   (f) Mercury is a liquid at 25°C.
   (g) Oxygen is not combustible.
   (h) Aluminum combines with oxygen to form a protective oxide coating.

2. Determine whether each of the following represents a physical property or a chemical property:
   (a) Chlorine gas has a greenish-yellow tint.
   (b) The density of water at 4°C is 1.000 g/mL.
   (c) Hydrogen gas is very flammable.
   (d) Aluminum is a solid at 25°C.
   (e) Water is colorless and odorless.
   (f) Lemon juice tastes sour.
   (g) Gold does not tarnish.
   (h) Copper cannot be decomposed.

3. Cite the evidence that indicates that only physical changes occur when a platinum wire is heated in a Bunsen burner flame.

4. Cite the evidence that indicates that both physical and chemical changes occur when a copper wire is heated in a Bunsen burner flame.

5. Identify the reactants and products when a copper wire is heated in air in a Bunsen burner flame.

6. Identify the reactants and products for the electrolysis of water.

7. State whether each of the following represents a chemical change or a physical change:
   (a) A steak is cooked on a grill until well done.
   (b) In the lab, students firepolish the end of a glass rod. The jagged edge of the glass has become smooth.
   (c) Chlorine bleach is used to remove a coffee stain on a white lab coat.
   (d) When two clear and colorless aqueous salt solutions are mixed together, the solution turns cloudy and yellow.
   (e) One gram of an orange crystalline solid is heated in a test tube, producing a green powdery solid whose volume is 10 times the volume of the original substance.
   (f) In the lab, a student cuts a 20-cm strip of magnesium metal into 1-cm pieces.

8. State whether each of the following represents a chemical change or a physical change:
   (a) A few grams of sucrose (table sugar) are placed in a small beaker of deionized water; the sugar crystals "disappear," and the liquid in the beaker remains clear and colorless.
   (b) A copper statue, over time, turns green.
   (c) When a teaspoon of baking soda (sodium bicarbonate) is placed into a few ounces of vinegar (acetic acid), volumes of bubbles (effervescence) are produced.
   (d) When a few grams of a blue crystalline solid are placed into a beaker of deionized water, the crystals "disappear" and the liquid becomes clear and blue in color.
   (e) In the lab, a student mixes 2 mL of sodium hydroxide with 2 mL of hydrochloric acid in a test tube. He notices that the test tube has become very warm to the touch.
   (f) A woman visits a hairdresser and has her hair colored a darker shade of brown. After several weeks the hair, even though washed several times, has not changed back to the original color.

9. Are the following examples of potential energy or kinetic energy?
   (a) fan blades spinning   (d) a person napping
   (b) a bird flying   (e) ocean waves rippling
   (c) sodium hydroxide in a sealed jar

10. Are the following examples of potential energy or kinetic energy?
    (a) a fish swimming   (d) water starting to boil
    (b) a skier at the top of a hill   (e) a leaf unfolding
    (c) a bird's egg in a nest

11. What happens to the kinetic energy of a speeding car when the car is braked to a stop?

12. What energy transformation is responsible for the fiery reentry of the space shuttle into Earth's atmosphere?

13. Indicate with a plus sign (+) any of these processes that require energy and a negative sign (−) any that release energy.
    (a) arctic ice melting   (d) dry ice changing to vapor
    (b) starting a car   (e) blowing up a balloon
    (c) flash of lightning

14. Indicate with a plus sign (+) any of these processes that require energy and a negative sign (−) any that release energy.
    (a) riding a bike   (d) tires deflating
    (b) fireworks bursting   (e) wood burning in a fireplace
    (c) water evaporating

15. How many joules of heat are required to heat 125 g aluminum from 19.0°C to 95.5°C? (Table 4.3)

16. How many joules of heat are required to heat 65 g lead from 22.0°C to 98.5°C? (Table 4.3)

17. How many joules of heat are required to heat 25.0 g of ethyl alcohol from the prevailing room temperature, 22.5°C, to its boiling point, 78.5°C?

18. How many joules of heat are required to heat 35.0 g of isopropyl alcohol from the prevailing room temperature, 21.2°C, to its boiling point, 82.4°C? (The specific heat of isopropyl alcohol is 2.604 J/g°C.)

19. A 135-g sample of a metal requires 2.50 kJ to change its temperature from 19.5°C to 100.0°C. What is the specific heat of this metal?

20. A 275-g sample of a metal requires 10.75 kJ to change its temperature from 21.2°C to its melting temperature, 327.5°C. What is the specific heat of this metal?

21. A 155-g sample of copper was heated to 150.0°C, then placed into 250.0 g water at 19.8°C. (The specific heat of copper is 0.385 J/g°C.) Calculate the final temperature of the mixture. (Assume no heat loss to the surroundings.)

22. A 225-g sample of aluminum was heated to 125.5°C, then placed into 500.0 g water at 22.5°C. (The specific heat of aluminum is 0.900 J/g°C). Calculate the final temperature of the mixture. (Assume no heat loss to the surroundings.)

ADDITIONAL EXERCISES

23. Read the following passage and identify at least two physical and two chemical properties of zeolites.

Zeolites are crystalline solids composed of silicon, aluminum, and oxygen with very porous structures. They are very useful in a variety of applications. Zeolites are formed naturally when volcanic rocks and ash react with alkaline groundwater. They may also be synthesized in the laboratory by allowing aluminosilicate crystals to form around small organic molecules. Zeolites generally have very low density due to their porous nature. They are often used as molecular sieves because they have large pores that can trap molecules of certain sizes and shapes while excluding others. Bandages used by the military sometimes contain zeolites, which absorb water from the blood, thus accelerating the natural clotting process. Zeolites have also replaced phosphates in many detergents, which is an ecological benefit, because zeolites do not encourage the growth of algae. Interestingly, the name *zeolite* comes from the fact that zeolites give off water in the form of steam when they are heated; thus, they are called boiling (zeo) stones (lithos).

24. A 110.0-g sample of a gray-colored, unknown, pure metal was heated to 92.0°C and put into a coffee-cup calorimeter containing 75.0 g of water at 21.0°C. When the heated metal was put into the water, the temperature of the water rose to a final temperature of 24.2°C. The specific heat of water is 4.184 J/g°C. (a) What is the specific heat of the metal. (b) Is it possible that the metal is either iron or lead? Explain.

25. The specific heat of a human is approximately 3.47 J/g°C. Given this information,
   (a) If a 165-lb man eats a candy bar containing 262 Cal, how much will his body temperature increase if all the calories from the candy bar are converted into heat energy? Remember that a food calorie (Cal) is equal to 1 kcal.
   (b) If a 165-lb man eats a roll of candy containing 25.0 Cal, how much will his body temperature increase if all the calories from the candy bar are converted into heat energy?

26. A sample of water at 23.0°C required an input of 1.69×10^4 J of heat to reach its boiling point, 100°C. What was the mass of the water? (Table 4.3)

27. A sample of gold required 3.25×10^4 J of heat to melt it from room temperature, 23.2°C, to its melting point, 1064.4°C. How many pounds of gold were in the sample?

28. If 40.0 kJ of energy are absorbed by 500.0 g of water at 10.0°C, what is the final temperature of the water?

29. A sample of iron was heated to 125.0°C, then placed into 375 g of water at 19.8°C. The temperature of the water rose to 25.6°C. How many grams of iron were in the sample? (Table 4.3)

30. A sample of copper was heated to 275.1°C and placed into 272 g of water at 21.0°C. The temperature of the water rose to 29.7°C. How many grams of copper were in the sample? (Table 4.3)

31. A 100.0-g sample of copper is heated from 10.0°C to 100.0°C.
   (a) Determine the number of calories needed. (The specific heat of copper is 0.0921 cal/g°C.)
   (b) The same amount of heat is added to 100.0 g of Al at 10.0°C. (The specific heat of Al is 0.215 cal/g°C.) Which metal gets hotter, the copper or the aluminum?

32. A 500.0-g piece of iron is heated in a flame and dropped into 400.0 g of water at 10.0°C. The temperature of the water rises to 90.0°C. How hot was the iron when it was first removed from the flame? (The specific heat of iron is 0.473 J/g°C.)

33. A 20.0-g piece of metal at 203°C is dropped into 100.0 g of water at 25.0°C. The water temperature rose to 29.0°C. Calculate the specific heat of the metal (J/g°C). Assume that all of the heat lost by the metal is transferred to the water and no heat is lost to the surroundings.

34. Assuming no heat loss by the system, what will be the final temperature when 50.0 g of water at 10.0°C are mixed with 10.0 g of water at 50.0°C?

35. Three 500.0-g pans of iron, aluminum, and copper are each used to fry an egg. Which pan fries the egg (105°C) the quickest? Explain.

36. At 6:00 P.M., you put a 300.0-g copper pan containing 800.0 g of water (all at room temperature, which is 25°C) on the stove. The stove supplies 628 J/s. When will the water reach the boiling point? (Assume no heat loss.)

37. Why does blowing gently across the surface of a cup of hot coffee help to cool it? Why does inserting a spoon into the coffee do the same thing?

38. If you are boiling some potatoes in a pot of water, will they cook faster if the water is boiling vigorously than if the water is only gently boiling? Explain your reasoning.

39. Homogenized whole milk contains 4% butterfat by volume. How many milliliters of fat are there in a glass (250 mL) of milk? How many grams of butterfat ($d = 0.8$ g/mL) are in this glass of milk?

40. Gloves are often worn to protect the hands from being burned when they come in contact with very hot or very cold objects. Gloves are often made of cotton or wool, but many of the newer heat-resistant gloves are made of silicon rubber. The specific heats of these materials are listed below:

Material	Specific heat (J/g°C)
wool felt	1.38
cotton	1.33
paper	1.33
rubber	3.65
silicon rubber	1.46

   (a) If a glove with a mass of 99.3 grams composed of cotton increases in temperature by 15.3°F, how much energy was absorbed by the glove?
   (b) A glove with a mass of 86.2 grams increases in temperature by 25.9°F when it absorbs 1.71 kJ of energy. Calculate the specific heat of the glove and predict its composition.
   (c) If a glove with a mass of 50.0 grams needs to absorb 1.65 kJ of energy, how much will the temperature of the glove increase for each of the materials listed above?
   (d) Which is the best material for a heat-resistant glove?
   (e) If you were designing a heat-resistant glove, what kind of specific heat would you look for?

41. If solar panels are placed in the Mojave Desert in California that generate 100 megawatts on a 1.3-square-mile site with an average of 7.50 hours of productive daylight each day, how many tons of coal are equivalent to the energy produced in one day by the solar panels? (One ton of coal will produce 26.6 gigajoules of energy, 1 Watt = 1 J/s.)

CHALLENGE EXERCISES

42. Suppose a ball is sitting at the top of a hill. At this point, the ball has potential energy. The ball rolls down the hill, so the potential energy is converted into kinetic energy. When the ball reaches the bottom of the hill, it goes halfway up the hill on the other side and stops. If energy is supposed to be conserved, then why doesn't the ball go up the other hill to the same level as it started from?

43. From the following molecular picture, determine whether a physical or a chemical change has occurred. Justify your answer.

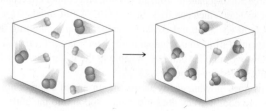

44. From the illustration below (a) describe the change that has occurred and (b) determine whether the change was a physical change or a chemical change. Justify your answer.

45. Describe the change(s) that you see in the following illustration. Was this a physical or a chemical change?

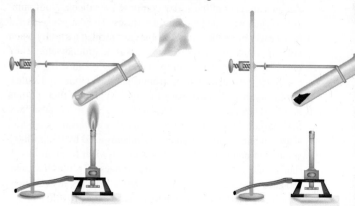

ANSWERS TO PRACTICE EXERCISES

4.1 Each compound has one atom of silver to one atom of the other elements. Using the periodic table, we find the atomic masses of these elements to be Ag (107.9), Cl (35.45), Br (79.90), and I (126.8). Each compound has the same amount of silver, but the percent of silver will be different. The highest percent of silver will be in AgCl, the middle will be in AgBr, and the lowest will be in AgI. So the order is AgCl, AgBr, AgI.

4.2 Physical change

4.3 $1.0 \times 10^2 \text{ J} = 24 \text{ cal}$

4.4 0.477 J/g°C

4.5 C_nH_{2n+2} where n is the number of carbons in each formula

PUTTING IT TOGETHER:
Review for Chapters 1–4

Answers for Putting It Together Reviews are found in Appendix VII.

Multiple Choice

Choose the correct answer to each of the following.

1. 1.00 cm is equal to how many meters?
   (a) 2.54   (b) 100.   (c) 10.0   (d) 0.0100

2. 1.00 cm is equal to how many inches?
   (a) 0.394   (b) 0.10   (c) 12   (d) 2.54

3. 4.50 ft is how many centimeters?
   (a) 11.4   (b) 21.3   (c) 454   (d) 137

4. The number 0.0048 contains how many significant figures?
   (a) 1   (b) 2   (c) 3   (d) 4

5. Express 0.00382 in scientific notation.
   (a) 3.82×10^3 (c) 3.82×10^{-2}
   (b) 3.8×10^{-3} (d) 3.82×10^{-3}

6. 42.0°C is equivalent to
   (a) 273 K   (c) 108°F
   (b) 5.55°F   (d) 53.3°F

7. 267°F is equivalent to
   (a) 404 K   (c) 540 K
   (b) 116°C   (d) 389 K

8. An object has a mass of 62 g and a volume of 4.6 mL. Its density is
   (a) 0.074 mL/g   (c) 7.4 g/mL
   (b) 285 g/mL   (d) 13 g/mL

9. The mass of a block is 9.43 g and its density is 2.35 g/mL. The volume of the block is
   (a) 4.01 mL   (c) 22.2 mL
   (b) 0.249 mL   (d) 2.49 mL

10. The density of copper is 8.92 g/mL. The mass of a piece of copper that has a volume of 9.5 mL is
    (a) 2.6 g   (c) 0.94 g
    (b) 85 g   (d) 1.1 g

11. An empty graduated cylinder has a mass of 54.772 g. When filled with 50.0 mL of an unknown liquid, it has a mass of 101.074 g. The density of the liquid is
    (a) 0.926 g/mL   (c) 2.02 g/mL
    (b) 1.08 g/mL   (d) 1.85 g/mL

12. The conversion factor to change grams to milligrams is
    (a) $\dfrac{100 \text{ mg}}{1 \text{ g}}$ (c) $\dfrac{1 \text{ g}}{1000 \text{ mg}}$
    (b) $\dfrac{1 \text{ g}}{100 \text{ mg}}$ (d) $\dfrac{1000 \text{ mg}}{1 \text{ g}}$

13. What Fahrenheit temperature is twice the Celsius temperature?
    (a) 64°F   (c) 200°F
    (b) 320°F   (d) 546°F

14. A gold alloy has a density of 12.41 g/mL and contains 75.0% gold by mass. The volume of this alloy that can be made from 255 g of pure gold is
    (a) 4.22×10^3 mL   (c) 27.4 mL
    (b) 2.37×10^3 mL   (d) 20.5 mL

15. A lead cylinder ($V = \pi r^2 h$) has radius 12.0 cm and length 44.0 cm and a density of 11.4 g/mL. The mass of the cylinder is
    (a) 2.27×10^5 g   (c) 1.78×10^3 g
    (b) 1.89×10^5 g   (d) 3.50×10^5 g

16. The following units can all be used for density *except*
    (a) g/cm^3 (b) kg/m^3 (c) g/L   (d) kg/m^2

17. 37.4 cm × 2.2 cm equals
    (a) 82.28 cm^2 (c) 82 cm^2
    (b) 82.3 cm^2 (d) 82.2 cm^2

18. The following elements are among the five most abundant by mass in the Earth's crust, seawater, and atmosphere *except*
    (a) oxygen   (c) silicon
    (b) hydrogen   (d) aluminum

19. Which of the following is a compound?
    (a) lead   (c) potassium
    (b) wood   (d) water

20. Which of the following is a mixture?
    (a) water   (c) wood
    (b) chromium   (d) sulfur

21. How many atoms are represented in the formula Na_2CrO_4?
    (a) 3   (b) (5)   (c) (7)   (d) 8

22. Which of the following is a characteristic of metals?
    (a) ductile   (c) extremely strong
    (b) easily shattered   (d) dull

23. Which of the following is a characteristic of nonmetals?
    (a) always a gas
    (b) poor conductor of electricity
    (c) shiny
    (d) combines only with metals

24. When a pure substance was analyzed, it was found to contain carbon and chlorine. This substance must be classified as
    (a) an element
    (b) a mixture
    (c) a compound
    (d) both a mixture and a compound

25. Chromium, fluorine, and magnesium have the symbols
    (a) Ch, F, Ma   (c) Cr, F, Mg
    (b) Cr, Fl, Mg   (d) Cr, F, Ma

26. Sodium, carbon, and sulfur have the symbols
    (a) Na, C, S   (c) Na, Ca, Su
    (b) So, C, Su   (d) So, Ca, Su

27. Coffee is an example of
    (a) an element   (c) a homogeneous mixture
    (b) a compound   (d) a heterogeneous mixture

28. The number of oxygen atoms in $Al(C_2H_3O_2)_3$ is
    (a) 2   (b) 3   (c) 5   (d) 6

29. Which of the following is a mixture?
    (a) water   (c) sugar solution
    (b) iron(II) oxide   (d) iodine

30. Which is the most compact state of matter?
    (a) solid   (c) gas
    (b) liquid   (d) amorphous

31. Which is not characteristic of a solution?
    (a) a homogeneous mixture
    (b) a heterogeneous mixture
    (c) one that has two or more substances
    (d) one that has a variable composition

32. A chemical formula is a combination of
    (a) symbols   (c) elements
    (b) atoms   (d) compounds

33. The number of nonmetal atoms in $Al_2(SO_3)_3$ is
    (a) 5   (b) 7   (c) 12   (d) 14

34. Which of the following is not a physical property?
    (a) boiling point   (c) bleaching action
    (b) physical state   (d) color

35. Which of the following is a physical change?
    (a) A piece of sulfur is burned.
    (b) A firecracker explodes.
    (c) A rubber band is stretched.
    (d) A nail rusts.

36. Which of the following is a chemical change?
    (a) Water evaporates.
    (b) Ice melts.
    (c) Rocks are ground to sand.
    (d) A penny tarnishes.

37. The changing of liquid water to ice is known as a
    (a) chemical change
    (b) heterogeneous change
    (c) homogeneous change
    (d) physical change

38. Which of the following does not represent a chemical change?
    (a) heating of copper in air
    (b) combustion of gasoline
    (c) cooling of red-hot iron
    (d) digestion of food

39. Heating 30. g of water from 20.°C to 50.°C requires
    (a) 30. cal   (c) 3.8×10^3 J
    (b) 50. cal   (d) 6.3×10^3 J

40. The specific heat of aluminum is 0.900 J/g°C. How many joules of energy are required to raise the temperature of 20.0 g of Al from 10.0°C to 15.0°C?
    (a) 79 J   (b) 90. J   (c) 100. J   (d) 112 J

41. A 100.-g iron ball (specific heat = 0.473 J/g°C) is heated to 125°C and is placed in a calorimeter holding 200. g of water at 25.0°C. What will be the highest temperature reached by the water?
    (a) 43.7°C   (c) 65.3°C
    (b) 30.4°C   (d) 35.4°C

42. Which has the highest specific heat?
    (a) ice   (b) lead   (c) water   (d) aluminum

43. When 20.0 g of mercury are heated from 10.0°C to 20.0°C, 27.6 J of energy are absorbed. What is the specific heat of mercury?
    (a) 0.726 J/g°C   (c) 2.76 J/g°C
    (b) 0.138 J/g°C   (d) no correct answer given

44. Changing hydrogen and oxygen into water is a
    (a) physical change
    (b) chemical change
    (c) conservation reaction
    (d) no correct answer given

Free Response Questions

Answer each of the following. Be sure to include your work and explanations in a clear, logical form.

1. You decide to go sailing in the tropics with some friends. Once there, you listen to the marine forecast, which predicts in-shore wave heights of 1.5 m, offshore wave heights of 4 m, and temperature of 27°C. Your friend the captain is unfamiliar with the metric system, and he needs to know whether it is safe for your small boat and if it will be warm. He asks you to convert the measurements to feet and degrees Fahrenheit, respectively.

2. Jane is melting butter in a copper pot on the stove. If she knows how much heat her stove releases per minute, what other measurements does she need to determine how much heat the butter absorbed? She assumes that the stove does not lose any heat to the surroundings.

3. Julius decided to heat 75 g $CaCO_3$ to determine how much carbon dioxide is produced. (*Note:* When $CaCO_3$ is heated, it produces CaO and carbon dioxide.) He collected the carbon dioxide in a balloon. Julius found the mass of the CaO remaining was 42 g. If 44 g of carbon dioxide take up 24 dm^3 of space, how many *liters* of gas were trapped in the balloon?

Use these pictures to answer Question 4.

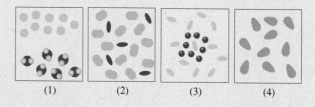

(1)   (2)   (3)   (4)

4. (a) Which picture best describes a homogeneous mixture?
   (b) How would you classify the contents of the other containers?
   (c) Which picture contains a compound? Explain how you made your choice.

Use these pictures to answer Question 5.

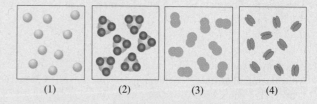

(1)   (2)   (3)   (4)

5. (a) Which picture best represents fluorine gas? Why?
   (b) Which other elements could that picture also represent?
   (c) Which of the pictures could represent SO_3 gas?

6. Sue and Tim each left a one-quart bowl outside one night. Sue's bowl was full of water and covered, while Tim's was empty and open. The next day there was a huge snowstorm that filled Tim's bowl with snow. The temperature that night went down to 12°F.
   (a) Which bowl would require less energy to bring its contents to room temperature (25°C)? Why?
   (b) What temperature change (°C) is required to warm the bowls to 25°C?
   (c) How much heat (in kJ) is required to raise the temperature of the contents of Sue's bowl to 0°C (without converting the ice to water)?
   (d) Did the water in Sue's bowl undergo chemical or physical changes or both?

7. One cup of Raisin Bran provides 60.% of the U.S. recommended daily allowance (RDA) of iron.
   (a) If the cereal provides 11 mg of iron, what is the U.S. RDA for Fe?
   (b) When the iron in the cereal is extracted, it is found to be the pure element. What is the volume of iron in a cup of the cereal?

8. Absent-minded Alfred put down a bottle containing silver on the table. When he went to retrieve it, he realized he had forgotten to label the bottle. Unfortunately, there were two full bottles of the same size side-by-side. Alfred realized he had placed a bottle of mercury on the same table last week. State two ways Alfred can determine which bottle contains silver without opening the bottles.

9. Suppose 25 g of solid sulfur and 35 g of oxygen gas are placed in a sealed container.
   (a) Does the container hold a mixture or a compound?
   (b) After heating, the container was weighed. From a comparison of the total mass before heating to the total mass after heating, can you tell whether a reaction took place? Explain.
   (c) After the container is heated, all the contents are gaseous. Has the density of the container including its contents changed? Explain.

Lightning occurs when electrons move to neutralize a charge difference between the clouds and the Earth.

CHAPTER **5**

EARLY ATOMIC THEORY AND STRUCTURE

Pure substances are classified as elements or compounds, but just what makes a substance possess its unique properties? How small a piece of salt will still taste salty? Carbon dioxide puts out fires, is used by plants to produce oxygen, and forms dry ice when solidified. But how small a mass of this material still behaves like carbon dioxide? Substances are in their simplest identifiable form at the atomic, ionic, or molecular level. Further division produces a loss of characteristic properties.

What particles lie within an atom or ion? How are these tiny particles alike? How do they differ? How far can we continue to divide them? Alchemists began the quest, early chemists laid the foundation, and modern chemists continue to build and expand on models of the atom.

CHAPTER OUTLINE

5.1 Dalton's Model of the Atom
5.2 Electric Charge
5.3 Subatomic Parts of the Atom
5.4 The Nuclear Atom
5.5 Isotopes of the Elements
5.6 Atomic Mass

5.1 DALTON'S MODEL OF THE ATOM

Describe Dalton's model of the atom and compare it to the earlier concepts of matter.

KEY TERM

Dalton's atomic model

The structure of matter has long intrigued and engaged us. The earliest models of the atom were developed by the ancient Greek philosophers. About 440 B.C. Empedocles stated that all matter was composed of four "elements"—earth, air, water, and fire. Democritus (about 470–370 B.C.) thought that all forms of matter were composed of tiny indivisible particles, which he called atoms, derived from the Greek word *atomos*, meaning "indivisible." He held that atoms were in constant motion and that they combined with one another in various ways. This hypothesis was not based on scientific observations. Shortly thereafter, Aristotle (384–322 B.C.) opposed the theory of Democritus and instead endorsed and advanced the Empedoclean theory. So strong was the influence of Aristotle that his theory dominated the thinking of scientists and philosophers until the beginning of the seventeenth century.

More than 2000 years after Democritus, the English schoolmaster John Dalton (1766–1844) revived the concept of atoms and proposed an atomic model based on facts and experimental evidence (**Figure 5.1**). His theory, described in a series of papers published from 1803 to 1810, rested on the idea of a different kind of atom for each element. The essence of **Dalton's atomic model** may be summed up as follows:

1. Elements are composed of minute, indivisible particles called atoms.

2. Atoms of the same element are alike in mass and size.

3. Atoms of different elements have different masses and sizes.

4. Chemical compounds are formed by the union of two or more atoms of different elements.

5. Atoms combine to form compounds in simple numerical ratios, such as one to one, one to two, two to three, and so on.

6. Atoms of two elements may combine in different ratios to form more than one compound.

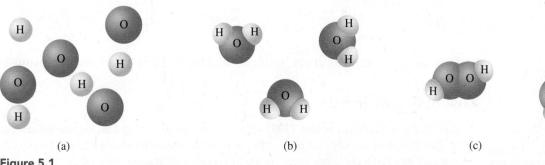

     (a)   (b)   (c)

Figure 5.1

(a) Dalton's atoms were individual particles, the atoms of each element being alike in mass and size but different in mass and size from other elements; (b) and (c) Dalton's atoms combine in specific ratios to form compounds.

Dalton's atomic model stands as a landmark in the development of chemistry. The major premises of his model are still valid, but some of his statements must be modified or qualified because later investigations have shown that (1) atoms are composed of subatomic particles; (2) not all the atoms of a specific element have the same mass; and (3) atoms, under special circumstances, can be decomposed.

In chemistry we use models (theories) such as Dalton's atomic model to explain the behavior of atoms, molecules, and compounds. Models are modified to explain new information. We frequently learn the most about a system when our models (theories) fail. That is the time when we must rethink our explanation and determine whether we need to modify our model or propose a new or different model to explain the behavior.

WileyPLUS

ENHANCED EXAMPLE

EXAMPLE 5.1

Which statement of Dalton's model of the atom does this series of compounds represent: NO_2, N_2O, N_2O_3, N_2O_5?

SOLUTION

Atoms combine to form compounds in simple numerical ratios. Atoms of two elements combine in different ratios to form more than one compound.

PRACTICE 5.1

List another series of compounds that illustrates the statements in Example 5.1.

5.2 ELECTRIC CHARGE

LEARNING OBJECTIVE   ● Recognize the force between particles and distinguish between a cation and an anion.

You've probably received a shock after walking across a carpeted area on a dry day. You may have also experienced the static electricity associated with combing your hair and have had your clothing cling to you. These phenomena result from an accumulation of *electric charge*. This charge may be transferred from one object to another. The properties of electric charge are as follows:

1. Charge may be of two types, positive and negative.
2. Unlike charges attract (positive attracts negative), and like charges repel (negative repels negative and positive repels positive).
3. Charge may be transferred from one object to another by contact or induction.
4. The less the distance between two charges, the greater the force of attraction between unlike charges (or repulsion between identical charges). The force of attraction (F) can be expressed using the following equation:

$$F = \frac{kq_1q_2}{r^2}$$

where q_1 and q_2 are the charges, r is the distance between the charges, and k is a constant.

Discovery of Ions

English scientist Michael Faraday (1791–1867) made the discovery that certain substances when dissolved in water conduct an electric current. He also noticed that certain compounds decompose into their elements when an electric current is passed through the compound. Atoms of some elements are attracted to the positive electrode, while atoms of other elements are attracted to the negative electrode. Faraday concluded that these atoms are electrically charged. He called them *ions* after the Greek word meaning "wanderer."

Any moving charge is an electric current. The electrical charge must travel through a substance known as a conducting medium. The most familiar conducting media are metals formed into wires.

The Swedish scientist Svante Arrhenius (1859–1927) extended Faraday's work. Arrhenius reasoned that an ion is an atom (or a group of atoms) carrying a positive or negative charge. When a compound such as sodium chloride (NaCl) is melted, it conducts electricity. Water is unnecessary. Arrhenius's explanation of this conductivity was that upon melting, the sodium chloride dissociates, or breaks up, into charged ions, Na^+ and Cl^-. The Na^+ ions move toward the negative electrode (cathode), whereas the Cl^- ions migrate toward the positive electrode (anode). Thus positive ions are called *cations*, and negative ions are called *anions*.

From Faraday's and Arrhenius's work with ions, Irish physicist G. J. Stoney (1826–1911) realized there must be some fundamental unit of electricity associated with atoms.

Richard Megna/Fundamental Photographs

When ions are present in a solution of salt water and an electric current is passed through the solution, the light bulb glows.

He named this unit the electron in 1891. Unfortunately, he had no means of supporting his idea with experimental proof. Evidence remained elusive until 1897, when English physicist J. J. Thomson (1856–1940) was able to show experimentally the existence of the electron.

EXAMPLE 5.2

What is a cation? What is an anion?

SOLUTION

A cation is a positively charged ion. An anion is a negatively charged ion.

PRACTICE 5.2

Which of the following ions are cations and which are anions?

Na^+, Cl^-, NH_4^+, Ca^{2+}, Br^-, Al^{3+}, NO_3^-, SO_4^{2-}, PO_4^{3-}, K^+, Ag^+, Fe^{2+}, Fe^{3+}, S^{2-}

WileyPLUS
ENHANCED EXAMPLE

5.3 SUBATOMIC PARTS OF THE ATOM

Describe the three basic subatomic particles and how they changed Dalton's model of the atom.

● **LEARNING OBJECTIVE**

KEY TERMS
subatomic particles
electron
proton
Thomson model of the atom
neutron

The concept of the atom—a particle so small that until recently it could not be seen even with the most powerful microscope—and the subsequent determination of its structure stand among the greatest creative intellectual human achievements.

Any visible quantity of an element contains a vast number of identical atoms. But when we refer to an atom of an element, we isolate a single atom from the multitude in order to present the element in its simplest form. **Figure 5.2** shows individual atoms as we can see them today.

What is this tiny particle we call the atom? The diameter of a single atom ranges from 0.1 to 0.5 nanometer (1 nm $= 1 \times 10^{-9}$ m). Hydrogen, the smallest atom, has a diameter of about 0.1 nm. To arrive at some idea of how small an atom is, consider this dot (•), which has a diameter of about 1 mm, or 1×10^6 nm. It would take 10 million hydrogen atoms to form a line of atoms across this dot. As inconceivably small as atoms are, they contain even smaller particles, the **subatomic particles**, including electrons, protons, and neutrons.

The development of atomic theory was helped in large part by the invention of new instruments. For example, the Crookes tube, developed by Sir William Crookes (1832–1919) in 1875, opened the door to the subatomic structure of the atom (**Figure 5.3**). The emissions generated in a Crookes tube are called *cathode rays*. J. J. Thomson demonstrated in 1897 that cathode rays (1) travel in straight lines, (2) are negative in charge, (3) are deflected by electric and magnetic fields, (4) produce sharp shadows, and (5) are capable of moving a small paddle wheel. This was the experimental discovery of the fundamental unit of charge—the electron.

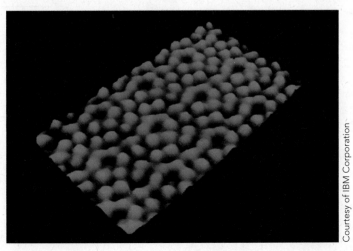

Courtesy of IBM Corporation

Figure 5.2

A scanning tunneling microscope shows an array of copper atoms.

Figure 5.3

Cathode ray tube. A stream of electrons passes between electrodes. The fast-moving particles excite the gas inside the tube, creating a greenish glow between the electrodes.

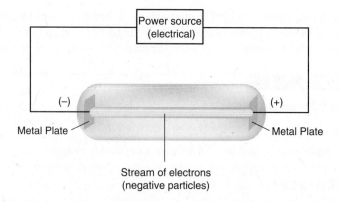

Power source (electrical)

(−)   (+)

Metal Plate   Metal Plate

Stream of electrons
(negative particles)

The **electron** (e^-) is a particle with a negative electrical charge and a mass of 9.110×10^{-28} g. This mass is 1/1837 the mass of a hydrogen atom. Although the actual charge of an electron is known, its value is too cumbersome for practical use and has therefore been assigned a relative electrical charge of −1. The size of an electron has not been determined exactly, but its diameter is believed to be less than 10^{-12} cm.

Protons were first observed by German physicist Eugen Goldstein (1850–1930) in 1886. However, it was Thomson who discovered the nature of the proton. He showed that the proton is a particle, and he calculated its mass to be about 1837 times that of an electron. The **proton** (p) is a particle with actual mass of 1.673×10^{-24} g. Its relative charge (+1) is equal in magnitude, but opposite in sign, to the charge on the electron. The mass of a proton is only very slightly less than that of a hydrogen atom.

Thomson had shown that atoms contain both negatively and positively charged particles. Clearly, the Dalton model of the atom was no longer acceptable. Atoms are not indivisible but are instead composed of smaller parts. Thomson proposed a new model of the atom.

In the **Thomson model of the atom**, the electrons are negatively charged particles embedded in the positively charged atomic sphere (see **Figure 5.4**). A neutral atom could become an ion by gaining or losing electrons.

Positive ions were explained by assuming that a neutral atom loses electrons. An atom with a net charge of +1 (for example, Na^+ or Li^+) has lost one electron. An atom with a net charge of +3 (for example, Al^{3+}) has lost three electrons (**Figure 5.5a**).

Negative ions were explained by assuming that additional electrons can be added to atoms. A net charge of −1 (for example, Cl^- or F^-) is produced by the addition of one electron. A net charge of −2 (for example, O^{2-} or S^{2-}) requires the addition of two electrons (**Figure 5.5b**).

The third major subatomic particle was discovered in 1932 by James Chadwick (1891–1974). This particle, the **neutron** (n), has neither a positive nor a negative charge and

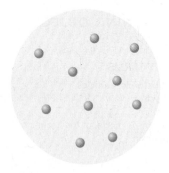

Figure 5.4

Thomson model of the atom. In this early model of the atom, negative particles (electrons) were thought to be embedded in a positively charged sphere. It is sometimes called the plum pudding model.

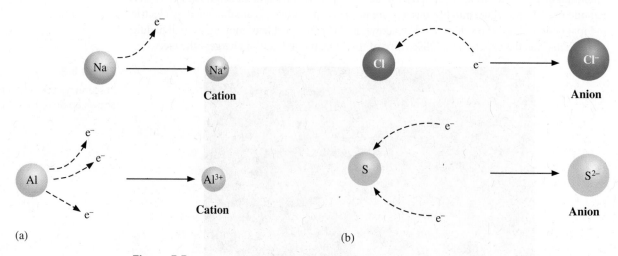

(a)   (b)

Figure 5.5

(a) When one or more electrons are lost from a neutral atom, a cation is formed. (b) When one or more electrons are added to a neutral atom, an anion is formed.

TABLE 5.1	Electrical Charge and Relative Mass of Electrons, Protons, and Neutrons		
Particle	Symbol	Relative electrical charge	Actual mass (g)
Electron	e^-	-1	9.110×10^{-28}
Proton	p	$+1$	1.673×10^{-24}
Neutron	n	0	1.675×10^{-24}

has an actual mass (1.675×10^{-24} g) that is only very slightly greater than that of a proton. The properties of these three subatomic particles are summarized in Table 5.1.

Nearly all the ordinary chemical properties of matter can be explained in terms of atoms consisting of electrons, protons, and neutrons. The discussion of atomic structure that follows is based on the assumption that atoms contain only these principal subatomic particles. Many other subatomic particles, such as mesons, positrons, neutrinos, and antiprotons, have been discovered, but it is not yet clear whether all these particles are actually present in the atom or whether they are produced by reactions occurring within the nucleus. The fields of atomic and high-energy physics have produced a long list of subatomic particles. Descriptions of the properties of many of these particles are to be found in physics textbooks.

WileyPLUS

ENHANCED EXAMPLE

EXAMPLE 5.3

The mass of a helium atom is 6.65×10^{-24} g. How many atoms are in a 4.0-g sample of helium?

SOLUTION

$$(4.0 \text{ g})\left(\frac{1 \text{ atom He}}{6.65 \times 10^{-24} \text{ g}}\right) = 6.0 \times 10^{23} \text{ atoms He}$$

PRACTICE 5.3

The mass of an atom of hydrogen is 1.673×10^{-24} g. How many atoms are in a 10.0-g sample of hydrogen?

5.4 THE NUCLEAR ATOM

Explain how the nuclear model of the atom differs from Dalton's and Thomson's models.

● **LEARNING OBJECTIVE**

KEY TERMS
nucleus
atomic number

The discovery that positively charged particles are present in atoms came soon after the discovery of radioactivity by Henri Becquerel (1852–1908) in 1896. Radioactive elements spontaneously emit alpha particles, beta particles, and gamma rays from their nuclei (see Chapter 18).

By 1907 Ernest Rutherford (1871–1937) had established that the positively charged alpha particles emitted by certain radioactive elements are ions of the element helium. Rutherford used these alpha particles to establish the nuclear nature of atoms. In experiments performed in 1911, he directed a stream of positively charged helium ions (alpha particles) at a very thin sheet of gold foil (about 1000 atoms thick). See Figure 5.6a. He observed that most of the alpha particles passed through the foil with little or no deflection; but a few of the particles were deflected at large angles, and occasionally one even bounced back from the foil (Figure 5.6b). It was known that like charges repel each other and that an electron with a mass of 1/1837 that of a proton could not possibly have an appreciable effect on the path of an alpha particle, which is about 7350 times more massive than an electron. Rutherford therefore reasoned that each gold atom must contain a positively charged mass occupying a relatively tiny volume and that, when an alpha particle approaches close enough to this positive mass, it is deflected. Rutherford spoke of this positively charged mass as the *nucleus* of the atom. Because alpha particles have relatively high masses, the extent of the deflections (some actually

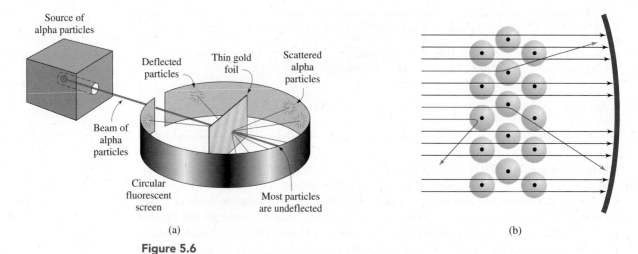

Figure 5.6

(a) Rutherford's experiment on alpha-particle scattering, where positive alpha particles (α), emanating from a radioactive source, were directed at a thin gold foil. (b) Deflection (red) and scattering (blue) of the positive alpha particles by the positive nuclei of the gold atoms.

bounced back) indicated to Rutherford that the nucleus is very heavy and dense. (The density of the nucleus of a hydrogen atom is about 10^{12} g/cm^3—about 1 trillion times the density of water.) Because most of the alpha particles passed through the thousand or so gold atoms without any apparent deflection, he further concluded that most of an atom consists of empty space.

When we speak of the mass of an atom, we are referring primarily to the mass of the nucleus. The nucleus contains all the protons and neutrons, which represent more than 99.9% of the total mass of any atom. By way of illustration, the largest number of electrons known to exist in an atom is 118. The mass of even 118 electrons is only about 1/17 of the mass of a single proton or neutron. The mass of an atom therefore is primarily determined by the combined masses of its protons and neutrons.

General Arrangement of Subatomic Particles

The alpha-particle scattering experiments of Rutherford established that the atom contains a dense, positively charged nucleus. The later work of Chadwick demonstrated that the atom contains neutrons, which are particles with mass, but no charge. Rutherford also noted that light, negatively charged electrons are present and offset the positive charges in the nucleus. Based on this experimental evidence, a model of the atom and the location of its subatomic particles was devised in which each atom consists of a **nucleus** surrounded by electrons (see **Figure 5.7**). The nucleus contains protons and neutrons but does not contain electrons. In a neutral atom the positive charge of the nucleus (due to protons) is exactly offset by the negative electrons. Because the charge of an electron is equal to, but of opposite sign than, the charge of a proton, a neutral atom must contain exactly the same number of electrons as protons. However, this model of atomic structure provides no information on the arrangement of electrons within the atom.

A neutral atom contains the same number of protons and electrons.

Figure 5.7

In the nuclear model of the atom, protons and neutrons are located in the nucleus. The electrons are found in the remainder of the atom (which is mostly empty space because electrons are very tiny).

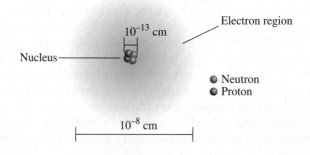

EXAMPLE 5.4

What are the major subatomic particles and where are they located in an atom?

SOLUTION

Protons, neutrons, and electrons are the three major subatomic particles in an atom. The protons and neutrons are located in the nucleus and the electrons are located outside the nucleus and around the nucleus.

WileyPLUS

ENHANCED EXAMPLE

Atomic Numbers of the Elements

The **atomic number** (Z) of an element is the number of protons in the nucleus of an atom of that element. The atomic number determines the identity of an atom. For example, every atom with an atomic number of 1 is a hydrogen atom; it contains one proton in its nucleus. Every atom with an atomic number of 6 is a carbon atom; it contains 6 protons in its nucleus. Every atom with an atomic number of 92 is a uranium atom; it contains 92 protons in its nucleus. The atomic number tells us not only the number of positive charges in the nucleus but also the number of electrons in the neutral atom, since a neutral atom contains the same number of electrons and protons.

atomic number = number of protons in the nucleus

You don't need to memorize the atomic numbers of the elements because a periodic table is usually provided in texts, in laboratories, and on examinations. The atomic numbers of all elements are shown in the periodic table on the inside front cover of this book and are also listed in the table of atomic masses on the inside front endpapers.

PRACTICE 5.4

Look in the periodic table and determine how many protons and electrons are in a Cu atom and an Xe atom.

5.5 ISOTOPES OF THE ELEMENTS

Define the terms *atomic number, mass number,* and *isotope.*

● **LEARNING OBJECTIVE**

KEY TERMS

isotope
mass number

Shortly after Rutherford's conception of the nuclear atom, experiments were performed to determine the masses of individual atoms. These experiments showed that the masses of nearly all atoms were greater than could be accounted for by simply adding up the masses of all the protons and electrons that were known to be present in an atom. This fact led to the concept of the neutron, a particle with no charge but with a mass about the same as that of a proton. Because this particle has no charge, it was very difficult to detect, and the existence of the neutron was not proven experimentally until 1932. All atomic nuclei except that of the simplest hydrogen atom contain neutrons.

All atoms of a given element have the same number of protons. Experimental evidence has shown that, in most cases, all atoms of a given element do not have identical masses. This is because atoms of the same element may have different numbers of neutrons in their nuclei.

Atoms of an element having the same atomic number but different atomic masses are called **isotopes** of that element. Atoms of the various isotopes of an element therefore have the same number of protons and electrons but different numbers of neutrons.

Three isotopes of hydrogen (atomic number 1) are known. Each has one proton in the nucleus and one electron. The first isotope (protium), without a neutron, has a mass number of 1; the second isotope (deuterium), with one neutron in the nucleus, has a mass number of 2; the third isotope (tritium), with two neutrons, has a mass number of 3 (see **Figure 5.8**).

The three isotopes of hydrogen may be represented by the symbols ^1_1H, ^2_1H and ^3_1H, indicating an atomic number of 1 and mass numbers of 1, 2, and 3, respectively. This method

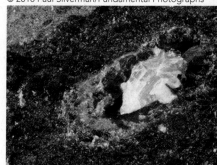

An uncut diamond from a Colorado mine. Diamonds are composed of carbon.

Figure 5.8

The isotopes of hydrogen. The number of protons (purple) and neutrons (blue) are shown within the nucleus. The electron (e⁻) exists outside the nucleus.

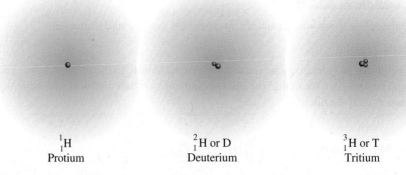

$^1_1$H
Protium

$^2_1$H or D
Deuterium

$^3_1$H or T
Tritium

of representing atoms is called *isotopic notation*. The subscript (Z) is the atomic number; the superscript (A) is the **mass number**, which is the sum of the number of protons and the number of neutrons in the nucleus. The hydrogen isotopes may also be referred to as hydrogen-1, hydrogen-2, and hydrogen-3.

The mass number of an element is the sum of the protons and neutrons in the nucleus.

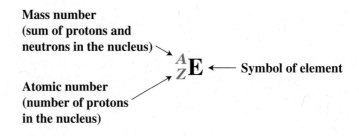

Mass number (sum of protons and neutrons in the nucleus) $\longrightarrow$ ^A_Z**E** $\longleftarrow$ **Symbol of element**

Atomic number (number of protons in the nucleus)

Most of the elements occur in nature as mixtures of isotopes. However, not all isotopes are stable; some are radioactive and are continuously decomposing to form other elements. For example, of the seven known isotopes of carbon, only two, carbon-12 and carbon-13, are stable. Of the seven known isotopes of oxygen, only three—$^{16}_8$O, $^{17}_8$O, and $^{18}_8$O—are stable. Of the fifteen known isotopes of arsenic, $^{75}_{33}$As is the only one that is stable.

The relationship between mass number and atomic number is such that if we subtract the atomic number from the mass number of a given isotope, we obtain the number of neutrons in the nucleus of an atom of that isotope. Table 5.2 shows this method of determining the number of neutrons. For example, the fluorine atom ($^{19}_9$F), atomic number 9, having a mass of 19 contains 10 neutrons:

mass number	−	atomic number	=	number of neutrons
19	−	9	=	10

Use four significant figures for atomic masses in this text.

The atomic masses given in the table on the front endpapers of this book are values accepted by international agreement. You need not memorize atomic masses. In the calculations in this book, the use of atomic masses rounded to four significant figures will give results of sufficient accuracy. (See periodic table.)

TABLE 5.2   Determination of the Number of Neutrons in an Atom by Subtracting Atomic Number from Mass Number

	Hydrogen ($^1_1$H)	Oxygen ($^{16}_8$O)	Sulfur ($^{32}_{16}$S)	Fluorine ($^{19}_9$F)	Iron ($^{56}_{26}$Fe)
Mass number	1	16	32	19	56
Atomic number	(−)1	(−)8	(−)16	(−)9	(−)26
Number of neutrons	0	8	16	10	30

EXAMPLE 5.5

WileyPLUS
ENHANCED EXAMPLE

How many protons, neutrons, and electrons are found in an atom of $^{14}_{6}C$?

SOLUTION

The element is carbon, atomic number 6. The number of protons or electrons equals the atomic number and is 6. The number of neutrons is determined by subtracting the atomic number from the mass number: $14 - 6 = 8$.

PRACTICE 5.5

How many protons, neutrons, and electrons are in each of these isotopes?

(a) $^{16}_{8}O$ (b) $^{80}_{35}Br$ (c) $^{235}_{92}U$ (d) $^{64}_{29}Cu$

PRACTICE 5.6

What are the atomic number and the mass number of the elements that contain

(a) 9 electrons   (b) 24 protons and 28 neutrons   (c) $^{197}_{79}X$

What are the names of these elements?

>CHEMISTRY *IN ACTION*

Isotope Detectives

Scientists are learning to use isotopes to determine the origin of drugs and gems. It turns out that isotope ratios similar to those used in carbon dating can also identify the source of cocaine or the birthplace of emeralds.

Researchers with the Drug Enforcement Agency (DEA) have created a database of the origin of coca leaves that pinpoints the origin of the leaves with a 90% accuracy. Cocaine keeps a chemical signature of the environment where it grew. Isotopes of carbon and nitrogen are found in a particular ratio based on climatic conditions in the growing region. These ratios correctly identified the source of 90% of the samples tested, according to James Ehleringer of the University of Utah, Salt Lake City. This new method can trace drugs a step further back than current techniques, which mainly look at chemicals introduced by processing practices in different locations. This could aid in tracking the original exporters and stopping production at the source.

It turns out that a similar isotopic analysis of oxygen has led researchers in France to be able to track the birthplace of emeralds. Very high quality emeralds have few inclusions (microscopic cavities). Gemologists use these inclusions and the material trapped in them to identify the source of the gem. High-quality gems can now also be identified by using an oxygen isotope ratio. These tests use an ion microscope that blasts a few atoms from the gems' surface (with virtually undetectable damage). The tiny sample is analyzed for its oxygen isotope ratio and then compared to a database from emerald mines around the world. Using the information, gemologists can determine the mine from which the emerald came. Since emeralds from Colombian mines are valued much more highly than those from other countries, this technique can be used to help collectors know just what they are paying for, as well as to identify the history of treasured emeralds.

Raw and cut emeralds

© PjrStudio/Alamy

5.6 ATOMIC MASS

LEARNING OBJECTIVE ● Explain the relationship between the atomic mass of an element and the masses of its isotopes.

KEY TERMS

atomic mass unit (amu)
atomic mass

$1 \text{ amu} = 1.6606 \times 10^{-24} \text{ g}$

Tetra Images /Getty Images, Inc.

The copper used in casting the Liberty Bell contains a mixture of the isotopes of copper.

The mass of a single atom is far too small to measure on a balance, but fairly precise determinations of the masses of individual atoms can be made with an instrument called a *mass spectrometer*. The mass of a single hydrogen atom is 1.673×10^{-24} g. However, it is neither convenient nor practical to compare the actual masses of atoms expressed in grams; therefore, a table of relative atomic masses using *atomic mass units* was devised. (The term *atomic weight* is sometimes used instead of *atomic mass*.) The carbon isotope having six protons and six neutrons and designated carbon-12, or $^{12}_{6}$C, was chosen as the standard for atomic masses. This reference isotope was assigned a value of exactly 12 atomic mass units (amu). Thus, 1 **atomic mass unit (amu)** is defined as equal to exactly 1/12 of the mass of a carbon-12 atom. The actual mass of a carbon-12 atom is 1.9927×10^{-23} g and that of one atomic mass unit is 1.6606×10^{-24} g. In the table of atomic masses, all elements then have values that are relative to the mass assigned to the reference isotope, carbon-12.

A table of atomic masses is given on the inside front cover of this book. Hydrogen atoms, with a mass of about 1/12 that of a carbon atom, have an average atomic mass of 1.00794 amu on this relative scale. Magnesium atoms, which are about twice as heavy as carbon, have an average mass of 24.305 amu. The average atomic mass of oxygen is 15.9994 amu.

Since most elements occur as mixtures of isotopes with different masses, the atomic mass determined for an element represents the average relative mass of all the naturally occurring isotopes of that element. The atomic masses of the individual isotopes are approximately whole numbers, because the relative masses of the protons and neutrons are approximately 1.0 amu each. Yet we find that the atomic masses given for many of the elements deviate considerably from whole numbers.

For example, the atomic mass of rubidium is 85.4678 amu, that of copper is 63.546 amu, and that of magnesium is 24.305 amu. The deviation of an atomic mass from a whole number is due mainly to the unequal occurrence of the various isotopes of an element.

The two principal isotopes of copper are $^{63}_{29}$Cu and $^{65}_{29}$Cu. Copper used in everyday objects and the Liberty Bell contains a mixture of these two isotopes. It is apparent that copper-63 atoms are the more abundant isotope, since the atomic mass of copper, 63.546 amu, is closer to 63 than to 65 amu (see **Figure 5.9**). The actual values of the copper isotopes observed by mass spectra determination are shown in the following table:

Isotope	Isotopic mass (amu)	Abundance (%)	Average atomic mass (amu)
$^{63}_{29}$Cu	62.9298	69.09	63.55
$^{65}_{29}$Cu	64.9278	30.91	

The average atomic mass is a weighted average of the masses of all the isotopes present in the sample.

The average atomic mass can be calculated by multiplying the atomic mass of each isotope by the fraction of each isotope present and adding the results. The calculation for copper is

$$(62.9298 \text{ amu})(0.6909) = 43.48 \text{ amu}$$

$$(64.9278 \text{ amu})(0.3091) = \underline{20.07 \text{ amu}}$$
$$63.55 \text{ amu}$$

The **atomic mass** of an element is the average relative mass of the isotopes of that element compared to the atomic mass of carbon-12 (exactly 12.0000... amu).

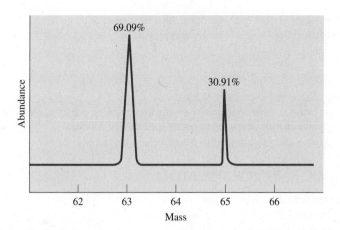

Figure 5.9
A typical reading from a mass spectrometer. The two principal isotopes of copper are shown with the abundance (%) given.

EXAMPLE 5.6

Chlorine is found in nature as two isotopes, $^{37}_{17}Cl$ (24.47%) and $^{35}_{17}Cl$ (75.53%). The atomic masses are 36.96590 and 34.96885 amu, respectively. Determine the average atomic mass of chlorine.

SOLUTION

Multiply each mass by its percentage and add the results to find the average:

$(0.2447)(36.96590 \text{ amu}) + (0.7553)(34.96885 \text{ amu})$

$= 35.4575 \text{ amu}$

$= 35.46 \text{ amu}$ (four significant figures)

PRACTICE 5.7

Silver occurs as two isotopes with atomic masses 106.9041 and 108.9047 amu, respectively. The first isotope represents 51.82% and the second 48.18%. Determine the average atomic mass of silver.

CHAPTER 5 REVIEW

5.1 DALTON'S MODEL OF THE ATOM

- Greek model of matter:
  - Four elements—earth, air, water, fire
  - Democritus—atoms (indivisible particles) make up matter
  - Aristotle—opposed atomic ideas
- Summary of Dalton's model of the atom:
  - Elements are composed of atoms.
  - Atoms of the same element are alike (mass and size).
  - Atoms of different elements are different in mass and size.
  - Compounds form by the union of two or more atoms of different elements.
  - Atoms form compounds in simple numerical ratios.
  - Atoms of two elements may combine in different ratios to form different compounds.

KEY TERM
Dalton's atomic model

5.2 ELECTRIC CHARGE

- The properties of electric charge:
  - Charges are one of two types—positive or negative.
  - Unlike charges attract and like charges repel.

- Charge is transferred from one object to another by contact or by induction.
- The force of attraction between charges is expressed by

$$F = \frac{kq_1q_2}{r^2}$$

- Michael Faraday discovered electrically charged ions.
- Svante Arrhenius explained that conductivity results from the dissociation of compounds into ions:
  - Cation—positive charge—attracted to negative electrode (cathode)
  - Anion—negative charge—attracted to positive electrode (anode)

5.3 SUBATOMIC PARTS OF THE ATOM

KEY TERMS

subatomic particles
electron
proton
Thomson model of the atom
neutron

- Atoms contain smaller subatomic particles:
  - Electron—negative charge, 1/1837 mass of proton
  - Proton—positive charge, 1.673×10^{-24} g
  - Neutron—no charge, 1.675×10^{-24} g
- Thomson model of the atom:
  - Negative electrons are embedded in a positive atomic sphere.
- Number of protons = number of electrons in a neutral atom.
- Ions are formed by losing or gaining electrons.

5.4 THE NUCLEAR ATOM

KEY TERMS

nucleus
atomic number

- Rutherford gold foil experiment modified the Thomson model to a nuclear model of the atom:
  - Atoms are composed of a nucleus containing protons and/or neutrons surrounded by electrons, which occupy mostly empty space.
  - Neutral atoms contain equal numbers of protons and electrons.

5.5 ISOTOPES OF THE ELEMENTS

KEY TERMS

isotope
mass number

- The mass number of an element is the sum of the protons and neutrons in the nucleus.

**Mass number
(sum of protons and
neutrons in the nucleus)** ↘

$$^A_Z\mathbf{E} \longleftarrow \textbf{Symbol of element}$$

**Atomic number
(number of protons
in the nucleus)**

- The number of neutrons in an atom is determined by

mass number − atomic number = number of neutrons
       A − Z = neutrons

5.6 ATOMIC MASS

KEY TERMS

atomic mass unit (amu)
atomic mass

- The average atomic mass is a weighted average of the masses of all the isotopes present in the sample.

REVIEW QUESTIONS

1. List the main features of Dalton's model of the atom.
2. What elements of Democritus' theory did Dalton incorporate into his model of the atom?
3. How does Dalton's theory explain why chemical formulas are always written with whole-number values?
4. Distinguish between an atom and an ion.
5. Does the force of attraction increase, decrease, or stay the same as two oppositely charged particles approach each other?
6. How do cations and anions differ?

7. A neutron is approximately how many times heavier than an electron?
8. From the chemist's point of view, what are the essential differences among a proton, a neutron, and an electron?
9. What are the atomic numbers of (a) copper, (b) nitrogen, (c) phosphorus, (d) radium, and (e) zinc?
10. What letters are used to designate atomic number and mass number in isotopic notation of atoms?
11. In what ways are isotopes alike? In what ways are they different?
12. Explain why the mass number of an element is always a whole number.

Most of the exercises in this chapter are available for assignment via the online homework management program, WileyPLUS (www.wileyplus. com). All exercises with blue numbers have answers in Appendix VI.

PAIRED EXERCISES

1. Identify the ions in a mixture containing Cu^{2+}, N^{3-}, O^{2-}, Ca^{2+}, Al^{3+}, Te^{3-}, and H^+ that would be attracted to the positive end of an electric field.

2. Identify the ions in a mixture containing Mg^{2+}, Br^-, Cr^{3+}, Ba^{2+}, P^{3-}, Ca^{2+}, Se^{2-}, and Y^{3+} that would be attracted to the anode of a battery.

3. Explain why, in Rutherford's experiments, some alpha particles were scattered at large angles by the gold foil or even bounced back.

4. What experimental evidence led Rutherford to conclude the following?
   (a) The nucleus of the atom contains most of the atomic mass.
   (b) The nucleus of the atom is positively charged.
   (c) The atom consists of mostly empty space.

5. Describe the general arrangement of subatomic particles in the atom.

6. What part of the atom contains practically all its mass?

7. What contribution did these scientists make to atomic models of the atom?
   (a) Dalton   (b) Thomson   (c) Rutherford

8. Consider the following models of the atom: (a) Dalton, (b) Thomson, (c) Rutherford. How does the location of the electrons in an atom vary? How does the location of the atom's positive matter compare?

9. Explain why the atomic masses of elements are not whole numbers.

10. Is the isotopic mass of a given isotope ever an exact whole number? Is it always? (Consider the masses of $^{12}_{6}C$ and $^{63}_{29}Cu$.)

11. What special names are given to the isotopes of hydrogen?

12. List the similarities and differences in the three isotopes of hydrogen.

13. What is the nuclear composition of the five naturally occurring isotopes of germanium having mass numbers 70, 72, 73, 74, and 76?

14. What is the nuclear composition of the five naturally occurring isotopes of zinc having mass numbers 64, 66, 67, 68, and 70?

15. Write isotopic notation symbols for each of the following:
   (a) $Z = 29$, $A = 65$
   (b) $Z = 20$, $A = 45$
   (c) $Z = 36$, $A = 84$

16. Write isotopic notation symbols for each of the following:
   (a) $Z = 47$, $Z = 109$
   (b) $Z = 8$, $A = 18$
   (c) $Z = 26$, $A = 57$

17. Give the number of protons, neutrons, and electrons in the following isotopes.
   (a) ^{63}Cu
   (b) ^{32}S
   (c) manganese-55 (^{55}Mn)
   (d) potassium-39 (^{39}K)

18. Tell the number of protons, neutrons, and electrons in the following isotopes.
   (a) ^{54}Fe
   (b) ^{23}Na
   (c) bromine-79 (^{79}Br)
   (d) phosphorus-31 (^{31}P)

19. An unknown element contains 33 protons, 36 electrons, and has a mass number 76. Answer the following questions:
   (a) What is the atomic number of this element?
   (b) What is the name and symbol of this element?
   (c) How many neutrons does it contain?
   (d) What is the charge on this ion? Is it a cation or an anion?
   (e) Write the symbolic notation for this element.

20. An unknown element contains 56 protons, 54 electrons, and has a mass number 135. Answer the following questions:
   (a) What is the atomic number of this element?
   (b) What is the name and symbol of this element?
   (c) How many neutrons does it contain?
   (d) What is the charge on this ion? Is it a cation or an anion?
   (e) Write the symbolic notation for this element.

21. Naturally occurring zirconium exists as five stable isotopes: ^{90}Zr with a mass of 89.905 amu (51.45%); ^{91}Zr with a mass of 90.906 amu (11.22%); ^{92}Zr with a mass of 91.905 amu (17.15%); ^{94}Zr with a mass of 93.906 amu (17.38%); and ^{96}Zr with a mass of 95.908 amu (2.80%). Calculate the average mass of zirconium.

22. Naturally occurring titanium exists as five stable isotopes. Four of the isotopes are ^{46}Ti with a mass of 45.953 amu (8.0%); ^{47}Ti with a mass of 46.952 amu (7.3%); ^{48}Ti with a mass of 47.948 amu (73.8%); and ^{49}Ti with a mass of 48.948 amu (5.5%). The average mass of an atom of titanium is 47.9 amu. Determine the mass of the fifth isotope of titanium.

23. A particular element exists in two stable isotopic forms. One isotope has a mass of 62.9296 amu (69.17% abundance). The other isotope has a mass of 64.9278 amu. Calculate the average mass of the element and determine its identity.

24. A particular element exists in two stable isotopic forms. One isotope has a mass of 34.9689 amu (75.77% abundance). The other isotope has a mass of 36.9659 amu. Calculate the average mass of the element and determine its identity.

25. An average dimension for the radius of an atom is 1.0×10^{-8} cm, and the average radius of the nucleus is 1.0×10^{-13} cm. Determine the ratio of atomic volume to nuclear volume. Assume that the atom is spherical [$V = (4/3)\pi r^3$ for a sphere].

26. An aluminum atom has an average diameter of about 3.0×10^{-8} cm. The nucleus has a diameter of about 2.0×10^{-13} cm. Calculate the ratio of the atom's diameter to its nucleus.

ADDITIONAL EXERCISES

27. What experimental evidence supports these statements?
    (a) The nucleus of an atom is small.
    (b) The atom consists of both positive and negative charges.
    (c) The nucleus of the atom is positive.

28. Given the following information, determine which, if any, elements are isotopes.
    (a) mass number = 32; number of neutrons = 17
    (b) mass number = 32; number of protons = 16
    (c) number of protons = 15; number of neutrons = 16

29. The diameter of a silicon atom is 2.34×10^{-8} cm. If the silicon atoms were placed next to each other in a straight line across an 8.5-in. width of paper, how many silicon atoms would be there?

30. How is it possible for there to be more than one kind of atom of the same element?

31. Place each of the following elements in order of increasing number of neutrons (least to most); ^{156}Dy, ^{160}Gd, ^{162}Er, ^{165}Ho. Does this order differ from the elements' positions on the periodic table?

32. An unknown element Q has two known isotopes: ^{60}Q and ^{63}Q. If the average atomic mass is 61.5 amu, what are the relative percentages of the isotopes?

33. The actual mass of one atom of an element is 3.27×10^{-19} mg.
    (a) Calculate the atomic mass of the element.
    (b) Identify the element, giving the symbol and the name. (There is only one stable isotope of this element.)

34. The mass of an atom of silver is 1.79×10^{-22} g. How many atoms would be in a 0.52-lb sample of silver?

35. Using the periodic table inside the front cover of the book, determine which of the first 20 elements have isotopes that you would expect to have the same number of protons, neutrons, and electrons.

36. Solar winds that send streams of charged particles toward the Earth cause the aurora borealis or the northern lights. As these charged particles hit the upper atmosphere of the Earth or the ionosphere (so named because it is composed of a plasma or gaseous soup of ions), they collide with ions located in this region. These high-energy ions then lose energy in the form of light, which is visible in the northernmost and southernmost regions of the planet as ribbons of light in the night sky. Some of the ions responsible for the colors seen are C^+, O^+, and O^{2+}. Determine the number of protons and electrons in each of these ions.

37. Often the minerals we need in our diet are provided to us in vitamin tablets as cations. Following is a list of ingredients for a vitamin and mineral supplement. The table below highlights some of these minerals and identifies the ion provided for each mineral. Determine the number of protons and electrons in each of these ions.

Mineral supplement	Mineral use	Ion provided	Number of protons	Number of electrons
Calcium carbonate	Bones and Teeth	Ca^{2+}		
Iron(II) sulfate	Hemoglobin	Fe^{2+}		
Chromium(III) nitrate	Insulin	Cr^{3+}		
Magnesium sulfate	Bones	Mg^{2+}		
Zinc sulfate	Cellular metabolism	Zn^{2+}		
Potassium iodide	Thyroid function	I^-		

38. Complete the following table with the appropriate data for each isotope given (all are neutral atoms):

Element	Symbol	Atomic number	Mass number	Number of protons	Number of neutrons	Number of electrons
	^{36}Cl					
Gold			197	•		
		56			79	
					20	18
			58	28		

39. Complete the following table with the appropriate data for each isotope given (all are neutral atoms):

Element	Symbol	Atomic number	Mass number	Number of protons	Number of neutrons	Number of electrons
	^{134}Xe					
Silver			107			
				9		
		92			143	92
			41	19		

40. Draw diagrams similar to those shown in Figure 5.5 for the following ions:
(a) O^{2-} (b) Na^+ (c) P^{3-} (d) Ca^{2+}

41. Draw pictures similar to those in Figure 5.8 for the following isotopes:
(a) ^3He and ^4He (c) ^{10}B and ^{11}B
(b) ^6Li and ^7Li (d) ^{12}C and ^{13}C

42. What percent of the total mass of one atom of each of the following elements comes from electrons?
(a) aluminum (mass of one atom = 4.480×10^{-23} g)
(b) phosphorus (mass of one atom = 5.143×10^{-23} g)
(c) krypton (mass of one atom = 1.392×10^{-22} g)
(d) platinum (mass of one atom = 3.240×10^{-22} g)

43. What percent of the total mass of one atom of each of the following elements comes from protons?
(a) selenium (mass of one atom = 1.311×10^{-22} g)
(b) xenon (mass of one atom = 2.180×10^{-22} g)
(c) chlorine (mass of one atom = 5.887×10^{-23} g)
(d) barium (mass of one atom = 2.280×10^{-22} g)

44. Figure 5.7 is a representation of the nuclear model of the atom. The area surrounding the nucleus is labeled as the electron region. What is the electron region?

CHALLENGE EXERCISE

45. You have discovered a new element and are trying to determine where on the periodic table it would fit. You decide to do a mass spectrometer analysis of the sample and discover that it contains three isotopes with masses of 270.51 amu, 271.23 amu, and 269.14 amu and relative abundances of 34.07%, 55.12%, and 10.81%, respectively. Sketch the mass spectrometer reading, determine the average atomic mass of the element, estimate its atomic number, and determine its approximate location on the periodic table.

ANSWERS TO PRACTICE EXERCISES

5.1 CH_4, C_2H_2, C_2H_4, C_2H_6

5.2   cations: Na^+, NH_4^+, Ca^{2+}, Al^{3+}, K^+, Ag^+, Fe^{2+}, Fe^{3+}
anions: Cl^-, Br^-, NO_3^-, S^{2-}, SO_4^{2-}, PO_4^{3-}

5.3 5.98×10^{24} atoms

5.4   Cu, 29 protons and 29 electrons; Xe, 36 protons and 36 electrons

5.5

	protons	neutrons	electrons
(a)	8	8	8
(b)	35	45	35
(c)	92	143	92
(d)	29	35	29

5.6
(a) atomic number	9
mass number	19
name	fluorine
(b) atomic number	24
mass number	52
name	chromium
(c) atomic number	79
mass number	179
name	gold

5.7   107.9 amu

This seashell is formed from the chemical calcium carbonate, commonly called limestone. It is the same chemical used in many calcium supplements for our diets.

CHAPTER **6**

NOMENCLATURE OF INORGANIC COMPOUNDS

s children, we begin to communicate with other people in our lives by learning the names of objects around us. As we continue to develop, we learn to speak and use language to complete a wide variety of tasks. As we enter school, we begin to learn of other languages—the languages of mathematics, of other cultures, of computers. In each case, we begin by learning the names of the building blocks and then proceed to more abstract concepts. Chemistry has a language all its own—a whole new way of describing the objects so familiar to us in our daily lives. Once we learn the language, we can begin to understand the macroscopic and microscopic

CHAPTER OUTLINE

6.1 Common and Systematic Names

6.2 Elements and Ions

6.3 Writing Formulas from Names of Ionic Compounds

6.4 Naming Binary Compounds

6.5 Naming Compounds Containing Polyatomic Ions

6.6 Acids

6.1 COMMON AND SYSTEMATIC NAMES

Distinguish between the common and systematic names of chemical substances.   ● **LEARNING OBJECTIVE**

Chemical nomenclature is the system of names that chemists use to identify compounds. When a new substance is formulated, it must be named in order to distinguish it from all other substances (see **Figure 6.1**). In this chapter, we will restrict our discussion to the nomenclature of inorganic compounds—compounds that do not generally contain carbon.

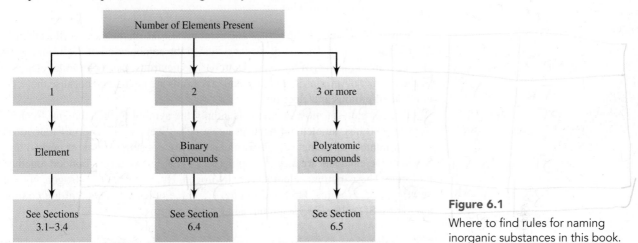

Figure 6.1
Where to find rules for naming inorganic substances in this book.

Common names are arbitrary names that are not based on the chemical composition of compounds. Before chemistry was systematized, a substance was given a name that generally associated it with one of its outstanding physical or chemical properties. For example, *quicksilver* is a common name for mercury, and nitrous oxide (N_2O), used as an anesthetic in dentistry, has been called *laughing gas* because it induces laughter when inhaled. Water and ammonia are also common names because neither provides any information about the chemical composition of the compounds. If every substance were assigned a common name, the amount of memorization required to learn over 50 million names would be astronomical.

Common names have distinct limitations, but they remain in frequent use. Common names continue to be used because the systematic name is too long or too technical for everyday use. For example, calcium oxide (CaO) is called *lime* by plasterers; photographers refer to *hypo* rather than sodium thiosulfate ($Na_2S_2O_3$); and nutritionists use the name *vitamin D_3*, instead of 9,10-secocholesta-5,7,10(19)-trien-3-ß-ol ($C_{27}H_{44}O$). **Table 6.1** lists the common names, formulas, and systematic names of some familiar substances.

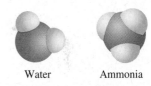

Water   Ammonia

Water (H_2O) and ammonia (NH_3) are almost always referred to by their common names.

TABLE 6.1   Common Names, Formulas, and Chemical Names of Familiar Substances

Common names	Formula	Chemical names	Common names	Formula	Chemical names
Acetylene	C_2H_2	ethyne	Gypsum	$CaSO_4 \cdot 2\,H_2O$	calcium sulfate dihydrate
Lime	CaO	calcium oxide	Grain alcohol	C_2H_5OH	ethanol, ethyl alcohol
Slaked lime	$Ca(OH)_2$	calcium hydroxide	Hypo	$Na_2S_2O_3$	sodium thiosulfate
Water	H_2O	water	Laughing gas	N_2O	dinitrogen monoxide
Galena	PbS	lead(II) sulfide	Lye, caustic soda	NaOH	sodium hydroxide
Alumina	Al_2O_3	aluminum oxide	Milk of magnesia	$Mg(OH)_2$	magnesium hydroxide
Baking soda	$NaHCO_3$	sodium hydrogen carbonate	Muriatic acid	HCl	hydrochloric acid
Cane or beet sugar	$C_{12}H_{22}O_{11}$	sucrose	Plaster of paris	$CaSO_4 \cdot {}^1\!/_2\,H_2O$	calcium sulfate hemihydrate
Borax	$Na_2B_4O_7 \cdot 10\,H_2O$	sodium tetraborate decahydrate	Potash	K_2CO_3	potassium carbonate
			Pyrite (fool's gold)	FeS_2	iron disulfide
Brimstone	S	sulfur	Quicksilver	Hg	mercury
Calcite, marble, limestone	$CaCO_3$	calcium carbonate	Saltpeter (chile)	$NaNO_3$	sodium nitrate
			Table salt	NaCl	sodium chloride
Cream of tartar	$KHC_4H_4O_6$	potassium hydrogen tartrate	Vinegar	$HC_2H_3O_2$	acetic acid
Epsom salts	$MgSO_4 \cdot 7\,H_2O$	magnesium sulfate heptahydrate	Washing soda	$Na_2CO_3 \cdot 10\,H_2O$	sodium carbonate decahydrate
			Wood alcohol	CH_3OH	methanol, methyl alcohol

Chemists prefer systematic names that precisely identify the chemical composition of chemical compounds. The system for inorganic nomenclature was devised by the International Union of Pure and Applied Chemistry (IUPAC), which was founded in 1921. The IUPAC meets regularly and constantly reviews and updates the system.

6.2 ELEMENTS AND IONS

LEARNING OBJECTIVE ● Discuss the formation, charge, and naming of simple ions.

In Chapter 3, we studied the names and symbols for the elements as well as their location on the periodic table. In Chapter 5, we investigated the composition of the atom and learned that all atoms are composed of protons, electrons, and neutrons; that a particular element is defined by the number of protons it contains; and that atoms are uncharged because they contain equal numbers of protons and electrons.

The formula for most elements is simply the symbol of the element. In chemical reactions or mixtures, an element behaves as though it were a collection of individual particles. A small number of elements have formulas that are not single atoms at normal temperatures. Seven of the elements are *diatomic* molecules—that is, two atoms bonded together to form a molecule. These diatomic elements are hydrogen, H_2, oxygen, O_2, nitrogen, N_2, fluorine, F_2, chlorine, Cl_2, bromine, Br_2, and iodine, I_2. Three other elements that are commonly polyatomic are S_8, sulfur, Se_8, selenium and P_4, phosphorus.

Elements occurring as polyatomic molecules					
Hydrogen	H_2	Chlorine	Cl_2	Sulfur	S_8
Oxygen	O_2	Fluorine	F_2	Phosphorus	P_4
Nitrogen	N_2	Bromine	Br_2	Selenium	Se_8
		Iodine	I_2		

We have learned that a charged particle, known as an *ion*, can be produced by adding or removing one or more electrons from a neutral atom. For example, potassium atoms contain 19 protons and 19 electrons. To make a potassium ion, we remove one electron, leaving 19 protons and only 18 electrons. This gives an ion with a positive one (+1) charge:

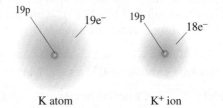

K atom   K⁺ ion

Written in the form of an equation, $K \rightarrow K^+ + e^-$. A positive ion is called a *cation*. Any neutral atom that *loses* an electron will form a cation. Sometimes an atom may lose one electron, as in the potassium example. Other atoms may lose more than one electron:

$$Mg \rightarrow Mg^{2+} + 2e^-$$

or

$$Al \rightarrow Al^{3+} + 3e^-$$

Cations are named the same as their parent atoms, as shown here:

Atom		Ion	
K	potassium	K^+	potassium ion
Mg	magnesium	Mg^{2+}	magnesium ion
Al	aluminum	Al^{3+}	aluminum ion

>CHEMISTRY *IN ACTION*

What's in a Name?

When a scientist discovered a new element in the early days of chemistry, he or she had the honor of naming it. Now researchers must submit their choices for a name to an international committee called the International Union of Pure and Applied Chemistry (IUPAC) before the name can be placed on the periodic table. Since 1997 the IUPAC has decided on names for 11 new elements from 104 through 118. The new elements are called rutherfordium (Rf), dubnium (Db), seaborgium (Sg), bohrium (Bh), hessium (Hs), meitnerium (Mt), darmstadtium (Ds), roentgenium (Rg), copernicium (Cn), flerovium (Fv), and livermorium (Lv).

The new names are compromises among the choices presented by different research teams. Some names given to the elements recognize places where much of the research to find these elements is done. The Russians gained recognition for work done at a laboratory in Dubna, the Germans for their laboratories in Darmstadt and Hesse, and the Americans for the Lawrence Livermore National Laboratory (LLNL).

Other names of elements recognize the contributions of great scientists and researchers such as the American Glenn Seaborg, the first living scientist to have an element named after him. The British recognized Ernest Rutherford, who discovered the atom's nucleus. The Germans recognized Lise Meitner who co-discovered atomic fusion. Both the Germans and the Russians won recognition for Niels Bohr whose model of the atom led to modern ideas about atomic structure. Wilhelm Roentgen was honored for discovering X-rays and winning the first Nobel Prize in physics. Most recently, the astronomer Copernicus was honored for changing our model of the solar system to a heliocentric model and to highlight the link between astronomy and nuclear science. The Russians honored Georgy Flerov, who developed the Soviet nuclear program, with the name for element 114.

Elements 113, 115, 117, and 118 do not yet have official names and are currently shown on the periodic table using three letter symbols representing their numbers.

Niels Bohr

The Image Works

Lise Meitner

© Corbis

Copernicus

Science Source/PhotoResearchers, Inc.

© Corbis

Ernest Rutherford

104	105	106	107	108	109	110	111	112	113	114
Rf	Db	Sg	Bh	Hs	Mt	Ds	Rg	Cn		Fl

Science Source/Photo Researchers, Inc.

Glenn Seaborg

© Hulton/Getty Images, Inc.

Wilhelm Roentgen

RIA Novosti/Photo Researchers, Inc.

Flerov

Ions can also be formed by adding electrons to a neutral atom. For example, the chlorine atom contains 17 protons and 17 electrons. The equal number of positive charges and negative charges results in a net charge of zero for the atom. If one electron is added to the chlorine atom, it now contains 17 protons and 18 electrons, resulting in a net charge of negative one (-1) on the ion:

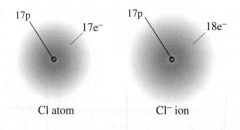

In a chemical equation, this process is summarized as $Cl + e^- \rightarrow Cl^-$. A negative ion is called an *anion*. Any neutral atom that *gains* an electron will form an anion. Atoms may gain more than one electron to form anions with different charges:

$$O + 2e^- \rightarrow O^{2-}$$
$$N + 3e^- \rightarrow N^{3-}$$

Anions are named differently from cations. To name an anion consisting of only one element, use the stem of the parent element name and change the ending to *-ide*. For example, the Cl^- ion is named by using the stem *chlor-* from chlorine and adding *-ide* to form chloride ion. Here are some examples:

Symbol	Name of atom	Ion	Name of ion
F	fluorine	F^-	fluoride ion
Br	bromine	Br^-	bromide ion
Cl	chlorine	Cl^-	chloride ion
I	iodine	I^-	iodide ion
O	oxygen	O^{2-}	oxide ion
N	nitrogen	N^{3-}	nitride ion

WileyPLUS

ENHANCED EXAMPLE

EXAMPLE 6.1

What is the general basis for naming a negative ion in a binary compound?

SOLUTION

Establish an identifying stem from the name of the negative element; then add the suffix *-ide* to that stem. For example, the stem for chlorine is *chlor-*. Add the suffix *-ide* to get the name of the ion, *chloride*. The stem for oxygen is *ox-*. Add the suffix *-ide* to get the name of the ion, *oxide*.

PRACTICE 6.1

Derive the names of these ions when used as the negative ion in a binary compound: P, N, B, C.

Ions are formed by adding or removing electrons from an atom. Atoms do not form ions on their own. Most often ions are formed when metals combine with nonmetals.

The charge on an ion can often be predicted from the position of the element on the periodic table. **Figure 6.2** shows the charges of selected ions from several groups on the periodic table. Notice that all the metals and hydrogen in the far left column (Group 1A) are ($1+$), all those in the next column (Group 2A) are ($2+$), and the metals in the next tall column (Group 3A) form ($3+$) ions. The elements in the lower center part of the table are called *transition metals*. These elements tend to form cations with various positive charges. There is no easy way to predict the charges on these cations. All metals lose electrons to form positive ions.

1A																	
H^+	2A											3A	4A	5A	6A	7A	
Li^+	Be^{2+}													N^{3-}	O^{2-}	F^-	
Na^+	Mg^{2+}											Al^{3+}		P^{3-}	S^{2-}	Cl^-	
K^+	Ca^{2+}			Cr^{2+} Cr^{3+}		Fe^{2+} Fe^{3+}		Cu^+ Cu^{2+}	Zn^{2+}							Br^-	
Rb^+	Sr^{2+}			Transition metals				Ag^+	Cd^{2+}							I^-	
Cs^+	Ba^{2+}																

Figure 6.2

Charges of selected ions in the periodic table.

In contrast, the nonmetals form anions by gaining electrons. On the right side of the periodic table in Figure 6.2, you can see that the atoms in Group 7A form $(1-)$ ions. The nonmetals in Group 6A form $(2-)$ ions. It's important to learn the charges on the ions shown in Figure 6.2 and their relationship to the group number at the top of the column. For nontransition metals the charge is equal to the group number. For nonmetals, the charge is equal to the group number minus 8. We will learn more about why these ions carry their particular charges later in the course.

6.3 WRITING FORMULAS FROM NAMES OF IONIC COMPOUNDS

Write the chemical formula for an ionic compound from the name of the compound.

LEARNING OBJECTIVE

In Chapters 3 and 5, we learned that compounds can be composed of ions. These substances are called *ionic compounds* and will conduct electricity when dissolved in water. An excellent example of an ionic compound is ordinary table salt composed of crystals of sodium chloride. When dissolved in water, sodium chloride conducts electricity very well, as shown in Figure 6.3.

A chemical compound must have a net charge of zero. If it contains ions, the charges on the ions must add up to zero in the formula for the compound. This is relatively easy in the case of sodium chloride. The sodium ion $(1+)$ and the chloride ion $(1-)$ add to zero, resulting in the formula NaCl. Now consider an ionic compound containing calcium (Ca^{2+}) and fluoride (F^-) ions. How can we write a formula with a net charge of zero? To do this we need one Ca^{2+} and two F^- ions. The correct formula is CaF_2. The subscript 2 indicates that two fluoride ions are needed for each calcium ion. Aluminum oxide is a bit more complicated because it consists of Al^{3+} and O^{2-} ions. Since 6 is the least common multiple of 3 and 2, we have $2(3+) + 3(2-) = 0$, or a formula containing 2 Al^{3+} ions and 3 O^{2-} ions for Al_2O_3. Here are a few more examples of formula writing for ionic compounds:

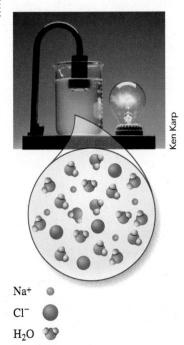

Ken Karp

Na^+ ●
Cl^- ●
H_2O ●

Figure 6.3

A solution of salt water contains Na^+ and Cl^- ions in addition to water molecules. The ions cause the solution to conduct electricity, lighting the bulb.

Name of compound	Ions	Least common multiple	Sum of charges on ions	Formula
Sodium bromide	Na^+, Br^-	1	$(+1) + (-1) = 0$	NaBr
Potassium sulfide	K^+, S^{2-}	2	$2(+1) + (-2) = 0$	K_2S
Zinc sulfate	Zn^{2+}, SO_4^{2-}	2	$(+2) + (-2) = 0$	$ZnSO_4$
Ammonium phosphate	NH_4^+, PO_4^{3-}	3	$3(+1) + (-3) = 0$	$(NH_4)_3PO_4$
Aluminum chromate	Al^{3+}, CrO_4^{2-}	6	$2(+3) + 3(-2) = 0$	$Al_2(CrO_4)_3$

RULES FOR WRITING FORMULAS FOR IONIC COMPOUNDS

1. Write the formula for the metal ion followed by the formula for the nonmetal ion.
2. Combine the smallest numbers of each ion needed to give the charge sum equal to zero.
3. Write the formula for the compound as the symbol for the metal and nonmetal, each followed by a subscript of the number determined in **2**.

WileyPLUS

ENHANCED EXAMPLE

EXAMPLE 6.2

Write formulas for (a) calcium chloride, (b) magnesium oxide, and (c) barium phosphide.

SOLUTION

(a) Use the following steps for calcium chloride.

1. From the name, we know that calcium chloride is composed of calcium and chloride ions. First write down the formulas of these ions:

$$Ca^{2+} \qquad Cl^-$$

2. To write the formula of the compound, combine the smallest numbers of Ca^{2+} and Cl^- ions to give the charge sum equal to zero. In this case the lowest common multiple of the charges is 2:

$$(Ca^{2+}) + 2(Cl^-) = 0$$
$$(2+) + 2(1-) = 0$$

3. Therefore, the formula is $CaCl_2$.

(b) Use the same procedure for magnesium oxide:

1. From the name, we know that magnesium oxide is composed of magnesium and oxide ions. First write down the formulas of these ions:

$$Mg^{2+} \qquad O^{2-}$$

2. To write the formula of the compound, combine the smallest numbers of Mg^{2+} and O^{2-} ions to give the charge sum equal to zero:

$$(Mg^{2+}) + (O^{2-}) = 0$$
$$(2+) + (2-) = 0$$

3. The formula is MgO.

(c) Use the same procedure for barium phosphide:

1. From the name, we know that barium phosphide is composed of barium and phosphide ions. First write down the formulas of these ions:

$$Ba^{2+} \qquad P^{3-}$$

2. To write the formula of this compound, combine the smallest numbers of Ba^{2+} and P^{3-} ions to give the charge sum equal to zero. In this case the lowest common multiple of the charges is 6:

$$3(Ba^{2+}) + 2(P^{3-}) = 0$$
$$3(2+) + 2(3-) = 0$$

3. The formula is Ba_3P_2.

PRACTICE 6.2

Write formulas for compounds containing the following ions:

(a) K^+ and F^-

(b) Ca^{2+} and Br^-

(c) Mg^{2+} and N^{3-}

(d) Na^+ and S^{2-}

(e) Ba^{2+} and O^{2-}

(f) As^{5+} and CO_3^{2-}

6.4 NAMING BINARY COMPOUNDS

Name binary ionic and nonionic compounds.

● LEARNING OBJECTIVE

KEY TERMS

binary compounds
Stock System

Binary compounds contain only two different elements. Many binary compounds are formed when a metal combines with a nonmetal to form a *binary ionic compound*. The metal loses one or more electrons to become a cation while the nonmetal gains one or more electrons to become an anion. The cation is written first in the formula, followed by the anion.

Binary Ionic Compounds Containing a Metal Forming Only One Type of Cation

The chemical name is composed of the name of the metal followed by the name of the non-metal, which has been modified to an identifying stem plus the suffix *-ide*.

For example, sodium chloride, NaCl, is composed of one atom each of sodium and chlorine. The name of the metal, sodium, is written first and is not modified. The second part of the name is derived from the nonmetal, chlorine, by using the stem *chlor-* and adding the ending *-ide*; it is named *chloride*. The compound name is sodium chloride.

NaCl	
Elements:	Sodium (metal)
	Chlorine (nonmetal)
	name modified to the stem *chlor-* + *-ide*
Name of compound:	Sodium chloride

Stems of the more common negative-ion-forming elements are shown in **Table 6.2**.

TABLE 6.2   Examples of Elements Forming Anions

Symbol	Element	Stem	Anion name
Br	bromine	brom	bromide
Cl	chlorine	chlor	chloride
F	fluorine	fluor	fluoride
H	hydrogen	hydr	hydride
I	iodine	iod	iodide
N	nitrogen	nitr	nitride
O	oxygen	ox	oxide
P	phosphorus	phosph	phosphide
S	sulfur	sulf	sulfide

RULES FOR NAMING BINARY IONIC COMPOUNDS OF METAL FORMING ONE TYPE OF CATION

1. Write the name of the cation.
2. Write the stem for the anion and add the suffix *-ide*.

Compounds may contain more than one atom of the same element, but as long as they contain only two different elements and only one compound of these two elements exists, the name follows the rules for binary compounds:

Examples:

$CaBr_2$ Mg_3N_2 Li_2O
calcium bromide   magnesium nitride   lithium oxide

Table 6.3 lists some compounds with names ending in -ide.

TABLE 6.3	Examples of Compounds with Names Ending in -ide		
Formula	**Name**	**Formula**	**Name**
$AlCl_3$	aluminum chloride	BaS	barium sulfide
Al_2O_3	aluminum oxide	LiI	lithium iodide
CaC_2	calcium carbide	$MgBr_2$	magnesium bromide
HCl	hydrogen chloride	NaH	sodium hydride
HI	hydrogen iodide	Na_2O	sodium oxide

WileyPLUS

ENHANCED EXAMPLE

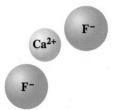

EXAMPLE 6.3

Name the compound CaF_2.

SOLUTION

From the formula it is a two-element compound and follows the rules for binary compounds.

1. The compound is composed of a metal, Ca, and a nonmetal, F. Elements in the 2A column of the periodic table form only one type of cation. Thus, we name the positive part of the compound *calcium*.

2. Modify the name of the second element to the stem *fluor-* and add the binary ending *-ide* to form the name of the negative part, *fluoride*.

Therefore, the name of the compound is *calcium fluoride*.

PRACTICE 6.3

Write formulas for these compounds:

(a) strontium chloride   (d) calcium sulfide
(b) potassium iodide   (e) sodium oxide
(c) aluminum nitride

Binary Ionic Compounds Containing a Metal That Can Form Two or More Types of Cations

The metals in the center of the periodic table (including the transition metals) often form more than one type of cation. For example, iron can form Fe^{2+} and Fe^{3+} ions, and copper can form Cu^+ and Cu^{2+} ions. This can be confusing when you are naming compounds. For example, copper chloride could be $CuCl_2$ or CuCl. To resolve this difficulty the IUPAC devised a system, known as the **Stock System**, to name these compounds. This system is currently recognized as the official system to name these compounds, although another older system is sometimes used. In the Stock System, when a compound contains a metal that can form more than one type of cation, the charge on the cation of the metal is designated by a Roman numeral placed in parentheses immediately following the name of the metal. The negative element is treated in the usual manner for binary compounds:

The Roman numeral is part of the cation name.

Cation charge	+1	+2	+3	+4	+5
Roman numeral	(I)	(II)	(III)	(IV)	(V)

Examples:			
$FeCl_2$	iron(II) chloride	Fe^{2+}	
$FeCl_3$	iron(III) chloride	Fe^{3+}	
CuCl	copper(I) chloride	Cu^+	
$CuCl_2$	copper(II) chloride	Cu^{2+}	

The fact that $FeCl_2$ has two chloride ions, each with a -1 charge, establishes that the charge on Fe is $+2$. To distinguish between the two iron chlorides, $FeCl_2$ is named iron(II) chloride and $FeCl_3$ is named iron(III) chloride:

	iron(II) chloride		iron(III) chloride	
	$FeCl_2$		$FeCl_3$	
charge:	$+2$	-1	$+3$	-1
name:	iron(II)	chloride	iron(III)	chloride

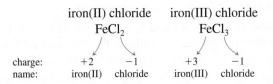

When a metal forms only one possible cation, we need not distinguish one cation from another, so Roman numerals are not needed. Thus we do not say calcium(II) chloride for $CaCl_2$, but rather calcium chloride, since the charge of calcium is understood to be $+2$.

RULES FOR NAMING BINARY IONIC COMPOUNDS OF METAL FORMING TWO OR MORE TYPES OF CATIONS (STOCK SYSTEM)

1. Write the name of the cation.
2. Write the charge on the cation as a Roman numeral in parentheses.
3. Write the stem of the anion and add the suffix *-ide*.

In classical nomenclature, when the metallic ion has only two cation types, the name of the metal (usually the Latin name) is modified with the suffixes *-ous* and *-ic* to distinguish between the two. The lower-charge cation is given the *-ous* ending, and the higher one, the *-ic* ending.

Examples:	$FeCl_2$	ferrous chloride	Fe^{2+}	(lower-charge cation)
	$FeCl_3$	ferric chloride	Fe^{3+}	(higher-charge cation)
	$CuCl$	cuprous chloride	Cu^{+}	(lower-charge cation)
	$CuCl_2$	cupric chloride	Cu^{2+}	(higher-charge cation)

Table 6.4 lists some common metals that have more than one type of cation.

TABLE 6.4 Names and Charges of Some Common Metal Ions That Have More Than One Type of Cation

Formula	Stock System name	Classical name	Formula	Stock System name	Classical name
Cu^{1+}	copper(I)	cuprous	Sn^{4+}	tin(IV)	stannic
Cu^{2+}	copper(II)	cupric	Pb^{2+}	lead(II)	plumbous
Hg^{1+} $(Hg_2)^{2+}$	mercury(I)	mercurous	Pb^{4+}	lead(IV)	plumbic
Hg^{2+}	mercury(II)	mercuric	As^{3+}	arsenic(III)	arsenous
Fe^{2+}	iron(II)	ferrous	As^{5+}	arsenic(V)	arsenic
Fe^{3+}	iron(III)	ferric	Ti^{3+}	titanium(III)	titanous
Sn^{2+}	tin(II)	stannous	Ti^{4+}	titanium(IV)	titanic

Notice that the *ous–ic* naming system does not give the charge of the cation of an element but merely indicates that at least two types of cations exist. The Stock System avoids any possible uncertainty by clearly stating the charge on the cation.

In this book we will use mainly the Stock System.

EXAMPLE 6.4

Name the compound FeS.

SOLUTION

This compound follows the rules for a binary compound.

1, 2. It is a compound of Fe, a metal, and S, a nonmetal, and Fe is a transition metal that has more than one type of cation. In sulfides, the charge on the S is −2. Therefore, the charge on Fe must be +2, and the name of the positive part of the compound is *iron(II)*.

3. We have already determined that the name of the negative part of the compound will be *sulfide*.

The name of FeS is *iron(II) sulfide*.

PRACTICE 6.4

Write the name for each of the following compounds using the Stock System:

(a) PbI_2 (b) SnF_4 (c) Fe_2O_3 (d) CuO

PRACTICE 6.5

Write formulas for the following compounds:

(a) tin(IV) chromate   (c) tin(II) fluoride
(b) chromium(III) bromide   (d) copper(I) oxide

Prefix	Number
mono	1
di	2
tri	3
tetra	4
penta	5
hexa	6
hepta	7
octa	8
nona	9
deca	10

Binary Compounds Containing Two Nonmetals

Compounds between nonmetals are molecular, not ionic. Therefore, a different system for naming them is used. In a compound formed between two nonmetals, the element that occurs first in this series is written and named first:

Si, B, P, H, C, S, I, Br, N, Cl, O, F

The name of the second element retains the *-ide* ending as though it were an anion. A Latin or Greek prefix (*mono-, di-, tri-,* and so on) is attached to the name of each element to indicate the number of atoms of that element in the molecule. The prefix *mono-* is rarely used for naming the first element. Some common prefixes and their numerical equivalences are as shown in the margin table.

RULES FOR NAMING BINARY COMPOUNDS CONTAINING TWO NONMETALS

1. Write the name for the first element using a prefix if there is more than one atom of this element.

2. Write the stem of the second element and add the suffix *-ide*. Use a prefix to indicate the number of atoms for the second element.

Here are some examples of compounds that illustrate this system:

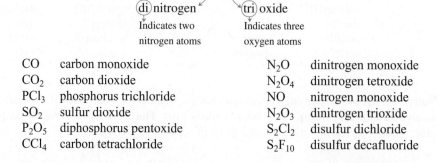

CO	carbon monoxide
CO_2	carbon dioxide
PCl_3	phosphorus trichloride
SO_2	sulfur dioxide
P_2O_5	diphosphorus pentoxide
CCl_4	carbon tetrachloride

N_2O	dinitrogen monoxide
N_2O_4	dinitrogen tetroxide
NO	nitrogen monoxide
N_2O_3	dinitrogen trioxide
S_2Cl_2	disulfur dichloride
S_2F_{10}	disulfur decafluoride

These examples illustrate that we sometimes drop the final *o* (mono) or *a* (penta) of the prefix when the second element is oxygen. This avoids creating a name that is awkward to pronounce. For example, CO is carbon monoxide instead of carbon monooxide.

EXAMPLE 6.5

Name the compound PCl_5.

SOLUTION

1. Phosphorus and chlorine are nonmetals, so the rules for naming binary compounds containing two nonmetals apply. Phosphorus is named first. Therefore, the compound is a chloride.

2. No prefix is needed for phosphorus because each molecule has only one atom of phosphorus. The prefix *penta-* is used with chloride to indicate the five chlorine atoms. (PCl_3 is also a known compound.)

The name for PCl_5 is *phosphorus pentachloride*.

Phosphorus pentachloride

WileyPLUS

ENHANCED EXAMPLE

PRACTICE 6.6

Name these compounds:

(a) Cl_2O (b) SO_2 (c) CBr_4 (d) N_2O_5 (e) NH_3 (f) ICl_3

PRACTICE 6.7

Name these compounds:

(a) KBr   (b) Ca_3N_2 (c) SO_3 (d) SnF_2 (e) $CuCl_2$ (f) N_2O_4

6.5 NAMING COMPOUNDS CONTAINING POLYATOMIC IONS

Recognize names, formulas, and charges of polyatomic ions, name compounds containing polyatomic ions, and write formulas from names of compounds containing polyatomic ions.

● LEARNING OBJECTIVE

A **polyatomic ion** is an ion that contains two or more elements. Compounds containing polyatomic ions are composed of three or more elements and usually consist of one or more cations combined with a negative polyatomic ion. In general, naming compounds containing polyatomic ions is similar to naming binary compounds. The cation is named first, followed by the name for the negative polyatomic ion.

To name these compounds, you must learn to recognize the common polyatomic ions (Table 6.5) and know their charges. Consider the formula $KMnO_4$. You must be able to recognize that it consists of two parts $KMnO_4$. These parts are composed of a K^+ ion and a MnO_4^-

KEY TERM

polyatomic ion

TABLE 6.5   Names, Formulas, and Charges of Some Common Polyatomic Ions

Name	Formula	Charge	Name	Formula	Charge
Acetate	$C_2H_3O_2^-$	−1	Cyanide	CN^-	−1
Ammonium	NH_4^+	+1	Dichromate	$Cr_2O_7^{2-}$	−2
Arsenate	AsO_4^{3-}	−3	Hydroxide	OH^-	−1
Hydrogen carbonate	HCO_3^-	−1	Nitrate	NO_3^-	−1
Hydrogen sulfate	HSO_4^-	−1	Nitrite	NO_2^-	−1
Bromate	BrO_3^-	−1	Permanganate	MnO_4^-	−1
Carbonate	CO_3^{2-}	−2	Phosphate	PO_4^{3-}	−3
Chlorate	ClO_3^-	−1	Sulfate	SO_4^{2-}	−2
Chromate	CrO_4^{2-}	−2	Sulfite	SO_3^{2-}	−2

More ions are listed on the back endpapers.

Ken Karp

Potassium permanganate crystals are dark purple.

ion. The correct name for this compound is potassium permanganate. Many polyatomic ions that contain oxygen are called *oxy-anions* and generally have the suffix *-ate* or *-ite*. Unfortunately, the suffix doesn't indicate the number of oxygen atoms present. The *-ate* form contains more oxygen atoms than the *-ite* form. Examples include sulfate (SO_4^{2-}), sulfite (SO_3^{2-}), nitrate (NO_3^-), and nitrite (NO_2^-).

RULES FOR NAMING COMPOUNDS CONTAINING POLYATOMIC IONS

1. Write the name of the cation.
2. Write the name of the anion.

Only four of the common negatively charged polyatomic ions do not use the *ate–ite* system. These exceptions are hydroxide (OH^-) hydrogen sulfide (HS^-) peroxide (O_2^{2-}), and cyanide (CN^-). Care must be taken with these, as their endings can easily be confused with the *-ide* ending for binary compounds (Section 6.4).

There are three common positively charged polyatomic ions as well—the ammonium, the mercury(I) (Hg_2^{2+}), and the hydronium (H_3O^+) ions. The ammonium ion (NH_4^+) is frequently found in polyatomic compounds, whereas the hydronium ion (H_3O^+) is usually associated with aqueous solutions of acids (Chapter 15).

WileyPLUS

ENHANCED EXAMPLE

EXAMPLE 6.6

What is the minimum number of different elements in a compound that contains a polyatomic ion?

SOLUTION

Most polyatomic ions are anions and contain two different elements. Adding a cation to a polyatomic ion to form a compound means the minimum is three different elements in a polyatomic compound.

PRACTICE 6.8

Name these compounds:
(a) $NaNO_3$ (b) $Ca_3(PO_4)_2$ (c) KOH   (d) Li_2CO_3 (e) $NaClO_3$

TABLE 6.6 Names of Selected Compounds That Contain More Than One Kind of Positive Ion

Formula	Name of compound
$KHSO_4$	potassium hydrogen sulfate
$Ca(HSO_3)_2$	calcium hydrogen sulfite
NH_4HS	ammonium hydrogen sulfide
$MgNH_4PO_4$	magnesium ammonium phosphate
NaH_2PO_4	sodium dihydrogen phosphate
Na_2HPO_4	sodium hydrogen phosphate
KHC_2O_4	potassium hydrogen oxalate
$KAl(SO_4)_2$	potassium aluminum sulfate
$Al(HCO_3)_3$	aluminum hydrogen carbonate

Inorganic compounds are also formed from more than three elements (see Table 6.6). In these cases, one or more of the ions is often a polyatomic ion. Once you have learned to recognize the polyatomic ions, naming these compounds follows the patterns we have already learned. First identify the ions. Name the cations in the order given and follow them with the names of the anions. Study the following examples:

Compound	Ions	Name
$NaHCO_3$	Na^+; HCO_3^-	sodium hydrogen carbonate
NaHS	Na^+; HS^-	sodium hydrogen sulfide
$MgNH_4PO_4$	Mg^{2+}; NH_4^+; PO_4^{3-}	magnesium ammonium phosphate
$NaKSO_4$	Na^+; K^+; SO_4^{2-}	sodium potassium sulfate

PRACTICE 6.9

Name these compounds:
(a) $KHCO_3$ (b) $NaHC_2O_4$ (c) $BaNH_4PO_4$ (d) $NaAl(SO_4)_2$

6.6 ACIDS

Use the rules to name an acid from its formula and to write the formula for an acid ● **LEARNING OBJECTIVE** from its name.

While we will learn much more about acids later (see Chapter 15), it is helpful to be able to recognize and name common acids both in the laboratory and in class. The simplest way to recognize many acids is to know that acid formulas often begin with hydrogen.

Binary Acids

Certain binary hydrogen compounds, when dissolved in water, form solutions that have *acid* properties. Because of this property, these compounds are given acid names in addition to their regular *-ide* names. For example, HCl is a gas and is called *hydrogen chloride*, but its water solution is known as *hydrochloric acid*. Binary acids are composed of hydrogen and one other nonmetallic element. However, not all binary hydrogen compounds are acids. To express the formula of a binary acid, it's customary to write the symbol of hydrogen first, followed by the symbol of the second element (e.g., HCl, HBr, or H_2S). When we see formulas such as CH_4 or NH_3, we understand that these compounds are not normally considered to be acids.

To name a binary acid, place the prefix *hydro-* in front of, and the suffix *-ic* after, the stem of the nonmetal name. Then add the word *acid*:

	HCl	H_2S
Examples:	*Hydro-*chlor-*ic acid*	*Hydro-*sulfur-*ic acid*
	(hydrochloric acid)	(hydrosulfuric acid)

RULES FOR NAMING BINARY ACIDS

1. Write the prefix *hydro-* followed by the stem of the second element and add the suffix *-ic*.

2. Add the word *acid*.

EXAMPLE 6.7

Name the acid HI.

SOLUTION

The compound follows the Rules for Naming Binary Acids.

1. Write the prefix *hydro-* and the stem of the second element with the suffix *-ic*: *hydro*iodic.

2. Add the word *acid*: hydroiodic *acid*.

PRACTICE 6.10

Name these binary acids:
(a) HCl   (b) HBr

TABLE 6.7 Names and Formulas of Selected Binary Acids

Formula	Acid name
HF	Hydrofluoric acid
HCl	Hydrochloric acid
HBr	Hydrobromic acid
HI	Hydroiodic acid
H_2S	Hydrosulfuric acid
H_2Se	Hydroselenic acid

Acids are hydrogen-containing substances that liberate hydrogen ions when dissolved in water. The same formula is often used to express binary hydrogen compounds, such as HCl, regardless of whether or not they are dissolved in water. **Table 6.7** shows several examples of binary acids.

Figure 6.4 summarizes how to name binary compounds.

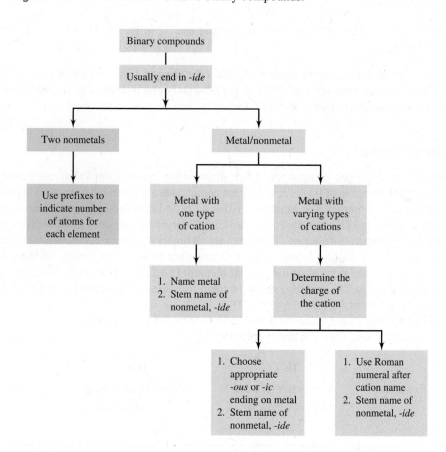

Figure 6.4

Flow diagram for naming binary compounds.

Naming Oxy-Acids

Inorganic compounds containing hydrogen, oxygen, and one other element are called *oxy-acids*. The first step in naming these acids is to determine that the compound in question is really an oxy-acid. The keys to identification are (1) hydrogen is the first element in the compound's formula and (2) the second part of the formula consists of a polyatomic ion containing oxygen.

Hydrogen in an oxy-acid is not specifically designated in the acid name. The presence of hydrogen in the compound is indicated by the use of the word *acid* in the name of the substance. To determine the particular type of acid, the polyatomic ion following hydrogen must be examined. The name of the polyatomic ion is modified in the following manner: (1) *-ate* changes to an *-ic* ending; (2) *-ite* changes to an *-ous* ending. (See Table 6.8.) The compound with the *-ic* ending contains more oxygen than the one with the *-ous* ending.

TABLE 6.8   Comparison of Acid and Anion Names for Selected Oxy-Acids

Acid	Anion	Acid	Anion	Acid	Anion
H_2SO_4	SO_4^{2-}	H_2CO_3	CO_3^{2-}	HIO_3	IO_3^-
Sulfuric acid	Sulfate ion	Carbonic acid	Carbonate ion	Iodic acid	Iodate ion
H_2SO_3	SO_3^{2-}	H_3BO_3	BO_3^{3-}	$HC_2H_3O_2$	$C_2H_3O_2^-$
Sulfurous acid	Sulfite ion	Boric acid	Borate ion	Acetic acid	Acetate ion
HNO_3	NO_3^-	H_3PO_4	PO_4^{3-}	$H_2C_2O_4$	$C_2O_4^{2-}$
Nitric acid	Nitrate ion	Phosphoric acid	Phosphate ion	Oxalic acid	Oxalate ion
HNO_2	NO_2^-	H_3PO_3	PO_3^{3-}	$HBrO_3$	BrO_3^-
Nitrous acid	Nitrite ion	Phosphorous acid	Phosphite ion	Bromic acid	Bromate ion

EXAMPLE 6.8

Name the acids H_2SO_4 and H_2SO_3.

SOLUTION

The names are written by modifying the name of the polyatomic ion in the acid.

For H_2SO_4 the polyatomic ion is the SO_4^{2-} sulfate ion. The *-ate* ending changes to an *-ic* ending in the acid. The stem of the sulfate ion is *sulfur* so the name for the compound is *sulfuric* and the word *acid* is added to show the presence of hydrogen in the compound: *sulfuric acid*.

For H_2SO_3 the polyatomic ion is the SO_3^{2-} sulfite ion. The *-ite* ending changes to an *-ous* ending in the acid, *sulfurous*, and the word *acid* is added to show the presence of hydrogen in the compound: *sulfurous acid*.

PRACTICE 6.11

Name these polyatomic acids:

(a) H_2CO_3 (b) $HC_2H_3O_2$ (c) HNO_3 (d) HNO_2

Some elements form more than two different polyatomic ions containing oxygen. To name these ions, prefixes are used in addition to a suffix. To indicate more oxygen than in the *-ate* form we add the prefix *per-*, which is a short form of *hyper-* meaning "more." The prefix *hypo-*, meaning "less" (oxygen in this case), is used for the ion containing less oxygen than the *-ite* form. An example of this system is shown for the oxy-acids containing chlorine and oxygen in Table 6.9. The prefixes are also used with other similar ions, such as iodate (IO_3^-), bromate (BrO_3^-).

We have now looked at ways of naming a variety of inorganic compounds—binary compounds consisting of a metal and a nonmetal and of two nonmetals, acids, and polyatomic compounds. (See Figure 6.5.) These compounds are just a small part of the classified chemical compounds. Most of the remaining classes are in the broad field of organic chemistry under such categories as hydrocarbons, alcohols, ethers, aldehydes, ketones, phenols, and carboxylic acids.

TABLE 6.9 Oxy-Acids of Chlorine

Acid formula	Anion formula	Anion name	Acid name
HClO	ClO^-	*hypo*chlor*ite*	*hypo*chlor*ous* acid
$HClO_2$	ClO_2^-	chlor*ite*	chlor*ous* acid
$HClO_3$	ClO_3^-	chlor*ate*	chlor*ic* acid
$HClO_4$	ClO_4^-	*per*chlor*ate*	*per*chlor*ic* acid

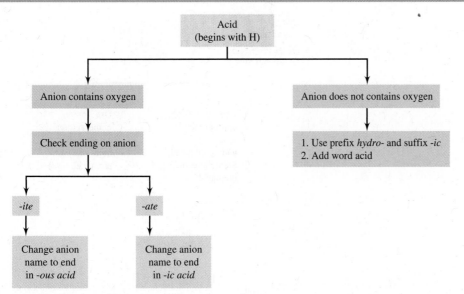

Figure 6.5

Flow diagram for naming acids.

CHAPTER **6 REVIEW**

6.1 COMMON AND SYSTEMATIC NAMES

- Common names are arbitrary and do not describe the chemical composition of compounds.
  - Examples—ammonia, water
- A standard system of nomenclature was devised and is maintained by the IUPAC.

6.2 ELEMENTS AND IONS

- Diatomic elements—H_2, O_2, N_2, F_2, Cl_2, Br_2, I_2.
- Polyatomic elements—P_4, S_8.
- Ions form through the gain or loss of electrons from neutral atoms.
- The charge on ions can often be predicted from the periodic table.

6.3 WRITING FORMULAS FROM NAMES OF IONIC COMPOUNDS

- Compounds must have a net charge of zero.
- To write a formula from a name:
  - Identify and write the symbols for the elements in the compound.
  - Combine the smallest number of ions required to produce a net charge of zero.

6.4 NAMING BINARY COMPOUNDS

KEY TERMS

binary compounds
Stock System

- The rules for naming binary compounds can be summarized in the following chart.

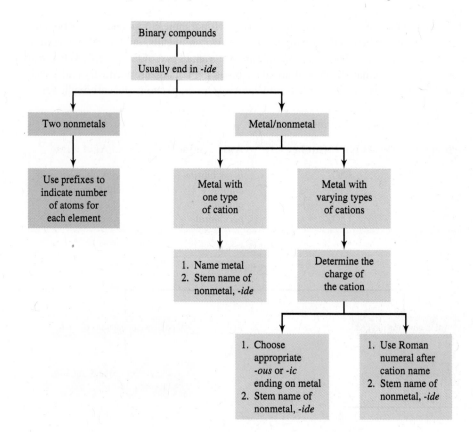

6.5 NAMING COMPOUNDS CONTAINING POLYATOMIC IONS

- The rules for naming these compounds:
  1. Write the name of the cation.
  2. Write the name of the anion.

6.6 ACIDS

- The rules for naming acids can be summarized in the chart that follows:

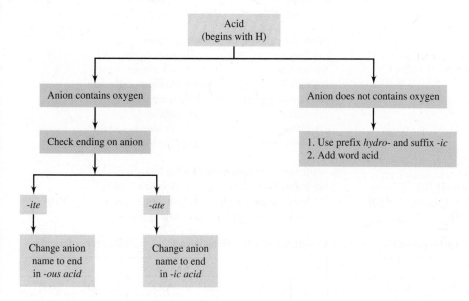

REVIEW QUESTIONS

1. List two compounds almost always referred to by their common names. What would be their systematic names?
2. What must happen in order to convert an atom into an anion? Into a cation?
3. Does the fact that two elements combine in a one-to-one atomic ratio mean that the charges on their ions are both 1? Explain.
4. When naming a binary ionic compound, how do you determine which element to name first?
5. Use the common ion table on the back endpapers of your text to determine the formulas for compounds composed of the following ions:
   (a) potassium and sulfide
   (b) cobalt(II) and bromate
   (c) ammonium and nitrate
   (d) hydrogen and phosphate
   (e) iron(III) oxide
   (f) magnesium and hydroxide
6. Explain why P_2O_5 is named dinitrogen pentoxide, using prefixes, but Al_2O_3 is named aluminum oxide, without prefixes.

7. Explain why $FeCl_2$ is named iron(II) chloride, using a Roman numeral, and $BaCl_2$ is named barium chloride, using no Roman numeral. Why does neither use prefixes?
8. The names of the nitrogen oxides are nitrogen dioxide (NO_2) and dinitrogen monoxide (N_2O). Why is the *mono-* prefix used only on the second molecule?
9. What is the advantage of using the Stock System for naming metals with variable charge instead of using the common name?
10. Write formulas for the compounds formed when a nickel(II) ion is combined with
    (a) sulfate   (f) acetate
    (b) phosphide   (g) dichromate
    (c) chromate   (h) bromide
    (d) hydroxide   (i) nitrate
    (e) iodite   (j) hypochlorite
    (Use the common ion table on the back endpapers of your text.)
11. Write the names and formulas for the four oxy-acids containing (a) bromine, (b) iodine.
12. How do you differentiate between different anions formed from a polyprotic acid?

Most of the exercises in this chapter are available for assignment via the online homework management program, WileyPLUS (www.wileyplus.com). All exercises with blue numbers have answers in Appendix VII.

PAIRED EXERCISES

1. Write the formula of the compound that will be formed between these elements:
   (a) Ba and S   (d) Mg and N
   (b) Cs and P   (e) Ca and I
   (c) Li and Br   (f) H and Cl

2. Write the formula of the compound that will be formed between these elements:
   (a) Al and S   (d) Sr and O
   (b) H and F   (e) Cs and P
   (c) K and N   (f) Al and Cl

3. Write formulas for the following cations:
   (a) potassium   (h) calcium
   (b) ammonium   (i) lead(II)
   (c) copper(I)   (j) zinc
   (d) titanium(IV)   (k) silver
   (e) nickel(III)   (l) hydrogen
   (f) cesium   (m) tin(II)
   (g) mercury(II)   (n) iron(III)

4. Write formulas for the following anions:
   (a) fluoride   (h) oxide
   (b) acetate   (i) dichromate
   (c) iodide   (j) hydrogen carbonate
   (d) carbonate   (k) phosphate
   (e) sulfide   (l) sulfate
   (f) nitrate   (m) nitride
   (g) phosphide   (n) chloride

5. Write the systematic names for the following:
   (a) baking soda ($NaHCO_3$)   (d) vinegar ($HC_2H_3O_2$)
   (b) quicksilver (Hg)   (e) Epsom salts ($MgSO_4 \cdot 7\ H_2O$)
   (c) lime (CaO)   (f) lye (NaOH)

6. Write the systematic names for the following:
   (a) hypo ($Na_2S_2O_3$)   (d) table salt (NaCl)
   (b) laughing gas (N_2O)   (e) milk of magnesia ($Mg(OH)_2$)
   (c) alumina (Al_2O_3)   (f) galena (PbS)

7. Complete the table, filling in each box with the proper formula.

<table>
<tr><th colspan="6">Anions</th></tr>
<tr><th></th><th>Br^-</th><th>O^{2-}</th><th>NO_3^-</th><th>PO_4^{3-}</th><th>CO_3^{2-}</th></tr>
<tr><td>K^+</td><td>KBr</td><td></td><td></td><td></td><td></td></tr>
<tr><td>Mg^{2+}</td><td></td><td></td><td></td><td></td><td></td></tr>
<tr><td>Al^{3+}</td><td></td><td></td><td></td><td></td><td></td></tr>
<tr><td>Zn^{2+}</td><td></td><td></td><td></td><td>$Zn_3(PO_4)_2$</td><td></td></tr>
<tr><td>H^+</td><td></td><td></td><td></td><td></td><td></td></tr>
</table>

Cations

8. Complete the table, filling in each box with the proper formula.

<table>
<tr><th colspan="6">Anions</th></tr>
<tr><th></th><th>SO_4^{2-}</th><th>OH^-</th><th>AsO_4^{3-}</th><th>$C_2H_3O_2^-$</th><th>CrO_4^{2-}</th></tr>
<tr><td>NH_4^+</td><td></td><td></td><td>$(NH_4)_3AsO_4$</td><td></td><td></td></tr>
<tr><td>Ca^{2+}</td><td></td><td></td><td></td><td></td><td></td></tr>
<tr><td>Fe^{3+}</td><td>$Fe_2(SO_4)_3$</td><td></td><td></td><td></td><td></td></tr>
<tr><td>Ag^+</td><td></td><td></td><td></td><td></td><td></td></tr>
<tr><td>Cu^{2+}</td><td></td><td></td><td></td><td></td><td></td></tr>
</table>

Cations

9. Write the names of each of the compounds formed in Question 7.

10. Write the names of each of the compounds formed in Question 8.

11. Write formulas for each of the following binary compounds, all of which are composed of nonmetals:
    (a) diphosphorus pentoxide   (e) carbon tetrachloride
    (b) carbon dioxide   (f) dichlorine heptoxide
    (c) tribromine octoxide   (g) boron trifluoride
    (d) sulfur hexachloride   (h) tetranitrogen hexasulfide

12. Write formulas for each of the following binary compounds, all of which are composed of a metal and nonmetal:
    (a) potassium nitride   (e) calcium nitride
    (b) barium oxide   (f) cesium bromide
    (c) iron(II) oxide   (g) manganese(III) iodide
    (d) strontium phosphide   (h) sodium selenide

13. Name each of the following binary compounds, all of which are composed of a metal and a nonmetal:
    (a) BaO   (e) Al_2O_3
    (b) K_2S (f) $CaBr_2$
    (c) $CaCl_2$ (g) SrI_2
    (d) Cs_2S (h) Mg_3N_2

14. Name each of the following binary compounds, all of which are composed of nonmetals:
    (a) PBr_5 (e) $SiCl_4$
    (b) I_4O_9 (f) ClO_2
    (c) N_2S_5 (g) P_4S_7
    (d) S_2F_{10} (h) IF_6

15. Name these compounds by the Stock System (IUPAC):
    (a) $CuCl_2$ (d) $FeCl_3$
    (b) $FeCl_2$ (e) SnF_2
    (c) $Fe(NO_3)_2$ (f) VPO_4

16. Write formulas for these compounds:
    (a) tin(IV) bromide   (d) mercury(II) nitrite
    (b) copper(I) sulfate   (e) cobalt(III) carbonate
    (c) nickel(II) borate   (f) iron(II) acetate

17. Write formulas for these acids:
    (a) hydrochloric acid   (d) carbonic acid
    (b) chloric acid   (e) sulfurous acid
    (c) nitric acid   (f) phosphoric acid

18. Write formulas for these acids:
    (a) acetic acid   (d) boric acid
    (b) hydrofluoric acid   (e) nitrous acid
    (c) hydrosulfuric acid (f) hypochlorous acid

19. Name these acids:
  (a) HNO_2 (d) HBr   (g) HF
  (b) H_2SO_4 (e) H_3PO_3 (h) $HBrO_3$
  (c) $H_2C_2O_4$ (f) $HC_2H_3O_2$ (i) HIO_4

20. Name these acids:
  (a) H_3PO_4 (d) HCl   (g) HI
  (b) H_2CO_3 (e) HClO   (h) $HClO_4$
  (c) HIO_3 (f) HNO_3 (i) H_2SO_3

21. Write formulas for these compounds:
  (a) silver sulfite
  (b) cobalt(II) bromide
  (c) tin(II) hydroxide
  (d) aluminum sulfate
  (e) lead(II) chloride
  (f) ammonium carbonate
  (g) chromium(III) oxide
  (h) copper(II) chloride
  (i) potassium permanganate
  (j) arsenic(V) sulfite
  (k) sodium peroxide
  (l) iron(II) sulfate
  (m) potassium dichromate
  (n) bismuth(III) chromate

22. Write formulas for these compounds:
  (a) sodium chromate
  (b) magnesium hydride
  (c) nickel(II) acetate
  (d) calcium chlorate
  (e) magnesium bromate
  (f) potassium dihydrogen phosphate
  (g) manganese(II) hydroxide
  (h) cobalt(II) hydrogen carbonate
  (i) sodium hypochlorite
  (j) barium perchlorate
  (k) chromium(III) sulfite
  (l) antimony(III) sulfate
  (m) sodium oxalate
  (n) potassium thiocyanate

23. Write the name of each compound:
  (a) $ZnSO_4$ (f) CoF_2
  (b) Hg_2S (g) $Cr(ClO_3)_3$
  (c) $CuCO_3$ (h) Ag_3PO_4
  (d) $Cd(NO_3)_2$ (i) MnS
  (e) $Al(C_2H_3O_2)_3$ (j) $BaCrO_4$

24. Write the name of each compound:
  (a) $Ca(HSO_4)_2$ (f) $BiAsO_4$
  (b) $As_2(SO_3)_3$ (g) $(NH_4)_2CO_3$
  (c) $Sn(NO_2)_2$ (h) $(NH_4)_2HPO_4$
  (d) CuI   (i) NaClO
  (e) $KHCO_3$ (j) $KMnO_4$

25. Write the chemical formulas for these substances:
  (a) slaked lime   (e) cane sugar
  (b) fool's gold   (f) borax
  (c) washing soda   (g) wood alcohol
  (d) calcite   (h) acetylene

26. Write the chemical formulas for these substances:
  (a) grain alcohol   (e) muriatic acid
  (b) cream of tartar   (f) plaster of paris
  (c) gypsum   (g) lye
  (d) brimstone   (h) laughing gas

ADDITIONAL EXERCISES

27. Write equations similar to those found in Section 6.2 for the formation of
  (a) potassium ion   (d) iron(II) ion
  (b) iodide ion   (e) calcium ion
  (c) bromide ion   (f) oxide ion

28. Name each of the following polyatomic ions:
  (a) $\left[\quad\right]^{2-}$ = S   = O
  (d) $\left[\quad\right]^{-}$ = Cl   = O

  (b) $\left[\quad\right]^{3-}$ = P   = O
  (e) $\left[\quad\right]^{-}$ = O   = H

  (c) $\left[\quad\right]^{-}$ = N   = O
  (f) $\left[\quad\right]^{2-}$ = C   = O

29. Write formulas for all possible compounds formed between the calcium ion and the anions shown in Question 28.

30. Write formulas for all possible compounds formed between the potassium ion and the anions shown in Question 28.

31. Write the formula and name for each of the following compounds:

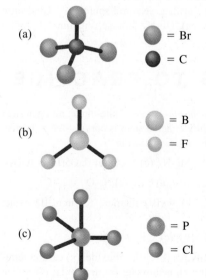

(a)
(b)
(c)

= Br
= C
= B
= F
= P
= Cl

32. Write the formula and name for the following compound:

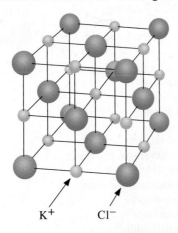

K⁺   Cl⁻

33. State how each of the following is used in naming inorganic compounds: *ide, ous, ic, hypo, per, ite, ate,* Roman numerals.

34. Several excerpts from newspaper articles follow. Tell the chemical formula of the chemical spilled and indicate whether this spill should raise concern in the local area.

(a) A mistake at the Columbia Water and Light Department's West Ash Street pumping station sent an estimated 7200 gallons of calcium-carbonate-laden water into Harmony Creek early yesterday, forcing crews to spend hours cleaning up the mess.

(b) A tanker truck carrying 6000 gallons of acetic acid overturned on the I-5 freeway in San Diego, California.

(c) A university in New Hampshire was discovered dumping dihydrogen oxide into the local sewer system.

35. How many of each type of subatomic particle (protons and electrons) is in

(a) an atom of tin?

(b) a Sn^{2+} ion?

(c) a Sn^{4+} ion?

36. The compound X_2Y_3 is a stable solid. What ionic charge do you expect for X and Y? Explain.

37. The ferricyanide ion has the formula $Fe(CN)_6^{3-}$. Write the formula for the compounds that ferricyanide would form with the cations of elements 3, 13, and 30.

38. Compare and contrast the formulas of

(a) nitride with nitrite

(b) nitrite with nitrate

(c) nitrous acid with nitric acid

CHALLENGE EXERCISE

39. After studying chemistry, you should be able to recognize more of the substances listed on consumer products. A list of ingredients for dog food follows:

Chicken By-Product Meal (Natural source of Chondroitin Sulfate and Glucosamine), Corn Meal, Ground Whole Grain Sorghum, Ground Whole Grain Barley, Fish Meal (source of fish oil), Chicken, Chicken Fat (preserved with mixed Tocopherols, a source of Vitamin E), Dried Beet Pulp, Chicken Flavor, Dried Egg Product, Potassium Chloride, Brewers Dried Yeast, Salt, Sodium Hexametaphosphate, Fructooligosaccharides, Fish Oil (preserved with mixed Tocopherols, a source of Vitamin E), Calcium Carbonate, Flax Meal, Choline Chloride, Minerals (Ferrous Sulfate, Zinc Oxide, Manganese Sulfate, Copper Sulfate, Manganous Oxide, Potassium Iodide, Cobalt Carbonate), Vitamin E Supplement, Dried Chicken Cartilage (Natural source of Chondroitin Sulfate and Glucosamine), DL-Methionine, Vitamins (Ascorbic Acid, Vitamin A Acetate, Calcium Pantothenate, Biotin, Thiamine Mononitrate (source of Vitamin B1), Vitamin B12 Supplement, Niacin, Riboflavin Supplement (source of Vitamin B2), Inositol, Pyridoxine Hydrochloride (source of Vitamin B6), Vitamin D3 Supplement, Folic Acid), Beta-Carotene, L-Carnitine, Marigold, Citric Acid, Rosemary Extract.

Many of the substances in this bag of dog food are ionic compounds that you should be able to recognize. The manufacturers of this dog food did not completely identify some of the compounds. Write the chemical formula of the following compounds found in this ingredient list.

(a) potassium chloride   (f) copper sulfate

(b) calcium carbonate   (g) manganous oxide

(c) ferrous sulfate   (h) potassium iodide

(d) zinc oxide   (i) cobalt carbonate

(e) manganese sulfate   (j) sodium chloride

ANSWERS TO PRACTICE EXERCISES

6.1   phosphorus stem is *phosph-* + *-ide* = phosphide; nitrogen stem is *nitr-* + *-ide* = nitride; boron stem is *bor-* + *-ide* = boride; carbon stem is *carb-* + *-ide* = carbide

6.2   (a) KF; (b) $CaBr_2$; (c) Mg_3N_2; (d) Na_2S; (e) BaO; (f) $As_2(CO_3)_5$

6.3   (a) $SrCl_2$; (b) KI; (c) AlN; (d) CaS; (e) Na_2O

6.4   (a) lead(II) iodide; (b) tin(IV) fluoride; (c) iron(III) oxide; (d) copper(II) oxide

6.5   (a) $Sn(CrO_4)_2$; (b) $CrBr_3$; (c) SnF_2; (d) Cu_2O

6.6   (a) dichlorine monoxide; (b) sulfur dioxide; (c) carbon tetrabromide; (d) dinitrogen pentoxide; (e) ammonia; (f) iodine trichloride

6.7   (a) potassium bromide; (b) calcium nitride; (c) sulfur trioxide; (d) tin(II) fluoride; (e) copper(II) chloride; (f) dinitrogen tetroxide

6.8   (a) sodium nitrate; (b) calcium phosphate; (c) potassium hydroxide; (d) lithium carbonate; (e) sodium chlorate

6.9   (a) potassium hydrogen carbonate; (b) sodium hydrogen oxalate; (c) barium ammonium phosphate; (d) sodium aluminum sulfate

6.10   (a) hydrochloric acid; (b) hydrobromic acid

6.11   (a) carbonic acid; (b) acetic acid; (c) nitric acid; (d) nitrous acid

PUTTING IT TOGETHER:
Review for Chapters 5–6

Multiple Choice

Choose the correct answer to each of the following.

1. The concept of positive charge and a small, "heavy" nucleus surrounded by electrons was the contribution of
   (a) Dalton   (c) Thomson
   (b) Rutherford   (d) Chadwick

2. The neutron was discovered in 1932 by
   (a) Dalton   (c) Thomson
   (b) Rutherford   (d) Chadwick

3. An atom of atomic number 53 and mass number 127 contains how many neutrons?
   (a) 53   (c) 127
   (b) 74   (d) 180

4. How many electrons are in an atom of $^{40}_{18}Ar$?
   (a) 20   (c) 40
   (b) 22   (d) no correct answer given

5. The number of neutrons in an atom of $^{139}_{56}Ba$ is
   (a) 56   (c) 139
   (b) 83   (d) no correct answer given

6. The name of the isotope containing one proton and two neutrons is
   (a) protium   (c) deuterium
   (b) tritium   (d) helium

7. Each atom of a specific element has the same
   (a) number of protons   (c) number of neutrons
   (b) atomic mass   (d) no correct answer given

8. Which pair of symbols represents isotopes?
   (a) $^{23}_{11}Na$ and $^{23}_{12}Na$ (c) $^{63}_{29}Cu$ and $^{29}_{64}Cu$
   (b) $^{7}_{3}Li$ and $^{6}_{3}Li$ (d) $^{12}_{24}Mg$ and $^{12}_{26}Mg$

9. Two naturally occurring isotopes of an element have masses and abundance as follows: 54.00 amu (20.00%) and 56.00 amu (80.00%). What is the relative atomic mass of the element?
   (a) 54.20   (c) 54.80
   (b) 54.40   (d) 55.60

10. Substance X has 13 protons, 14 neutrons, and 10 electrons. Determine its identity.
    (a) ^{27}Mg (c) $^{27}Al^{3+}$
    (b) ^{27}Ne (d) ^{27}Al

11. The mass of a chlorine atom is 5.90×10^{-23}g. How many atoms are in a 42.0-g sample of chlorine?
    (a) 2.48×10^{-21} (c) 1.40×10^{-24}
    (b) 7.12×10^{23} (d) no correct answer given

12. The number of neutrons in an atom of $^{108}_{47}Ag$ is
    (a) 47   (c) 155
    (b) 108   (d) no correct answer given

13. The number of electrons in an atom of $^{27}_{13}Al$ is
    (a) 13   (c) 27
    (b) 14   (d) 40

14. The number of protons in an atom of $^{65}_{30}Zn$ is
    (a) 65   (c) 30
    (b) 35   (d) 95

15. The number of electrons in the nucleus of an atom of $^{24}_{12}Mg$ is
    (a) 12   (c) 36
    (b) 24   (d) no correct answer given

Names and Formulas

In which of the following is the formula correct for the name given?

1. copper(II) sulfate, $CuSO_4$

2. ammonium hydroxide, NH_4OH

3. mercury(I) carbonate, $HgCO_3$

4. phosphorus triiodide, PI_3

5. calcium acetate, $Ca(C_2H_3O_2)_2$

6. hypochlorous acid, $HClO$

7. dichlorine heptoxide, Cl_2O_7

8. magnesium iodide, MgI

9. sulfurous acid, H_2SO_3

10. potassium manganate, $KMnO_4$

11. lead(II) chromate, $PbCrO_4$

12. ammonium hydrogen carbonate, NH_4HCO_3

13. iron(II) phosphate, $FePO_4$

14. calcium hydrogen sulfate, $CaHSO_4$

15. mercury(II) sulfate, $HgSO_4$

16. dinitrogen pentoxide, N_2O_5

17. sodium hypochlorite, $NaClO$

18. sodium dichromate, $Na_2Cr_2O_7$

19. cadmium cyanide, $Cd(CN)_2$

20. bismuth(III) oxide, Bi_3O_2

21. carbonic acid, H_2CO_3

22. silver oxide, Ag_2O

23. ferric iodide, FeI_2

24. tin(II) fluoride, TiF_2

25. carbon monoxide, CO

26. phosphoric acid, H_3PO_3

27. sodium bromate, Na_2BrO_3

28. hydrosulfuric acid, H_2S

29. potassium hydroxide, POH

30. sodium carbonate, Na_2CO_3

31. zinc sulfate, $ZnSO_3$

32. sulfur trioxide, SO_3

33. tin(IV) nitrate, $Sn(NO_3)_4$

34. ferrous sulfate, $FeSO_4$

35. chloric acid, HCl

36. aluminum sulfide, Al_2S_3

37. cobalt(II) chloride, $CoCl_2$

38. acetic acid, $HC_2H_3O_2$

39. zinc oxide, ZnO_2

40. stannous fluoride, SnF_2

Free Response Questions

Answer each of the following. Be sure to include your work and explanations in a clear, logical form.

1. (a) What is an ion?
   (b) The average mass of a calcium atom is 40.08 amu. Why do we also use 40.08 amu as the average mass of a calcium ion (Ca^{2+})?

2. Congratulations! You discover a new element you name wyzzlebium (Wz). The average atomic mass of Wz was found to be 303.001 amu, and its atomic number is 120.
   (a) If the masses of the two isotopes of wyzzlebium are 300.9326 amu and 303.9303 amu, what is the relative abundance of each isotope?
   (b) What are the isotopic notations of the two isotopes? (e.g., $_Z^A$Wz)
   (c) How many neutrons are in one atom of the more abundant isotope?

3. How many protons are in one molecule of dichlorine heptoxide? Is it possible to determine precisely how many electrons and neutrons are in a molecule of dichlorine heptoxide? Why or why not?

4. An unidentified metal forms an ionic compound with phosphate. The metal forms a 2+ cation. If the minimum ratio of protons in the metal to the phosphorus is 6:5, what metal is it? (*Hint:* First write the formula for the ionic compound formed with phosphate anion.)

5. For each of the following compounds, indicate what is wrong with the name and why. If possible, fix the name.
   (a) iron hydroxide
   (b) dipotassium dichromium heptoxide
   (c) sulfur oxide

6. Sulfur dioxide is a gas formed as a by-product of burning coal. Sulfur trioxide is a significant contributor to acid rain. Does the existence of these two substances violate the law of multiple proportions? Explain.

7. (a) Which subatomic particles are not in the nucleus?
   (b) What happens to the size of an atom when it becomes an anion?
   (c) What do an ion of Ca and an atom of Ar have in common?

8. An unidentified atom is found to have an atomic mass 7.18 times that of the carbon-12 isotope.
   (a) What is the mass of the unidentified atom?
   (b) What are the possible identities of this atom?
   (c) Why are you unable to positively identify the element based on the atomic mass and the periodic table?
   (d) If the element formed a compound M_2O, where M is the unidentified atom, identify M by writing the isotopic notation for the atom.

9. Scientists such as Dalton, Thomson, and Rutherford proposed important models, which were ultimately challenged by later technology. What do we know to be false in Dalton's atomic model? What was missing in Thomson's model of the atom? What was Rutherford's experiment that led to the current model of the atom?

These black pearls are made of layers of calcium carbonate. They can be measured by counting or weighing.

© Hemis/Alamy

CHAPTER **7**

QUANTITATIVE COMPOSITION OF COMPOUNDS

Cereals, cleaning products, and pain remedies all list their ingredients on the package label. The ingredients are listed in order from most to least, but the amounts are rarely given. However, it is precisely these amounts that give products their desired properties and distinguish them from the competition. Understandably, manufacturers carefully regulate the amounts of ingredients to maintain quality and hopefully their customers' loyalty. In the medicines we purchase, these quantities are especially important for safety reasons—for example, they determine whether a medicine is given to children or is safe only for adults.

The composition of compounds is an important concept in chemistry. Determining numerical relationships among the elements in compounds and measuring exact quantities of particles are fundamental tasks that chemists routinely perform in their daily work.

CHAPTER OUTLINE

7.1 The Mole

7.2 Molar Mass of Compounds

7.3 Percent Composition of Compounds

7.4 Calculating Empirical Formulas

7.5 Calculating the Molecular Formula from the Empirical Formula

7.1 THE MOLE

LEARNING OBJECTIVE ● Apply the concepts of the mole, molar mass, and Avogadro's number to solve chemistry problems.

KEY TERMS

Avogadro's number
mole
molar mass

PhotoDisc, Inc./Getty Images

Oranges can be "counted" by weighing them in the store.

The atom is an incredibly tiny object. Its mass is far too small to measure on an ordinary balance. In Chapter 5 (Section 5.6), we learned to compare atoms using a table of atomic mass units. These units are valuable when we compare the masses of individual atoms (mentally), but they have no practical use in the laboratory. The mass in grams for an "average" carbon atom (atomic mass 12.01 amu) would be 2.00×10^{-23} g, which is much too tiny for the best laboratory balance.

So how can we confidently measure masses for these very tiny atoms? We increase the number of atoms in a sample until we have an amount large enough to measure on a laboratory balance. The problem then is how to count our sample of atoms.

Consider for a moment the produce in a supermarket. Frequently, apples and oranges are sorted by size and then sold by weight, not by the piece of fruit. The grocer is counting by weighing. To do this, he needs to know the mass of an "average" apple (235 g) and the mass of an "average" orange (186 g). Now suppose he has an order from the local college for 275 apples and 350 oranges. It would take a long time to count and package this order. The grocer can quickly count fruit by weighing. For example,

$$(275 \text{ apples})\left(\frac{235 \text{ g}}{\text{apple}}\right) = 6.46 \times 10^4 \text{ g} = 64.6 \text{ kg}$$

$$(350 \text{ oranges})\left(\frac{186 \text{ g}}{\text{orange}}\right) = 6.51 \times 10^4 \text{ g} = 65.1 \text{ kg}$$

He can now weigh 64.6 kg of apples and 65.1 kg of oranges and pack them without actually counting them. Manufacturers and suppliers often count by weighing. Other examples of counting by weighing include nuts, bolts, and candy.

Chemists also count atoms by weighing. We know the "average" masses of atoms, so we can count atoms by defining a unit to represent a larger number of atoms. Chemists have chosen the mole (mol) as the unit for counting atoms. The mole is a unit for counting just as a dozen or a ream or a gross is used to count:

$$1 \text{ dozen} = 12 \text{ objects}$$
$$1 \text{ ream} = 500 \text{ objects}$$
$$1 \text{ gross} = 144 \text{ objects}$$
$$1 \text{ mole} = 6.022 \times 10^{23} \text{ objects}$$

Note that we use a unit only when it is appropriate. A dozen eggs is practical in our kitchen, a gross might be practical for a restaurant, but a ream of eggs would not be very practical. Chemists can't use dozens, grosses, or reams because atoms are so tiny that a dozen, gross, or ream of atoms still couldn't be measured in the laboratory.

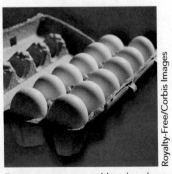

Royalty-Free/Corbis Images

Eggs are measured by the dozen.

Copyright John Wiley & Sons, Inc.

Paper is measured by the ream (500 sheets).

Tom Pantages

Pencils are measured by the gross (144).

Units of measurement need to be appropriate for the object being measured.

The number represented by 1 mol, 6.022×10^{23}, is called **Avogadro's number**, in honor of Amadeo Avogadro (1776–1856), who investigated several quantitative aspects in chemistry. It's difficult to imagine how large Avogadro's number really is, but this example may help: If 10,000 people started to count Avogadro's number, and each counted at the rate of 100 numbers per minute each minute of the day, it would take them over 1 trillion (10^{12}) years to count the total number. So even the tiniest amount of matter contains extremely large numbers of atoms.

Symbol name	C (carbon)	Al (aluminum)	Pb (lead)
Element	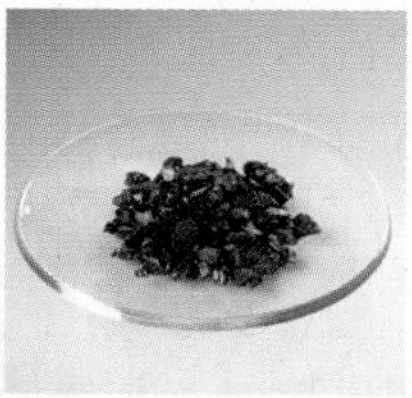	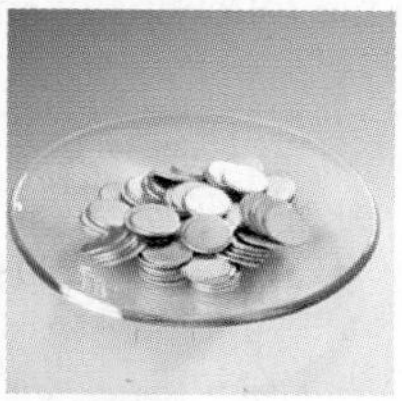	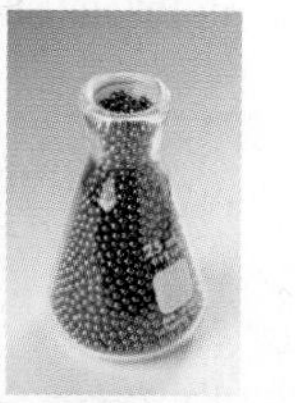
Average atomic mass	12.01 amu	26.98 amu	207.2 amu
Mass of sample	12.01 g	26.98 g	207.2 g
Number of atoms in sample	6.022×10^{23} atoms	6.022×10^{23} atoms	6.022×10^{23} atoms

Avogadro's number has been experimentally determined by several methods. How does it relate to atomic mass units? Remember that the atomic mass for an element is the average relative mass of all the isotopes for the element. The atomic mass (expressed in grams) of 1 mole of any element contains the same number of particles (Avogadro's number) as there are in exactly 12 g of ^{12}C. Thus, 1 **mole** of anything is the amount of the substance that contains the same number of items as there are atoms in exactly 12 g of ^{12}C.

Remember that ^{12}C is the reference isotope for atomic masses.

$$1 \text{ mole} = 6.022 \times 10^{23} \text{ items}$$

From the definition of mole, we can see that the atomic mass in grams of any element contains 1 mol of atoms. The term *mole* is so commonplace in chemistry that chemists use it as freely as the words *atom* or *molecule*. A mole of atoms, molecules, ions, or electrons represents Avogadro's number of these particles. If we can speak of a mole of atoms, we can also speak of a mole of molecules, a mole of electrons, or a mole of ions, understanding that in each case we mean 6.022×10^{23} particles:

The abbreviation for mole is mol, both singular and plural.

$$1 \text{ mol of atoms} = 6.022 \times 10^{23} \text{ atoms}$$
$$1 \text{ mol of molecules} = 6.022 \times 10^{23} \text{ molecules}$$
$$1 \text{ mol of ions} = 6.022 \times 10^{23} \text{ ions}$$

PRACTICE 7.1

How many atoms are in 1.000 mole of the following?
(a) Fe   (b) H_2 (c) H_2SO_4

The atomic mass of an element in grams contains Avogadro's number of atoms and is defined as the **molar mass** of that element. To determine molar mass, we change the units of the atomic mass (found in the periodic table) from atomic mass units to grams. For example, sulfur has an atomic mass of 32.07 amu, so 1 mol of sulfur has a molar mass of 32.07 g and contains 6.022×10^{23} atoms of sulfur. Here are some other examples:

Element	Atomic mass	Molar mass	Number of atoms
H	1.008 amu	1.008 g	6.022×10^{23}
Mg	24.31 amu	24.31 g	6.022×10^{23}
Na	22.99 amu	22.99 g	6.022×10^{23}

In this text molar masses of elements are given to four significant figures.

To summarize:

1. The atomic mass expressed in grams is the *molar mass* of an element. It is different for each element. In this text, molar masses are expressed to four significant figures.

$$1 \text{ molar mass} = \text{atomic mass of an element in grams}$$

2. One mole of any element contains Avogadro's number of atoms.

$$1 \text{ mol of atoms} = 6.022 \times 10^{23} \text{ atoms}$$

We can use these relationships to make conversions between number of atoms, mass, and moles, as shown in the following examples.

Sulfur	**Iodine**	**Iron**	**Mercury**
32.07 g 6.022×10^{23} atoms	126.9 g 6.022×10^{23} atoms	55.85 g 6.022×10^{23} atoms	200.6 g 6.022×10^{23} atoms

WileyPLUS

ENHANCED EXAMPLE

EXAMPLE 7.1

How many moles of iron does 25.0 g of iron (Fe) represent?

SOLUTION

Plan •   **Solution map:** g Fe → moles Fe

1 mole Fe = 55.85 g (from periodic table)

The conversion factor needed is

$$\frac{1 \text{ mole Fe}}{55.85 \text{ g Fe}} \quad \text{or} \quad \frac{55.85 \text{ g Fe}}{1 \text{ mole Fe}}$$

Calculate • $(25.0 \text{ g Fe})\left(\dfrac{1 \text{ mole Fe}}{55.85 \text{ g Fe}}\right) = 0.448 \text{ mol Fe}$

Check •   Since 25.0 g has three significant figures, the number of significant figures allowed in the answer is three. The answer is reasonable since 0.448 mol is about ½ mole and 25.0 g is about ½ of 55.85 g (one mole Fe).

EXAMPLE 7.2

How many magnesium atoms are contained in 5.00 g Mg?

SOLUTION

Plan •   **Solution map:** g Mg → atoms Mg

1 mol Mg = 24.31 g (from periodic table)

The conversion factor needed is

$$\frac{6.022 \times 10^{23} \text{ atoms Mg}}{24.31 \text{ g Mg}} \quad \text{or} \quad \frac{24.31 \text{ g Mg}}{6.022 \times 10^{23} \text{ atoms Mg}}$$

Calculate • $(5.00 \text{ g Mg})\left(\dfrac{6.022 \times 10^{23} \text{ atoms Mg}}{24.31 \text{ g Mg}}\right) = 1.24 \times 10^{23} \text{ atoms Mg}$

Check •   Since 5.00 g has three significant figures, the number of significant figures allowed in the answer is three.

EXAMPLE 7.3

What is the mass, in grams, of one atom of carbon (C)?

SOLUTION

Plan •   **Solution map:** atoms C → grams C

1 mol C = 12.01 g (from periodic table)

The conversion factor needed is

$$\frac{6.022 \times 10^{23} \text{ atoms C}}{12.01 \text{ g C}} \quad \text{or} \quad \frac{12.01 \text{ g C}}{6.022 \times 10^{23} \text{ atoms C}}$$

Calculate • $(1 \text{ atom C})\left(\dfrac{12.01 \text{ g C}}{6.022 \times 10^{23} \text{ atoms C}}\right) = 1.994 \times 10^{-23} \text{ g C}$

Check •   Since 1 atom is an exact number, the number of significant figures allowed in the answer is four based on the number of significant figures in the molar mass of C.

EXAMPLE 7.4

What is the mass of 3.01×10^{23} atoms of sodium (Na)?

SOLUTION

Plan •   **Solution map:** atoms Na → grams Na

1 mol Na = 22.99 g (from periodic table)

The conversion factor needed is

$$\frac{6.022 \times 10^{23} \text{ atoms Na}}{22.99 \text{ g Na}} \quad \text{or} \quad \frac{22.99 \text{ g Na}}{6.022 \times 10^{23} \text{ atoms Na}}$$

Calculate • $(3.01 \times 10^{23} \text{ atoms Na})\left(\dfrac{22.99 \text{ g Na}}{6.022 \times 10^{23} \text{ atoms Na}}\right) = 11.5 \text{ g Na}$

Check •   Since 3.01×10^{23} has three significant figures, the answer (mass) should have three as well.

EXAMPLE 7.5

What is the mass of 0.252 mol of copper (Cu)?

SOLUTION

Plan •   **Solution map:** mol Cu → grams Cu

1 mol Cu = 63.55 g (from periodic table)

The conversion factor needed is

$$\frac{1 \text{ mol Cu}}{63.55 \text{ g Cu}} \quad \text{or} \quad \frac{63.55 \text{ g Cu}}{1 \text{ mol Cu}}$$

Calculate • $(0.252 \text{ mol Cu})\left(\dfrac{63.55 \text{ g Cu}}{1 \text{ mol Cu}}\right) = 16.0 \text{ g Cu}$

Check •   The number of significant figures in both the question and answer is three.

EXAMPLE 7.6

How many oxygen atoms are present in 1.00 mol of oxygen molecules?

SOLUTION

Plan •   Oxygen is a diatomic molecule with the formula O_2. Therefore, a molecule of oxygen contains 2 oxygen atoms: $\dfrac{2 \text{ atoms O}}{1 \text{ molecule } O_2}$.

Solution map: moles $O_2 \rightarrow$ molecules $O_2 \rightarrow$ atoms O

The conversion factors needed are:

$$\frac{6.022 \times 10^{23} \text{ molecules } O_2}{1 \text{ mole } O_2} \quad \text{and} \quad \frac{2 \text{ atoms O}}{1 \text{ molecule } O_2}$$

Calculate • $(1.00 \text{ mol } O_2)\left(\dfrac{6.022 \times 10^{23} \text{ molecules } O_2}{1 \text{ mole } O_2}\right)\left(\dfrac{2 \text{ atoms O}}{1 \text{ molecule } O_2}\right)$

$$= 1.20 \times 10^{24} \text{ atoms O}$$

PRACTICE 7.2

What is the mass of 2.50 mol of helium (He)?

PRACTICE 7.3

How many atoms are present in 0.025 mol of iron?

7.2 MOLAR MASS OF COMPOUNDS

LEARNING OBJECTIVE •   Calculate the molar mass of a compound.

A formula unit is indicated by the formula, for example, Mg, MgS, H_2O, NaCl.

One mole of a compound contains Avogadro's number of *formula units* of that compound. The terms *molecular weight, molecular mass, formula weight,* and *formula mass* have been used in the past to refer to the mass of 1 mol of a compound. However, the term *molar mass* is more inclusive because it can be used for all types of compounds.

If the formula of a compound is known, its molar mass can be determined by adding the molar masses of all the atoms in the formula. If more than one atom of any element is present, its mass must be added as many times as it appears in the compound.

EXAMPLE 7.7

The formula for water is H_2O. What is its molar mass?

SOLUTION

Read •   **Known:** H_2O

Plan •   The formula contains 1 atom O and 2 atoms H. We need to use the atomic masses from the inside front cover of the text.

Calculate • $O = 16.00 \text{ g}   = 16.00 \text{ g}$

$2\,H = 2(1.008 \text{ g}) = \underline{2.016 \text{ g}}$

$\phantom{2\,H = 2(1.008 \text{ g}) = }\ 18.02 \text{ g} = \text{molar mass of } H_2O$

EXAMPLE 7.8

Calculate the molar mass of calcium hydroxide, $Ca(OH)_2$.

SOLUTION

Read •   **Known:** $Ca(OH)_2$

Plan •   The formula contains 1 atom Ca and 2 atoms each of O and H. We need to use the atomic masses from the inside front cover of the text.

Calculate •

$$
\begin{aligned}
1\ Ca &= 1(40.08\ g) = 40.08\ g \\
2\ O &= 2(16.00\ g) = 32.00\ g \\
2\ H &= 2(1.008\ g) = \underline{\ 2.016\ g} \\
&\qquad\qquad\qquad\quad 74.10\ \ g = \text{molar mass of } Ca(OH)_2
\end{aligned}
$$

PRACTICE 7.4

Calculate the molar mass of KNO_3.

In this text we round all molar masses to four significant figures, although you may need to use a different number of significant figures for other work (in the lab).

The mass of 1 mol of a compound contains Avogadro's number of formula units. Consider the compound hydrogen chloride (HCl). One atom of H combines with one atom of Cl to form HCl. When 1 mol of H (1.008 g of H or 6.022×10^{23} H atoms) combines with 1 mol of Cl (35.45 g of Cl or 6.022×10^{23} Cl atoms), 1 mol of HCl (36.46 g of HCl or 6.022×10^{23} HCl molecules) is produced. These relationships are summarized in the following table:

H	Cl	HCl
6.022×10^{23} H *atoms*	6.022×10^{23} Cl *atoms*	6.022×10^{23} HCl *molecules*
1 mol H *atoms*	1 mol Cl *atoms*	1 mol HCl *molecules*
1.008 g H	34.45 g Cl	36.46 g HCl
1 molar mass H *atoms*	1 molar mass Cl *atoms*	1 molar mass HCl *molecules*

In dealing with diatomic elements (H_2, O_2, N_2, F_2, Cl_2, Br_2, and I_2), we must take special care to distinguish between a mole of atoms and a mole of molecules. For example, consider 1 mol of oxygen molecules, which has a mass of 32.00 g. This quantity is equal to 2 mol of

A mole of table salt (in front of a salt shaker) and a mole of water (in the plastic container) have different sizes but both contain Avogadro's number of formula units.

Tom Pantages

oxygen atoms. Remember that 1 mol represents Avogadro's number of the particular chemical entity that is under consideration:

$$1 \text{ mol } H_2O = 18.02 \text{ g } H_2O \quad = 6.022 \times 10^{23} \text{ molecules}$$
$$1 \text{ mol } NaCl = 58.44 \text{ g } NaCl \quad = 6.022 \times 10^{23} \text{ formula units}$$
$$1 \text{ mol } H_2 = 2.016 \text{ g } H_2 \quad = 6.022 \times 10^{23} \text{ molecules}$$
$$1 \text{ mol } HNO_3 = 63.02 \text{ g } HNO_3 = 6.022 \times 10^{23} \text{ molecules}$$
$$1 \text{ mol } K_2SO_4 = 174.3 \text{ g } K_2SO_4 = 6.022 \times 10^{23} \text{ formula units}$$
$$1 \text{ mol} = 6.022 \times 10^{23} \text{ formula unit or molecules}$$
$$= 1 \text{ molar mass of a compound}$$

Formula units are often used in place of molecules for substances that contain ions.

WileyPLUS

ENHANCED EXAMPLE

EXAMPLE 7.9

What is the mass of 1 mol of sulfuric acid (H_2SO_4)?

SOLUTION

Read •   **Known:** H_2SO_4

Plan •   We need to use the atomic masses from the inside front cover of the text.

Calculate • $2 \text{ H} = 2(1.008 \text{ g}) = 2.016 \text{ g}$

$1 \text{ S} = 32.07 \text{ g} = 32.07 \text{ g}$

$4 \text{ O} = 4(16.00 \text{ g}) = \underline{64.00 \text{ g}}$
$98.09 \text{ g} = \text{mass of 1 mol of } H_2SO_4$

EXAMPLE 7.10

How many moles of sodium hydroxide (NaOH) are there in 1.00 kg of sodium hydroxide?

SOLUTION

Read •   **Known:**   1.00 kg NaOH

Plan •   We need to use the atomic masses from the inside front cover of the text to find the molar mass of NaOH.

$$\begin{array}{ccc} \text{Na} & \text{O} & \text{H} \\ 22.99 \text{ g} + 16.00 \text{ g} + 1.008 \text{ g} = 40.00 \text{ g NaOH} = \text{Molar mass NaOH} \end{array}$$

Solution map: kg NaOH → g NaOH → mol NaOH

Create the appropriate conversion factor by placing the unit desired in the numerator and the unit to be eliminated in the denominator.

We need two conversion factors: $\dfrac{1000 \text{ g}}{1 \text{ kg}}$ and $\dfrac{1 \text{ mol NaOH}}{40.00 \text{ g NaOH}}$.

Calculate • $(1.00 \text{ kg NaOH})\left(\dfrac{1000 \text{ g NaOH}}{1 \text{ kg NaOH}}\right)\left(\dfrac{1 \text{ mol NaOH}}{40.00 \text{ g NaOH}}\right)$

$= 25.00 \text{ mol NaOH}$

Therefore, 1.00 kg NaOH = 25.00 mol NaOH.

EXAMPLE 7.11

What is the mass of 5.00 mol water?

SOLUTION

Read •   **Known:**   5.00 mol H_2O

Plan • We need to use the atomic masses from the inside front cover of the text to find the molar mass of H_2O.

Solution map: mol H_2O → g H_2O

$$\frac{18.02 \text{ g } H_2O}{1 \text{ mol } H_2O} \text{ (conversion factor)}$$

Calculate • $(5.00 \text{ mol } H_2O)\left(\dfrac{18.02 \text{ g } H_2O}{1 \text{ mol } H_2O}\right) = 90.1 \text{ g } H_2O$

Therefore, 5.00 mol H_2O = 90.1 g H_2O.

EXAMPLE 7.12

How many molecules of hydrogen chloride (HCl) are there in 25.0 g of hydrogen chloride?

SOLUTION

Read • **Known:** 25.0 g HCl

Plan • We need to use the atomic masses from the inside front cover of the text to find the molar mass of HCl.

$$\begin{array}{cc} H & Cl \\ 1.008 \text{ g} + 34.45 \text{ g} = 36.46 \text{ g} = \text{molar mass HCl} \end{array}$$

Solution map: g HCl → mol HCl → molecules HCl

$$\frac{1 \text{ mol HCl}}{36.46 \text{ g HCl}} \text{ and } \frac{6.022 \times 10^{23} \text{ molecules HCl}}{1 \text{ mol HCl}} \text{ (conversion factors)}$$

Calculate • $(25.0 \text{ g HCl})\left(\dfrac{1 \text{ mol HCl}}{36.46 \text{ g HCl}}\right)\left(\dfrac{6.022 \times 10^{23} \text{ molecules HCl}}{1 \text{ mol HCl}}\right)$

$= 4.13 \times 10^{23}$ molecules HCl

PRACTICE 7.5

What is the mass of 0.150 mol of Na_2SO_4?

PRACTICE 7.6

How many moles and molecules are there in 500.0 g of $HC_2H_3O_2$?

7.3 PERCENT COMPOSITION OF COMPOUNDS

Calculate the percent composition of a compound from its chemical formula and from experimental data.

● LEARNING OBJECTIVE

KEY TERM

percent composition of a compound

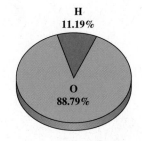

Percent means parts per 100 parts. Just as each piece of pie is a percent of the whole pie, each element in a compound is a percent of the whole compound. The **percent composition of a compound** is the *mass percent* of each element in the compound. The molar mass represents the total mass, or 100%, of the compound. Thus, the percent composition of water, H_2O, is 11.19% H and 88.79% O by mass. According to the law of definite composition, the percent composition must be the same no matter what size sample is taken.

The percent composition of a compound can be determined (1) from knowing its formula or (2) from experimental data.

Percent Composition from Formula

If the formula of a compound is known, a two-step process is needed to determine the percent composition:

PROBLEM-SOLVING STRATEGY: For Calculating Percent Composition from Formula

1. Calculate the molar mass (Section 7.2).

2. Divide the total mass of each element in the formula by the molar mass and multiply by 100. This gives the percent composition:

$$\frac{\text{total mass of the element}}{\text{molar mass}} \times 100 = \text{percent of the element}$$

WileyPLUS

ENHANCED EXAMPLE

EXAMPLE 7.13

Calculate the percent composition of sodium chloride (NaCl).

SOLUTION

Read •   **Known:**   NaCl

Plan •   Use the Problem-Solving Strategy for Calculating Percent Composition from Formula.

Calculate • **1.** We need to use the atomic masses from the inside front cover of the text to find the molar mass of NaCl.

$$\begin{array}{cc} \text{Na} & \text{Cl} \\ 22.99 \text{ g} + 34.45 \text{ g} = 58.44 \text{ g} = \text{molar mass NaCl} \end{array}$$

2. Calculate the percent composition for each element.

$$\text{Na:} \quad \left(\frac{22.99 \text{ g Na}}{58.44 \text{ g}}\right)(100) = 39.34\% \text{ Na}$$

$$\text{Cl:} \quad \left(\frac{35.45 \text{ g Cl}}{58.44 \text{ g}}\right)(100) = \underline{60.66\% \text{ Cl}} \atop 100.00\% \text{ total}$$

NOTE   In any two-component system, if the percent of one component is known, the other is automatically defined by difference; that is, if Na is 39.43%, then Cl is 100% − 39.34% = 60.66%. However, the calculation of the percent should be carried out for each component, since this provides a check against possible error. The percent composition should add to 100% ± 0.2%.

EXAMPLE 7.14

Calculate the percent composition of potassium chloride (KCl).

SOLUTION

Read •   **Known:**   KCl

Plan •   Use the Problem-Solving Strategy for Calculating Percent Composition from Formula.

Calculate • **1.** We need to use the atomic masses from the inside front cover of the text to find the molar mass of KCl.

$$\begin{array}{cc} \text{K} & \text{Cl} \\ 39.10 \text{ g} + 35.45 \text{ g} = 74.55 \text{ g} = \text{molar mass KCl} \end{array}$$

2. Calculate the percent composition for each element.

K: $\left(\dfrac{39.10 \text{ g K}}{74.55 \text{ g}}\right)(100) = 52.45\% \text{ K}$

Cl: $\left(\dfrac{35.45 \text{ g Cl}}{74.55 \text{ g}}\right)(100) = \underline{47.55\% \text{ Cl}}$

$100.00\% \text{ total}$

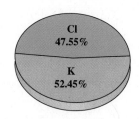

Comparing the results calculated for NaCl and for KCl, we see that NaCl contains a higher percentage of Cl by mass, although each compound has a one-to-one atom ratio of Cl to Na and Cl to K. The reason for this mass percent difference is that Na and K do not have the same atomic masses.

It is important to realize that when we compare *1 mol of NaCl with 1 mol of KCl*, each quantity contains the same number of Cl atoms—namely, 1 mol of Cl atoms. However, if we compare *equal masses* of NaCl and KCl, there will be more Cl atoms in the mass of NaCl, since NaCl has a higher mass percent of Cl.

1 mol NaCl contains	100.00 g NaCl contains	1 mol KCl contains	100.00 g KCl contains
1 mol Na	39.34 g Na	1 mol K	52.45 g K
1 mol Cl	60.66 g Cl	1 mol Cl	47.55 g Cl
	60.66% Cl		47.55% Cl

EXAMPLE 7.15

Calculate the percent composition of potassium sulfate (K_2SO_4).

SOLUTION

Read •   **Known:** K_2SO_4

Plan •   Use the Problem-Solving Strategy for Calculating Percent Composition from Formula.

Calculate •   **1.** We need to use the atomic masses from the inside front cover of the text to find the molar mass of K_2SO_4.

$$\begin{array}{ccc} \text{K} & \text{S} & \text{O} \\ 2(39.10) \text{ g} + 32.07 \text{ g} + 4(16.00) \end{array} = 174.3 \text{ g} = \text{molar mass } K_2SO_4$$

2. Calculate the percent composition for each element.

K: $\left(\dfrac{78.20 \text{ g K}}{174.3 \text{ g}}\right)(100) = 44.87\% \text{ K}$

S: $\left(\dfrac{32.07 \text{ g S}}{174.3 \text{ g}}\right)(100) = 18.40\% \text{ S}$

O: $\left(\dfrac{64.00 \text{ g O}}{174.3 \text{ g}}\right)(100) = \underline{36.72\% \text{ O}}$

$99.99\% \text{ total}$

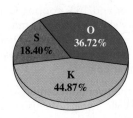

PRACTICE 7.7

Calculate the percent composition of $Ca(NO_3)_2$.

PRACTICE 7.8

Calculate the percent composition of K_2CrO_4.

PROBLEM-SOLVING STRATEGY: For Percent Composition from Experimental Data

1. Calculate the mass of the compound formed.

2. Divide the mass of each element by the total mass of the compound and multiply by 100.

EXAMPLE 7.16

Zinc oxide is a compound with many uses from preventing sunburn to a pigment in white paint. When heated in air, 1.63 g of zinc (Zn) combines with 0.40 g of oxygen (O_2) to form zinc oxide. Calculate the percent composition of the compound formed.

SOLUTION

Read •   **Knowns:**   1.63 g Zn

                      0.40 g O_2

Plan •   Use the Problem-Solving Strategy for Calculating Percent Composition from Experimental Data.

Calculate •   **1.** We need to calculate the mass of the product formed.

$$1.63 \text{ g Zn} + 0.40 \text{ g } O_2 = 2.03 \text{ g product}$$

2. Calculate the percent for each element.

$$\text{Zn:} \quad \left(\frac{1.63 \text{ g Zn}}{2.03 \text{ g}} \right)(100) = \quad 80.3\% \text{ Zn}$$

$$\text{O:} \quad \left(\frac{0.40 \text{ g O}}{2.03 \text{ g}} \right)(100) = \frac{20. \ \% \text{ O}}{100.3\% \text{ total}}$$

PRACTICE 7.9

Aluminum chloride is formed by reacting 13.43 g aluminum with 53.18 g chlorine. What is the percent composition of the compound?

> CHEMISTRY *IN ACTION*

Vanishing Coins?

Modern technology is changing our coins in some pretty interesting ways. The first U.S. coins were produced during the late 1700s from "coin silver," an alloy of 90% silver and 10% copper. Coins were made following a very simple rule: The mass of the coin reflects its relative value. Therefore, a half dollar weighed half as much as a dollar coin, and a quarter weighed ¼ as much as a dollar coin.

In the twentieth century our society began using machines to collect coins. We use vending machines, parking meters, toll baskets, laundromats, slot machines, and video games to make our lives more convenient and fun. This change produced a huge demand for coins. At the same time, the price of silver rose rapidly.

The solution? Make coins from a different alloy that would still work in all of our machines. This was a tricky process since the machines require a specific mass and electric "resistivity." In the mid-1960s new coins that sandwiched a layer of copper between layers of an alloy of copper and nickel began to replace quarters and dimes. And a new Susan B. Anthony dollar appeared briefly in 1979. Finally, in 1997 Congress decided to revive the dollar coin once more.

Congress specified that the new dollar coin had to look and feel different to consumers, be golden in color, and have a distinctive edge. At the same time, it needed to fool vending machines so that both the Susan B. Anthony dollar and the new coin would work. Designers decided on the portrait of Sacagawea, a Shoshone woman and her child. She was a guide on the Lewis and Clark expedition. On the back is an eagle. The chemists struggled to find a golden alloy for the coin. It is slightly larger than a quarter and weighs 8.1 grams. Vending machines determine a coin's value by its size, weight, and electromagnetic signature. This electromagnetic signature is hard to duplicate. All the golden coins tested had three times the conductivity of the silver Anthony coin. Metallurgists finally tried adding manganese and zinc to the copper core and found a "golden" alloy that fooled the vending machines! Sacagawea dollars are made of an alloy that is 77% copper, 12% zinc, 7% manganese, and 4% nickel. And on top of it all, the new coins only cost 12 cents each to make, leaving an 88 cent profit for the mint on each coin!

The future of coins is becoming less certain, however, as our machines are converted to electronic devices that use magnetic strips and a swipe of a card in place of "coins" and scan our cars for toll pass information. Some day soon, coins may vanish altogether from our pockets into collectors' albums.

7.4 CALCULATING EMPIRICAL FORMULAS

Determine the empirical formula for a compound from its percent composition.

LEARNING OBJECTIVE

KEY TERM
empirical formula

The **empirical formula**, or *simplest formula*, gives the smallest whole-number ratio of atoms present in a compound. This formula gives the relative number of atoms of each element in the compound.

We can establish empirical formulas because (1) individual atoms in a compound are combined in whole-number ratios and (2) each element has a specific atomic mass.

To calculate an empirical formula, we need to know (1) the elements that are combined, (2) their atomic masses, and (3) the ratio by mass or percentage in which they are combined. If elements A and B form a compound, we may represent the empirical formula as $A_x B_y$, where x and y are small whole numbers that represent the atoms of A and B. To write the empirical formula, we must determine x and y:

PROBLEM-SOLVING STRATEGY: For Calculating an Empirical Formula

1. Assume a definite starting quantity (usually 100.0 g) of the compound, if not given, and express the mass of each element in grams.

2. Convert the grams of each element into moles using each element's molar mass. This conversion gives the number of moles of atoms of each element in the quantity assumed in Step 1. At this point, these numbers will usually not be whole numbers.

3. Divide each value obtained in Step 2 by the smallest of these values. If the numbers obtained are whole numbers, use them as subscripts and write the empirical formula. If the numbers obtained are not whole numbers, go on to Step 4.

4. Multiply the values obtained in Step 3 by the smallest number that will convert them to whole numbers. Use these whole numbers as the subscripts in the empirical formula. For example, if the ratio of A to B is 1.0 : 1.5, multiply both numbers by 2 to obtain a ratio of 2 : 3. The empirical formula then is A_2B_3.

In many of these calculations, results will vary somewhat from an exact whole number; this can be due to experimental errors in obtaining the data or from rounding off numbers. Calculations that vary by no more than ± 0.1 from a whole number usually are rounded off to the nearest whole number. Deviations greater than about 0.1 unit usually mean that the calculated ratios need to be multiplied by a factor to make them all whole numbers. For example, an atom ratio of 1 : 1.33 should be multiplied by 3 to make the ratio 3 : 4. Let's do a few examples to see how it works.

Some common fractions and their decimal equivalents are

$$1/4 = 0.25$$
$$1/3 = 0.333\ldots$$
$$2/3 = 0.666\ldots$$
$$1/2 = 0.5$$
$$3/4 = 0.75$$

Multiply the decimal equivalent by the number in the denominator of the fraction to get a whole number: $4(0.75) = 3$.

WileyPLUS

ENHANCED EXAMPLE

EXAMPLE 7.17

Calculate the empirical formula of a compound containing 11.19% hydrogen (H) and 88.79% oxygen (O).

SOLUTION

Read • **Knowns:** 11.19% H
88.79% O

Plan • Use the Problem-Solving Strategy for Calculating an Empirical Formula.

Calculate • **1.** Assume 100.0 g of material. We know that the percent of each element equals the grams of each element:

11.19 g H
88.79 g O

2. Convert grams of each element to moles:

$$H: (11.19 \text{ g})\left(\frac{1 \text{ mol H atoms}}{1.008 \text{ g H}}\right) = 11.10 \text{ mol H atoms}$$

$$O: (88.79 \text{ g})\left(\frac{1 \text{ mol O atoms}}{16.00 \text{ g O}}\right) = 5.549 \text{ mol O atoms}$$

The formula could be expressed as $H_{11.10}O_{5.549}$. However, it's customary to use the smallest whole-number ratio of atoms.

3. Change the number of moles to whole numbers by dividing by the smallest number.

$$H = \frac{11.10 \text{ mol}}{5.549 \text{ mol}} = 2.000$$

$$O = \frac{5.549 \text{ mol}}{5.549 \text{ mol}} = 1.000$$

The simplest ratio of H to O is 2:1.

Empirical formula = H_2O

EXAMPLE 7.18

The analysis of a salt shows that it contains 56.58% potassium (K), 8.68% carbon (C), and 34.73% oxygen (O). Calculate the empirical formula for this substance.

SOLUTION

Read •   **Knowns:** 56.58% K
                              8.68% C
                              34.73% O

Plan •   Use the Problem-Solving Strategy for Calculating an Empirical Formula.

Calculate • **1.** Assume 100.0 g of material. We know that the percent of each element equals the grams of each element:

   56.58 g K
   8.68 g C
   34.73 g O

2. Convert the grams of each element to moles:

$$K: (56.58 \text{ g K})\left(\frac{1 \text{ mol K atoms}}{39.10 \text{ g K}}\right) = 1.447 \text{ mol K atoms}$$

$$C: (8.68 \text{ g C})\left(\frac{1 \text{ mol C atoms}}{12.01 \text{ g C}}\right) = 0.723 \text{ mol C atoms}$$

$$O: (34.73 \text{ g O})\left(\frac{1 \text{ mol O atoms}}{16.00 \text{ g O}}\right) = 2.171 \text{ mol O atoms}$$

3. Change the number of moles to whole numbers by dividing by the smallest number.

$$K = \frac{1.447 \text{ mol}}{0.723 \text{ mol}} = 2.00$$

$$C = \frac{0.723 \text{ mol}}{0.723 \text{ mol}} = 1.00$$

$$O = \frac{2.171 \text{ mol}}{0.723 \text{ mol}} = 3.00$$

The simplest ratio of K:C:O is 2:1:3.

Empirical formula = K_2CO_3

EXAMPLE 7.19

A sulfide of iron was formed by combining 2.233 g of iron (Fe) with 1.926 g of sulfur (S). What is the empirical formula for the compound?

SOLUTION

Read •   **Knowns:**   2.233 g Fe
                           1.926 g S

Plan •   Use the Problem-Solving Strategy for Calculating an Empirical Formula.

Calculate • **1.** The mass of each element is known, so we use it directly.

2. Convert grams of each element to moles.

$$\text{Fe:}\quad (2.233\ \cancel{\text{g Fe}})\left(\frac{1\ \text{mol Fe atoms}}{55.85\ \cancel{\text{g Fe}}}\right) = 0.03998\ \text{mol Fe atoms}$$

$$\text{S:}\quad (1.926\ \cancel{\text{g S}})\left(\frac{1\ \text{mol S atoms}}{32.07\ \cancel{\text{g S}}}\right) = 0.06006\ \text{mol S atoms}$$

3. Change the number of moles to whole numbers by dividing by the smallest number.

$$\text{Fe} = \frac{0.03998\ \text{mol}}{0.03998\ \text{mol}} = 1.000$$

$$\text{S} = \frac{0.06006\ \text{mol}}{0.03998\ \text{mol}} = 1.502$$

4. We still have not reached a ratio that gives whole numbers in the formula, so we multiply by a number that will give us whole numbers.

Fe: (1.000)2 = 2.000
S: (1.502)2 = 3.004

Empirical formula = Fe_2S_3

PRACTICE 7.10

Calculate the empirical formula of a compound containing 52.14% C, 13.12% H, and 34.73% O.

PRACTICE 7.11

Calculate the empirical formula of a compound that contains 43.7% phosphorus and 56.3% O by mass.

7.5 CALCULATING THE MOLECULAR FORMULA FROM THE EMPIRICAL FORMULA

Compare an empirical formula to a molecular formula and calculate a molecular formula from the empirical formula of the compound and its molar mass.

• **LEARNING OBJECTIVE**

KEY TERM
molecular formula

The **molecular formula** is the true formula, representing the total number of atoms of each element present in one molecule of a compound. It is entirely possible that two or more substances will have the same percent composition yet be distinctly different compounds. For example, acetylene (C_2H_2) is a common gas used in welding; benzene (C_6H_6) is an important solvent obtained from coal tar and is used in the synthesis of styrene and nylon. Both acetylene and benzene contain 92.3% C and 7.7% H. The smallest ratio of C and H corresponding to these percentages is CH (1:1) Therefore, the *empirical* formula for both acetylene and benzene is CH, even though the *molecular* formulas are C_2H_2 and

C_6H_6, respectively. Often the molecular formula is the same as the empirical formula. If the molecular formula is not the same, it will be an integral (whole number) multiple of the empirical formula. For example,

$$CH = \text{empirical formula}$$

$$(CH)_2 = C_2H_2 = \text{acetylene} \quad \text{(molecular formula)}$$

$$(CH)_6 = C_6H_6 = \text{benzene} \quad \text{(molecular formula)}$$

Table 7.1 compares the formulas of these substances. Table 7.2 shows empirical and molecular formula relationships of other compounds.

TABLE 7.1 Molecular Formulas of Two Compounds Having an Empirical Formula with a 1:1 Ratio of Carbon and Hydrogen Atoms

Formula	Composition		Molar mass
	% C	% H	
CH (empirical)	92.3	7.7	13.02 (empirical)
C_2H_2 (acetylene)	92.3	7.7	26.04 (2 × 13.02)
C_6H_6 (benzene)	92.3	7.7	78.12 (6 × 13.02)

TABLE 7.2 Some Empirical and Molecular Formulas

Substance	Empirical formula	Molecular formula	Substance	Empirical formula	Molecular formula
Acetylene	CH	C_2H_2	Diborane	BH_3	B_2H_6
Benzene	CH	C_6H_6	Hydrazine	NH_2	N_2H_4
Ethylene	CH_2	C_2H_4	Hydrogen	H	H_2
Formaldehyde	CH_2O	CH_2O	Chlorine	Cl	Cl_2
Acetic acid	CH_2O	$C_2H_4O_2$	Bromine	Br	Br_2
Glucose	CH_2O	$C_6H_{12}O_6$	Oxygen	O	O_2
Hydrogen chloride	HCl	HCl	Nitrogen	N	N_2
Carbon dioxide	CO_2	CO_2	Iodine	I	I_2

The molecular formula can be calculated from the empirical formula if the molar mass is known. The molecular formula will be equal either to the empirical formula or some multiple of it. For example, if the empirical formula of a compound of hydrogen and fluorine is HF, the molecular formula can be expressed as $(HF)_n$, where $n = 1, 2, 3, 4, \ldots$. This n means that the molecular formula could be HF, H_2F_2, H_3F_3, H_4F_4, and so on. To determine the molecular formula, we must evaluate n:

$$n = \frac{\text{molar mass}}{\text{mass of empirical formula}} = \text{number of empirical formula units}$$

What we actually calculate is the number of units of the empirical formula contained in the molecular formula.

WileyPLUS

ENHANCED EXAMPLE

EXAMPLE 7.20

A compound of nitrogen and oxygen with a molar mass of 92.00 g was found to have an empirical formula of NO_2. What is its molecular formula?

SOLUTION

Read •   **Knowns:** NO_2 = empirical formula

Molar mass = 92.00

Plan • Let n be the number of NO_2 units in a molecule; then the molecular formula is $(NO_2)_n$. Find the molar mass of an NO_2 unit and determine the numbers of units needed for the molar mass of the compound.

Calculate • The molar mass of NO_2 is 14.01 g + 2(16.00 g) = 46.01 g.

$$n = \frac{92.00 \text{ g}}{46.01 \text{ g}} = 2 \text{ (empirical formula units)}$$

The molecular formula is $(NO_2)_2$ or N_2O_4.

N_2O_4

EXAMPLE 7.21

Propylene is a compound frequently polymerized to make polypropylene, which is used for rope, laundry bags, blankets, carpets, and textile fibers. Propylene has a molar mass of 42.08 g and contains 14.3% H and 85.7% C. What is its molecular formula?

SOLUTION

Read • **Knowns:** Molar mass = 42.08 g

14.3% H

85.7% C

Plan • First find the empirical formula using the Problem-Solving Strategy for Calculating an Empirical Formula. Then use n to determine the molecular formula from the empirical formula.

Calculate • **1.** Assume 100.0 g of material. We know that the percent of each element equals the grams of each element:

14.3 g H

85.7 g C

2. Convert the grams of each element to moles:

$$\text{H:} \quad (14.3 \text{ g H})\left(\frac{1 \text{ mol H atoms}}{1.008 \text{ g H}}\right) = 14.2 \text{ mol H atoms}$$

$$\text{C:} \quad (85.7 \text{ g C})\left(\frac{1 \text{ mol C atoms}}{12.01 \text{ g C}}\right) = 7.14 \text{ mol C atoms}$$

3. Change the number of moles to whole numbers by dividing by the smallest number.

$$\text{H} = \frac{14.2 \text{ mol}}{7.14 \text{ mol}} = 1.99$$

$$\text{C} = \frac{7.14 \text{ mol}}{7.14 \text{ mol}} = 1.00$$

Empirical formula = CH_2

We determine the molecular formula from the empirical formula and the molar mass:

Molecular formula = $(CH_2)_n$

CH_2 molar mass = 12.01 g + 2(1.008 g) = 14.03 g

$$n = \frac{42.08 \text{ g}}{14.03 \text{ g}} = 3 \text{ (empirical formula units)}$$

The molecular formula is $(CH_2)_3 = C_3H_6$.

When propylene is polymerized into polypropylene, it can be used for rope and other recreational products.

C_3H_6

PRACTICE 7.12

Calculate the empirical and molecular formulas of a compound that contains 80.0% C, 20.0% H, and has a molar mass of 30.00 g.

CHAPTER **7 REVIEW**

7.1 THE MOLE

- We count atoms by weighing them since they are so tiny.
- 1 mole $= 6.022 \times 10^{23}$ items.
- Avogadro's number is 6.022×10^{23}.

7.2 MOLAR MASS OF COMPOUNDS

- One mole of a compound contains Avogadro's number of formula units of that compound.
- The mass (grams) of one mole of a compound is the molar mass.
- Molar mass is determined by adding the molar masses of all the atoms in a formula.
- Molar masses are given to four significant figures in this text.

7.3 PERCENT COMPOSITION OF COMPOUNDS

- To determine the percent composition from a formula:
  - Calculate the molar mass.
  - For each element in the formula

$$\frac{\text{total mass of the element}}{\text{molar mass of the compound}} \times 100 = \text{percent of the element}$$

- To determine percent composition from experimental data:
  - Calculate the mass of the compound formed.
  - For each element in the formula

$$\frac{\text{mass of the element}}{\text{mass of the compound formed}} \times 100 = \text{percent of the element}$$

7.4 CALCULATING EMPIRICAL FORMULAS

- The empirical formula is the simplest formula giving the smallest whole-number ratio of atoms present in a compound.
- To determine the empirical formula for a compound you need to know:
  - The elements that are combined
  - Their atomic masses
  - The ratio of masses or percentage in which they are combined
- Empirical formulas are represented in the form $A_x B_y$. To determine the empirical formula of this compound:
  - Assume a starting quantity (100.0 g is a good choice).
  - Convert mass (g) to moles for each element.
  - Divide each element's moles by the smallest number of moles.
    - If the ratios are whole numbers, use them as subscripts and write the empirical formula.
  - If the ratios are not whole numbers, multiply them all by the smallest number, which will convert them to whole numbers.
  - Use the whole numbers to write the empirical formula.

7.5 CALCULATING THE MOLECULAR FORMULA FROM THE EMPIRICAL FORMULA

- The molecular formula is the true formula representing the total number of atoms of each element present in one molecule of the compound.
- Two or more substances may have the same empirical formulas but different molecular formulas.
- The molecular formula is calculated from the empirical formula when the molar mass is known:

- $n = \dfrac{\text{molar mass}}{\text{mass of empirical formula}} = \text{number of empirical formula units}$

- The molecular formula is $(A_x B_y)_n$.

REVIEW QUESTIONS

1. What is a mole?
2. Which would have a higher mass: a mole of K atoms or a mole of Au atoms?
3. Which would contain more atoms: a mole of K atoms or a mole of Au atoms?
4. Which would contain more electrons: a mole of K atoms or a mole of Au atoms?
5. What is the numerical value of Avogadro's number?
6. What is the relationship between Avogadro's number and the mole?
7. Which has the greater number of oxygen atoms, 1 mole of oxygen gas (O_2) or 1 mole of ozone gas (O_3)?
8. What is molar mass?
9. If the atomic mass scale had been defined differently, with an atom of $^{12}_{6}C$ defined as a mass of 50 amu, would this have any effect on the value of Avogadro's number? Explain.
10. Complete these statements, supplying the proper quantity.
    (a) A mole of O atoms contains _____ atoms.
    (b) A mole of O_2 molecules contains _____ molecules.
    (c) A mole of O_2 molecules contains _____ atoms.
    (d) A mole of O atoms has a mass of _____ grams.
    (e) A mole of O_2 molecules has a mass of _____ grams.

11. How many molecules are present in 1 molar mass of sulfuric acid (H_2SO_4)? How many atoms are present?
12. What are the two types of information that can be used to calculate the percent composition of a compound?
13. A compound is composed of only nitrogen and oxygen. If the compound contains 43.8% nitrogen, what is the percent oxygen in the compound?
14. Write the steps used to calculate percent composition of an element in a compound.
15. In calculating the empirical formula of a compound from its percent composition, why do we choose to start with 100.0 g of the compound?
16. Which of the following compounds have the same empirical formula? What is the empirical formula? CH_4, C_2H_4, C_4H_6, C_4H_{10}, C_8H_{12}, C_8H_{18}
17. What is the difference between an empirical formula and a molecular formula?
18. What data beyond the percent composition are generally used to determine the molecular formula of a compound?
19. What information does the ratio $\dfrac{\text{molar mass}}{\text{mass of emprical formula}}$ provide?

Most of the exercises in this chapter are available for assignment via the online homework management program, WileyPLUS (www.wileyplus.com). All exercises with blue numbers have answers in Appendix VI.

PAIRED EXERCISES

1. Determine the molar masses of these compounds:
    (a) KBr
    (b) Na_2SO_4
    (c) $Pb(NO_3)_2$
    (d) C_2H_5OH
    (e) $HC_2H_3O_2$
    (f) Fe_3O_4
    (g) $C_{12}H_{22}O_{11}$
    (h) $Al_2(SO_4)_3$
    (i) $(NH_4)_2HPO_4$

2. Determine the molar masses of these compounds:
    (a) NaOH
    (b) Ag_2CO_3
    (c) Cr_2O_3
    (d) $(NH_4)_2CO_3$
    (e) $Mg(HCO_3)_2$
    (f) C_6H_5COOH
    (g) $C_6H_{12}O_6$
    (h) $K_4Fe(CN)_6$
    (i) $BaCl_2 \cdot 2H_2O$

3. How many moles of atoms are contained in the following?
    (a) 22.5 g Zn
    (b) 0.688 g Mg
    (c) 4.5×10^{22} atoms Cu
    (d) 382 g Co
    (e) 0.055 g Sn
    (f) 8.5×10^{24} molecules N_2

4. How many moles of atoms are contained in the following?
    (a) 25.0 g NaOH
    (b) 44.0 g Br_2
    (c) 0.684 g $MgCl_2$
    (d) 14.8 g CH_3OH
    (e) 2.88 g Na_2SO_4
    (f) 4.20 lb ZnI_2

5. Calculate the number of grams in each of the following:
    (a) 0.550 mol Au
    (b) 15.8 mol H_2O
    (c) 12.5 mol Cl_2
    (d) 3.15 mol NH_4NO_3

6. Calculate the number of grams in each of the following:
    (a) 4.25×10^{-4} mol H_2SO_4
    (b) 4.5×10^{22} molecules CCl_4
    (c) 0.00255 mol Ti
    (d) 1.5×10^{16} atoms S

7. How many molecules are contained in each of the following?
    (a) 2.5 mol S_8
    (b) 7.35 mol NH_3
    (c) 17.5 g C_2H_5OH
    (d) 225 g Cl_2

8. How many molecules are contained in each of the following?
    (a) 9.6 mol C_2H_4
    (b) 2.76 mol N_2O
    (c) 23.2 g CH_3OH
    (d) 32.7 g CCl_4

9. How many atoms are contained in each of the following?
    (a) 25 molecules P_2O_5
    (b) 3.62 mol O_2
    (c) 12.2 mol CS_2
    (d) 1.25 g Na
    (e) 2.7 g CO_2
    (f) 0.25 g CH_4

10. How many atoms are contained in each of the following?
    (a) 2 molecules CH_3COOH
    (b) 0.75 mol C_2H_6
    (c) 25 mol H_2O
    (d) 92.5 g Au
    (e) 75 g PCl_3
    (f) 15 g $C_6H_{12}O_6$

11. Calculate the mass in grams of each of the following:
    (a) 1 atom He
    (b) 15 atoms C
    (c) 4 molecules N_2O_5
    (d) 11 molecules $C_6H_5NH_2$

12. Calculate the mass in grams of each of the following:
    (a) 1 atom Xe
    (b) 22 atoms Cl
    (c) 9 molecules CH_3COOH
    (d) 15 molecules $C_4H_4O_2(NH_2)_2$

13. Make the following conversions:
    (a) 25 kg CO_2 to mol CO_2
    (b) 5 atoms Pb to mol Pb
    (c) 6 mol O_2 to atoms oxygen
    (d) 25 molecules P_4 to g P_4

14. Make the following conversions:
    (a) 275 atoms W to mol W
    (b) 95 atoms H_2O to kg H_2O
    (c) 12 molecules SO_2 to g SO_2
    (d) 25 mol Cl_2 to atoms chlorine

15. 1 molecule of tetraphosphorus decoxide contains
    (a) how many moles?
    (b) how many grams?
    (c) how many phosphorus atoms?
    (d) how many oxygen atoms?
    (e) how many total atoms?

16. 125 grams of disulfur decafluoride contain
    (a) how many moles?
    (b) how many molecules?
    (c) how many total atoms?
    (d) how many atoms of sulfur?
    (e) how many atoms of fluorine?

17. How many atoms of hydrogen are contained in each of the following?
    (a) 25 molecules $C_6H_5CH_3$ (c) 36 g CH_3CH_2OH
    (b) 3.5 mol H_2CO_3

18. How many atoms of hydrogen are contained in each of the following?
    (a) 23 molecules CH_3CH_2COOH (c) 57 g $C_6H_5ONH_2$
    (b) 7.4 mol H_3PO_4

19. Calculate the number of
    (a) grams of silver in 25.0 g AgBr
    (b) grams of nitrogen in 6.34 mol $(NH_4)_3PO_4$
    (c) grams of oxygen in 8.45×10^{22} molecules SO_3

20. Calculate the number of
    (a) grams of chlorine in 5.0 g $PbCl_2$
    (b) grams of hydrogen in 4.50 mol H_2SO_4
    (c) grams of hydrogen in 5.45×10^{22} molecules NH_3

21. Calculate the percent composition by mass of these compounds:
    (a) NaBr   (d) $SiCl_4$
    (b) $KHCO_3$ (e) $Al_2(SO_4)_3$
    (c) $FeCl_3$ (f) $AgNO_3$

22. Calculate the percent composition by mass of these compounds:
    (a) $ZnCl_2$ (d) $(NH_4)_2SO_4$
    (b) $NH_4C_2H_3O_2$ (e) $Fe(NO_3)_3$
    (c) MgP_2O_7 (f) ICl_3

23. Calculate the percent of iron in the following compounds:
    (a) FeO   (c) Fe_3O_4
    (b) Fe_2O_3 (d) $K_4Fe(CN)_6$

24. Which of the following chlorides has the highest and which has the lowest percentage of chlorine, by mass, in its formula?
    (a) KCl   (c) $SiCl_4$
    (b) $BaCl_2$ (d) LiCl

25. A 73.16-g sample of an interesting barium silicide reported to have superconducting properties was found to contain 33.62 g barium and the remainder silicon. Calculate the percent composition of the compound.

26. One of the compounds that give garlic its characteristic odor is ajoene. A 7.52-g sample of this compound contains 3.09 g sulfur, 0.453 g hydrogen, 0.513 g oxygen, and the remainder is carbon. Calculate the percent composition of ajoene.

27. Examine the following formulas. Which compound has the
    (a) higher percent by mass of hydrogen: H_2O or H_2O_2?
    (b) lower percent by mass of nitrogen: NO or N_2O_3?
    (c) higher percent by mass of oxygen: NO_2 or N_2O_4?

    Check your answers by calculation if you wish.

28. Examine the following formulas. Which compound has the
    (a) lower percent by mass of chlorine: $NaClO_3$ or $KClO_3$?
    (b) higher percent by mass of sulfur: $KHSO_4$ or K_2SO_4?
    (c) lower percent by mass of chromium: Na_2CrO_4 or $Na_2Cr_2O_7$?

    Check your answers by calculation if you wish.

29. Calculate the empirical formula of each compound from the percent compositions given:
    (a) 63.6% N, 36.4% O   (d) 43.4% Na, 11.3% C, 45.3% O
    (b) 46.7% N, 53.3% O   (e) 18.8% Na, 29.0% Cl, 52.3% O
    (c) 25.9% N, 74.1% O   (f) 72.02% Mn, 27.98% O

30. Calculate the empirical formula of each compound from the percent compositions given:
    (a) 64.1% Cu, 35.9% Cl   (d) 55.3% K, 14.6% P, 30.1% O
    (b) 47.2% Cu, 52.8% Cl   (e) 38.9% Ba, 29.4% Cr, 31.7% O
    (c) 51.9% Cr, 48.1% S   (f) 3.99% P, 82.3% Br, 13.7% Cl

31. Determine the empirical formula for each of the following, using the given masses:
    (a) a compound containing 26.08 g zinc, 4.79 g carbon, and 19.14 g oxygen
    (b) a 150.0-g sample of a compound containing 57.66 g carbon, 7.26 g hydrogen, and the rest, chlorine
    (c) a 75.0-g sample of an oxide of vanadium, containing 42.0 g V and the rest, oxygen
    (d) a compound containing 67.35 g nickel, 43.46 g oxygen, and 23.69 g phosphorus

32. Determine the empirical formula for each of the following, using the given masses:
    (a) a compound containing 55.08 g carbon, 3.85 g hydrogen, and 61.07 g bromine
    (b) a 65.2-g sample of a compound containing 36.8 g silver, 12.1 g chlorine, and the rest, oxygen
    (c) a 25.25-g sample of a sulfide of vanadium, containing 12.99 g V, and the rest, sulfur
    (d) a compound containing 38.0 g zinc and 12.0 g phosphorus

33. A 15.267-g sample of a sulfide of gold was found to contain 12.272 g gold, and the rest, sulfur. Calculate the empirical formula of this sulfide of gold.

34. A 10.724-g sample of an oxide of titanium was found to contain 7.143 g titanium, and the rest, oxygen. Calculate the empirical formula of this oxide of titanium.

35. Lenthionine, a naturally occurring sulfur compound, gives shitake mushrooms their interesting flavor. This compound is composed of carbon, sulfur, and hydrogen. A 5.000-g sample of lenthionine contains 0.6375 g carbon and 0.1070 g hydrogen, and the remainder is sulfur. The molar mass of lenthionine is 188.4 g/mol. Determine the empirical formula for lenthionine.

36. A 5.276-g sample of an unknown compound was found to contain 3.989 g of mercury and the rest, chlorine. Calculate the empirical formula of this chloride of mercury.

37. Traumatic acid is a plant hormone that induces injured cells to begin to divide and repair the trauma they have experienced. Traumatic acid contains 63.13% carbon, 8.830% hydrogen, and 28.03% oxygen. Its molar mass is 228 g/mol. Determine the empirical and molecular formulas of traumatic acid.

38. Dixanthogen, an herbicide, is composed of 29.73% carbon, 4.16% hydrogen, 13.20% oxygen, and 52.91% sulfur. It has a molar mass of 242.4 g/mol. Determine the empirical and the molecular formulas.

39. Ethanedioic acid, a compound that is present in many vegetables, has a molar mass of 90.04 g/mol and a composition of 26.7% C, 2.2% H, and 71.1% O. What is the molecular formula for this substance?

40. Butyric acid is a compound that is present in butter. It has a molar mass of 88.11 g/mol and is composed of 54.5% C, 9.2% H, and 36.3% O. What is the molecular formula for this substance?

41. Calculate the percent composition and determine the molecular formula and the empirical formula for the nitrogen-oxygen compound that results when 12.04 g of nitrogen are reacted with enough oxygen to produce 39.54 g of product. The molar mass of the product is 92.02 g.

42. Calculate the percent composition and determine the molecular formula and the empirical formula for the carbon-hydrogen-oxygen compound that results when 30.21 g of carbon, 40.24 g of oxygen, and 5.08 g of hydrogen are reacted to produce a product with a molar mass of 180.18 g.

43. The compound XYZ_3 has a molar mass of 100.09 g and a percent composition (by mass) of 40.04% X, 12.00% Y, and 47.96% Z. What is the formula of the compound?

44. The compound $X_2(YZ_3)_3$ has a molar mass of 282.23 g and a percent composition (by mass) of 19.12% X, 29.86% Y, and 51.02% Z. What is the formula of the compound?

ADDITIONAL EXERCISES

45. White phosphorus is one of several forms of phosphorus and exists as a waxy solid consisting of P_4 molecules. How many atoms are present in 0.350 mol of P_4?

46. How many grams of sodium contain the same number of atoms as 10.0 g of potassium?

47. One molecule of an unknown compound is found to have a mass of 3.27×10^{-22} g. What is the molar mass of this compound?

48. How many molecules of sugar are in a 5-lb bag of sugar? The formula for table sugar or sucrose is $C_{12}H_{22}O_{11}$.

49. If a stack of 500 sheets of paper is 4.60 cm high, what will be the height, in meters, of a stack of Avogadro's number of sheets of paper?

50. There are about 7.0 billion (7.0×10^9) people on Earth. If exactly 1 mol of dollars were distributed equally among these people, how many dollars would each person receive?

51. If 20. drops of water equal 1.0 mL (1.0 cm^3),
    (a) how many drops of water are there in a cubic mile of water?
    (b) what would be the volume in cubic miles of a mole of drops of water?

52. Silver has a density of 10.5 g/cm^3. If 1.00 mol of silver were shaped into a cube,
    (a) what would be the volume of the cube?
    (b) what would be the length of one side of the cube?

53. Given 1.00-g samples of each of the compounds CO_2, O_2, H_2O, and CH_3OH,
    (a) which sample will contain the largest number of molecules?
    (b) which sample will contain the largest number of atoms? Show proof for your answers.

54. How many grams of Fe_2S_3 will contain a total number of atoms equal to Avogadro's number?

55. How many grams of calcium must be combined with 1 g phosphorus to form the compound Ca_3P_2?

56. An iron ore contains 5% $FeSO_4$ by mass. How many grams of iron could be obtained from 1.0 ton of this ore?

57. How many grams of lithium will combine with 20.0 g of sulfur to form the compound Li_2S?

58. Calculate the percentage of
    (a) mercury in $HgCO_3$
    (b) oxygen in $Ca(ClO_3)_2$
    (c) nitrogen in $C_{10}H_{14}N_2$ (nicotine)
    (d) Mg in $C_{55}H_{72}MgN_4O_5$ (chlorophyll)

59. Zinc and sulfur react to form zinc sulfide, ZnS. If we mix 19.5 g of zinc and 9.40 g of sulfur, have we added sufficient sulfur to fully react all the zinc? Show evidence for your answer.

60. Vomitoxin is produced by some fungi that grow on wheat and barley. It derives its name from the fact that it causes pigs that eat contaminated wheat to vomit. The chemical formula for vomitoxin is $C_{15}H_{20}O_6$. What is the percent composition of this compound?

61. Diphenhydramine hydrochloride, a drug used commonly as an antihistamine, has the formula $C_{17}H_{21}NO \cdot HCl$. What is the percent composition of each element in this compound?

62. What is the percent composition of each element in sucrose, $C_{12}H_{22}O_{11}$?

63. Aspirin is well known as a pain reliever (analgesic) and as a fever reducer (antipyretic). It has a molar mass of 180.2 g/mol and a composition of 60.0% C, 4.48% H, and 35.5% O. Calculate the molecular formula of aspirin.

64. How many grams of oxygen are contained in 8.50 g of $Al_2(SO_4)_3$?

65. Gallium arsenide is one of the newer materials used to make semiconductor chips for use in supercomputers. Its composition is 48.2% Ga and 51.8% As. What is the empirical formula?

66. Calcium tartrate is used as a preservative for certain foods and as an antacid. It contains 25.5% C, 2.1% H, 21.3% Ca, and 51.0% O. What is the empirical formula for calcium tartrate?

67. The compositions of four different compounds of carbon and chlorine follow. Determine the empirical formula and the molecular formula for each compound.

Percent C	Percent Cl	Molar mass (g)
(a) 7.79	92.21	153.8
(b) 10.13	89.87	236.7
(c) 25.26	74.74	284.8
(d) 11.25	88.75	319.6

68. How many years is a mole of seconds?

69. A normal penny has a mass of about 2.5 g. If we assume the penny to be pure copper (which means the penny is very old since newer pennies are a mixture of copper and zinc), how many atoms of copper does it contain?

70. What would be the mass (in grams) of one thousand trillion molecules of glycerin ($C_3H_8O_3$)?

71. If we assume that there are 7.0 billion people on the Earth, how many moles of people is this?

72. An experimental catalyst used to make polymers has the following composition: Co, 23.3%; Mo, 25.3%; and Cl, 51.4%. What is the empirical formula for this compound?

73. If a student weighs 18 g of aluminum and needs twice as many atoms of magnesium as she has of aluminum, how many grams of Mg does she need?

74. If 10.0 g of an unknown compound composed of carbon, hydrogen, and nitrogen contains 17.7% N and 3.8×10^{23} atoms of hydrogen, what is its empirical formula?

75. A substance whose formula is A_2O (A is a mystery element) is 60.0% A and 40.0% O. Identify the element A.

76. For the following compounds whose molecular formulas are given, indicate the empirical formula:
    (a) $C_6H_{12}O_6$ glucose
    (b) C_8H_{18} octane
    (c) $C_3H_6O_3$ lactic acid
    (d) $C_{25}H_{52}$ paraffin
    (e) $C_{12}H_4Cl_4O_2$ dioxin (a powerful poison)

77. Copper is often added to paint for ocean-going ships and boats to discourage the growth of marine organisms on their hulls. Unfortunately the copper dissolves into the water and is toxic to the marine life in harbors where these boats are docked. When copper(II) ions reach a concentration of 9.0 microgram/L, the growth of some marine life slows. How many copper ions are found in a liter of water containing 9.0 micrograms of Cu^{2+}/L?

78. Molecules of oxygen gas have been very difficult to find in space. Recently the European Space Agency's Herschel space observatory has found molecular oxygen in the Orion star-forming complex. It is present at very low concentrations, 1 molecule of oxygen per every million (1.0×10^6) molecules of hydrogen, but it does exist. How many molecules of oxygen would exist in a 3.0-g sample of hydrogen gas?

79. Scientists from the European Space Agency have been studying geysers located around the southern pole of one of Saturn's moons. These geysers send 250.0 kg of water into the atmosphere every second. Scientists trying to discover what happens to this water have found that it forms a very thin vapor ring around Saturn. How many moles of water are shot into the atmosphere in a Saturn day? (One Saturn day is equal to 10 hours and 45 minutes.)

80. Researchers from the University of Cape Town in South Africa have discovered that penguins are able to detect a chemical released by plankton as they are being consumed by schools of fish. By following this scent they are able to find these schools of fish and feast on them. The chemical they smell is composed of 38.65% carbon, 9.74% hydrogen, and 51.61% sulfur. Determine the empirical formula of this compound.

CHALLENGE EXERCISES

81. The compound $A(BC)_3$ has a molar mass of 78.01 g and a percent composition of 34.59% A, 61.53% B, and 3.88% C. Determine the identity of the elements A, B, and C. What is the percent composition by mass of the compound A_2B_3?

82. A 2.500-g sample of an unknown compound containing only C, H, and O was burned in O_2. The products were 4.776 g of CO_2 and 2.934 g of H_2O.
    (a) What is the percent composition of the original compound?
    (b) What is the empirical formula of the compound?

83. Researchers at Anna Gudmundsdottir's laboratory at the University of Cincinnati have been studying extremely reactive chemicals known as radicals. One of the interesting phenomena they have discovered is that these radicals can be chemically attached to fragrance molecules, effectively tethering them to a solution. When light strikes these tethered molecules, the fragrance is released. This property would allow us to produce perfumes, cleansers, and other consumer products that release fragrance only when exposed to light. If limonene, $C_{10}H_{16}$, the molecule that gives fruits their citrus scent, were able to be tethered to one of these radicals and every photon of light would release one molecule of limonene, calculate the time in seconds required to release 1.00 picogram of limonene if ambient light releases 2.64×10^{18} photons/sec.

ANSWERS TO PRACTICE EXERCISES

7.1   (a) 6.022×10^{23} atoms Fe
      (b) 1.204×10^{24} atoms H
      (c) 4.215×10^{24} atoms

7.2   10.0 g helium

7.3 1.5×10^{22} atoms

7.4   101.1 g KNO_3

7.5   21.3 g Na_2SO_4

7.6   8.326 mol and 5.014×10^{24} molecules $HC_2H_3O_2$

7.7   24.42% Ca; 17.07% N; 58.50% O

7.8   40.27% K; 26.78% Cr; 32.96% O

7.9   20.16% Al; 79.84% Cl

7.10 C_2H_6O

7.11 P_2O_5

7.12 The empirical formula is CH_3; the molecular formula is C_2H_6.

Flames and sparks result when aluminum foil is dropped into liquid bromine.

CHAPTER **8**

CHEMICAL EQUATIONS

I n the world today, we continually strive to express information in a concise, useful manner. From early childhood, we are taught to translate our ideas and desires into sentences. In mathematics, we learn to describe numerical relationships and situations through mathematical expressions and equations. Historians describe thousands of years of history in 500-page textbooks. Filmmakers translate entire events, such as the Olympics, into a few hours of entertainment.

Chemists use chemical equations to describe reactions they observe in the laboratory or in nature. Chemical equations provide us with the means to (1) summarize the reaction, (2) display the substances that are reacting, (3) show the products, and (4) indicate the amounts of all component substances in a reaction.

CHAPTER OUTLINE

8.1 The Chemical Equation

8.2 Writing and Balancing Chemical Equations

8.3 Types of Chemical Equations

8.4 Heat in Chemical Reactions

8.5 Global Warming: The Greenhouse Effect

8.1 THE CHEMICAL EQUATION

LEARNING OBJECTIVE • Describe the information present in a chemical equation.

KEY TERMS

reactants
products
chemical equation
law of conservation of mass

Molten iron falls from the reaction container during the thermite reaction.

Chemical reactions always involve change. Atoms, molecules, or ions rearrange to form different substances, sometimes in a spectacular manner. For example, the thermite reaction is a reaction between aluminum metal and iron(III) oxide, which produces molten iron and aluminum oxide. The substances entering the reaction are called the **reactants**, and the substances formed are called the **products**. In our example,

reactants	aluminum
	iron(III) oxide
products	iron
	aluminum oxide

During reactions, chemical bonds are broken and new bonds are formed. The reactants and products may be present as solids, liquids, gases, or in solution.

> In a chemical reaction atoms are neither created nor destroyed. All atoms present in the reactants must also be present in the products.

A **chemical equation** is a shorthand expression for a chemical change or reaction. A chemical equation uses the chemical symbols and formulas of the reactants and products and other symbolic terms to represent a chemical reaction. The equations are written according to this general format:

1. Reactants are separated from products by an arrow ($\longrightarrow$) that indicates the direction of the reaction. The reactants are placed to the left and the products to the right of the arrow. A plus sign (+) is placed between reactants and between products when needed.

$$Al + Fe_2O_3 \longrightarrow Fe + Al_2O_3$$
$$\underset{\text{reactants}}{\qquad\qquad} \underset{\text{products}}{\qquad\qquad}$$

2. Coefficients (whole numbers) are placed in front of substances to balance the equation and to indicate the number of units (atoms, molecules, moles, ions) of each substance reacting or being produced. When no number is shown, it is understood that one unit of the substance is indicated.

$$2\,Al + Fe_2O_3 \longrightarrow 2\,Fe + Al_2O_3$$

3. Conditions required to carry out the reaction may, if desired, be placed above or below the arrow or equality sign. For example, a delta sign placed over the arrow ($\xrightarrow{\Delta}$) indicates that heat is supplied to the reaction.

$$2\,Al + Fe_2O_3 \xrightarrow{\Delta} 2\,Fe + Al_2O_3$$

4. The physical state of a substance is indicated by the following symbols: (s) for solid state; (l) for liquid state; (g) for gaseous state; and (aq) for substances in aqueous solution. States are not always given in chemical equations.

$$2\,Al(s) + Fe_2O_3(s) \xrightarrow{\Delta} Fe(l) + Al_2O_3(s)$$

Symbols commonly used in chemical equations are given in Table 8.1.

TABLE 8.1 Symbols Commonly Used in Chemical Equations

Symbol	Meaning
+	Plus or added to (placed between substances)
$\longrightarrow$	Yields; produces (points to products)
(s)	Solid state (written after a substance)
(l)	Liquid state (written after a substance)
(g)	Gaseous state (written after a substance)
(aq)	Aqueous solution (substance dissolved in water)
Δ	Heat is added (when written above or below arrow)

EXAMPLE 8.1

Classify each formula in the chemical reactions below as a reactant or a product.

(a) $MnO_2 + 4\,HCl \longrightarrow MnCl_2 + Cl_2 + 2\,H_2O$
(b) $Na_2SO_3 + H_2SO_4 \longrightarrow Na_2SO_4 + SO_2 + H_2O$

SOLUTION

(a) Reactants: MnO_2 and HCl
Products: $MnCl_2$, Cl_2, and H_2O
(b) Reactants: Na_2SO_3, and H_2SO_4
Products: Na_2SO_4, SO_2, and H_2O

PRACTICE 8.1

How is a substance in a chemical equation identified as a solid, a gas, or in a water solution?

Conservation of Mass

The **law of conservation of mass** states that no change is observed in the total mass of the substances involved in a chemical change. This law, tested by extensive laboratory experimentation, is the basis for the quantitative mass relationships among reactants and products.

The decomposition of water into hydrogen and oxygen illustrates this law. Thus, 100.0 g of water decompose into 11.2 g of hydrogen and 88.8 g of oxygen:

$$\text{water} \longrightarrow \text{hydrogen} + \text{oxygen}$$
$$100.0 \text{ g} \qquad\qquad 11.2 \text{ g} \qquad\quad 88.8 \text{ g}$$
$$100.0 \text{ g} \longrightarrow \qquad\qquad 100.0 \text{ g}$$
$$\text{reactant} \qquad\qquad\qquad \text{products}$$

In a chemical reaction

$$\textbf{mass of reactants = mass of products}$$

Combustion of gasoline is a chemical change.

8.2 WRITING AND BALANCING CHEMICAL EQUATIONS

Write and balance chemical equations.

● **LEARNING OBJECTIVE**

KEY TERM
balanced equation

To represent the quantitative relationships of a reaction, the chemical equation must be balanced. A **balanced equation** contains the same number of each kind of atom on each side of the equation. The balanced equation therefore obeys the law of conservation of mass.

Every chemistry student must learn to *balance* equations. Many equations are balanced by trial and error, but care and attention to detail are still required. The way to balance an equation is to adjust the number of atoms of each element so that they are the same on each side of the equation, but a correct formula is never changed in order to balance an equation. The general procedure for balancing equations is as follows:

Correct formulas are not changed to balance an equation.

PROBLEM-SOLVING STRATEGY: For Writing and Balancing a Chemical Equation

1. **Identify the reaction.** Write a description or word equation for the reaction. For example, let's consider mercury(II) oxide decomposing into mercury and oxygen.

$$\text{mercury(II) oxide} \xrightarrow{\Delta} \text{mercury} + \text{oxygen}$$

2. **Write the unbalanced (skeleton) equation.** Make sure that the formula for each substance is correct and that reactants are written to the left and products to the right of the arrow. For our example,

$$\text{HgO} \xrightarrow{\Delta} \text{Hg} + \text{O}_2$$

The correct formulas must be known or determined from the periodic table, lists of ions, or experimental data.

3. **Balance the equation.** Use the following process as necessary:

(a) Count and compare the number of atoms of each element on each side of the equation and determine those that must be balanced:

Hg is balanced   (1 on each side)

O needs to be balanced   (1 on reactant side, 2 on product side)

(b) Balance each element, one at a time, by placing whole numbers (coefficients) in front of the formulas containing the unbalanced element. It is usually best to balance metals first, then nonmetals, then hydrogen and oxygen. Select the smallest coefficients that will give the same number of atoms of the element on each side. A coefficient placed in front of a formula multiplies every atom in the formula by that number (e.g., $2 \text{ H}_2\text{SO}_4$

Study this procedure carefully and refer to it when you work examples.

Leave elements that are in two or more formulas (on the same side of the equation) unbalanced until just before balancing hydrogen and oxygen.

means two units of sulfuric acid and also means four H atoms, two S atoms, and eight O atoms). Place a 2 in front of HgO to balance O:

$$2\,HgO \xrightarrow{\Delta} Hg + O_2$$

(c) Check all other elements after each individual element is balanced to see whether, in balancing one element, other elements have become unbalanced. Make adjustments as needed. Now Hg is not balanced. To adjust this, we write a 2 in front of Hg:

$$2\,HgO \xrightarrow{\Delta} 2\,Hg + O_2 \quad \text{(balanced)}$$

(d) Do a final check, making sure that each element and/or polyatomic ion is balanced and that the smallest possible set of whole-number coefficients has been used:

$$2\,HgO \xrightarrow{\Delta} 2\,Hg + O_2 \quad \text{(correct form)}$$
$$4\,HgO \xrightarrow{\Delta} 4\,Hg + 2\,O_2 \quad \text{(incorrect form)}$$

Not all chemical equations can be balanced by the simple method of inspection just described. The following examples show *stepwise* sequences leading to balanced equations. Study each one carefully.

WileyPLUS

ENHANCED EXAMPLE

EXAMPLE 8.2

Write the balanced equation for the reaction that takes place when magnesium metal is burned in air to produce magnesium oxide.

SOLUTION

Use the Problem-Solving Strategy for Writing and Balancing a Chemical Equation.

1. *Word equation:*

$$\text{magnesium} + \text{oxygen} \longrightarrow \text{magnesium oxide}$$
$$\text{reactants (R)} \qquad\qquad \text{product (P)}$$

Elements shown in color are balanced.

R = reactant
P = product

2. *Skeleton equation:*

$$Mg + O_2 \longrightarrow MgO \quad \text{(unbalanced)}$$

3. *Balance:*

(a) Mg is balanced.
Oxygen is not balanced. Two O atoms appear on the left side and one on the right side.

(a)
R	1 Mg	2 O
P	1 Mg	1 O

(b) Place the coefficient 2 in front of MgO to balance oxygen:

$$Mg + O_2 \longrightarrow 2\,MgO \quad \text{(unbalanced)}$$

(b)
R	1 Mg	2 O
P	2 Mg	2 O

(c) Now Mg is not balanced. One Mg atom appears on the left side and two on the right side. Place a 2 in front of Mg:

$$2\,Mg + O_2 \longrightarrow 2\,MgO \quad \text{(balanced)}$$

(c)
R	2 Mg	2 O
P	2 Mg	2 O

(d) *Check:* Each side has two Mg and two O atoms.

EXAMPLE 8.3

When methane, CH_4, undergoes complete combustion, it reacts with oxygen to produce carbon dioxide and water. Write the balanced equation for this reaction.

SOLUTION

Use the Problem-Solving Strategy for Writing and Balancing a Chemical Equation.

1. *Word equation:*

$$\text{methane} + \text{oxygen} \longrightarrow \text{carbon dioxide} + \text{water}$$

2. *Skeleton equation:*

$$CH_4 + O_2 \longrightarrow CO_2 + H_2O \quad \text{(unbalanced)}$$

3. *Balance:*

(a) Carbon is balanced.
Hydrogen and oxygen are not balanced.

(a)
R	1 C	4 H	2 O
P	1 C	2 H	3 O

(b) Balance H atoms by placing a 2 in front of H_2O:

$$CH_4 + O_2 \longrightarrow CO_2 + 2\,H_2O \quad \text{(unbalanced)}$$

Each side of the equation has four H atoms; oxygen is still not balanced. Place a 2 in front of O_2 to balance the oxygen atoms:

$$CH_4 + 2\,O_2 \longrightarrow CO_2 + 2\,H_2O \quad \text{(balanced)}$$

(c) The other atoms remain balanced.

(d) *Check:* The equation is correctly balanced; it has one C, four O, and four H atoms on each side.

(b)			
R	1 C	4 H	2 O
P	1 C	4 H	4 O

(c)			
R	1 C	4 H	4 O
P	1 C	4 H	4 O

EXAMPLE 8.4

Oxygen and potassium chloride are formed by heating potassium chlorate. Write a balanced equation for this reaction.

SOLUTION

Use the Problem-Solving Strategy for Writing and Balancing a Chemical Equation.

1. *Word equation:*

$$\text{potassium chlorate} \xrightarrow{\Delta} \text{potassium chloride} + \text{oxygen}$$

2. *Skeleton equation:*

$$KClO_3 \xrightarrow{\Delta} KCl + O_2 \quad \text{(unbalanced)}$$

3. *Balance:*

(a) Potassium and chlorine are balanced.
Oxygen is unbalanced (three O atoms on the left and two on the right side).

(b) How many oxygen atoms are needed? The subscripts of oxygen (3 and 2) in $KClO_3$ and O_2 have a least common multiple of 6. Therefore, coefficients for $KClO_3$ and O_2 are needed to get six O atoms on each side. Place a 2 in front of $KClO_3$ and a 3 in front of O_2 to balance oxygen:

$$2\,KClO_3 \xrightarrow{\Delta} KCl + 3\,O_2 \quad \text{(unbalanced)}$$

(c) Now K and Cl are not balanced. Place a 2 in front of KCl, which balances both K and Cl at the same time:

$$2\,KClO_3 \xrightarrow{\Delta} 2\,KCl + 3\,O_2 \quad \text{(balanced)}$$

(d) *Check:* Each side now contains two K, two Cl, and six O atoms.

(a)			
R	1 K	1 Cl	3 O
P	1 K	1 Cl	2 O

(b)			
R	2 K	2 Cl	6 O
P	1 K	1 Cl	6 O

(c)			
R	2 K	2 Cl	6 O
P	2 K	2 Cl	6 O

EXAMPLE 8.5

Silver nitrate reacts with hydrogen sulfide to produce silver sulfide and nitric acid. Write a balanced equation for this reaction.

SOLUTION

Use the Problem-Solving Strategy for Writing and Balancing a Chemical Equation.

1. *Word equation:*

$$\text{silver nitrate} + \text{hydrogen sulfide} \longrightarrow \text{silver sulfide} + \text{nitric acid}$$

2. *Skeleton equation:*

$$AgNO_3 + H_2S \longrightarrow Ag_2S + HNO_3 \quad \text{(unbalanced)}$$

3. *Balance:*

(a) Ag and H are unbalanced.

(b) Place a 2 in front of $AgNO_3$ to balance Ag:

$$2\,AgNO_3 + H_2S \longrightarrow Ag_2S + HNO_3 \quad \text{(unbalanced)}$$

(c) H and NO_3^- are still unbalanced. Balance by placing a 2 in front of HNO_3:

$$2\,AgNO_3 + H_2S \longrightarrow Ag_2S + 2\,HNO_3 \quad \text{(balanced)}$$

In this example, N and O atoms are balanced by balancing the NO_3^- ion as a unit.

(d) The other atoms remain balanced.

(e) *Check:* Each side has two Ag, two H, and one S atom. Also, each side has two NO_3^- ions.

(a)				
R	1 Ag	2 H	1 S	1 NO₃
P	2 Ag	1 H	1 S	1 NO₃

(b)				
R	2 Ag	2 H	1 S	2 NO₃
P	2 Ag	1 H	1 S	1 NO₃

(c)				
R	2 Ag	2 H	1 S	2 NO₃
P	2 Ag	2 H	1 S	2 NO₃

EXAMPLE 8.6

When aluminum hydroxide is mixed with sulfuric acid, the products are aluminum sulfate and water. Write a balanced equation for this reaction.

SOLUTION

Use the Problem-Solving Strategy for Writing and Balancing a Chemical Equation.

1. *Word equation:*

aluminum hydroxide + sulfuric acid $\longrightarrow$ aluminum sulfate + water

2. *Skeleton equation:*

$$Al(OH)_3 + H_2SO_4 \longrightarrow Al_2(SO_4)_3 + H_2O \qquad \text{(unbalanced)}$$

3. *Balance:*

(a) All elements are unbalanced.

(b) Balance Al by placing a 2 in front of $Al(OH)_3$. Treat the unbalanced SO_4^{2-} ion as a unit and balance by placing a 3 in front of H_2SO_4:

$$2\,Al(OH)_3 + 3\,H_2SO_4 \longrightarrow Al_2(SO_4)_3 + H_2O \quad \text{(unbalanced)}$$

Balance the unbalanced H and O by placing a 6 in front of H_2O:

$$2\,Al(OH)_3 + 3\,H_2SO_4 \longrightarrow Al_2(SO_4)_3 + 6\,H_2O \quad \text{(balanced)}$$

(c) The other atoms remain balanced.

(d) *Check:* Each side has 2 Al, 12 H, 3 S, and 18 O atoms.

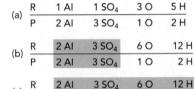

(a)	R	1 Al	1 SO₄	3 O	5 H
	P	2 Al	3 SO₄	1 O	2 H
(b)	R	2 Al	3 SO₄	6 O	12 H
	P	2 Al	3 SO₄	1 O	2 H
(c)	R	2 Al	3 SO₄	6 O	12 H
	P	2 Al	3 SO₄	6 O	12 H

EXAMPLE 8.7

When the fuel in a butane gas stove undergoes complete combustion, it reacts with oxygen to form carbon dioxide and water. Write the balanced equation for this reaction.

SOLUTION

Use the Problem-Solving Strategy for Writing and Balancing a Chemical Equation.

1. *Word equation:*

butane + oxygen $\longrightarrow$ carbon dioxide + water

2. *Skeleton equation:*

$$C_4H_{10} + O_2 \longrightarrow CO_2 + H_2O \qquad \text{(unbalanced)}$$

3. *Balance:*

(a) All elements are unbalanced.

(b) Balance C by placing a 4 in front of CO_2:

$$C_4H_{10} + O_2 \longrightarrow 4\,CO_2 + H_2O \qquad \text{(unbalanced)}$$

Balance H by placing a 5 in front of H_2O:

$$C_4H_{10} + O_2 \longrightarrow 4\,CO_2 + 5\,H_2O \qquad \text{(unbalanced)}$$

Oxygen remains unbalanced. The oxygen atoms on the right side are fixed because 4 CO_2 and 5 H_2O are derived from the single C_4H_{10} molecule on the left. When we try to balance the O atoms, we find that there is no whole number that can be placed in front of O_2 to bring about a balance, so we double the coefficients of each substance, and then balance the oxygen:

$$2\,C_4H_{10} + 13\,O_2 \longrightarrow 8\,CO_2 + 10\,H_2O \quad \text{(balanced)}$$

(c) The other atoms remain balanced.

(d) *Check:* Each side now has 8 C, 20 H, and 26 O atoms.

Butane
C_4H_{10}

(a)	R	4 C	10 H	2 O
	P	1 C	2 H	3 O
(b)	R	4 C	10 H	2 O
	P	4 C	10 H	13 O
(c)	R	8 C	20 H	26 O
	P	8 C	20 H	26 O

PRACTICE 8.2

Write a balanced formula equation for:

aluminum + oxygen $\longrightarrow$ aluminum oxide

PRACTICE 8.3

Write a balanced formula equation for:

  magnesium hydroxide + phosphoric acid ⟶ magnesium phosphate + water

Information in a Chemical Equation

Depending on the particular context in which it is used, a formula can have different meanings. A formula can refer to an individual chemical entity (atom, ion, molecule, or formula unit) or to a mole of that chemical entity. For example, the formula H_2O can mean any of the following:

- 2 H atoms and 1 O atom
- 1 molecule of water
- 1 *mol of water*
- 6.022×10^{23} molecules of water
- 18.02 g of water

Formulas used in equations can represent units of individual chemical entities or moles, the latter being more commonly used. For example, in the reaction of hydrogen and oxygen to form water,

$$2\ H_2 \quad + \quad O_2 \quad \longrightarrow \quad 2\ H_2O$$

| 2 molecules hydrogen | 1 molecule oxygen | 2 molecules water |
| 2 mol hydrogen | 1 mol oxygen | 2 mol water |

We generally use moles in equations because molecules are so small.

As indicated earlier, a chemical equation is a shorthand description of a chemical reaction. Interpretation of a balanced equation gives us the following information:

1. What the reactants are and what the products are
2. The formulas of the reactants and products
3. The number of molecules or formula units of reactants and products in the reaction
4. The number of atoms of each element involved in the reaction
5. The number of moles of each substance

Consider the reaction that occurs when propane gas (C_3H_8) is burned in air; the products are carbon dioxide (CO_2) and water (H_2O). The balanced equation and its interpretation are as follows:

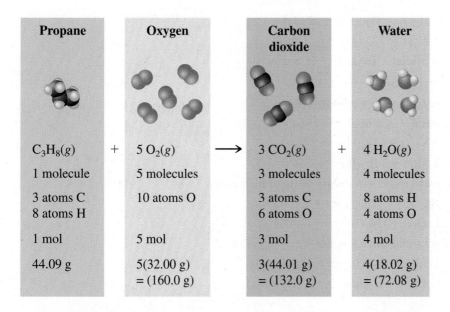

Propane		**Oxygen**		**Carbon dioxide**		**Water**
$C_3H_8(g)$	+	$5\ O_2(g)$	⟶	$3\ CO_2(g)$	+	$4\ H_2O(g)$
1 molecule		5 molecules		3 molecules		4 molecules
3 atoms C		10 atoms O		3 atoms C		8 atoms H
8 atoms H				6 atoms O		4 atoms O
1 mol		5 mol		3 mol		4 mol
44.09 g		5(32.00 g) = (160.0 g)		3(44.01 g) = (132.0 g)		4(18.02 g) = (72.08 g)

PRACTICE 8.4

Consider the reaction that occurs when hydrogen gas reacts with chlorine gas to produce gaseous hydrogen chloride:

(a) Write a word equation for this reaction.

(b) Write a balanced formula equation including the state for each substance.

(c) Label each reactant and product to show the relative amounts of each substance.

   (1) number of molecules   (3) number of moles
   (2) number of atoms   (4) mass

(d) What mass of HCl would be produced if you reacted 2 mol hydrogen gas with 2 mol chlorine gas?

The quantities involved in chemical reactions are important when working in industry or the laboratory. We will study the relationship among quantities of reactants and products in the next chapter.

8.3 TYPES OF CHEMICAL EQUATIONS

LEARNING OBJECTIVE ● Give examples of a combination reaction, decomposition reaction, single-displacement reaction, and double-displacement reaction.

KEY TERMS

combination reaction
decomposition reaction
single-displacement reaction
double-displacement reaction

Chemical equations represent chemical changes or reactions. Reactions are classified into types to assist in writing equations and in predicting other reactions. Many chemical reactions fit one or another of the four principal reaction types that we discuss in the following paragraphs. Reactions are also classified as oxidation–reduction. Special methods are used to balance complex oxidation–reduction equations. (See Chapter 17.)

Combination Reaction

In a **combination reaction**, two reactants combine to give one product. The general form of the equation is

$$A + B \longrightarrow AB$$

in which A and B are either elements or compounds and AB is a compound. The formula of the compound in many cases can be determined from knowledge of the ionic charges of the reactants in their combined states. Some reactions that fall into this category are given here:

1. metal + oxygen ⟶ metal oxide:

   $2\,Mg(s) + O_2(g) \xrightarrow{\Delta} 2\,MgO(s)$

   $4\,Al(s) + 3\,O_2(g) \xrightarrow{\Delta} 2\,Al_2O_3(s)$

2. nonmetal + oxygen ⟶ nonmetal oxide:

   $S(s) + O_2(g) \xrightarrow{\Delta} SO_2(g)$

   $N_2(g) + O_2(g) \xrightarrow{\Delta} 2\,NO(g)$

3. metal + nonmetal ⟶ salt:

   $2\,Na(s) + Cl_2(g) \longrightarrow 2\,NaCl(s)$

   $2\,Al(s) + 3\,Br_2(l) \longrightarrow 2\,AlBr_3(s)$

4. metal oxide + water ⟶ metal hydroxide:

   $Na_2O(s) + H_2O(l) \longrightarrow 2\,NaOH(aq)$

   $CaO(s) + H_2O(l) \longrightarrow Ca(OH)_2(aq)$

5. nonmetal oxide + water ⟶ oxy-acid:

   $SO_3(g) + H_2O(l) \longrightarrow H_2SO_4(aq)$

   $N_2O_5(s) + H_2O(l) \longrightarrow 2\,HNO_3(aq)$

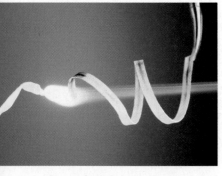

A strip of Mg burning in air produces MgO (left of the flame).

>CHEMISTRY *IN ACTION*

CO Poisoning—A Silent Killer

Carbon monoxide (CO) is one possible product when carbon-containing fuels such as coal, natural gas, propane, heating oil, and gasoline are burned. For example, when methane, the main component of natural gas, burns, the reaction is

$$CH_4(g) + 2\ O_2(g) \longrightarrow 2\ H_2O(g) + CO_2(g)$$

However, if not enough oxygen is available, carbon monoxide will be produced instead of carbon dioxide:

$$2\ CH_4(g) + 3\ O_2(g) \longrightarrow 4\ H_2O(g) + 2\ CO(g)$$

Carbon monoxide is colorless and has no flavor or smell. It often can poison several people at once, since it can build up undetected in enclosed spaces. The symptoms of carbon monoxide poisoning are so easily missed that people often don't realize that they have been poisoned. Common symptoms are head-ache, ringing in ears, nausea, dizziness, weakness, confusion, and drowsiness. A low concentration of CO in the air may go completely unnoticed while the toxicity level increases in the blood.

Carbon monoxide in the blood is poisonous because of its ability to bond to a hemoglobin molecule. Both CO and O_2 bond to the same place on the hemoglobin molecule. The CO bonds more tightly than O_2. We use hemoglobin to transport oxygen to the various tissues from the lungs. Unfortunately, CO also bonds to hemo-globin, forming a molecule called carboxyhemoglobin. In fact, hemoglobin prefers CO to O_2. Once a CO is bound to a hemoglobin molecule, it cannot carry an oxygen molecule. So, until the CO is released, that molecule of hemoglobin is lost as an O_2 carrier. If you breathe normal air (about 20% oxygen) containing 0.1% CO, in just one hour the CO bonds to 50% of the hemoglobin molecules. To keep CO from being a silent killer, there are three possible actions:

1. Immediate treatment of victims to restore the hemoglobin
2. Detection of CO levels before poisoning occurs
3. Conversion of CO to CO_2 to eliminate the threat

If a person already has CO poisoning, the treatment must focus on getting CO released from the hemoglobin. This can be done by giving the victim oxygen or by placing the victim in a hyperbaric chamber (at about 3 atm pressure). These treatments reduce the time required for carboxy-hemoglobin to release CO from between 4 to 6 hours to 30 to 90 min. This treatment also supplies more oxygen to the system to help keep the victim's brain functioning.

A CO detector can sound an alarm before the CO level becomes toxic in a home. Current CO detectors sound an alarm when levels increase rapidly or when lower concen-trations are present over a long time. Inside a CO detector is a beam of infrared light that shines on a chromophore (a substance that turns darker with increasing CO levels). If the infrared light passing through the chromophore is too low, the alarm sounds. Typically, these alarms ring when CO levels reach 70 ppm, or greater than 30 ppm for 30 days.

The final option is to convert CO to CO_2 before toxicity occurs. David Schreyer from NASA's Langley Research Center has devised a catalyst (tin hydroxide with small amounts of Pd) that speeds up the conversion of CO to CO_2 at room temperature. He and his colleagues have adapted their catalyst to home ventilation systems to eliminate even low levels of CO before it has a chance to build up. Using all of these approaches can help to eliminate 500–10,000 cases of CO poisoning each year.

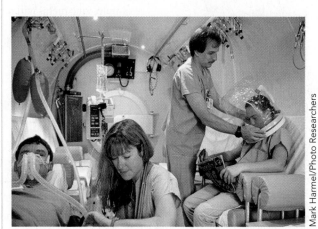

People in a hyperbaric chamber.

Decomposition Reaction

In a **decomposition reaction**, a single substance is decomposed, or broken down, to give two or more different substances. This reaction may be considered the reverse of combination. The starting material must be a compound, and the products may be elements or compounds. The general form of the equation is

$$AB \longrightarrow A + B$$

Predicting the products of a decomposition reaction can be difficult and requires an under-standing of each individual reaction. Heating oxygen-containing compounds often results in decomposition. The following reactions fall into this category.

Hydrogen peroxide decomposes to steam ($H_2O(g)$) and oxygen.

1. Some metal oxides decompose to yield the free metal plus oxygen; others give another oxide, and some are very stable, resisting decomposition by heating:

$$2 \, HgO(s) \xrightarrow{\Delta} 2 \, Hg(l) + O_2(g)$$

$$2 \, PbO_2(s) \xrightarrow{\Delta} 2 \, PbO(s) + O_2(g)$$

2. Carbonates and hydrogen carbonates decompose to yield CO_2 when heated:

$$CaCO_3(s) \xrightarrow{\Delta} CaO(s) + CO_2(g)$$

$$2 \, NaHCO_3(s) \xrightarrow{\Delta} Na_2CO_3(s) + H_2O(g) + CO_2(g)$$

3. Miscellaneous reactions in this category:

$$2 \, KClO_3(s) \xrightarrow{\Delta} 2 \, KCl(s) + 3 \, O_2(g)$$

$$2 \, NaNO_3(s) \xrightarrow{\Delta} 2 \, NaNO_2(s) + O_2(g)$$

$$2 \, H_2O_2(l) \xrightarrow{\Delta} 2 \, H_2O(l) + O_2(g)$$

Single-Displacement Reaction

In a **single-displacement reaction**, one element reacts with a compound to replace one of the elements of that compound, yielding a different element and a different compound. The general forms of the equation follow.

If A is a *metal*, A will replace B to form AC, provided that A is a more reactive metal than B.

$$A + BC \longrightarrow B + AC$$

If A is a *halogen*, it will replace C to form BA, provided that A is a more reactive halogen than C.

$$A + BC \longrightarrow C + BA$$

A brief activity series of selected metals (and hydrogen) and halogens is shown in Table 8.2. This series is listed in descending order of chemical activity, with the most active metals and halogens at the top. Many chemical reactions can be predicted from an activity series because the atoms of any element in the series will replace the atoms of those elements below it. For example, zinc metal will replace hydrogen from a hydrochloric acid solution. But copper metal, which is below hydrogen on the list and thus less reactive than hydrogen, will not replace hydrogen from a hydrochloric acid solution. Here are some reactions that fall into this category:

When pieces of zinc metal are placed in hydrochloric acid, hydrogen bubbles form immediately.

1. metal + acid $\longrightarrow$ hydrogen + salt

$$Zn(s) + 2 \, HCl(aq) \longrightarrow H_2(g) + ZnCl_2(aq)$$

$$2 \, Al(s) + 3 \, H_2SO_4(aq) \longrightarrow 3 \, H_2(g) + Al_2(SO_4)_3(aq)$$

2. metal + water $\longrightarrow$ hydrogen + metal hydroxide or metal oxide

$$2 \, Na(s) + 2 \, H_2O \longrightarrow H_2(g) + 2 \, NaOH(aq)$$

$$Ca(s) + 2 \, H_2O \longrightarrow H_2(g) + Ca(OH)_2(aq)$$

$$3 \, Fe(s) + 4 \, \underset{\text{steam}}{H_2O(g)} \longrightarrow 4 \, H_2(g) + Fe_3O_4(s)$$

3. metal + salt $\longrightarrow$ metal + salt

$$Fe(s) + CuSO_4(aq) \longrightarrow Cu(s) + FeSO_4(aq)$$

$$Cu(s) + 2 \, AgNO_3(aq) \longrightarrow 2 \, Ag(s) + Cu(NO_3)_2(aq)$$

4. halogen + halide salt $\longrightarrow$ halogen + halide salt

$$Cl_2(g) + 2 \, NaBr(aq) \longrightarrow Br_2(l) + 2 \, NaCl(aq)$$

$$Cl_2(g) + 2 \, KI(aq) \longrightarrow I_2(s) + 2 \, KCl(aq)$$

A common chemical reaction is the displacement of hydrogen from water or acids (shown in 1 and 2 above). This reaction is a good illustration of the relative reactivity of metals and the use of the activity series. For example,

- K, Ca, and Na displace hydrogen from cold water, steam (H_2O), and acids.
- Mg, Al, Zn, and Fe displace hydrogen from steam and acids.
- Ni, Sn, and Pb displace hydrogen only from acids.
- Cu, Ag, Hg, and Au do not displace hydrogen.

TABLE 8.2 Activity Series

Metals	Halogens
K	F_2
Ca	Cl_2
Na	Br_2
Mg	I_2
Al	
Zn	
Fe	
Ni	
Sn	
Pb	
H	
Cu	
Ag	
Hg	
Au	

increasing activity ↑

EXAMPLE 8.8

Will a reaction occur between (a) nickel metal and hydrochloric acid and (b) tin metal and a solution of aluminum chloride? Write balanced equations for the reactions.

SOLUTION

(a) Nickel is more reactive than hydrogen, so it will displace hydrogen from hydrochloric acid. The products are hydrogen gas and a salt of Ni^{2+} and Cl^- ions:

$$Ni(s) + 2\ HCl(aq) \longrightarrow H_2(g) + NiCl_2(aq)$$

(b) According to the activity series, tin is less reactive than aluminum, so no reaction will occur:

$$Sn(s) + AlCl_3(aq) \longrightarrow \text{no reaction}$$

WileyPLUS

ENHANCED EXAMPLE

PRACTICE 8.5

Write balanced equations for these reactions:
(a) iron metal and a solution of magnesium chloride
(b) zinc metal and a solution of lead(II) nitrate

Double-Displacement Reaction

In a **double-displacement reaction**, two compounds exchange partners with each other to produce two different compounds. The general form of the equation is

$$AB + CD \longrightarrow AD + CB$$

This reaction can be thought of as an exchange of positive and negative groups, in which A combines with D and C combines with B. In writing formulas for the products, we must account for the charges of the combining groups.

It's also possible to write an equation in the form of a double-displacement reaction when a reaction has not occurred. For example, when solutions of sodium chloride and potassium nitrate are mixed, the following equation can be written:

$$NaCl(aq) + KNO_3(aq) \longrightarrow NaNO_3(aq) + KCl(aq)$$

When the procedure is carried out, no physical changes are observed, indicating that no chemical reaction has taken place.

A double-displacement reaction is accompanied by evidence of reaction such as:

1. The evolution of heat
2. The formation of an insoluble precipitate
3. The production of gas bubbles

Let's look at some of these reactions more closely:

Neutralization of an acid and a base The production of a molecule of water from an H^+ and an OH^- ion is accompanied by a *release of heat*, which can be detected by touching the reaction container or by using a thermometer. For neutralization reactions, $H^+ + OH^- \longrightarrow H_2O$:

$$\text{acid} + \text{base} \longrightarrow \text{salt} + \text{water} + \text{heat}$$
$$HCl(aq) + NaOH(aq) \longrightarrow NaCl(aq) + H_2O(l) + \text{heat}$$
$$H_2SO_4(aq) + Ba(OH)_2(aq) \longrightarrow BaSO_4(s) + 2\ H_2O(l) + \text{heat}$$

Metal oxide + acid *Heat is released* by the production of a molecule of water.

$$\text{metal oxide} + \text{acid} \longrightarrow \text{salt} + \text{water} + \text{heat}$$
$$CuO(s) + 2\ HNO_3(aq) \longrightarrow Cu(NO_3)_2(aq) + H_2O(l) + \text{heat}$$
$$CaO(s) + 2\ HCl(aq) \longrightarrow CaCl_2(aq) + H_2O(l) + \text{heat}$$

A double-displacement reaction results from pouring a clear, colorless solution of Pb(NO₃)₂ into a clear, colorless solution of KI, forming a yellow precipitate of PbI₂.
$Pb(NO_3)_2 + 2KI \longrightarrow PbI_2 + 2KNO_3.$

Formation of an insoluble precipitate The solubilities of the products can be determined by consulting the solubility table in Appendix V to see whether one or both of the products are insoluble in water. An insoluble product (precipitate) is indicated by placing (s) after its formula in the equation.

$$BaCl_2(aq) + 2\,AgNO_3(aq) \longrightarrow 2\,AgCl(s) + Ba(NO_3)_2(aq)$$

$$Pb(NO_3)_2(aq) + 2\,KI(aq) \longrightarrow PbI_2(s) + 2\,KNO_3(aq)$$

Formation of a gas A gas such as HCl or H_2S may be produced directly, as in these two examples:

$$H_2SO_4(l) + NaCl(s) \longrightarrow NaHSO_4(s) + HCl(g)$$

$$2\,HCl(aq) + ZnS(s) \longrightarrow ZnCl_2(aq) + H_2S(g)$$

A gas can also be produced indirectly. Some unstable compounds formed in a double-displacement reaction, such as H_2CO_3, H_2SO_3, and NH_4OH, will decompose to form water and a gas:

$$2\,HCl(aq) + Na_2CO_3(aq) \longrightarrow 2\,NaCl(aq) + H_2CO_3(aq) \longrightarrow 2\,NaCl(aq) + H_2O(l) + CO_2(g)$$

$$2\,HNO_3(aq) + K_2SO_3(aq) \longrightarrow 2\,KNO_3(aq) + H_2SO_3(aq) \longrightarrow 2\,KNO_3(aq) + H_2O(l) + SO_2(g)$$

$$NH_4Cl(aq) + NaOH(aq) \longrightarrow NaCl(aq) + NH_4OH(aq) \longrightarrow NaCl(aq) + H_2O(l) + NH_3(g)$$

EXAMPLE 8.9

Write the equation for the reaction between aqueous solutions of hydrobromic acid and potassium hydroxide.

SOLUTION

First we write the formulas for the reactants. They are HBr and KOH. Then we classify the type of reaction that would occur between them. Because the reactants are compounds, one an acid and the other a base, the reaction will be of the neutralization type:

$$acid + base \longrightarrow salt + water$$

Now rewrite the equation using the formulas for the known substances:

$$HBr(aq) + KOH(aq) \longrightarrow salt + H_2O$$

In this reaction, which is a double-displacement type, the H^+ from the acid combines with the OH^- from the base to form water. The ionic compound must be composed of the other two ions, K^+ and Br^-. We determine the formula of the ionic compound to be KBr from the fact that K is a +1 cation and Br is a −1 anion. The final balanced equation is

$$HBr(aq) + KOH(aq) \longrightarrow KBr(aq) + H_2O(l)$$

EXAMPLE 8.10

Complete and balance the equation for the reaction between aqueous solutions of barium chloride and sodium sulfate.

SOLUTION

First determine the formulas for the reactants. They are $BaCl_2$ and Na_2SO_4. Then classify these substances as acids, bases, or ionic compounds. Both substances are ionic compounds. Since both substances are compounds, the reaction will be of the double-displacement type. Start writing the equation with the reactants:

$$BaCl_2(aq) + Na_2SO_4(aq) \longrightarrow$$

If the reaction is double-displacement, Ba^{2+} will be written combined with SO_4^{2-}, and Na^+ with Cl^- as the products. The balanced equation is

$$BaCl_2(aq) + Na_2SO_4(aq) \longrightarrow BaSO_4(s) + 2\,NaCl(aq)$$

When barium chloride is poured into a solution of sodium sulfate, a white precipitate of barium sulfate forms.

The final step is to determine the nature of the products, which controls whether or not the reaction will take place. If both products are soluble, we have a mixture of all the ions in solution. But if an insoluble precipitate is formed, the reaction will definitely occur. We know from experience that NaCl is fairly soluble in water, but what about $BaSO_4$? Consulting the solubility table in Appendix V, we see that $BaSO_4$ is insoluble in water, so it will be a precipitate in the reaction. Thus, the reaction will occur, forming a precipitate. The equation is

$$BaCl_2(aq) + Na_2SO_4(aq) \longrightarrow BaSO_4(s) + 2\ NaCl(aq)$$

PRACTICE 8.6

Complete and balance the equations for these reactions in water:
(a) potassium phosphate + barium chloride
(b) hydrochloric acid + nickel carbonate
(c) ammonium chloride + sodium nitrate

Some of the reactions you attempt may fail because the substances are not reactive or because the proper conditions for reaction are not present. For example, mercury(II) oxide does not decompose until it is heated; magnesium does not burn in air or oxygen until the temperature reaches a certain point. When silver is placed in a solution of copper(II) sulfate, no reaction occurs. When copper wire is placed in a solution of silver nitrate, a single-displacement reaction takes place because copper is a more reactive metal than silver. (See **Figure 8.1**.)

$$Cu(s) + 2\ AgNO_3(aq) \longrightarrow 2Ag(s) + Cu(NO_3)_2(aq)$$

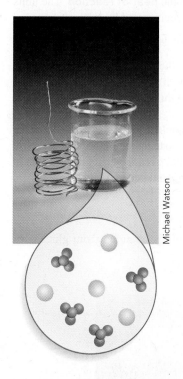

Ag$^+$

NO$_3^-$

Cu^{2+}

A solution of silver nitrate contains Ag$^+$ ions and NO$_3^-$ ions.

A copper wire is placed in a solution of silver nitrate. After 24 hours, crystals of silver are seen hanging on the copper wire and the solution has turned blue, indicating copper ions are present there.

The silver metal clings to the copper wire, and the blue solution contains Cu^{2+} and NO$_3^-$ ions.

Figure 8.1

Reaction of a copper wire and silver nitrate solution.

The successful prediction of the products of a reaction is not always easy. The ability to predict products correctly comes with knowledge and experience. Although you may not be able to predict many reactions at this point, as you continue to experiment you will find that reactions can be categorized and that prediction of the products becomes easier, if not always certain.

8.4 HEAT IN CHEMICAL REACTIONS

LEARNING OBJECTIVE • Explain the following terms and how they relate to a chemical reaction: *exothermic reaction, endothermic reaction, heat of reaction,* and *activation energy.*

KEY TERMS

exothermic reactions
endothermic reactions
heat of reaction
hydrocarbons
activation energy

Energy changes always accompany chemical reactions. One reason reactions occur is that the products attain a lower, more stable energy state than the reactants. When the reaction leads to a more stable state, energy is released to the surroundings as heat and/or work. When a solution of a base is neutralized by the addition of an acid, the liberation of heat energy is signaled by an immediate rise in the temperature of the solution. When an automobile engine burns gasoline, heat is certainly liberated; at the same time, part of the liberated energy does the work of moving the automobile.

Reactions are either exothermic or endothermic. **Exothermic reactions** liberate heat; **endothermic reactions** absorb heat. In an exothermic reaction, heat is a product and may be written on the right side of the equation for the reaction. In an endothermic reaction, heat can be regarded as a reactant and is written on the left side of the equation. Here are two examples:

$$H_2(g) + Cl_2(g) \longrightarrow 2\ HCl(g) + 185\ kJ \quad (exothermic)$$
$$N_2(g) + O_2(g) + 181\ kJ \longrightarrow 2\ NO(g) \quad (endothermic)$$

The quantity of heat produced by a reaction is known as the **heat of reaction**. The units used can be kilojoules or kilocalories. Consider the reaction represented by this equation:

$$C(s) + O_2(g) \longrightarrow CO_2(g) + 393\ kJ$$

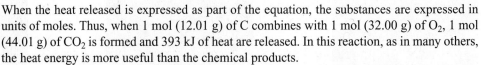

When the heat released is expressed as part of the equation, the substances are expressed in units of moles. Thus, when 1 mol (12.01 g) of C combines with 1 mol (32.00 g) of O_2, 1 mol (44.01 g) of CO_2 is formed and 393 kJ of heat are released. In this reaction, as in many others, the heat energy is more useful than the chemical products.

Aside from relatively small amounts of energy from nuclear processes, the sun is the major provider of energy for life on Earth. The sun maintains the temperature necessary for life and also supplies light energy for the endothermic photosynthetic reactions of green plants. In photosynthesis, carbon dioxide and water are converted to free oxygen and glucose:

$$6\ CO_2 + 6\ H_2O + 2519\ kJ \longrightarrow C_6H_{12}O_6 + 6\ O_2$$
glucose

Nearly all of the chemical energy used by living organisms is obtained from glucose or compounds derived from glucose.

Glucose
$C_6H_{12}O_6$

$\bigcirc$ = O
● = C
$\bigcirc$ = H

This cornfield is a good example of the endothermic reactions happening through photosynthesis in plants.

The major source of energy for modern technology is fossil fuel—coal, petroleum, and natural gas. The energy is obtained from the combustion (burning) of these fuels, which are converted to carbon dioxide and water. Fossil fuels are mixtures of **hydrocarbons**, compounds containing only hydrogen and carbon.

Natural gas is primarily methane, CH_4. Petroleum is a mixture of hydrocarbons (compounds of carbon and hydrogen). Liquefied petroleum gas (LPG) is a mixture of propane (C_3H_8) and butane (C_4H_{10}).

Here are some examples:

$$CH_4(g) + 2\,O_2(g) \longrightarrow CO_2(g) + 2\,H_2O(g) + 890\ kJ$$

$$C_3H_8(g) + 5\,O_2(g) \longrightarrow 3\,CO_2(g) + 4\,H_2O(g) + 2200\ kJ$$

Combustion of Mg inside a block of dry ice (CO_2) makes a glowing lantern.

The combustion of these fuels releases a tremendous amount of energy, but reactions won't occur to a significant extent at ordinary temperatures. A spark or a flame must be present before methane will ignite. The amount of energy that must be supplied to start a chemical reaction is called the **activation energy**. In an exothermic reaction, once this activation energy is provided, enough energy is then generated to keep the reaction going.

Be careful not to confuse an exothermic reaction that requires heat (activation energy) to get it *started* with an endothermic process that requires energy to keep it going. The combustion of magnesium, for example, is highly exothermic, yet magnesium must be heated to a fairly high temperature in air before combustion begins. Once started, however, the combustion reaction goes very vigorously until either the magnesium or the available supply of oxygen is exhausted. The electrolytic decomposition of water to hydrogen and oxygen is highly endothermic. If the electric current is shut off when this process is going on, the reaction stops instantly. The relative energy levels of reactants and products in exothermic and endothermic processes are presented graphically in Figures 8.2 and 8.3.

Examples of endothermic and exothermic processes can be easily demonstrated. In Figure 8.2, the products are at a higher potential energy than the reactants. Energy has therefore been absorbed, and the reaction is endothermic. An endothermic reaction takes place when you apply a cold pack to an injury. When a cold pack is activated, ammonium chloride (NH_4Cl) dissolves in water. For example, temperature changes from 24.5°C to 18.1°C result when 10 g of NH_4Cl are added to 100 mL of water. Energy, in the form of heat, is taken from the immediate surroundings (water), causing the salt solution to become cooler.

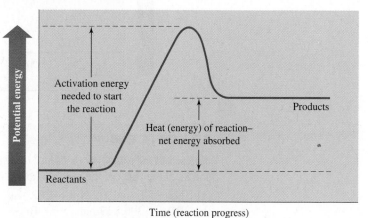

Endothermic reaction

Figure 8.2

An endothermic process occurs when a cold pack containing an ampule of solid NH_4Cl is broken, releasing the crystals into water in the pack. The dissolving of NH_4Cl in water is endothermic, producing a salt solution that is cooler than the surroundings. This process is represented graphically to show the energy changes between the reactants and the products.

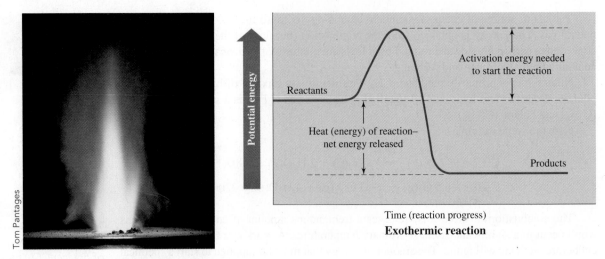

Figure 8.3

Exothermic reaction between $KClO_3$ and sugar. A sample of $KClO_3$ and sugar is well mixed and placed on a fireproof pad. Several drops of concentrated H_2SO_4 are used to ignite the mixture.

EXAMPLE 8.11

Label the graph in Figure 8.2 showing the specific reactants and products as well as the heat of reaction for the reaction.

SOLUTION

For Figure 8.2 the chemical equation is $NH_4Cl(s) + H_2O(l) \longrightarrow NH_4^+(aq) + Cl^-(aq) + H_2O(l)$. The reactants are $NH_4Cl(s) + H_2O(l)$ and the products are $NH_4^+(aq) + Cl^-(aq) + H_2O(l)$. The graph is labeled and shown here:

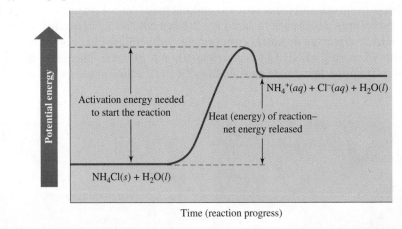

PRACTICE 8.7

For each of the following reactions, draw a reaction graph (like those in Figure 8.2 or 8.3) and label the parts of the graph with the information from the chemical equation.

(a) $2 Na + Cl_2 \longrightarrow 2 NaCl + 822$ kJ

(b) $2 BrF_3 + 300.8$ kJ/mol $\longrightarrow Br_2 + 3 F_2$

In Figure 8.3, the products are at a lower potential energy than the reactants. Energy (heat) is given off, producing an exothermic reaction. Here, potassium chlorate ($KClO_3$) and sugar are mixed and placed into a pile. A drop of concentrated sulfuric acid is added, creating a spectacular exothermic reaction.

8.5 GLOBAL WARMING: THE GREENHOUSE EFFECT

Discuss the possible causes and results of the greenhouse effect.

Fossil fuels, derived from coal and petroleum, provide the energy we use to power our industries, heat and light our homes and workplaces, and run our cars. As these fuels are burned, they produce carbon dioxide and water, releasing over 50 billion tons of carbon dioxide into our atmosphere each year.

The concentration of CO_2 has been monitored by scientists since 1958. Each week a scientist walks outdoors at a particular location and, carefully holding his breath, opens a basketball-sized glass sphere and collects 5 liters of the atmosphere. At these same remote sites all over the globe (and at dozens of others), instruments sniff the air adding more measurements of the atmosphere to the data. These data are summarized in **Figure 8.4**. Several trends have been discovered in these data.

1. In the Northern Hemisphere the concentration of CO_2 rises and falls 7 ppm over a year, peaking in May and dropping as plants use the CO_2 to produce growth till October, when a second peak occurs as a result of the fallen leaves decaying.

2. The 7-ppm annual variation is superimposed on an average increase in the concentration of CO_2. Our current average is more then 380 ppm, or an increase of more than 65 ppm over the last 50 years. It continues to rise about 2 ppm annually.

Carbon dioxide is a minor component in our atmosphere and is not usually considered to be a pollutant. The concern expressed by scientists arises from the dramatic increase occurring in the Earth's atmosphere. Without the influence of humans in the environment, the exchange of carbon dioxide between plants and animals would be relatively balanced. Our continued use of fossil fuels led to an increase of 7.4% in carbon dioxide between 1900 and 1970 and an additional 3.5% increase during the 1980s. Continued increases were observed in the 1990s.

Besides our growing consumption of fossil fuels, other factors contribute to increased carbon dioxide levels in the atmosphere: Rain forests are being destroyed by cutting and burning to make room for increased population and agricultural needs. Carbon dioxide is added to the atmosphere during the burning, and the loss of trees diminishes the uptake of carbon dioxide by plants.

About half of all the carbon dioxide released into our atmosphere each year remains there, thus increasing its concentration. The other half is absorbed by plants during photosynthesis or is dissolved in the ocean to form hydrogen carbonates and carbonates.

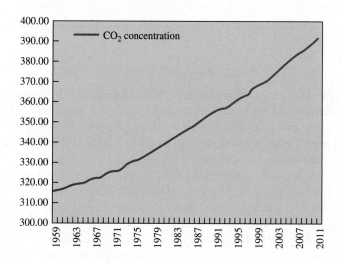

Figure 8.4

Concentration of CO_2 in the atmosphere.

Source: Data from the National Oceanic and Atmospheric Administration, http://www.esrl.noaa.gov/gmd/ccgg/trends/#mol-full, February 28, 2012.

>CHEMISTRY *IN ACTION*

Decreasing Carbon Footprints

In order to slow down global warming, we need to stop the increase of carbon dioxide (CO_2) in the air. We can, of course, continue to try to limit our dependence on fossil fuels. But scientists are also working to capture carbon dioxide emissions. Capturing CO_2 presents two challenges: how to remove the CO_2 from the air and what to do with it once it has been collected.

One way of capturing and storing carbon dioxide is to let nature work. Plants capture 505 million tons of carbon dioxide in North America through photosynthesis by storing it as carbon in woody plants and trees or as organic material in the soil and wetlands or sediments in rivers and lakes. The difficulty with this method is that as the decay process continues in an oxygen-rich environment, the carbon is returned to the atmosphere as CO_2. Ning Zing, from the University of Maryland, suggests that we could prevent the release of this CO_2 from forests by burying wood in an oxygen-poor environment. The carbon would then remain in the ground and not in the atmosphere. He notes that it is possible to accomplish this process at a cost of $14 per ton of wood, but that would mean collecting wood and burying it in trenches, which is not an environment-friendly method of capture and storage. Other scientists have suggested seeding the oceans to increase the algae and phytoplankton. This would also pull CO_2 from the atmosphere, but many scientists think that this method would not have much effect on climate since the oceans could only capture a small fraction of the CO_2 emitted by humans.

Coal and petroleum each account for about 40% of global CO_2 emissions. Coal is the resource that is the greatest threat to our climate since it produces more CO_2 per unit of energy than any other fossil fuel. Each year coal-fired power plants produce 8 billion tons of CO_2 (containing 2.2 billion tons of C). Future coal plants (planned in the next 25 years) will generate 660 billion tons of CO_2. This is 25% more than all the CO_2 that humans have produced by burning coal since 1751. How can we reduce these numbers? The current goal is to capture 90% of the CO_2 emissions from a power plant but only increase the cost of electricity by 20%. Scientists are developing materials that act as CO_2 sponges. These materials, called zeolites, are very porous and have large surface areas where CO_2 molecules can attach themselves.

The question here is that once the CO_2 is captured, how can it be stored? Many scientists propose to lock it underground or in deep ocean. Since CO_2 is a liquid at high pressure (as in ocean depths below 500 meters) and does not mix very well with water, it could be stored as large pool on the ocean floor. This storage method raises a number of environmental concerns. In some places there are saline aquifers underground. The water from these aquifers is not drinkable, and so some of them might be a suitable storage place for CO_2. Another possibility is using volcanic rock as a storage site. Tests indicate that liquid CO_2 will react with basalt to produce minerals such as calcium carbonate. These deposits contain many holes and cracks where the CO_2 might be injected. It is thought that in the United States these saline aquifers and geologic formations could store 150 years' worth of power plant emissions.

We are now looking seriously at ways to slow down the rise in atmospheric CO_2 and to decrease our carbon footprints without losing our fossil fuels as one of our energy sources.

Steven Raniszewski /Design Pics/Newscom

Zeolite (a basaltic volcanic rock) could be used as a storage place for carbon dioxide from power plants.

Methane is another important greenhouse gas. Its concentration in the atmosphere has also increased significantly since the 1850s, as shown in **Figure 8.5**. Methane is produced by cows, termites, agriculture, and anaerobic bacteria. Coal mining and oil wells also release methane into the atmosphere. The greenhouse effect of methane is 20 times more than that of CO_2, but there is less methane in the atmosphere.

Carbon dioxide and other greenhouse gases, such as methane and water, act to warm our atmosphere by trapping heat near the surface of the Earth. Solar radiation strikes the Earth and warms the surface. The warmed surface then reradiates this energy as heat. The greenhouse gases absorb some of this heat energy from the surface, which then warms our atmosphere. (See **Figure 8.6**.) A similar principle is illustrated in a greenhouse, where sunlight comes through the glass yet heat cannot escape. The air in the greenhouse warms, producing a climate considerably different than that in nature.

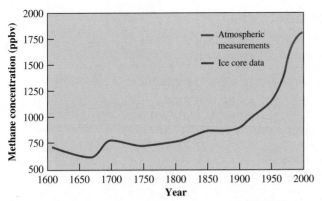

Figure 8.5
Concentration of methane in the atmosphere.

The long-term effects of global warming are still a matter of debate. One consideration is whether the polar ice caps will melt; this would cause a rise in sea level and lead to major flooding on the coasts of our continents. Further effects could include shifts in rainfall patterns, producing droughts and extreme seasonal change in such major agricultural regions as California. To reverse these trends will require major efforts in the following areas:

• The development of new energy sources to cut our dependence on fossil fuels

• An end to deforestation worldwide

• Intense efforts to improve conservation

On an individual basis each of us play a significant role. Recycling, switching to more fuel-efficient cars, and using energy-efficient appliances, heaters, and air conditioners all would result in decreased energy consumption.

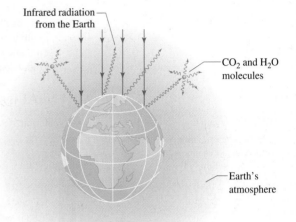

Figure 8.6
Visible light from the sun (red arrows) penetrates the atmosphere and strikes the Earth. Part of this light is changed to infrared radiation. Molecules of CO_2 and H_2O in the atmosphere absorb infrared radiation, acting like the glass of a greenhouse by trapping the energy.

CHAPTER **8 REVIEW**

8.1 THE CHEMICAL EQUATION

• A chemical equation is shorthand for expressing a chemical change or reaction.
• In a chemical reaction atoms are neither created nor destroyed.
• All atoms in the reactants must be present in the products.

KEY TERMS
reactants
products
chemical equation
law of conservation of mass

8.2 WRITING AND BALANCING CHEMICAL EQUATIONS

• To balance a chemical equation:
  • Identify the reaction.
  • Write the unbalanced (skeleton) equation.
  • Balance the equation:
    • Count the number of atoms of each element on each side and determine which need to be balanced.
    • Balance each element (one at a time) by placing whole numbers (coefficients) in front of the formulas containing the unbalanced element:
      • Begin with metals, then nonmetals, and then H and O.
    • Check the other elements to see if they have become unbalanced in the process of balancing the chosen element. If so, rebalance as needed.
    • Do a final check to make sure all elements are balanced.
• The following information can be found in a chemical equation:
  • Identity of reactants and products
  • Formulas for reactants and products
  • Number of formula units for reactants and products
  • Number of atoms of each element in the reaction
  • Number of moles of each substance

KEY TERM
balanced equation

8.3 TYPES OF CHEMICAL EQUATIONS

KEY TERMS

combination reaction
decomposition reaction
single-displacement reaction
double-displacement reaction

- Combination reactions $A + B \rightarrow AB$
- Decomposition reactions $AB \rightarrow A + B$
- Single-displacement reactions:
  - In which A is a metal $A + BC \rightarrow B + AC$
  - In which A is a halogen $A + BC \rightarrow C + BA$
- Double-displacement reactions $AB + CD \rightarrow AD + CB$
- Evidence for a chemical reaction:
  - Evolution of heat
  - Formation of an insoluble precipitate
  - Production of a gas

8.4 HEAT IN CHEMICAL REACTIONS

KEY TERMS

exothermic reactions
endothermic reactions
heat of reaction
hydrocarbons
activation energy

- Exothermic reactions release heat.

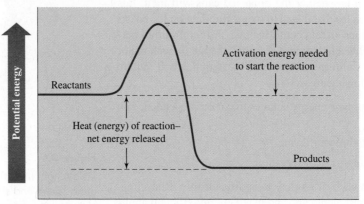

- Endothermic reactions absorb heat.

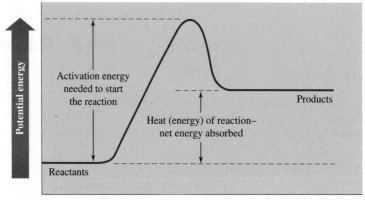

- The amount of heat released or absorbed in a chemical reaction is called the heat of reaction:
  - It can be written as a reactant or product in the chemical equation.
  - Units are joules (J) or kilojoules (kJ).
- The major source of energy for modern technology is hydrocarbon combustion.
- To initiate a chemical reaction, activation energy is required:
  - In exothermic reactions, this energy is returned and more is released, which allows the reaction to continue on its own.
  - In endothermic reactions, energy must be added to start and continue to be added to sustain the reaction.

8.5 GLOBAL WARMING: THE GREENHOUSE EFFECT

- Carbon dioxide levels are now increasing on our planet every year.
- Carbon dioxide and other greenhouse gases warm the atmosphere by trapping heat near the surface of the Earth.

REVIEW QUESTIONS

1. What is represented by the numbers (coefficients) that are placed in front of the formulas in a balanced equation?
2. What is meant by the physical state of a substance? What symbols are used to represent these physical states, and what does each symbol mean?
3. What is the purpose of balancing chemical equations?
4. In a balanced chemical equation:
   (a) Are atoms conserved?
   (b) Are molecules conserved?
   (c) Are moles conserved?
   Explain your answers briefly.
5. In Section 8.2, there are small charts in color along the margins. What information is represented in those charts?
6. Why is it incorrect to balance chemical reactions by changing the subscripts on the reactants and products?

7. What is a combustion reaction? How can you identify whether a reaction is a combustion reaction?
8. What information does the activity series shown in Table 8.2 give?
9. What major types of chemical reactions are studied in this chapter?
10. What three observations indicate a chemical reaction has taken place?
11. Explain how endothermic reactions differ from exothermic reactions.
12. Why does an exothermic reaction require heat to occur?
13. Name several greenhouse gases.
14. What factors are responsible for the annual rise and fall in carbon dioxide levels?

Most of the exercises in this chapter are available for assignment via the online homework management program, WileyPLUS (www.wileyplus.com). All exercises with blue numbers have answers in Appendix VI.

PAIRED EXERCISES

1. Classify the following as an endothermic or exothermic reaction:
   (a) freezing water
   (b) the reaction inside an ice pack
   (c) burning wood
   (d) combustion of Mg in dry ice
   (e) melting ice

2. Classify the following as an endothermic or exothermic reaction:
   (a) making popcorn in a microwave oven
   (b) a burning match
   (c) boiling water
   (d) burning rocket fuel
   (e) the reaction inside a heat pack

3. Balance each of the following equations. Classify each reaction as combination, decomposition, single-displacement, or double-displacement.
   (a) $H_2 + O_2 \longrightarrow H_2O$
   (b) $N_2H_4(l) \longrightarrow NH_3(g) + N_2(g)$
   (c) $H_2SO_4 + NaOH \longrightarrow H_2O + Na_2SO_4$
   (d) $Al_2(CO_3)_3 \xrightarrow{\Delta} Al_2O_3 + CO_2$
   (e) $NH_4I + Cl_2 \longrightarrow NH_4Cl + I_2$

4. Balance each of the following equations. Classify each reaction as combination, decomposition, single-displacement, or double-displacement.
   (a) $H_2 + Br_2 \longrightarrow HBr$
   (b) $BaO_2(s) + H_2SO_4(aq) \longrightarrow BaSO_4(s) + H_2O_2(aq)$
   (c) $Ba(ClO_3)_2 \xrightarrow{\Delta} BaCl_2 + O_2$
   (d) $CrCl_3 + AgNO_3 \longrightarrow Cr(NO_3)_3 + AgCl$
   (e) $H_2O_2 \longrightarrow H_2O + O_2$

5. What reactant(s) is (are) required to form an oxide product?

6. What reactant(s) is (are) required to form a salt?

7. Balance the following equations:
   (a) $MnO_2 + CO \longrightarrow Mn_2O_3 + CO_2$
   (b) $Cu_2O(s) + C(s) \longrightarrow Cu(s) + CO(g)$
   (c) $C_3H_5(NO_3)_3 \longrightarrow CO_2 + H_2O + N_2 + O_2$
   (d) $FeS + O_2 \longrightarrow Fe_2O_3 + SO_2$
   (e) $Cu(NO_3)_2 \longrightarrow CuO + NO_2 + O_2$
   (f) $NO_2 + H_2O \longrightarrow HNO_3 + NO$
   (g) $Fe(s) + S(l) \longrightarrow Fe_2S_3(s)$
   (h) $HCN + O_2 \longrightarrow N_2 + CO_2 + H_2O$
   (i) $B_5H_9 + O_2 \longrightarrow B_2O_3 + H_2O$

8. Balance the following equations:
   (a) $SO_2 + O_2 \longrightarrow SO_3$
   (b) $Li_2O(s) + H_2O(l) \longrightarrow LiOH(aq)$
   (c) $Na + H_2O \longrightarrow NaOH + H_2$
   (d) $AgNO_3 + Ni \longrightarrow Ni(NO_3)_2 + Ag$
   (e) $Bi_2S_3 + HCl \longrightarrow BiCl_3 + H_2S$
   (f) $PbO_2 \xrightarrow{\Delta} PbO + O_2$
   (g) $Hg_2(C_2H_3O_2)_2(aq) + KCl(aq) \longrightarrow Hg_2Cl_2(s) + KC_2H_3O_2(aq)$
   (h) $KI + Br_2 \longrightarrow KBr + I_2$
   (i) $K_3PO_4 + BaCl_2 \longrightarrow KCl + Ba_3(PO_4)_2$

9. Change these word equations into formula equations and balance them. Be sure to use the proper symbols to indicate the state of each substance, as given.
   (a) Magnesium metal is placed into hydrobromic acid solution, forming hydrogen gas and aqueous magnesium bromide.
   (b) When heated, solid calcium chlorate decomposes into calcium chloride solid, releasing oxygen gas.
   (c) Lithium metal reacts with oxygen gas to form solid lithium oxide.
   (d) Solutions of barium bromate and sodium phosphate combine to form solid barium phosphate and aqueous sodium bromate.
   (e) Solutions of acetic acid and sodium carbonate are mixed together, forming a solution of sodium acetate, along with carbon dioxide gas and liquid water.
   (f) Solutions of silver nitrate and aluminum iodide are mixed together, forming solid silver iodide and aqueous aluminum nitrate.

10. Change these word equations into formula equations and balance them. Be sure to use the proper symbols to indicate the state of each substance, as given.
    (a) Upon heating, solid magnesium carbonate decomposes into solid magnesium oxide and carbon dioxide gas.
    (b) Solid calcium hydroxide reacts with aqueous chloric acid to form a solution of calcium chlorate along with liquid water.
    (c) Solutions of iron(III) sulfate and sodium hydroxide are mixed together, forming solid iron(III) hydroxide and a solution of sodium sulfate.
    (d) Zinc metal is placed into a solution of acetic acid, producing hydrogen gas and aqueous zinc acetate.
    (e) Gaseous sulfur trioxide reacts with liquid water to form a solution of sulfuric acid.
    (f) Solutions of sodium carbonate and cobalt(II) chloride react to form solid cobalt(II) carbonate and a solution of sodium chloride.

11. For each of the following reactions, predict the products, converting each to a balanced formula equation:
    (a) Aqueous solutions of sulfuric acid and sodium hydroxide are mixed together. (Heat is released during the reaction.)
    (b) Aqueous solutions of lead(II) nitrate and potassium bromide are mixed together. (The solution turns cloudy white during the reaction.)
    (c) Aqueous solutions of ammonium chloride and silver nitrate are mixed together. (The solution turns cloudy white during the reaction.)
    (d) Solid calcium carbonate is mixed with acetic acid. (Bubbles of gas are formed during the reaction.)

12. For each of the following reactions, predict the products, converting each to a balanced formula equation:
    (a) Aqueous solutions of copper(II) sulfate and potassium hydroxide are mixed together. (The solution turns cloudy and light blue during the reaction.)
    (b) Aqueous solutions of phosphoric acid and sodium hydroxide are mixed together. (Heat is produced during the reaction.)
    (c) Solid sodium bicarbonate is mixed with phosphoric acid. (Bubbles of gas are formed during the reaction.)
    (d) Aqueous solutions of aluminum chloride and lead(II) nitrate are mixed together. (The solution turns cloudy white during the reaction.)

13. Use the activity series to predict which of the following reactions will occur. Complete and balance the equations. If no reactions occurs, write "no reaction" as the product.
    (a) $Ca(s) + H_2O(l) \rightarrow$
    (b) $Br_2(l) + KI(aq) \rightarrow$
    (c) $Cu(s) + HCl(aq) \rightarrow$
    (d) $Al(s) + H_2SO_4(aq) \rightarrow$

14. Use the activity series to predict which of the following reactions will occur. Complete and balance the equations. If no reaction occurs, write "no reaction" as the product.
    (a) $Cu(s) + NiCl_2(aq) \rightarrow$
    (b) $Rb(s) + H_2O(l) \rightarrow$
    (c) $I_2(s) + CaCl_2(aq) \rightarrow$
    (d) $Mg(s) + Al(NO_3)_3(aq) \rightarrow$

15. Complete and balance the equations for each of the following reactions. All yield products.
    (a) $Sr(s) + H_2O(l) \rightarrow$
    (b) $BaCl_2(aq) + AgNO_3(aq) \rightarrow$
    (c) $Mg(s) + ZnBr_2(aq) \rightarrow$
    (d) $K(s) + Cl_2(g) \rightarrow$

16. Complete and balance the equations for each of the following reactions. All yield products.
    (a) $Li_2O(s) + H_2O(l) \rightarrow$
    (b) $Na_2SO_4(aq) + Pb(NO_3)_2(aq) \rightarrow$
    (c) $Zn(s) + CuSO_4(aq) \rightarrow$
    (d) $Al(s) + O_2(g) \rightarrow$

17. Complete and balance the equations for these reactions. All reactions yield products.
    (a) $Ba + O_2 \longrightarrow$
    (b) $NaHCO_3 \xrightarrow{\Delta} Na_2CO_3 +$
    (c) $Ni + CuSO_4 \longrightarrow$
    (d) $MgO + HCl \longrightarrow$
    (e) $H_3PO_4 + KOH \longrightarrow$

18. Complete and balance the equations for these reactions. All reactions yield products.
    (a) $C + O_2 \longrightarrow$
    (b) $Al(ClO_3)_3 \xrightarrow{\Delta} O_2 +$
    (c) $CuBr_2 + Cl_2 \longrightarrow$
    (d) $SbCl_3 + (NH_4)_2S \longrightarrow$
    (e) $NaNO_3 \longrightarrow NaNO_2 +$

19. Interpret these chemical reactions in terms of the number of moles of each reactant and product:
    (a) $MgBr_2 + 2 AgNO_3 \longrightarrow Mg(NO_3)_2 + 2 AgBr$
    (b) $N_2 + 3 H_2 \longrightarrow 2 NH_3$
    (c) $2 C_3H_7OH + 9 O_2 \longrightarrow 6 CO_2 + 8 H_2O$

20. Interpret these equations in terms of the relative number of moles of each substance involved and indicate whether the reaction is exothermic or endothermic:
    (a) $2 Na + Cl_2 \longrightarrow 2 NaCl + 822 kJ$
    (b) $PCl_5 + 92.9 kJ \longrightarrow PCl_3 + Cl_2$
    (c) $S(s) + 2 CO(g) \longrightarrow SO_2(g) + 2 C(s) + 76 kJ$

21. Write balanced equations for each of these reactions, including the heat term:
    (a) Solid mercury(II) oxide decomposes into liquid mercury and oxygen gas upon the absorption of 90.8 kJ for each mole of mercury(II) oxide decomposed.
    (b) Hydrogen gas reacts with oxygen gas to form liquid water. The reaction produces 285.8 kJ of heat for each mole of water formed

22. Write balanced equations for each of these reactions, including the heat term:
    (a) Calcium metal reacts with water to produce calcium hydroxide and hydrogen gas, releasing 635.1 kJ of heat for every mole of calcium that reacts.
    (b) Bromine trifluoride decomposes into bromine and fluorine upon the absorption of 300.8 kJ of heat for each mole of bromine trifluoride decomposed.

23. Determine what reactants would form the given products. Give a balanced equation for each and classify the reaction as a combination, decomposition, single-displacement, or double-displacement reaction.
    (a) $AgCl(s) + O_2(g)$
    (b) $H_2(g) + FeSO_4(aq)$
    (c) $ZnCl_2(s)$
    (d) $KBr(aq) + H_2O(l)$

24. Determine what reactants would form the given products. Give a balanced equation for each and classify the reaction as a combination, decomposition, single-displacement, or double-displacement reaction.
    (a) $Pb(s) + Ni(NO_3)_2(aq)$
    (b) $Mg(OH)_2(s)$
    (c) $Hg(l) + O_2(g)$
    (d) $PbCO_3(s) + NH_4Cl(aq)$

ADDITIONAL EXERCISES

25. An aqueous mixture of the nitrate salts of sodium, calcium, and silver is mixed with ammonium salts of chloride, sulfate, and carbonate. Write balanced equations for any combination that will form a precipitate.

26. Balance this equation, using the smallest possible whole numbers. Then determine how many atoms of oxygen appear on each side of the equation:

$$P_4O_{10} + HClO_4 \longrightarrow Cl_2O_7 + H_3PO_4$$

27. Suppose that in a balanced equation the term 5 $Ni_3(PO_4)_2$ appears.
(a) How many atoms of nickel are represented?
(b) How many atoms of phosphorus are represented?
(c) How many atoms of oxygen are represented?
(d) How many atoms of all kinds are represented?

28. In a lab experiment, a student places a few stones, composed mainly of calcium carbonate, into a beaker and then adds some acetic acid. The student observes that effervescence occurs immediately. Explain this observation by writing a balanced formula equation.

29. Make a drawing to show the combustion reaction of one molecule of methane, CH_4.

30. A variety of iron-containing ores can be converted into iron and steel in a blast furnace. The reactions that take place are outlined below. Write balanced chemical equations for each reaction.
(a) Pure carbon is reacted with oxygen gas to form carbon dioxide.
(b) Carbon dioxide is reacted with more carbon to form monoxide.
(c) Carbon monoxide reacts with the iron ore to form elemental iron and carbon dioxide. There are two main types of ore that may be used: hematite (Fe_2O_3) and magnetite (Fe_3O_4).

31. The box at the left represents the reactant atoms in blue (B) and molecules in green (G_2). After reacting, these reactants from the compound pictured in the box at the right. Write a balanced chemical reaction for this process.

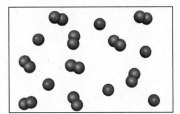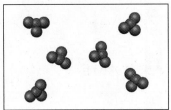

32. Students in a lab exercise were asked to choose among a sample of different metals to react with a solution of nickel(II) chloride. Which of the following metals should they choose? Write balanced equations for all that will react and explain why each will or will not react.
(a) copper   (d) lead
(b) zinc   (e) calcium
(c) aluminum

33. A student does an experiment to determine where titanium metal should be placed on the activity series chart. He places newly cleaned pieces of titanium into solutions of nickel(II) nitrate, lead(II) nitrate, and magnesium nitrate. He finds that the titanium reacts with the nickel(II) nitrate and lead(II) nitrate solutions, but not with the magnesium nitrate solution. From this information, place titanium in the activity series in a position relative to these ions.

34. Predict the products of each of the following combination reactions in words. Then change each equation into a balanced formula equation.
(a) Cesium reacts with oxygen.
(b) Aluminum metal reacts with sulfur.
(c) Sulfur trioxide reacts with water.
(d) Sodium oxide reacts with water.

35. Automobile designers are currently working on cars that will run on hydrogen. Hydrogen-fueled cars have the advantage that they produce only water as a waste product. In these cars, hydrogen gas reacts with oxygen gas to form gaseous water.
(a) Write a balanced chemical equation for this reaction.
(b) There are many ways to generate hydrogen to fuel these cars. The most commonly used method of making hydrogen gas is to react methane gas (CH_4) with gaseous water at 1000°C to form gaseous carbon monoxide and hydrogen gas. The carbon monoxide gas can then be reacted with more gaseous water to produce more hydrogen gas and carbon dioxide gas. Write balanced chemical equations for these two reactions.
(c) One of the products of burning fossil fuels to provide energy is carbon dioxide. This carbon dioxide may be increasing the temperature of the planet. Methane can also be used to provide energy by reacting it with oxygen gas to form carbon dioxide and water. Write the balanced chemical equation for this reaction.
(d) Compare the number of molecules of carbon dioxide produced out of one molecule of methane by burning it directly to produce energy and by converting it to hydrogen gas and using this hydrogen gas to produce energy. Is there an advantage to using hydrogen gas as the fuel source? Explain.

36. Many paint pigments are composed of ionic compounds. For each of the pigments listed below, write a complete balanced equation to synthesize the pigment using the reagents listed in parentheses. Write the correct IUPAC name of each pigment.
(a) chrome yellow, $PbCrO_4$ ($Pb(NO_3)_2(aq) + K_2CrO_4(aq)$)
(b) cadmium yellow, CdS ($CdCl_2(aq) + Li_2S(aq)$)
(c) white lead, $PbCO_3$ ($Pb(C_2H_3O_2)_2(aq) + H_2CO_3(aq)$)
(d) vermillion, HgS ($Hg(NO_3)_2(aq) + Na_2S(aq)$)
(e) titanium black, Ti_2O_3 ($Ti(s) + O_2(g)$)
(f) iron oxide red, Fe_2O_3 ($Fe(s) + O_2(g)$)

37. Predict the products of each of the following decomposition reactions in words. Then change each equation into a balanced formula equation.
(a) Zinc oxide decomposes upon heating strongly.
(b) Tin(IV) oxide decomposes into two products when heated.
(c) Sodium carbonate decomposes when heated.
(d) Magnesium chlorate decomposes when heated.

38. Predict the products of each of the following single-displacement reactions in words. Then change each equation into a balanced formula equation. All reactions occur in aqueous solution.
(a) Magnesium metal reacts with hydrochloric acid.
(b) Sodium bromide reacts with chlorine.
(c) Zinc metal reacts with iron(III) nitrate.
(d) Aluminum reacts with copper(II) sulfate.

39. Predict the products of each of the following double-displacement reactions in words. Then change each equation into a balanced formula equation. All reactants are in aqueous solution.
(a) Ammonium phosphate reacts with barium nitrate.
(b) Sodium sulfide reacts with lead(II) acetate.
(c) Copper(II) sulfate reacts with calcium chlorate.
(d) Barium hydroxide reacts with oxalic acid.
(e) Phosphoric acid reacts with potassium hydroxide.
(f) Sulfuric acid reacts with sodium carbonate.

40. Predict which of the following double-displacement reactions will occur. For those that do, write a balanced formula equation. For any

reaction that does not occur, write "no reaction" as the product. All reactants are in aqueous solution.
(a) potassium sulfate and barium acetate
(b) sulfuric acid and lithium hydroxide
(c) ammonium phosphate and sodium bromide
(d) calcium iodide and silver nitrate
(e) nitric acid and strontium hydroxide
(f) cesium nitrate and calcium hydroxide

41. Write balanced equations for each of the following combustion reactions:

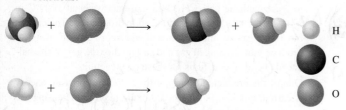

42. Write balanced equations for the complete combustion of the following hydrocarbons:
(a) CH_4 (b) C_3H_6 (c) $C_6H_5CH_3$
43. List the factors that contribute to an increase in carbon dioxide in our atmosphere.
44. List three gases considered to be greenhouse gases. Explain why they are given this name.
45. How can the effects of global warming be reduced?
46. What happens to carbon dioxide released into our atmosphere?
47. In the Northern Hemisphere, the concentration of CO_2 peaks twice a year. When do these peaks occur, and what causes them?

CHALLENGE EXERCISE

48. You are given a solution of the following cations in water: Ag^+, Co^{2+}, Ba^{2+}, Zn^{2+}, and Sn^{2+}. You want to separate the ions out *one at a time* by precipitation using the following reagents: NaF, NaI, Na_2SO_4, and NaCl.
(a) In what order would you add the reagents to ensure that *only one cation* is precipitating out at a time? Use the solubility table in Appendix V to help you. (*Note:* A precipitate that is slightly soluble in water is still considered to precipitate out of the solution.)
(b) Why are the anionic reagents listed above all sodium salts?

ANSWERS TO PRACTICE EXERCISES

8.1 To show that a substance is a solid, write (*s*) after the formula of the substance. When the substance is a gas, write (*g*) after the formula, and when the substance is in an aqueous solution write (*aq*) after the formula.

8.2 $4\,Al + 3\,O_2 \longrightarrow 2\,Al_2O_3$

8.3 $3\,Mg(OH)_2 + 2\,H_3PO_4 \longrightarrow Mg_3(PO_4)_2 + 6\,H_2O$

8.4 (a) hydrogen gas + chlorine gas $\longrightarrow$ hydrogen chloride gas
(b) $H_2(g) + Cl_2(g) \longrightarrow 2\,HCl(g)$
(c)

$H_2(g)$	+	$Cl_2(g)$	$\longrightarrow$	$2\,HCl(g)$
1 molecule		1 molecule		2 molecules
2 atoms H		2 atoms Cl		2 atoms H 2 atoms Cl
1 mol 2.016 g		1 mol 70.90 g		2 mol 2 (36.46 g) (72.92 g)

(d) $2\,mol\,H_2 + 2\,mol\,Cl_2 \longrightarrow 4\,mol\,HCl(145.8\,g)$

8.5 (a) $Fe + MgCl_2 \longrightarrow$ no reaction
(b) $Zn(s) + Pb(NO_3)_2(aq) \longrightarrow Pb(s) + Zn(NO_3)_2(aq)$

8.6 (a) $2\,K_3PO_4(aq) + 3\,BaCl_2(aq) \longrightarrow$
$$Ba_3(PO_4)_2(s) + 6\,KCl(aq)$$
(b) $2\,HCl(aq) + NiCO_3(aq) \longrightarrow$
$$NiCl_2(aq) + H_2O(l) + CO_2(g)$$
(c) $NH_4Cl(aq) + NaNO_3(aq) \longrightarrow$ no reaction

8.7 (a)

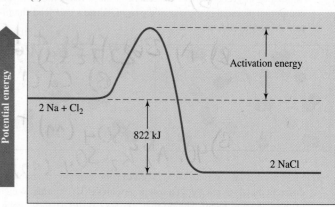

(b)

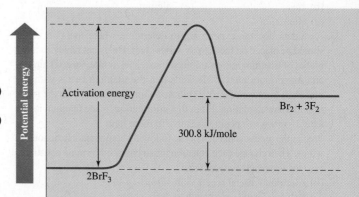

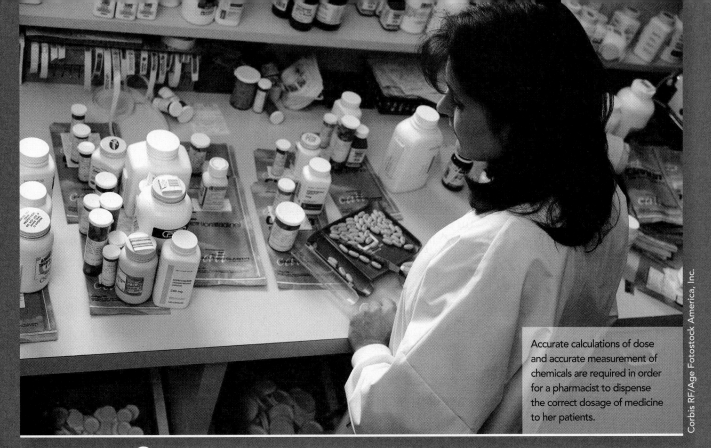

Accurate calculations of dose and accurate measurement of chemicals are required in order for a pharmacist to dispense the correct dosage of medicine to her patients.

Corbis RF/Age Fotostock America, Inc.

CHAPTER **9**

CALCULATIONS FROM CHEMICAL EQUATIONS

The old adage "waste not, want not" is appropriate in our daily life and in the laboratory. Determining correct amounts comes into play in almost all professions. A seamstress determines the amount of material, lining, and trim necessary to produce a gown for her client by relying on a pattern or her own experience to guide the selection. A carpet layer determines the correct amount of carpet and padding necessary to recarpet a customer's house by calculating the floor area. The IRS determines the correct deduction for federal income taxes from your paycheck based on your expected annual income.

CHAPTER OUTLINE

9.1 Introduction to Stoichiometry

9.2 Mole–Mole Calculations

9.3 Mole–Mass Calculations

9.4 Mass–Mass Calculations

9.5 Limiting Reactant and Yield Calculations

9.1 INTRODUCTION TO STOICHIOMETRY

LEARNING OBJECTIVE ● Define stoichiometry and describe the strategy required to solve problems based on chemical equations.

KEY TERMS
stoichiometry
molar mass
mole ratio

We often need to calculate the amount of a substance that is either produced from, or needed to react with, a given quantity of another substance. The area of chemistry that deals with quantitative relationships among reactants and products is known as **stoichiometry** (*stoy-key-ah-meh-tree*). Solving problems in stoichiometry requires the use of *moles* in the form of *mole ratios*.

The chemist also finds it necessary to calculate amounts of products or reactants by using a balanced chemical equation. With these calculations, the chemist can control the amount of product by scaling the reaction up or down to fit the needs of the laboratory and can thereby minimize waste or excess materials formed during the reaction.

A Short Review

Molar mass The sum of the atomic masses of all the atoms in an element or compound is called **molar mass**. The term *molar mass* also applies to the mass of a mole of any formula unit—atoms, molecules, or ions; it is the atomic mass of an atom or the sum of the atomic masses in a molecule or an ion (in grams).

Relationship between molecule and mole A molecule is the smallest unit of a molecular substance (e.g., Br_2), and a mole is Avogadro's number (6.022×10^{23}) of molecules of that substance. A mole of bromine (Br_2) has the same number of molecules as a mole of carbon dioxide, a mole of water, or a mole of any other molecular substance. When we relate molecules to molar mass, 1 molar mass is equivalent to 1 mol, or 6.022×10^{23} molecules.

The term *mole* also refers to any chemical species. It represents a quantity (6.022×10^{23} particles) and may be applied to atoms, ions, electrons, and formula units of nonmolecular substances. In other words,

$$1 \text{ mole} = \begin{cases} 6.022 \times 10^{23} \text{ molecules} \\ 6.022 \times 10^{23} \text{ formula units} \\ 6.022 \times 10^{23} \text{ atoms} \\ 6.022 \times 10^{23} \text{ ions} \end{cases}$$

Other useful mole relationships are

$$\text{molar mass} = \frac{\text{grams of a substance}}{\text{number of moles of the substance}}$$

$$\text{molar mass} = \frac{\text{grams of a monatomic element}}{\text{number of moles of the element}}$$

$$\text{number of moles} = \frac{\text{number of molecules}}{6.022 \times 10^{23} \text{ molecules/mole}}$$

A mole of water, salt, and any gas all have the same number of particles (6.022×10^{23}).

Balanced equations When using chemical equations for calculations of mole–mass–volume relationships between reactants and products, the equations must be balanced. *Remember:* The number in front of a formula in a balanced chemical equation represents the number of moles of that substance in the chemical reaction.

A **mole ratio** is a ratio between the number of moles of any two species involved in a chemical reaction. For example, in the reaction

$$2\,H_2 + O_2 \longrightarrow 2\,H_2O$$
$$\text{2 mol} \quad \text{1 mol} \qquad \text{2 mol}$$

six mole ratios can be written:

$$\frac{2 \text{ mol } H_2}{1 \text{ mol } O_2} \qquad \frac{2 \text{ mol } H_2}{2 \text{ mol } H_2O} \qquad \frac{1 \text{ mol } O_2}{2 \text{ mol } H_2O}$$

$$\frac{1 \text{ mol } O_2}{2 \text{ mol } H_2} \qquad \frac{2 \text{ mol } H_2O}{2 \text{ mol } H_2} \qquad \frac{2 \text{ mol } H_2O}{1 \text{ mol } O_2}$$

We use the mole ratio to convert the number of moles of one substance to the corresponding number of moles of another substance in a chemical reaction. For example, if we want to calculate the number of moles of H_2O that can be obtained from 4.0 mol of O_2, we use the mole ratio 2 mol H_2O/1 mol O_2:

$$(4.0 \text{ mol } O_2)\left(\frac{2 \text{ mol } H_2O}{1 \text{ mol } O_2}\right) = 8.0 \text{ mol } H_2O$$

EXAMPLE 9.1

Use the following equation $CO_2 + 4 H_2 \rightarrow CH_4 + 2 H_2O$ to write the mole ratio needed to calculate

(a) the number of moles of water that can be produced from 3 moles of CO_2.
(b) the number of moles of hydrogen required to make 3 moles of water.

SOLUTION

In order to write the mole ratio we need to determine the relationship between the desired substance and the starting substance.

(a) In this case the desired substance is water. The moles of water belong in the numerator of the mole ratio. The starting substance is CO_2. The moles of CO_2 then belong in the denominator of the mole ratio. In the mole ratio the coefficients of the balanced equation are used. Therefore, the mole ratio is $\dfrac{2 \text{ mol } H_2O}{1 \text{ mol } CO_2}$.

(b) In this case the desired substance is hydrogen. The moles of hydrogen belong in the numerator of the mole ratio. The starting substance is water. The moles of water then belong in the denominator of the mole ratio. In the mole ratio the coefficients of the balanced equation are used. Therefore, the mole ratio is $\dfrac{4 \text{ mol } H_2}{2 \text{ mol } H_2O}$.

PRACTICE 9.1

Use the equation $2 KClO_3 \rightarrow 2 KCl + 3 O_2$ to write the mole ratio needed to calculate

(a) moles of KCl produced from 3.5 moles of $KClO_3$.
(b) moles of KCl produced when 4.5 moles of O_2 are produced.

PROBLEM-SOLVING STRATEGY: For Stoichiometry Problems

1. Convert the quantity of starting substance to moles (if it is not given in moles).
2. Convert the moles of starting substance to moles of desired substance.
3. Convert the moles of desired substance to the units specified in the problem.

Write and balance the equation before you begin the problem.

Like balancing chemical equations, making stoichiometric calculations requires practice. Several worked examples follow. Study this material and practice on the problems at the end of this chapter.

PROBLEM-SOLVING STRATEGY: For Stoichiometry Problems

Use a balanced equation.

1. **Determine the number of moles of starting substance.** Identify the starting substance from the data given in the problem statement. Convert the quantity of the starting substance to moles, if it is not already done:

$$\text{moles} = (\text{grams})\left(\frac{1 \text{ mole}}{\text{molar mass}}\right)$$

As in all problems with units, the desired quantity is in the numerator, and the quantity to be eliminated is in the denominator.

2. **Determine the mole ratio of the desired substance to the starting substance.** The number of moles of each substance in the balanced equation is indicated by the coefficient in front of each substance. Use these coefficients to set up the mole ratio:

$$\text{mole ratio} = \frac{\text{moles of desired substance in the equation}}{\text{moles of starting substance in the equation}}$$

Multiply the number of moles of starting substance (from Step 1) by the mole ratio to obtain the number of moles of desired substance:

Units of moles of starting substance cancel in the numerator and denominator.

$$\underset{\text{substance}}{\text{moles of desired}} = \left(\underset{\text{substance}}{\overset{\text{From Step 1}}{\text{moles of starting}}}\right)\left(\frac{\text{moles of desired substance in the balanced equation}}{\text{moles of starting substance in the balanced equation}}\right)$$

3. **Calculate the desired substance in the units specified in the problem.** If the answer is to be in moles, the calculation is complete. If units other than moles are wanted, multiply the moles of the desired substance (from Step 2) by the appropriate factor to convert moles to the units required. For example, if grams of the desired substance are wanted:

$$\text{grams} = \overset{\text{From Step 2}}{(\text{moles})}\left(\frac{\text{molar mass}}{1\ \text{mol}}\right)$$

If moles $\longrightarrow$ atoms, use $\dfrac{6.022 \times 10^{23}\ \text{atoms}}{1\ \text{mol}}$

If moles $\longrightarrow$ molecules, use $\dfrac{6.022 \times 10^{23}\ \text{molecules}}{1\ \text{mol}}$

The steps for converting the mass of a starting substance A to either the mass, atoms, or molecules of desired substance B are summarized in **Figure 9.1**.

Figure 9.1

Steps for converting starting substance A to mass, atoms, or molecules of desired substance B.

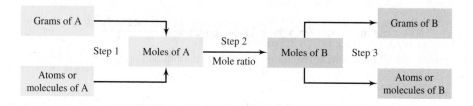

9.2 MOLE–MOLE CALCULATIONS

LEARNING OBJECTIVE ● Solve problems in which the reactants and products are both in moles.

Let's solve stoichiometric problems for mole–mole calculations. The quantity of starting substance is given in moles, and the quantity of desired substance is requested in moles.

WileyPLUS

ENHANCED EXAMPLE

EXAMPLE 9.2

How many moles of carbon dioxide will be produced by the complete reaction of 2.0 mol of glucose ($C_6H_{12}O_6$) according to the following equation?

$$\underset{\text{1 mol}}{C_6H_{12}O_6} + \underset{\text{6 mol}}{6\,O_2} \rightarrow \underset{\text{6 mol}}{6\,CO_2} + \underset{\text{6 mol}}{6\,H_2O}$$

SOLUTION

Read ●   **Knowns:**   2.0 mol of glucose
$$C_6H_{12}O_6 + 6\,O_2 \rightarrow 6\,CO_2 + 6\,H_2O$$

Solving for: moles of carbon dioxide

Plan • The balanced equation states that 6 mol CO_2 will be produced from 1 mol $C_6H_{12}O_6$. Even though we can readily see that 12 mol CO_2 will be formed, let's use the Problem-Solving Strategy for Stoichiometry Problems.

Solution map: moles of $C_6H_{12}O_6$ → moles of CO_2

The mole ratio from the balanced equation is $\dfrac{6 \text{ mol } CO_2}{1 \text{ mol } C_6H_{12}O_6}$.

The mole ratio is exact and does not affect the number of significant figures in the answer.

Calculate • $(2.0 \text{ mol } C_6H_{12}O_6)\left(\dfrac{6 \text{ mol } CO_2}{1 \text{ mol } C_6H_{12}O_6}\right) = 12 \text{ mol } CO_2$

Check • Note that the moles $C_6H_{12}O_6$ cancel and that the answer agrees with our reasoned answer.

EXAMPLE 9.3

How many moles of ammonia can be produced from 8.00 mol of hydrogen reacting with nitrogen? The balanced equation is

$$3 H_2 + N_2 \rightarrow 2 NH_3$$

SOLUTION

Read • **Knowns:** 8.00 mol of hydrogen
$3 H_2 + N_2 \rightarrow 2 NH_3$

Solving for: moles of ammonia

Plan • The balanced equation states that we get 2 mol NH_3 for every 3 mol H_2 that react. Let's use the Problem-Solving Strategy for Stoichiometry Problems.

Solution map: moles of H_2 → moles of NH_3

The mole ratio from the balanced equation is $\dfrac{2 \text{ mol } NH_3}{3 \text{ mol } H_2}$.

Calculate • $(8.00 \text{ mol } H_2)\left(\dfrac{2 \text{ mol } NH_3}{3 \text{ mol } H_2}\right) = 5.33 \text{ mol } NH_3$

Check • Note that the moles H_2 cancel and that the answer is less than the starting 8.00 mol, which makes sense since our mole ratio is 2:3.

Ammonia
NH_3

EXAMPLE 9.4

Given the balanced equation

$$\underset{\substack{1 \text{ mol} \quad 6 \text{ mol}}}{K_2Cr_2O_7 + 6 KI + 7 H_2SO_4} \rightarrow \underset{3 \text{ mol}}{Cr_2(SO_4)_3 + 4 K_2SO_4 + 3 I_2} + 7 H_2O$$

Calculate (a) the number of moles of potassium dichromate ($K_2Cr_2O_7$) that will react with 2.0 mol of potassium iodide (KI) and (b) the number of moles of iodine (I_2) that will be produced from 2.0 mol of potassium iodide.

SOLUTION

Read • **Knowns:**

$$K_2Cr_2O_7 + 6 KI + 7 H_2SO_4 \rightarrow Cr_2(SO_4)_3 + 4 K_2SO_4 + 3 I_2 + 7 H_2O$$

(a) 2.0 mol KI
(b) 2.0 mol KI

(a) **Solving for:** moles of $K_2Cr_2O_7$

Plan •   Use the Problem-Solving Strategy for Stoichiometry Problems.

Solution map: moles of KI → moles of $K_2Cr_2O_7$

The mole ratio from the balanced equation is $\dfrac{1 \text{ mol } K_2Cr_2O_7}{6 \text{ mol KI}}$.

Calculate • $(2.0 \text{ mol KI})\left(\dfrac{1 \text{ mol } K_2Cr_2O_7}{6 \text{ mol KI}}\right) = 0.33 \text{ mol } K_2Cr_2O_7$

Check •   Note that the moles KI cancel and that the answer is less than the starting 2.0 mol, which makes sense since our mole ratio is 1:6.

(b) **Solving for:** moles of I_2

Plan •   Use the Problem-Solving Strategy for Stoichiometry Problems.

Solution map: mole of KI → moles of I_2

The mole ratio from the balanced equation is $\dfrac{3 \text{ mol } I_2}{6 \text{ mol KI}}$.

Calculate • $(2.0 \text{ mol KI})\left(\dfrac{3 \text{ mol } I_2}{6 \text{ mol KI}}\right) = 1.0 \text{ mol } I_2$

Check •   Note that the moles KI cancel and that the answer is half the starting 2.0 mol, which makes sense since our mole ratio is 1:2.

EXAMPLE 9.5

How many molecules of water can be produced by reacting 0.010 mol of oxygen with hydrogen? The balanced equation is $2 H_2 + O_2 \rightarrow 2 H_2O$.

SOLUTION

Read •   **Knowns:** $0.010 \text{ mol } O_2$
$2 H_2 + O_2 \rightarrow 2 H_2O$

Solving for: molecules H_2O

Plan •   Use the Problem-Solving Strategy for Stoichiometry Problems.

Solution map: moles of O_2 → moles of H_2O → molecules of H_2O

The mole ratio from the balanced equation is $\dfrac{2 \text{ mol } H_2O}{1 \text{ mol } O_2}$.

Calculate • $(0.010 \text{ mol } O_2)\left(\dfrac{2 \text{ mol } H_2O}{1 \text{ mol } O_2}\right) = 0.020 \text{ mol } H_2O$

Conversion factor for moles H_2O → molecules H_2O

$$\dfrac{6.022 \times 10^{23} \text{ molecules}}{1 \text{ mol}}$$

$$(0.020 \text{ mol } H_2O)\left(\dfrac{6.022 \times 10^{23} \text{ molecules}}{1 \text{ mol}}\right) = 1.2 \times 10^{22} \text{ molecules } H_2O$$

NOTE   Note that 0.020 mol is still quite a large number of water molecules.

PRACTICE 9.2

How many moles of aluminum oxide will be produced from 0.50 mol of oxygen?

$$4\ Al + 3\ O_2 \longrightarrow 2\ Al_2O_3$$

PRACTICE 9.3

How many moles of aluminum hydroxide are required to produce 22.0 mol of water?

$$2\ Al(OH)_3 + 3\ H_2SO_4 \longrightarrow Al_2(SO_4)_3 + 6\ H_2O$$

9.3 MOLE–MASS CALCULATIONS

Solve problems in which mass is given and the answer is to be determined in moles or the moles are given and mass is to be determined. • **LEARNING OBJECTIVE**

The object of this type of problem is to calculate the mass of one substance that reacts with or is produced from a given number of moles of another substance in a chemical reaction. If the mass of the starting substance is given, we need to convert it to moles. We use the mole ratio to convert moles of starting substance to moles of desired substance. We can then change moles of desired substance to mass.

WileyPLUS
ENHANCED EXAMPLE

EXAMPLE 9.6

What mass of hydrogen can be produced by reacting 6.0 mol of aluminum with hydrochloric acid? The balanced equation is $2\ Al(s) + 6\ HCl(aq) \rightarrow 2\ AlCl_3(aq) + H_2(g)$.

SOLUTION

Read • **Knowns:** 6.0 mol Al

$2\ Al(s) + 6\ HCl(aq) \rightarrow 2\ AlCl_3(aq) + 3\ H_2(g)$

Solving for: mass H_2

Plan • Use the Problem-Solving Strategy for Stoichiometry Problems.

Solution map: moles of Al $\rightarrow$ moles of H_2 $\rightarrow$ grams H_2

The mole ratio from the balanced equation is $\dfrac{3\text{ mol } H_2}{2\text{ mol Al}}$.

Calculate • $(6.0\ \cancel{\text{mol Al}})\left(\dfrac{3\text{ mol } H_2}{2\ \cancel{\text{mol Al}}}\right) = 9.0\text{ mol } H_2$

Conversion factor for moles $H_2 \rightarrow$ grams H_2 $\qquad \dfrac{2.016\text{ g } H_2}{1\text{ mol } H_2}$

$$(9.0\ \cancel{\text{mol } H_2})\left(\frac{2.016\text{ g } H_2}{1\ \cancel{\text{mol } H_2}}\right) = 18\text{ g } H_2$$

NOTE You can keep the answer for the first calculation (9.0 mol H_2) on your calculator and continue the calculation to the answer of 18 g H_2 in one continuous calculation as shown below:

moles of Al $\rightarrow$ moles of H_2 $\rightarrow$ grams H_2

$$(6.0\ \cancel{\text{mol Al}})\left(\frac{3\ \cancel{\text{mol } H_2}}{2\ \cancel{\text{mol Al}}}\right)\left(\frac{2.016\text{ g } H_2}{1\ \cancel{\text{mol } H_2}}\right) = 18\text{ g } H_2$$

EXAMPLE 9.7

How many moles of water can be produced by burning 325 g of octane (C_8H_{18})? The balanced equation is $2\,C_8H_{18}(l) + 25\,O_2(g) \rightarrow 16\,CO_2(g) + 18\,H_2O(g)$.

SOLUTION

Read • **Knowns:** 325 g C_8H_{18}

$$2\,C_8H_{18}(l) + 25\,O_2(g) \rightarrow 16\,CO_2(g) + 18\,H_2O(g)$$

Solving for: mol H_2O

Plan • Use the Problem-Solving Strategy for Stoichiometry Problems.

Solution map: grams of $C_8H_{18} \rightarrow$ moles of $C_8H_{18} \rightarrow$ moles of H_2O

Conversion factor for grams $C_8H_{18} \rightarrow$ moles C_8H_{18} $\dfrac{1\ \text{mol}\ C_8H_{18}}{114.2\ \text{g}\ C_8H_{18}}$

The mole ratio from the balanced equation is $\dfrac{18\ \text{mol}\ H_2O}{2\ \text{mol}\ C_8H_{18}}$.

Calculate • $(325\ \text{g}\ \cancel{C_8H_{18}})\left(\dfrac{1\ \text{mol}\ C_8H_{18}}{114.2\ \text{g}\ \cancel{C_8H_{18}}}\right) = 2.85\ \text{mol}\ C_8H_{18}$

$(2.85\ \cancel{\text{mol}\ C_8H_{18}})\left(\dfrac{18\ \text{mol}\ H_2O}{2\ \cancel{\text{mol}\ C_8H_{18}}}\right) = 25.7\ \text{mol}\ H_2O$

Check • On your calculator in continuous calculation you might get 25.6 mol H_2O for your answer. The difference in the last digit is due to rounding off at different points in the calculation. You should check with your instructor to find out the appropriate rules for your course. In our examples we round at the end of each step to show the proper number of significant figures for each calculation.

Octane
C_8H_{18}

PRACTICE 9.4

How many moles of potassium chloride and oxygen can be produced from 100.0 g of potassium chlorate? The balanced equation is:

$$2\,KClO_3 \longrightarrow 2\,KCl + 3\,O_2$$

PRACTICE 9.5

How many grams of silver nitrate are required to produce 0.25 mol of silver sulfide? The balanced equation is:

$$2\,AgNO_3 + H_2S \longrightarrow Ag_2S + 2\,HNO_3$$

9.4 MASS–MASS CALCULATIONS

LEARNING OBJECTIVE • Solve problems in which mass is given and the answer is to be determined as mass.

Solving mass–mass stoichiometry problems requires all the steps of the mole-ratio method. The mass of starting substance is converted to moles. The mole ratio is then used to determine moles of desired substance, which, in turn, is converted to mass of desired substance.

EXAMPLE 9.8

What mass of carbon dioxide is produced by the complete combustion of 100. g of the hydrocarbon pentane, C_5H_{12}? The balanced equation is
$C_5H_{12}(l) + 8\,O_2(g) \rightarrow 5\,CO_2(g) + 6\,H_2O(g)$.

SOLUTION

Read • **Knowns:** 100. g C_5H_{12}

$C_5H_{12}(l) + 8\,O_2(g) \rightarrow 5\,CO_2(g) + 6\,H_2O(g)$

Solving for: mass CO_2

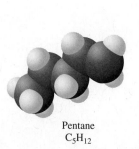

Pentane
C_5H_{12}

Plan •   Use the Problem-Solving Strategy for Stoichiometry Problems.

Solution map: grams of C_5H_{12} → moles of C_5H_{12} → moles CO_2 → g CO_2

Conversion factor for grams C_5H_{12} → moles C_5H_{12} $\dfrac{1\ \text{mol}\ C_5H_{12}}{72.15\ \text{g}\ C_5H_{12}}$

The mole ratio from the balanced equation is $\dfrac{5\ \text{mol}\ CO_2}{1\ \text{mol}\ C_5H_{12}}$.

Conversion factor for moles CO_2 → g CO_2 $\dfrac{44.01\ \text{g}\ CO_2}{1\ \text{mol}\ CO_2}$

Calculate •   grams of C_5H_{12} → moles of C_5H_{12}

$$(100.\ \text{g}\ C_5H_{12})\left(\frac{1\ \text{mol}\ C_5H_{12}}{72.15\ \text{g}\ C_5H_{12}}\right) = 1.39\ \text{mol}\ C_5H_{12}$$

moles of C_5H_{12} → moles CO_2

$$(1.39\ \text{mol}\ C_5H_{12})\left(\frac{5\ \text{mol}\ CO_2}{1\ \text{mol}\ C_5H_{12}}\right) = 6.95\ \text{mol}\ CO_2$$

moles of CO_2 → g CO_2

$$(6.95\ \text{mol}\ CO_2)\left(\frac{44.01\ \text{g}\ CO_2}{1\ \text{mol}\ CO_2}\right) = 306\ \text{g}\ CO_2$$

EXAMPLE 9.9

How many grams of nitric acid (HNO_3) are required to produce 8.75 g of dinitrogen monoxide (N_2O) according to the following equation?

$$4\ Zn(s) + 10\ HNO_3(aq) \rightarrow 4\ Zn(NO_3)_2(aq) + N_2O(g) + 5\ H_2O(l)$$

SOLUTION

Read •   **Knowns:**   8.75 g N_2O

$$4\ Zn(s) + 10\ HNO_3(aq) \rightarrow 4\ Zn(NO_3)_2(aq) + N_2O(g) + 5\ H_2O(l)$$

Solving for: g HNO_3

Plan •   Use the Problem-Solving Strategy for Stoichiometry Problems.

Solution map: grams of N_2O → moles N_2O → moles HNO_3 → g HNO_3

Conversion factor for grams N_2O → moles N_2O $\dfrac{1\ \text{mol}\ N_2O}{44.02\ \text{g}\ N_2O}$

The mole ratio from the balanced equation is $\dfrac{10\ \text{mol}\ HNO_3}{1\ \text{mol}\ N_2O}$.

Conversion factor for moles HNO_3 → g HNO_3 $\dfrac{63.02\ \text{g}\ HNO_3}{1\ \text{mol}\ HNO_3}$

Calculate •   grams of N_2O → moles of N_2O

$$(8.75\ \text{g}\ N_2O)\left(\frac{1\ \text{mol}\ N_2O}{44.02\ \text{g}\ N_2O}\right) = 0.199\ \text{mol}\ N_2O$$

moles N_2O → moles HNO_3

$$(0.199\ \text{mol}\ N_2O)\left(\frac{10\ \text{mol}\ HNO_3}{1\ \text{mol}\ N_2O}\right) = 1.99\ \text{mol}\ HNO_3$$

moles HNO_3 → g HNO_3

$$(1.99\ \text{mol}\ HNO_3)\left(\frac{63.02\ \text{g}\ HNO_3}{1\ \text{mol}\ HNO_3}\right) = 125\ \text{g}\ HNO_3$$

9.5 LIMITING REACTANT AND YIELD CALCULATIONS

LEARNING OBJECTIVE ● Solve problems involving limiting reactants and yield.

KEY TERMS

limiting reactant
theoretical yield
actual yield
percent yield

In many chemical processes, the quantities of the reactants used are such that one reactant is in excess. The amount of the product(s) formed in such a case depends on the reactant that is not in excess. This reactant is called the **limiting reactant**—it limits the amount of product that can be formed.

Consider the case illustrated in **Figure 9.2**. How many bicycles can be assembled from the parts shown? The limiting part in this case is the number of pedal assemblies; only three bicycles can be built because there are only three pedal assemblies. The wheels and frames are parts in excess.

Let's consider a chemical example at the molecular level in which seven molecules of H_2 are combined with four molecules of Cl_2 (**Figure 9.3** before reaction). How many molecules of HCl can be produced according to this reaction?

$$H_2 + Cl_2 \longrightarrow 2\,HCl$$

Figure 9.2

The number of bicycles that can be built from these parts is determined by the "limiting reactant" (the pedal assemblies).

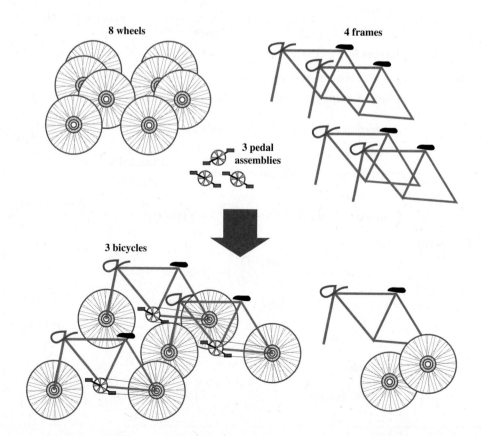

8 wheels   4 frames

3 pedal assemblies

3 bicycles

>CHEMISTRY *IN ACTION*

A Shrinking Technology

The microchip has revolutionized the field of electronics. Engineers at Bell Laboratories, Massachusetts Institute of Technology, the University of California, and Stanford University are racing to produce parts for tiny machines and robots. New techniques now produce gears smaller than a grain of sand and motors lighter than a speck of dust.

To produce ever smaller computers, calculators, and even microbots (microsized robots), precise quantities of chemicals in exact proportions are required. The secret to producing minute circuits is to print the entire circuit or blueprint at one time. Computers are used to draw a chip. This image is then transferred onto a pattern, or mask, with details finer than a human hair. In a process similar to photography, light is then shined through the mask onto a silicon-coated surface. The areas created on the silicon exhibit high or low resistance to chemical etching. Chemicals then etch away the silicon.

Micromachinery is produced in the same way. First a thin layer of silicon dioxide is applied (sacrificial material); then a layer of polysilicon is carefully applied (structural material). Next, a mask is applied, and the whole structure is covered with plasma (excited gas). The plasma acts as a tiny sandblaster, removing everything the mask doesn't protect. This process is repeated as the entire machine is constructed. When the entire assembly is complete, the whole machine is placed in hydrofluoric acid, which dissolves all the sacrificial material and permits the various parts of the machine to move.

To turn these micromachines into true microbots, current research is focusing on locomotion and sensing imaging systems. Possible uses for these microbots include "smart" pills, which could contain sensors or drug reservoirs (currently used in birth control). Tiny pumps, once inside the body, will dispense the proper amount of medication at precisely the correct site. These microbots are currently in production for the treatment of diabetes (to release insulin).

These nanorobots attach themselves to red blood cells.

If the molecules of H_2 and Cl_2 are taken apart and recombined as HCl (Figure 9.3 after reaction), we see that eight molecules of HCl can be formed before we run out of Cl_2. Therefore, the Cl_2 is the limiting reactant and H_2 is in excess—three molecules of H_2 remain unreacted.

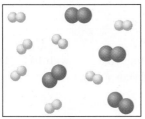

Before reaction

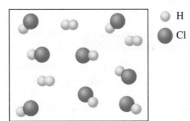

After reaction

○ H
● Cl

Figure 9.3

Reaction of H_2 with Cl_2. Eight molecules of HCl are formed, and three molecules of H_2 remain. The limiting reactant is Cl_2.

When problem statements give the amounts of two reactants, one of them is usually a limiting reactant. We can identify the limiting reactant using the following method:

PROBLEM-SOLVING STRATEGY: For Limiting Reactant Problems

1. Calculate the amount of product (moles or grams, as needed) formed from each reactant.

2. Determine which reactant is limiting. (The reactant that gives the least amount of product is the limiting reactant; the other reactant is in excess.)

3. Once we know the limiting reactant, the amount of product formed can be determined. It is the amount determined by the limiting reactant.

4. If we need to know how much of the other reactant remains, we calculate the amount of the other reactant required to react with the limiting reactant, then subtract this amount from the starting quantity of the reactant. This gives the amount of that substance that remains unreacted.

EXAMPLE 9.10

How many moles of HCl can be produced by reacting 4.0 mol H_2 and 3.5 mol Cl_2? Which compound is the limiting reactant? The equation is $H_2(g) + Cl_2(g) \rightarrow 2\ HCl(g)$.

SOLUTION

Read •   **Knowns:**   4.0 mol H_2

3.5 mol Cl_2

$H_2(g) + Cl_2(g) \rightarrow 2\ HCl(g)$

Solving for: mol HCl

Plan •   Use the Problem-Solving Strategy for Limiting Reactant Problems.

Calculate the mol HCl formed from each reactant.

The mole ratios from the balanced equation are

$$\frac{2\ \text{mol HCl}}{1\ \text{mol}\ H_2} \qquad \frac{2\ \text{mol HCl}}{1\ \text{mol}\ Cl_2}$$

Calculate •   moles $H_2 \rightarrow$ moles HCl

$$(4.0\ \text{mol}\ H_2)\left(\frac{2\ \text{mol HCl}}{1\ \text{mol}\ H_2}\right) = 8.0\ \text{mol HCl}$$

moles of $Cl_2^- \rightarrow$ moles HCl

$$(3.5\ \text{mol}\ Cl_2)\left(\frac{2\ \text{mol HCl}}{1\ \text{mol}\ Cl_2}\right) = 7.0\ \text{mol HCl}$$

Determine the limiting reactant. The limiting reactant is Cl_2 because it produces less HCl than H_2. The H_2 is in excess. The yield of HCl is 7.0 mol.

Check •   In this reaction we can decide on the limiting reactant by inspection. From the equation you can see that 1 mol H_2 reacts with 1 mol Cl_2. Therefore, when we react 4.0 mol H_2 with 3.5 mol Cl_2, Cl_2 is the reactant that limits the amount of HCl produced, since it is present in smaller amount.

EXAMPLE 9.11

How many moles of Fe_3O_4 can be obtained by reacting 16.8 g Fe with 10.0 g H_2O? Which substance is the limiting reactant? Which substance is in excess?

$$3\ Fe(s) + 4\ H_2O(g) \xrightarrow{\Delta} Fe_3O_4(s) + 4\ H_2(g)$$

SOLUTION

Read •   **Knowns:**   16.5 g Fe

10.0 g H_2O

$3\ Fe(s) + 4\ H_2O(g) \xrightarrow{\Delta} Fe_3O_4(s) + 4\ H_2(g)$

Solving for: mol Fe_3O_4

Plan •   Use the Problem-Solving Strategy for Limiting Reactant Problems.

Calculate the mol Fe_3O_4 formed from each reactant:

g reactant $\rightarrow$ mol reactant $\rightarrow$ mol Fe_3O_4

Calculate •   For Fe

$$(16.8\ \text{g Fe})\left(\frac{1\ \text{mol Fe}}{55.85\ \text{g Fe}}\right)\left(\frac{1\ \text{mol}\ Fe_3O_4}{3\ \text{mol Fe}}\right) = 0.100\ \text{mol}\ Fe_3O_4$$

For H_2O

$$(10.0 \text{ g } H_2O)\left(\frac{1 \text{ mol } H_2O}{18.02 \text{ g } H_2O}\right)\left(\frac{1 \text{ mol } Fe_3O_4}{4 \text{ mol } H_2O}\right) = 0.139 \text{ mol } Fe_3O_4$$

Determine the limiting reactant. The limiting reactant is Fe because it produces less Fe_3O_4 than H_2O. The H_2O is in excess. The yield of Fe_3O_4 is 0.100 mol.

EXAMPLE 9.12

How many grams of silver bromide (AgBr) can be formed when solutions containing 50.0 g of $MgBr_2$ and 100.0 g of $AgNO_3$ are mixed together? How many grams of excess reactant remain unreacted?

SOLUTION

Read •   **Knowns:**   50.0 g $MgBr_2$

                             100.0 g $AgNO_3$

          $MgBr_2(aq) + 2 \, AgNO_3(aq) \rightarrow 2 \, AgBr(s) + Mg(NO_3)_2(aq)$

          Solving for: g AgBr

Plan •   Use the Problem-Solving Strategy for Limiting Reactant Problems.

          Calculate the g AgBr formed from each reactant:

              g reactant $\rightarrow$ mol reactant $\rightarrow$ mol AgBr $\rightarrow$ g AgBr

Calculate •   For $MgBr_2$

$$(50.0 \text{ g } MgBr_2)\left(\frac{1 \text{ mol } MgBr_2}{184.1 \text{ g } MgBr_2}\right)\left(\frac{2 \text{ mol } AgBr}{1 \text{ mol } MgBr_2}\right)\left(\frac{187.8 \text{ g } AgBr}{1 \text{ mol } AgBr}\right)$$

        $= 102 \text{ g } AgBr$

        For $AgNO_3$

$$(100.0 \text{ g } AgNO_3)\left(\frac{1 \text{ mol } AgNO_3}{169.9 \text{ g } AgNO_3}\right)\left(\frac{2 \text{ mol } AgBr}{2 \text{ mol } AgNO_3}\right)\left(\frac{187.8 \text{ g } AgBr}{1 \text{ mol } AgBr}\right)$$

        $= 110.5 \text{ g } AgBr$

Determine the limiting reactant. The limiting reactant is $MgBr_2$ because it produces less AgBr than $AgNO_3$. The $AgNO_3$ is in excess. The yield of AgBr is 102 g.

Calculate the mass of unreacted $AgNO_3$. We must first determine the g $AgNO_3$ that react with 50.0 g $MgBr_2$.

        g $MgBr_2$ $\rightarrow$ mol $MgBr_2$ $\rightarrow$ mol $AgNO_3$ $\rightarrow$ g $AgNO_3$

$$(50.0 \text{ g } MgBr_2)\left(\frac{1 \text{ mol } MgBr_2}{184.1 \text{ g } MgBr_2}\right)\left(\frac{2 \text{ mol } AgNO_3}{1 \text{ mol } MgBr_2}\right)\left(\frac{169.9 \text{ g } AgNO_3}{1 \text{ mol } AgNO_3}\right)$$

        $= 92.3 \text{ g } AgNO_3$

Unreacted $AgNO_3 = 100.0 \text{ g } AgNO_3 - 92.3 \text{ g } AgNO_3 = 7.7 \text{ g } AgNO_3$

The final mixture will contain:

        102 g $AgBr(s)$ 7.7 g $AgNO_3$

PRACTICE 9.8

How many grams of hydrogen chloride can be produced from 0.490 g of hydrogen and 50.0 g of chlorine? The balanced equation is:

$$H_2(g) + Cl_2(g) \longrightarrow 2 \, HCl(g)$$

PRACTICE 9.9

How many grams of barium sulfate will be formed from 200.0 g of barium nitrate and 100.0 g of sodium sulfate? The balanced equation is:

$$Ba(NO_3)_2(aq) + Na_2SO_4(aq) \longrightarrow BaSO_4(s) + 2\,NaNO_3(aq)$$

The quantities of the products we have been calculating from chemical equations represent the maximum yield (100%) of product according to the reaction represented by the equation. Many reactions, especially those involving organic substances, fail to give a 100% yield of product. The main reasons for this failure are the side reactions that give products other than the main product and the fact that many reactions are reversible. In addition, some product may be lost in handling and transferring from one vessel to another. The **theoretical yield** of a reaction is the calculated amount of product that can be obtained from a given amount of reactant, according to the chemical equation. The **actual yield** is the amount of product that we finally obtain.

The **percent yield** is the ratio of the actual yield to the theoretical yield multiplied by 100. Both the theoretical and the actual yields must have the same units to obtain a percent:

$$\frac{\text{actual yield}}{\text{theoretical yield}} \times 100 = \text{percent yield}$$

For example, if the theoretical yield calculated for a reaction is 14.8 g, and the amount of product obtained is 9.25 g, the percent yield is

Round off as appropriate for your particular course.

$$\text{percent yield} = \left(\frac{9.25\ \cancel{g}}{14.8\ \cancel{g}}\right)(100) = 62.5\%\ \text{yield}$$

WileyPLUS

ENHANCED EXAMPLE

EXAMPLE 9.13

Carbon tetrachloride (CCl_4) was prepared by reacting 100. g of carbon disulfide and 100. g of chlorine. Calculate the percent yield if 65.0 g of CCl_4 were obtained from the reaction $CS_2 + 3\,Cl_2 \rightarrow CCl_4 + S_2Cl_2$.

SOLUTION

Read •   **Knowns:**   100. g CS_2

100. g Cl_2

65.0 g CCl_4 obtained

$$CS_2 + 3\,Cl_2 \rightarrow CCl_4 + S_2Cl_2$$

Solving for: % yield of CCl_4

Plan •   Use the equation for percent yield

$$\frac{\text{actual yield}}{\text{theoretical yield}} \times 100 = \text{percent yield}$$

Determine the theoretical yield. Calculate the g CCl_4 formed from each reactant.

$$\text{g reactant} \rightarrow \text{mol reactant} \rightarrow \text{mol } CCl_4 \rightarrow \text{g } CCl_4$$

Calculate •   For CS_2

$$(100.\ \text{g } CS_2)\left(\frac{1\ \text{mol } \cancel{CS_2}}{76.15\ \text{g } \cancel{CS_2}}\right)\left(\frac{1\ \text{mol } \cancel{CCl_4}}{1\ \text{mol } \cancel{CS_2}}\right)\left(\frac{153.8\ \text{g } CCl_4}{1\ \text{mol } \cancel{CCl_4}}\right)$$

$$= 202\ \text{g } CCl_4$$

For Cl_2

$$(100. \text{ g } Cl_2)\left(\frac{1 \text{ mol } Cl_2}{70.90 \text{ g } Cl_2}\right)\left(\frac{1 \text{ mol } CCl_4}{3 \text{ mol } Cl_2}\right)\left(\frac{153.8 \text{ g } CCl_4}{1 \text{ mol } CCl_4}\right)$$

$$= 72.3 \text{ g } CCl_4$$

Determine the limiting reactant. The limiting reactant is Cl_2 because it produces less CCl_4 than CS_2. The CS_2 is in excess. The theoretical yield of CCl_4 is 72.3 g.

Calculate the percent yield

$$\frac{\text{actual yield}}{\text{theoretical yield}} \times 100 = \left(\frac{65.0 \text{ g}}{72.3 \text{ g}}\right)(100) = 89.9\% \text{ yield}$$

EXAMPLE 9.14

Silver bromide was prepared by reacting 200.0 g of magnesium bromide and an adequate amount of silver nitrate. Calculate the percent yield if 375.0 g of silver bromide were obtained from the reaction:

$$MgBr_2 + 2 AgNO_3 \rightarrow Mg(NO_3)_2 + 2 AgBr$$

SOLUTION

Read •   **Knowns:**   200.0 g $MgBr_2$

375.0 g AgBr

$MgBr_2 + 2 AgNO_3 \rightarrow Mg(NO_3)_2 + 2 AgBr$

Solving for: % yield AgBr

Plan •   Use the equation for percent yield

$$\frac{\text{actual yield}}{\text{theoretical yield}} \times 100 = \text{percent yield}$$

Determine the theoretical yield. Calculate the grams of AgBr formed:

$$\text{g } MgBr_2 \rightarrow \text{mol } MgBr_2 \rightarrow \text{mol } AgBr \rightarrow \text{g } AgBr$$

Calculate • $(200.0 \text{ g } MgBr_2)\left(\frac{1 \text{ mol } MgBr_2}{184.1 \text{ g } MgBr_2}\right)\left(\frac{2 \text{ mol } AgBr}{1 \text{ mol } MgBr_2}\right)\left(\frac{187.8 \text{ g } AgBr}{1 \text{ mol } AgBr}\right)$

$$= 408.0 \text{ g } AgBr$$

The theoretical yield is 408.0 g AgBr.

Calculate the percent yield.

$$\frac{\text{actual yield}}{\text{theoretical yield}} \times 100 = \left(\frac{375.0 \text{ g}}{408.0 \text{ g}}\right)(100) = 91.91\% \text{ yield}$$

PRACTICE 9.10

Aluminum oxide was prepared by heating 225 g of chromium(II) oxide with 125 g of aluminum. Calculate the percent yield if 100.0 g of aluminum oxide were obtained. The balanced equation is:

$$2 Al + 3 CrO \longrightarrow Al_2O_3 + 3 Cr$$

CHAPTER 9 REVIEW

9.1 INTRODUCTION TO STOICHIOMETRY

- Solving stoichiometry problems requires the use of moles and mole ratios.
- To solve a stoichiometry problem:
  - Convert the quantity of starting substance to moles.
  - Convert the moles of starting substance to moles of desired substance.
  - Convert the moles of desired substance to the appropriate unit.

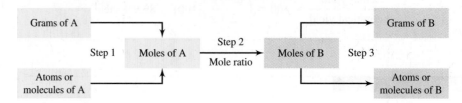

9.2 MOLE–MOLE CALCULATIONS

- If the starting substance is given in moles and the unit specified for the answer is moles:
  - Convert the moles of starting substance to moles of desired substance.

9.3 MOLE–MASS CALCULATIONS

- If the starting substance is given in moles and the unit specified for the answer is grams:
  - Convert the moles of starting substance to moles of desired substance.
  - Convert the moles of desired substance to grams.

9.4 MASS–MASS CALCULATIONS

- If the starting substance is given in grams and the unit specified for the answer is grams:
  - Convert the quantity of starting substance to moles.
  - Convert the moles of starting substance to moles of desired substance.
  - Convert the moles of desired substance to grams.

9.5 LIMITING REACTANT AND YIELD CALCULATIONS

- To identify the limiting reactant in a reaction:
  - Calculate the amount of product formed from each reactant.
  - Determine the limiting reactant by selecting the one that gives the least amount of product.
- To determine the actual amount of product formed in a limiting reactant situation:
  - Use the result calculated that is the least amount of product.
- To determine the amount of the other reactants required to react with the limiting reactant:
  - Calculate the amount of the other reactant needed to react with the limiting reactant.
  - Subtract this amount from the original amount of the other reactant to find the amount of excess reactant (unreacted).
- The theoretical yield of a chemical reaction is the calculated amount of product from a given amount of reactant (the limiting reactant).
- The actual yield is the amount of product actually obtained experimentally.
- The percent yield is

$$\frac{\text{actual yield}}{\text{theoretical yield}} \times 100 = \text{percent yield}$$

REVIEW QUESTIONS

1. What is a mole ratio?
2. What piece of information is needed to convert grams of a compound to moles of the same compound?
3. Phosphine (PH_3) can be prepared by the reaction of calcium phosphide, Ca_3P_2:

$$Ca_3P_2 + 6 H_2O \longrightarrow 3 Ca(OH)_2 + 2 PH_3$$

Based on this equation, which of the following statements are correct? Show evidence to support your answer.
(a) One mole of Ca_3P_2 produces 2 mol of PH_3.
(b) One gram of Ca_3P_2 produces 2 g of PH_3.
(c) Three moles of $Ca(OH)_2$ are produced for each 2 mol of PH_3 produced.
(d) The mole ratio between phosphine and calcium phosphide is

$$\frac{2 \text{ mol } PH_3}{1 \text{ mol } Ca_3P_2}$$

(e) When 2.0 mol of Ca_3P_2 and 3.0 mol of H_2O react, 4.0 mol of PH_3 can be formed.
(f) When 2.0 mol of Ca_3P_2 and 15.0 mol of H_2O react, 6.0 mol of $Ca(OH)_2$ can be formed.
(g) When 200. g of Ca_3P_2 and 100. g of H_2O react, Ca_3P_2 is the limiting reactant.
(h) When 200. g of Ca_3P_2 and 100. g of H_2O react, the theoretical yield of PH_3 is 57.4 g.

4. The equation representing the reaction used for the commercial preparation of hydrogen cyanide is

$$2 CH_4 + 3 O_2 + 2 NH_3 \longrightarrow 2 HCN + 6 H_2O$$

Based on this equation, which of the following statements are correct? Rewrite incorrect statements to make them correct.
(a) Three moles of O_2 are required for 2 mol of NH_3.
(b) Twelve moles of HCN are produced for every 16 mol of O_2 that react.
(c) The mole ratio between H_2O and CH_4 is

$$\frac{6 \text{ mol } H_2O}{2 \text{ mol } CH_4}$$

(d) When 12 mol of HCN are produced, 4 mol of H_2O will be formed.
(e) When 10 mol of CH_4, 10 mol of O_2, and 10 mol of NH_3 are mixed and reacted, O_2 is the limiting reactant.
(f) When 3 mol each of CH_4, O_2, and NH_3 are mixed and reacted, 3 mol of HCN will be produced.

5. What information beyond the mole ratio is needed to convert from moles of reactant to grams of product?
6. Draw a flowchart showing the conversions necessary to determine grams of ammonia formed from 10 g of hydrogen gas.

$$3 H_2 + 2 N_2 \longrightarrow 2 NH_3$$

7. What is the difference between the theoretical and the actual yield of a chemical reaction?
8. How is the percent yield of a reaction calculated?

Most of the exercises in this chapter are available for assignment via the online homework management program, WileyPLUS (www.wileyplus.com). All exercises with blue numbers have answers in Appendix VI.

PAIRED EXERCISES

1. Calculate the number of moles in these quantities:
   (a) 25.0 g KNO_3
   (b) 56 millimol NaOH
   (c) 5.4×10^2 g $(NH_4)_2C_2O_4$
   (d) 16.8 mL H_2SO_4 solution
       ($d = 1.727$ g/mL, 80.0% H_2SO_4 by mass)
3. Calculate the number of grams in these quantities:
   (a) 2.55 mol $Fe(OH)_3$
   (b) 125 kg $CaCO_3$
   (c) 10.5 mol NH_3
   (d) 72 millimol HCl
   (e) 500.0 mL of liquid Br_2 ($d = 3.119$ g/mL)

5. Which contains the larger number of molecules, 10.0 g H_2O or 10.0 g H_2O_2? Show evidence for your answer.

7. Balance the equation for the synthesis of sucrose

   $$CO_2 + H_2O \longrightarrow C_{12}H_{22}O_{11} + O_2$$

   and set up the mole ratio of

   (a) CO_2 to H_2O (d) $C_{12}H_{22}O_{11}$ to CO_2
   (b) H_2O to $C_{12}H_{22}O_{11}$ (e) H_2O to O_2
   (c) O_2 to CO_2 (f) O_2 to $C_{12}H_{22}O_{11}$

2. Calculate the number of moles in these quantities:
   (a) 2.10 kg $NaHCO_3$
   (b) 525 mg $ZnCl_2$
   (c) 9.8×10^{24} molecules CO_2
   (d) 250 mL ethyl alcohol, C_2H_5OH ($d = 0.789$ g/mL)
4. Calculate the number of grams in these quantities:
   (a) 0.00844 mol $NiSO_4$
   (b) 0.0600 mol $HC_2H_3O_2$
   (c) 0.725 mol Bi_2S_3
   (d) 4.50×10^{21} molecules glucose, $C_6H_{12}O_6$
   (e) 75 mL K_2CrO_4 solution
       ($d = 1.175$ g/mL, 20.0% K_2CrO_4 by mass)

6. Which contains the larger numbers of molecules, 25.0 g HCl or 85.0 g $C_6H_{12}O_6$? Show evidence for your answer.

8. Balance the equation for the combustion of butanol

   $$C_4H_9OH + O_2 \longrightarrow CO_2 + H_2O$$

   and set up the mole ratio of

   (a) O_2 to C_4H_9OH (d) C_4H_9OH to CO_2
   (b) H_2O to O_2 (e) H_2O to C_4H_9OH
   (c) CO_2 to H_2O (f) CO_2 to O_2

9. Given the unbalanced equation

$$CO_2 + H_2 \longrightarrow CH_4 + H_2O$$

(a) How many moles of water can be produced from 25 moles of carbon dioxide?
(b) How many moles of CH_4 will be produced along with 12 moles of water?

11. Given the equation

$$MnO_2(s) + HCl(aq) \longrightarrow Cl_2(g) + MnCl_2(aq) + H_2O(l)$$
(unbalanced)

(a) How many moles of HCl will react with 1.05 mol of MnO_2?
(b) How many moles of $MnCl_2$ will be produced when 1.25 mol of H_2O are formed?
(c) How many grams of Cl_2 will be produced when 3.28 mol of MnO_2 react with excess HCl?
(d) How many moles of HCl are required to produce 15.0 kg of $MnCl_2$?

13. Carbonates react with acids to form a salt, water, and carbon dioxide gas. When 50.0 g of calcium carbonate are reacted with sufficient hydrochloric acid, how many grams of calcium chloride will be produced? (Write a balanced equation first.)

15. In a blast furnace, iron(III) oxide reacts with coke (carbon) to produce molten iron and carbon monoxide:

$$Fe_2O_3 + 3\,C \xrightarrow{\Delta} 2\,Fe + 3\,CO$$

How many kilograms of iron would be formed from 125 kg of Fe_2O_3?

17. Polychlorinated biphenyls (PCBs), which were formerly used in the manufacture of electrical transformers, are environmental and health hazards. They break down very slowly in the environment. The decomposition of PCBs can be represented by the equation

$$2\,C_{12}H_4Cl_6 + 23\,O_2 + 2\,H_2O \longrightarrow 24\,CO_2 + 12\,HCl$$

(a) How many moles of water are needed to react 10.0 mol O_2?
(b) How many grams of HCl are produced when 15.2 mol H_2O react?
(c) How many moles of CO_2 are produced when 76.5 g HCl are produced?
(d) How many grams of $C_{12}H_4Cl_6$ are reacted when 100.25 g CO_2 are produced?
(e) How many grams of HCl are produced when 2.5 kg $C_{12}H_4Cl_6$ react?

19. Draw the molecules of product(s) formed for each of the following reactions; then determine which substance is the limiting reactant:

(a)

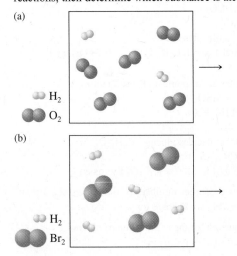

H₂
O₂

(b)

H₂
Br₂

21. Draw pictures similar to those in Question 19 and determine which is the limiting reactant in the following reactions:
(a) Eight atoms of potassium react with five molecules of chlorine to produce potassium chloride.
(b) Ten atoms of aluminum react with three molecules of oxygen to produce aluminum oxide.

10. Given the unbalanced equation

$$H_2SO_4 + NaOH \longrightarrow Na_2SO_4 + H_2O$$

(a) How many moles of NaOH will react with 17 mol of H_2SO_4?
(b) How many moles of Na_2SO_4 will be produced when 21 mol of NaOH react?

12. Given the equation

$$Al_4C_3 + 12\,H_2O \longrightarrow 4\,Al(OH)_3 + 3\,CH_4$$

(a) How many moles of water are needed to react with 100. g of Al_4C_3?
(b) How many moles of $Al(OH)_3$ will be produced when 0.600 mol of CH_4 is formed?
(c) How many moles of CH_4 will be formed by the reaction of 275 grams of Al_4C_3 with excess water?
(d) How many grams of water are required to produce 4.22 moles of $Al(OH)_3$?

14. Certain metals can displace hydrogen from acids to produce hydrogen gas and a salt. When 25.2 g of aluminum metal are placed into hydrobromic acid, how many grams of aluminum bromide will be produced? (Write the balanced equation first.)

16. How many grams of steam and iron must react to produce 375 g of magnetic iron oxide, Fe_3O_4?

$$3\,Fe(s) + 4\,H_2O(g) \xrightarrow{\Delta} Fe_3O_4(s) + 4\,H_2(g)$$

18. Mercury can be isolated from its ore by the reaction

$$4\,HgS + 4\,CaO \longrightarrow 4\,Hg + 3\,CaS + CaSO_4$$

(a) How many moles of CaS are produced from 2.5 mol CaO?
(b) How many grams of Hg are produced along with 9.75 mol of $CaSO_4$?
(c) How many moles of CaO are required to react with 97.25 g of HgS?
(d) How many grams of Hg are produced from 87.6 g of HgS?
(e) How many grams of CaS are produced along with 9.25 kg of Hg?

20. Draw the molecules of product(s) formed for the following reactions; then determine which substance is the limiting reactant:

(a)

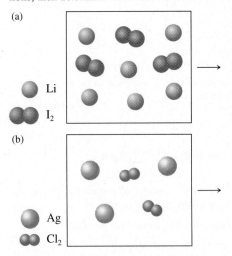

Li
I₂

(b)

Ag
Cl₂

22. Draw pictures similar to those in Question 20 and determine which is the limiting reactant in the following reactions:
(a) Eight molecules of nitrogen react with six molecules of oxygen to produce nitrogen dioxide.
(b) Fifteen atoms of iron react with twelve molecules of water to produce magnetic iron oxide (Fe_3O_4) and hydrogen.

23. In the following equations, determine which reactant is the limiting reactant and which reactant is in excess. The amounts used are given below each reactant. Show evidence for your answers.
 (a) $KOH + HNO_3 \longrightarrow KNO_3 + H_2O$
     16.0 g   12.0 g
 (b) $2\,NaOH + H_2SO_4 \longrightarrow Na_2SO_4 + H_2O$
     10.0 g   10.0 g

24. In the following equations, determine which reactant is the limiting reactant and which reactant is in excess. The amounts used are given below each reactant. Show evidence for your answers.
 (a) $2\,Bi(NO_3)_3 + 3\,H_2S \longrightarrow Bi_2S_3 + 6\,HNO_3$
     50.0 g   6.00 g
 (b) $3\,Fe + 4\,H_2O \longrightarrow Fe_3O_4 + 4\,H_2$
     40.0 g 16.0 g

25. Aluminum sulfate has many uses in industry. It can be prepared by reacting aluminum hydroxide and sulfuric acid:

$$2\,Al(OH)_3 + 3\,H_2SO_4 \longrightarrow Al_2(SO_4)_3 + 6\,H_2O$$

 (a) When 35.0 g of $Al(OH)_3$ and 35.0 g of H_2SO_4 are reacted, how many moles of $Al_2(SO_4)_3$ can be produced? Which substance is the limiting reactant?
 (b) When 45.0 g of H_2SO_4 and 25.0 g of $Al(OH)_3$ are reacted, how many grams of $Al_2(SO_4)_3$ are produced? Which reactant is in excess?
 (c) In a reaction vessel, 2.5 mol of $Al(OH)_3$ and 5.5 mol of H_2SO_4 react. Determine which substances will be present in the container when the reaction goes to completion and how many moles of each.

26. Phosphorous trichloride, a precursor for the manufacture of certain pesticides and herbicides, can be produced by the reaction of phosphorous and chlorine:

$$P_4 + 6\,Cl_2 \longrightarrow 4\,PCl_3$$

 (a) When 20.5 g of P_4 and 20.5 g of Cl_2 are reacted, how many moles of PCl_3 can be produced?
 (b) When 55 g of Cl_2 are reacted with 25 g of P_4, how many grams of PCl_3 can be produced? Which reactant is in excess?
 (c) In a reaction vessel, 15 mol of P_4 and 35 mol of Cl_2 react. Determine which substances will be present in the container after the reaction goes to completion and how many moles of each substance is present.

27. When a certain nonmetal whose formula is X_8 burns in air, XO_3 forms. Write a balanced equation for this reaction. If 120.0 g of oxygen gas are consumed completely, along with 80.0 g of X_8, identify the element X.

28. When a particular metal X reacts with HCl, the resulting products are XCl_2 and H_2. Write and balance the equation. When 78.5 g of the metal react completely, 2.42 g of hydrogen gas result. Identify the element X.

29. Elemental silicon can be produced by the reduction of silicon dioxide, or sand, with carbon:

$$SiO_2 + 2\,C \longrightarrow Si + 2\,CO$$

When 35.0 kg of SiO_2 reacts with 25.3 kg of carbon, and 14.4 kg of silicon is recovered, what is the percent yield for the reaction?

30. Chromium metal can be produced by the reduction of Cr_2O_3 with elemental silicon:

$$2\,Cr_2O_3 + 3\,Si \longrightarrow 4\,Cr + 3\,SiO_2$$

If 350.0 grams of Cr_2O_3 are reacted with 235.0 grams of elemental silicon and 213.2 grams of chromium metal are recovered, what is the percent yield?

31. In a laboratory experiment, 27.5 g of Cu were reacted with 125 g of HNO_3:

$$3\,Cu + 8\,HNO_3 \longrightarrow 3\,Cu(NO_3)_2 + 4\,H_2O + 2\,NO$$

 (a) How many grams of $Cu(NO_3)_2$ can be theoretically produced?
 (b) How many grams of the excess reactant remain?
 (c) If the percent yield of $Cu(NO_3)_2$ is 87.3%, what is the actual yield in grams?

32. In aqueous solution, metal oxides can react with acids to form a salt and water:

$$Fe_2O_3(s) + 6\,HCl(aq) \longrightarrow 2\,FeCl_3(aq) + 3\,H_2O(l)$$

 (a) How many moles of each product will be formed when 35 g of Fe_2O_3 react with 35 g of HCl?
 (b) From part (a), how many grams of the excess reactant remain?
 (c) If the percent yield of $FeCl_3$ is 92.5%, what is the actual yield in grams?

ADDITIONAL EXERCISES

33. A tool set contains 6 wrenches, 4 screwdrivers, and 2 pliers. The manufacturer has 1000 pliers, 2000 screwdrivers, and 3000 wrenches in stock. Can an order for 600 tool sets be filled? Explain briefly.
34. Write a balanced equation for the fermentation of glucose, $C_6H_{12}O_6$, into ethyl alcohol, C_2H_5OH and CO_2. Determine how many liters of ethyl alcohol ($d = 0.789$ g/mL) can be produced from 575 lb of glucose.
35. For mass–mass calculation, why is it necessary to convert the grams of starting material to moles of starting material, then determine the moles of product from the moles of starting material, and convert the moles of product to grams of product? Why can't you calculate the grams of product directly from the grams of starting material?
36. Oxygen masks for producing O_2 in emergency situations contain potassium superoxide (KO_2). It reacts according to this equation:

$$4\,KO_2 + 2\,H_2O + 4\,CO_2 \longrightarrow 4\,KHCO_3 + 3\,O_2$$

 (a) If a person wearing such a mask exhales 0.85 g of CO_2 every minute, how many moles of KO_2 are consumed in 10.0 minutes?
 (b) How many grams of oxygen are produced in 1.0 hour?

37. In the presence of concentrated H_2SO_4, sucrose, $C_{12}H_{22}O_{11}$, undergoes dehydration, forming carbon and water:

$$C_{12}H_{22}O_{11} \xrightarrow{\;H_2SO_4\;} 12\,C + 11\,H_2O$$

 (a) How many grams of carbon and how many grams of water can be produced from 2.0 lb of sucrose?
 (b) If 25.2 g sucrose are dehydrated, how many grams of liquid water with a density of 0.994 g/mL can be collected?

38. Alfred Nobel extended Sobrero's work on nitroglycerin and discovered that when mixed with diatomaceous earth it would form a stable mixture. This mixture is now known as dynamite. The reaction of nitroglycerin is

$$4\,C_3H_5(NO_3)_3(l) \longrightarrow 12\,CO_2(g) + 10\,H_2O(g) + 6\,N_2(g) + O_2(g)$$

 (a) Calculate the number of grams of each product that are produced by the decomposition of 45.0 g of nitroglycerin.
 (b) Calculate the total number of moles of gas produced.

39. The methyl alcohol (CH_3OH) used in alcohol burners combines with oxygen gas to form carbon dioxide and water. How many grams of oxygen are required to burn 60.0 mL of methyl alcohol ($d = 0.787$ g/mL)?

40. Hydrazine (N_2H_4) and hydrogen peroxide (H_2O_2) have been used as rocket propellants. They react according to the following equation:

$$7\ H_2O_2 + N_2H_4 \longrightarrow 2\ HNO_3 + 8\ H_2O$$

(a) When 75 kg of hydrazine react, how many grams of nitric acid can be formed?

(b) When 250 L of hydrogen peroxide ($d = 1.41$ g/mL) react, how many grams of water can be formed?

(c) How many grams of hydrazine will be required to react with 725 g hydrogen peroxide?

(d) How many grams of water can be produced when 750 g hydrazine combine with 125 g hydrogen peroxide?

(e) How many grams of the excess reactant in part (d) are left unreacted?

41. Chlorine gas can be prepared according to the reaction

$$16\ HCl + 2\ KMnO_4 \longrightarrow 5\ Cl_2 + 2\ KCl + 2\ MnCl_2 + 8\ H_2O$$

(a) How many moles of $MnCl_2$ can be produced when 25 g $KMnO_4$ are mixed with 85 g HCl?

(b) How many grams of water will be produced when 75 g KCl are produced?

(c) What is the percent yield of Cl_2 if 150 g HCl are reacted, producing 75 g of Cl_2?

(d) When 25 g HCl react with 25 g $KMnO_4$, how many grams of Cl_2 can be produced?

(e) How many grams of the excess reactant in part (d) are left unreacted?

42. Silver tarnishes in the presence of hydrogen sulfide (which smells like rotten eggs) and oxygen because of the reaction

$$4\ Ag + 2\ H_2S + O_2 \longrightarrow 2\ Ag_2S + 2\ H_2O$$

(a) How many grams of silver sulfide can be formed from a mixture of 1.1 g Ag, 0.14 g H_2S, and 0.080 g O_2?

(b) How many more grams of H_2S would be needed to completely react all of the Ag?

43. Humans long ago discovered that they could make soap by cooking fat with ashes from a fire. This happens because the ash contains potassium hydroxide that reacts with the triglycerides in the fat to form soap. The chemical reaction for this process is

$$C_3H_5(C_{15}H_{31}CO_2)_3 + 3\ KOH \longrightarrow C_3H_5(OH)_3 + 3\ C_{15}H_{31}CO_2K$$
(triglyceride)   (soap)

If 500.0 g of this triglyceride are mixed with 500.0 g of potassium hydroxide and 361.7 g of soap are produced, what is the percent yield for the reaction?

44. Group 2A metal oxides react with water to produce hydroxides. When 35.55 g of CaO are placed into 125 mL water ($d = 1.000$ g/mL), how many grams of $Ca(OH)_2$ can be produced?

45. When a solution of lead(II) nitrate is mixed with a solution of sodium chromate, a yellow precipitate forms.

(a) Write the balanced equation for the reaction, including the states of all the substances.

(b) When 26.41 g of lead(II) nitrate are mixed with 18.33 g of sodium chromate, what is the percent yield of the solid if 21.23 g are recovered?

46. Ethyl alcohol burns, producing carbon dioxide and water:

$$C_2H_5OH + 3\ O_2 \longrightarrow 2\ CO_2 + 3\ H_2O$$

(a) If 2.5 mol of C_2H_5OH react with 7.5 mol of O_2 in a closed container, what substances and how many moles of each will be present in the container when the reaction has gone to completion?

(b) When 225 g of C_2H_5OH are burned, how many grams of CO_2 and how many grams of H_2O will be produced?

47. After 180.0 g of zinc were dropped into a beaker of hydrochloric acid and the reaction ceased, 35 g of unreacted zinc remained in the beaker:

$$Zn + HCl \longrightarrow ZnCl_2 + H_2$$

(a) How many grams of hydrogen gas were produced?

(b) How many grams of HCl were reacted?

(c) How many more grams of HCl would be required to completely react with the original sample of zinc?

48. Use this equation to answer (a) and (b):

$$Fe(s) + CuSO_4(aq) \longrightarrow Cu(s) + FeSO_4(aq)$$

(a) When 2.0 mol of Fe and 3.0 mol of $CuSO_4$ are reacted, what substances will be present when the reaction is over? How many moles of each substance are present?

(b) When 20.0 g of Fe and 40.0 g of $CuSO_4$ are reacted, what substances will be present when the reaction is over? How many grams of each substance are present?

49. Methyl alcohol (CH_3OH) is made by reacting carbon monoxide and hydrogen in the presence of certain metal oxide catalysts. How much alcohol can be obtained by reacting 40.0 g of CO and 10.0 g of H_2? How many grams of excess reactant remain unreacted?

$$CO(g) + 2\ H_2(g) \longrightarrow CH_3OH(l)$$

50. Ethyl alcohol (C_2H_5OH), also called grain alcohol, can be made by the fermentation of glucose:

$$C_6H_{12}O_6 \longrightarrow 2\ C_2H_5OH + 2\ CO_2$$
glucose   ethyl alcohol

If an 84.6% yield of ethyl alcohol is obtained,

(a) what mass of ethyl alcohol will be produced from 750 g of glucose?

(b) what mass of glucose should be used to produce 475 g of C_2H_5OH?

51. Both $CaCl_2$ and $MgCl_2$ react with $AgNO_3$ to precipitate AgCl. When solutions containing equal masses of $CaCl_2$ and $MgCl_2$ are reacted with $AgNO_3$, which salt solution will produce the larger amount of AgCl? Show proof.

52. An astronaut excretes about 2500 g of water a day. If lithium oxide (Li_2O) is used in the spaceship to absorb this water, how many kilograms of Li_2O must be carried for a 30-day space trip for three astronauts?

$$Li_2O + H_2O \longrightarrow 2\ LiOH$$

53. Much commercial hydrochloric acid is prepared by the reaction of concentrated sulfuric acid with sodium chloride:

$$H_2SO_4 + 2\ NaCl \longrightarrow Na_2SO_4 + 2\ HCl$$

How many kilograms of concentrated H_2SO_4, 96% H_2SO_4 by mass, are required to produce 20.0 L of concentrated hydrochloric acid ($d = 1.20$ g/mL, 42.0% HCl by mass)?

54. Silicon carbide or carborundum is a widely used ceramic material in car brakes and bullet-proof vests because of its hardness. Carborundum is found naturally in the mineral moissanite, but the demand for it exceeds the supply. For this reason it is generally synthesized in the laboratory. The reaction is

$$SiO_2 + 3\ C \longrightarrow SiC + 2\ CO$$

How many kg of silicon carbide can be formed by the reaction of 250.0 kg of sand (SiO_2) with excess carbon?

55. Three chemical reactions that lead to the formation of sulfuric acid are

$$S + O_2 \longrightarrow SO_2$$
$$2\,SO_2 + O_2 \longrightarrow 2\,SO_3$$
$$SO_3 + H_2O \longrightarrow H_2SO_4$$

Starting with 100.0 g of sulfur, how many grams of sulfuric acid will be formed, assuming a 10% loss in each step? What is the percent yield of H_2SO_4?

56. A 10.00-g mixture of $NaHCO_3$ and Na_2CO_3 was heated and yielded 0.0357 mol H_2O and 0.1091 mol CO_2. Calculate the percent composition of the mixture:

$$NaHCO_3 \longrightarrow Na_2O + CO_2 + H_2O$$
$$Na_2CO_3 \longrightarrow Na_2O + CO_2$$

57. Draw a picture showing the reaction of 2 molecules CH_4 with 3 molecules of O_2 to form water and carbon dioxide gas. Also show any unreacted CH_4 and O_2. Does one of the reactants limit the amount of products formed? If so, which one?

CHALLENGE EXERCISES

58. When 12.82 g of a mixture of $KClO_3$ and NaCl are heated strongly, the $KClO_3$ reacts according to this equation:

$$2\,KClO_3(s) \longrightarrow 2\,KCl(s) + 3\,O_2(g)$$

The NaCl does not undergo any reaction. After the heating, the mass of the residue (KCl and NaCl) is 9.45 g. Assuming that all the loss of mass represents loss of oxygen gas, calculate the percent of $KClO_3$ in the original mixture.

59. Gastric juice contains about 3.0 g HCl per liter. If a person produces about 2.5 L of gastric juice per day, how many antacid tablets, each containing 400 mg of $Al(OH)_3$, are needed to neutralize all the HCl produced in one day?

$$Al(OH)_3(s) + 3\,HCl(aq) \longrightarrow AlCl_3(aq) + 3\,H_2O(l)$$

60. Phosphoric acid, H_3PO_4, can be synthesized from phosphorus, oxygen, and water according to these two reactions:

$$4\,P + 5\,O_2 \longrightarrow P_4O_{10}$$
$$P_4O_{10} + 6\,H_2O \longrightarrow 4\,H_3PO_4$$

Starting with 20.0 g P, 30.0 g O_2, and 15.0 g H_2O, what is the mass of phosphoric acid that can be formed?

61. You have just finished your first year of college in South Carolina and decide to take a cross-country trip to visit California. You begin your drive west, but upon arriving in Kansas you discover to your dismay that your funds have dwindled. You meet up with the manager of a feed mill who has just lost his formulations chemist and needs help determining how many bags of feed supplement he can produce with the oats, molasses, alfalfa, and apples that are arriving on Monday. Luckily you just finished your chemistry class and remember learning about limiting reactants. If you can use these ideas to solve the following problem your money worries may be over! Here is the problem: A new customer just ordered a new formulation of supplement containing 25.0 lb of oats, 17.5 lb of molasses, 30.0 lb of alfalfa, and 28.5 lb of apples per bag. If four train cars of materials are due in the plant on Monday containing the components listed below, how much money will you make for the remainder of your trip if you earn $2.25 for every bag of supplement produced?

Raw materials

1 boxcar of oats containing 4.20 tons of oats

1 tanker of molasses containing 2490 gallons of molasses (molasses has a density of 1.46 g/mL)

250.0 crates of apples, each containing 12.5 kg of apples

350.0 bales of alfalfa, each weighing 62.8 kg

ANSWERS TO PRACTICE EXERCISES

9.1 (a) $\dfrac{2\text{ mol KCl}}{2\text{ mol KClO}_3}$ (b) $\dfrac{2\text{ mol KCl}}{3\text{ mol O}_2}$

9.2   0.33 mol Al_2O_3

9.3   7.33 mol $Al(OH)_3$

9.4   0.8157 mol KCl, 1.223 mol O_2

9.5   85 g $AgNO_3$

9.6   27.6 g $CrCl_3$

9.7   348.6 g H_2O

9.8   17.7 g HCl

9.9   164.3 g $BaSO_4$

9.10 89.1% yield

PUTTING IT TOGETHER:
Review for Chapters 7–9

Multiple Choice

Choose the correct answer to each of the following.

1. 4.0 g of oxygen contain
   (a) 1.5×10^{23} atoms of oxygen
   (b) 4.0 molar masses of oxygen
   (c) 0.50 mol of oxygen
   (d) 6.022×10^{23} atoms of oxygen

2. One mole of hydrogen atoms contains
   (a) 2.016 g of hydrogen
   (b) 6.022×10^{23} atoms of hydrogen
   (c) 1 atom of hydrogen
   (d) 12 g of carbon-12

3. The mass of one atom of magnesium is
   (a) 24.31 g   (c) 12.00 g
   (b) 54.94 g   (d) 4.037×10^{-23} g

4. Avogadro's number of magnesium atoms
   (a) has a mass of 1.0 g
   (b) has the same mass as Avogadro's number of sulfur atoms
   (c) has a mass of 12.0 g
   (d) is 1 mol of magnesium atoms

5. Which of the following contains the largest number of moles?
   (a) 1.0 g Li   (c) 1.0 g Al
   (b) 1.0 g Na   (d) 1.0 g Ag

6. The number of moles in 112 g of acetylsalicylic acid (aspirin), $C_9H_8O_4$, is
   (a) 1.61   (c) 112
   (b) 0.622   (d) 0.161

7. How many moles of aluminum hydroxide are in one antacid tablet containing 400 mg of $Al(OH)_3$?
   (a) 5.13×10^{-3} (c) 5.13
   (b) 0.400   (d) 9.09×10^{-3}

8. How many grams of Au_2S can be obtained from 1.17 mol of Au?
   (a) 182 g   (c) 364 g
   (b) 249 g   (d) 499 g

9. The molar mass of $Ba(NO_3)_2$ is
   (a) 199.3   (c) 247.3
   (b) 261.3   (d) 167.3

10. A 16-g sample of oxygen
    (a) is 1 mol of O_2
    (b) contains 6.022×10^{23} molecules of O_2
    (c) is 0.50 molecule of O_2
    (d) is 0.50 molar mass of O_2

11. What is the percent composition for a compound formed from 8.15 g of zinc and 2.00 g of oxygen?
    (a) 80.3% Zn, 19.7% O   (c) 70.3% Zn, 29.7% O
    (b) 80.3% O, 19.7% Zn   (d) 65.3% Zn, 34.7% O

12. Which of these compounds contains the largest percentage of oxygen?
    (a) SO_2 (c) N_2O_3
    (b) SO_3 (d) N_2O_5

13. 2.00 mol of CO_2
    (a) have a mass of 56.0 g
    (b) contain 1.20×10^{24} molecules
    (c) have a mass of 44.0 g
    (d) contain 6.00 molar masses of CO_2

14. In Ag_2CO_3, the percent by mass of
    (a) carbon is 43.5%   (c) oxygen is 17.4%
    (b) silver is 64.2%   (d) oxygen is 21.9%

15. The empirical formula of the compound containing 31.0% Ti and 69.0% Cl is
    (a) TiCl   (c) $TiCl_3$
    (b) $TiCl_2$ (d) $TiCl_4$

16. A compound contains 54.3% C, 5.6% H, and 40.1% Cl. The empirical formula is
    (a) CH_3Cl (c) $C_2H_4Cl_2$
    (b) C_2H_5Cl (d) C_4H_5Cl

17. A compound contains 40.0% C, 6.7% H, and 53.3% O. The molar mass is 60.0 g/mol. The molecular formula is
    (a) $C_2H_3O_2$ (c) C_2HO
    (b) C_3H_8O (d) $C_2H_4O_2$

18. How many chlorine atoms are in 4.0 mol of PCl_3?
    (a) 3   (c) 12
    (b) 7.2×10^{24} (d) 2.4×10^{24}

19. What is the mass of 4.53 mol of Na_2SO_4?
    (a) 142.1 g   (c) 31.4 g
    (b) 644 g   (d) 3.19×10^{-2} g

20. The percent composition of Mg_3N_2 is
    (a) 72.2% Mg, 27.8% N   (c) 83.9% Mg, 16.1% N
    (b) 63.4% Mg, 36.6% N   (d) no correct answer given

21. How many grams of oxygen are contained in 0.500 mol of Na_2SO_4?
    (a) 16.0 g   (c) 64.0 g
    (b) 32.0 g   (d) no correct answer given

22. The empirical formula of a compound is CH. If the molar mass of this compound is 78.11, then the molecular formula is
    (a) C_2H_2 (c) C_6H_6
    (b) C_5H_{18} (d) no correct answer given

23. The reaction

    $$BaCl_2 + (NH_4)_2CO_3 \longrightarrow BaCO_3 + 2\,NH_4Cl$$

    is an example of
    (a) combination   (c) single displacement
    (b) decomposition   (d) double displacement

24. When the equation

$$Al + O_2 \longrightarrow Al_2O_3$$

is properly balanced, which of the following terms appears?
(a) 2 Al   (b) 2 Al_2O_3 (c) 3 Al   (d) 2 O_2

25. Which equation is *incorrectly* balanced?

(a) 2 $KNO_3 \xrightarrow{\Delta} 2 KNO_2 + O_2$
(b) $H_2O_2 \longrightarrow H_2O + O_2$
(c) 2 $Na_2O_2 + 2 H_2O \longrightarrow 4 NaOH + O_2$
(d) 2 $H_2O \xrightarrow[H_2SO_4]{electrical\ energy} 2 H_2 + O_2$

26. The reaction

$$2 Al + 3 Br_2 \longrightarrow 2 AlBr_3$$

is an example of
(a) combination   (c) decomposition
(b) single displacement   (d) double displacement

27. When the equation

$$PbO_2 \xrightarrow{\Delta} PbO + O_2$$

is balanced, one term in the balanced equation is
(a) PbO_2 (c) 3 PbO
(b) 3 O_2 (d) O_2

28. When the equation

$$Cr_2S_3 + HCl \longrightarrow CrCl_3 + H_2S$$

is balanced, one term in the balanced equation is
(a) 3 HCl   (c) 3 H_2S
(b) $CrCl_3$ (d) 2 Cr_2S_3

29. When the equation

$$F_2 + H_2O \longrightarrow HF + O_2$$

is balanced, one term in the balanced equation is
(a) 2 HF   (c) 4 HF
(b) 3 O_2 (d) 4 H_2O

30. When the equation

$$NH_4OH + H_2SO_4 \longrightarrow$$

is completed and balanced, one term in the balanced equation is
(a) NH_4SO_4 (c) H_2OH
(b) 2 H_2O (d) 2 $(NH_4)_2SO_4$

31. When the equation

$$H_2 + V_2O_5 \longrightarrow V +$$

is completed and balanced, one term in the balanced equation is
(a) 2 V_2O_5 (c) 2 V
(b) 3 H_2O (d) 8 H_2

32. When the equation

$$Al(OH)_3 + H_2SO_4 \longrightarrow Al_2(SO_4)_3 + H_2O$$

is balanced, the sum of the coefficients will be
(a) 9   (c) 12
(b) 11   (d) 15

33. When the equation

$$H_3PO_4 + Ca(OH)_2 \longrightarrow H_2O + Ca_3(PO_4)_2$$

is balanced, the proper sequence of coefficients is
(a) 3, 2, 1, 6   (c) 2, 3, 1, 6
(b) 2, 3, 6, 1   (d) 2, 3, 3, 1

34. When the equation

$$Fe_2(SO_4)_3 + Ba(OH)_2 \longrightarrow$$

is completed and balanced, one term in the balanced equation is
(a) $Ba_2(SO_4)_3$ (c) 2 $Fe_2(SO_4)_3$
(b) 2 $Fe(OH)_2$ (d) 2 $Fe(OH)_3$

35. For the reaction

$$2 H_2 + O_2 \longrightarrow 2 H_2O + 572.4\ kJ$$

which of the following is not true?
(a) The reaction is exothermic.
(b) 572.4 kJ of heat are liberated for each mole of water formed.
(c) 2 mol of hydrogen react with 1 mol of oxygen.
(d) 572.4 kJ of heat are liberated for each 2 mol of hydrogen reacted.

36. How many moles are 20.0 g of Na_2CO_3?
(a) 1.89 mol   (c) 212 mol
(b) 2.12×10^3 mol   (d) 0.189 mol

37. What is the mass in grams of 0.30 mol of $BaSO_4$?
(a) 7.0×10^3 g   (c) 70. g
(b) 0.13 g   (d) 700.2 g

38. How many molecules are in 5.8 g of acetone, C_3H_6O?
(a) 0.10 molecule
(b) 6.0×10^{22} molecules
(c) 3.5×10^{24} molecules
(d) 6.0×10^{23} molecules

Problems 39–45 refer to the reaction

$$2 C_2H_4 + 6 O_2 \longrightarrow 4 CO_2 + 4 H_2O$$

39. If 6.0 mol of CO_2 are produced, how many moles of O_2 were reacted?
(a) 4.0 mol   (c) 9.0 mol
(b) 7.5 mol   (d) 15.0 mol

40. How many moles of O_2 are required for the complete combustion of 45 g of C_2H_4?
(a) 1.3×10^2 mol   (c) 112.5 mol
(b) 0.64 mol   (d) 4.8 mol

41. If 18.0 g of CO_2 are produced, how many grams of H_2O are produced?
(a) 7.37 g   (c) 9.00 g
(b) 3.68 g   (d) 14.7 g

42. How many moles of CO_2 can be produced by the reaction of 5.0 mol of C_2H_4 and 12.0 mol of O_2?
(a) 4.0 mol   (c) 8.0 mol
(b) 5.0 mol   (d) 10. mol

43. How many moles of CO_2 can be produced by the reaction of 0.480 mol of C_2H_4 and 1.08 mol of O_2?
(a) 0.240 mol   (c) 0.720 mol
(b) 0.960 mol   (d) 0.864 mol

44. How many grams of CO_2 can be produced from 2.0 g of C_2H_4 and 5.0 g of O_2?
(a) 5.5 g   (c) 7.6 g
(b) 4.6 g   (d) 6.3 g

45. If 14.0 g of C_2H_4 are reacted and the actual yield of H_2O is 7.84 g, the percent yield in the reaction is
(a) 0.56%   (c) 87.1%
(b) 43.6%   (d) 56.0%

Problems 46–48 refer to the equation

$$H_3PO_4 + MgCO_3 \longrightarrow Mg_3(PO_4)_2 + CO_2 + H_2O$$

46. The sequence of coefficients for the balanced equation is
(a) 2, 3, 1, 3, 3   (c) 2, 2, 1, 2, 3
(b) 3, 1, 3, 2, 3   (d) 2, 3, 1, 3, 2

47. If 20.0 g of carbon dioxide are produced, the number of moles of magnesium carbonate used is
(a) 0.228 mol   (c) 0.910 mol
(b) 1.37 mol   (d) 0.454 mol

48. If 50.0 g of magnesium carbonate react completely with H_3PO_4, the number of grams of carbon dioxide produced is
(a) 52.2 g   (c) 13.1 g
(b) 26.1 g   (d) 50.0 g

49. When 10.0 g of $MgCl_2$ and 10.0 g of Na_2CO_3 are reacted in

$$MgCl_2 + Na_2CO_3 \longrightarrow MgCO_3 + 2\, NaCl$$

the limiting reactant is
(a) $MgCl_2$ (c) $MgCO_3$
(b) Na_2CO_3 (d) NaCl

50. When 50.0 g of copper are reacted with silver nitrate solution

$$Cu + 2\, AgNO_3 \longrightarrow Cu(NO_3)_2 + 2\, Ag$$

148 g of silver are obtained. What is the percent yield of silver obtained?
(a) 87.1%   (c) 55.2%
(b) 84.9%   (d) no correct answer given

Free Response Questions

Answer each of the following. Be sure to include your work and explanations in a clear, logical form.

1. Compound X requires 104 g of O_2 to produce 2 moles of CO_2 and 2.5 moles of H_2O.
(a) What is the empirical formula for X?
(b) What additional information would you need to determine the molecular formula for X?

2. Consider the reaction of sulfur dioxide with oxygen to form sulfur trioxide taking place in a closed container.

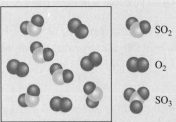

(a) Draw what you would expect to see in the box at the completion of the reaction.
(b) If you begin with 25 g of SO_2 and 5 g of oxygen gas, which is the limiting reagent?
(c) Is the following statement true or false? "When SO_2 is converted into SO_3, the percent composition of S in the compounds changes from 33% to 25%." Explain.

3. The percent composition of compound Z is 63.16% C and 8.77% H. When compound Z burns in air, the only products are carbon dioxide and water. The molar mass for Z is 114.
(a) What is the molecular formula for compound Z?
(b) What is the balanced reaction for Z burning in air?

4. 16% of the U.S. RDA for Ca is 162 mg.
(a) What mass of calcium phosphate, $Ca_3(PO_4)_2$, provides 162 mg of calcium?

(b) Is $Ca_3(PO_4)_2$ an element, a mixture, or a compound?
(c) Milk is a good source of calcium in the diet. If 120 mL of skim milk provide 13% of the U.S. RDA for Ca, how many cups of milk should you drink per day if that is your only source of calcium?

5. Compound A decomposes at room temperature in an exothermic reaction, while compound B requires heating before it will decompose in an endothermic reaction.
(a) Draw reaction profiles (potential energy vs. reaction progress) for both reactions.
(b) The decomposition of 0.500 mol of $NaHCO_3$ to form sodium carbonate, water, and carbon dioxide requires 85.5 kJ of heat. How many grams of water could be collected, and how many kJ of heat are absorbed when 24.0 g of CO_2 are produced?
(c) Could $NaHCO_3$ be compound A or compound B? Explain your reasoning.

6. Aqueous ammonium hydroxide reacts with aqueous cobalt(II) sulfate to produce aqueous ammonium sulfate and solid cobalt(II) hydroxide. When 38.0 g of one of the reactants is fully reacted with enough of the other reactant, 8.09 g of ammonium sulfate is obtained, which corresponded to a 25.0% yield.
(a) What type of reaction took place?
(b) Write the balanced chemical equation.
(c) What is the theoretical yield of ammonium sulfate?
(d) Which reactant was the limiting reagent?

7. $C_6H_{12}O_6 \longrightarrow 2\, C_2H_5OH + 2\, CO_2(g)$
(a) If 25.0 g of $C_6H_{12}O_6$ were used, and only 11.2 g of C_2H_5OH were produced, how much reactant was left over and what volume of gas was produced? (*Note:* One mole of gas occupies 24.0 L of space at room temperature.)
(b) What was the yield of the reaction?
(c) What type of reaction took place?

8. When solutions containing 25 g each of lead(II) nitrate and potassium iodide were mixed, a yellow precipitate resulted.
(a) What type of reaction occurred?
(b) What are the name and formula for the solid product?
(c) If after filtration and drying, the solid product weighed 7.66 g, what was the percent yield for the reaction?

9. Consider the following unbalanced reaction:

$$XNO_3 + CaCl_2 \longrightarrow XCl + Ca(NO_3)_2$$

(a) If 30.8 g of $CaCl_2$ produced 79.6 g of XCl, what is X?
(b) Would X be able to displace hydrogen from an acid?

10. Consider the following reaction: $H_2O_2 \longrightarrow H_2O + O_2$
(a) If at the end of the reaction there are eight water molecules and eight oxygen molecules, what was in the flask at the start of the reaction?
(b) Does the following reaction profile indicate the reaction is exothermic or endothermic?

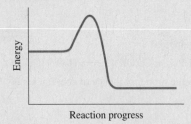

(c) What type of reaction is given above?
(d) What is the empirical formula for hydrogen peroxide?

The amazing colors in these fireworks explosions are the result of electrons transferring between energy levels in atoms.

© Hannu Viitanen/iStockphoto

CHAPTER **10**

MODERN ATOMIC THEORY AND THE PERIODIC TABLE

How do we go about studying an object that is too small to see? Think back to that birthday present you could look at but not yet open. Judging from the wrapping and size of the box was not very useful, but shaking, turning, and lifting the package all gave indirect clues to its contents. After all your experiments were done, you could make a fairly good guess about the contents. But was your guess correct? The only way to know for sure would be to open the package.

Chemists have the same dilemma when they study the atom. Atoms are so very small that it isn't possible to use the normal senses to describe them. We are essentially working in the dark with this package we call the atom. However, our improvements in instruments (X-ray machines and scanning tunneling microscopes) and measuring devices (spectrophotometers and magnetic resonance imaging, MRI) as well as in our mathematical skills are bringing us closer to revealing the secrets of the atom.

CHAPTER OUTLINE

10.1 Electromagnetic Radiation

10.2 The Bohr Atom

10.3 Energy Levels of Electrons

10.4 Atomic Structures of the First 18 Elements

10.5 Electron Structures and the Periodic Table

10.1 ELECTROMAGNETIC RADIATION

LEARNING OBJECTIVE ● List the three basic characteristics of electromagnetic radiation.

KEY TERMS

wavelength
frequency
speed
photons

In the last 200 years, vast amounts of data have been accumulated to support the atomic theory. When atoms were originally suggested by the early Greeks, no physical evidence existed to support their ideas. Early chemists did a variety of experiments, which culminated in Dalton's model of the atom. Because of the limitations of Dalton's model, modifications were proposed first by Thomson and then by Rutherford, which eventually led to our modern concept of the nuclear atom. These early models of the atom work reasonably well—in fact, we continue to use them to visualize a variety of chemical concepts. There remain questions that these models cannot answer, including an explanation of how atomic structure relates to the periodic table. In this chapter, we will present our modern model of the atom; we will see how it varies from and improves upon the earlier atomic models.

Electromagnetic Radiation

Scientists have studied energy and light for centuries, and several models have been proposed to explain how energy is transferred from place to place. One way energy travels through space is by *electromagnetic radiation*. Examples of electromagnetic radiation include light from the sun, X-rays in your dentist's office, microwaves from your microwave oven, radio and television waves, and radiant heat from your fireplace. While these examples seem quite different, they are all similar in some important ways. Each shows wavelike behavior, and all travel at the same speed in a vacuum (3.00×10^8 m/s).

The study of wave behavior is a topic for another course, but we need some basic terminology to understand atoms. Waves have three basic characteristics: wavelength, frequency, and speed. **Wavelength** (lambda, λ) is the distance between consecutive peaks (or troughs) in a wave, as shown in **Figure 10.1**. **Frequency** (nu, ν) tells how many waves pass a particular point per second. **Speed** (ν) tells how fast a wave moves through space.

Surfers judge the wavelength, frequency, and speed of waves to get the best ride.

Tony Freeman/PhotoEdit

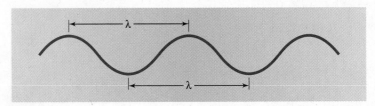

Figure 10.1

The wavelength of this wave is shown by λ. It can be measured from peak to peak or trough to trough.

Light is one form of electromagnetic radiation and is usually classified by its wavelength, as shown in **Figure 10.2**. Visible light, as you can see, is only a tiny part of the electromagnetic spectrum. Some examples of electromagnetic radiation involved in energy transfer outside the visible region are hot coals in your backyard grill, which transfer infrared radiation to cook your food, and microwaves, which transfer energy to water molecules in the food, causing them to move more quickly and thus raise the temperature of the food.

Figure 10.2

The electromagnetic spectrum.

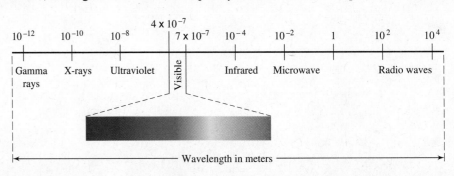

>CHEMISTRY *IN ACTION*

You Light Up My Life

Birds in the parrot family have an unusual way to attract their mates—their feathers glow in the dark! This phenomenon is called fluorescence. It results from the absorption of ultraviolet (UV) light, which is then reemitted at longer wavelengths that both birds and people can see. In everyday life this happens in a fluorescent bulb or in the many glow-in-the-dark products such as light sticks.

Kathleen Arnold from the University of Glasgow, Scotland, discovered that the feathers of parrots that fluoresced were only those used in display or those shown off during courtship. She decided to experiment using budgerigars with their natural colors. The researchers offered birds a choice of two companion birds, which were smeared with petroleum jelly. One of the potential companions also had a UV blocker in the petroleum jelly. The birds clearly preferred companions without the UV blocker, leading the researchers to conclude that the parrots prefer to court radiant partners. The researchers also tested same-sex companions and discovered they did not prefer radiant companions.

Perhaps we may find that the glow of candlelight really does add to the radiance of romance. Ultraviolet light certainly does for parrots!

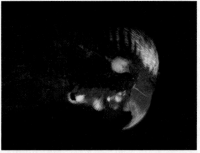

Budgerigars under normal light and under UV light showing the glow used to attract mates.

We have evidence for the wavelike nature of light. We also know that a beam of light behaves like a stream of tiny packets of energy called **photons**. So what is light exactly? Is it a particle? Is it a wave? Scientists have agreed to explain the properties of electromagnetic radiation by using both wave and particle properties. Neither explanation is ideal, but currently these are our best models.

EXAMPLE 10.1

What is a wavelength?

SOLUTION

A wavelength is the distance between consecutive peaks or troughs of a wave. The symbol used for wavelength is lambda (λ).

PRACTICE 10.1

Draw a wave function and indicate the distance of two wavelengths (2 lambda).

10.2 THE BOHR ATOM

Explain the relationship between the line spectrum and the quantized energy levels of an electron in an atom.

● **LEARNING OBJECTIVE**

As scientists struggled to understand the properties of electromagnetic radiation, evidence began to accumulate that atoms could radiate light. At high temperatures, or when subjected to high voltages, elements in the gaseous state give off colored light. Brightly colored neon signs illustrate this property of matter very well. When the light emitted by a gas is passed through a prism or diffraction grating, a set of brightly colored lines called

KEY TERMS

line spectrum
quanta
ground state
orbital

410   434   486   656

(in nanometers)

Figure 10.3

Line spectrum of hydrogen. Each line corresponds to the wavelength of the energy emitted when the electron of a hydrogen atom, which has absorbed energy, falls back to a lower principal energy level.

a **line spectrum** results (**Figure 10.3**). These colored lines indicate that the light is being emitted only at certain wavelengths, or frequencies, that correspond to specific colors. Each element possesses a unique set of these spectral lines that is different from the sets of all the other elements.

In 1912–1913, while studying the line spectrum of hydrogen, Niels Bohr (1885–1962), a Danish physicist, made a significant contribution to the rapidly growing knowledge of atomic structure. His research led him to believe that electrons exist in specific regions at various distances from the nucleus. He also visualized the electrons as revolving in orbits around the nucleus, like planets rotating around the sun, as shown in **Figure 10.4**.

Bohr's first paper in this field dealt with the hydrogen atom, which he described as a single electron revolving in an orbit about a relatively heavy nucleus. He applied the concept of energy quanta, proposed in 1900 by the German physicist Max Planck (1858–1947), to the observed line spectrum of hydrogen. Planck stated that energy is never emitted in a continuous stream but only in small, discrete packets called **quanta** (Latin, *quantus*, "how much"). From this, Bohr theorized that electrons have several possible energies corresponding to several possible orbits at different distances from the nucleus. Therefore, an electron has to be in one specific energy level; it cannot exist between energy levels. In other words, the energy of the electron is said to be quantized. Bohr also stated that when a hydrogen atom absorbed one or more quanta of energy, its electron would "jump" to a higher energy level.

Bohr was able to account for spectral lines of hydrogen this way. A number of energy levels are available, the lowest of which is called the **ground state**. When an electron falls from a high energy level to a lower one (say, from the fourth to the second), a quantum of energy is emitted as light at a specific frequency, or wavelength (**Figure 10.5**). This light corresponds to one of the lines visible in the hydrogen spectrum (Figure 10.3). Several lines are visible in this spectrum, each one corresponding to a specific electron energy-level shift within the hydrogen atom.

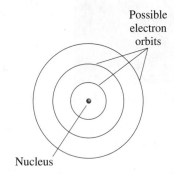

Possible electron orbits

Nucleus

Figure 10.4

The Bohr model of the hydrogen atom described the electron revolving in certain allowed circular orbits around the nucleus.

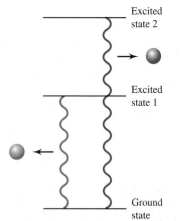

Excited state 2

Excited state 1

Ground state

Figure 10.5

When an excited electron returns to the ground state, energy emitted as a photon is released. The color (wavelength) of the light is determined by the difference in energy between the two states (excited and ground).

EXAMPLE 10.2

What does the line spectrum of a hydrogen atom illustrate?

SOLUTION

Each line shows the frequency of light energy emitted when an electron that has absorbed energy falls from a higher energy level to a lower energy level.

PRACTICE 10.2

What is a photon?

The chemical properties of an element and its position in the periodic table depend on electron behavior within the atoms. In turn, much of our knowledge of the behavior of electrons within atoms is based on spectroscopy. Niels Bohr contributed a great deal to our knowledge of atomic structure by (1) suggesting quantized energy levels for electrons and (2) showing that spectral lines result from the radiation of small increments of energy (Planck's quanta) when electrons shift from one energy level to another. Bohr's calculations succeeded very well in correlating the experimentally observed spectral lines with electron energy levels for the hydrogen atom. However, Bohr's methods of calculation did not succeed for heavier atoms. More theoretical work on atomic structure was needed.

In 1924, the French physicist Louis de Broglie (1892–1957) suggested a surprising hypothesis: All objects have wave properties. Louis de Broglie used sophisticated mathematics to show that the wave properties for an object of ordinary size, such as a baseball, are too small to be observed. But for smaller objects, such as an electron, the wave properties become significant. Other scientists confirmed de Broglie's hypothesis, showing that electrons do exhibit wave properties. In 1926, Erwin Schrödinger (1887–1961), an Austrian physicist, created a mathematical model that described electrons as waves. Using Schrödinger's wave mechanics, we can determine the *probability* of finding an electron in a certain region around the nucleus of the atom.

This treatment of the atom led to a new branch of physics called *wave mechanics* or *quantum mechanics*, which forms the basis for our modern understanding of atomic structure. Although the wave-mechanical description of the atom is mathematical, it can be translated, at least in part, into a visual model. It is important to recognize that we cannot locate an electron precisely within an atom; however, it is clear that electrons are not revolving around the nucleus in orbits as Bohr postulated. The electrons are instead found in *orbitals*. An **orbital** is pictured in **Figure 10.6** as a region in space around the nucleus where there is a high probability of finding a given electron.

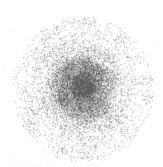

Figure 10.6
An orbital for a hydrogen atom. The intensity of the dots shows that the electron spends more time closer to the nucleus.

10.3 ENERGY LEVELS OF ELECTRONS

Describe the principal energy levels, sublevels, and orbitals of an atom.

● **LEARNING OBJECTIVE**

KEY TERMS
principal energy levels
sublevels
spin
Pauli exclusion principle

One of the ideas Bohr contributed to the modern concept of the atom was that the energy of the electron is quantized—that is, the electron is restricted to only certain allowed energies. The wave-mechanical model of the atom also predicts discrete **principal energy levels** within the atom. These energy levels are designated by the letter n, where n is a positive integer (**Figure 10.7**). The lowest principal energy level corresponds to $n = 1$, the next to $n = 2$, and so on. As n increases, the energy of the electron increases, and the electron is found on average farther from the nucleus.

Each principal energy level is divided into **sublevels**, which are illustrated in **Figure 10.8**. The first principal energy level has one sublevel. The second principal energy level has two sublevels, the third energy level has three sublevels, and so on. Each of these sublevels contains spaces for electrons called orbitals.

In each sublevel the electrons are found within specified orbitals (s, p, d, f). Let's consider each principal energy level in turn. The first principal energy level ($n = 1$) has one

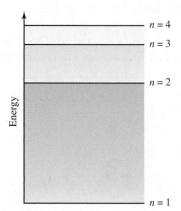

Figure 10.7
The first four principal energy levels in the hydrogen atom. Each level is assigned a principal quantum number n.

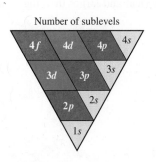

Figure 10.8
The types of orbitals on each of the first four principal energy levels.

Figure 10.9

Perspective representation of the p_x, p_y, and p_z atomic orbitals.

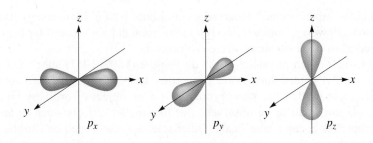

sublevel or type of orbital. It is spherical in shape and is designated as 1s. It is important to understand what the spherical shape of the 1s orbital means. The electron does *not* move around on the surface of the sphere, but rather the surface encloses a space where there is a 90% probability where the electron may be found. It might help to consider these orbital shapes in the same way we consider the atmosphere. There is no distinct dividing line between the atmosphere and "space." The boundary is quite fuzzy. The same is true for atomic orbitals. Each has a region of highest density roughly corresponding to its shape. The probability of finding the electron outside this region drops rapidly but never quite reaches zero. Scientists often speak of orbitals as electron "clouds" to emphasize the fuzzy nature of their boundaries.

How many electrons can fit into a 1s orbital? To answer this question, we need to consider one more property of electrons. This property is called **spin**. Each electron appears to be spinning on an axis, like a globe. It can only spin in two directions. We represent this spin with an arrow: ↑ *or* ↓. In order to occupy the same orbital, electrons must have *opposite* spins. That is, two electrons with the same spin cannot occupy the same orbital. This gives us the answer to our question: An atomic orbital can hold a maximum of two electrons, which must have opposite spins. This rule is called the **Pauli exclusion principle**. The first principal energy level contains one type of orbital (1s) that holds a maximum of two electrons.

What happens with the second principal energy level ($n = 2$)? Here we find two sublevels, 2s and 2p. Like 1s in the first principal energy level, the 2s orbital is spherical in shape but is larger in size and higher in energy. It also holds a maximum of two electrons. The second type of orbital is designated by 2p. The 2p sublevel consists of three orbitals: $2p_x$, $2p_y$, and $2p_z$. The shape of p orbitals is quite different from the s orbitals, as shown in **Figure 10.9**.

Each p orbital has two "lobes." Remember, the space enclosed by these surfaces represents the regions of probability for finding the electrons 90% of the time. There are three separate p orbitals, each oriented in a different direction, and each p orbital can hold a maximum of two electrons. Thus the total number of electrons that can reside in all three p orbitals is six. To summarize our model, the first principal energy level of an atom has a 1s orbital. The second principal energy level has a 2s and three 2p orbitals labeled $2p_x$, $2p_y$, and $2p_z$, as shown in **Figure 10.10**.

Notice that there is a correspondence between the energy level and number of sublevels.

The third principal energy level has three sublevels labeled 3s, 3p, and 3d. The 3s orbital is spherical and larger than the 1s and 2s orbitals. The $3p_x$, $3p_y$, and $3p_z$ orbitals are shaped

Figure 10.10

Orbitals on the second principal energy level are one 2s and three 2p orbitals.

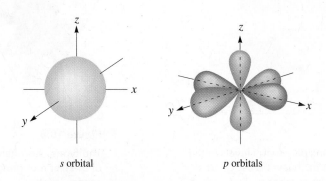

s orbital   p orbitals

>CHEMISTRY *IN ACTION*

Atomic Clocks

Imagine a clock that keeps time to within 1 second over a million years. The National Institute of Standards and Technology in Boulder, Colorado, has an atomic clock that does just that—a little better than your average alarm, grandfather, or cuckoo clock! This atomic clock serves as the international standard for time and frequency. How does it work?

Within the case are several layers of magnetic shielding. In the heart of the clock is a small oven that heats cesium metal to release cesium atoms, which are collected into a narrow beam (1 mm wide). The beam of atoms passes down a long evacuated tube while being excited by a laser until all the cesium atoms are in the same electron state.

The atoms then enter another chamber filled with re-flecting microwaves. The frequency of the microwaves (9,192,631,770 cycles per second) is exactly the same frequency required to excite a cesium atom from its ground state to the next higher energy level. These excited cesium atoms then release electromagnetic radia-tion in a process known as fluorescence. Electronic circuits maintain the microwave frequency at precisely the right level to keep the cesium atoms moving from one level to the next. One second is equal to 9,192,631,770 of these vibrations. The clock is set to this frequency and can keep accurate time for over a million years.

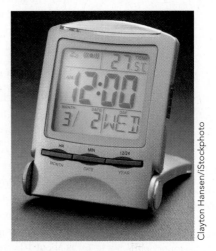

Clayton Hansen/iStockphoto

This clock automatically updates itself by comparing time with an atomic clock by radio signal.

like those of the second level, only larger. The five $3d$ orbitals have the shapes shown in **Figure 10.11.** You don't need to memorize these shapes, but notice that they look different from the s or p orbitals.

Each time a new principal energy level is added, we also add a new sublevel. This makes sense because each energy level corresponds to a larger average distance from the nucleus, which provides more room on each level for new sublevels containing more orbitals.

The pattern continues with the fourth principal energy level. It has $4s$, $4p$, $4d$, and $4f$ orbitals. There are one $4s$, three $4p$, five $4d$, and seven $4f$ orbitals. The shapes of the s, p, and d orbitals are the same as those for lower levels, only larger. We will not consider the shapes of the f orbitals. Remember that for all s, p, d, and f orbitals, the maximum number of electrons per orbital is two. We summarize each principal energy level:

$n = 1$ $1s$
$n = 2$ $2s$ $2p\ 2p\ 2p$
$n = 3$ $3s$ $3p\ 3p\ 3p$ $3d\ 3d\ 3d\ 3d\ 3d$
$n = 4$ $4s$ $4p\ 4p\ 4p$ $4d\ 4d\ 4d\ 4d\ 4d$ $4f\ 4f\ 4f\ 4f\ 4f\ 4f\ 4f$

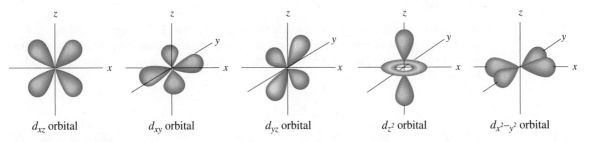

d_{xz} orbital d_{xy} orbital d_{yz} orbital d_{z^2} orbital $d_{x^2-y^2}$ orbital

Figure 10.11
The five d orbitals are found in the third principal energy level along with one $3s$ orbital and three $3p$ orbitals.

EXAMPLE 10.3

Energy levels of the atoms are designated as principal energy levels and energy sublevels.

(a) What labels are used for these energy levels?
(b) How many of each of these labels are there?

SOLUTION

(a) The letter n is used to designate the principal energy levels. The letters s, p, d, and f are used to designate the energy sublevels.
(b) According to the periodic table, there are seven (7) principal energy levels and $2s$, $6p$, $10d$, and $14f$ energy sublevels.

PRACTICE 10.3

(a) How many electrons can occupy a single subenergy level?
(b) What is the maximum number of electrons that can occupy the third principal energy level? List them.

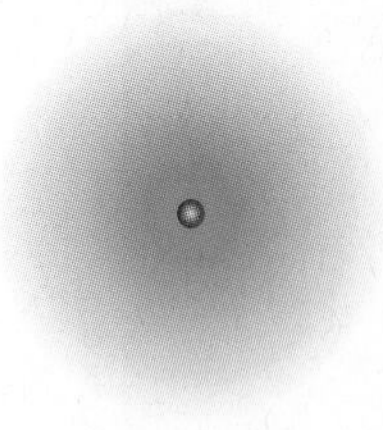

Figure 10.12

The modern concept of a hydrogen atom consists of a proton and an electron in an s orbital. The shaded area represents a region where the electron may be found with 90% probability.

The hydrogen atom consists of a nucleus (containing one proton) and one electron occupying a region outside of the nucleus. In its ground state, the electron occupies a $1s$ orbital, but by absorbing energy the electron can become *excited* and move to a higher energy level.

The hydrogen atom can be represented as shown in **Figure 10.12**. The diameter of the nucleus is about 10^{-13} cm, and the diameter of the electron orbital is about 10^{-8} cm. The diameter of the electron cloud of a hydrogen atom is about 100,000 times greater than the diameter of the nucleus.

10.4 ATOMIC STRUCTURES OF THE FIRST 18 ELEMENTS

LEARNING OBJECTIVE ● Use the guidelines to write electron configurations.

KEY TERMS

electron configuration
orbital diagram
valence electrons

We have seen that hydrogen has one electron that can occupy a variety of orbitals in different principal energy levels. Now let's consider the structure of atoms with more than one electron. Because all atoms contain orbitals similar to those found in hydrogen, we can describe the structures of atoms beyond hydrogen by systematically placing electrons in these hydrogen-like orbitals. We use the following guidelines:

1. No more than two electrons can occupy one orbital.

2. Electrons occupy the lowest energy orbitals available. They enter a higher energy orbital only when the lower orbitals are filled. For the atoms beyond hydrogen, orbital energies vary as $s < p < d < f$ for a given value of n.

3. Each orbital in a sublevel is occupied by a single electron before a second electron enters. For example, all three p orbitals must contain one electron before a second electron enters a p orbital.

We can use several methods to represent the atomic structures of atoms, depending on what we are trying to illustrate. When we want to show both the nuclear makeup and the electron structure of each principal energy level (without orbital detail), we can use a diagram such as **Figure 10.13**.

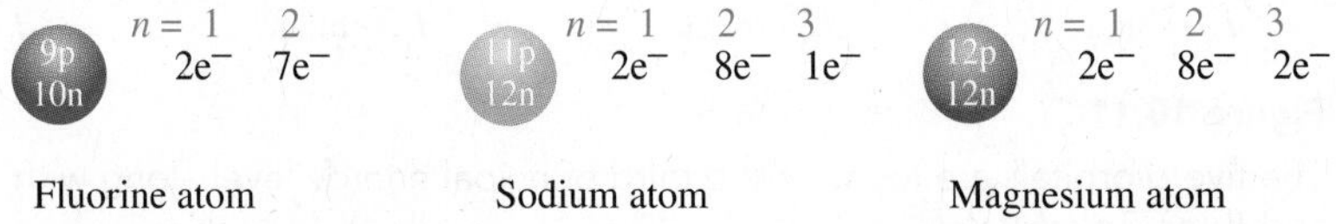

Figure 10.13

Atomic structure diagrams of fluorine, sodium, and magnesium atoms. The number of protons and neutrons is shown in the nucleus. The number of electrons is shown in each principal energy level outside the nucleus.

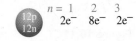

Fluorine atom   Sodium atom   Magnesium atom

Often we are interested in showing the arrangement of the electrons in an atom in their orbitals. There are two ways to do this. The first method is called the **electron configuration**. In this method, we list each type of orbital, showing the number of electrons in it as an exponent. An electron configuration is read as follows:

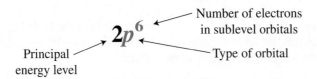

We can also represent this configuration with an **orbital diagram** in which boxes represent the orbitals (containing small arrows indicating the electrons). When the orbital contains one electron, an arrow, pointing upward ($\uparrow$), is placed in the box. A second arrow, pointing downward ($\downarrow$), indicates the second electron in that orbital.

Let's consider each of the first 18 elements on the periodic table in turn. The order of filling for the orbitals in these elements is $1s$, $2s$, $2p$, $3s$, $3p$, and $4s$. Hydrogen, the first element, has only one electron. The electron will be in the $1s$ orbital because this is the most favorable position (where it will have the greatest attraction for the nucleus). Both representations are shown here:

$$H \quad \boxed{\uparrow} \quad 1s^1$$

Orbital   Electron
diagram   configuration

Helium, with two electrons, can be shown as

$$He \quad \boxed{\uparrow\downarrow} \quad 1s^2$$

Orbital   Electron
diagram   configuration

The first energy level, which can hold a maximum of two electrons, is now full. An atom with three electrons will have its third electron in the second principal energy level. Thus, in lithium (atomic number 3), the first two electrons are in the $1s$ orbital, and the third electron is in the $2s$ orbital of the second energy level.

Lithium has the following structure:

$$Li \quad \boxed{\uparrow\downarrow} \quad \boxed{\uparrow} \quad 1s^2 2s^1$$
$$\qquad\quad 1s \qquad 2s$$

All four electrons of beryllium are s electrons:

$$Be \quad \boxed{\uparrow\downarrow} \quad \boxed{\uparrow\downarrow} \quad 1s^2 2s^2$$
$$\qquad\quad 1s \qquad 2s$$

The next six elements illustrate the filling of the p orbitals. Boron has the first p electron. Because p orbitals all have the same energy, it doesn't matter which of these orbitals fills first:

$$B \quad \boxed{\uparrow\downarrow} \quad \boxed{\uparrow\downarrow} \quad \boxed{\uparrow\ \ |\ \ } \quad 1s^2 2s^2 2p^1$$
$$\quad\ 1s \qquad 2s \qquad\ \ 2p$$

Carbon is the sixth element. It has two electrons in the $1s$ orbital, two electrons in the $2s$ orbital, and two electrons to place in the $2p$ orbitals. Because it is more difficult for the p electrons to pair up than to occupy a second p orbital, the second p electron is located in a different p orbital. We could show this by writing $2p_x^1 2p_y^1$, but we usually write it as $2p^2$; it is *understood* that the electrons are in different p orbitals. The spins on these electrons are alike, for reasons we will not explain here.

$$C \quad \boxed{\uparrow\downarrow} \quad \boxed{\uparrow\downarrow} \quad \boxed{\uparrow\ |\ \uparrow\ |\ } \quad 1s^2 2s^2 2p^2$$
$$\quad\ 1s \qquad 2s \qquad\ \ 2p$$

Nitrogen has seven electrons. They occupy the $1s$, $2s$, and $2p$ orbitals. The third p electron in nitrogen is still unpaired and is found in the $2p_z$ orbital:

$$N \quad \boxed{\uparrow\downarrow}_{1s} \quad \boxed{\uparrow\downarrow}_{2s} \quad \boxed{\uparrow\,|\,\uparrow\,|\,\uparrow}_{2p} \quad 1s^2 2s^2 2p^3$$

Oxygen is the eighth element. It has two electrons in both the $1s$ and $2s$ orbitals and four electrons in the $2p$ orbitals. One of the $2p$ orbitals is now occupied by a second electron, which has a spin opposite the electron already in that orbital:

$$O \quad \boxed{\uparrow\downarrow}_{1s} \quad \boxed{\uparrow\downarrow}_{2s} \quad \boxed{\uparrow\downarrow\,|\,\uparrow\,|\,\uparrow}_{2p} \quad 1s^2 2s^2 2p^4$$

The next two elements are fluorine with nine electrons and neon with ten electrons:

$$F \quad \boxed{\uparrow\downarrow}_{1s} \quad \boxed{\uparrow\downarrow}_{2s} \quad \boxed{\uparrow\downarrow\,|\,\uparrow\downarrow\,|\,\uparrow}_{2p} \quad 1s^2 2s^2 2p^5$$

$$Ne \quad \boxed{\uparrow\downarrow}_{1s} \quad \boxed{\uparrow\downarrow}_{2s} \quad \boxed{\uparrow\downarrow\,|\,\uparrow\downarrow\,|\,\uparrow\downarrow}_{2p} \quad 1s^2 2s^2 2p^6$$

With neon, the first and second energy levels are filled as shown in Table 10.1. The second energy level can hold a maximum of eight electrons, $2s^2 2p^6$.

Sodium, element 11, has two electrons in the first energy level and eight electrons in the second energy level, with the remaining electron occupying the $3s$ orbital in the third energy level:

$$Na \quad \boxed{\uparrow\downarrow}_{1s} \quad \boxed{\uparrow\downarrow}_{2s} \quad \boxed{\uparrow\downarrow\,|\,\uparrow\downarrow\,|\,\uparrow\downarrow}_{2p} \quad \boxed{\uparrow}_{3s} \quad 1s^2 2s^2 2p^6 3s^1$$

TABLE 10.1   Orbital Filling for the First Ten Elements*

Atomic number	Element	Orbitals			Electron configuration
		$1s$	$2s$	$2p$	
1	H	$\boxed{\uparrow}$			$1s^1$
2	He	$\boxed{\uparrow\downarrow}$			$1s^2$
3	Li	$\boxed{\uparrow\downarrow}$	$\boxed{\uparrow}$		$1s^2 2s^1$
4	Be	$\boxed{\uparrow\downarrow}$	$\boxed{\uparrow\downarrow}$		$1s^2 2s^2$
5	B	$\boxed{\uparrow\downarrow}$	$\boxed{\uparrow\downarrow}$	$\boxed{\uparrow\,|\,\,|\,\,}$	$1s^2 2s^2 2p^1$
6	C	$\boxed{\uparrow\downarrow}$	$\boxed{\uparrow\downarrow}$	$\boxed{\uparrow\,|\,\uparrow\,}$	$1s^2 2s^2 2p^2$
7	N	$\boxed{\uparrow\downarrow}$	$\boxed{\uparrow\downarrow}$	$\boxed{\uparrow\,|\,\uparrow\,|\,\uparrow}$	$1s^2 2s^2 2p^3$
8	O	$\boxed{\uparrow\downarrow}$	$\boxed{\uparrow\downarrow}$	$\boxed{\uparrow\downarrow\,|\,\uparrow\,|\,\uparrow}$	$1s^2 2s^2 2p^4$
9	F	$\boxed{\uparrow\downarrow}$	$\boxed{\uparrow\downarrow}$	$\boxed{\uparrow\downarrow\,|\,\uparrow\downarrow\,|\,\uparrow}$	$1s^2 2s^2 2p^5$
10	Ne	$\boxed{\uparrow\downarrow}$	$\boxed{\uparrow\downarrow}$	$\boxed{\uparrow\downarrow\,|\,\uparrow\downarrow\,|\,\uparrow\downarrow}$	$1s^2 2s^2 2p^6$

*Boxes represent the orbitals grouped by sublevel. Electrons are shown by arrows.

TABLE 10.2   Orbital Diagrams and Electron Configurations for Elements 11–18

Atomic number	Element	1s	2s	2p			3s	3p			Electron configuration
11	Na	↑↓	↑↓	↑↓	↑↓	↑↓	↑				$1s^22s^22p^63s^1$
12	Mg	↑↓	↑↓	↑↓	↑↓	↑↓	↑↓				$1s^22s^22p^63s^2$
13	Al	↑↓	↑↓	↑↓	↑↓	↑↓	↑↓	↑			$1s^22s^22p^63s^23p^1$
14	Si	↑↓	↑↓	↑↓	↑↓	↑↓	↑↓	↑	↑		$1s^22s^22p^63s^23p^2$
15	P	↑↓	↑↓	↑↓	↑↓	↑↓	↑↓	↑	↑	↑	$1s^22s^22p^63s^23p^3$
16	S	↑↓	↑↓	↑↓	↑↓	↑↓	↑↓	↑↓	↑	↑	$1s^22s^22p^63s^23p^4$
17	Cl	↑↓	↑↓	↑↓	↑↓	↑↓	↑↓	↑↓	↑↓	↑	$1s^22s^22p^63s^23p^5$
18	Ar	↑↓	↑↓	↑↓	↑↓	↑↓	↑↓	↑↓	↑↓	↑↓	$1s^22s^22p^63s^23p^6$

Magnesium (12), aluminum (13), silicon (14), phosphorus (15), sulfur (16), chlorine (17), and argon (18) follow in order. Table 10.2 summarizes the filling of the orbitals for elements 11–18.

The electrons in the outermost (highest) energy level of an atom are called the **valence electrons**. For example, oxygen, which has the electron configuration of $1s^22s^22p^4$, has electrons in the first and second energy levels. Therefore, the second principal energy level is the valence level for oxygen. The 2s and 2p electrons are the valence electrons. Valence electrons are involved in bonding atoms together to form compounds and are of particular interest to chemists, as we will see in Chapter 11.

The number of the column in 1A–7A of the periodic table gives the number of valence electrons of the element in those columns.

EXAMPLE 10.4

What is the electron configuration for the valence electrons in magnesium?

SOLUTION

Magnesium has the electron configuration of $1s^22s^22p^63s^2$. The electrons are in the first, second, and third energy levels. This means the valence electrons are in the third energy level. The 3s electrons are the valence electrons for magnesium. The electron configuration for the valence electrons in $3s^2$.

PRACTICE 10.4

Give the electron configuration for the valence electrons in these elements.

(a) B   (b) N   (c) Na   (d) Cl

WileyPLUS

ENHANCED EXAMPLE

10.5 ELECTRON STRUCTURES AND THE PERIODIC TABLE

Describe how the electron configurations of the atoms relate to their position on the periodic table and write electron configurations for elements based on their position on the periodic table.

LEARNING OBJECTIVE

KEY TERMS

period
groups or families (of elements)
representative elements
transition elements

We have seen how the electrons are assigned for the atoms of elements 1–18. How do the electron structures of these atoms relate to their position on the periodic table? To answer this question, we need to look at the periodic table more closely.

>CHEMISTRY *IN ACTION*
Collecting the Elements

Theodore Gray loves collecting samples of elements. He had an idea several years ago to design a conference table for his company, Wolfram Research in Champaign, Illinois. He built the table in the shape of the periodic table and made little boxes below each element to house a sample of the element. As he was collecting the element samples, he began to realize that he needed to keep track of where each sample had come from before he forgot. So he began to keep notes on each element sample, where he obtained it, what it was, and why it was there. Gray says, "Once I had collected three or four dozen such descriptions, I thought, "'Well I had better put this on a website, because that's what you do when you have anything these days—you put it on your own website." His element collection is ever expanding and contains more than 1400 samples.

Theodore Gray's websites, http://www.theodoregray.com/PeriodicTable and periodictable.com, both include elemental data and many stories that are both entertaining and very distracting. Gray also has created video for each element from all angles in stunning high-resolution photography, formed into video loops. He used to take all the photos himself, but the project has grown so big he now has an assistant and utilizes Mathematica, a computer program from Wolfram Research, to help him. He has also created a beautiful periodic table poster from the element images.

Take a couple of minutes and spend some time with Gray on his websites enjoying the photos and stories of the elements. He has some truly amazing examples of the elements.

© Mike Walker Photography

Theodore Gray and his periodic table conference table containing samples of each element.

The periodic table represents the efforts of chemists to organize the elements logically. Chemists of the early nineteenth century had sufficient knowledge of the properties of elements to recognize similarities among groups of elements. In 1869, Dimitri Mendeleev (1834–1907) of Russia and Lothar Meyer (1830–1895) of Germany independently published periodic arrangements of the elements based on increasing atomic masses. Mendeleev's arrangement is the precursor to the modern periodic table, and his name is associated with it. The modern periodic table is shown on the inside front cover of this book.

Each horizontal row in the periodic table is called a **period**, as shown in **Figure 10.14**. There are seven periods of elements. The number of each period corresponds to the outermost

Figure 10.14

The periodic table of the elements.

Figure 10.15

Valence electron configurations for the first 18 elements.

1A							Noble gases
1 **H** $1s^1$							2 **He** $1s^2$
	2A	3A	4A	5A	6A	7A	
3 **Li** $2s^1$	4 **Be** $2s^2$	5 **B** $2s^22p^1$	6 **C** $2s^22p^2$	7 **N** $2s^22p^3$	8 **O** $2s^22p^4$	9 **F** $2s^22p^5$	10 **Ne** $2s^22p^6$
11 **Na** $3s^1$	12 **Mg** $3s^2$	13 **Al** $3s^23p^1$	14 **Si** $3s^23p^2$	15 **P** $3s^23p^3$	16 **S** $3s^23p^4$	17 **Cl** $3s^23p^5$	18 **Ar** $3s^23p^6$

energy level that contains electrons for elements in that period. Those in Period 1 contain electrons only in energy level 1, while those in Period 2 contain electrons in levels 1 and 2. In Period 3, electrons are found in levels 1, 2, 3, and so on.

Elements that behave in a similar manner are found in **groups** or **families**. These form the vertical columns on the periodic table. Two systems exist for numbering the groups. In one system, the columns are numbered from left to right using the numbers 1–18. However, we use a system that numbers the columns with numbers and the letters A and B, as shown in Figure 10.14. The A groups are known as the **representative elements**. The B groups are called the **transition elements**. In this book we will focus on the representative elements. The groups (columns) of the periodic table often have family names. For example, the group on the far right side of the periodic table (He, Ne, Ar, Kr, Xe, and Rn) is called the *noble gases*. Group 1A is called the *alkali metals*, Group 2A the *alkaline earth metals*, and Group 7A the *halogens*.

How is the structure of the periodic table related to the atomic structures of the elements? We've just seen that the periods of the periodic table are associated with the energy level of the outermost electrons of the atoms in that period. Look at the valence electron configurations of the elements we have just examined (**Figure 10.15**). Do you see a pattern? The valence electron configuration for the elements in each column is the same. The chemical behavior and properties of elements in a particular family must therefore be associated with the electron configuration of the elements. The number for the principal energy level is different. This is expected since each new period is associated with a different energy level for the valence electrons.

The electron configurations for elements beyond these first 18 become long and tedious to write. We often abbreviate the electron configuration using the following notation:

$$\text{Na} \qquad [\text{Ne}]3s^1$$

Look carefully at Figure 10.15 and you will see that the p orbitals are full at the noble gases. By placing the symbol for the noble gas in square brackets, we can abbreviate the complete electron configuration and focus our attention on the valence electrons (the electrons we will be interested in when we discuss bonding in Chapter 11). To write the abbreviated electron configuration for any element, go back to the previous noble gas and place its symbol in square brackets. Then list the valence electrons. Here are some examples:

B $1s^22s^22p^1$ $[\text{He}]\,2s^22p^1$

Cl $1s^22s^22p^63s^23p^5$ $[\text{Ne}]\,3s^23p^5$

Na $1s^22s^22p^63s^1$ $[\text{Ne}]\,3s^1$

The sequence for filling the orbitals is exactly as we would expect up through the 3p orbitals. The third energy level might be expected to fill with 3d electrons before electrons enter the 4s orbital, but this is not the case. The behavior and properties of the next two elements, potassium (19) and calcium (20), are very similar to the elements in Groups 1A and 2A, respectively. They clearly belong in these groups. The other elements in Group 1A and Group 2A have electron configurations that indicate valence electrons in the s orbitals. For example, since the electron configuration is connected to the element's properties,

Both numbering systems are shown in the periodic table on the inside front cover.

we should place the last electrons for potassium and calcium in the $4s$ orbital. Their electron configurations are

K $1s^2 2s^2 2p^6 3s^2 3p^6 4s^1$ or $[Ar] 4s^1$

Ca $1s^2 2s^2 2p^6 3s^2 3p^6 4s^2$ or $[Ar] 4s^2$

PRACTICE 10.5

Write the abbreviated electron configuration for the following elements:
(a) Br   (b) Sr   (c) Ba   (d) Te

Elements 21–30 belong to the elements known as *transition elements*. Electrons are placed in the $3d$ orbitals for each of these elements. When the $3d$ orbitals are full, the electrons fill the $4p$ orbitals to complete the fourth period. Let's consider the overall relationship between orbital filling and the periodic table. **Figure 10.16** illustrates the type of orbital filling and its location on the periodic table. The tall columns on the table (labeled 1A–7A, and noble gases) are often called the *representative elements*. Valence electrons in these elements occupy s and p orbitals. The period number corresponds to the energy level of the valence electrons. The elements in the center of the periodic table (shown in ▨) are the transition elements where the d orbitals are being filled. Notice that the number for the d orbitals is one less than the period number. The two rows shown at the bottom of the table in Figure 10.16 are called the *inner transition elements* or the *lanthanide* and *actinide series*. The last electrons in these elements are placed in the f orbitals. The number for the f orbitals is always two less than that of the s and p orbitals. A periodic table is almost always available to you, so if you understand the relationship between the orbitals and the periodic table, you can write the electron configuration for any element. There are several minor variations to these rules, but we won't concern ourselves with them in this course.

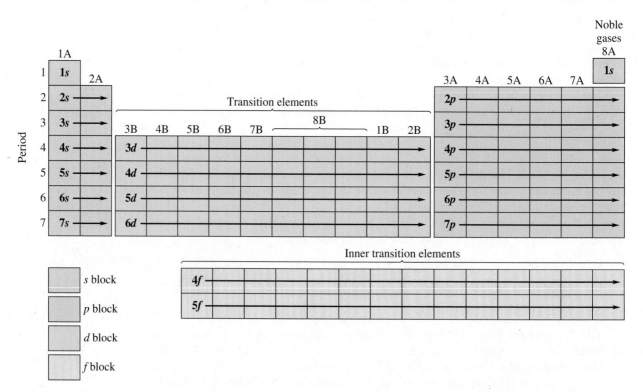

Figure 10.16
Arrangement of elements according to the electron sublevel being filled.

EXAMPLE 10.5

Use the periodic table to write the electron configuration for phosphorus and tin.

WileyPLUS

ENHANCED EXAMPLE

SOLUTION

Phosphorus is element 15 and is located in Period 3, Group 5A. The electron configuration must have a full first and second energy level:

P $1s^2 2s^2 2p^6 3s^2 3p^3$ or $[Ne]3s^2 3p^3$

You can determine the electron configuration by looking across the period and counting the element blocks.

Tin is element 50 in Period 5, Group 4A, two places after the transition metals. It must have two electrons in the $5p$ series. Its electron configuration is

Sn $1s^2 2s^2 2p^6 3s^2 3p^6 4s^2 3d^{10} 4p^6 5s^2 4d^{10} 5p^2$ or $[Kr]5s^2 4d^{10} 5p^2$

Notice that the d series of electrons always has a principal energy level number one less than its period.

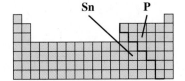

PRACTICE 10.6

Use the periodic table to write the complete electron configuration for (a) O, (b) Ca, and (c) Ti.

The early chemists classified the elements based only on their observed properties, but modern atomic theory gives us an explanation for why the properties of elements vary periodically. For example, as we "build" atoms by filling orbitals with electrons, the same orbitals occur on each energy level. This means that the same electron configuration reappears regularly for each level. Groups of elements show similar chemical properties because of the similarity of these outermost electron configurations.

In Figure 10.17, only the electron configuration of the outermost electrons is given. This periodic table illustrates these important points:

1. The number of the period corresponds with the highest energy level occupied by electrons in that period.

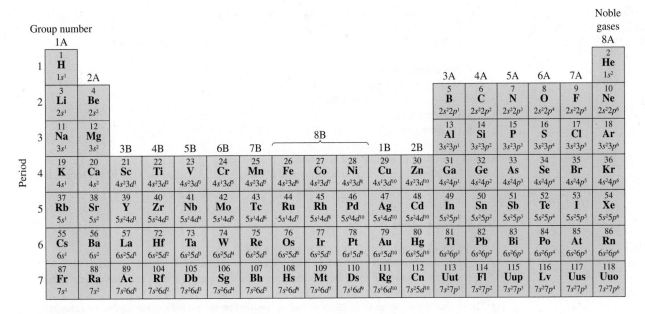

Figure 10.17

Outermost electron configurations.

2. The group numbers for the representative elements are equal to the total number of outermost electrons in the atoms of the group. For example, elements in Group 7A always have the electron configuration ns^2np^5. The d and f electrons are always in a lower energy level than the highest principal energy level and so are not considered as outermost (valence) electrons.

3. The elements of a family have the same outermost electron configuration except that the electrons are in different principal energy levels.

4. The elements within each of the s, p, d, f blocks are filling the s, p, d, f orbitals, as shown in Figure 10.16.

5. Within the transition elements, some discrepancies in the order of filling occur. (Explanation of these discrepancies and similar ones in the inner transition elements is beyond the scope of this book.)

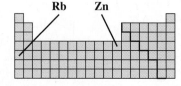

EXAMPLE 10.6

Write the complete electron configuration of a zinc atom and a rubidium atom.

SOLUTION

The atomic number of zinc is 30; it therefore has 30 protons and 30 electrons in a neutral atom. Using Figure 10.14, we see that the electron configuration of a zinc atom is $1s^22s^22p^63s^23p^64s^23d^{10}$. Check by adding the superscripts, which should equal 30.

The atomic number of rubidium is 37; therefore, it has 37 protons and 37 electrons in a neutral atom. With a little practice using a periodic table, you can write the electron configuration directly. The electron configuration of a rubidium atom is $1s^22s^22p^63s^23p^64s^23d^{10}4p^65s^1$. Check by adding the superscripts, which should equal 37.

PRACTICE 10.7

Write the complete electron configuration for a gallium atom and a lead atom.

CHAPTER **10 REVIEW**

10.1 ELECTROMAGNETIC RADIATION

KEY TERMS

wavelength
frequency
speed
photons

- Atomic theory has changed over the past 200 years:
  - Early Greek ideas
  - Dalton's model
  - Thomson's model
  - Rutherford's model
  - Modern atomic theory
- Basic wave characteristics include
  - Wavelength (λ)
  - Frequency (ν)
  - Speed (v)
- Light is a form of electromagnetic radiation.
- Light can also be considered to be composed of energy packets called photons, which behave like particles.

10.2 THE BOHR ATOM

KEY TERMS

line spectrum
quanta
ground state
orbital

- The study of atomic spectra led Bohr to propose that
  - Electrons are found in quantized energy levels in an atom.
  - Spectral lines result from the radiation of quanta of energy when the electron moves from a higher level to a lower level.
- The chemical properties of an element and its position on the periodic table depend on electrons.
- Louis de Broglie suggested that all objects have wave properties.
- Schrödinger created a mathematical model to describe electrons as waves.
  - Electrons are located in orbitals, or regions of probability, around the nucleus of an atom.

10.3 ENERGY LEVELS OF ELECTRONS

- Modern atomic theory predicts that
  - Electrons are found in discrete principal energy levels ($n = 1, 2, 3 \ldots$):
    - Energy levels contain sublevels.

Number of sublevels

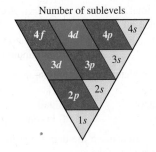

- Two electrons fit into each orbital but must have opposite spin to do so.

10.4 ATOMIC STRUCTURES OF THE FIRST 18 ELEMENTS

- Guidelines for writing electron configurations:
  - Not more than two electrons per orbital
  - Electrons fill lowest energy levels first:
    - $s < p < d < f$ for a given value of n
  - Orbitals on a given sublevel are each filled with a single electron before pairing of electrons begins to occur
- For the representative elements, only electrons in the outermost energy level (valence electrons) are involved in bonding.

10.5 ELECTRON STRUCTURES AND THE PERIODIC TABLE

- Elements in horizontal rows on the periodic table contain elements whose valence electrons (s and p) are generally on the same energy level as the number of the row.
- Elements that are chemically similar are arranged in columns (groups) on the periodic table.
- The valence electron configurations of elements in a group or family are the same, but they are located in different principal energy levels.

REVIEW QUESTIONS

1. Define the terms *wavelength* and *frequency*. What symbols do chemists generally use to represent these quantities?
2. What is the wavelength range for visible light? Which has a longer wavelength, red light or blue light?
3. What is the name given to a packet of energy?
4. What is an orbital?
5. What is meant when we say the electron structure of an atom is in its ground state?
6. What is the major difference between an orbital and a Bohr orbit?
7. Explain how and why Bohr's model of the atom was modified to include the cloud model of the atom.
8. How do 1s and 2s orbitals differ? How are they alike?
9. What letters are used to designate the types of orbitals?
10. List the following orbitals in order of increasing energy: 2s, 2p, 4s, 1s, 3d, 3p, 4p, 3s.
11. How many s electrons, p electrons, and d electrons are possible in any energy level?
12. Sketch the s, p_x, p_y, and p_z orbitals.
13. Under what conditions can a second electron enter an orbital already containing one electron?
14. What is a valence shell?
15. What are valence electrons, and why are they important?

16. A lanthanide element has the designation $4f^3$ in its electron structure. What is the significance of the 4, the f, and the 3?
17. Of the elements Ir, Pb, Xe, Zr, and Ag, which ones are not contained in the groups of representative elements?
18. From the standpoint of electron structure, what do the elements in the p block have in common?
19. Write symbols for the elements with atomic numbers 6, 7, 8, 15, and 33. Which of those elements have something in common? Explain.
20. Write the symbols of the first three elements that have six electrons in their outermost energy level.
21. What is the greatest number of elements to be found in any period? Which periods have this number?
22. From the standpoint of energy level, how does the placement of the last electron in the Group A elements differ from that of the Group B elements?
23. Find the places in the periodic table where elements are not in proper sequence according to atomic mass. (See inside of front cover for periodic table.)
24. What are the names of the two scientists who independently published results that have led to the establishment of the periodic table?
25. Which scientist is credited with being the author of the modern periodic table?

Most of the exercises in this chapter are available for assignment via the online homework management program, WileyPLUS (www.wileyplus.com). All exercises with blue numbers have answers in Appendix VI.

PAIRED EXERCISES

1. How many total and how many valence electrons are in each of the following neutral atoms?
(a) lithium   (c) calcium
(b) magnesium   (d) fluorine

2. How many total and how many valence electrons are in each of the following neutral atoms?
(a) sodium   (c) phosphorus
(b) arsenic   (d) aluminum

3. Write the complete electron configuration for each of the following elements:
(a) scandium   (c) bromine
(b) rubidium   (d) sulfur

4. Write the complete electron configuration for each of the following elements:
(a) manganese   (c) gallium
(b) krypton   (d) boron

5. Explain how the spectral lines of hydrogen occur.

6. Explain how Bohr used the data from the hydrogen spectrum to support his model of the atom.

7. How many orbitals exist in the fourth principal energy level? What are they, and in what periods can they be found?

8. How many total electrons can be present in the third principal energy level? In what orbitals are they found?

9. Write orbital diagrams for these elements:
(a) N   (b) Cl   (c) Zn   (d) Zr   (e) I

10. Write orbital diagrams for these elements:
(a) Si   (b) S   (c) Ar   (d) V   (e) P

11. For each of the orbital diagrams given, write out the corresponding electron configurations.

(a) O   [↑↓] [↑↓] [↑↓|↑|↑]

(b) Ca   [↑↓] [↑↓] [↑↓|↑↓|↑↓] [↑↓] [↑↓|↑↓|↑↓] [↑↓]

(c) Ar   [↑↓] [↑↓] [↑↓|↑↓|↑↓] [↑↓] [↑↓|↑↓|↑↓]

(d) Br   [↑↓] [↑↓] [↑↓|↑↓|↑↓] [↑↓] [↑↓|↑↓|↑↓] [↑↓] [↑↓|↑↓|↑↓|↑↓|↑↓] [↑↓|↑↓|↑]

(e) Fe   [↑↓] [↑↓] [↑↓|↑↓|↑↓] [↑↓] [↑↓|↑↓|↑↓] [↑↓] [↑↓|↑|↑|↑|↑]

12. For each of the orbital diagrams given, write out the corresponding electron configurations.

(a) Li   [↑↓] [↑]

(b) P   [↑↓] [↑↓] [↑↓|↑↓|↑↓] [↑↓] [↑|↑|↑]

(c) Zn   [↑↓] [↑↓] [↑↓|↑↓|↑↓] [↑↓] [↑↓|↑↓|↑↓] [↑↓] [↑↓|↑↓|↑↓|↑↓|↑↓]

(d) Na   [↑↓] [↑↓] [↑↓|↑↓|↑↓] [↑]

(e) K   [↑↓] [↑↓] [↑↓|↑↓|↑↓] [↑↓] [↑↓|↑↓|↑↓] [↑]

13. Identify the correct orbital diagrams below. For the incorrect diagrams redraw the correct orbital diagram.

    1s   2s   2p   3s   3p   4s   3d   4p

(a) [↑↓] [↑↓] [↑↓|↑↓|↑] [↑]

(b) [↑↓] [↑↓] [↑↓|↑↓|↑↓] [↑↓] [↑|↑|↑]

(c) [↑↓] [↑↓] [↑↓|↑↓|↑↓] [↑↓] [↑↓|↑↓|↑↓] [↑↓] [↑↓|↑↓|↑↓|↑↓|↑↓] [↑| |]

(d) [↑↓] [↑↓] [↑↓|↑↓|↑↓] [↑↓] [↑↓|↑↓|↑↓] [↑↓] [↑↓|↑↓|↑| |]

14. Identify the correct orbital diagrams below. For the incorrect diagrams redraw the correct orbital diagram.

    1s   2s   2p   3s   3p   4s   4p   3d

(a) [↑↓] [↑↓] [↑|↑|↑]

(b) [↑↓] [↑↓] [↑↓|↑↓|↑↓] [↑↓] [↑↓|↑↓|↑↓] [↑↓] [↑↓|↑↓|↑↓] [↑|↑| |]

(c) [↑↓] [↑↓] [↑↓|↑↓|↑↓] [↑↓] [↑↓|↑↓|↑↓] [↑↑]

(d) [↑↓] [↑↓] [↑↓|↑↓|↑↓] [↑↓] [↑↓|↑|↑]

15. Identify the element represented by each of the orbital diagrams in Problem 13.

16. Identify the element represented by each of the orbital diagrams in Problem 14.

17. Which elements have these electron configurations? Give both the symbol and the name:
(a) $1s^2 2s^2 2p^5$ (c) $[Ne]3s^2 3p^4$
(b) $1s^2 2s^2 2p^6 3s^1$ (d) $[Ar]4s^2 3d^8$

18. Which of the elements have these electron configurations? Give both the symbol and the name:
(a) $1s^2 2s^2 2p^1$ (c) $[Xe]6s^2 4f^{14} 5d^{10} 6p^2$
(b) $1s^2 2s^2 2p^6 3s^2 3p^2$ (d) $[Kr]5s^2 4d^{10} 5p^4$

19. Identify the element and draw orbital diagrams for elements with these atomic numbers:
(a) 22   (b) 18   (c) 33   (d) 35   (e) 25

20. Identify the element and draw orbital diagrams for elements with these atomic numbers:
(a) 15   (b) 30   (c) 20   (d) 34   (e) 19

21. For each of the electron configurations given, write the corresponding orbital diagrams.
(a) F $1s^2 2s^2 2p^5$
(b) S $1s^2 2s^2 2p^6 3s^2 3p^4$
(c) Co $1s^2 2s^2 2p^6 3s^2 3p^6 4s^2 3d^7$
(d) Kr $1s^2 2s^2 2p^6 3s^2 3p^6 4s^2 3d^{10} 4p^6$
(e) Ru $1s^2 2s^2 2p^6 3s^2 3p^6 4s^2 3d^{10} 4p^6 5s^2 4d^6$

22. For each of the electron configuration given, write the corresponding orbital diagrams.
(a) Cl $1s^2 2s^2 2p^6 3s^2 3p^5$
(b) Mg $1s^2 2s^2 2p^6 3s^2$
(c) Ni $1s^2 2s^2 2p^6 3s^2 3p^6 4s^2 3d^8$
(d) Cu $1s^2 2s^2 2p^6 3s^2 3p^6 4s^2 3d^9$
(e) Ba $1s^2 2s^2 2p^6 3s^2 3p^6 4s^2 3d^{10} 4p^6 5s^2 4d^{10} 5p^6 6s^2$

23. Identify these elements from their atomic structure diagrams:

(a) $\left(\begin{array}{c}16p\\16n\end{array}\right)$ $2e^-$ $8e^-$ $6e^-$

(b) $\left(\begin{array}{c}28p\\32n\end{array}\right)$ $2e^-$ $8e^-$ $16e^-$ $2e^-$

24. Diagram the atomic structures (as you see in Exercise 23) for these elements:
(a) $^{27}_{13}Al$
(b) $^{48}_{22}Ti$

25. Why is the eleventh electron of the sodium atom located in the third energy level rather than in the second energy level?

26. Why is the last electron in potassium located in the fourth energy level rather than in the third energy level?

27. What electron structure do the noble gases have in common?

28. What is the unique about the noble gases from an electron point of view?

29. How are elements in a period related to one another?

30. How are elements in a group related to one another?

31. How many valence electrons do each of the following elements have?
(a) C   (b) S   (c) K   (d) I   (e) B

32. How many valence electrons do each of the following elements have?
(a) N   (b) P   (c) O   (d) Ba   (e) Al

33. What do the electron structures of the alkali metals have in common?

34. Why would you expect elements zinc, cadmium, and mercury to be in the same chemical family?

35. Pick the electron structures that represent elements in the same chemical family:
(a) $1s^2 2s^1$ (e) $1s^2 2s^2 2p^6 3s^2 3p^6$
(b) $1s^2 2s^2 2p^4$ (f) $1s^2 2s^2 2p^6 3s^2 3p^6 4s^2$
(c) $1s^2 2s^2 2p^2$ (g) $1s^2 2s^2 2p^6 3s^2 3p^6 4s^1$
(d) $1s^2 2s^2 2p^6 3s^2 3p^4$ (h) $1s^2 2s^2 2p^6 3s^2 3p^6 4s^2 3d^1$

36. Pick the electron structures that represent elements in the same chemical family:
(a) $[He]2s^2 2p^6$ (e) $[Ar]4s^2 3d^{10}$
(b) $[Ne]3s^1$ (f) $[Ar]4s^2 3d^{10} 4p^6$
(c) $[Ne]3s^2$ (g) $[Ar]4s^2 3d^5$
(d) $[Ne]3s^2 3p^3$ (h) $[Kr]5s^2 4d^{10}$

37. In the periodic table, calcium, element 20, is surrounded by elements 12, 19, 21, and 38. Which of these have physical and chemical properties most resembling calcium?

38. In the periodic table, phosphorus, element 15, is surrounded by elements 14, 7, 16, and 33. Which of these have physical and chemical properties most resembling phosphorus?

39. Classify the following elements as metals, nonmetals, or metalloids (review Chapter 3 if you need help):
(a) potassium   (c) sulfur
(b) plutonium   (d) antimony

40. Classify the following elements as metals, nonmetals, or metalloids (review Chapter 3 if you need help):
(a) iodine   (c) molybdenum
(b) tungsten   (d) germanium

41. In which period and group does an electron first appear in an f orbital?

42. In which period and group does an electron first appear in a d orbital?

43. How many electrons occur in the valence level of Group 7A and 7B elements? Why are they different?

44. How many electrons occur in the valence level of Group 3A and 3B elements? Why are they different?

45. Determine the identity of the element that contains exactly
(a) three $4p$ electrons in the ground state
(b) seven $3d$ electrons in the ground state
(c) one $2s$ electrons in the ground state
(d) five $3p$ electrons in the ground state

46. Determine the identity of the element that contains exactly
(a) two $6p$ electrons in the ground state
(b) five $4f$ electrons in the ground state
(c) one $4p$ electrons in the ground state
(d) seven $5d$ electrons in the ground state

ADDITIONAL EXERCISES

47. Using only the periodic table, explain how the valence energy level and the number of valence electrons could be determined.

48. Using only a periodic table, write the full electron configuration for each of the following elements. Circle the valence electrons.
  (a) Mg (b) P (c) K (d) F (e) Se (f) N

49. Which of the following would have the same number of valence electrons?
  (a) Na^+ (b) O (c) Li (d) F^- (e) Ne

50. Name the group in which each of the following elements appear:
  (a) $1s^2 2s^2 2p^5$
  (b) $1s^2 2s^2 2p^6 3s^2 3p^6 4s^2$
  (c) $1s^2 2s^2 2p^6 3s^1$
  (d) $1s^2 2s^2 2p^6 3s^2 3p^6 4s^2 3d^{10} 4p^6$
  (e) $1s^2$
  (f) $1s^2 2s^2 2p^6 3s^2 3p^6 4s^2 3d^{10} 4p^6 5s^1$

51. Chromium is a lustrous silver-colored metal that has been used to prevent corrosion for centuries. Bronze swords and other weapons discovered in burial pits from the Qin Dynasty were coated with chromium and had not corroded at all since their entombing. Today many items are coated with a layer of chromium as a decorative and protective covering.

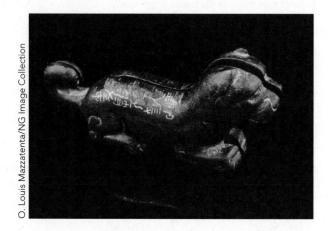

O. Louis Mazzatenta/NG Image Collection

  (a) Sometimes the actual electron configurations of the elements differ from those predicted by the periodic table. The experimentally determined electron configuration for chromium is $1s^2 2s^2 2p^6 3s^2 3p^6 4s^1 3d^5$. Is this the electronic configuration you would predict based on the periodic table? If not, what is the configuration predicted by the periodic table for chromium?
  (b) Chromium has a density of 7.19 g/cm^3. How many atoms of chromium are contained in a 5.00-cm^3 sample of chromium?
  (c) If the radius of a chromium atom is 1.40×10^{-8} cm, what is the volume $\left(V = \dfrac{4}{3}\pi r^3 \right)$ of a single chromium atom?
  (d) How many chromium atoms occupy a volume of 5.00 cm^3?

52. Anyone watching fireworks enjoys the beautiful colors produced. The colors are produced by adding different elements to the mix of fuels in the fireworks. Red is produced by adding calcium, yellow by adding sodium, and blue by adding copper. Explain why adding different elements might give fireworks their characteristic colors.

53. There has long been speculation that Wolfgang Amadeus Mozart may have been poisoned by a rival. There is some evidence that he may have been poisoned by an antimony compound. This compound would have been prescribed by his doctor, however, suggesting that there was no foul play. What is the electronic configuration of antimony.

54. A dog is suspected of attacking a runner in the park. This runner was wearing a pearlescent lip gloss containing bismuth hypochlorite, a compound often used in cosmetics to give this pearlescent effect. Upon analyzing a residue on the dog's fur, this compound is detected giving additional support to the victim's claim. Write the complete and shorthand electronic configurations for bismuth.

55. Write the electron configuration of each of the following neutral atoms. (You may need to refer to Figure 3.2.)
  (a) the four most abundant elements in the Earth's crust, seawater, and air
  (b) the five most abundant elements in the human body

56. Give the maximum number of electrons that can reside in the following:
  (a) a p orbital
  (b) a d sublevel
  (c) the third principal energy level
  (d) an s orbital
  (e) an f sublevel

57. Give the names of each of the following elements based on the information given:
  (a) the second element in Period 3
  (b) $[Ne]3s^2 3p^3$
  (c)

↑↓	↑↓	↑↓ ↑↓ ↑↓	↑↓	↑↓ ↑↓ ↑↓
$1s$	$2s$	$2p$	$3s$	$3p$

58. Why does the emission spectrum for nitrogen reveal many more spectral lines than that for hydrogen?

59. List the first element on the periodic table that satisfies each of these conditions:
  (a) a completed set of p orbitals
  (b) two $4p$ electrons
  (c) seven valence electrons
  (d) three unpaired electrons

60. In Group 7A, the first two elements are gases, the next one is a liquid, and the fourth one is a solid. What do they have in common that causes them to be in the same group?

61. In which groups are the transition elements located?

62. How do the electron structures of the transition elements differ from those of the representative elements?

63. The atomic numbers of the noble gases are 2, 10, 18, 36, 54 and 86. Without looking at a periodic table, determine the atomic numbers for the elements that have five electrons in their valence shell.

64. What is the family name for
  (a) Group 1A?   (c) Group 7A?
  (b) Group 2A?

65. What sublevel is being filled in
  (a) Period 3, Group 3A to 7A?
  (b) Period 5, transition elements?
  (c) the lanthanide series?

66. Classify each of the following as a noble gas, a representative element, or a transition metal. Also indicate whether the element is a metal, nonmetal, or metalloid.
(a) Na   (d) Ra
(b) N   (e) As
(c) Mo   (f) Ne

67. If element 36 is a noble gas, in which groups would you expect elements 35 and 37 to occur?

68. Write a paragraph describing the general features of the periodic table.

69. What is the relationship between two elements if
(a) one of them has 10 electrons, 10 protons, and 10 neutrons and the other has 10 electrons, 10 protons, and 12 neutrons?
(b) one of them has 23 electrons, 23 protons, and 27 neutrons and the other has 24 electrons, 24 protons, and 26 neutrons?

70. Is there any pattern for the location of gases on the periodic table? for the location of liquids? for the location of solids?

CHALLENGE EXERCISES

71. A valence electron in an atom of sulfur is excited by heating a sample. The electron jumps from the s orbital to the p orbital. What is the electron configuration of the excited sulfur atom, and what would the orbital diagram look like?

72. Element 87 is in Group 1A, Period 7. In how many principal energy levels are electrons located? Describe its outermost energy level.

73. Use the periodic table to explain why metals tend to lose electrons and nonmetals tend to gain electrons.

74. Show how the periodic table helps determine the expected electron configuration for any element.

ANSWERS TO PRACTICE EXERCISES

10.1

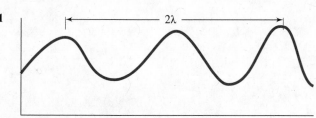

10.2 A photon is a little packet of electromagnetic radiation energy and has the property of both a particle and a wave. It travels at the speed of light, 3.00×10^8 meters per second.

10.3 (a) Two electrons. Wolfgang Pauli's exclusion principle.
(b) $1s^2 2s^2 2p^6 3s^2 3p^6$ for a total of 18 electrons.

10.4 (a) $2s^2 2p^1$ (c) $3s^1$
(b) $2s^2 2p^3$ (d) $3s^2 3p^5$

10.5 (a) $[Ar]4s^2 4p^5$ (c) $[Xe]6s^2$
(b) $[Kr]5s^2$ (d) $[Kr]5s^2 5p^4$

10.6 (a) O $1s^2 2s^2 2p^4$
(b) Ca $1s^2 2s^2 2p^6 3s^2 3p^6 4s^2$
(c) Ti $1s^2 2s^2 2p^6 3s^2 3p^6 4s^2 3d^2$

10.7 Ga, $1s^2 2s^2 2p^6 3s^2 3p^6 4s^2 3d^{10} 4p^1$
Pb, $1s^2 2s^2 2p^6 3s^2 3p^6 4s^2 3d^{10} 4p^6 5s^2 4d^{10} 5p^6 6s^2 4f^{14} 5d^{10} 6p^2$

The atoms in tartaric acid bond together in a very specific orientation to form the shape of the molecule. The molecules collect together to form a crystal, photographed here in a polarized light micrograph. Tartaric acid is used in baking powder and as a food additive.

Pasieka/Photo Researchers, Inc.

CHAPTER **11**

CHEMICAL BONDS: THE FORMATION OF COMPOUNDS FROM ATOMS

For centuries we've been aware that certain metals cling to a magnet. We've seen balloons sticking to walls. Why? High-speed levitation trains are heralded to be the wave of the future. How do they function? In each case, forces of attraction and repulsion are at work.

Human interactions also suggest that "opposites attract" and "likes repel." Attractions draw us into friendships and significant relationships, whereas repulsive forces may produce debate and antagonism. We form and break apart interpersonal bonds throughout our lives.

In chemistry, we also see this phenomenon. Substances form chemical bonds as a result of electrical attractions. These bonds provide the tremendous diversity of compounds found in nature. This chapter is one of the most significant and useful chapters in the book—chemical bonding between atoms. This is what chemistry is really all about. Study it carefully.

CHAPTER OUTLINE

11.1 Periodic Trends in Atomic Properties

11.2 Lewis Structures of Atoms

11.3 The Ionic Bond: Transfer of Electrons from One Atom to Another

11.4 Predicting Formulas of Ionic Compounds

11.5 The Covalent Bond: Sharing Electrons

11.6 Electronegativity

11.7 Lewis Structures of Compounds

11.8 Complex Lewis Structures

11.9 Compounds Containing Polyatomic Ions

11.10 Molecular Shape

11.1 PERIODIC TRENDS IN ATOMIC PROPERTIES

LEARNING OBJECTIVE

Discuss the atomic trends for metals and nonmetals, atomic radius, and ionization energy.

KEY TERM
ionization energy

Although atomic theory and electron configuration help us understand the arrangement and behavior of the elements, it's important to remember that the design of the periodic table is based on observing properties of the elements. Before we use the concept of atomic structure to explain how and why atoms combine to form compounds, we need to understand the characteristic properties of the elements and the trends that occur in these properties on the periodic table. These trends allow us to use the periodic table to accurately predict properties and reactions of a wide variety of substances.

Metals and Nonmetals

In Section 3.5, we classified elements as metals, nonmetals, or metalloids. The heavy stair-step line beginning at boron and running diagonally down the periodic table separates the elements into metals and nonmetals. Metals are usually lustrous, malleable, and good conductors of heat and electricity. Nonmetals are just the opposite—nonlustrous, brittle, and poor conductors. Metalloids are found bordering the heavy diagonal line and may have properties of both metals and nonmetals.

Most elements are classified as metals (see **Figure 11.1**). Metals are found on the left side of the stair-step line, while the nonmetals are located toward the upper right of the table. Note that hydrogen does not fit into the division of metals and nonmetals. It displays nonmetallic properties under normal conditions, even though it has only one outermost electron like the alkali metals. Hydrogen is considered to be a unique element.

It is the chemical properties of metals and nonmetals that interest us most. Metals tend to lose electrons and form positive ions, while nonmetals tend to gain electrons and form negative ions. When a metal reacts with a nonmetal, electrons are often transferred from the metal to the nonmetal.

Metals
Metalloids
Nonmetals

1 H																	2 He
3 Li	4 Be											5 B	6 C	7 N	8 O	9 F	10 Ne
11 Na	12 Mg											13 Al	14 Si	15 P	16 S	17 Cl	18 Ar
19 K	20 Ca	21 Sc	22 Ti	23 V	24 Cr	25 Mn	26 Fe	27 Co	28 Ni	29 Cu	30 Zn	31 Ga	32 Ge	33 As	34 Se	35 Br	36 Kr
37 Rb	38 Sr	39 Y	40 Zr	41 Nb	42 Mo	43 Tc	44 Ru	45 Rh	46 Pd	47 Ag	48 Cd	49 In	50 Sn	51 Sb	52 Te	53 I	54 Xe
55 Cs	56 Ba	57 La*	72 Hf	73 Ta	74 W	75 Re	76 Os	77 Ir	78 Pt	79 Au	80 Hg	81 Tl	82 Pb	83 Bi	84 Po	85 At	86 Rn
87 Fr	88 Ra	89 Ac†	104 Rf	105 Db	106 Sg	107 Bh	108 Hs	109 Mt	110 Ds	111 Rg	112 Cn	113 Uut	114 Fl	115 Uup	116 Lv	117 Uus	118 Uuo

*	58 Ce	59 Pr	60 Nd	61 Pm	62 Sm	63 Eu	64 Gd	65 Tb	66 Dy	67 Ho	68 Er	69 Tm	70 Yb	71 Lu
†	90 Th	91 Pa	92 U	93 Np	94 Pu	95 Am	96 Cm	97 Bk	98 Cf	99 Es	100 Fm	101 Md	102 No	103 Lr

Figure 11.1
The elements are classified as metals, nonmetals, and metalloids.

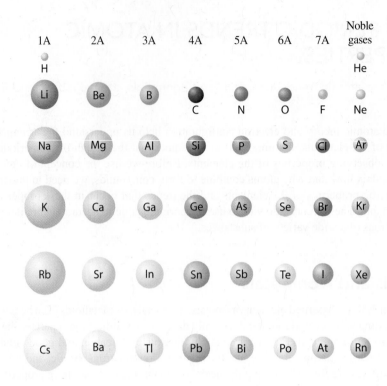

Figure 11.2

Relative atomic radii for the representative elements. Atomic radius decreases across a period and increases down a group in the periodic table.

Atomic Radius

The relative radii of the representative elements are shown in **Figure 11.2**. Notice that the radii of the atoms tend to increase down each group and that they tend to decrease from left to right across a period.

The increase in radius down a group can be understood if we consider the electron structure of the atoms. For each step down a group, an additional energy level is added to the atom. The average distance from the nucleus to the outside edge of the atom must increase as each new energy level is added. The atoms get bigger as electrons are placed in these new higher-energy levels.

Understanding the decrease in atomic radius across a period requires more thought, however. As we move from left to right across a period, electrons within the same block are being added to the same principal energy level. Within a given energy level, we expect the orbitals to have about the same size. We would then expect the atoms to be about the same size across the period. But each time an electron is added, a proton is added to the nucleus as well. The increase in positive charge (in the nucleus) pulls the electrons closer to the nucleus, which results in a gradual decrease in atomic radius across a period.

Ionization Energy

The **ionization energy** of an atom is the energy required to remove an electron from the atom. For example,

$$Na + \text{ionization energy} \longrightarrow Na^+ + e^-$$

The first ionization energy is the amount of energy required to remove the first electron from an atom, the second is the amount required to remove the second electron from that atom, and so on.

Table 11.1 gives the ionization energies for the removal of one to five electrons from several elements. The table shows that even higher amounts of energy are needed to remove the second, third, fourth, and fifth electrons. This makes sense because removing electrons leaves fewer electrons attracted to the same positive charge in the nucleus. The data in Table 11.1 also show that an extra-large ionization energy (blue) is needed when an electron is removed

TABLE 11.1   Ionization Energies for Selected Elements*

Element	Required amounts of energy (kJ/mol)				
	1st e⁻	2nd e⁻	3rd e⁻	4th e⁻	5th e⁻
H	1,314				
He	2,372	5,247			
Li	520	7,297	11,810		
Be	900	1,757	14,845	21,000	
B	800	2,430	3,659	25,020	32,810
C	1,088	2,352	4,619	6,222	37,800
Ne	2,080	3,962	6,276	9,376	12,190
Na	496	4,565	6,912	9,540	13,355

*Values are expressed in kilojoules per mole, showing energies required to remove 1 to 5 electrons per atom.
Blue type indicates the energy needed to remove an electron from a noble gas electron structure.

from a noble gas-like structure, clearly showing the stability of the electron structure of the noble gases.

First ionization energies have been experimentally determined for most elements. **Figure 11.3** plots these energies for representative elements in the first four periods. Note these important points:

1. Ionization energy in Group A elements decreases from top to bottom in a group. For example, in Group 1A the ionization energy changes from 520 kJ/mol for Li to 419 kJ/mol for K.

2. Ionization energy gradually increases from left to right across a period. Noble gases have a relatively high value, confirming the nonreactive nature of these elements.

Metals don't behave in exactly the same manner. Some metals give up electrons much more easily than others. In the alkali metal family, cesium gives up its 6s electron much more easily than the metal lithium gives up its 2s electron. This makes sense when we consider that the size of the atoms increases down the group. The distance between the nucleus and the outer electrons increases and the ionization energy decreases. The most chemically active metals are located at the lower left of the periodic table.

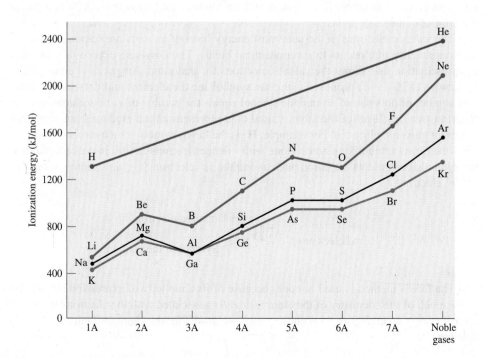

Figure 11.3

Periodic relationship of the first ionization energy for representative elements in the first four periods.

Nonmetals have relatively large ionization energies compared to metals. Nonmetals tend to gain electrons and form anions. Since the nonmetals are located at the right side of the periodic table, it is not surprising that ionization energies tend to increase from left to right across a period. The most active nonmetals are found in the *upper* right corner of the periodic table (excluding the noble gases).

EXAMPLE 11.1

Why is there such a large increase in the ionization energy to remove the fifth electron from a carbon atom? (See Table 11.1.)

SOLUTION

Two main factors apply here. (1) The remaining carbon atom has six protons attracting only two electrons and (2) these two electrons constitute the stable noble gas helium. Both factors make it much more difficult to withdraw the fifth electron.

PRACTICE 11.1

(a) What are the trends in the first ionization energy for the groups of elements and for the periods of elements in the periodic table?

(b) Consider your answer to part (a) and state why this is the case for the groups of elements.

11.2 LEWIS STRUCTURES OF ATOMS

LEARNING OBJECTIVE ● Draw the Lewis structure for a given atom.

KEY TERM
Lewis structure

Metals tend to form cations (positively charged ions) and nonmetals form anions (negatively charged ions) in order to attain a stable valence electron structure. For many elements this stable valence level contains eight electrons (two s and six p), identical to the valence electron configuration of the noble gases. Atoms undergo rearrangements of electron structure to lower their chemical potential energy (or to become more stable). These rearrangements are accomplished by losing, gaining, or sharing electrons with other atoms. For example, a hydrogen atom could accept a second electron and attain an electron structure the same as the noble gas helium. A fluorine atom could gain an electron and attain an electron structure like neon. A sodium atom could lose one electron to attain an electron structure like neon.

The valence electrons in the outermost energy level of an atom are responsible for the electron activity that occurs to form chemical bonds. The **Lewis structure** of an atom is a representation that shows the valence electrons for that atom. American chemist Gilbert N. Lewis (1875–1946) proposed using the symbol for the element and dots for electrons. The number of dots placed around the symbol equals the number of s and p electrons in the outermost energy level of the atom. Paired dots represent paired electrons; unpaired dots represent unpaired electrons. For example, **H·** is the Lewis symbol for a hydrogen atom, $1s^1$; **:B** is the Lewis symbol for a boron atom, with valence electrons $2s^2 2p^1$. In the case of boron, the symbol B represents the boron nucleus and the $1s^2$ electrons; the dots represent only the $2s^2 2p^1$ electrons.

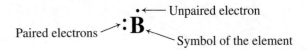

Paired electrons $\longrightarrow$:**B** $\longleftarrow$ Symbol of the element

Unpaired electron

The Lewis method is used not only because of its simplicity of expression but also because much of the chemistry of the atom is directly associated with the electrons in the outermost energy level. **Figure 11.4** shows Lewis structures for the elements hydrogen through calcium.

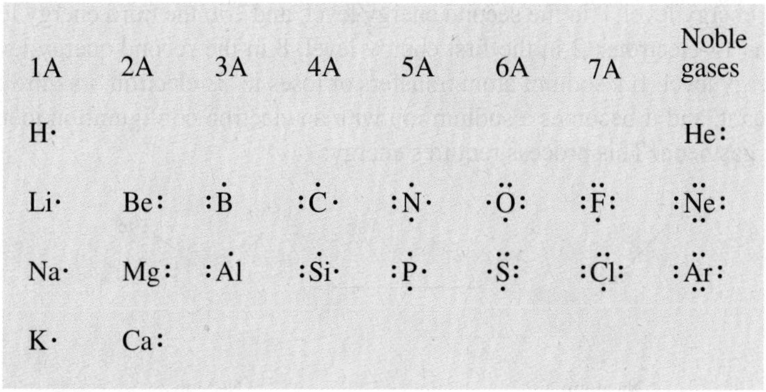

							Noble gases
1A	2A	3A	4A	5A	6A	7A	
H·							He:
Li·	Be:	:B	:C·	:N·	·Ö·	:F:	:Ne:
Na·	Mg:	·Al	:Si·	:P·	·S·	:Cl:	:Ar:
K·	Ca:						

Figure 11.4

Lewis structures of the first 20 elements. Dots represent electrons in the outermost s and p energy levels only.

EXAMPLE 11.2

Write the Lewis structure for a phosphorus atom.

SOLUTION

First establish the electron structure for a phosphorus atom, which is $1s^2 2s^2 2p^6 3s^2 3p^3$. Note that there are five electrons in the outermost energy level; they are $3s^2 3p^3$. Write the symbol for phosphorus and place the five electrons as dots around it.

The $3s^2$ electrons are paired and are represented by the paired dots. The $3p^3$ electrons, which are unpaired, are represented by the single dots.

PRACTICE 11.2

Write the Lewis structure for the following elements:
(a) N   (b) Al   (c) Sr   (d) Br

A quick way to determine the correct number of dots (electrons) for a Lewis structure is to use the group number. For the A groups on the periodic table, the group number is the same as the number of electrons in the Lewis structure.

11.3 THE IONIC BOND: TRANSFER OF ELECTRONS FROM ONE ATOM TO ANOTHER

Discuss the formation of an ionic bond and the chemical change that results from the bond.

LEARNING OBJECTIVE

The chemistry of many elements, especially the representative ones, is to attain an outer electron structure like that of the chemically stable noble gases. With the exception of helium, this stable structure consists of eight electrons in the outermost energy level (see Table 11.2).

KEY TERM

ionic bond

Let's look at the electron structures of sodium and chlorine to see how each element can attain a structure of 8 electrons in its outermost energy level. A sodium atom has 11 electrons:

TABLE 11.2   Arrangement of Electrons in the Noble Gases*

Noble gas	Symbol	Electron structure					
		$n = 1$	2	3	4	5	6
Helium	He	$1s^2$					
Neon	Ne	$1s^2$	$2s^2 2p^6$				
Argon	Ar	$1s^2$	$2s^2 2p^6$	$3s^2 3p^6$			
Krypton	Kr	$1s^2$	$2s^2 2p^6$	$3s^2 3p^6 3d^{10}$	$4s^2 4p^6$		
Xenon	Xe	$1s^2$	$2s^2 2p^6$	$3s^2 3p^6 3d^{10}$	$4s^2 4p^6 4d^{10}$	$5s^2 5p^6$	
Radon	Rn	$1s^2$	$2s^2 2p^6$	$3s^2 3p^6 3d^{10}$	$4s^2 4p^6 4d^{10} 4f^{14}$	$5s^2 5p^6 5d^{10}$	$6s^2 6p^6$

*Each gas except helium has eight electrons in its outermost energy level.

2 in the first energy level, 8 in the second energy level, and 1 in the third energy level. A chlorine atom has 17 electrons: 2 in the first energy level, 8 in the second energy level, and 7 in the third energy level. If a sodium atom transfers or loses its $3s$ electron, its third energy level becomes vacant, and it becomes a sodium ion with an electron configuration identical to that of the noble gas neon. This process requires energy:

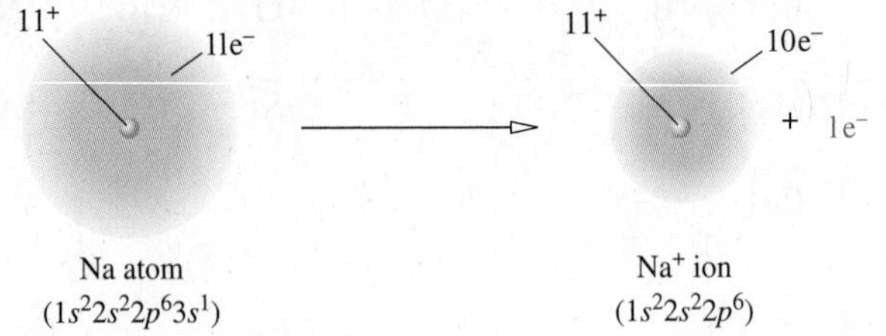

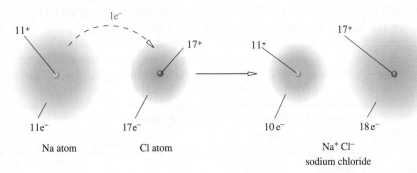

An atom that has lost or gained electrons will have a positive or negative charge, depending on which particles (protons or electrons) are in excess. Remember that a charged particle or group of particles is called an *ion*.

By losing a negatively charged electron, the sodium atom becomes a positively charged particle known as a sodium ion. The charge, $+1$, results because the nucleus still contains 11 positively charged protons, and the electron orbitals contain only 10 negatively charged electrons. The charge is indicated by a plus sign ($+$) and is written as a superscript after the symbol of the element (Na^+).

A chlorine atom with seven electrons in the third energy level needs one electron to pair up with its one unpaired $3p$ electron to attain the stable outer electron structure of argon. By gaining one electron, the chlorine atom becomes a chloride ion (Cl^-), a negatively charged particle containing 17 protons and 18 electrons. This process releases energy:

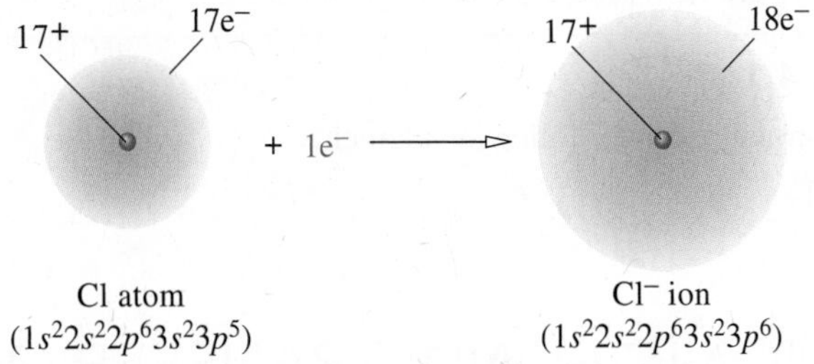

Consider sodium and chlorine atoms reacting with each other. The $3s$ electron from the sodium atom transfers to the half-filled $3p$ orbital in the chlorine atom to form a positive sodium ion and a negative chloride ion. The compound sodium chloride results because the Na^+ and Cl^- ions are strongly attracted to each other by their opposite electrostatic charges. The force holding the oppositely charged ions together is called an ionic bond:

The Lewis representation of sodium chloride formation is

$$Na \cdot + \cdot \ddot{\underset{\cdot\cdot}{Cl}} : \longrightarrow [Na]^+ \left[: \ddot{\underset{\cdot\cdot}{Cl}} : \right]^-$$

The chemical reaction between sodium and chlorine is a very vigorous one, producing considerable heat in addition to the salt formed. When energy is released in a chemical reaction, the products are more stable than the reactants. Note that in NaCl both atoms attain a noble gas electron structure.

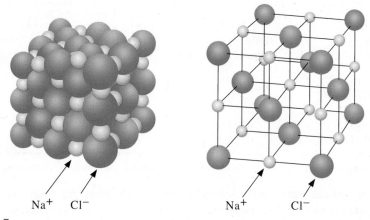

These tiny NaCl crystals show the cubic structure illustrated in Figure 11.5.

Figure 11.5

Sodium chloride crystal. Diagram represents a small fragment of sodium chloride, which forms cubic crystals. Each sodium ion is surrounded by six chloride ions, and each chloride ion is surrounded by six sodium ions. The tiny NaCl crystals show the cubic crystal structure of salt.

Sodium chloride is made up of cubic crystals in which each sodium ion is surrounded by six chloride ions and each chloride ion by six sodium ions, except at the crystal surface. A visible crystal is a regularly arranged aggregate of millions of these ions, but the ratio of sodium to chloride ions is 1:1, hence the formula NaCl. The cubic crystalline lattice arrangement of sodium chloride is shown in **Figure 11.5**.

Figure 11.6 contrasts the relative sizes of sodium and chlorine atoms with those of their ions. The sodium ion is smaller than the atom due primarily to two factors: (1) The sodium atom has lost its outermost electron, thereby reducing its size; and (2) the 10 remaining electrons are now attracted by 11 protons and are thus drawn closer to the nucleus. Conversely, the chloride ion is larger than the atom because (1) it has 18 electrons but only 17 protons and (2) the nuclear attraction on each electron is thereby decreased, allowing the chlorine atom to expand as it forms an ion.

We've seen that when sodium reacts with chlorine, each atom becomes an ion. Sodium chloride, like all ionic substances, is held together by the attraction existing between positive and negative charges. An **ionic bond** is the attraction between oppositely charged ions.

Ionic bonds are formed whenever one or more electrons are transferred from one atom to another. Metals, which have relatively little attraction for their valence electrons, tend to form ionic bonds when they combine with nonmetals.

It's important to recognize that substances with ionic bonds do not exist as molecules. In sodium chloride, for example, the bond does not exist solely between a single sodium ion and a single chloride ion. Each sodium ion in the crystal attracts six near-neighbor negative chloride ions; in turn, each negative chloride ion attracts six near-neighbor positive sodium ions (see Figure 11.5).

A metal will usually have one, two, or three electrons in its outer energy level. In reacting, metal atoms characteristically lose these electrons, attain the electron structure of a noble gas, and become positive ions. A nonmetal, on the other hand, is only a few electrons short of having a noble gas electron structure in its outer energy level and thus has a tendency to gain electrons. In reacting with metals, nonmetal atoms characteristically gain one to four electrons; attain the electron structure of a noble gas; and become negative ions. The ions

Remember: A cation is always smaller than its parent atom, whereas an anion is always larger than its parent atom.

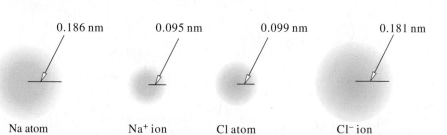

Figure 11.6

Relative radii of sodium and chlorine atoms and their ions.

TABLE 11.3   Change in Atomic Radii (nm) of Selected Metals and Nonmetals*

Atomic radius		Ionic radius		Atomic radius		Ionic radius	
Li	0.152	Li^+	0.060	F	0.071	F^-	0.136
Na	0.186	Na^+	0.095	Cl	0.099	Cl^-	0.181
K	0.227	K^+	0.133	Br	0.114	Br^-	0.195
Mg	0.160	Mg^{2+}	0.065	O	0.074	O^{2-}	0.140
Al	0.143	Al^{3+}	0.050	S	0.103	S^{2-}	0.184

*The metals shown lose electrons to become positive ions. The nonmetals gain electrons to become negative ions.

formed by loss of electrons are much smaller than the corresponding metal atoms; the ions formed by gaining electrons are larger than the corresponding nonmetal atoms. The dimensions of the atomic and ionic radii of several metals and nonmetals are given in Table 11.3.

PRACTICE 11.3

What noble gas structure is formed when an atom of each of these metals loses all its valence electrons? Write the formula for the metal ion formed.

(a) K   (b) Mg   (c) Al   (d) Ba

Study the following examples. Note the loss and gain of electrons between atoms; also note that the ions in each compound have a noble gas electron structure.

EXAMPLE 11.3

Explain how magnesium and chlorine combine to form magnesium chloride, $MgCl_2$.

SOLUTION

A magnesium atom of electron structure $1s^2 2s^2 2p^6 3s^2$ must lose two electrons or gain six to reach a stable electron structure. If magnesium reacts with chlorine and each chlorine atom can accept only one electron, two chlorine atoms will be needed for the two electrons from each magnesium atom. The compound formed will contain one magnesium ion and two chloride ions. The magnesium atom, having lost two electrons, becomes a magnesium ion with a +2 charge. Each chloride ion will have a −1 charge. The transfer of electrons from a magnesium atom to two chlorine atoms is shown in the following illustration:

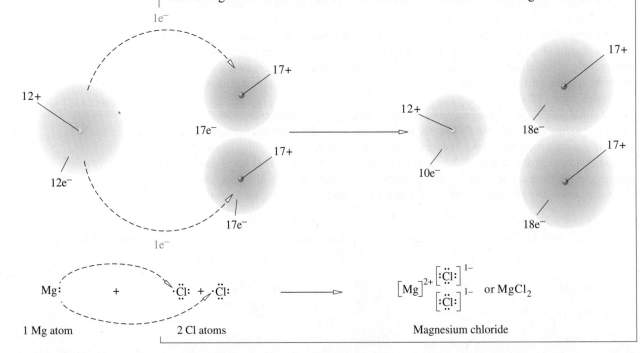

EXAMPLE 11.4

Explain the formation of sodium fluoride (NaF) from its elements.

SOLUTION

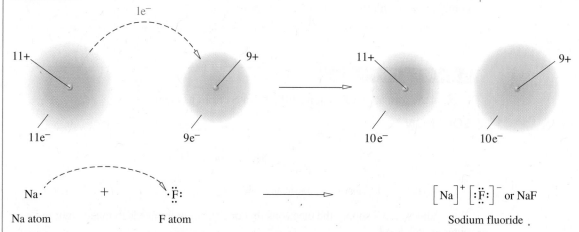

$$\left[\text{Na} \right]^+ \left[:\ddot{\text{F}}: \right]^- \text{ or NaF}$$

The fluorine atom, with seven electrons in its outer energy level, behaves similarly to the chlorine atom.

EXAMPLE 11.5

Explain the formation of aluminum fluoride (AlF_3) from its elements.

SOLUTION

$$\ddot{\text{Al}} \cdot \ + \ \begin{array}{c} \cdot \ddot{\text{F}}: \\ \cdot \ddot{\text{F}}: \\ \cdot \ddot{\text{F}}: \end{array} \longrightarrow [\text{Al}]^{3+} \left[:\ddot{\text{F}}: \right]^- \quad \begin{array}{c} \left[:\ddot{\text{F}}: \right]^- \\ \left[:\ddot{\text{F}}: \right]^- \\ \left[:\ddot{\text{F}}: \right]^- \end{array} \quad \text{or} \quad AlF_3$$

    1 Al atom   3 F atoms   Aluminium fluoride

Each fluorine atom can accept only one electron. Therefore, three fluorine atoms are needed to combine with the three valence electrons of one aluminum atom. The aluminum atom has lost three electrons to become an aluminum ion (Al^{3+}) with a +3 charge.

EXAMPLE 11.6

Explain the formation of magnesium oxide (MgO) from its elements.

SOLUTION

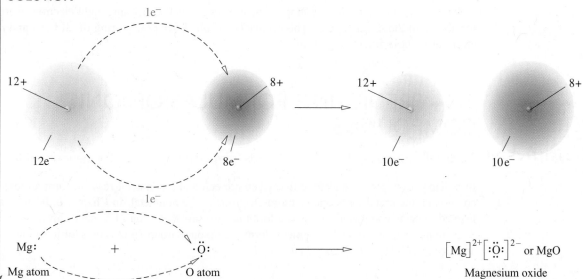

$$\left[\text{Mg} \right]^{2+} \left[:\ddot{\text{O}}: \right]^{2-} \text{ or MgO}$$
Magnesium oxide

The magnesium atom, with two electrons in the outer energy level, exactly fills the need of one oxygen atom for two electrons. The resulting compound has a ratio of one magnesium atom to one oxygen atom. The oxygen (oxide) ion has a -2 charge, having gained two electrons. In combining with oxygen, magnesium behaves the same way as when it combines with chlorine—it loses two electrons.

EXAMPLE 11.7

Explain the formation of sodium sulfide (Na_2S) from its elements.

SOLUTION

$$Na\cdot \quad \overset{}{} \quad \ddot{S}: \longrightarrow \begin{matrix} [Na]^+ \\ [Na]^+ \end{matrix} \left[:\ddot{S}:\right]^{2-} \quad or \quad Na_2S$$

2 Na atoms   1 S atom   Sodium sulfide

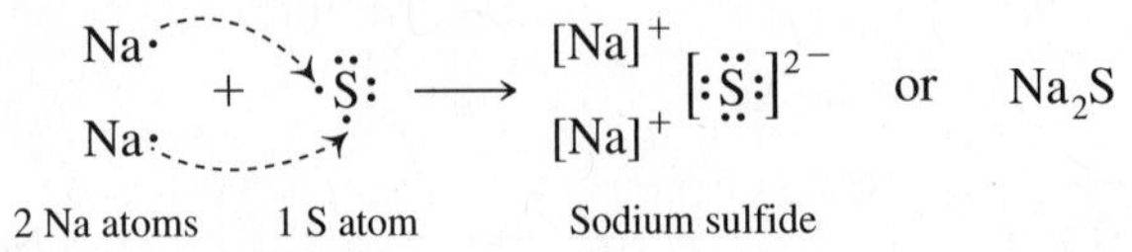

Two sodium atoms supply the electrons that one sulfur atom needs to make eight in its outer energy level.

EXAMPLE 11.8

Explain the formation of aluminum oxide (Al_2O_3) from its elements.

SOLUTION

$$\begin{matrix} \ddot{Al}\cdot \\ + \\ \ddot{Al}\cdot \end{matrix} \quad \begin{matrix} \cdot\ddot{O}: \\ \cdot\ddot{O}: \\ \cdot\ddot{O}: \end{matrix} \longrightarrow \begin{matrix} [Al]^{3+} \\ [Al]^{3+} \end{matrix} \begin{matrix} \left[:\ddot{O}:\right]^{2-} \\ \left[:\ddot{O}:\right]^{2-} \\ \left[:\ddot{O}:\right]^{2-} \end{matrix} \quad or \quad Al_2O_3$$

2 Al atoms   3 O atoms   Aluminium oxide

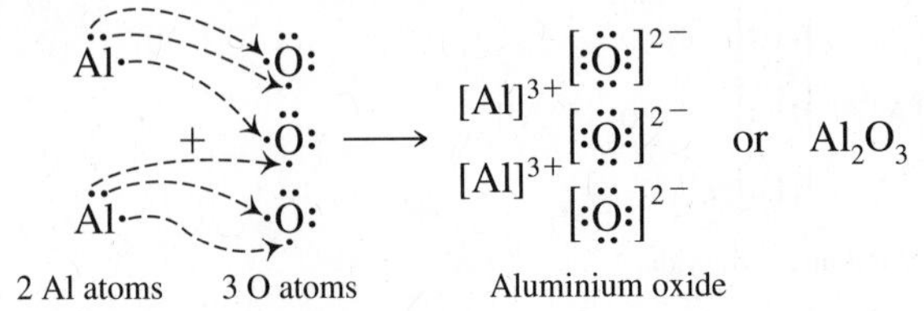

One oxygen atom, needing two electrons, cannot accommodate the three electrons from one aluminum atom. One aluminum atom falls one electron short of the four electrons needed by two oxygen atoms. A ratio of two atoms of aluminum to three atoms of oxygen, involving the transfer of six electrons (two to each oxygen atom), gives each atom a stable electron configuration.

Note that in each of these examples, outer energy levels containing eight electrons were formed in all the negative ions. This formation resulted from the pairing of all the s and p electrons in these outer energy levels.

11.4 PREDICTING FORMULAS OF IONIC COMPOUNDS

LEARNING OBJECTIVE   •   Predict the formulas of ionic compounds from their position on the periodic table.

In previous examples, we learned that when a metal and a nonmetal react to form an ionic compound, the metal loses one or more electrons to the nonmetal. In Chapter 6, where we learned to name compounds and write formulas, we saw that Group 1A metals always form $+1$ cations, whereas Group 2A elements form $+2$ cations. Group 7A elements form -1 anions and Group 6A elements form -2 anions.

It stands to reason, then, that this pattern is directly related to the stability of the noble gas configuration. Metals lose electrons to attain the electron configuration of a noble gas (the previous one on the periodic table). A nonmetal forms an ion by gaining enough electrons to achieve the electron configuration of the noble gas following it on the periodic table. These observations lead us to an important chemical principle:

> In almost all stable chemical compounds of representative elements, each atom attains a noble gas electron configuration. This concept forms the basis for our understanding of chemical bonding.

We can apply this principle in predicting the formulas of ionic compounds. To predict the formula of an ionic compound, we must recognize that chemical compounds are always electrically neutral. In addition, the metal will lose electrons to achieve noble gas configuration and the nonmetal will gain electrons to achieve noble gas configuration. Consider the compound formed between barium and sulfur. Barium has two valence electrons, whereas sulfur has six valence electrons:

$$\text{Ba} \quad [Xe]6s^2 \qquad \text{S} \quad [Ne]3s^23p^4$$

If barium loses two electrons, it will achieve the configuration of xenon. By gaining two electrons, sulfur achieves the configuration of argon. Consequently, a pair of electrons is transferred between atoms. Now we have Ba^{2+} and S^{2-}. Since compounds are electrically neutral, there must be a ratio of one Ba to one S, giving the formula BaS.

The same principle works for many other cases. Since the key lies in the electron configuration, the periodic table can be used to extend the prediction even further. Because of similar electron structures, the elements in a family generally form compounds with the same atomic ratios. In general, if we know the atomic ratio of a particular compound—say, NaCl—we can predict the atomic ratios and formulas of the other alkali metal chlorides. These formulas are LiCl, KCl, RbCl, CsCl, and FrCl (see **Table 11.4**).

Similarly, if we know that the formula of the oxide of hydrogen is H_2O, we can predict that the formula of the sulfide will be H_2S, because sulfur has the same valence electron structure as oxygen. Recognize, however, that these are only predictions; it doesn't necessarily follow that every element in a group will behave like the others or even that a predicted compound will actually exist. For example, knowing the formulas for potassium chlorate, bromate, and iodate to be $KClO_3$, $KBrO_3$, and KIO_3, we can correctly predict the corresponding sodium compounds to have the formulas $NaClO_3$, $NaBrO_3$, and $NaIO_3$. Fluorine belongs to the same family of elements (Group 7A) as chlorine, bromine, and iodine, so it would appear that fluorine should combine with potassium and sodium to give fluorates with the formulas KFO_3 and $NaFO_3$. However, potassium and sodium fluorates are not known to exist.

In the discussion in this section, we refer only to representative metals (Groups 1A, 2A, and 3A). The transition metals (Group B) show more complicated behavior (they form multiple ions), and their formulas are not as easily predicted.

TABLE 11.4   Formulas of Compounds Formed by Alkali Metals

Lewis structure	Oxides	Chlorides	Bromides	Sulfates
Li·	Li_2O	LiCl	LiBr	Li_2SO_4
Na·	Na_2O	NaCl	NaBr	Na_2SO_4
K·	K_2O	KCl	KBr	K_2SO_4
Rb·	Rb_2O	RbCl	RbBr	Rb_2SO_4
Cs·	Cs_2O	CsCl	CsBr	Cs_2SO_4

EXAMPLE 11.9

The formula for calcium sulfide is CaS and that for lithium phosphide is Li_3P. Predict formulas for (a) magnesium sulfide, (b) potassium phosphide, and (c) magnesium selenide.

SOLUTION

(a) Look up calcium and magnesium in the periodic table; they are both in Group 2A. The formula for calcium sulfide is CaS, so it's reasonable to predict that the formula for magnesium sulfide is MgS.

(b) Find lithium and potassium in the periodic table; they are in Group 1A. Since the formula for lithium phosphide is Li_3P, it's reasonable to predict that K_3P is the formula for potassium phosphide.

(c) Find selenium in the periodic table; it is in Group 6A just below sulfur. Therefore, it's reasonable to assume that selenium forms selenide in the same way that sulfur forms sulfide. Since MgS was the predicted formula for magnesium sulfide in part (a), we can reasonably assume that the formula for magnesium selenide is MgSe.

PRACTICE 11.4

The formula for sodium oxide is Na_2O. Predict the formula for

(a) sodium sulfide   (b) rubidium oxide

PRACTICE 11.5

The formula for barium phosphide is Ba_3P_2. Predict the formula for

(a) magnesium nitride   (b) barium arsenide

11.5 THE COVALENT BOND: SHARING ELECTRONS

LEARNING OBJECTIVE • Draw the electron structure of a covalent bond.

KEY TERMS

covalent bond
polar covalent bond

Some atoms do not transfer electrons from one atom to another to form ions. Instead they form a chemical bond by sharing pairs of electrons between them. A **covalent bond** consists of a pair of electrons shared between two atoms. This bonding concept was introduced in 1916 by G. N. Lewis. In the millions of known compounds, the covalent bond is the predominant chemical bond.

True molecules exist in substances in which the atoms are covalently bonded. It is proper to refer to molecules of such substances as hydrogen, chlorine, hydrogen chloride, carbon dioxide, water, or sugar (**Figure 11.7**). These substances contain only covalent bonds and exist as aggregates of molecules. We don't use the term *molecule* when talking about ionically bonded compounds, such as sodium chloride, because such substances exist as large aggregates of positive and negative ions, not as molecules (Figure 11.7).

A study of the hydrogen molecule gives us an insight into the nature of the covalent bond and its formation. The formation of a hydrogen molecule (H_2) involves the overlapping and pairing of $1s$ electron orbitals from two hydrogen atoms, shown in **Figure 11.8**. Each atom contributes one electron of the pair that is shared jointly by two hydrogen nuclei. The orbital of the electrons now includes both hydrogen nuclei, but probability factors show that the most

O   C   Dry ice

Na$^+$ Cl$^-$

Sodium chloride

Figure 11.7

Solid carbon dioxide (dry ice) is composed of individual covalently bonded molecules of CO_2 closely packed together. Table salt is a large aggregate of Na$^+$ and Cl$^-$ ions instead of molecules.

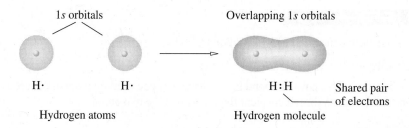

1s orbitals   Overlapping 1s orbitals

H·   H·   H:H   Shared pair of electrons

Hydrogen atoms   Hydrogen molecule

Figure 11.8

The formation of a hydrogen molecule from two hydrogen atoms. The two 1s orbitals overlap, forming the H_2 molecule. In this molecule the two electrons are shared between the atoms, forming a covalent bond.

likely place to find the electrons (the point of highest electron density) is between the two nuclei. The two nuclei are shielded from each other by the pair of electrons, allowing the two nuclei to be drawn very close to each other.

The formula for chlorine gas is Cl_2. When the two atoms of chlorine combine to form this molecule, the electrons must interact in a manner similar to that shown in the hydrogen example. Each chlorine atom would be more stable with eight electrons in its outer energy level. But chlorine atoms are identical, and neither is able to pull an electron away from the other. What happens is this: The unpaired $3p$ electron orbital of one chlorine atom overlaps the unpaired $3p$ electron orbital of the other atom, resulting in a pair of electrons that is mutually shared between the two atoms. Each atom furnishes one of the pair of shared electrons. Thus, each atom attains a stable structure of eight electrons by sharing an electron pair with the other atom. The pairing of the p electrons and the formation of a chlorine molecule are illustrated in **Figure 11.9**. Neither chlorine atom has a positive or negative charge because both contain the same number of protons and have equal attraction for the pair of electrons being shared. Other examples of molecules in which electrons are equally shared between two atoms are hydrogen (H_2), oxygen (O_2), nitrogen (N_2), fluorine (F_2), bromine (Br_2), and iodine (I_2). Note that more than one pair of electrons may be shared between atoms:

H:H   :F̈:F̈:   :B̈r:B̈r: :Ï:Ï: :Ö::Ö: :N⫶⫶N:

hydrogen   fluorine   bromine   iodine   oxygen   nitrogen

The Lewis structure given for oxygen does not adequately account for all the properties of the oxygen molecule. Other theories explaining the bonding in oxygen molecules have been advanced, but they are complex and beyond the scope of this book.

In writing structures, we commonly replace the pair of dots used to represent a shared pair of electrons with a dash (—). One dash represents a single bond; two dashes, a double bond; and three dashes, a triple bond. The six structures just shown may be written thus:

H—H   :F̈—F̈:   :B̈r—B̈r: :Ï—Ï: :Ö=Ö: :N≡N:

The ionic bond and the covalent bond represent two extremes. In ionic bonding the atoms are so different that electrons are transferred between them, forming a charged pair of ions. In covalent bonding, two identical atoms share electrons equally. The bond is the mutual attraction of the two nuclei for the shared electrons. Between these extremes lie many cases in which the atoms are not different enough for a transfer of electrons but are different enough that the electron pair cannot be shared equally. This unequal sharing of electrons results in the formation of a **polar covalent bond**.

Tom Pantages

Molecular models for F_2 (green, single bond), O_2 (black, double bond), and N_2 (blue, triple bond).

Remember: A dash represents a shared pair of electrons.

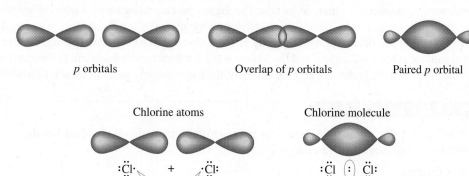

p orbitals   Overlap of p orbitals   Paired p orbital

Chlorine atoms   Chlorine molecule

:C̈l·   +   ·C̈l: :C̈l (:) C̈l:

Unshared p orbitals   Shared pair of p electrons

Figure 11.9

Pairing of p electrons in the formation of a chlorine molecule.

EXAMPLE 11.10

What is the basic concept of a covalent bond?

SOLUTION

The basic concept of a covalent bond is the sharing of one or more pairs of electrons between two atoms to form a covalent bond between the two atoms.

PRACTICE 11.6

The atoms in the two molecules HBr and C_2H_6 are bonded by covalent bonds. Draw Lewis structures showing these covalent bonds.

11.6 ELECTRONEGATIVITY

LEARNING OBJECTIVE ● Explain how electronegativities of component atoms in a molecule determine the polarity of the molecule.

KEY TERMS

electronegativity
nonpolar covalent bond
dipole

When two *different* kinds of atoms share a pair of electrons, a bond forms in which electrons are shared unequally. One atom assumes a partial positive charge and the other a partial negative charge with respect to each other. This difference in charge occurs because the two atoms exert unequal attraction for the pair of shared electrons. The attractive force that an atom of an element has for shared electrons in a molecule or polyatomic ion is known as its **electronegativity**. Elements differ in their electronegativities. For example, both hydrogen and chlorine need one electron to form stable electron configurations. They share a pair of electrons in hydrogen chloride (HCl). Chlorine is more electronegative and therefore has a greater attraction for the shared electrons than does hydrogen. As a result, the pair of electrons is displaced toward the chlorine atom, giving it a partial negative charge and leaving the hydrogen atom with a partial positive charge. Note that the electron is not transferred entirely to the chlorine atom (as in the case of sodium chloride) and that no ions are formed. The entire molecule, HCl, is electrically neutral. A partial charge is usually indicated by the Greek letter delta, δ. Thus, a partial positive charge is represented by $\delta+$ and a partial negative charge by $\delta-$.

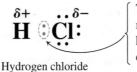

Hydrogen chloride

The pair of shared electrons in HCl is closer to the more electronegative chlorine atom than to the hydrogen atom, giving chlorine a partial negative charge with respect to the hydrogen atom.

A scale of relative electronegativities, in which the most electronegative element, fluorine, is assigned a value of 4.0, was developed by the Nobel Laureate (1954 and 1962) Linus Pauling (1901–1994). Table 11.5 shows that the relative electronegativity of the nonmetals is high and that of the metals is low. These electronegativities indicate that atoms of metals have a greater tendency to lose electrons than do atoms of nonmetals and that nonmetals have a greater tendency to gain electrons than do metals. The higher the electronegativity value, the greater the attraction for electrons. Note that electronegativity generally increases from left to right across a period and decreases down a group for the representative elements. The highest electronegativity is 4.0 for fluorine, and the lowest is 0.7 for francium and cesium. It's important to remember that the higher the electronegativity, the more strongly an atom attracts electrons.

EXAMPLE 11.11

What are the electronegativity trends for the groups of representative elements and the periods of representative elements in the periodic table?

SOLUTION

In general, electronegativity decreases from top to bottom for a group of representative elements and increases from left to right across a period of representative elements.

TABLE 11.5   Three-Dimensional Representation of Electronegativity

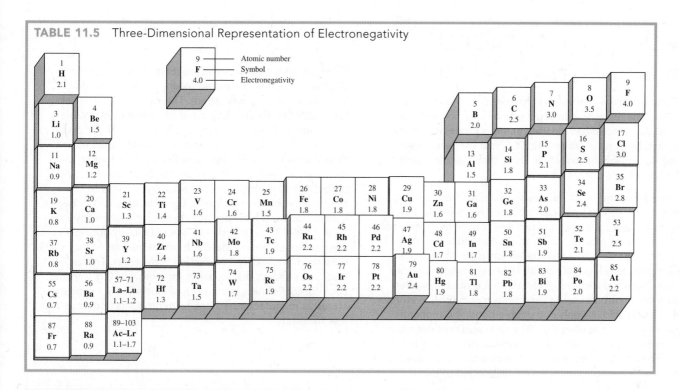

PRACTICE 11.7

Explain the term *electronegativity* and how it relates to the elements in the periodic table.

The polarity of a bond is determined by the difference in electronegativity values of the atoms forming the bond (see **Figure 11.10**). If the electronegativities are the same, the bond is **nonpolar covalent** and the electrons are shared equally. If the atoms have greatly different electronegativities, the bond is very *polar*. At the extreme, one or more electrons are actually transferred and an ionic bond results.

A **dipole** is a molecule that is electrically asymmetrical, causing it to be oppositely charged at two points. A dipole is often written as ⊕⊖. A hydrogen chloride molecule is polar and behaves as a small dipole. The HCl dipole may be written as H⟵⟶Cl. The arrow points toward the negative end of the dipole. Molecules of H_2O, HBr, and ICl are polar:

$$H\longleftrightarrow Cl \qquad H\longleftrightarrow Br \qquad I\longleftrightarrow Cl \qquad H\overset{O}{\underset{}{\diagup\hspace{-0.3em}\diagdown}}H$$

How do we know whether a bond between two atoms is ionic or covalent? The difference in electronegativity between the two atoms determines the character of the bond formed between them. As the difference in electronegativity increases, the polarity of the bond (or percent ionic character) increases.

In general, if the electronegativity difference between two bonded atoms is greater than 1.7–1.9, the bond will be more ionic than covalent.

If the electronegativity difference is greater than 2.0, the bond is strongly ionic. If the electro-negativity difference is less than 1.5, the bond is strongly covalent.

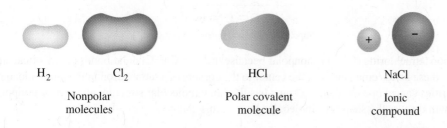

H_2	Cl_2	HCl	NaCl
Nonpolar molecules		Polar covalent molecule	Ionic compound

Figure 11.10

Nonpolar, polar covalent, and ionic compounds.

>CHEMISTRY *IN ACTION*

Trans-forming Fats

Trans fats are virtually everywhere in the American diet. These fats remain when vegetable oils are converted into solid substances that are used in many processed foods. In fact, beginning January 1, 2006, manufacturers were required to label products to show their *trans* fat content. So just what is a *trans* fat?

Different categories of fat can be identified from the pattern of bonds and hydrogen atoms in the molecule. Fatty acids are one component of fats and contain long chains of carbon atoms with hydrogen atoms bonded to some or all of the carbon atoms. Unsaturated fats (such as corn or soybean oil) contain double bonds between some of the carbon atoms in the chains. A carbon that is doubly bonded to another carbon usually also bonds to a hydrogen atom. If the fatty acid has only one double bond, it is monounsaturated (such as olive oil). If the fatty acid contains more than one double bond, it is polyunsaturated. All fat with no double bonds is saturated—all the carbon atoms have the maximum possible hydrogen atoms bonded to them.

Trans fats contain a particular kind of unsaturated fatty acid. When there is a double bond between carbon atoms, the molecule bends in one of two ways: the *cis* or the *trans* direction. In cis configuration, the carbon chain on both sides of the double bond bends in to the same side of the double bond (see structure). In the *trans* configuration, the chain on either side of the double bond bends toward opposite sides of the double bond (see structure). Most *trans* fats come from processing oils for prepared foods and from solid fats such as margarine.

When an oil is converted into a solid fat, some of the double bonds are converted to single bonds by adding hydrogen (hydrogenation). This process is easier at *cis* double bonds and, therefore, the remaining double bonds are mainly in the *trans* configuration. These *trans* fatty acids tend to stack together, making a solid easier than the *cis* forms. Studies have linked diets high in *trans* fats to poor health, high cholesterol, heart disease, and diabetes. Food producers are working on ways to lower the *trans* fat content of foods. Gary List at USDA in Peoria, Illinois, has used high-pressure hydrogen gas on soybean oil at 140°–170°C to hydrogenate the oil, producing a soft margarine containing 5%–6% *trans* fat instead of the ~40% from the standard hydrogenation techniques. This could lead to a product that qualifies for a label of 0 g *trans* fat. Look for lots of new products and labels in your grocery stores as manufacturers *trans*-form products into ones with less *trans* fat.

cis fatty acid

trans fatty acid

PRACTICE 11.8

Which of these compounds would you predict to be ionic and which would be covalent?
(a) $SrCl_2$ (d) RbBr
(b) PCl_3 (e) LiCl
(c) NH_3 (f) CS_2

Care must be taken to distinguish between polar bonds and polar molecules. A covalent bond between different kinds of atoms is always polar. But a molecule containing different kinds of atoms may or may not be polar, depending on its shape or geometry. Molecules of HF, HCl, HBr, HI, and ICl are all polar because each contains a single polar bond. However, CO_2, CH_4, and CCl_4 are nonpolar molecules despite the fact that all three contain polar bonds. The carbon dioxide molecule O=C=O is nonpolar because the carbon–oxygen dipoles cancel each other by acting in opposite directions.

$$O=C=O$$

dipoles in equal and opposite directions

Spacefilling molecular model CCl_4 (top) and methane (CH_4).

Carbon tetrachloride (CCl_4) is nonpolar because the four C—Cl polar bonds are identical, and since these bonds emanate from the center to the corners of a tetrahedron in the molecule, their polarities cancel one another. Methane has the same molecular structure and is also nonpolar. We will discuss the shapes of molecules later in this chapter.

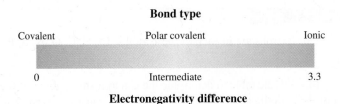

Figure 11.11

Relating bond type to electronegativity difference between atoms.

We have said that water is a polar molecule. If the atoms in water were linear like those in carbon dioxide, the two O—H dipoles would cancel each other, and the molecule would be nonpolar. However, water is definitely polar and has a nonlinear (bent) structure with an angle of 105° between the two O—H bonds.

The relationships among types of bonds are summarized in **Figure 11.11**. It is important to realize that bonding is a continuum; that is, the difference between ionic and covalent is a gradual change.

11.7 LEWIS STRUCTURES OF COMPOUNDS

Draw the Lewis structure of a covalent compound.   ● **LEARNING OBJECTIVE**

As we have seen, Lewis structures are a convenient way of showing the covalent bonds in many molecules or ions of the representative elements. In writing Lewis structures, the most important consideration for forming a stable compound is that the atoms attain a noble gas configuration.

The most difficult part of writing Lewis structures is determining the arrangement of the atoms in a molecule or an ion. In simple molecules with more than two atoms, one atom will be the central atom surrounded by the other atoms. Thus, Cl_2O has two possible arrangements, Cl—Cl—O or Cl—O—Cl. Usually, but not always, the single atom in the formula (except H) will be the central atom.

Although Lewis structures for many molecules and ions can be written by inspection, the following procedure is helpful for learning to write them:

PROBLEM-SOLVING STRATEGY: Writing a Lewis Structure

1. Obtain the total number of valence electrons to be used in the structure by adding the number of valence electrons in all the atoms in the molecule or ion. If you are writing the structure of an ion, add one electron for each negative charge or subtract one electron for each positive charge on the ion.

2. Write the skeletal arrangement of the atoms and connect them with a single covalent bond (two dots or one dash). Hydrogen, which contains only one bonding electron, can form only one covalent bond. Oxygen atoms are not normally bonded to each other, except in compounds known to be peroxides. Oxygen atoms normally have a maximum of two covalent bonds (two single bonds or one double bond).

3. Subtract two electrons for each single bond you used in Step 2 from the total number of electrons calculated in Step 1. This gives you the net number of electrons available for completing the structure.

4. Distribute pairs of electrons (pairs of dots) around each atom (except hydrogen) to give each atom a noble gas structure.

5. If there are not enough electrons to give these atoms eight electrons, change single bonds between atoms to double or triple bonds by shifting unbonded pairs of electrons as needed. Check to see that each atom has a noble gas electron structure (two electrons for hydrogen and eight for the others). A double bond counts as four electrons for each atom to which it is bonded.

Remember: The number of valence electrons of Group A elements is the same as their group number in the periodic table.

EXAMPLE 11.12

How many valence electrons are in each of these atoms: Cl, H, C, O, N, S, P, I?

SOLUTION

You can look at the periodic table to determine the electron structure, or, if the element is in Group A of the periodic table, the number of valence electrons is equal to the group number:

Atom	Group	Valence electrons
Cl	7A	7
H	1A	1
C	4A	4
O	6A	6
N	5A	5
S	6A	6
P	5A	5
I	7A	7

EXAMPLE 11.13

Use the Writing a Lewis Structure Problem-Solving Strategy for water (H_2O).

SOLUTION

1. The total number of valence electrons is eight, two from the two hydrogen atoms and six from the oxygen atom.

2. The two hydrogen atoms are connected to the oxygen atom. Write the skeletal structure:

   H O   or   H O H
     H

Place two dots between the hydrogen and oxygen atoms to form the covalent bonds:

   H:O   or   H:O:H
     Ḧ

3. Subtract the four electrons used in Step 2 from eight to obtain four electrons yet to be used.

4. Distribute the four electrons in pairs around the oxygen atom. Hydrogen atoms cannot accommodate any more electrons:

   H—Ö: or   H—Ö—H
      |
      H

These arrangements are Lewis structures because each atom has a noble gas electron structure. Note that the shape of the molecule is not shown by the Lewis structure.

EXAMPLE 11.14

Use the Writing a Lewis Structure Problem-Solving Strategy for a molecule of methane (CH_4).

SOLUTION

1. The total number of valence electrons is eight, one from each hydrogen atom and four from the carbon atom.

2. The skeletal structure contains four H atoms around a central C atom. Place two electrons between the C and each H.

    H   H
  H C H   H:C:H
    H   Ḧ

3. Subtract the eight electrons used in Step 2 from eight (obtained in Step 1) to obtain zero electrons yet to be placed. Therefore, the Lewis structure must be as written in Step 2:

$$H\!:\!\ddot{C}\!:\!H \quad \text{or} \quad H\!-\!\underset{\underset{\textstyle H}{|}}{\overset{\overset{\textstyle H}{|}}{C}}\!-\!H$$

EXAMPLE 11.15

Use the Writing a Lewis Structure Problem-Solving Strategy for a molecule of carbon tetrachloride (CCl_4).

SOLUTION

1. The total number of valence electrons to be used is 32, 4 from the carbon atom and 7 from each of the four chlorine atoms.

2. The skeletal structure contains the four Cl atoms around a central C atom. Place 2 electrons between the C and each Cl:

$$\begin{array}{ccc} & Cl & \\ Cl & C & Cl \\ & Cl & \end{array} \qquad Cl\!:\!\ddot{C}\!:\!Cl$$

3. Subtract the 8 electrons used in Step 2 from 32 (obtained in Step 1) to obtain 24 electrons yet to be placed.

4. Distribute the 24 electrons (12 pairs) around the Cl atoms so that each Cl atom has 8 electrons around it:

$$:\!\ddot{\underset{..}{C}}\!l\!: \quad \text{or} \quad :\!\ddot{C}\!l\!-\!C\!-\!\ddot{C}\!l\!:$$

This arrangement is the Lewis structure; CCl_4 contains four covalent bonds.

EXAMPLE 11.16

Use the Writing a Lewis Structure Problem-Solving Strategy for CO_2.

SOLUTION

1. The total number of valence electrons is 16, 4 from the C atom and 6 from each O atom.

2. The two O atoms are bonded to a central C atom. Write the skeletal structure and place 2 electrons between the C and each O atom.

$$O : C : O$$

3. Subtract the 4 electrons used in Step 2 from 16 (found in Step 1) to obtain 12 electrons yet to be placed.

4. Distribute the 12 electrons (six pairs) around the C and O atoms. Several possibilities exist:

$$:\!\ddot{O}\!:\!C\!:\!\ddot{O}\!: \qquad :\!\ddot{O}\!:\!\ddot{C}\!:\!\ddot{O}\!: \qquad :\!\ddot{O}\!:\!\ddot{C}\!:\!\ddot{O}\!:$$
$$\quad I \qquad\qquad II \qquad\qquad III$$

5. Not all the atoms have 8 electrons around them (noble gas structure). Remove one pair of unbonded electrons from each O atom in structure I and place one pair between each O and the C atom, forming two double bonds:

$$:\!\ddot{O}\!:\!:\!C\!:\!:\!\ddot{O}\!: \quad \text{or} \quad :\!\ddot{O}\!=\!C\!=\!\ddot{O}\!:$$

Each atom now has 8 electrons around it. The carbon is sharing four pairs of electrons, and each oxygen is sharing two pairs. These bonds are known as double bonds because each involves sharing two pairs of electrons.

>CHEMISTRY *IN ACTION*

Strong Enough to Stop a Bullet?

What do color-changing pens, bullet-resistant vests, and calculators have in common? The chemicals that make each of them work are liquid crystals. These chemicals find numerous applications; you are probably most familiar with liquid crystal displays (LCDs) and color-changing products, but these chemicals are also used to make superstrong synthetic fibers.

Molecules in a normal crystal remain in an orderly arrangement, but in a liquid crystal the molecules can flow *and* maintain an orderly arrangement at the same time. Liquid crystal molecules are linear and polar. Since the atoms tend to lie in a relatively straight line, the molecules are generally much longer than they are wide. These polar molecules are attracted to each other and are able to line up in an orderly fashion without solidifying.

Liquid crystals with twisted arrangements of molecules give us novelty color-changing products. In these liquid crystals the molecules lie side by side in a nearly flat layer. The next layer is similar but at an angle to the one below. The closely packed flat layers have a special effect on light. As the light strikes the surface, some of it is reflected from the top layer and some from lower layers. When the same wavelength is reflected from many layers, we see a color. (This is similar to the rainbow of colors formed by oil in a puddle on the street or the film of a soap bubble.) As the temperature is increased, the molecules move faster, causing a change in the angle and the space between the layers. This results in a color change in the reflected light. Different compounds change color within different temperature ranges, allowing a variety of practical and amusing applications.

Liquid crystal (nematic) molecules that lie parallel to one another are used to manufacture very strong synthetic fibers. Perhaps the best example of these liquid crystals is Kevlar, a synthetic fiber used in bullet-resistant vests, canoes, and parts of the space shuttle. Kevlar is a synthetic polymer, like nylon or polyester, that gains strength by passing through a liquid crystal state during its manufacture.

In a typical polymer, the long molecular chains are jumbled together, somewhat like spaghetti. The strength of the material is limited by the disorderly arrangement. The trick is to get the molecules to line up parallel to each other. Once the giant molecules have been synthesized, they are dissolved in sulfuric acid. At the proper concentration the molecules align, and the solution is forced through tiny holes in a nozzle and further aligned. The sulfuric acid is removed in a water bath, thereby forming solid fibers in near-perfect alignment. One strand of Kevlar is stronger than an equal-sized strand of steel. It has a much lower density as well, making it a material of choice in bullet-resistant vests.

Kevlar is used to make protective vests for police.

PRACTICE 11.9

Write the Lewis structures for the following:
(a) PBr_3 (b) $CHCl_3$ (c) HF   (d) H_2CO (e) N_2

Although many compounds attain a noble gas structure in covalent bonding, there are numerous exceptions. Sometimes it's impossible to write a structure in which each atom has 8 electrons around it. For example, in BF_3 the boron atom has only 6 electrons around it, and in SF_6 the sulfur atom has 12 electrons around it.

Although there are exceptions, many molecules can be described using Lewis structures where each atom has a noble gas electron configuration. This is a useful model for understanding chemistry.

11.8 COMPLEX LEWIS STRUCTURES

LEARNING OBJECTIVE ● Draw the resonance structures for a polyatomic ion.

KEY TERM

resonance structure

Most Lewis structures give bonding pictures that are consistent with experimental information on bond strength and length. There are some molecules and polyatomic ions for which no single Lewis structure consistent with all characteristics and bonding information can be written. For example, consider the nitrate ion, NO_3^-. To write a Lewis structure for this polyatomic ion, we use the following steps.

1. The total number of valence electrons is 24, 5 from the nitrogen atom, 6 from each oxygen atom, and 1 from the −1 charge.

2. The three O atoms are bonded to a central N atom. Write the skeletal structure and place two electrons between each pair of atoms. Since we have an extra electron in this ion, resulting in a −1 charge, we enclose the group of atoms in square brackets and add a − charge as shown.

$$\left[\begin{array}{c} O \\ O\!:\!\ddot{N}\!:\!O \end{array}\right]^{-}$$

3. Subtract the 6 electrons used in Step 2 from 24 (found in Step 1) to obtain 18 electrons yet to be placed.

4. Distribute the 18 electrons around the N and O atoms:

$$\begin{array}{c} :\ddot{O} \longleftarrow \text{electron deficient} \\ :\ddot{O}\!:\!\ddot{N}\!:\!\ddot{O}: \end{array}$$

5. One pair of electrons is still needed to give all the N and O atoms a noble gas structure. Move the unbonded pair of electrons from the N atom and place it between the N and the electron-deficient O atom, making a double bond.

$$\left[\begin{array}{c} :\ddot{O} \\ \| \\ :\ddot{O}-N-\ddot{O}: \end{array}\right]^{-} \quad \text{or} \quad \left[\begin{array}{c} :\ddot{O}: \\ | \\ :\ddot{O}-N=\ddot{O}: \end{array}\right]^{-} \quad \text{or} \quad \left[\begin{array}{c} :\ddot{O}: \\ | \\ \ddot{O}=N-\ddot{O}: \end{array}\right]^{-}$$

Are these all valid Lewis structures? Yes, so there really are three possible Lewis structures for NO_3^-.

A molecule or ion that has multiple correct Lewis structures shows *resonance*. Each of these Lewis structures is called a **resonance structure**. In this book, however, we will not be concerned with how to choose the correct resonance structure for a molecule or ion. Therefore, any of the possible resonance structures may be used to represent the ion or molecule.

EXAMPLE 11.17

Write the Lewis structure for a carbonate ion (CO_3^{2-}).

SOLUTION

1. These four atoms have 22 valence electrons plus 2 electrons from the −2 charge, which makes 24 electrons to be placed.

2. In the carbonate ion, the carbon is the central atom surrounded by the three oxygen atoms. Write the skeletal structure and place 2 electrons (or a single line) between each C and O:

$$\begin{array}{c} O \\ | \\ C-O \\ | \\ O \end{array}$$

3. Subtract the 6 electrons used in Step 2 from 24 (from Step 1) to give 18 electrons yet to be placed.

4. Distribute the 18 electrons around the three oxygen atoms and indicate that the carbonate ion has a −2 charge:

$$\left[\begin{array}{c} :\ddot{O}: \\ | \\ C \\ :\ddot{O} \quad \ddot{O}: \end{array}\right]^{2-}$$

The difficulty with this structure is that the carbon atom has only six electrons around it instead of a noble gas octet.

5. Move one of the nonbonding pairs of electrons from one of the oxygens and place them between the carbon and the oxygen. Three Lewis structures are possible:

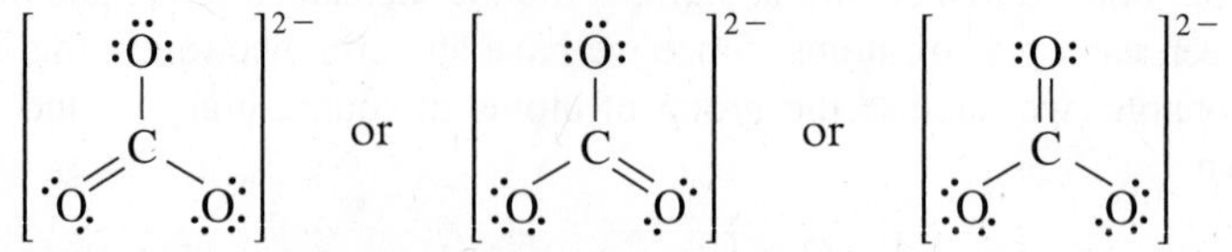

PRACTICE 11.10

Write the Lewis structure for each of the following:
(a) NH_3 (b) H_3O^+ (c) NH_4^+ (d) HCO_3^-

11.9 COMPOUNDS CONTAINING POLYATOMIC IONS

LEARNING OBJECTIVE   • Describe a compound that contains both ionic and covalent bonds.

A polyatomic ion is a stable group of atoms that has either a positive or a negative charge and behaves as a single unit in many chemical reactions. Sodium carbonate, Na_2CO_3, contains two sodium ions and a carbonate ion. The carbonate ion (CO_3^{2-}) is a polyatomic ion composed of one carbon atom and three oxygen atoms and has a charge of -2. One carbon and three oxygen atoms have a total of 22 electrons in their outer energy levels. The carbonate ion contains 24 outer electrons and therefore has a charge of -2. In this case, the 2 additional electrons come from the two sodium atoms, which are now sodium ions:

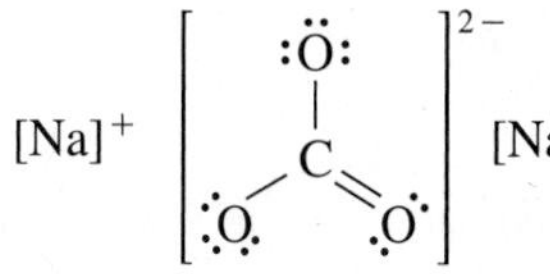

Sodium carbonate 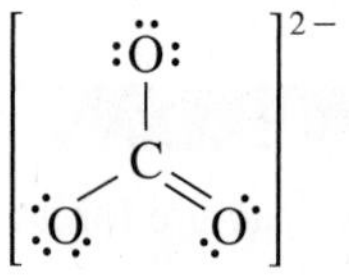   Carbonate ion

Sodium carbonate has both ionic and covalent bonds. Ionic bonds exist between each of the sodium ions and the carbonate ion. Covalent bonds are present between the carbon and oxygen atoms within the carbonate ion. One important difference between the ionic and covalent bonds in this compound can be demonstrated by dissolving sodium carbonate in water. It dissolves in water, forming three charged particles—two sodium ions and one carbonate ion—per formula unit of Na_2CO_3:

$$Na_2CO_3(s) \xrightarrow{water} 2\,Na^+(aq) + CO_3^{2-}(aq)$$

sodium carbonate   sodium ions   carbonate ion

The CO_3^{2-} ion remains as a unit, held together by covalent bonds; but where the bonds are ionic, dissociation of the ions takes place. Do not think, however, that polyatomic ions are so stable that they cannot be altered. Chemical reactions by which polyatomic ions can be changed to other substances do exist.

PRACTICE 11.11

Write Lewis structures for the calcium compounds of the following anions: nitrate and sulfate.

11.10 MOLECULAR SHAPE

Determine the shape of a compound by using VSEPR method.

● **LEARNING OBJECTIVE**

KEY TERMS
linear structure
trigonal planar structure
tetrahedral structure
bent structure

So far in our discussion of bonding we have used Lewis structures to represent valence electrons in molecules and ions, but they don't indicate anything regarding the molecular or geometric shape of a molecule. The three-dimensional arrangement of the atoms within a molecule is a significant feature in understanding molecular interactions. Let's consider several examples illustrated in **Figure 11.12**.

Water is known to have the geometric shape known as "bent" or "V-shaped." Carbon dioxide exhibits a linear shape. BF_3 forms a third molecular shape called *trigonal planar* since all the atoms lie in one plane in a triangular arrangement. One of the more common molecular shapes is the tetrahedron, illustrated by the molecule methane (CH_4).

How do we predict the geometric shape of a molecule? We will now study a model developed to assist in making predictions from the Lewis structure.

The Valence Shell Electron Pair Repulsion (VSEPR) Model

The chemical properties of a substance are closely related to the structure of its molecules. A change in a single site on a large biomolecule can make a difference in whether or not a particular reaction occurs.

Figure 11.12
Geometric shapes of common molecules. Each molecule is shown as a ball-and-stick model (showing the bonds) and as a spacefilling model (showing the shape).

Water	Carbon dioxide	Boron trifluoride	Methane
H_2O	CO_2	BF_3	CH_4
(V-shaped)	(linear shape)	(trigonal planar)	(tetrahedral)

Instrumental analysis can be used to determine exact spatial arrangements of atoms. Quite often, though, we only need to be able to predict the approximate structure of a molecule. A relatively simple model has been developed to allow us to make predictions of shape from Lewis structures.

The VSEPR model is based on the idea that electron pairs will repel each other electrically and will seek to minimize this repulsion. To accomplish this minimization, the electron pairs will be arranged around a central atom as far apart as possible. Consider $BeCl_2$, a molecule with only two pairs of electrons surrounding the central atom. These electrons are arranged 180° apart for maximum separation:

$$Cl \overset{180°}{\underset{}{\longleftrightarrow}} Be \longleftrightarrow Cl$$

This molecular structure can now be labeled as a **linear structure**. When only two pairs of electrons surround a central atom, they should be placed 180° apart to give a linear structure.

What occurs when there are only three pairs of electrons around the central atom? Consider the BF_3 molecule. The greatest separation of electron pairs occurs when the angles between atoms are 120°:

This arrangement of atoms is flat (planar) and, as noted earlier, is called **trigonal planar structure**. When three pairs of electrons surround an atom, they should be placed 120° apart to show the trigonal planar structure.

Now consider the most common situation (CH_4), with four pairs of electrons on the central carbon atom. In this case the central atom exhibits a noble gas electron structure. What arrangement best minimizes the electron pair repulsions? At first, it seems that an obvious choice is a 90° angle with all the atoms in a single plane:

However, we must consider that molecules are three-dimensional. This concept results in a structure in which the electron pairs are actually 109.5° apart:

In this diagram the wedged line seems to protrude from the page, whereas the dashed line recedes. Two representations of this arrangement, known as **tetrahedral structure**, are illustrated in **Figure 11.13**. When four pairs of electrons surround a central atom, they should be placed 109.5° apart to give them a tetrahedral structure.

Figure 11.13

Ball-and-stick models of methane and carbon tetrachloride. Methane and carbon tetrachloride are nonpolar molecules because their polar bonds cancel each other in the tetrahedral arrangement of their atoms. The carbon atoms are located in the centers of the tetrahedrons.

Nonbonding pairs of electrons are not shown here, so you can focus your attention on the shapes, not electron arrangement.

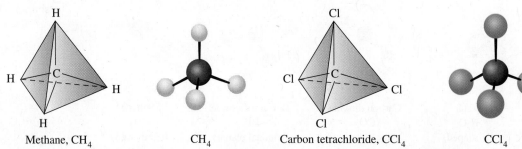

Methane, CH_4 CH_4 Carbon tetrachloride, CCl_4 CCl_4

Figure 11.14

(a) The tetrahedral arrangement of electron pairs around the N atom in the NH_3 molecule. (b) Three pairs are shared and one is unshared. (c) The NH_3 molecule is pyramidal.

The VSEPR model is based on the premise that we are counting electron pairs. It's quite possible that one or more of these electron pairs may be nonbonding (lone) pairs. What happens to the molecular structure in these cases?

Consider the ammonia molecule. First we draw the Lewis structure to determine the number of electron pairs around the central atom:

$$H\!:\!\ddot{N}\!:\!H$$
$$\overset{..}{H}$$

Since there are four pairs of electrons, the arrangement of electrons around the central atom will be tetrahedral (**Figure 11.14a**). However, only three of the pairs are bonded to another atom, so the molecule itself is pyramidal. It is important to understand that the placement of the electron pairs determines the shape but the name for the molecule is determined by the position of the atoms themselves. Therefore, ammonia is pyramidal. See Figure 11.14c.

Now consider the effect of two unbonded pairs of electrons in the water molecule. The Lewis structure for water is

$$H\!-\!\ddot{O}\!:$$
$$|$$
$$H$$

The four electron pairs indicate that a tetrahedral electron arrangement is necessary (see **Figure 11.15a**). The molecule is not called tetrahedral because two of the electron pairs are unbonded pairs. The water molecule displays a **bent structure**.

Let's summarize the VSEPR model.

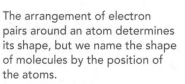

PROBLEM-SOLVING STRATEGY: Determining Molecular Shape Using VSEPR

1. Draw the Lewis structure for the molecule.

2. Count the electron pairs around the central atom and arrange them to minimize repulsions (as far apart as possible). This determines the electron pair arrangement.

3. Determine the positions of the atoms.

4. Name the molecular structure from the position of the atoms.

The arrangement of electron pairs around an atom determines its shape, but we name the shape of molecules by the position of the atoms.

It is important to recognize that the placement of the electron pairs determines the structure but the name of the molecular structure is determined by the position of the atoms. **Table 11.6** shows the results of this process. Note that when the number of electron pairs is the same as the number of atoms, the electron pair arrangement and the molecular structure are the same. But when the number of atoms and the number of electron pairs are not the same, the molecular structure is different from the electron pair arrangement. This is illustrated when the number of electron pairs is four (a tetrahedral arrangement) in Table 11.6.

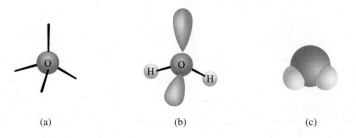

Figure 11.15

(a) The tetrahedral arrangement of the four electron pairs around oxygen in the H_2O molecule. (b) Two of the pairs are shared and two are unshared. (c) The H_2O molecule is bent.

TABLE 11.6   Arrangement of Electron Pairs and Molecular Structure

Number of electron pairs	Electron pair arrangement	Ball-and-stick model	Bonds	Molecular structure	Molecular structure model
2	Linear	 180°	2	Linear	
3	Trigonal planar	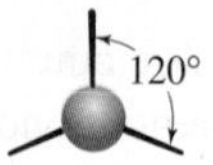 120°	3	Trigonal planar	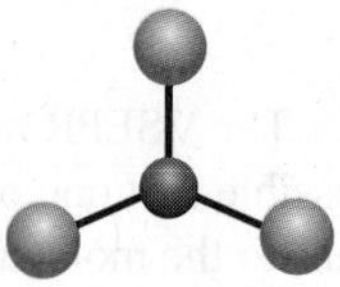
4	Tetrahedral	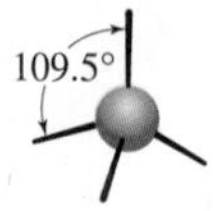 109.5°	4	Tetrahedral	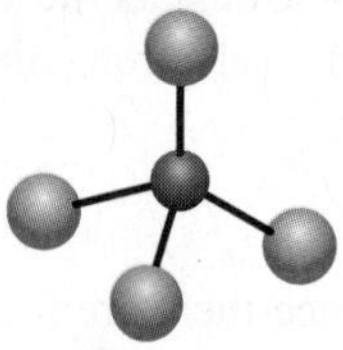
4	Tetrahedral	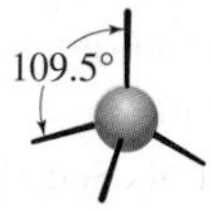 109.5°	3	Pyramidal	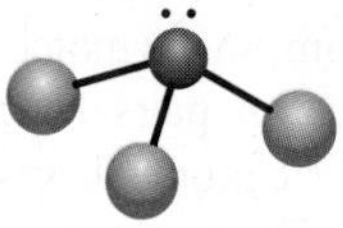
4	Tetrahedral	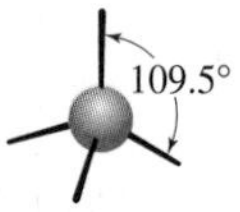 109.5°	2	Bent	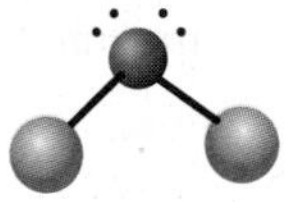

WileyPLUS

ENHANCED EXAMPLE

EXAMPLE 11.18

Predict the molecular shape for these molecules: H_2S, CCl_4, $AlCl_3$.

SOLUTION

1. Draw the Lewis structure.
2. Count the electron pairs around the central atom and determine the electron arrangement that will minimize repulsions.
3. Determine the positions of the atoms and name the shape of the molecule.

Molecule	Lewis structure	Number of electron pairs	Electron pair arrangement	Molecular shape
H_2S	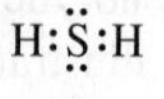 H:S̈:H	4	tetrahedral	bent
CCl_4	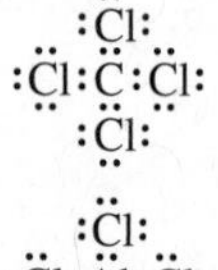 :Cl̈:C̈:Cl̈: :Cl̈: (top) :Cl̈: (bottom)	4	tetrahedral	tetrahedral
$AlCl_3$	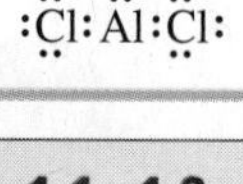 :Cl̈: :Cl̈:Al:Cl̈:	3	trigonal planar	trigonal planar

PRACTICE 11.12

Predict the shape for CF_4, NF_3, and BeI_2.

CHAPTER 11 REVIEW

11.1 PERIODIC TRENDS IN ATOMIC PROPERTIES

- Metals and nonmetals
- Atomic radius:
  - Increases down a group
  - Decreases across a row
- Ionization energy:
  - Energy required to remove an electron from an atom
  - Decreases down a group
  - Increases across a row

KEY TERM

ionization energy

11.2 LEWIS STRUCTURES OF ATOMS

- A Lewis structure is a representation of the atom where the symbol represents the element and dots around the symbol represent the valence electrons.
- To determine a Lewis structure for representative elements, use the group number as the number of electrons to place around the symbol for the element.

KEY TERM

Lewis structure

11.3 THE IONIC BOND: TRANSFER OF ELECTRONS FROM ONE ATOM TO ANOTHER

- The goal of bonding is to achieve stability:
  - For representative elements, this stability can be achieved by attaining a valence electron structure of a noble gas.
- In an ionic bond stability is attained by transferring an electron from one atom to another:
  - The atom that loses an electron becomes a cation:
    - Positive ions are smaller than their parent atoms.
    - Metals tend to form cations.
  - The atom gaining an electron becomes an anion:
    - Negative ions are larger than their parent atoms.
    - Nonmetals tend to form anions.
  - Ionic compounds do not exist as molecules:
    - Ions are attracted by multiple ions of the opposite charge to form a crystalline structure.

KEY TERM

ionic bond

11.4 PREDICTING FORMULAS OF IONIC COMPOUNDS

- Chemical compounds are always electrically neutral.
- Metals lose electrons and nonmetals gain electrons to form compounds.
- Stability is achieved (for representative elements) by attaining a noble gas electron configuration.

11.5 THE COVALENT BOND: SHARING ELECTRONS

- Covalent bonds are formed when two atoms share a pair of electrons between them:
  - This is the predominant type of bonding in compounds.
  - True molecules exist in covalent compounds.
  - Overlap of orbitals forms a covalent bond.
- Unequal sharing of electrons results in a polar covalent bond.

KEY TERMS

covalent bond
polar covalent bond

11.6 ELECTRONEGATIVITY

- Electronegativity is the attractive force an atom has for shared electrons in a molecule or polyatomic ion.
- Electrons spend more time closer to the more electronegative atom in a bond forming a polar bond.
- The polarity of a bond is determined by the electronegativity difference between the atoms involved in the bond:
  - The greater the difference, the more polar the bond is.
  - At the extremes:
    - Large differences result in ionic bonds.
    - Tiny differences (or no difference) result(s) in a nonpolar covalent bond.

KEY TERMS

electronegativity
nonpolar covalent bond
dipole

• A molecule that is electrically asymmetrical has a dipole, resulting in charged areas within the molecule.

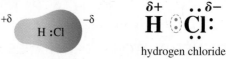

hydrogen chloride

• If the electronegativity difference between two bonded atoms is greater than 1.7–1.9, the bond will be more ionic than covalent.
• Polar bonds do not always result in polar molecules.

11.7 LEWIS STRUCTURES OF COMPOUNDS

> **PROBLEM-SOLVING STRATEGY: Writing a Lewis Structure**
>
> 1. Obtain the total number of valence electrons to be used in the structure by adding the number of valence electrons in all the atoms in the molecule or ion. If you are writing the structure of an ion, add one electron for each negative charge or subtract one electron for each positive charge on the ion.
> 2. Write the skeletal arrangement of the atoms and connect them with a single covalent bond (two dots or one dash). Hydrogen, which contains only one bonding electron, can form only one covalent bond. Oxygen atoms are not normally bonded to each other, except in compounds known to be peroxides. Oxygen atoms normally have a maximum of two covalent bonds (two single bonds or one double bond).
> 3. Subtract two electrons for each single bond you used in Step 2 from the total number of electrons calculated in Step 1. This gives you the net number of electrons available for completing the structure.
> 4. Distribute pairs of electrons (pairs of dots) around each atom (except hydrogen) to give each atom a noble gas structure.
> 5. If there are not enough electrons to give these atoms eight electrons, change single bonds between atoms to double or triple bonds by shifting unbonded pairs of electrons as needed. Check to see that each atom has a noble gas electron structure (two electrons for hydrogen and eight for the others). A double bond counts as four electrons for each atom to which it is bonded.

11.8 COMPLEX LEWIS STRUCTURES

KEY TERM

resonance structure

• When a single unique Lewis structure cannot be drawn for a molecule, resonance structures (multiple Lewis structures) are used to represent the molecule.

11.9 COMPOUNDS CONTAINING POLYATOMIC IONS

• Polyatomic ions behave like a single unit in many chemical reactions.
• The bonds within a polyatomic ion are covalent.

11.10 MOLECULAR SHAPE

KEY TERMS

linear structure
trigonal planar structure
tetrahedral structure
bent structure

• Lewis structures do not indicate the shape of a molecule.

> **PROBLEM-SOLVING STRATEGY: Determining Molecular Shape Using VSEPR**
>
> 1. Draw the Lewis structure for the molecule.
> 2. Count the electron pairs around the central atom and arrange them to minimize repulsions (as far apart as possible). This determines the electron pair arrangement.
> 3. Determine the positions of the atoms.
> 4. Name the molecular structure from the position of the atoms.

REVIEW QUESTIONS

1. Rank these elements according to the radii of their atoms, from smallest to largest: Na, Mg, Cl, K, and Rb. (Figure 11.2)
2. Explain why much more ionization energy is required to remove the first electron from neon than from sodium. (Table 11.1)
3. Explain the large increase in ionization energy needed to remove the third electron from beryllium compared with that needed for the second electron. (Table 11.1)

4. Does the first ionization energy increase or decrease from top to bottom in the periodic table for the alkali metal family? Explain. (Figure 11.3)

5. Does the first ionization energy increase or decrease from top to bottom in the periodic table for the noble gas family? Explain. (Figure 11.3)

6. Explain the reason that an atom of helium has a much higher first ionization energy than does an atom of hydrogen. (Table 11.1)

7. Why is there such a large increase in the energy required to remove a second electron from an atom of lithium? (Table 11.1)

8. Which element has the larger radius?
   (a) Li or Be   (d) F or Cl
   (b) K or Rb   (e) O or Se
   (c) Al or P   (f) As or Kr

9. In Groups 1A–7A, which element in each group has the smallest atomic radius? (Figure 11.2)

10. Why does the atomic size increase in going down any family of the periodic table?

11. Explain why potassium usually forms a K^+ ion but not a K^{2+} ion.

12. Why does an aluminum ion have a +3 charge?

13. Why are only valence electrons represented in a Lewis structure?

14. All the atoms within each Group A family of elements can be represented by the same Lewis structure. Complete the following table, expressing the Lewis structure for each group. (Use E to represent the elements.) (Figure 11.4)

Group	1A	2A	3A	4A	5A	6A	7A
	E·						

15. Draw the Lewis structure for Cs, Ba, Tl, Pb, Po, At, and Rn. How do these structures correlate with the group in which each element occurs?

16. What are valence electrons?

17. Why do metals tend to lose electrons and nonmetals tend to gain electrons when forming ionic bonds?

18. State whether the elements in each group gain or lose electrons in order to achieve a noble gas configuration. Explain.
   (a) Group 1A   (c) Group 6A
   (b) Group 2A   (d) Group 7A

19. What is the overall charge on an ionic compound?

20. Which family of elements tends to form ionic compounds with a 1 : 1 ratio of cation to anion with sulfur?

21. If the formula for calcium phosphate is $Ca_3(PO_4)_2$, predict the formulas of magnesium phosphate, beryllium phosphate, strontium phosphate, and barium phosphate.

22. Explain why magnesium tends to lose two electrons when forming an ionic compound.

23. Why is the term *molecule* used to describe covalent compounds but not ionic compounds?

24. How many shared electrons make up a single covalent bond? What is the maximum number of covalent bonds that can be formed between any two atoms?

25. Explain in simple terms how a chemist describes the formation of a covalent bond.

26. What does the dash between atoms represent in a Lewis electron dot structure?

27. Are all molecules that contain polar bonds polar molecules? Explain.

28. In a polar covalent bond, how do you determine which atom has a partial negative charge ($\delta-$) and which has a partial positive charge ($\delta+$)?

29. In which general areas of the periodic table are the elements with (a) the highest and (b) the lowest electronegativities located?

30. What is the purpose of a Lewis structure?

31. In a Lewis structure, what do the dots represent and what do the lines represent?

32. Can there be more than one correct Lewis structure for a compound? Explain.

33. If a molecule has more than one correct Lewis structure, what is the term used to describe these structures?

34. When drawing a Lewis structure for an ion, how is the charge represented?

35. Write two examples of molecules that have both ionic and covalent bonds.

36. What is the difference between electron pair arrangement and molecular shape?

Most of the exercises in this chapter are available for assignment via the online homework management program, WileyPLUS (www.wileyplus.com). All exercises with blue numbers have answers in Appendix VI.

PAIRED EXERCISES

1. An atom of potassium and a K^+ ion are drawn below. Identify each and explain your choice.

1   2

2. An atom of chlorine and a Cl^- ion are drawn below. Identify each and explain your choice.

1   2

3. Which one in each pair has the larger radius? Explain.
   (a) a calcium atom or a calcium ion
   (b) a chlorine atom or a chloride ion
   (c) a magnesium ion or an aluminum ion
   (d) a sodium atom or a silicon atom
   (e) a potassium ion or a bromide ion

4. Which one in each pair has the larger radius? Explain.
   (a) Fe^{2+} or Fe^{3+}
   (b) a potassium atom or a potassium ion
   (c) a sodium ion or a chloride ion
   (d) a strontium atom or an iodine atom
   (e) a rubidium ion or a strontium ion

5. Using the table of electronegativity values (Table 11.5), indicate which element is more positive and which is more negative in these compounds:
   (a) H_2O (d) PbS
   (b) RbCl   (e) PF_3
   (c) NH_3 (f) CH_4

6. Using the table of electronegativity values (Table 11.5), indicate which element is more positive and which is more negative in these compounds:
   (a) HCl   (d) IBr
   (b) SO_2 (e) CsI
   (c) CCl_4 (f) OF_2

7. Classify the bond between these pairs of elements as principally ionic or principally covalent (use Table 11.5):
   (a) sulfur and oxygen
   (b) barium and nitrogen
   (c) potassium and bromine
   (d) carbon and chlorine

8. Classify the bond between these pairs of elements as principally ionic or principally covalent (use Table 11.5):
   (a) sodium and oxygen
   (b) nitrogen and hydrogen
   (c) oxygen and hydrogen
   (d) phosphorus and chlorine

9. Write an equation representing each of the following:
   (a) the change of a magnesium atom to a magnesium ion
   (b) the change of a bromine atom to a bromide ion

10. Write an equation representing each of the following:
    (a) the change of a potassium atom to a potassium ion
    (b) the change of a sulfur atom to a sulfide ion

11. Use Lewis structures to show the electron transfer that enables these ionic compounds to form:
    (a) Li_2O (b) K_3N

12. Use Lewis structures to show the electron transfer that enables these ionic compounds to form:
    (a) K_2S (b) Ca_3N_2

13. State the number of valence electrons in an atom of each of the following elements:
    (a) selenium   (d) magnesium
    (b) phosphorus   (e) helium
    (c) bromine   (f) arsenic

14. State the number of valence electrons in an atom of each of the following elements:
    (a) lead   (d) cesium
    (b) lithium   (e) gallium
    (c) oxygen   (f) argon

15. How many electrons must be gained or lost for the following to achieve a noble gas electron configuration?
    (a) a potassium atom   (c) a bromine atom
    (b) an aluminum ion   (d) a selenium atom

16. How many electrons must be gained or lost for the following to achieve a noble gas electron structure?
    (a) a sulfur atom   (c) a nitrogen atom
    (b) a calcium atom   (d) an iodide ion

17. Determine whether the following atoms will form an ionic compound or a molecular compound, and give the formula of the compound.
    (a) sodium and chlorine
    (b) carbon and 4 hydrogen
    (c) magnesium and bromine
    (d) 2 bromine
    (e) carbon and 2 oxygen

18. Determine whether each of the following atoms will form a nonpolar covalent compound or a polar covalent compound, and give the formula of the compound.
    (a) 2 oxygen
    (b) hydrogen and bromine
    (c) oxygen and 2 hydrogen
    (d) 2 iodine

19. Let E be any representative element. Following the pattern in the table, write formulas for the hydrogen and oxygen compounds of the following:
    (a) Na   (c) Al
    (b) Ca   (d) Sn

20. Let E be any representative element. Following the pattern in the table, write formulas for the hydrogen and oxygen compounds of the following:
    (a) Sb   (c) Cl
    (b) Se   (d) C

Group						
1A	2A	3A	4A	5A	6A	7A
EH	EH_2	EH_3	EH_4	EH_3	H_2E	HE
E_2O	EO	E_2O_3	EO_2	E_2O_5	EO_3	E_2O_7

Group						
1A	2A	3A	4A	5A	6A	7A
EH	EH_2	EH_3	EH_4	EH_3	H_2E	HE
E_2O	EO	E_2O_3	EO_2	E_2O_5	EO_3	E_2O_7

21. The formula for sodium sulfate is Na_2SO_4. Write the names and formulas for the other alkali metal sulfates.

22. The formula for calcium bromide is $CaBr_2$. Write the names and formulas for the other alkaline earth metal bromides.

23. Write Lewis structures for the following:
    (a) Na   (b) Br^- (c) O^{2-}

24. Write Lewis structures for the following:
    (a) Ga   (b) Ga^{3+} (c) Ca^{2+}

25. Classify the bonding in each compound as ionic or covalent:
    (a) H_2O (c) MgO
    (b) NaCl   (d) Br_2

26. Classify the bonding in each compound as ionic or covalent:
    (a) HCl   (c) NH_3
    (b) $BaCl_2$ (d) SO_2

27. Predict the type of bond that would be formed between the following pairs of atoms:
    (a) P and I   (b) N and S   (c) Br and I

28. Predict the type of bond that would be formed between the following pairs of atoms:
    (a) H and Si   (b) O and F   (c) Si and Br

29. Draw Lewis structures for the following:
    (a) H_2 (b) N_2 (c) Cl_2

30. Draw Lewis structures for the following:
    (a) O_2 (b) Br_2 (c) I_2

31. Draw Lewis structures for the following:
    (a) NCl_3 (c) C_2H_6
    (b) H_2CO_3 (d) $NaNO_3$

32. Draw Lewis structures for the following:
    (a) H_2S (c) NH_3
    (b) CS_2 (d) NH_4Cl

33. Draw Lewis structures for the following:
    (a) Ba^{2+} (d) CN^-
    (b) Al^{3+} (e) HCO_3^-
    (c) SO_3^{2-}

34. Draw Lewis structures for the following:
    (a) I^- (d) ClO_3^-
    (b) S^{2-} (e) NO_3^-
    (c) CO_3^{2-}

35. Classify each of the following molecules as polar or nonpolar:
    (a) CH_3Cl (c) OF_2
    (b) Cl_2 (d) PBr_3

36. Classify each of the following molecules as polar or nonpolar:
    (a) H_2 (c) CH_3OH
    (b) NI_3 (d) CS_2

37. Give the number and arrangement of the electron pairs around the central atom:
    (a) C in CCl_4 (c) Al in AlH_3
    (b) S in H_2S

38. Give the number and arrangement of the electron pairs around the central atom:
    (a) Ga in $GaCl_3$ (c) Cl in ClO_3^-
    (b) N in NF_3

39. Use VSEPR theory to predict the structure of these polyatomic ions:
    (a) sulfate ion   (c) periodate ion
    (b) chlorate ion

40. Use VSEPR theory to predict the structure of these polyatomic ions:
    (a) ammonium ion   (c) phosphate ion
    (b) sulfite ion

41. Use VSEPR theory to predict the shape of these molecules:
    (a) SiH_4 (b) PH_3 (c) SeF_2

42. Use VSEPR theory to predict the shape of these molecules:
    (a) SiF_4 (b) OF_2 (c) Cl_2O

43. Element X reacts with sodium to form the compound Na_2X and is in the second period on the periodic table. Identify this element.

44. Element Y reacts with oxygen to form the compound Y_2O and has the lowest ionization energy of any Period 4 element on the periodic table. Identify this element.

ADDITIONAL EXERCISES

45. Atoms of sodium, cesium, and argon are drawn below. Identify each and explain your choice.

    1   2   3

46. Spheres representing Br^-, Se^{2-}, Rb^+, Sr^{2+}, and Kr are drawn below. Identify each and explain your choice.

  1   2   3   4   5

47. Write Lewis structures for hydrazine (N_2H_4) and hydrazoic acid (HN_3).

48. Draw Lewis structures and give the shape of each of the following compounds:
    (a) NO_2^- (c) $SOCl_2$
    (b) SO_4^{2-} (d) Cl_2O

49. Draw Lewis structures for each of the following compounds:
    (a) ethane (C_2H_6)
    (b) ethylene (C_2H_4)
    (c) acetylene (C_2H_2)

50. Using the periodic table, identify each element given the description:
    (a) the most electronegative element of the alkaline earth metals
    (b) the noble gas with the highest first ionization energy
    (c) the element in Period 4 with the lowest first ionization energy

(d) the element with the highest electronegativity
(e) the alkali metal with the largest radius
(f) the element in Period 2 with the smallest radius

51. Choose the element that fits each description:
    (a) the higher electronegativity   Cl or Br
    (b) the higher first ionization energy   O or S
    (c) the lower electronegativity   Ca or C

52. Why do you think there are no electronegativity values given for the noble gases in Table 11.5?

53. When one electron is removed from an atom of Li, it has two left. Helium atoms also have two electrons. Why is more energy required to remove the second electron from Li than to remove the first from He?

54. Group 1B elements (see the periodic table on the inside cover of your book) have one electron in their outer energy level, as do Group 1A elements. Would you expect them to form compounds such as CuCl, AgCl, and AuCl? Explain.

55. The formula for lead(II) bromide is $PbBr_2$. Predict formulas for tin(II) and germanium(II) bromides.

56. Why is it not proper to speak of sodium chloride molecules?

57. What is a covalent bond? How does it differ from an ionic bond?

58. Briefly comment on the structure Na:$\ddot{O}$:Na for the compound Na_2O.

59. What are the four most electronegative elements?

60. Rank these elements from highest electronegativity to lowest: Mg, S, F, H, O, Cs.

61. Is it possible for a molecule to be nonpolar even though it contains polar covalent bonds? Explain.

62. Why is CO_2 a nonpolar molecule, whereas CO is a polar molecule?

63. Rank the following bonds in order of increasing polarity:
    (a) NaO, CO, NO
    (b) CO, SiO, GeO
    (c) BO, BS, BSe

64. Estimate the bond angle between atoms in these molecules:
    (a) H_2S
    (b) NH_3
    (c) NH_4^+
    (d) $SiCl_4$

65. Consider the two molecules BF_3 and NF_3. Compare and contrast them in terms of the following:
    (a) valence-level orbitals on the central atom that are used for bonding
    (b) shape of the molecule
    (c) number of lone electron pairs on the central atom
    (d) type and number of bonds found in the molecule

66. With respect to electronegativity, why is fluorine such an important atom? What combination of atoms on the periodic table results in the most ionic bond?

67. Why does the Lewis structure of each element in a given group of representative elements on the periodic table have the same number of dots?

68. A sample of an air pollutant composed of sulfur and oxygen was found to contain 1.40 g sulfur and 2.10 g oxygen. What is the empirical formula for this compound? Draw a Lewis structure to represent it.

69. A dry-cleaning fluid composed of carbon and chlorine was found to haye the composition 14.5% carbon and 85.5% chlorine. Its known molar mass is 166 g/mol. Draw a Lewis structure to represent the compound.

70. Identify each of the following drawings as representing ionic compounds, covalent compounds, or compounds composed of both ionic and covalent portions.

(a)
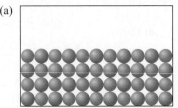

(b)
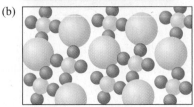

(c)
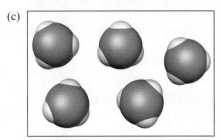

(d)
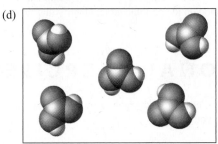

71. Identify each of the following compounds as ionic, covalent, or both. Give the correct IUPAC name for each compound.
    (a) Na_3P (c) SO_2 (e) $Cu(NO_3)_2$
    (b) NH_4I (d) H_2S (f) MgO

CHALLENGE EXERCISES

72. Draw the following Lewis structures, showing any full charges where they exist on ions:
    (a) H_2SO_4 (d) HCN
    (b) $NaNO_3$ (e) $Al_2(SO_4)_3$
    (c) K_2CO_3 (f) CH_3COOH

73. The first ionization energy for lithium is 520 kJ/mol. How much energy would be required to change 25 g of lithium atoms to lithium ions? (Refer to Table 11.1.)

74. What is the total amount of energy required to remove the first two electrons from 15 moles of sodium atoms? (Refer to Table 11.1.)

75. Sulfur trioxide gas reacts with water to form sulfuric acid. Draw the Lewis structures of each substance to represent the reaction.

76. Compounds found in garlic have been shown to have significant biological effects including immune system enhancement, anticancer activity, and cardiovascular activity. Garlic clearly seems to have more value than as just a great tasting addition to spaghetti!

Some of the compounds in garlic that are currently being studied are shown in the following figures.

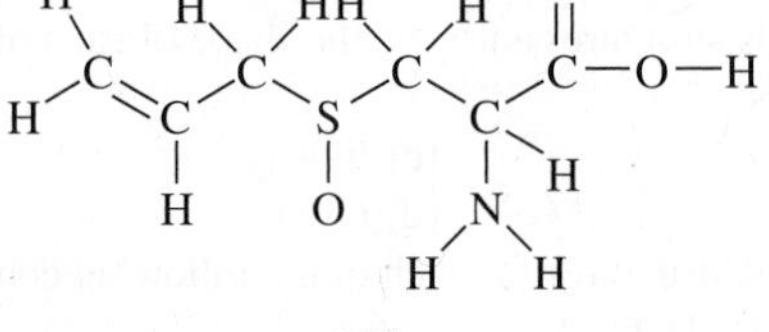

Alliin

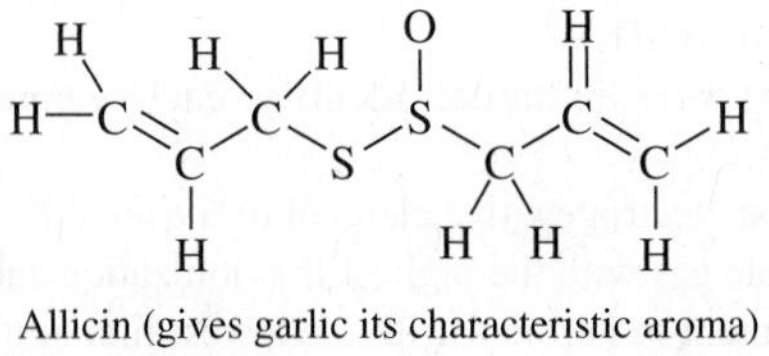

Allicin (gives garlic its characteristic aroma)

Redraw these molecules adding in dots to represent any lone pairs present on atoms. Draw a circle around atoms with tetrahedral geometry, draw a square around atoms with trigonal planar geometry, draw a pentagon around atoms with a trigonal pyramidal geometry, and draw a hexagon around atoms with a bent geometry. Draw a triangle around any atoms that have an unexpected number of bonds.

77. Two of the most complex organic compounds found outside of our solar system have been detected in a massive star-forming region near the heart of the Milky Way known as Sagittarius B2. The skeleton structures of these two compounds, ethyl formate (the compound that gives raspberries their flavor) and *n*-propyl cyanide, are drawn below. Draw the complete Lewis electron dot structures for these two compounds, identify the electron pair and molecular geometry around each atom, and predict the bond angles.

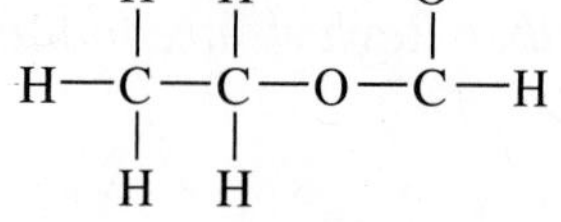

Ethyl formate 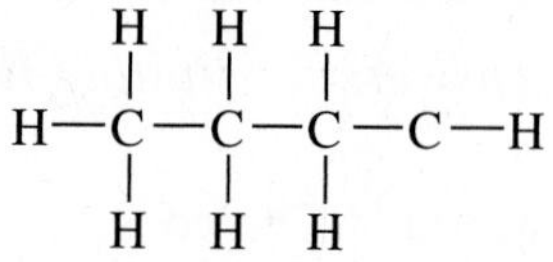   *n*-propyl cyanide

ANSWERS TO PRACTICE EXERCISES

11.1 (a) Ionization energy decreases from top to bottom for a group of elements and increases from left to right across a period of elements.

(b) For a group of elements the outer shell of electrons is increasingly further away from the nucleus as you go from top to bottom of the group. Thus, they are not held as strongly by the positive nucleus and are easier to ionize.

11.2 (a) $:\dot{\ddot{N}}\cdot$

(b) $:\dot{\ddot{Al}}$

(c) $Sr:$

(d) $:\dot{\ddot{Br}}\cdot$

11.3 (a) Ar; K^+
(b) Ne; Mg^{2+}
(c) Ne; Al^{3+}
(d) Xe; Ba^{2+}

11.4 (a) Na_2S
(b) Rb_2O

11.5 (a) Mg_3N_2
(b) Ba_3As_2

11.6 $H:\ddot{\ddot{Br}}:$ or $H—Br$

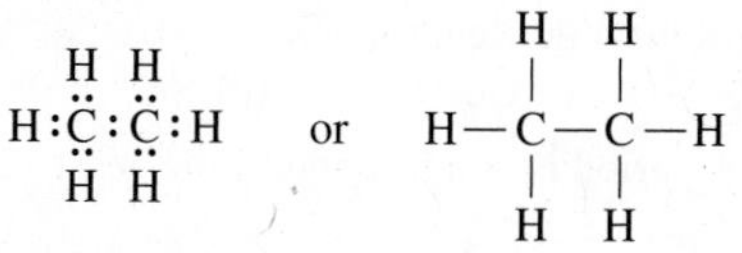

11.7 Electronegativity of an atom gives information about the strength that the atom has for attracting a pair of shared electrons in a covalent bond. The higher the electronegativity value,

the greater the attraction for a pair of shared electrons. Consequently, when two different atoms with different electronegativities share a pair of electrons, one atom becomes partially negative and the other atom becomes partially positive due to the attractive forces for the pair of electrons. The bond formed is said to be polar covalent.

11.8 ionic: (a), (d), (e)
covalent: (b), (c), (f)

11.9 (a) 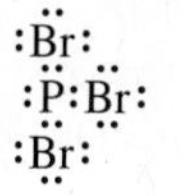 $:\ddot{P}:\ddot{Br}:$ (with $:\ddot{Br}:$ above and $:\ddot{Br}:$ below)   (c) $H:\ddot{F}:$ (e) $:N:::N:$

(b) 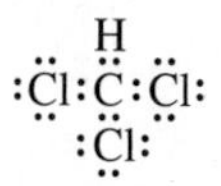 $:\ddot{Cl}:\ddot{C}:\ddot{Cl}:$ with H above and $:\ddot{Cl}:$ below   (d) 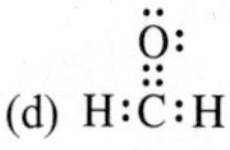 $H:\ddot{C}:H$ with $\ddot{\ddot{O}}:$ above

11.10 (a) 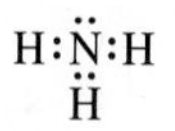 $H:\ddot{N}:H$ with H below   (c) 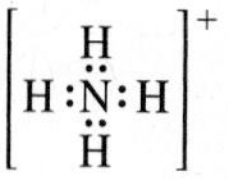 $\left[H:\ddot{N}:H \right]^+$ with H above and H below

(b) 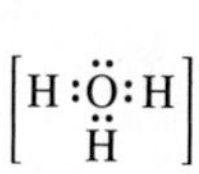 $\left[H:\ddot{O}:H \right]^+$ with H below   (d) 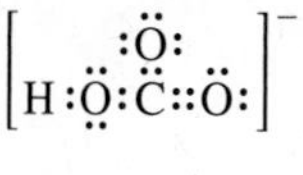 $\left[H:\ddot{O}:\ddot{C}::\ddot{O}: \right]^-$ with $:\ddot{O}:$ above

11.11

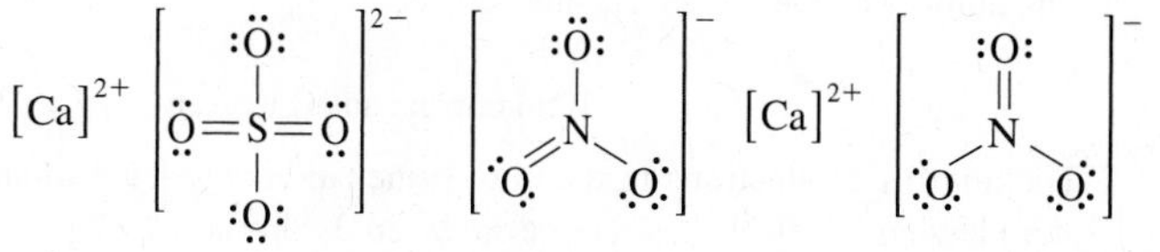

11.12 CF_4, tetrahedral; NF_3, pyramidal; BeI_2, linear

PUTTING IT TOGETHER:
Review for Chapters 10–11

Answers for Putting It Together Reviews are found in Appendix VII.

Multiple Choice

Choose the correct answer to each of the following.

1. The concept of electrons existing in specific orbits around the nucleus was the contribution of
   (a) Thomson (c) Bohr
   (b) Rutherford (d) Schrödinger

2. The correct electron structure for a fluorine atom (F) is
   (a) $1s^2 2s^2 2p^5$ (c) $1s^2 2s^2 2p^4 3s^1$
   (b) $1s^2 2s^2 2p^3 3s^2 3p^1$ (d) $1s^2 2s^2 2p^3$

3. The correct electron structure for $_{48}$Cd is
   (a) $1s^2 2s^2 2p^6 3s^2 3p^6 4s^2 3d^{10}$
   (b) $1s^2 2s^2 2p^6 3s^2 3p^6 4s^2 3d^{10} 4p^6 5s^2 4d^{10}$
   (c) $1s^2 2s^2 2p^6 3s^2 3p^6 4s^2 3d^{10} 4p^6 4d^4$
   (d) $1s^2 2s^2 2p^6 3s^2 3p^6 4s^2 4p^6 4d^{10} 5s^2 5d^{10}$

4. The correct electron structure of $_{23}$V is
   (a) $[Ar]4s^2 3d^3$ (c) $[Ar]4s^2 4d^3$
   (b) $[Ar]4s^2 4p^3$ (d) $[Kr]4s^2 3d^3$

5. Which of the following is the correct atomic structure for $_{22}^{48}$Ti?

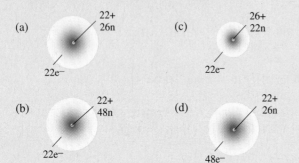

6. The number of orbitals in a d sublevel is
   (a) 3 (c) 7
   (b) 5 (d) no correct answer given

7. The number of electrons in the third principal energy level in an atom having the electron structure $1s^2 2s^2 2p^6 3s^2 3p^2$ is
   (a) 2 (b) 4 (c) 6 (d) 8

8. The total number of orbitals that contain at least one electron in an atom having the structure $1s^2 2s^2 2p^6 3s^2 3p^2$ is
   (a) 5 (c) 14
   (b) 8 (d) no correct answer given

9. Which of these elements has two s and six p electrons in its outer energy level?
   (a) He (c) Ar
   (b) O (d) no correct answer given

10. Which element is not a noble gas?
    (a) Ra (b) Xe (c) He (d) Ar

11. Which element has the largest number of unpaired electrons?
    (a) F (b) S (c) Cu (d) N

12. How many unpaired electrons are in the electron structure of $_{24}$Cr, $[Ar]4s^1 3d^5$?
    (a) 2 (b) 4 (c) 5 (d) 6

13. Groups 3A–7A plus the noble gases form the area of the periodic table where the electron sublevels being filled are
    (a) p sublevels (c) d sublevels
    (b) s and p sublevels (d) f sublevels

14. In moving down an A group on the periodic table, the number of electrons in the outermost energy level
    (a) increases regularly
    (b) remains constant
    (c) decreases regularly
    (d) changes in an unpredictable manner

15. Which of the following is an incorrect formula?
    (a) NaCl (b) K_2O (c) AlO (d) BaO

16. Elements of the noble gas family
    (a) form no compounds at all
    (b) have no valence electrons
    (c) have an outer electron structure of $ns^2 np^6$ (helium excepted), where n is the period number
    (d) no correct answer given

17. The lanthanide and actinide series of elements are
    (a) representative elements
    (b) transition elements
    (c) filling in d-level electrons
    (d) no correct answer given

18. The element having the structure $1s^2 2s^2 2p^6 3s^2 3p^2$ is in Group
    (a) 2A (b) 2B (c) 4A (d) 4B

19. In Group 5A, the element having the smallest atomic radius is
    (a) Bi (b) P (c) As (d) N

20. In Group 4A, the most metallic element is
    (a) C (b) Si (c) Ge (d) Sn

21. Which group in the periodic table contains the least reactive elements?
    (a) 1A (b) 2A (c) 3A (d) noble gases

22. Which group in the periodic table contains the alkali metals?
    (a) 1A (b) 2A (c) 3A (d) 4A

23. An atom of fluorine is smaller than an atom of oxygen. One possible explanation is that, compared to oxygen, fluorine has
    (a) a larger mass number
    (b) a smaller atomic number
    (c) a greater nuclear charge
    (d) more unpaired electrons

24. If the size of the fluorine atom is compared to the size of the fluoride ion,
    (a) they would both be the same size
    (b) the atom is larger than the ion
    (c) the ion is larger than the atom
    (d) the size difference depends on the reaction

25. Sodium is a very active metal because
   (a) it has a low ionization energy
   (b) it has only one outermost electron
   (c) it has a relatively small atomic mass
   (d) all of the above

26. Which of the following formulas is not correct?
   (a) Na^+ (b) S^- (c) Al^{3+} (d) F^-

27. Which of the following molecules does not have a polar covalent bond?
   (a) CH_4 (b) H_2O (c) CH_3OH (d) Cl_2

28. Which of the following molecules is a dipole?
   (a) HBr   (b) CH_4 (c) H_2 (d) CO_2

29. Which of the following has bonding that is ionic?
   (a) H_2 (b) MgF_2 (c) H_2O (d) CH_4

30. Which of the following is a correct Lewis structure?
   (a) $:\ddot{O}:C:\ddot{O}:$ (c) $\ddot{C}l::\ddot{C}l$
   (b) $:\ddot{C}l:C:\ddot{C}l:$ with $:\ddot{C}l:$ above and $:\ddot{C}l:$ below   (d) $:\ddot{N}:\ddot{N}:$

31. Which of the following is an incorrect Lewis structure?
   (a) $H:\ddot{N}:H$ (c) $H:\ddot{C}:H$ with H above and H below
   (b) $:\ddot{O}:H$ with H below   (d) $:N:::N:$

32. The correct Lewis structure for SO_2 is
   (a) $:\ddot{O}:\ddot{S}:\ddot{O}:$ (c) $:\ddot{O}::S::\ddot{O}:$
   (b) $:\ddot{O}:S::\ddot{O}:$ (d) $:\ddot{O}:\ddot{S}:\ddot{O}:$

33. Carbon dioxide (CO_2) is a nonpolar molecule because
   (a) oxygen is more electronegative than carbon
   (b) the two oxygen atoms are bonded to the carbon atom
   (c) the molecule has a linear structure with the carbon atom in the middle
   (d) the carbon–oxygen bonds are polar covalent

34. When a magnesium atom participates in a chemical reaction, it is most likely to
   (a) lose 1 electron   (c) lose 2 electrons
   (b) gain 1 electron   (d) gain 2 electrons

35. If X represents an element of Group 3A, what is the general formula for its oxide?
   (a) X_3O_4 (b) X_3O_2 (c) XO   (d) X_2O_3

36. Which of the following has the same electron structure as an argon atom?
   (a) Ca^{2+} (b) Cl^0 (c) Na^+ (d) K^0

37. As the difference in electronegativity between two elements decreases, the tendency for the elements to form a covalent bond
   (a) increases
   (b) decreases
   (c) remains the same
   (d) sometimes increases and sometimes decreases

38. Which compound forms a tetrahedral molecule?
   (a) NaCl   (b) CO_2 (c) CH_4 (d) $MgCl_2$

39. Which compound has a bent (V-shaped) molecular structure?
   (a) NaCl   (b) CO_2 (c) CH_4 (d) H_2O

40. Which compound has double bonds within its molecular structure?
   (a) NaCl   (b) CO_2 (c) CH_4 (d) H_2O

41. The total number of valence electrons in a nitrate ion, NO_3^-, is
   (a) 12   (b) 18   (c) 23   (d) 24

42. The number of electrons in a triple bond is
   (a) 3   (b) 4   (c) 6   (d) 8

43. The number of unbonded pairs of electrons in H_2O is
   (a) 0   (b) 1   (c) 2   (d) 4

44. Which of the following does not have a noble gas electron structure?
   (a) Na   (b) Sc^{3+} (c) Ar   (d) O^{2-}

Free Response Questions

Answer each of the following. Be sure to include your work and explanations in a clear, logical form.

1. An alkaline earth metal, M, combines with a halide, X. Will the resulting compound be ionic or covalent? Why? What is the Lewis structure for the compound?

2. "All electrons in atoms with even atomic numbers are paired." Is this statement true or false? Explain your answer using an example.

3. Discuss whether the following statement is true or false: "All non-metals have two valence electrons in an *s* sublevel with the exception of the noble gases, which have at least one unpaired electron in a *p* sublevel."

4. The first ionization energy (IE) of potassium is lower than the first IE for calcium, but the second IE of calcium is lower than the second IE of potassium. Use an electron configuration or size argument to explain this trend in ionization energies.

5. Chlorine has a very large first ionization energy, yet it forms a chloride ion relatively easily. Explain.

6. Three particles have the same electron configuration. One is a cation of an alkali metal, one is an anion of the halide in the third period, and the third particle is an atom of a noble gas. What are the identities of the three particles (including charges)? Which particle should have the smallest atomic/ionic radius, which should have the largest, and why?

7. Why is the Lewis structure of AlF_3 not written as

$$:\ddot{F}: \diagup \atop :\ddot{F}-Al \atop \diagdown :\ddot{F}:$$

What is the correct Lewis structure, and which electrons are shown in a Lewis structure?

8. Why does carbon have a maximum of four covalent bonds?

9. Both NCl_3 and BF_3 have a central atom bonded to three other atoms, yet one is pyramidal and the other is trigonal planar. Explain.

10. Draw the Lewis structure of the atom whose electron configuration is $1s^2 2s^2 2p^6 3s^2 3p^6 4s^2 3d^{10} 4p^5$. Would you expect this atom to form an ionic, nonpolar covalent, or polar covalent bond with sulfur?

The air in a hot air balloon expands when it is heated. Some of the air escapes from the top of the balloon, lowering the air density inside the balloon making the balloon buoyant.

Kees Van DenBerg/Photo Researchers, Inc.

CHAPTER **12**

THE GASEOUS STATE OF MATTER

Our atmosphere is composed of a mixture of gases, including nitrogen, oxygen, argon, carbon dioxide, ozone, and trace amounts of others. These gases are essential to life, yet they can also create hazards for us. For example, carbon dioxide is valuable when it is taken in by plants and converted to carbohydrates, but it is also associated with the potentially hazardous greenhouse effect. Ozone surrounds the Earth at high altitudes and protects us from harmful ultraviolet rays, but it also destroys rubber and plastics. We require air to live, yet scuba divers must be concerned about oxygen poisoning and the "bends."

In chemistry, the study of the behavior of gases allows us to understand our atmosphere and the effects that gases have on our lives.

CHAPTER OUTLINE

12.1 Properties of Gases
12.2 Boyle's Law
12.3 Charles' Law
12.4 Avogadro's Law
12.5 Combined Gas Laws
12.6 Ideal Gas Law
12.7 Dalton's Law of Partial Pressures
12.8 Density of Gases
12.9 Gas Stoichiometry

12.1 PROPERTIES OF GASES

Explain atmospheric pressure and how it is measured. Be able to convert among the various units of pressure.

Gases are the least dense and most mobile of the three states of matter (see **Figure 12.1**). A solid has a rigid structure, and its particles remain in essentially fixed positions. When a solid absorbs sufficient heat, it melts and changes into a liquid. Melting occurs because the molecules (or ions) have absorbed enough energy to break out of the rigid structure of the solid. The molecules or ions in the liquid are more energetic than they were in the solid, as indicated by their increased mobility. Molecules in the liquid state cling to one another. When the liquid absorbs additional heat, the more energetic molecules break away from the liquid surface and go into the gaseous state—the most mobile state of matter. Gas molecules move at very high velocities and have high kinetic energy. The average velocity of hydrogen molecules at 0°C is over 1600 meters (1 mile) per second. Mixtures of gases are uniformly distributed within the container in which they are confined.

The same quantity of a substance occupies a much greater volume as a gas than it does as a liquid or a solid (see **Figure 12.2**). For example, 1 mol of water (18.02 g) has a volume of 18 mL at 4°C. This same amount of liquid water would occupy about 22,400 mL in the gaseous state—more than a 1200-fold increase in volume. We may assume from this difference in volume that (1) gas molecules are relatively far apart, (2) gases can be greatly compressed, and (3) the volume occupied by a gas is mostly empty space.

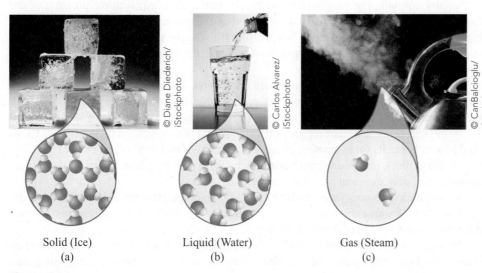

Solid (Ice)
(a)

Liquid (Water)
(b)

Gas (Steam)
(c)

© Diane Diederich/iStockphoto
© Carlos Alvarez/iStockphoto
© CanBalcioglu/iStockphoto

Figure 12.1

The three states of matter. (a) Solid—water molecules are held together rigidly and are very close to each other. (b) Liquid—water molecules are close together but are free to move around and slide over each other. (c) Gas—water molecules are far apart and move freely and randomly.

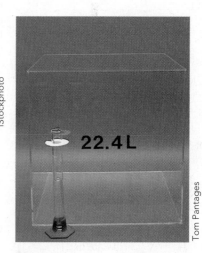

22.4 L

Tom Pantages

Figure 12.2

A mole of water occupies 18 mL as a liquid but would fill this box (22.4 L) as a gas at the same temperature.

Measuring the Pressure of a Gas

Pressure is defined as force per unit area. When a rubber balloon is inflated with air, it stretches and maintains its larger size because the pressure on the inside is greater than that on the outside. Pressure results from the collisions of gas molecules with the walls of the balloon (see **Figure 12.3**). When the gas is released, the force or pressure of the air escaping from the small neck propels the balloon in a rapid, irregular flight. If the balloon is inflated until it bursts, the gas escaping all at once causes an explosive noise.

The effects of pressure are also observed in the mixture of gases in the atmosphere, which is composed of about 78% nitrogen, 21% oxygen, 1% argon, and other minor constituents by volume (see **Table 12.1**). The outer boundary of the atmosphere is not known precisely, but more than 99% of the atmosphere is below an altitude of 32 km (20 miles). Thus, the concentration of gas molecules in the atmosphere decreases with altitude, and at about

© Jon Helgason/iStockphoto

Figure 12.3

The pressure resulting from the collisions of gas molecules with the walls of the balloon keeps the balloon inflated.

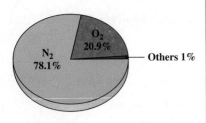

TABLE 12.1	Average Composition of Dry Air		
Gas	**Percent by volume**	**Gas**	**Percent by volume**
N_2	78.08	He	0.0005
O_2	20.95	CH_4	0.0002
Ar	0.93	Kr	0.0001
CO_2	0.033	Xe, H_2, and N_2O	Trace
Ne	0.0018		

6.4 km (4 miles) the amount of oxygen is insufficient to sustain human life. The gases in the atmosphere exert a pressure known as **atmospheric pressure**. The pressure exerted by a gas depends on the number of molecules of gas present, the temperature, and the volume in which the gas is confined. Gravitational forces hold the atmosphere relatively close to Earth and prevent air molecules from flying off into space. Thus, the atmospheric pressure at any point is due to the mass of the atmosphere pressing downward at that point.

The pressure of the gases in the atmosphere can be measured with a **barometer**. A mercury barometer may be prepared by completely filling a long tube with pure, dry mercury and inverting the open end into an open dish of mercury. If the tube is longer than 760 mm, the mercury level will drop to a point at which the column of mercury in the tube is just supported by the pressure of the atmosphere. If the tube is properly prepared, a vacuum will exist above the mercury column. The weight of mercury, per unit area, is equal to the pressure of the atmosphere. The column of mercury is supported by the pressure of the atmosphere, and the height of the column is a measure of this pressure (see **Figure 12.4**). The mercury barometer was invented in 1643 by the Italian physicist E. Torricelli (1608–1647), for whom the unit of pressure *torr* was named.

Air pressure is measured and expressed in many units. The standard atmospheric pressure, or simply **1 atmosphere** (atm), is the pressure exerted by a column of mercury 760 mm high at a temperature of 0°C. The normal pressure of the atmosphere at sea level is 1 atm, or 760 torr, or 760 mm Hg. The SI unit for pressure is the pascal (Pa), where 1 atm = 101,325 Pa, or 101.3 kPa. Other units for expressing pressure are inches of mercury, centimeters of mercury, the millibar (mbar), and pounds per square inch (lb/in.² or psi). The values of these units equivalent to 1 atm are summarized in **Table 12.2**.

Atmospheric pressure varies with altitude. The average pressure at Denver, Colorado, 1.61 km (1 mile) above sea level, is 630 torr (0.83 atm). Atmospheric pressure is 380 torr (0.5 atm) at about 5.5 km (3.4 miles) altitude.

Pressure is often measured by reading the height of a mercury column in millimeters on a barometer. Thus pressure may be recorded as mm Hg (torr). In problems dealing with gases,

TABLE 12.2   Pressure Units Equivalent to 1 Atmosphere
1 atm
760 torr
760 mm Hg
76 cm Hg
101.325 kPa
1013 mbar
29.9 in. Hg
14.7 lb/in.²

Figure 12.4

Preparation of a mercury barometer. The full tube of mercury at the left is inverted and placed in a dish of mercury.

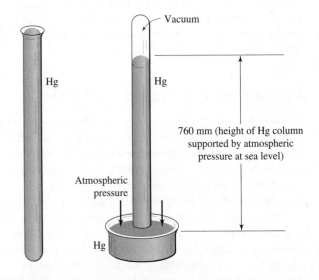

it is necessary to make conversions among the various pressure units. Since atm, torr, and mm Hg are common pressure units, we give examples involving all three of these units:

$$1 \text{ atm} = 760 \text{ torr} = 760 \text{ mm Hg}$$

EXAMPLE 12.1

The average atmospheric pressure at Walnut, California, is 740. mm Hg. Calculate this pressure in (a) torr and (b) atmospheres.

SOLUTION

Let's use conversion factors that relate one unit of pressure to another.

(a) To convert mm Hg to torr, use the conversion factor

760 torr/760 mm Hg (1 torr/1 mm Hg):

$$(740. \text{ mm Hg})\left(\frac{1 \text{ torr}}{1 \text{ mm Hg}}\right) = 740. \text{ torr}$$

(b) To convert mm Hg to atm, use the conversion factor

1 atm/760. mm Hg:

$$(740. \text{ mm Hg})\left(\frac{1 \text{ atm}}{760. \text{ mm Hg}}\right) = 0.974 \text{ atm}$$

PRACTICE 12.1

A barometer reads 1.12 atm. Calculate the corresponding pressure in (a) torr, (b) mm Hg, and (c) kilopascals.

WileyPLUS

ENHANCED EXAMPLE

Pressure Dependence on the Number of Molecules and the Temperature

Pressure is produced by gas molecules colliding with the walls of a container. At a specific temperature and volume, the number of collisions depends on the number of gas molecules present. The number of collisions can be increased by increasing the number of gas molecules present. If we double the number of molecules, the frequency of collisions and the pressure should double. We find, for an ideal gas, that this doubling is actually what happens. When the temperature and mass are kept constant, the pressure is directly proportional to the number of moles or molecules of gas present. **Figure 12.5** illustrates this concept.

A good example of this molecule–pressure relationship may be observed in an ordinary cylinder of compressed gas equipped with a pressure gauge. When the valve is opened, gas escapes from the cylinder. The volume of the cylinder is constant, and the decrease in quantity (moles) of gas is registered by a drop in pressure indicated on the gauge.

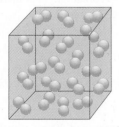
2 mol H_2
$P = 2$ atm

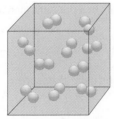
1 mol H_2
$P = 1$ atm

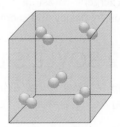
0.5 mol H_2
$P = 0.5$ atm

Figure 12.5

The pressure exerted by a gas is directly proportional to the number of molecules present. In each case shown, the volume is 22.4 L and the temperature is 0°C.

>CHEMISTRY *IN ACTION*

What the Nose Knows

The sense of smell is one of the most intriguing senses of animals. Through the sense of smell dogs can detect the presence of many different drugs, explosives, and other substances. In some cases they may even be able to detect certain human diseases. The air around us is filled with traces of the chemicals composing our world. Much work has gone into finding ways to identify the chemicals present in the air. Developing methods to identify chemicals at low concentrations will have tremendous benefits.

BETTER COFFEE? Kenneth Suslick at the University of Illinois, Urbana-Champaign, and his colleagues have developed an electronic nose. Molecules placed in polymer films change color when exposed to the aromatic compounds found in coffee. By comparing the relative amounts of each of these compounds to the amounts found in an exceptional cup of coffee, coffee roasters are able to easily identify problems with their coffee beans and thereby ensure a consistently good cup of coffee.

NO MORE STINKY FEET? If you have been flying recently, you have probably been required to remove your shoes. If researchers at the University of Illinois are successful, you may no longer have to remove your shoes at the airport. A handheld sensor is able to quickly and accurately detect the presence of a commonly used explosive in shoes. Now as passengers walk through security, chemical sensors will look for triacetone triperoxide or TATP vapors emanating from their shoes. This technology holds the promise that other harmful compounds may also be detected in this way.

THE DOCTOR NOSE YOU'RE SICK! Many human diseases produce volatile compounds that are exhaled with every breath you take. Physicians can often diagnose diabetes simply by sniffing the breath of their patients for the odor of acetone. While most diseases do not produce easily detectable odors, physicians can take advantage of technology to detect these compounds at the low concentrations in which they are produced. For example, lung cancer patients emit specific metabolic by-products in their breath that can be detected by dogs. Researchers are actively looking for techniques to detect these same compounds so that physicians can have a quick, easy screening tool for lung cancer.

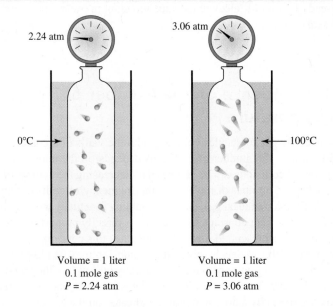

Figure 12.6

The pressure of a gas in a fixed volume increases with increasing temperature. The increased pressure is due to more frequent and more energetic collisions of the gas molecules with the walls of the container at the higher temperature.

2.24 atm

3.06 atm

0°C

100°C

Volume = 1 liter
0.1 mole gas
P = 2.24 atm

Volume = 1 liter
0.1 mole gas
P = 3.06 atm

The pressure of a gas in a fixed volume also varies with temperature. When the temperature is increased, the kinetic energy of the molecules increases, causing more frequent and more energetic collisions of the molecules with the walls of the container. This increase in collision frequency and energy results in a pressure increase (see **Figure 12.6**).

12.2 BOYLE'S LAW

LEARNING OBJECTIVE Use Boyle's law to calculate changes in pressure or volume of a sample of gas at a constant temperature.

KEY TERM

Boyle's law

Through a series of experiments, Robert Boyle (1627–1691) determined the relationship between the pressure (P) and volume (V) of a particular quantity of a gas. This relationship of P and V is known as **Boyle's law**:

At constant temperature (T), the volume (V) of a fixed mass of a gas is inversely proportional to the pressure (P), which may be expressed as

$$V \propto \frac{1}{P} \quad \text{or} \quad P_1V_1 = P_2V_2$$

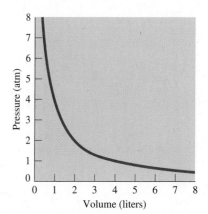

This law says that the volume varies ($\propto$) inversely with the pressure at constant mass and temperature. When the pressure on a gas is increased, its volume will decrease, and vice versa. The inverse relationship of pressure and volume is graphed in **Figure 12.7**.

When Boyle doubled the pressure on a specific quantity of a gas, keeping the temperature constant, the volume was reduced to one-half the original volume; when he tripled the pressure on the system, the new volume was one-third the original volume; and so on. His work showed that the product of volume and pressure is constant if the temperature is not changed:

$$PV = \text{constant} \quad \text{or} \quad PV = k \quad \text{(mass and temperature are constant)}$$

Figure 12.7

Graph of pressure versus volume showing the inverse *PV* relationship of an ideal gas.

Let's demonstrate this law using a cylinder with a movable piston so that the volume of gas inside the cylinder may be varied by changing the external pressure (see **Figure 12.8**). Assume that the temperature and the number of gas molecules do not change. We start with a volume of 1000 mL and a pressure of 1 atm. When we change the pressure to 2 atm, the gas molecules are crowded closer together, and the volume is reduced to 500 mL. When we increase the pressure to 4 atm, the volume becomes 250 mL.

Note that the product of the pressure times the volume is the same number in each case, verifying Boyle's law. We may then say that

$$P_1V_1 = P_2V_2$$

where P_1V_1 is the pressure–volume product at one set of conditions, and P_2V_2 is the product at another set of conditions. In each case, the new volume may be calculated by multiplying the starting volume by a ratio of the two pressures involved.

$$V_2 = V_1\left(\frac{P_1}{P_2}\right)$$

Of course, the ratio of pressures used must reflect the direction in which the volume should change. When the pressure is changed from 1 atm to 2 atm, the ratio to be used is 1atm/2atm. Now we can verify the volumes shown in Figure 12.8:

1. Starting volume, 1000 mL; pressure change, 1 atm $\longrightarrow$ 2 atm
   $P_1 P_2$

$$(1000 \text{ mL})\left(\frac{1 \text{ atm}}{2 \text{ atm}}\right) = 500 \text{ mL}$$

P = 1 atm

P = 2 atm

P = 4 atm

Figure 12.8

The effect of pressure on the volume of a gas.

$V = 1000$ mL	=	$V = 500$ mL	=	$V = 250$ mL
PV = 1 atm x 1000 mL	=	2 atm x 500 mL	=	4 atm x 250 mL

2. Starting volume, 1000 mL; pressure change, $\underset{P_1}{1 \text{ atm}} \longrightarrow \underset{P_2}{4 \text{ atm}}$

$$(1000 \text{ mL})\left(\frac{1 \text{ atm}}{4 \text{ atm}}\right) = 250 \text{ mL}$$

3. Starting volume, 500 mL; pressure change, $\underset{P_1}{2 \text{ atm}} \longrightarrow \underset{P_2}{4 \text{ atm}}$

$$(500 \text{ mL})\left(\frac{2 \text{ atm}}{4 \text{ atm}}\right) = 250 \text{ mL}$$

In summary, a change in the volume of a gas due to a change in pressure can be calculated by multiplying the original volume by a ratio of the two pressures. If the pressure is increased, the ratio should have the smaller pressure in the numerator and the larger pressure in the denominator. If the pressure is decreased, the larger pressure should be in the numerator and the smaller pressure in the denominator:

$$\text{new volume} = \text{original volume} \times \text{ratio of pressures} = V_1\left(\frac{P_1}{P_2}\right)$$

We use Boyle's law in the following examples. If no mention is made of temperature, assume that it remains constant.

WileyPLUS

ENHANCED EXAMPLE

EXAMPLE 12.2

What volume will 2.50 L of a gas occupy if the pressure is changed from 760. mm Hg to 630. mm Hg?

SOLUTION

Read •   **Knowns:** $V_1 = 2.50 \text{ L}$

$P_1 = 760. \text{ mm Hg}$ $P_2 = 630. \text{ mm Hg}$

Solving for: new volume (V_2)

Plan •   Since the problem is concerned only with pressure and volume, Boyle's law can be used: $P_1V_1 = P_2V_2$

Solution map: pressure decreases → volume increases

$$P_1V_1 = P_2V_2 \text{ solving for } V_2 \text{ gives } V_2 = \frac{P_1V_1}{P_2}$$

Calculate • $V_2 = \dfrac{(2.50 \text{ L})(760. \text{ mm Hg})}{630. \text{ mm Hg}} = 3.20 \text{ L}$

Another way to think about this is to use a ratio of pressures that will result in an increase in volume.

$$V_2 = (2.50 \text{ L})\left(\frac{760. \text{ mm Hg}}{630. \text{ mm Hg}}\right) = 3.20 \text{ L}$$

Note this produces the same equation as we used above to solve the problem.

Check •   The final volume is larger than the original as predicted by Boyle's law and the solution map.

EXAMPLE 12.3

A given mass of hydrogen occupies 40.0 L at 700. torr. What volume will it occupy at 5.00 atm pressure?

SOLUTION

Read • **Knowns:** $P_1 = 700. \text{ torr} = 700. \text{ torr} \left(\dfrac{1 \text{ atm}}{760 \text{ torr}} \right) = 0.921 \text{ atm}$

$P_2 = 5.00 \text{ atm}$

$V_1 = 40.0 \text{ L}$

Solving for: new volume (V_2)

Plan • Since the problem is concerned only with pressure and volume, Boyle's law can be used: $P_1V_1 = P_2V_2$

Solution map: pressure increases → volume decreases

$P_1V_1 = P_2V_2$ solving for V_2 gives $V_2 = \dfrac{P_1V_1}{P_2}$

Calculate • $V_2 = \dfrac{(0.921 \text{ atm}) (40.0 \text{ L})}{5.00 \text{ atm}} = 7.37 \text{ L}$

We can also use a ratio of pressures to solve this problem.

$V_2 = (40.0 \text{ L}) \left(\dfrac{0.921 \text{ atm}}{5.00 \text{ atm}} \right) = 7.37 \text{ L}$

Check • The final volume is smaller than the original as predicted by Boyle's law and the solution map.

EXAMPLE 12.4

A gas occupies a volume of 200. mL at 400. torr pressure. To what pressure must the gas be subjected in order to change the volume to 75.0 mL?

SOLUTION

Read • **Knowns:** $P_1 = 400. \text{ torr}$ $V_1 = 200. \text{ mL}$

$V_2 = 75.0 \text{ mL}$

Solving for: new pressure (P_2)

Plan • Since the problem is concerned only with pressure and volume, Boyle's law can be used: $P_1V_1 = P_2V_2$

Solution map: volume decreases → pressure increases

Since $P_1V_1 = P_2V_2$ solving for P_2 gives $P_2 = \dfrac{P_1V_1}{V_2}$

Calculate • $P_2 = \dfrac{(400. \text{ torr}) (200. \text{ mL})}{75.0 \text{ mL}} = 1.07 \times 10^3 \text{ torr}$

We can also think about this by using a ratio of volumes that will result in an increase in pressure.

$V_2 = (400. \text{ torr}) \left(\dfrac{200. \text{ mL}}{75.0 \text{ mL}} \right) = 1.07 \times 10^3 \text{ torr}$

This result is the same as the one used above to solve the problem.

Check • The final pressure is larger than the original as predicted by Boyle's law and the solution map.

PRACTICE 12.2

A gas occupies a volume of 3.86 L at 0.750 atm. At what pressure will the volume be 4.86 L?

12.3 CHARLES' LAW

Use Charles' law to calculate changes in temperature or volume of a sample of gas at constant pressure.

KEY TERMS

absolute zero
Charles' law

The effect of temperature on the volume of a gas was observed in about 1787 by the French physicist J. A. C. Charles (1746–1823). Charles found that various gases expanded by the same fractional amount when they underwent the same change in temperature. Later it was found that if a given volume of any gas initially at 0°C was cooled by 1°C, the volume decreased by $\frac{1}{273}$; if cooled by 2°C, it decreased by $\frac{2}{273}$; if cooled by 20°C, by $\frac{20}{273}$; and so on. Since each degree of cooling reduced the volume by $\frac{1}{273}$, it was apparent that any quantity of any gas would have zero volume if it could be cooled to −273°C. Of course, no real gas can be cooled to −273°C for the simple reason that it would liquefy before that temperature is reached. However, −273°C (more precisely, −273.15°C) is referred to as **absolute zero**; this temperature is the zero point on the Kelvin (absolute) temperature scale—the temperature at which the volume of an ideal, or perfect, gas would become zero.

The volume–temperature relationship for methane is shown graphically in **Figure 12.9**. Experimental data show the graph to be a straight line that, when extrapolated, crosses the temperature axis at −273.15°C, or absolute zero. This is characteristic for all gases.

Figure 12.9

Volume–temperature relationship of methane (CH_4). Extrapolated portion of the graph is shown by the broken line.

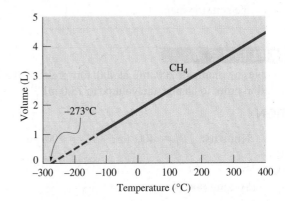

In modern form, **Charles' law** is as follows:

At *constant pressure* the volume of a fixed mass of any gas is directly proportional to the absolute temperature, which may be expressed as

$$V \propto T \quad \text{or} \quad \frac{V_1}{T_1} = \frac{V_2}{T_2}$$

A capital *T* is usually used for absolute temperature (K) and a small *t* for °C.

Mathematically, this states that the volume of a gas varies directly with the absolute temperature when the pressure remains constant. In equation form, Charles' law may be written as

$$V = kT \quad \text{or} \quad \frac{V}{T} = k \quad \text{(at constant pressure)}$$

where *k* is a constant for a fixed mass of the gas. If the absolute temperature of a gas is doubled, the volume will double.

To illustrate, let's return to the gas cylinder with the movable or free-floating piston (see **Figure 12.10**). Assume that the cylinder labeled (a) contains a quantity of gas and the pressure on it is 1 atm. When the gas is heated, the molecules move faster, and their kinetic

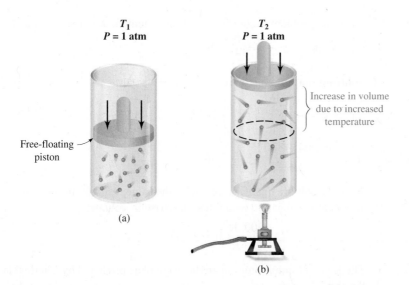

T_1
$P = 1$ atm

T_2
$P = 1$ atm

Increase in volume
due to increased
temperature

Free-floating
piston

(a)

(b)

Figure 12.10
The effect of temperature on the volume of a gas. The gas in cylinder (a) is heated from T_1 to T_2. With the external pressure constant at 1 atm, the free-floating piston rises, resulting in an increased volume, shown in cylinder (b).

energy increases. This action should increase the number of collisions per unit of time and therefore increase the pressure. However, the increased internal pressure will cause the piston to rise to a level at which the internal and external pressures again equal 1 atm, as we see in cylinder (b). The net result is an increase in volume due to an increase in temperature.

Another equation relating the volume of a gas at two different temperatures is

$$\frac{V_1}{T_1} = \frac{V_2}{T_2} \qquad (\text{constant } P)$$

where V_1 and T_1 are one set of conditions and V_2 and T_2 are another set of conditions.

A simple experiment showing the variation of the volume of a gas with temperature is illustrated in **Figure 12.11**. A balloon is placed in a beaker, and liquid N_2 is poured over it. The volume is reduced, as shown by the collapse of the balloon; when the balloon is removed from the liquid N_2, the gas expands as it warms back to room temperature and the balloon increases in size.

(a)

(b)

(c)

Charles D. Winters/Photo Researchers, Inc.

Figure 12.11
The air-filled balloons (a) are being placed into the beaker containing liquid nitrogen. Notice that the blue and red balloons on the left in (a) have been placed in the beaker in (b) and are much smaller. In (c) all of the balloons now fit inside the beaker.

EXAMPLE 12.5

Three liters of hydrogen at $-20.°C$ are allowed to warm to a room temperature of $27°C$. What is the volume at room temperature if the pressure remains constant?

SOLUTION

Read •   **Knowns:** $V_1 = 3.00\,\text{L}$ $T_1 = -20.°C + 273 = 253\,\text{K}$

$T_2 = 27°C + 273 = 300.\,\text{K}$

Solving for: new volume (V_2)

WileyPLUS

ENHANCED EXAMPLE

Remember: Temperature must be changed to Kelvin in gas law problems. Note that we use 273 to convert instead of 273.15 since our original measurements are to the nearest degree.

Plan • Since the problem is concerned only with temperature and volume, Charles' law can be used: $\dfrac{V_1}{T_1} = \dfrac{V_2}{T_2}$

Solution map: temperature increases → volume increases

Since $\dfrac{V_1}{T_1} = \dfrac{V_2}{T_2}$, $V_2 = \dfrac{V_1 T_2}{T_1}$

Calculate • $V_2 = \dfrac{(3.00\ \text{L})(300\ \cancel{K})}{253\ \cancel{K}} = 3.56\ \text{L}$

Note the solution map also produces the same result by using a ratio of temperatures that will result in an increase in volume.

$$V_2 = (3.00\ \text{L})\left(\dfrac{300.\ \cancel{K}}{253\ \cancel{K}}\right) = 3.56\ \text{L}$$

Check • The final volume is larger than the original as predicted by Charles' law and the solution map.

EXAMPLE 12.6

If 20.0 L of oxygen are cooled from 100.°C to 0.°C, what is the new volume?

SOLUTION

Read • Since no mention is made of pressure, we assume the pressure does not change. Remember that temperature must be in kelvins.

Knowns: $V_1 = 20.0\ \text{L}$ $T_1 = 100.°\text{C} = 373\ \text{K}$

$T_2 = 0.°\text{C} = 273\ \text{K}$

Solving for: new volume (V_2)

Plan • Since the problem is concerned only with temperature and volume, Charles' law can be used: $\dfrac{V_1}{T_1} = \dfrac{V_2}{T_2}$

Solution map: temperature decreases → volume decreases

Since $\dfrac{V_1}{T_1} = \dfrac{V_2}{T_2}$, $V_2 = \dfrac{V_1 T_2}{T_1}$

Calculate • $V_2 = \dfrac{(20.0\ \text{L})(273\ \cancel{K})}{373\ \cancel{K}} = 14.6\ \text{L}$

Or we can use the solution map to multiply the original volume by a ratio of temperatures that will result in a decrease in volume.

$$V_2 = (20.0\ \text{L})\left(\dfrac{273\ \cancel{K}}{373\ \cancel{K}}\right) = 14.6\ \text{L}$$

Check • The final volume is smaller than the original as predicted by Charles' law and the solution map.

PRACTICE 12.3

A 4.50-L container of nitrogen gas at 28.0°C is heated to 56.0°C. Assuming that the volume of the container can vary, what is the new volume of the gas?

12.4 AVOGADRO'S LAW

Solve problems using the relationship among moles, mass, and volume of gases.

● LEARNING OBJECTIVE

KEY TERMS
Gay-Lussac's law of combining
volumes
Avogadro's law

Early in the nineteenth century, a French chemist, J.L. Gay-Lussac (1778–1850) studied the volume relationships of reacting gases. His results, published in 1809, were summarized in a statement known as **Gay-Lussac's law of combining volumes**:

When measured at the same temperature and pressure, the ratios of the volumes of reacting gases are small whole numbers.

Thus, H_2 and O_2 combine to form water vapor in a volume ratio of 2:1 (**Figure 12.12**); H_2 and Cl_2 react to form HCl in a volume ratio of 1:1 and H_2, and N_2 react to form NH_3 in a volume ratio of 3:1.

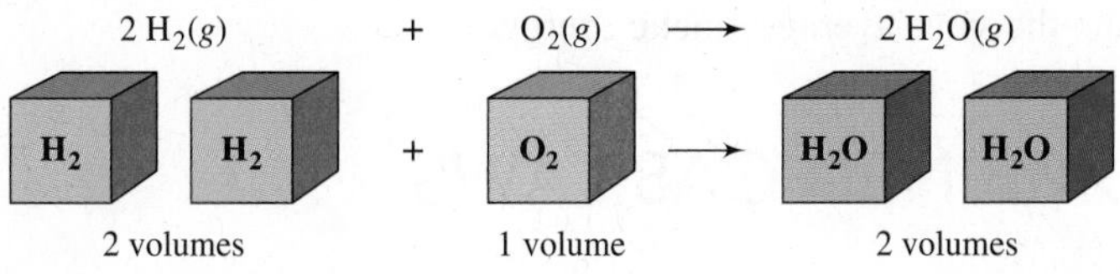

$$2\,H_2(g) \quad + \quad O_2(g) \longrightarrow 2\,H_2O(g)$$

H₂ + H₂ + O₂ ⟶ H₂O + H₂O

2 volumes   1 volume   2 volumes

Figure 12.12

Gay-Lussac's law of combining volumes of gases applied to the reaction of hydrogen and oxygen. When measured at the same temperature and pressure, hydrogen and oxygen react in a volume ratio of 2:1.

Two years later, in 1811, Amedeo Avogadro (1776–1856) used the law of combining volumes of gases to make a simple but significant and far-reaching generalization concerning gases. **Avogadro's law** states:

Equal volumes of different gases at the same temperature and pressure contain the same number of molecules.

This law was a real breakthrough in understanding the nature of gases.

1. It offered a rational explanation of Gay-Lussac's law of combining volumes of gases and indicated the diatomic nature of such elemental gases as hydrogen, chlorine, and oxygen.
2. It provided a method for determining the molar masses of gases and for comparing the densities of gases of known molar mass (see Section 12.8).
3. It afforded a firm foundation for the development of the kinetic-molecular theory.

By Avogadro's law, equal volumes of hydrogen and chlorine at the same temperature and pressure contain the same number of molecules. On a volume basis, hydrogen and chlorine react thus (see **Figure 12.13**):

hydrogen + chlorine ⟶ hydrogen chloride
1 volume   1 volume   2 volumes

Therefore, hydrogen molecules react with chlorine molecules in a 1:1 ratio. Since two volumes of hydrogen chloride are produced, one molecule of hydrogen and one molecule of chlorine must produce two molecules of hydrogen chloride. Therefore, each hydrogen molecule and

Figure 12.13

Avogadro's law proved the concept of diatomic molecules for hydrogen and chlorine.

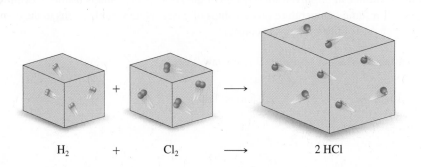

H_2 + Cl_2 ⟶ 2 HCl

each chlorine molecule must consist of two atoms. The coefficients of the balanced equation for the reaction give the correct ratios for volumes, molecules, and moles of reactants and products:

$$H_2 \quad + \quad Cl_2 \quad \longrightarrow \quad 2HCl$$

1 volume	1 volume	2 volumes
1 molecule	1 molecule	2 molecules
1 mol	1 mol	2 mol

By like reasoning, oxygen molecules also must contain at least two atoms because one volume of oxygen reacts with two volumes of hydrogen to produce two volumes of water vapor.

The volume of a gas depends on the temperature, the pressure, and the number of gas molecules. Different gases at the same temperature have the same average kinetic energy. Hence, if two different gases are at the same temperature, occupy equal volumes, and exhibit equal pressures, each gas must contain the same number of molecules. This statement is true because systems with identical PVT properties can be produced only by equal numbers of molecules having the same average kinetic energy.

12.5 COMBINED GAS LAWS

LEARNING OBJECTIVE ● Use the Combined gas law to calculate changes in pressure temperature, or volume of a sample of gas.

KEY TERMS

standard conditions
standard temperature and pressure (STP)
molar volume

In this text we'll use 273 K for temperature conversions and calculations. Check with your instructor for rules in your class.

Remember that you determine the correct ratios by thinking about the final result. For example an increase in pressure should decrease the volume, and so on.

To compare volumes of gases, common reference points of temperature and pressure were selected and called **standard conditions** or **standard temperature and pressure** (abbreviated **STP**). Standard temperature is 273.15 K (0°C), and standard pressure is 1 atm (760 torr, 760 mm Hg, or 101.325 kPa). For purposes of comparison, volumes of gases are often changed to STP conditions:

> standard temperature $=$ 273.15 K or 0.00°C
>
> standard pressure $=$ 1 atm or 760 torr, 760 mm Hg, or 101.325 kPa

When temperature and pressure change at the same time, the new volume may be calculated by multiplying the initial volume by the correct ratios of both pressure and temperature as follows:

$$\text{final volume} = (\text{initial volume})\left(\frac{\text{ratio of}}{\text{pressures}}\right)\left(\frac{\text{ratio of}}{\text{temperatures}}\right)$$

The P, V, and T relationships for a given mass of any gas, in fact, may be expressed as a single equation, $PV/T = k$. For problem solving, this equation is usually written

$$\frac{P_1 V_1}{T_1} = \frac{P_2 V_2}{T_2}$$

where P_1, V_1, and T_1 are the initial conditions and P_2, V_2, and T_2 are the final conditions.

This equation can be solved for any one of the six variables and is useful in dealing with the pressure–volume–temperature relationships of gases. Note what happens to the combined gas law when one of the variables is constant:

- T constant $\rightarrow P_1 V_1 = P_2 V_2$ Boyle's law

- P constant $\rightarrow \dfrac{V_1}{T_1} = \dfrac{V_2}{T_2}$ Charles' law

EXAMPLE 12.7

Given 20.0 L of ammonia gas at 5°C and 730. torr, calculate the volume at 50.°C and 800. torr.

SOLUTION

Read • Remember that temperature must be in kelvins.

Knowns: $P_1 = 730.$ torr $P_2 = 800.$ torr

  $T_1 = 5°C = 278$ K $T_2 = 50.°C = 323$ K

  $V_1 = 20.0$ L

Solving for: new volume (V_2)

Plan • Since P, V, and T are all changing, we must use the combined gas law:

$$\frac{P_1 V_1}{T_1} = \frac{P_2 V_2}{T_2}$$

$$V_2 = \frac{V_1 P_1 T_2}{P_2 T_1}$$

Calculate • $V_2 = \dfrac{(20.0 \text{ L})(730. \text{ torr})(323 \text{ K})}{(800. \text{ torr})(278 \text{ K})} = 21.2$ L

Check • The number of significant figures should be three, and it is, as indicated by the significant figures, in the knowns.

EXAMPLE 12.8

To what temperature (°C) must 10.0 L of nitrogen at 25°C and 700. torr be heated in order to have a volume of 15.0 L and a pressure of 760. torr?

SOLUTION

Read • Remember that temperature must be in kelvins.

Knowns: $P_1 = 700.$ torr $P_2 = 760.$ torr
  $V_1 = 10.0$ L $V_2 = 15.0$ L
  $T_1 = 25°C = 298$ K

Solving for: new temperature (T_2)

Plan • Since P, V, and T are all changing, we use the combined gas law:

$$\frac{P_1 V_1}{T_1} = \frac{P_2 V_2}{T_2}$$

$$T_2 = \frac{T_1 P_2 V_2}{P_1 V_1}$$

Calculate • $T_2 = \dfrac{(298 \text{ K})(760. \text{ torr})(15.0 \text{ L})}{(700. \text{ torr})(10.0 \text{ L})} = 485$ K

Since the problem asks for °C, we must convert our answer:

485 K – 273 = 212°C

EXAMPLE 12.9

The volume of a gas-filled balloon is 50.0 L at 20.°C and 742 torr. What volume will it occupy at standard temperature and pressure (STP)?

SOLUTION

Read •   Remember that temperature must be in kelvins.

Knowns: $P_1 = 742$ torr $P_2 = 760.$ torr (standard pressure)

$T_1 = 20.°C = 293$ K $T_2 = 273$ K (standard temperature)

$V_1 = 50.0$ L

Solving for: new volume (V_2)

Plan •   Since P, V, and T are all involved, we use the combined gas law:

$$\frac{P_1 V_1}{T_1} = \frac{P_2 V_2}{T_2}$$

$$V_2 = \frac{P_1 V_1 T_2}{P_2 T_1}$$

Calculate • $V_2 = \dfrac{(742 \text{ torr})(50.0 \text{ L})(273 \text{ K})}{(760. \text{ torr})(293 \text{ K})} = 45.5$ L

PRACTICE 12.4

15.00 L of gas at 45.0°C and 800. torr are heated to 400.°C, and the pressure is changed to 300. torr. What is the new volume?

PRACTICE 12.5

To what temperature must 5.00 L of oxygen at 50.°C and 600. torr be heated in order to have a volume of 10.0 L and a pressure of 800. torr?

Mole–Mass–Volume Relationships of Gases

As with many constants, the molar volume is known more exactly to be 22.414 L. We use 22.4 L in our calculations, since the extra figures don't often affect the result, given the other measurements in the calculation.

Because a mole contains 6.022×10^{23} molecules (Avogadro's number), a mole of any gas will have the same volume as a mole of any other gas at the same temperature and pressure. It has been experimentally determined that the volume occupied by a mole of any gas is 22.4 L at STP. This volume, 22.4 L, is known as the **molar volume** of a gas. The molar volume is a cube about 28.2 cm (11.1 in.) on a side. The molar masses of several gases, each occupying 22.4 L at STP, are shown in **Table 12.3** and **Figure 12.14**:

One mole of a gas occupies 22.4 L at STP.

The molar volume is useful for determining the molar mass of a gas or of substances that can be easily vaporized. If the mass and the volume of a gas at STP are known, we can calculate its molar mass. For example, 1.00 L of pure oxygen at STP has a mass of 1.429 g. The molar mass of oxygen may be calculated by multiplying the mass of 1.00 L by 22.4 L/mol:

$$\left(\frac{1.429 \text{ g}}{1.00 \text{ L}}\right)\left(\frac{22.4 \text{ L}}{1 \text{ mol}}\right) = 32.0 \text{ g/mol} \qquad \text{(molar mass)}$$

Standard conditions apply only to pressure, temperature, and volume. Mass is not affected.

If the mass and volume are at other than standard conditions, we change the volume to STP and then calculate the molar mass.

The molar volume, 22.4 L/mol, is used as a conversion factor to convert grams per liter to grams per mole (molar mass) and also to convert liters to moles. The two conversion factors are

$$\frac{22.4\text{ L}}{1\text{ mol}} \quad \text{and} \quad \frac{1\text{ mol}}{22.4\text{ L}}$$

These conversions must be done at STP except under certain special circumstances.

EXAMPLE 12.10

If 2.00 L of a gas measured at STP has a mass of 3.23 g, what is the molar mass of the gas?

SOLUTION

Read • **Knowns:** $V = 2.00$ L

$T = 273$ K

$P = 1.00$ atm

$m = 3.23$ g

Plan • The unit for molar mass is g/mol.

Solution map: g/L → g/mol

1 mol = 22.4 L, so the conversion factor is 22.4 L/1 mol.

Calculate • $\left(\dfrac{3.23\text{ g}}{2.00\text{ L}}\right)\left(\dfrac{22.4\text{ L}}{1\text{ mol}}\right) = 36.2$ g/mol (molar mass)

EXAMPLE 12.11

Measured at 40.°C and 630. torr, the mass of 691 mL of diethyl ether is 1.65 g. Calculate the molar mass of diethyl ether.

SOLUTION

Read • To find the molar mass we must first convert to STP.

Knowns: $P_1 = 630.$ torr $P_2 = 760$ torr

$T_1 = 40.°C = 313$ K $T_2 = 273$ K

$V_1 = 691$ mL

Plan • Since P, V, and T are all changing, we must use the combined gas law:

$$\frac{P_1 V_1}{T_1} = \frac{P_2 V_2}{T_2}$$

$$V_2 = \frac{V_1 P_1 T_2}{P_2 T_1}$$

Calculate • $V_2 = \dfrac{(691\text{ mL})(630.\text{ torr})(273\text{ K})}{(760.\text{ torr})(313\text{ K})} = 500.$ mL $= 0.500$ L (at STP)

$\left(\dfrac{1.65\text{ g}}{0.500\text{ L}}\right)\left(\dfrac{22.4\text{ L}}{1\text{ mol}}\right) = 73.9$ g/mol

PRACTICE 12.6

A gas with a mass of 86 g occupies 5.00 L at 25°C and 3.00 atm pressure. What is the molar mass of the gas?

TABLE 12.3 Mass of a Molar Volume (22.4 L) of Various Gases	
Gas	**Mass of 22.4 L**
Xe	131.3 g
O_2	32.00 g
CO_2	44.01 g
N_2	28.02 g
NH_3	17.03 g
Ar	39.95 g
H_2S	34.09 g
H_2	2.016 g
SO_2	64.07 g
HCl	36.46 g
CH_4	16.04 g
Cl_2	71.90 g

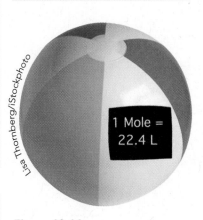

Lisa Thornberg/iStockphoto

1 Mole = 22.4 L

Figure 12.14
One mole of a gas occupies 22.4 L at STP.

Diethyl ether

WileyPLUS

ENHANCED EXAMPLE

12.6 IDEAL GAS LAW

LEARNING OBJECTIVE ● Use the ideal gas law to solve problems involving pressure, volume, temperature, and number of moles.

KEY TERMS

ideal gas law
kinectic-molecular theory (KMT)
ideal gas

Now that we have considered all of the relationships used in gas calculations, we can simplify our work by combining the relationships into a single equation describing a gas.
The four variables that characterize a gas are:

- volume (V)
- pressure (P)
- absolute temperature (T)
- number of molecules of moles (n)

Combining these variables, we obtain

$$V \propto \frac{nT}{P} \quad \text{or} \quad V = \frac{nRT}{P}$$

where R is a proportionality constant known as the *ideal gas constant*. The equation is commonly written as

$$PV = nRT$$

and is known as the **ideal gas law**. This law summarizes in a single expression what we have considered in our earlier discussions. The value and units of R depend on the units of P, V, and T. We can calculate one value of R by taking 1 mol of a gas at STP conditions. Solve the equation for R:

$$R = \frac{PV}{nT} = \frac{(1 \text{ atm})(22.4 \text{ L})}{(1 \text{ mol})(273 \text{ K})} = 0.0821 \frac{\text{L} \cdot \text{atm}}{\text{mol} \cdot \text{K}}$$

The units of R in this case are liter · atmospheres (L · atm) per mole kelvin (mol · K). When the value of $R = 0.0821$ L · atm/mol · K, P is in atmospheres, n is in moles, V is in liters, and T is in kelvins.

The ideal gas equation can be used to calculate any one of the four variables when the other three are known.

WileyPLUS

EXAMPLE 12.12

What pressure will be exerted by 0.400 mol of a gas in a 5.00-L container at 17°C?

SOLUTION

Read •   **Knowns:** $V = 5.00$ L

$T = 17°C = 290.$ K

$n = 0.400$ mol

Plan •   We need to use the ideal gas law:

$$PV = nRT$$

$$P = \frac{nRT}{V}$$

Calculate • $P = \dfrac{(0.400 \text{ mol})\left(0.0821 \dfrac{\text{L} \cdot \text{atm}}{\text{mol} \cdot \text{K}}\right)(290. \text{ K})}{5.00 \text{ L}} = 1.90 \text{ atm}$

EXAMPLE 12.13

How many moles of oxygen gas are in a 50.0-L tank at 22°C if the pressure gauge reads 2000. lb/in.²?

SOLUTION

Read • **Knowns:** $P = \left(\dfrac{2000.\ \cancel{lb}}{\cancel{in.^2}}\right)\left(\dfrac{1\ atm}{14.7\ \dfrac{\cancel{lb}}{\cancel{in.^2}}}\right) = 136.1\ atm$

$V = 50.0\ L$

$T = 22°C = 295\ K$

Plan •   We need to use the ideal gas law:

$$PV = nRT$$

$$n = \frac{PV}{RT}$$

Calculate • $n = \dfrac{(136.1\ \cancel{atm})(50.0\ \cancel{L})}{(0.0821\ \dfrac{\cancel{L}\cdot\cancel{atm}}{mol\cdot\cancel{K}})(295\ \cancel{K})} = 281\ mol\ O_2$

1 atm = 14.7 lb/in.² from Table 12.2.

PRACTICE 12.7

A 23.8-L cylinder contains oxygen gas at 20.0°C and 732 torr. How many moles of oxygen are in the cylinder?

The molar mass of a gaseous substance can also be determined using the ideal gas law. Since molar mass = g/mol it follows that mol = g/molar mass. Using M for molar mass and g for grams, we can substitute g/M for n (moles) in the ideal gas law to get

$$PV = \frac{g}{M}RT \quad \text{or} \quad M = \frac{gRT}{PV} \quad \text{(modified ideal gas law)}$$

which allows us to calculate the molar mass, M, for any substance in the gaseous state.

This form of the ideal gas law is most useful in problems containing mass instead of moles.

EXAMPLE 12.14

Calculate the molar mass of butane gas if 3.69 g occupy 1.53 L at 20.°C and 1.00 atm.

SOLUTION

Read • **Knowns:** $P = 1.00\ atm$

$V = 1.53\ L$

$T = 20.°C = 293\ K$

$m = 3.69\ g\ (mass)$

Plan •   Since we know m and not the number of moles (n) of the gas, we need to use the ideal gas law, modifying it since $n = g/M$, where M is the molar mass of the gas:

$$PV = \frac{g}{M}RT$$

$$M = \frac{gRT}{PV}$$

© Emil Schreiner/iStockphoto

Butane

Calculate • $M = \dfrac{(3.69 \text{ g})\left(0.0821 \dfrac{\text{L} \cdot \text{atm}}{\text{mol} \cdot \text{K}}\right)(293 \text{ K})}{(1.00 \text{ atm})(1.53 \text{ L})} = 58.0 \text{ g/mol}$

Check •   If we look up the formula for butane, we find it is C_4H_{10} with a molar mass of 58.0 g/mol, so our calculation is correct.

PRACTICE 12.8

A 0.286-g sample of a certain gas occupies 50.0 mL at standard temperature and 76.0 cm Hg. Determine the molar mass of the gas.

The Kinetic-Molecular Theory

The data accumulated from the gas laws allowed scientists to formulate a general theory to explain the behavior and properties of gases. This theory is called the **kinetic-molecular theory (KMT)**. The KMT has since been extended to cover, in part, the behavior of liquids and solids.

The KMT is based on the motion of particles, particularly gas molecules. A gas that behaves exactly as outlined by the theory is known as an **ideal gas**. No ideal gases exist, but under certain conditions of temperature and pressure, real gases approach ideal behavior or at least show only small deviations from it.

The principal assumptions of the kinetic-molecular theory are as follows:

1. Gases consist of tiny particles.
2. The distance between particles is large compared with the size of the particles themselves. The volume occupied by a gas consists mostly of empty space.
3. Gas particles have no attraction for one another.
4. Gas particles move in straight lines in all directions, colliding frequently with one another and with the walls of the container.
5. No energy is lost by the collision of a gas particle with another gas particle or with the walls of the container. All collisions are perfectly elastic.
6. The average kinetic energy for particles is the same for all gases at the same temperature, and its value is directly proportional to the Kelvin temperature.

The kinetic energy (KE) of a particle is expressed by the equation

$$\text{KE} = \frac{1}{2} mv^2$$

where m is the mass and v is the velocity of the particle.

All gases have the same kinetic energy at the same temperature. Therefore, from the kinetic energy equation we can see that, if we compare the velocities of the molecules of two gases, the lighter molecules will have a greater velocity than the heavier ones. For example, calculations show that the velocity of a hydrogen molecule is four times the velocity of an oxygen molecule.

Real Gases

All the gas laws are based on the behavior of an ideal gas—that is, a gas with a behavior that is described exactly by the gas laws for all possible values of P, V, and T. Most real gases actually do behave very nearly as predicted by the gas laws over a fairly wide range of temperatures and pressures. However, when conditions are such that the gas molecules are crowded closely together (high pressure and/or low temperature), they show marked deviations from ideal behavior. Deviations occur because molecules have finite volumes and also have intermolecular attractions, which result in less compressibility at high pressures and greater compressibility at low temperatures than predicted by the gas laws. Many gases become liquids at high pressure and low temperature.

>CHEMISTRY *IN ACTION*

Air Quality

Chemical reactions occur among the gases that are emitted into our atmosphere. In recent years, there has been growing concern over the effects these reactions have on our environment and our lives.

The outer portion (stratosphere) of the atmosphere plays a significant role in determining the conditions for life at the surface of the Earth. This stratosphere protects the surface from the intense radiation and particles bombarding our planet. Some of the high-energy radiation from the sun acts upon oxygen molecules, O_2, in the stratosphere, converting them into ozone, O_3.

$$O_2 \xrightarrow{\text{sunlight}} O + O$$
oxygen atoms

$$O_2 + O \longrightarrow O_3$$
ozone

Ultraviolet radiation from the sun is highly damaging to living tissues of plants and animals. The ozone layer, however, shields the Earth by absorbing ultraviolet radiation and thus prevents most of this lethal radiation from reaching the Earth's surface.

In the lower atmosphere ozone is a harmful pollutant. Ozone is formed in the atmosphere during electrical storms and by the photochemical action of ultraviolet radiation on a mixture of nitrogen dioxide and oxygen. Areas with high air pollution are subject to high atmospheric ozone concentrations. Ozone is not a desirable low-altitude constituent of the atmosphere because it is known to cause extensive plant damage, cracking of rubber, and the formation of eye-irritating substances.

In addition to ozone, the air in urban areas contains nitrogen oxides, which are components of smog. The term *smog* refers to air pollution in urban environments. Often the chemical reactions occur as part of a *photochemical process*. Nitrogen monoxide (NO) is oxidized in the air or in automobile engines to produce nitrogen dioxide (NO_2). In addition to nitrogen oxides, combustion of fossil fuels releases CO_2, CO, and sulfur oxides. Incomplete combustion releases unburned and partially burned hydrocarbons. Society is continually attempting to discover, understand, and control emissions that contribute to this sort of atmospheric chemistry.

Hong Kong skyline comparing poor and good air quality.

12.7 DALTON'S LAW OF PARTIAL PRESSURES

Use Dalton's law of partial pressures to calculate the total pressure from a mixture of gases or the pressure of a single gas in a mixture of gases.

● **LEARNING OBJECTIVE**

If gases behave according to the kinetic-molecular theory, there should be no difference in the pressure–volume–temperature relationships whether the gas molecules are all the same or different. This similarity in the behavior of gases is the basis for an understanding of **Dalton's law of partial pressures**:

KEY TERMS

Dalton's law of partial pressures
partial pressure

> The total pressure of a mixture of gases is the sum of the partial pressures exerted by each of the gases in the mixture.

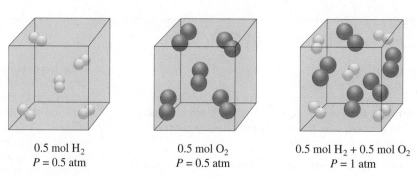

0.5 mol H₂
$P = 0.5$ atm

0.5 mol O₂
$P = 0.5$ atm

0.5 mol H₂ + 0.5 mol O₂
$P = 1$ atm

Figure 12.15
Each gas in a mixture of gases (H₂ and O₂) has a partial pressure which is the same as though the gas were alone in the container.

Each gas in the mixture exerts a pressure that is independent of the other gases present. These pressures are called **partial pressures** (see **Figure 12.15**). Thus, if we have a mixture of hydrogen and oxygen gases (H₂ and O₂), exerting partial pressures of 0.5 atm and 0.5 atm, respectively, the total pressure will be 1.0 atm:

$$P_{total} = P_{H_2} + P_{O_2}$$

$$P_{total} = 0.5 \text{ atm} + 0.5 \text{ atm} = 1.0 \text{ atm}$$

We can see an application of Dalton's law in the collection of insoluble gases over water. When prepared in the laboratory, oxygen is commonly collected by the downward displacement of water. Thus the oxygen is not pure but is mixed with water vapor (see **Figure 12.16**). When the water levels are adjusted to the same height inside and outside the bottle, the pressure of the oxygen plus water vapor inside the bottle is equal to the atmospheric pressure:

$$P_{atm} = P_{O_2} + P_{H_2O}$$

>CHEMISTRY *IN ACTION*

Getting High to Lose Weight?

Researchers in Germany have reported an interesting relationship between atmospheric pressure and weight loss. They invited 20 obese men with an average mass of 105 kg to spend a week at a research station 2650 meters above sea level. The men were instructed to maintain the same activity level as they normally would and wore pedometers to ensure that they were walking the same number of steps each day. At the end of the week the average weight loss was 1.5 kg, a half kilogram more than would have been expected based on their lowered appetites at this altitude.

The research group hypothesized that some of the extra weight loss could have been due to the low partial pressure of oxygen gas in the atmosphere. This would have required the subjects' hearts to work harder in order to draw in enough oxygen to survive. This increase in heart rate resulted in a higher energy consumption and net weight loss. Comparing this result to data from similar experiments with athletes training at high altitude, a similar weight loss was observed. In these experiments, however, the relative amounts of fat and muscle tissue metabolized to produce energy were compared. The percentage of fat burned decreased and the amount of muscle burned increased at high altitude. The result suggests that in an oxygen-poor environment the body chooses to metabolize the energy source requiring the least oxygen.

So, is moving to a high altitude the answer to your weight loss struggles? Probably not, considering that you most likely want to lose fat mass rather than muscle mass.

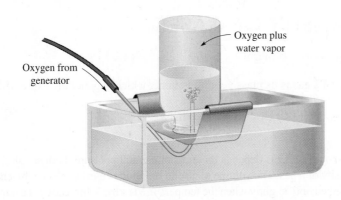

Oxygen plus
water vapor

Oxygen from
generator

Figure 12.16
Oxygen collected over water.

To determine the amount of O_2 or any other gas collected over water, we subtract the pressure of the water vapor (vapor pressure) from the total pressure of the gases:

$$P_{O_2} = P_{atm} - P_{H_2O}$$

The vapor pressure of water at various temperatures is tabulated in Appendix IV.

EXAMPLE 12.15

A 500.-mL sample of oxygen was collected over water at 23°C and 760. torr. What volume will the dry O_2 occupy at 23°C and 760. torr? The vapor pressure of water at 23°C is 21.2 torr.

SOLUTION

Read •   **Knowns:** $V_{oxygen\ and\ water\ vapor} = 500.$ mL
$P_{oxygen\ and\ water\ vapor} = 760$ torr
$P_{water\ vapor} = 21.2$ torr
Temperature constant at 23°C

Solving for: $V_{dry\ oxygen}$ (at 760 torr)

Plan •   First, we must find the pressure of the dry oxygen, and then determine the volume of the dry oxygen using Boyle's law (since the temperature is constant).

$$P_1V_1 = P_2V_2$$
$$P_{dry\ oxygen} = P_{oxygen\ and\ water\ vapor} - P_{water\ vapor}$$
$$= 760.\ torr - 21.2\ torr = 739\ torr$$
$$P_1 = 739\ torr \qquad P_2 = 760.\ torr$$
$$V_1 = 500.\ mL$$

Solution map: pressure increases → volume decreases

Since $P_1V_1 = P_2V_2,$ $V_2 = \dfrac{P_1V_1}{P_2}$

Calculate • $V_2 = \dfrac{(739\ torr)(500.\ mL)}{760.\ torr} = 486\ mL\ dry\ O_2$

Check •   The final volume is smaller than the original as predicted by Dalton's law.

PRACTICE 12.9

Hydrogen gas was collected by downward displacement of water. A volume of 600.0 mL of gas was collected at 25.0°C and 740.0 torr. What volume will the dry hydrogen occupy at STP?

12.8 DENSITY OF GASES

LEARNING OBJECTIVE • Calculate the density of a gas.

The density, d, of a gas is its mass per unit volume, which is generally expressed in grams per liter as follows:

$$d = \frac{mass}{volume} = \frac{g}{L}$$

Because the volume of a gas depends on temperature and pressure, both should be given when stating the density of a gas. The volume of a solid or liquid is hardly affected by changes in pressure and is changed only slightly when the temperature is varied. Increasing the temperature from 0°C to 50°C will reduce the density of a gas by about 18% if the gas is allowed to expand, whereas a 50°C rise in the temperature of water (0°C → 50°C) will change its density by less than 0.2%.

The density of a gas at any temperature and pressure can be determined by calculating the mass of gas present in 1 L. At STP, in particular, the density can be calculated by multiplying the molar mass of the gas by 1 mol/22.4 L:

$$d_{STP} = (molar\ mass)\left(\frac{1\ mol}{22.4\ L}\right)$$

$$molar\ mass = (d_{STP})\left(\frac{22.4\ L}{1\ mol}\right)$$

Table 12.4 lists the densities of some common gases.

TABLE 12.4 Density of Common Gases at STP

Gas	Molar mass (g/mol)	Density (g/L at STP)
H_2	2.016	0.0900
CH_4	16.04	0.716
NH_3	17.03	0.760
C_2H_2	26.04	1.16
HCN	27.03	1.21
CO	28.01	1.25
N_2	28.02	1.25
air	**(28.9)**	**(1.29)**
O_2	32.00	1.43
H_2S	34.09	1.52
HCl	36.46	1.63
F_2	38.00	1.70
CO_2	44.01	1.96
C_3H_8	44.09	1.97
O_3	48.00	2.14
SO_2	64.07	2.86
Cl_2	70.90	3.17

EXAMPLE 12.16

Calculate the density of Cl_2 at STP.

SOLUTION

Read • To find the density, we need the molar mass of Cl_2. Using the periodic table, we find it to be 70.90 g/mol.

Plan • The unit for density for gases is g/L.

  Solution map: g/mol → g/L

  1 mol = 22.4 L, so the conversion factor is 1 mol /22.4 L.

Calculate • $\left(\dfrac{70.90\ g}{1\ mol}\right)\left(\dfrac{1\ mol}{22.4\ L}\right) = 3.165\ g/L$

Check • Compare the answer 3.165 g/L to Cl_2 density in Table 12.4.

PRACTICE 12.10

The molor mass of a gas is 20. g/mol. Calculate the density of the gas at STP.

PRACTICE 12.11

List these gases in order of increasing density (same T and P): H_2, CO, C_3H_8, NO_2, O_2, N_2O, POF_3, and B_2H_6.

12.9 GAS STOICHIOMETRY

LEARNING OBJECTIVE • Solve stoichiometric problems involving gases.

Mole–Volume and Mass–Volume Calculations

Stoichiometric problems involving gas volumes can be solved by the general mole-ratio method outlined in Chapter 9 (summarized in **Figure 12.17**). The factors 1 mol/22.4 L and

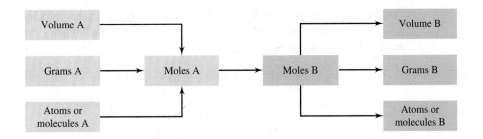

Figure 12.17

Summary of the primary conversions involved in stoichiometry. The conversion for volumes of gases is included.

22.4 L/1 mol are used for converting volume to moles and moles to volume, respectively. These conversion factors are used under the assumption that the gases are at STP and that they behave as ideal gases. In actual practice, gases are measured at other than STP conditions, and the volumes are converted to STP for stoichiometric calculations. The following are examples of typical problems involving gases in chemical reactions.

EXAMPLE 12.17

What volume of oxygen (at STP) can be formed from 0.500 mol of potassium chlorate?

SOLUTION

Read • **Knowns:** $P = 1.00$ atm

$T = 273$ K

$n = 0.500$ mol $KClO_3$

Plan • We can use the balanced equation to solve this problem since moles of a gas are directly convertible to L at STP.

$$2\ KClO_3(s) \longrightarrow 2\ KCl(s) + 3\ O_2(g)$$

Solution map: mol $KClO_3$ → mol O_2 → L O_2

We use our Problem-Solving Strategy for Stoichiometry Problems to find the moles of oxygen produced:

$$(0.500\ \text{mol } KClO_3) \left(\frac{3\ \text{mol } O_2}{2\ \text{mol } KClO_3} \right) = 0.750\ \text{mol } O_2$$

Calculate • Now we can convert moles of O_2 to liters of O_2.

$$(0.750\ \text{mol } O_2) \left(\frac{22.4\ \text{L}}{1\ \text{mol}} \right) = 16.8\ \text{L } O_2$$

WileyPLUS

ENHANCED EXAMPLE

See Section 9.1 if you have forgotten the Problem-Solving Strategy.

EXAMPLE 12.18

How many grams of aluminum must react with sulfuric acid to produce 1.25 L of hydrogen gas at STP?

SOLUTION

Read • **Knowns:** $V = 1.25$ L H_2

$T = 273$ K

$P = 1.00$ atm

Plan • We can use the balanced equation to solve this problem since moles of a gas are directly convertible to liters at STP.

$$2\ Al(s) + 3\ H_2SO_4(aq) \longrightarrow Al_2(SO_4)_3(aq) + 3\ H_2(g)$$

Solution map: L H_2 → mol H_2 → mol Al → g Al

We first use the conversion factor of 1 mol/22.4 L to convert the liters H_2 to moles. Then we use our Problem-Solving Strategy for Stoichiometry Problems to find the moles of aluminum produced:

$$(1.25 \text{ L } H_2)\left(\frac{1 \text{ mol}}{22.4 \text{ L}}\right)\left(\frac{2 \text{ mol Al}}{3 \text{ mol } H_2}\right) = 0.0372 \text{ mol Al}$$

Calculate • Now we can convert moles of Al to grams of Al.

$$(0.0372 \text{ mol Al})\left(\frac{26.98 \text{ g Al}}{\text{mol Al}}\right) = 1.00 \text{ g Al}$$

EXAMPLE 12.19

What volume of hydrogen, collected at 30.°C and 700. torr, will be formed by reacting 50.0 g of aluminum with hydrochloric acid?

SOLUTION

Read •   **Knowns:** $T = 30.°C = 303 \text{ K}$
                                $P = 700. \text{ torr} = 0.921 \text{ atm}$
                                $m = 50.0 \text{ g}$

Plan •   We need the balanced equation to determine the number of moles of Al and then the moles of H_2.

$$2 \text{ Al}(s) + 6 \text{ HCl}(aq) \rightarrow 2 \text{ AlCl}_3(aq) + 3 \text{ H}_2(g)$$

Solution map: $\text{g Al} \rightarrow \text{mol Al} \rightarrow \text{mol } H_2$

We first use the molar mass for Al to convert the g Al to moles. Then we use the mole ratio from the balanced equation to find the moles of hydrogen produced:

$$(50.0 \text{ g Al})\left(\frac{1 \text{ mol Al}}{26.98 \text{ g Al}}\right)\left(\frac{3 \text{ mol } H_2}{2 \text{ mol Al}}\right) = 2.78 \text{ mol } H_2$$

Now we can use the ideal gas law to complete the calculation:

$$PV = nRT$$

Calculate • $V = \dfrac{(2.78 \text{ mol } H_2)\left(0.0821\dfrac{\text{L} \cdot \text{atm}}{\text{mol} \cdot \text{K}}\right)(303 \text{ K})}{(0.921 \text{ atm})} = 75.1 \text{ L } H_2$

PRACTICE 12.12

If 10.0 g of sodium peroxide (Na_2O_2) react with water to produce sodium hydroxide and oxygen, how many liters of oxygen will be produced at 20.°C and 750. torr?

$$2 \text{ Na}_2O_2(s) + 2 \text{ H}_2O(l) \longrightarrow 4 \text{ NaOH}(aq) + O_2(g)$$

Volume–Volume Calculations

When all substances in a reaction are in the gaseous state, simplifications in the calculation can be made. These are based on Avogadro's law, which states that gases under identical conditions of temperature and pressure contain the same number of molecules and occupy the same volume. Under the standard conditions of temperature and pressure, the volumes of gases reacting are proportional to the numbers of moles of the gases in the balanced equation. Consider the reaction:

$H_2(g)$	+	$Cl_2(g)$	$\longrightarrow$	$2 \text{ HCl}(g)$
1 mol		1 mol		2 mol
1 volume		1 volume		2 volumes
Y volume		Y volume		2 Y volumes

This statement is true because these volumes are equivalent to the number of reacting moles in the equation. Therefore, Y volume of H_2 will combine with Y volume of Cl_2 to give $2Y$ volumes of HCl. For example, 100 L of H_2 react with 100 L of Cl_2 to give 200 L of HCl.

Remember: For gases at the same T and P, equal volumes contain equal numbers of particles.

For reacting gases at constant temperature and pressure, volume–volume relationships are the same as mole–mole relationships.

EXAMPLE 12.20

What volume of oxygen will react with 150. L of hydrogen to form water vapor? What volume of water vapor will be formed?

SOLUTION

Assume that both reactants and products are measured at standard conditions. Calculate by using reacting volumes:

$$2 H_2(g) \quad + \quad O_2(g) \quad \longrightarrow \quad 2 H_2O(g)$$

2 mol	1 mol	2 mol
2 volumes	1 volume	2 volumes
150. L	75 L	150. L

For every two volumes of H_2 that react, one volume of O_2 reacts and two volumes of $H_2O(g)$ are produced:

$$(150.\ L\ H_2)\left(\frac{1\ volume\ O_2}{2\ volumes\ H_2}\right) = 75.0\ L\ O_2$$

$$(150.\ L\ H_2)\left(\frac{2\ volumes\ H_2O}{2\ volumes\ H_2}\right) = 150.\ L\ H_2O$$

EXAMPLE 12.21

The equation for the preparation of ammonia is

$$3 H_2(g) + N_2(g) \xrightarrow{400°C} 2 NH_3(g)$$

Assuming that the reaction goes to completion, determine the following:

(a) What volume of H_2 will react with 50.0 L of N_2?
(b) What volume of NH_3 will be formed from 50.0 L of N_2?
(c) What volume of N_2 will react with 100. mL of H_2?
(d) What volume of NH_3 will be produced from 100. mL of H_2?
(e) If 600. mL of H_2 and 400. mL of N_2 are sealed in a flask and allowed to react, what amounts of H_2, N_2, and NH_3 are in the flask at the end of the reaction?

SOLUTION

The answers to parts (a)–(d) are shown in the boxes and can be determined from the equation by inspection, using the principle of reacting volumes:

$$3 H_2(g) \quad + \quad N_2(g) \quad \longrightarrow \quad 2 NH_3(g)$$

3 volume	1 volume	2 volumes

(a) 150. L   50.0 L
(b)   50.0 L   100. L
(c) 100. mL   33.3 mL
(d) 100. mL   66.7 mL

(e) Volume ratio from the equation $= \dfrac{3 \text{ volumes } H_2}{1 \text{ volume } N_2}$

Volume ratio used $= \dfrac{600. \text{ mL } H_2}{400. \text{ mL } N_2} = \dfrac{3 \text{ volumes } H_2}{2 \text{ volumes } N_2}$

Comparing these two ratios, we see that an excess of N_2 is present in the gas mixture. Therefore, the reactant limiting the amount of NH_3 that can be formed is H_2:

$$3 H_2(g) + N_2(g) \longrightarrow 2 NH_3(g)$$

| 600. mL | 200. mL | 400. mL |

To have a 3:1 ratio of volumes reacting, 600. mL of H_2 will react with 200. mL of N_2 to produce 400. mL of NH_3, leaving 200. mL of N_2 unreacted. At the end of the reaction, the flask will contain 400. mL of NH_3 and 200. mL of N_2.

PRACTICE 12.13

What volume of oxygen will react with 15.0 L of propane (C_3H_8) to form carbon dioxide and water? What volume of carbon dioxide will be formed? What volume of water vapor will be formed?

$$C_3H_8(g) + 5 O_2 \longrightarrow 3 CO_2(g) + 4 H_2O(g)$$

CHAPTER 12 REVIEW

12.1 PROPERTIES OF GASES

KEY TERMS

pressure
atmospheric pressure
barometer
1 atmosphere

- Gases:
    - Particles are relatively far apart.
    - Particles are very mobile.
    - Gases take the shape and volume of the container.
    - Gases are easily compressible.
- Pressure is force per unit area.
- Pressure of the atmosphere is measured by using a barometer:
    - Units of pressure include:
        - Atmosphere (atm) = 760 mm Hg.
        - Pascal, 1 atm = 101,325 Pa = 101.3 kPa.
        - Torr, 1 atm = 760 torr.
- Pressure is directly related to the number of molecules in the sample.
- Pressure is directly related to the Kelvin temperature of the sample.

12.2 BOYLE'S LAW

KEY TERM

Boyle's law

- At constant temperature, the volume of a gas is inversely proportional to the pressure of the gas:

$$V \propto \frac{1}{P} \quad \text{or} \quad P_1 V_1 = P_2 V_2$$

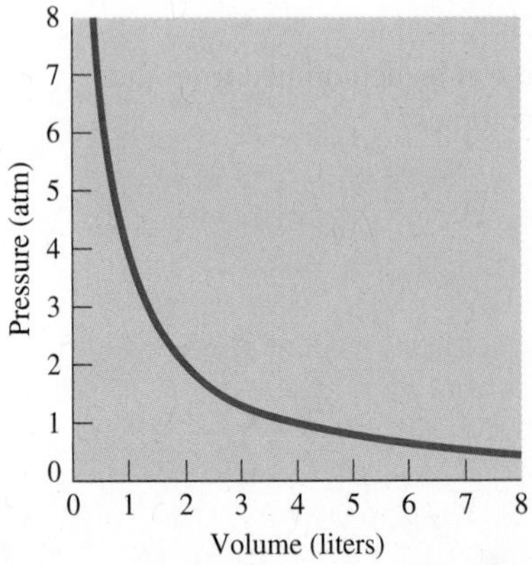

Pressure (atm) vs. Volume (liters)

12.3 CHARLES' LAW

• At constant pressure, the volume of a gas is directly proportional to the absolute temperature of the gas.

$$V \propto T \quad \text{or} \quad \frac{V_1}{T_1} = \frac{V_2}{T_2}$$

KEY TERMS

absolute zero
Charles' law

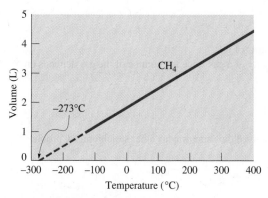

12.4 AVOGADRO'S LAW

• Gay-Lussac's law of combining volumes states that when measured at constant T and P, the ratios of the volumes of reacting gases are small whole numbers.
• Avogadro's law states that equal volumes of different gases at the same T and P contain the same number of particles.

KEY TERMS

Gay-Lussac's law of
    combining volumes
Avogadro's law

12.5 COMBINED GAS LAWS

• The $P\,V\,T$ relationship for gases can be expressed in a single equation known as the combined gas law:

$$\frac{P_1 V_1}{T_1} = \frac{P_2 V_2}{T_2}$$

KEY TERMS

standard conditions
standard temperature
    and pressure (STP)
molar volume

• One mole of any gas occupies 22.4 L at STP.

12.6 IDEAL GAS LAW

• The ideal gas law combines all the variables involving gases into a single expression:

$$PV = nRT$$

KEY TERMS

ideal gas law
kinetic-molecular theory (KMT)
ideal gas

• R is the ideal gas constant and can be expressed as

$$R = 0.0821 \frac{\text{L} \cdot \text{atm}}{\text{mol} \cdot \text{K}}$$

• Kinetic-molecular theory assumptions:
    • Gases are tiny particles with no attraction for each other.
    • The distance between particles is great compared to the size of the particles.
    • Gas particles move in straight lines.
    • No energy is lost in particle collisions (perfectly elastic collisions).
    • The average kinetic energy for particles is the same for all gases at the same temperature and pressure.
• A gas that follows the KMT is an ideal gas.
• The kinetic energy of a particle is expressed as $\text{KE} = \dfrac{1}{2} mv^2$.
• Real gases show deviation from the ideal gas law:
    • Deviations occur at:
        • High pressure
        • Low temperature
    • These deviations occur because:
        • Molecules have finite volumes.
        • Molecules have intermolecular attractions.

12.7 DALTON'S LAW OF PARTIAL PRESSURES

KEY TERMS

Dalton's law of partial
    pressures
partial pressure

- The total pressure of a mixture of gases is the sum of the partial pressures of the component gases in the mixture.
- When a gas is collected over water, the pressure of the collected gas is the difference between the atmospheric pressure and the vapor pressure of water at that temperature.

12.8 DENSITY OF GASES

- The density of a gas is usually expressed in units of g/L.
- Temperature and pressure are given for a density of a gas since the volume of the gas depends on these conditions.

12.9 GAS STOICHIOMETRY

- Stoichiometry problems involving gases are solved the same way as other stoichiometry problems. See graphic below.

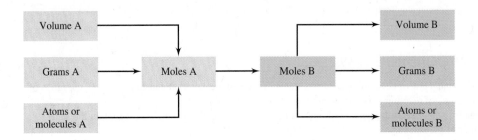

- For reacting gases (contant T and P) volume–volume relationships are the same as mole–mole relationships.

REVIEW QUESTIONS

1. What is meant by pressure (for a gas)?
2. How does the air pressure inside the balloon shown in Figure 12.3 compare with the air pressure outside the balloon? Explain.
3. According to Table 12.1, what two gases are the major constituents of dry air?
4. How does the pressure represented by 1 torr compare in magnitude to the pressure represented by 1 mm Hg? See Table 12.2.
5. In which container illustrated in Figure 12.6 are the molecules of gas moving faster? Assume both gases to be hydrogen.
6. In Figure 12.7, what gas pressure corresponds to a volume of 4 L?
7. How do the data illustrated in Figure 12.7 verify Boyle's law?
8. What effect would you observe in Figure 12.10 if T_2 were lower than T_1?
9. Explain how the reaction

$$N_2(g) + O_2(g) \xrightarrow{\Delta} 2\,NO(g)$$

proves that nitrogen and oxygen are diatomic molecules.
10. What is the reason for comparing gases to STP?
11. What are the four parameters used to describe the behavior of a gas?
12. What are the characteristics of an ideal gas?
13. How is Boyle's law related to the ideal gas law?
14. How is Charles' law related to the ideal gas law?
15. What are the basic assumptions of the kinetic-molecular theory?

16. Arrange the following gases, all at standard temperature, in order of increasing relative molecular velocities: H_2, CH_4, Rn, N_2, F_2, and He. What is your basis for determining the order?
17. List, in descending order, the average kinetic energies of the molecules in Question 16.
18. Under what condition of temperature, high or low, is a gas least likely to exhibit ideal behavior? Explain.
19. Under what condition of pressure, high or low, is a gas least likely to exhibit ideal behavior? Explain.
20. How does the kinetic-molecular theory account for the behavior of gases as described by
    (a) Boyle's law?
    (b) Charles' law?
    (c) Dalton's law of partial pressures?
21. Is the conversion of oxygen to ozone an exothermic or endothermic reaction? How do you know?
22. Write formulas for an oxygen atom, an oxygen molecule, and an ozone molecule. How many electrons are in an oxygen molecule?
23. In the diagram shown in Figure 12.16, is the pressure of the oxygen plus water vapor inside the bottle equal to, greater than, or less than the atmospheric pressure outside the bottle? Explain.
24. List five gases in Table 12.4 that are more dense than air. Explain the basis for your selection.

25. Compare, at the same temperature and pressure, equal volumes of H_2 and O_2 as to the following:
(a) number of molecules
(b) mass
(c) number of moles
(d) average kinetic energy of the molecules
(e) density

26. When constant pressure is maintained, what effect does heating a mole of N_2 gas have on
(a) its density?
(b) its mass?
(c) the average kinetic energy of its molecules?
(d) the average velocity of its molecules?
(e) the number of N_2 molecules in the sample?

Most of the exercises in this chapter are available for assignment via the online homework management program, WileyPLUS (www.wileyplus.com) All exercises with blue numbers have answers in Appendix VI.

PAIRED EXERCISES

1. Fill in the table below with the missing information:

	torr	in. Hg	kilopascals (kPa)
(a)		30.2	
(b)	752		
(c)			99.3

2. Fill in the table below with the missing information:

	mm Hg	lb/in.^{2}	atmospheres (atm)
(a)	789		
(b)		32	
(c)			1.4

3. Perform the following pressure conversions:
(a) Convert 953 torr to kPa.
(b) Convert 2.98 kPa to atm.
(c) Convert 2.77 atm to mm Hg.
(d) Convert 372 torr to atm.
(e) Convert 2.81 atm to cm Hg.

4. Perform the following pressure in conversions:
(a) Convert 649 torr to kPa.
(b) Convert 5.07 kPa to atm.
(c) Convert 3.64 atm to mm Hg.
(d) Convert 803 torr to atm.
(e) Convert 1.08 atm to cm Hg.

5. This scuba diver watch shows the air pressure in the diver's scuba tank as 1920 lb/in.^{2}. Convert this pressure to
(a) atm   (b) torr   (c) kPa

6. A pressure sensor shows tire pressure as 31 lb/in.^{2}. Convert this pressure to
(a) atm   (b) torr   (c) kPa

Courtesy Aeris

© Thomas Acop/iStockphoto

7. A sample of a gas occupies a volume of 725 mL at 825 torr. At constant temperature, what will be the new pressure (torr) when the volume changes to the following:
(a) 283 mL
(b) 2.87 L

8. A sample of a gas occupies a volume of 486 mL at 508 torr. At constant temperature, what will be the new pressure (torr) when the volume changes to the following:
(a) 185 mL
(b) 6.17 L

9. A sample of methane gas, CH_4, occupies a volume (L) of 58.2 L at a pressure of 7.25 atm. What volume will the gas occupy if the pressure is lowered to 2.03 atm?

10. A sample of nitrous oxide gas, N_2O, occupies a volume of 832 L at a pressure of 0.204 atm. What volume (L) will the gas occupy if the pressure is increased to 8.02 atm?

11. A sample of CO_2 gas occupies a volume of 125 mL at 21°C. If pressure remains constant, what will be the new volume if temperature changes to:
(a) −5°C   (b) 95°F   (c) 1095 K

12. A sample of CH_4 gas occupies a volume of 575 mL at −25°C. If pressure remains constant, what will be the new volume if temperature changes to:
(a) 298 K   (b) 32°F   (c) 45°C

13. A sample of a gas occupies a volume of 1025 mL at 75°C and 0.75 atm. What will be the new volume if temperature decreases to 35°C and pressure increases to 1.25 atm?

14. A sample of a gas occupies a volume of 25.6 L at 19°C and 678 torr. What will be the new volume if temperature increases to 35°C and pressure decreases to 595 torr?

15. An expandable balloon contains 1400. L of He at 0.950 atm pressure and 18°C. At an altitude of 22 miles (temperature 2.0°C and pressure 4.0 torr), what will be the volume of the balloon?

16. A gas occupies 22.4 L at 2.50 atm and 27°C. What will be its volume at 1.50 atm and −5.00°C?

17. A 775-mL sample of NO_2 gas is at STP. If the volume changes to 615 mL and the temperature changes to 25°C, what will be the new pressure?

18. A 2.5-L sample of SO_3 is at 19°C and 1.5 atm. What will be the new temperature in °C if the volume changes to 1.5 L and the pressure to 765 torr?

19. A sample of O_2 gas was collected over water at 23°C and 772 torr. What is the partial pressure of the O_2? (Refer to Appendix IV for the vapor pressure of water.)

20. A sample of CH_4 gas was collected over water at 29°C and 749 mm Hg. What is the partial pressure of the CH_4? (Refer to Appendix IV for the vapor pressure of water.)

21. A mixture contains H_2 at 600. torr pressure, N_2 at 200. torr pressure, and O_2 at 300. torr pressure. What is the total pressure of the gases in the system?

22. A mixture contains H_2 at 325 torr pressure, N_2 at 475 torr pressure, and O_2 at 650. torr pressure. What is the total pressure of the gases in the system?

23. A sample of methane gas, CH_4 was collected over water at 25.0°C and 720. torr. The volume of the wet gas is 2.50 L. What will be the volume of the dry methane at standard pressure?

24. A sample of propane gas, C_3H_8 was collected over water at 22.5°C and 745 torr. The volume of the wet gas is 1.25 L. What will be the volume of the dry propane at standard pressure?

25. Calculate the volume of 6.26 mol of nitrogen gas, N_2, at STP.

26. Calculate the volume of 5.89 mol of carbon dioxide, CO_2, at STP.

27. What volume will each of the following occupy at STP?
(a) 6.02×10^{23} molecules of CO_2 (c) 12.5 g oxygen
(b) 2.5 mol CH_4

28. What volume will each of the following occupy at STP?
(a) 1.80×10^{24} molecules of SO_3 (c) 25.2 g chlorine
(b) 7.5 mol C_2H_6

29. How many grams of NH_3 are present in 725 mL of the gas at STP?

30. How many grams of C_3H_6 are present in 945 mL of the gas at STP?

31. How many liters of CO_2 gas at STP will contain 1025 molecules?

32. How many molecules of CO_2 gas at STP are present in 10.5 L?

33. What volume would result if a balloon were filled with 10.0 grams of chlorine gas at STP?

34. How many grams of methane gas were used to fill a balloon to a volume of 3.0 L at STP?

35. At 22°C and 729 torr pressure, what will be the volume of 75 mol of NH_3 gas?

36. At 39°C and 1.5 atm, what will be the volume of 105 mol of CH_4?

37. How many moles of O_2 are contained in 5.25 L at 26°C and 1.2 atm?

38. How many moles of CO_2 are contained in 9.55 L at 45°C and 752 torr?

39. At what Kelvin temperature will 25.2 mol of Xe occupy a volume of 645 L at a pressure of 732 torr?

40. At what Kelvin temperature will 37.5 mol of Ar occupy a volume of 725 L at a pressure of 675 torr?

41. Calculate the density of each of the following gases at STP:
(a) He   (b) HF   (c) C_3H_6 (d) CCl_2F_2

42. Calculate the density of each of the following gases at STP:
(a) Rn   (b) NO_2 (c) SO_3 (d) C_2H_4

43. Calculate the density of each of the following gases:
(a) NH_3 at 25°C and 1.2 atm
(b) Ar at 75°C and 745 torr

44. Calculate the density of each of the following gases:
(a) C_2H_4 at 32°C and 0.75 atm
(b) He at 57°C and 791 torr

45. In the lab, students decomposed a sample of calcium carbonate by heating it over a Bunsen burner and collected carbon dioxide according to the following equation.

$$CaCO_3(s) \longrightarrow CaO(s) + CO_2(g)$$

(a) How many mL of carbon dioxide gas were generated by the decomposition of 6.24 g of calcium carbonate at STP?
(b) If 52.6 L of carbon dioxide at STP were needed, how many moles of calcium carbonate would be required?

46. In the lab, students generated and collected hydrogen gas according to the following equation:

$$Mg(s) + 2HCl(aq) \longrightarrow MgCl_2(aq) + H_2(g)$$

(a) How many mL of hydrogen gas at STP were generated from 42.9 g of magnesium metal?
(b) If 825 mL of hydrogen gas at STP were needed, how many moles of HCl would be required?

47. Consider the following equation:

$$4 NH_3(g) + 5 O_2(g) \longrightarrow 4 NO(g) + 6 H_2O(g)$$

(a) How many liters of oxygen are required to react with 2.5 L NH_3? Both gases are at STP.
(b) How many grams of water vapor can be produced from 25 L NH_3 if both gases are at STP?
(c) How many liters of NO can be produced when 25 L O_2 are reacted with 25 L NH_3? All gases are at the same temperature and pressure.

48. Consider the following equation:

$$C_3H_8(g) + 5 O_2(g) \longrightarrow 3 CO_2(g) + 4 H_2O(g)$$

(a) How many liters of oxygen are required to react with 7.2 L C_3H_8? Both gases are at STP.
(b) How many grams of CO_2 will be produced from 35 L C_3H_8 if both gases are at STP?
(c) How many liters of water vapor can be produced when 15 L C_3H_8 are reacted with 15 L O_2? All gases are at the same temperature and pressure.

49. Oxygen gas can be generated by the decomposition of potassium chlorate according to the following equation:

$$2 \, KClO_3(s) \longrightarrow 2 \, KCl(s) + 3 \, O_2(g)$$

How many liters of oxygen at STP will be produced when 0.525 kg of KCl is also produced?

50. When glucose is burned in a closed container, carbon dioxide gas and water are produced according to the following equation:

$$C_6H_{12}O_6(s) + 6 \, O_2(g) \longrightarrow 6 \, CO_2(g) + 6 \, H_2O(l)$$

How many liters of CO_2 at STP can be produced when 1.50 kg of glucose are burned?

ADDITIONAL EXERCISES

51. How is it possible that 1 mole of liquid water occupies a volume of 18 mL, but 1 mole of gaseous water occupies a volume of 22.4 L? (See Figure 12.2.)

52. Explain why it is necessary to add air to a car's tires during the winter.

53. Look at the three pictures below. If each of these pictures represents a gas at 85°C, which of the gases is exerting the greatest pressure?

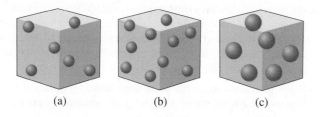

(a)   (b)   (c)

54. The balloon below is filled with helium at a temperature of 37°C. If the balloon is placed into a freezer with a temperature of −20°C, which diagram best represents the molecules inside the balloon?

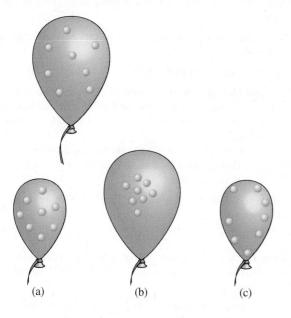

(a)   (b)   (c)

55. You have a 10-L container filled with 0.5 mol of O_2 gas at a temperature of 30.°C with a pressure of 945 torr.
   (a) What will happen to the pressure if the container size is doubled while keeping the temperature and number of moles constant?
   (b) What will happen to the pressure when the temperature is doubled while keeping the size of the container and the number of moles constant?
   (c) What will happen to the pressure when the amount of O_2 gas is cut in half while keeping the size of the container and the temperature constant?

   (d) What will happen to the pressure if 1 mole of N_2 gas is added to the container while keeping the temperature and size of the container the same?

56. Sketch a graph to show each of the following relationships:
   (a) P vs. V at constant temperature and number of moles
   (b) T vs. V at constant pressure and number of moles
   (c) n vs. V at constant temperature and pressure

57. For gases the relationship among the factors P, V, T, and n can be used to solve many different problems. Using the factors and the graphs below, complete the table.

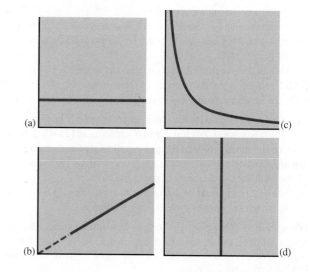

(a)   (c)
(b)   (d)

Law	Factors that are constant	Factors that are variable	Graph showing the relationship of variable factors
Boyle's law			
Charles' law			
Avogadro's law			

58. Why is it dangerous to incinerate an aerosol can?

59. What volume does 1 mol of an ideal gas occupy at standard conditions?

60. Which of these occupies the greatest volume?
   (a) 0.2 mol of chlorine gas at 48°C and 80 cm Hg
   (b) 4.2 g of ammonia at 0.65 atm and −11°C
   (c) 21 g of sulfur trioxide at 55°C and 110 kPa

61. Which of these contains the largest number of molecules?
   (a) 1.00 L of CH_4 at STP
   (b) 3.29 L of N_2 at 952 torr and 235°F
   (c) 5.05 L of Cl_2 at 0.624 atm and 0°C

62. Which of these has the greatest density?
    (a) SF_6 at STP
    (b) C_2H_6 at room conditions
    (c) He at $-80°C$ and 2.15 atm

63. For each of the following pairs, predict which sample would have the greater density.

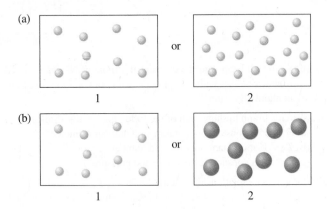

64. A chemist carried out a chemical reaction that produced a gas. It was found that the gas contained 80.0% carbon and 20.0% hydrogen. It was also noticed that 1500 mL of the gas at STP had a mass of 2.01 g.
    (a) What is the empirical formula of the compound?
    (b) What is the molecular formula of the compound?
    (c) What Lewis structure fits this compound?

65. Three gases were added to the same 2.0-L container. The total pressure of the gases was 790 torr at room temperature (25.0°C). If the mixture contained 0.65 g of oxygen gas, 0.58 g of carbon dioxide, and an unknown amount of nitrogen gas, determine the following:
    (a) the total number of moles of gas in the container
    (b) the number of grams of nitrogen in the container
    (c) the partial pressure of each gas in the mixture

66. When carbon monoxide and oxygen gas react, carbon dioxide results. If 500. mL of O_2 at 1.8 atm and 15°C are mixed with 500. mL of CO at 800 mm Hg and 60°C, how many milliliters of CO_2 at STP could possibly result?

67. One of the methods for estimating the temperature at the center of the sun is based on the ideal gas law. If the center is assumed to be a mixture of gases whose average molar mass is 2.0 g/mol, and if the density and pressure are 1.4 g/cm^3 and 1.3×10^9 atm, respectively, calculate the temperature.

68. A soccer ball of constant volume 2.24 L is pumped up with air to a gauge pressure of 13 $lb/in.^2$ at 20.0°C. The molar mass of air is about 29 g/mol.
    (a) How many moles of air are in the ball?
    (b) What mass of air is in the ball?
    (c) During the game, the temperature rises to 30.0°C. What mass of air must be allowed to escape to bring the gauge pressure back to its original value?

69. A balloon will burst at a volume of 2.00 L. If it is partially filled at 20.0°C and 65 cm Hg to occupy 1.75 L, at what temperature will it burst if the pressure is exactly 1 atm at the time that it bursts?

70. Given a sample of a gas at 27°C, at what temperature would the volume of the gas sample be doubled, the pressure remaining constant?

71. A gas sample at 22°C and 740 torr pressure is heated until its volume is doubled. What pressure would restore the sample to its original volume?

72. A gas occupies 250. mL at 700. torr and 22°C. When the pressure is changed to 500. torr, what temperature (°C) is needed to maintain the same volume?

73. The tires on an automobile were filled with air to 30. psi at 71.0°F. When driving at high speeds, the tires become hot. If the tires have a bursting pressure of 44 psi, at what temperature (°F) will the tires "blow out"?

74. What pressure will 800. mL of a gas at STP exert when its volume is 250. mL at 30°C?

75. How many L of SO_2 gas at STP will contain 9.14 g of sulfur dioxide gas?

76. How many gas molecules are present in 600. mL of N_2O at 40°C and 400. torr pressure? How many atoms are present? What would be the volume of the sample at STP?

77. An automobile tire has a bursting pressure of 60 $lb/in.^2$. The normal pressure inside the tire is 32 $lb/in.^2$. When traveling at high speeds, a tire can get quite hot. Assuming that the temperature of the tire is 25°C before running it, determine whether the tire will burst when the inside temperature gets to 212°F. Show your calculations.

78. If you prepared a barometer using water instead of mercury, how high would the column of water be at one atmosphere pressure? (Neglect the vapor pressure of water.)

79. How many moles of oxygen are in a 55-L cylinder at 27°C and 2.20×10^3 $lb/in.^2$?

80. How many moles of Cl_2 are in one cubic meter (1.00 m^3) of Cl_2 gas at STP?

81. At STP, 560. mL of a gas have a mass of 1.08 g. What is the molar mass of the gas?

82. A gas has a density at STP of 1.78 g/L. What is its molar mass?

83. Using the ideal gas law, $PV = nRT$, calculate the following:
    (a) the volume of 0.510 mol of H_2 at 47°C and 1.6 atm pressure
    (b) the number of grams in 16.0 L of CH_4 at 27°C and 600. torr pressure
    (c) the density of CO_2 at 4.00 atm pressure and 20.0°C
    (d) the molar mass of a gas having a density of 2.58 g/L at 27°C and 1.00 atm pressure.

84. Acetylene (C_2H_2) and hydrogen fluoride (HF) react to give difluoroethane:

$$C_2H_2(g) + 2\,HF(g) \longrightarrow C_2H_4F_2(g)$$

    When 1.0 mol of C_2H_2 and 5.0 mol of HF are reacted in a 10.0-L flask, what will be the pressure in the flask at 0°C when the reaction is complete?

85. What volume of hydrogen at STP can be produced by reacting 8.30 mol of Al with sulfuric acid? The equation is

$$2\,Al(s) + 3\,H_2SO_4(aq) \longrightarrow Al_2(SO_4)_3(aq) + 3\,H_2(g)$$

86. A gas has a percent composition by mass of 85.7% carbon and 14.3% hydrogen. At STP the density of the gas is 2.50 g/L. What is the molecular formula of the gas?

87. Assume that the reaction

$$2\,CO(g) + O_2(g) \longrightarrow 2\,CO_2(g)$$

goes to completion. When 10. mol of CO and 8.0 mol of O_2 react in a closed 10.-L vessel,

(a) how many moles of CO, O_2, and CO_2 are present at the end of the reaction?

(b) what will be the total pressure in the flask at 0°C?

CHALLENGE EXERCISES

89. Air has a density of 1.29 g/L at STP. Calculate the density of air on Pikes Peak, where the pressure is 450 torr and the temperature is 17°C.

90. Consider the arrangement of gases shown below. If the valve between the gases is opened and the temperature is held constant, determine the following:

(a) the pressure of each gas

(b) the total pressure in the system

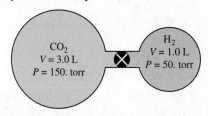

91. A steel cylinder contains 50.0 L of oxygen gas under a pressure of 40.0 atm and at a temperature of 25°C. What was the pressure in the cylinder during a storeroom fire that caused the temperature to rise 152°C? (Be careful!)

92. A balloon has a mass of 0.5 g when completely deflated. When it is filled with an unknown gas, the mass increases to 1.7 g. You notice on the canister of the unknown gas that it occupies a volume of 0.4478 L at a temperature of 50°C. You note the temperature in the room is 25°C. Identify the gas.

88. If 250 mL of O_2, measured at STP, are obtained by the decomposition of the $KClO_3$ in a 1.20-g mixture of KCl and $KClO_3$,

$$2\ KClO_3(s) \longrightarrow 2\ KCl(s) + 3\ O_2(g)$$

what is the percent by mass of $KClO_3$ in the mixture?

93. A baker is making strawberry cupcakes and wants to ensure they are very light. To do this, he must add enough baking soda to increase the volume of the cupcakes by 55.0%. All of the carbon dioxide produced by the decomposition of the baking soda is incorporated into the cupcakes. The baking soda or sodium bicarbonate will react with citric acid in the batter according to the reaction

$$3\ NaHCO_3 + H_3C_6H_5O_7 \longrightarrow Na_3C_6H_5O_7 + 3\ H_2O + 3\ CO_2$$

which occurs with a 63.7% yield. How many grams of baking soda must be added to 1.32 L of cupcake batter at a baking temperature of 325°F and a pressure of 738 torr to achieve the correct consistency?

ANSWERS TO PRACTICE EXERCISES

12.1 (a) 851 torr, (b) 851 mm Hg, (c) 113 kPa

12.2 0.596 atm

12.3 4.92 L

12.4 84.7 L

12.5 861 K (588°C)

12.6 1.4×10^2 g/mol

12.7 0.953 mol

12.8 128 g/mol

12.9 518 mL

12.10 0.89 g/L

12.11 $H_2 < B_2H_6 < CO < O_2 < N_2O < C_3H_8 < NO_2 < POF_3$

12.12 1.55 L O_2

12.13 75.0 L O_2, 45.0 L CO_2, 60.0 L H_2O

Liquid water provides the base for the recreation of windsurfing and also for our bodies. Water is a unique liquid and is the most common liquid on our blue planet.

Ingram Publishing/SuperStock

CHAPTER **13**

LIQUIDS

Planet Earth, that magnificent blue sphere we enjoy viewing from space, is spectacular. Over 75% of Earth is covered with water. We are born from it, drink it, bathe in it, cook with it, enjoy its beauty in waterfalls and rainbows, and stand in awe of the majesty of icebergs. Water supports and enhances life.

In chemistry, water provides the medium for numerous reactions. The shape of the water molecule is the basis for hydrogen bonds. These bonds determine the unique properties and reactions of water. The tiny water molecule holds the answers to many of the mysteries of chemical reactions.

CHAPTER OUTLINE

13.1 States of Matter: A Review
13.2 Properties of Liquids
13.3 Boiling Point and Melting Point
13.4 Changes of State
13.5 Intermolecular Forces
13.6 Hydrates
13.7 Water, a Unique Liquid

13.1 STATES OF MATTER: A REVIEW

In the last chapter, we found that gases contain particles that are far apart, in rapid random motion, and essentially independent of each other. The kinetic-molecular theory, along with the ideal gas law, summarizes the behavior of most gases at relatively high temperatures and low pressures.

Solids are obviously very different from gases. Solids contain particles that are very close together; solids have a high density, compress negligibly, and maintain their shape regardless of container. These characteristics indicate large attractive forces between particles. The model for solids is very different from the one for gases.

Liquids, on the other hand, lie somewhere between the extremes of gases and solids. Liquids contain particles that are close together; liquids are essentially incompressible and have definite volume. These properties are very similar to those of solids. But liquids also take the shape of their containers; this is closer to the model of a gas.

Although liquids and solids show similar properties, they differ tremendously from gases (see **Figure 13.1**). No simple mathematical relationship, like the ideal gas law, works well for liquids or solids. Instead, these models are directly related to the forces of attraction between molecules. With these general statements in mind, let's consider some specific properties of liquids.

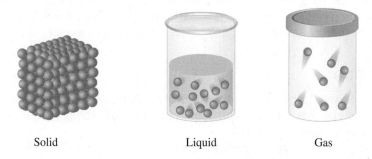

Solid   Liquid   Gas

Figure 13.1

The three states of matter. Solid—water molecules are held together rigidly and are very close to each other. Liquid—water molecules are close together but are free to move around and slide over each other. Gas—water molecules are far apart and move freely and randomly.

13.2 PROPERTIES OF LIQUIDS

Explain why liquids tend to form drops and explain the process of evaporation and its relationship to vapor pressure.

● LEARNING OBJECTIVE

KEY TERMS

surface tension
capillary action
meniscus
evaporation
vaporization
sublimation
condensation
vapor pressure
volatile

Surface Tension

Have you ever observed water and mercury in the form of small drops? These liquids form drops because liquids have *surface tension*. A droplet of liquid that is not falling or under the influence of gravity (as on the space shuttle) will form a sphere. Spheres minimize the ratio of surface area to volume. The molecules within the liquid are attracted to the surrounding liquid molecules, but at the liquid's surface, the attraction is nearly all inward. This pulls the surface into a spherical shape. The resistance of a liquid to an increase in its surface area is called the **surface tension** of the liquid. Substances with large attractive forces between molecules have high surface tensions. The effect of surface tension in water is illustrated by floating a needle on the surface of still water. Other examples include a water strider walking across a calm pond (see **Figure 13.2**) and water beading on a freshly waxed car. Surface tension is temperature dependent, decreasing with increasing temperature.

Liquids also exhibit a phenomenon called **capillary action**, the spontaneous rising of a liquid in a narrow tube. This action results from the *cohesive forces* within the liquid and the *adhesive forces* between the liquid and the walls of the container. If the forces between the liquid and the container are greater than those within the liquid itself, the liquid will climb the walls of the container. For example, consider the California sequoia, a tree that reaches over 200 feet in height. Although water rises only 33 feet in a glass tube (under atmospheric pressure), capillary action causes water to rise from the sequoia's roots to all its parts.

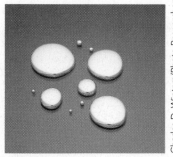

Mercury droplets.

Figure 13.2

A water strider skims the surface of the water as a result of surface tension. At the molecular level, the surface tension results from the net attraction of the water molecules toward the liquid below. In the interior of the water, the forces are balanced in all directions.

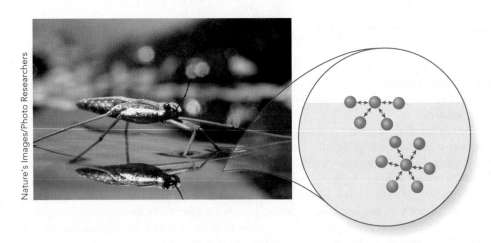

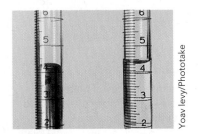

Figure 13.3

The meniscus of mercury (left) and water (right). The meniscus is the characteristic curve of the surface of a liquid in a narrow tube.

The meniscus in liquids is further evidence of cohesive and adhesive forces. When a liquid is placed in a glass cylinder, the surface of the liquid shows a curve called the **meniscus** (see **Figure 13.3**). The concave shape of water's meniscus shows that the adhesive forces between the glass and water are stronger than the cohesive forces within the water. In a nonpolar substance such as mercury, the meniscus is convex, indicating that the cohesive forces within mercury are greater than the adhesive forces between the glass wall and the mercury.

Evaporation

When beakers of water, ethyl ether, and ethyl alcohol (all liquids at room temperature) are allowed to stand uncovered, their volumes gradually decrease. The process by which this change takes place is called *evaporation.*

Attractive forces exist between molecules in the liquid state. Not all of these molecules, however, have the same kinetic energy. Molecules that have greater-than-average kinetic energy can overcome the attractive forces and break away from the surface of the liquid to become a gas (see **Figure 13.4**). **Evaporation**, or **vaporization**, is the escape of molecules from the liquid state to the gas or vapor state.

In evaporation, molecules of greater-than-average kinetic energy escape from a liquid, leaving it cooler than it was before they escaped. For this reason, evaporation of perspiration is one way the human body cools itself and keeps its temperature constant. When volatile liquids such as ethyl chloride (C_2H_5Cl) are sprayed on the skin, they evaporate rapidly, cooling the area by removing heat. The numbing effect of the low temperature produced by evaporation of ethyl chloride allows it to be used as a local anesthetic for minor surgery.

Solids such as iodine, camphor, naphthalene (moth balls), and, to a small extent, even ice will go directly from the solid to the gaseous state, bypassing the liquid state. This change is a form of evaporation and is called **sublimation**:

$$\text{liquid} \xrightarrow{\text{evaporation}} \text{vapor}$$

$$\text{solid} \xrightarrow{\text{sublimation}} \text{vapor}$$

Figure 13.4

The high-energy molecules escape from the surface of a liquid in a process known as evaporation.

EXAMPLE 13.1

Which will evaporate more quickly: 60 mL of water in a beaker with a diameter of 5 cm or 60 mL of water in a shallow dish with a diameter of 15 cm? Why?

SOLUTION

The water in the shallow dish will evaporate more quickly since there is more surface area in the dish than in the beaker.

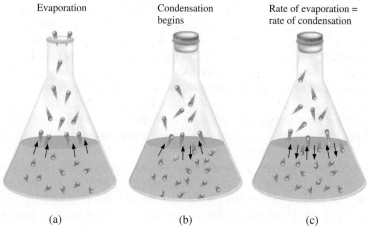

Figure 13.5
(a) Molecules in an open flask evaporate from the liquid and disperse into the atmosphere.
(b) When the flask is stoppered, the number of gaseous molecules increases inside the flask. Some gaseous molecules collide with the surface of the liquid and stick—condensation occurs.
(c) When the rate of evaporation equals the rate of condensation, an equilibrium between the liquid and vapor is established.

Vapor Pressure

When a liquid vaporizes in a closed system like that shown in **Figure 13.5b**, some of the molecules in the vapor or gaseous state strike the surface and return to the liquid state by the process of **condensation**. The rate of condensation increases until it's equal to the rate of vaporization. At this point, the space above the liquid is said to be saturated with vapor, and an equilibrium, or steady state, exists between the liquid and the vapor. The equilibrium equation is

$$\text{liquid} \underset{\text{condensation}}{\overset{\text{evaporation}}{\rightleftharpoons}} \text{vapor}$$

This equilibrium is dynamic; both processes—vaporization and condensation—are taking place, even though we cannot see or measure a change. The number of molecules leaving the liquid in a given time interval is equal to the number of molecules returning to the liquid.

At equilibrium, the molecules in the vapor exert a pressure like any other gas. The pressure exerted by a vapor in equilibrium with its liquid is known as the **vapor pressure** of the liquid. The vapor pressure may be thought of as a measure of the "escaping" tendency of molecules to go from the liquid to the vapor state. The vapor pressure of a liquid is independent of the amount of liquid and vapor present, but it increases as the temperature rises. We can measure the vapor pressure of a liquid by using a simple barometer as shown in **Figure 13.6**.

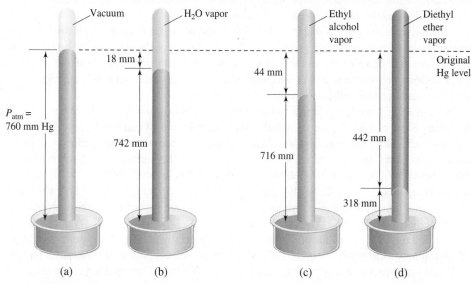

Figure 13.6

(a) A barometer measures atmospheric pressure. (Values given for 20°C)
  To measure the vapor pressure of a liquid, a tiny amount is injected into the barometer.
(b) The water vapor presses the Hg down 18 mm (vapor pressure = 18 mm).
(c) Ethyl alcohol presses the Hg level down 44 mm (vapor pressure = 44 mm).
(d) Ethyl ether has a vapor pressure of 442 mm Hg.

Mercury is so dense that if we inject a sample of liquid at the bottom of the Hg tube, the liquid rises to the top of the Hg column. It is then trapped in a closed space where it produces a vapor that pushes down the Hg column. When equilibrium between the liquid and vapor is reached, we can measure the vapor pressure by the change in height of the Hg column.

When equal volumes of water, ethyl ether, and ethyl alcohol are placed in beakers and allowed to evaporate at the same temperature, we observe that the ether evaporates faster than the alcohol, which evaporates faster than the water. This order of evaporation is consistent with the fact that ether has a higher vapor pressure at any particular temperature than ethyl alcohol or water. One reason for this higher vapor pressure is that the attraction is less between ether molecules than between alcohol or water molecules.

Substances that evaporate readily are said to be **volatile**. A volatile liquid has a relatively high vapor pressure at room temperature. Ethyl ether is a very volatile liquid; water is not too volatile; and mercury, which has a vapor pressure of 0.0012 torr at 20°C, is essentially a nonvolatile liquid. Most substances that are normally in a solid state are nonvolatile (solids that sublime are exceptions).

PRACTICE 13.1

How is the vapor pressure of a substance defined? How does it depend on temperature?

PRACTICE 13.2

A liquid in a closed container has a constant vapor pressure. What is the relationship between the rate of evaporation of the liquid and the rate of condensation of the vapor in the container?

13.3 BOILING POINT AND MELTING POINT

LEARNING OBJECTIVE ● Define boiling point and melting point and determine the boiling point of a liquid from a graph of temperature versus vapor pressure.

KEY TERMS

boiling point
normal boiling point
vapor pressure curve
melting point (or freezing point)

The boiling temperature of a liquid is related to its vapor pressure. We've seen that vapor pressure increases as temperature increases. When the internal or vapor pressure of a liquid becomes equal to the external pressure, the liquid boils. (By external pressure we mean the pressure of the atmosphere above the liquid.) The boiling temperature of a pure liquid remains constant as long as the external pressure does not vary.

The boiling point (bp) of water is 100°C at 1 atm pressure. **Figure 13.7** shows that the vapor pressure of water at 100°C is 760 torr. The significant fact here is that the boiling point is the temperature at which the vapor pressure of the water or other liquid is equal to standard, or atmospheric, pressure at sea level. **Boiling point** is the temperature at which the vapor pressure of a liquid is equal to the external pressure above the liquid.

We can readily see that a liquid has an infinite number of boiling points. When we give the boiling point of a liquid, we should also state the pressure. When we express the boiling point without stating the pressure, we mean it to be the **normal boiling point** at standard pressure (760 torr). Using Figure 13.7 again, we see that the normal boiling point of ethyl ether is between 30°C and 40°C, and for ethyl alcohol it is between 70°C and 80°C because for each compound 760 torr lies within these stated temperature ranges. At the normal boiling point, 1 g of a liquid changing to a vapor (gas) absorbs an amount of energy equal to its heat of vaporization (see Table 13.1).

The boiling point at various pressures can be evaluated by plotting the vapor pressures of various liquids on the graph in Figure 13.7, where temperature is plotted horizontally along the x-axis and vapor pressure is plotted vertically along the y-axis. The resulting curve is known as **vapor pressure curve**. Any point on this curve represents a vapor–liquid equilibrium at a particular temperature and pressure. We can find the boiling point at any pressure by tracing a horizontal line from the designated pressure to a point on the vapor pressure curve. From this point, we draw a vertical line to obtain the boiling point on the temperature axis. Three such points are shown in Figure 13.7; they represent the normal boiling points of the three compounds at 760 torr. By reversing this process, you can ascertain at what pressure a substance

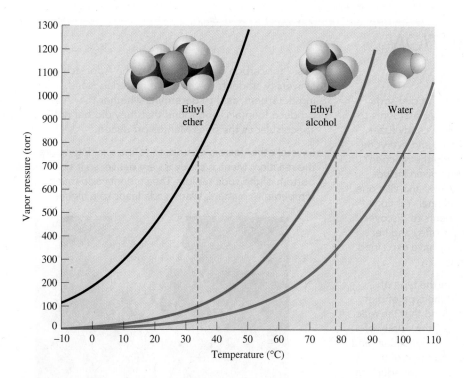

Figure 13.7

Vapor pressure–temperature curves for ethyl ether, ethyl alcohol, and water.

will boil at a specific temperature. The boiling point is one of the most commonly used physical properties for characterizing and identifying substances.

EXAMPLE 13.2

Determine the boiling point of ethyl alcohol at a vapor pressure of 500 torr.

SOLUTION

Use Figure 13.7. Trace a horizontal line from 500 torr to the green ethyl alcohol curve. Then trace a vertical line down from the curve to the *x*-axis. The boiling point is 69°C.

PRACTICE 13.3

Use the graph in Figure 13.7 to determine the boiling points of ethyl ether, ethyl alcohol, and water at 600 torr.

PRACTICE 13.4

The average atmospheric pressure in Denver is 0.83 atm. What is the boiling point of water in Denver?

As heat is removed from a liquid, the liquid becomes colder and colder, until a temperature is reached at which it begins to solidify. A liquid changing into a solid is said to be *freezing*, or *solidifying*. When a solid is heated continuously, a temperature is reached at which the solid begins to liquefy. A solid that is changing into a liquid is said to be *melting*. The temperature

WileyPLUS

ENHANCED EXAMPLE

TABLE 13.1 Physical Properties of Ethyl Chloride, Ethyl Ether, Ethyl Alcohol, and Water

Substance	Boiling point (°C)	Melting point (°C)	Heat of vaporization J/g (cal/g)	Heat of fusion J/g (cal/g)
Ethyl chloride	12.3	−139	387 (92.5)	—
Ethyl ether	34.6	−116	351 (83.9)	—
Ethyl alcohol	78.4	−112	855 (204.3)	104 (24.9)
Water	100.0	0	2259 (540)	335 (80)

>CHEMISTRY *IN ACTION*

Chemical Eye Candy

The chemical reaction between Mentos and diet soda has been seen everywhere from chemistry classrooms to David Letterman to YouTube. The Mentos candy fizzes, the diet soda sprays out in a tall fountain, and everyone laughs. But just what is happening during the reaction? Lots of informal explanations have been given including one in *Chemical and Engineering News* and also one on "MythBusters" on the Discovery Channel. Yet none of these have been systematic experiments to discover the details of the reaction. Now Tonya Coffey and her students at Appalachian State University have explained the details of the reaction's parameters.

Coffey and her students discovered that the type of Mentos did not matter and neither did the type of diet soda (caffeinated or clear). They also found that the reaction was not a simple acid–base reaction; the acidity of the soda was the same before and after the reaction, and Mentos candy is not basic. The key component of the reaction is the ability of some of the reactants to reduce the surface tension of the soda, allowing the carbon dioxide to escape more quickly, producing the fountain. It turns out that the gum Arabic in the Mentos's outer coating is an excellent surfactant (a substance that reduces the surface tension of a liquid). In addition, two chemicals in the diet soda (aspartame and potassium benzoate) also reduce surface tension and assist in the formation of bubbles. Coffey and her students also determined (using a scanning electron microscope) that the surface of the Mentos candy is very rough, providing many "nucleation

sites" for carbon dioxide bubbles. The Appalachian State University students tested other materials (sand, dishwashing detergent, salt) to determine whether they might produce fountains and found that they did but not as spectacular as the ones Mentos produced.

Several other factors contribute to the exciting nature of the reaction. Mentos candy is very dense, so it sinks to the bottom of the soda rapidly. The students also found that temperature matters: Warm soda leads to a higher fountain.

The Mentos in Diet Coke reactions

at which the solid phase of a substance is in equilibrium with its liquid phase is known as the **melting point** or **freezing point** of that substance. The equilibrium equation is

$$\text{solid} \underset{\text{freezing}}{\overset{\text{melting}}{\rightleftharpoons}} \text{liquid}$$

When a solid is slowly and carefully heated so that a solid–liquid equilibrium is achieved and then maintained, the temperature will remain constant as long as both phases are present. The energy is used solely to change the solid to the liquid. The melting point is another physical property that is commonly used for characterizing substances.

    The most common example of a solid–liquid equilibrium is ice and water. In a well-stirred system of ice and water, the temperature remains at 0°C as long as both phases are present. The melting point changes only slightly with pressure unless the pressure change is very large.

13.4 CHANGES OF STATE

LEARNING OBJECTIVE ● Calculate the amount of energy involved in a change of state.

KEY TERMS
heat of fusion
heat of vaporization

The majority of solids undergo two changes of state upon heating. A solid changes to a liquid at its melting point, and a liquid changes to a gas at its boiling point. This warming process can be represented by a graph called a *heating curve* (**Figure 13.8**). This figure shows ice being heated at a constant rate. As energy flows into the ice, the vibrations within the crystal increase and the temperature rises ($A \longrightarrow B$). Eventually, the molecules begin to break free

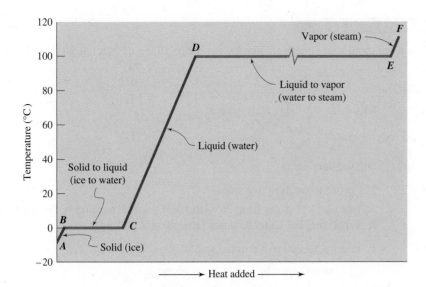

Figure 13.8
Heating curve for a pure substance—the absorption of heat by a substance from the solid state to the vapor state. Using water as an example, the *AB* interval represents the ice phase; *BC* interval, the melting of ice to water; *CD* interval, the elevation of the temperature of water from 0°C to 100°C; DE interval, the boiling of water to steam; and *EF* interval, the heating of steam.

from the crystal and melting occurs ($B \longrightarrow C$). During the melting process, all energy goes into breaking down the crystal structure; the temperature remains constant.

The energy required to change exactly one gram of a solid at its melting point into a liquid is called the **heat of fusion**. When the solid has completely melted, the temperature once again rises ($C \longrightarrow D$); the energy input is increasing the molecular motion within the water. At 100°C, the water reaches its boiling point; the temperature remains constant while the added energy is used to vaporize the water to steam ($D \longrightarrow E$). The **heat of vaporization** is the energy required to change exactly one gram of liquid to vapor at its normal boiling point. The attractive forces between the liquid molecules are overcome during vaporization. Beyond this temperature, all the water exists as steam and is being heated further ($E \longrightarrow F$).

WileyPLUS

ENHANCED EXAMPLE

EXAMPLE 13.3

How many joules of energy are needed to change 10.0 g of ice at 0.00°C to water at 20.0°C?

SOLUTION

Read • **Knowns:** $m = 10.0$ g ice
$t_{initial} = 0.00°C$
$t_{final} = 20.0°C$

Solving for: energy

Plan • The process involves two steps:
1. melting the ice (absorbing 335 J/g) and
2. warming the water from 0.00°C − 20.0°C, which requires 4.184 J/g°C.

Setup • 1. Energy needed to melt the ice

$$(10.0 \text{ g})\left(\frac{335 \text{ J}}{1 \text{ g}}\right) = 3350 \text{ J} = 3.35 \times 10^3 \text{ J}$$

2. Energy needed to warm the water (mass)(specific heat)(Δt) = energy

$$(10.0 \text{ g})\left(\frac{4.184 \text{ J}}{1 \text{ g°C}}\right)(20.0 °C) = 837 \text{ J}$$

Calculate • The total amount of energy required for the process is the sum of the two steps:

$$3350 \text{ J} + 837 \text{ J} = 4.19 \times 10^3 \text{ J}$$

EXAMPLE 13.4

How many kilojoules of energy are needed to change 20.0 g of water at 20.°C to steam at 100.°C?

SOLUTION

Read •   **Knowns:** $m = 20.0$ g water
                         $t_{initial} = 20.°C$
                         $t_{final} = 100.°C$

            Solving for: energy

Plan •   The process involves two steps:

            1. warming the water from 20°C to 100°C, which requires 4.184 J/g°C, and
            2. vaporizing the water to steam (absorbing 2.26 kJ/g).

Setup •   **1.** Energy needed to warm the water

$$(mass)(specific\ heat)\ (\Delta t) = energy$$

$$(20.0\ g)\left(\frac{4.184\ J}{1\ g°C}\right)(80.°C) = 6.7 \times 10^3\ J = 6.7\ kJ$$

            2. Energy needed to vaporize the water to steam

$$(20.0\ g)\left(\frac{2.26\ kJ}{1\ g}\right) = 45.2\ kJ$$

Calculate •   The total amount of energy required for the process is the sum of the two steps:

$$6.7\ kJ + 45.2\ kJ = 51.9\ kJ$$

PRACTICE 13.5

How many kilojoules of energy are required to change 50.0 g of ethyl alcohol from 60.0°C to vapor at 78.4°C? The specific heat of ethyl alcohol is 2.138 J/g°C.

13.5 INTERMOLECULAR FORCES

LEARNING OBJECTIVE • Describe the three types of intermolecular forces and explain their significance in liquids.

KEY TERMS

intermolecular forces
intramolecular forces
dipole–dipole attractions
hydrogen bond
London dispersion forces

Why do molecules of a liquid stay together instead of floating away from each other like molecules of a gas? **Intermolecular forces** are the forces of attraction between molecules. These forces hold molecules close together and allow for the formation of liquids and solids. These intermolecular forces can also help us to predict some of the properties of liquids. For example, if the molecules of a liquid are strongly held together, it will be difficult to pull the molecules apart, resulting in a low vapor pressure and a high boiling point. Molecules with strong intermolecular forces are also likely to have strong cohesive forces resulting in a high surface tension as well.

Molecules are held together by bonds, sometimes called **intramolecular forces**, that occur inside the molecules. These covalent bonding forces arise from sharing electrons between atoms. Intermolecular forces occur *between* molecules. How do these forces happen? There are several different kinds of intermolecular forces. Some of the most common are dipole-dipole attractions, hydrogen bonds, and London dispersion forces. Let's look at each of these types of intermolecular forces in more detail.

Dipole–Dipole Attractions

As we learned in Chapter 11, some covalent bonds do not share electrons equally because of differences in the electronegativity of the bonded atoms. When this happens, the polar bonds that form may create polar molecules or molecules with a positive end and a negative end. Some examples of polar molecules, such as HF, CO, and ICl, are easy to identify because they have only two atoms. Others are more difficult because of their complex molecular geometry.

EXAMPLE 13.5

Determine which of the following molecules are polar: HCl, CO_2, and NCl_3.

SOLUTION

The Lewis structures and three-dimensional drawings for these three molecules are shown here. Arrows represent the dipole moments of the individual bonds.

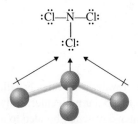

Chlorine is more electronegative than hydrogen so both the H—Cl bond and the molecule are polar.	The carbon dioxide molecule is nonpolar because the carbon–oxygen dipoles cancel each other by acting in equal and opposite directions.	The nitrogen–chlorine bonds are all polar and all three dipoles point toward the nitrogen atom resulting in a net dipole up. This is a polar molecule.

PRACTICE 13.6

Determine which of the following molecules are polar: SO_2, Br_2, and CH_3F.

As we saw in Chapter 11 water is a polar molecule—it has a dipole moment. When molecules with dipole moments are put together, they orient themselves to take advantage of their charge distribution. The positive ends of the molecules are attracted to the negative ends of other molecules. Because of this attraction, the molecules tend to stick together creating **dipole–dipole attraction**.

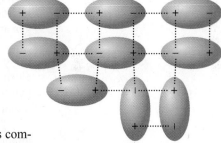

The Hydrogen Bond

Table 13.2 compares the physical properties of H_2O, H_2S, H_2Se, and H_2Te. From this comparison, it is apparent that four physical properties of water—melting point, boiling point, heat of fusion, and heat of vaporization—are extremely high and do not fit the trend relative to the molar masses of the four compounds. For example, if the properties of water followed the progression shown by the other three compounds, we would expect the melting point of water to be below −85°C and the boiling point to be below −60°C.

Why does water exhibit these anomalies? Because liquid water molecules are held together more strongly than other molecules in the same family. The intermolecular force acting

---

TABLE 13.2   Physical Properties of Water and Other Hydrogen Compounds of Group 6A Elements

Formula	Color	Molar mass (g/mol)	Melting point (°C)	Boiling point, 1 atm (°C)	Heat of fusion J/g (cal/g)	Heat of vaporization J/g (cal/g)
H_2O	Colorless	18.02	0.00	100.0	335 (80.1)	2.26×10^3 (540)
H_2S	Colorless	34.09	−85.5	−60.3	69.9 (16.7)	548 (131)
H_2Se	Colorless	80.98	−65.7	−41.3	31 (7.4)	238 (57.0)
H_2Te	Colorless	129.6	−49	−2	—	179 (42.8)

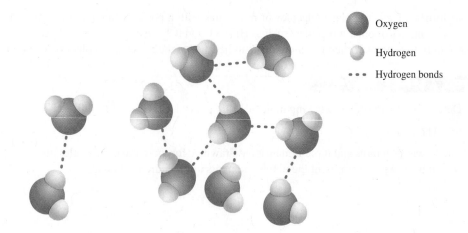

Oxygen

Hydrogen

– – – Hydrogen bonds

Figure 13.9

Hydrogen bonding. Water in the liquid and solid states exists as aggregates in which the water molecules are linked together by hydrogen bonds.

between water molecules is called a **hydrogen bond**, which acts like a very weak bond between two polar molecules. A hydrogen bond is formed between polar molecules that contain hydrogen covalently bonded to a small, highly electronegative atom such as fluorine, oxygen, or nitrogen (F—H, O—H, N—H). A hydrogen bond is actually the dipole–dipole attraction between polar molecules containing these three types of polar bonds.

> Compounds that have significant hydrogen-bonding ability are those that contain H covalently bonded to F, O, or N.

Because a hydrogen atom has only one electron, it forms only one covalent bond. When it is bonded to a strong electronegative atom such as oxygen, a hydrogen atom will also be attracted to an oxygen atom of another molecule, forming a dipole–dipole attraction (H bond) between the two molecules. Water has two types of bonds: covalent bonds that exist between hydrogen and oxygen atoms within a molecule and hydrogen bonds that exist between hydrogen and oxygen atoms in *different* water molecules.

Hydrogen bonds are *intermolecular* bonds; that is, they are formed between atoms in different molecules. They are somewhat ionic in character because they are formed by electrostatic attraction. Hydrogen bonds are much weaker than the ionic or covalent bonds that unite atoms to form compounds. Despite their weakness, they are of great chemical importance.

The oxygen atom in water can form two hydrogen bonds—one through each of the unbonded pairs of electrons. **Figure 13.9** shows two water molecules linked by a hydrogen bond and eight water molecules linked by hydrogen bonds. A dash (–) is used for the covalent bond and a dotted line (····) for the hydrogen bond. In water each molecule is linked to others through hydrogen bonds to form a three-dimensional aggregate of water molecules. This intermolecular hydrogen bonding effectively gives water the properties of a much larger, heavier molecule, explaining in part its relatively high melting point, boiling point, heat of fusion, and heat of vaporization. As water is heated and energy absorbed, hydrogen bonds are continually being broken until at 100°C, with the absorption of an additional 2.26 kJ/g, water separates into individual molecules, going into the gaseous state. Sulfur, selenium, and tellurium are not sufficiently electronegative for their hydrogen compounds to behave like water. The lack of hydrogen bonding is one reason H_2S is a gas, not a liquid, at room temperature.

Fluorine, the most electronegative element, forms the strongest hydrogen bonds. This bonding is strong enough to link hydrogen fluoride molecules together as *dimers*, H_2F_2, or as larger $(HF)_n$ molecular units. The dimer structure may be represented in this way:

>CHEMISTRY *IN ACTION*

How Sweet It Is!

Did you think artificial sweeteners were a product of the post–World War II chemical industry? Not so—many of them have been around a long time, and several of the important ones were discovered quite by accident. In 1878, Ira Remsen was working late in his laboratory and realized he was about to miss a dinner with friends. In his haste to leave the lab, he forgot to wash his hands. Later at dinner he broke a piece of bread and tasted it only to discover that it was very sweet. The sweet taste had to be the chemical he had been working with in the lab. Back at the lab, he isolated saccharin—the first of the artificial sweeteners.

In 1937, Michael Sveda was smoking a cigarette in his laboratory (a very dangerous practice to say the least!). He touched the cigarette to his lips and was surprised by the exceedingly sweet taste. The chemical on his hands turned out to be cyclamate, which soon became a staple of the artificial sweetener industry.

In 1965, James Schlatter was researching antiulcer drugs for the pharmaceutical firm G. D. Searle. In the course of his work, he accidentally ingested a small amount of a preparation and found to his surprise that it had an extremely sweet taste. He had discovered aspartame, a molecule consisting of two amino acids joined together. Since only very small quantities of aspartame are necessary to produce sweetness, it proved to be an excellent low-calorie artificial sweetener. More than 50 different molecules have a sweet taste, and it is difficult to find a single binding site that could interact with all of them.

Our taste receptors are composed of proteins that can form hydrogen bonds with other molecules. The proteins contain —NH and —OH groups (with hydrogen available to bond) as well as C=O groups (providing oxygen for hydrogen bonding). "Sweet molecules" also contain H-bonding groups including —OH, —NH$_2$, and O or N. These molecules not only must have the proper atoms to form hydrogen bonds, but they must also contain a hydrophobic region (repels H$_2$O). A new model for binding to a sweetness receptor has been developed at Senomyx in La Jolla, California. The model shows four binding sites that can act independently. Small molecules bind to a pocket on a subunit as shown in the model. Large molecules (such as proteins) bind to a different site above one of the pockets.

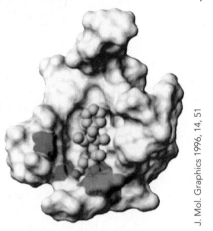

Sweet bondage. Model shows how the sweetener aspartame binds to a site on the sweetness receptor's T1R3 subunit. Red and blue are hydrogen-bond donor and acceptor residues, respectively; aspartame is in gold, except for its carboxylate (red) and ammonium (blue) groups. Model prepared with MOLMOL. Adapted from *J. Med. Chem.*

J. Mol. Graphics 1996, 14, 51

Hydrogen bonding can occur between two different atoms that are capable of forming H bonds. Thus, we may have an O····H—N or O—H····N linkage in which the hydrogen atom forming the H bond is between an oxygen and a nitrogen atom. This form of H bond exists in certain types of protein molecules and many biologically active substances.

EXAMPLE 13.6

WileyPLUS
ENHANCED EXAMPLE

Would you expect hydrogen bonding to occur between molecules of these substances?

(a) H—C—C—Ö—H
ethyl alcohol

(b) H—C—Ö—C—H
dimethyl ether

SOLUTION

(a) Hydrogen bonding should occur in ethyl alcohol because one hydrogen atom is bonded to an oxygen atom:

H—C—C—Ö—H····Ö—C—C—H

H bond

(b) There is no hydrogen bonding in dimethyl ether because all the hydrogen atoms are bonded only to carbon atoms.

Both ethyl alcohol and dimethyl ether have the same molar mass (46.07 g/mol). Although both compounds have the same molecular formula, C_2H_6O, ethyl alcohol has a much higher boiling point (78.4°C) than dimethyl ether (−23.7°C) because of hydrogen bonding between the alcohol molecules.

PRACTICE 13.7

Would you expect hydrogen bonding to occur between molecules of these substances?

London Dispersion Forces

Even molecules without dipole moments must exert forces on each other. We know this because all substances, even noble gases, exist in the liquid and solid states (at very low temperatures). The forces that exist between nonpolar molecules and also between noble gas molecules are called **London dispersion forces**. Let's consider a couple of noble gas atoms. We would assume the electrons are evenly distributed around the nucleus.

But the electrons are in motion around the nucleus, and so this even distribution is not true all of the time. Atoms can develop an instantaneous dipolar arrangement of charge. This instantaneous dipole can induce a similar dipole in a nearby atom.

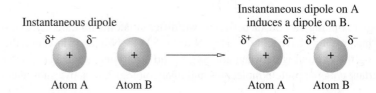

The interaction is weak and short-lived but can be important in large atoms and molecules. London dispersion forces become more significant as the atoms or molecules get larger. Larger size means more electrons are available to form dipoles.

The strength of this attraction depends on the molecular mass of the substance. Nonpolar molecules with very high molar masses such as paraffin wax, $C_{25}H_{52}$, have strong enough intermolecular forces to form soft solids.

EXAMPLE 13.7

Predict which of the following compounds will have the strongest London dispersion forces: C_4H_{10}, $C_{15}H_{32}$, and C_8H_{18}.

SOLUTION

London dispersion forces depend on molar mass. The molar masses of these compounds are 58.12 amu, 114.2 amu, and 212.4 amu, so $C_{15}H_{32}$ will have the strongest London dispersion forces.

Predict which of the following compounds will have the strongest London dispersion forces: F_2, Br_2, and I_2.

13.6 HYDRATES

Explain what hydrates are, write formulas for hydrates, and calculate the percent water in a hydrate.

● LEARNING OBJECTIVE

KEY TERMS

hydrate
water of hydration
water of crystallization

When certain solutions containing ionic compounds are allowed to evaporate, some water molecules remain as part of the crystalline compound that is left after evaporation is complete. Solids that contain water molecules as part of their crystalline structure are known as **hydrates**. Water in a hydrate is known as **water of hydration** or **water of crystallization**.

Formulas for hydrates are expressed by first writing the usual anhydrous (without water) formula for the compound and then adding a dot followed by the number of water molecules present. An example is $BaCl_2 \cdot 2\,H_2O$. This formula tells us that each formula unit of this compound contains one barium ion, two chloride ions, and two water molecules. A crystal of the compound contains many of these units in its crystalline lattice.

In naming hydrates, we first name the compound exclusive of the water and then add the term *hydrate*, with the proper prefix representing the number of water molecules in the formula. For example, $BaCl_2 \cdot 2\,H_2O$ is called *barium chloride dihydrate*. Hydrates are true compounds and follow the law of definite composition. The molar mass of $BaCl_2 \cdot 2\,H_2O$ is 244.2 g/mol; it contains 56.22% barium, 29.03% chlorine, and 14.76% water.

Water molecules in hydrates are bonded by electrostatic forces between polar water molecules and the positive or negative ions of the compound. These forces are not as strong as covalent or ionic chemical bonds. As a result, water of crystallization can be removed by moderate heating of the compound. A partially dehydrated or completely anhydrous compound may result. When $BaCl_2 \cdot 2\,H_2O$ is heated, it loses its water at about 100°C:

$$BaCl_2 \cdot 2\,H_2O(s) \xrightarrow{100°C} BaCl_2(s) + 2\,H_2O(g)$$

When a solution of copper(II) sulfate ($CuSO_4$) is allowed to evaporate, beautiful blue crystals containing 5 moles water per 1 mole $CuSO_4$ are formed (**Figure 13.10a**). The formula for

(b)

Figure 13.10

(a) When these blue crystals of $CuSO_4 \cdot 5\,H_2O$ are dissolved in water, a blue solution forms. (b) The anhydrous crystals of $CuSO_4$ are pale green. When water is added, they immediately change color to blue $CuSO_4 \cdot 5\,H_2O$ crystals.

(a)

this hydrate is $CuSO_4 \cdot 5\,H_2O$; it is called copper(II) sulfate pentahydrate. When $CuSO_4 \cdot 5\,H_2O$ is heated, water is lost, and a pale green-white powder, anhydrous $CuSO_4$, is formed:

$$CuSO_4 \cdot 5\,H_2O(s) \xrightarrow{150°C} CuSO_4(s) + 5\,H_2O(g)$$

When water is added to anhydrous copper(II) sulfate, the foregoing reaction is reversed, and the compound turns blue again (Figure 13.10b). Because of this outstanding color change, anhydrous copper(II) sulfate has been used as an indicator to detect small amounts of water. The formation of the hydrate is noticeably exothermic.

The formula for plaster of paris is $(CaSO_4)_2 \cdot H_2O$. When mixed with the proper quantity of water, plaster of paris forms a dihydrate and sets to a hard mass. It is therefore useful for making patterns for the production of art objects, molds, and surgical casts. The chemical reaction is

$$(CaSO_4)_2 \cdot H_2O(s) + 3\,H_2O(l) \longrightarrow 2\,CaSO_4 \cdot 2\,H_2O(s)$$

Table 13.3 lists a number of common hydrates.

TABLE 13.3   Selected Hydrates

Hydrate	Name	Hydrate	Name
$CaCl_2 \cdot 2\,H_2O$	calcium chloride dihydrate	$Na_2CO_3 \cdot 10\,H_2O$	sodium carbonate decahydrate
$Ba(OH)_2 \cdot 8\,H_2O$	barium hydroxide octahydrate	$(NH_4)_2C_2O_4 \cdot H_2O$	ammonium oxalate monohydrate
$MgSO_4 \cdot 7\,H_2O$	magnesium sulfate heptahydrate	$NaC_2H_3O_2 \cdot 3\,H_2O$	sodium acetate trihydrate
$SnCl_2 \cdot 2\,H_2O$	tin(II) chloride dihydrate	$Na_2B_4O_7 \cdot 10\,H_2O$	sodium tetraborate decahydrate
$CoCl_2 \cdot 6\,H_2O$	cobalt(II) chloride hexahydrate	$Na_2S_2O_3 \cdot 5\,H_2O$	sodium thiosulfate pentahydrate

WileyPLUS

ENHANCED EXAMPLE

EXAMPLE 13.8

Write the formula for calcium chloride hexahydrate.

SOLUTION

Begin by writing the formula for calcium chloride, $CaCl_2$. Remember the Ca is +2 and the Cl is −1 so we need to have two Cl atoms in the formula for the charge to be neutral. Next determine the number of water molecules in the hydrate. The prefix *hexa-* means 6, so the formula for the hydrate would be $CaCl_2 \cdot 6\,H_2O$.

PRACTICE 13.9

Write formulas for

(a) beryllium carbonate tetrahydrate
(b) cadmium permanganate hexahydrate
(c) chromium(III) nitrate nonahydrate
(d) platinum(IV) oxide trihydrate

EXAMPLE 13.9

Calculate the percent water in $CuSO_4 \cdot 5\,H_2O$.

SOLUTION

Using the periodic table, find the molar mass of $CuSO_4$. It is 159.6 g/mol. The molar mass for water is 18.02 g/mol. Since there are 5 mol of water associated with each mol of $CuSO_4$ in the hydrate, the water has a mass of $5 \times 18.02 = 90.10$ g. The percent water is

$$\frac{90.10\text{ g}}{159.6\text{ g} + 90.10\text{ g}} \times 100\% = 36.08\% \text{ water}$$

PRACTICE 13.10

Calculate the percent water in Epsom salts, $MgSO_4 \cdot 7\,H_2O$

13.7 WATER, A UNIQUE LIQUID

Describe the characteristics of water in terms of its structure and list the sources of drinking water.

Water is our most common natural resource. It covers about 75% of Earth's surface. Not only is it found in the oceans and seas, in lakes, rivers, streams, and glacial ice deposits, it is always present in the atmosphere and in cloud formations.

About 97% of Earth's water is in the oceans. This *saline* water contains vast amounts of dissolved minerals. More than 70 elements have been detected in the mineral content of seawater. Only four of these—chlorine, sodium, magnesium, and bromine—are now commercially obtained from the sea. The world's *fresh* water comprises the other 3%, of which about two-thirds is locked up in polar ice caps and glaciers. The remaining fresh water is found in groundwater, lakes, rivers, and the atmosphere.

Water is an essential constituent of all living matter. It is the most abundant compound in the human body, making up about 70% of total body mass. About 92% of blood plasma is water; about 80% of muscle tissue is water; and about 60% of a red blood cell is water. Water is more important than food in the sense that we can survive much longer without food than without water.

Physical Properties of Water

Water is a colorless, odorless, tasteless liquid with a melting point of 0°C and a boiling point of 100°C at 1 atm. The heat of fusion of water is 335 J/g (80 cal/g). The heat of vaporization of water is 2.26 kJ/g (540 cal/g). The values for water for both the heat of fusion and the heat of vaporization are high compared with those for other substances; this indicates strong attractive forces between the molecules.

Ice and water exist together in equilibrium at 0°C, as shown in **Figure 13.11**. When ice at 0°C melts, it absorbs 335 J/g in changing into a liquid; the temperature remains at 0°C. To refreeze the water, 335 J/g must be removed from the liquid at 0°C.

In Figure 13.11, both boiling water and steam are shown to have a temperature of 100°C. It takes 418 J to heat 1 g of water from 0°C to 100°C, but water at its boiling point absorbs 2.26 kJ/g in changing to steam. Although boiling water and steam are both at the same temperature, steam contains considerably more heat per gram and can cause more severe burns than hot water.

The maximum density of water is 1.000 g/mL at 4°C. Water has the unusual property of contracting in volume as it is cooled to 4°C and then expanding when cooled from 4°C to 0°C. Therefore, 1 g of water occupies a volume greater than 1 mL at all temperatures except 4°C. Although most liquids contract in volume all the way down to the point at which they solidify, a large increase (about 9%) in volume occurs when water changes from a liquid at 0°C to a solid (ice) at 0°C. The density of ice at 0°C is 0.917 g/mL, which means that ice, being less dense than water, will float in water.

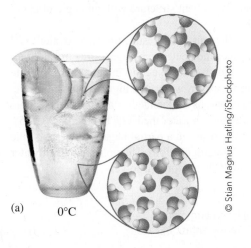

(a)   0°C

© Stian Magnus Hatling/iStockphoto

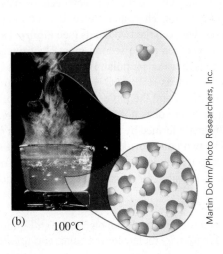

(b)   100°C

Martin Dohrn/Photo Researchers, Inc.

Figure 13.11

(a) Ice and water in equilibrium at 0°C. When ice melts, 335 J/g are needed to change it from solid to liquid. The same amount is released in freezing. (b) Boiling water and steam in equilibrium at 100°C. It takes 2.26 kJ/g to convert water to steam.

>CHEMISTRY *IN ACTION*

Reverse Osmosis?

If you have ever made a salad and been disappointed to find that the lettuce has become limp and the dressing has become watery, you have experienced osmosis. *Osmosis* is the process by which water flows through a membrane from a region of more pure water to a region of less pure water. (See Section 14.6.) In the case of your salad, lettuce contains lots of water in its cells. When the lettuce is surrounded by salad dressing (water with lots of things dissolved in it), the water in the lettuce leaf will flow through the cell membranes, resulting in diluted dressing and wilted lettuce.

This same process can be pushed in the reverse direction as well. Suppose you were to soak a raisin in water. The water would flow into the raisin to dilute the concentrated sugar solution inside the raisin, resulting in a plumper raisin. What would happen if you put the raisin into a box and put pressure on the sides so that the raisin could not expand? The water would not be able to enter the raisin. And if you increased the pressure, some of the water would even flow out of the raisin. This is *reverse osmosis.*

This is the process used in the reverse osmosis purification of water. A sample of water containing contaminant molecules and ions (such as seawater) is put into a vessel with a semiperme-able membrane. A **semipermeable membrane** allows the passage of water (solvent) molecules through it in either direction but prevents the passage of larger solute molecules or ions. As pressure is applied to the water sample, water molecules will pass through the membrane and pure water can be isolated.

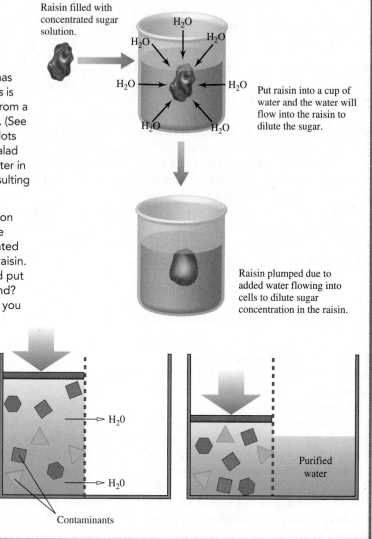

Raisin filled with concentrated sugar solution.

Put raisin into a cup of water and the water will flow into the raisin to dilute the sugar.

Raisin plumped due to added water flowing into cells to dilute sugar concentration in the raisin.

Contaminants

Purified water

Structure of the Water Molecule

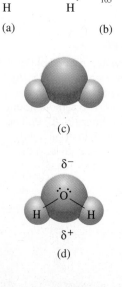

(a)

0.096 nm   105
(b)

(c)

δ^-

δ^+
(d)

A single water molecule consists of two hydrogen atoms and one oxygen atom. Each hydrogen atom is bonded to the oxygen atom by a single covalent bond. This bond is formed by the overlap of the $1s$ orbital of hydrogen with an unpaired $2p$ orbital of oxygen. The average distance between the two nuclei is known as the *bond length*. The O—H bond length in water is 0.096 nm. The water molecule is nonlinear and has a bent structure with an angle of about 105° between the two bonds (see **Figure 13.12**).

Oxygen is the second most electronegative element. As a result, the two covalent OH bonds in water are polar. If the three atoms in a water molecule were aligned in a linear structure, such as H +⟶ O ⟵+ H, the two polar bonds would be acting in equal and opposite directions and the molecule would be nonpolar. However, water is a highly polar molecule. It therefore does not have a linear structure. When atoms are bonded in a nonlinear fashion, the angle formed by the bonds is called the *bond angle*. In water the HOH bond angle is 105°. The two polar covalent bonds and the bent structure result in a partial negative charge on the oxygen atom and a partial positive charge on each hydrogen atom. The polar nature of water is responsible for many of its properties, including its behavior as a solvent.

Figure 13.12

Diagrams of a water molecule: (a) electron distribution, (b) bond angle and O—H bond length, (c) molecular structure, and (d) dipole representation.

Sources of Water for a Thirsty World

One of the greatest challenges facing our planet is finding sufficient potable water for the global population that now surpasses the 7 billion mark. A U.S. citizen, on average, consumes about 575 liters of water a day (United Nations Development Program—Human Development Report, 2006). Thus, it is not difficult to see why water is becoming a very precious resource. To satisfy the insatiable growing demand, fresh water is being depleted from streams, aquifers, and reservoirs faster than it can be replenished. Climate change also contributes to the water problem: Changes in weather patterns are predicted to cause severe drought in many parts of the globe. Reclamation of wastewater and desalination of seawater are currently being considered to augment the supplies of fresh water.

Bottled water provides a convenient source of drinking water for people on the go.

1. Reclamation of wastewater is actually an excellent source of fresh water. Although it may be very difficult to get comfortable with the idea of drinking reclaimed wastewater (the "ick factor" as termed by the reclamation industry), most of the water provided by municipalities is in fact reclaimed water from upstream cities. Many cities are using reclaimed water for irrigation and for industrial uses. Some cities are beginning to augment their water supplies using this reclaimed water. As an example, Singapore has been a leader in the challenge to reclaim wastewater for many uses. NEWater (provided by the Singapore water treatment department) is water that has been filtered, purified by reverse osmosis (see Chemistry in Action, Reverse Osmosis), and reused for agricultural and industrial use. Some of this water is also treated with UV light to further purify it and it is bottled for drinking. Thirty percent of Singapore's water is currently reclaimed and plans are to increase this to 50% by 2060. Orange County in California pumps reclaimed water into a local aquifer for storage and reuse. San Diego County has a plan to pump reclaimed water to upstream reservoirs to be mixed with "natural" water to be reused. These are excellent models for other cities that suffer from a shortage in available fresh water.

2. Desalination of seawater is a good source of fresh water for people living near the ocean. Unfortunately, desalination is very expensive and it is inefficient to extract fresh water from the ocean. On average it takes about 4 liters of seawater to make 1 liter of fresh water. The wastewater, which has a very high salt concentration, may also have an adverse impact on the environment. Generally, salt water is desalinated using the process of reverse osmosis in which a sample of salt water is placed in a vessel with a semipermeable membrane and pressure is applied to push the water molecules through the membrane. (See **Figure 13.13.**) This method of water purification has been championed by Israel, a nation with easy access to the sea but limited fresh water. Israel is currently building its fifth desalination plant and will produce over 75% of its water in this way by 2013.

Low temperature distillation and combustion of hydrogen are two other promising technologies being investigated as possible new sources of fresh water.

• In low-temperature distillation, water is subjected to a very low pressure environment. When the pressure is lowered, the boiling temperature is also lowered. This could possibly result in producing distilled water at low temperature and with a lower energy cost.

• To produce hydrogen gas for combustion, an electric current is run through a sample of water breaking it down into its component elements. In regions with lots of sunshine, solar panels can create electricity to decompose water. The hydrogen and oxygen gas can then be collected and recombined later to produce electricity and pure water. This provides both a source of fresh water and a storage method for energy produced from the sun.

Though these technologies are still in the experimental stage, hopefully they will become important sources of fresh water in the future.

Figure 13.13

Catalina Island, California, gets most of its water from a desalination plant.

CHAPTER **13 REVIEW**

13.1 STATES OF MATTER: A REVIEW

- Solids
    - Particles close together
    - High density
    - Incompressible
    - Maintain shape
- Liquids
    - Particles close together, yet free to move
    - Incompressible
    - Definite volume, but take the shape of the container
- Gases
    - Particles far apart
    - Low density
    - Compressible
    - Take volume and shape of container

13.2 PROPERTIES OF LIQUIDS

KEY TERMS

surface tension
capillary action
meniscus
evaporation
vaporization
sublimation
condensation
vapor pressure
volatile

- The resistance of a liquid to an increase in its surface area is the surface tension of the liquid.
- Capillary action is caused by the cohesive forces within the liquid and adhesive forces between the liquid and the walls of the container:
    - If the forces between the liquid and the container are greater than those within the liquid, the liquid will climb the walls of the container.
- A meniscus is evidence of cohesive and adhesive forces:
    - It is concave if the adhesive forces are stronger than the cohesive forces.
    - It is convex if the cohesive forces are stronger than the adhesive forces.
- During evaporation, molecules of greater-than-average kinetic energy escape from the liquid.
- Sublimation is the evaporation of a solid directly to a gas.
- Gaseous molecules return to the liquid state through condensation.
- At equilibrium the rate of evaporation equals the rate of condensation.
- The pressure exerted by the vapor in equilibrium with its liquid in a closed container is the vapor pressure of the liquid.
- A volatile substance evaporates readily.

13.3 BOILING POINT AND MELTING POINT

KEY TERMS

boiling point
normal boiling point
vapor pressure curve
melting point (or freezing point)

- The boiling point of a liquid is the temperature at which its vapor pressure equals the atmospheric pressure:
    - At 1 atm the boiling point is called the normal boiling point for a liquid.
- The temperature at which a solid is in equilibrium with its liquid phase is the freezing point or melting point.

13.4 CHANGES OF STATE

KEY TERMS

heat of fusion
heat of vaporization

- A graph of the warming of a liquid is called a heating curve:
    - Horizontal lines on the heating curve represent changes of state:
        - The energy required to change 1 g of solid to liquid at its melting point is the heat of fusion.
        - The energy required to change 1 g of liquid to gas at its normal boiling point is the heat of vaporization.
        - The energy required to change the phase of a sample (at its melting or boiling point) is
          energy = (mass)(heat of fusion (or vaporization))
- The energy required to heat molecules without a phase change is determined by
  energy = (mass)(sp ht)(Δt)

13.5 INTERMOLECULAR FORCES

KEY TERMS

intermolecular forces
intramolecular forces
dipole–dipole attractions
hydrogen bond
London dispersion forces

- Dipole–dipole attractions among polar molecules.

- A hydrogen bond is the dipole–dipole attraction between polar molecules containing any of these types of bonds: F—H, O—H, N—H.
- London dispersion forces cause attractions among nonpolar atoms and molecules.

13.6 HYDRATES

- Solids that contain water molecules as part of their crystalline structure are called hydrates.
- Formulas of hydrates are given by writing the formula for the anhydrous compound followed by a dot and then the number of water molecules present.
- Water molecules in hydrates are bonded by electrostatic forces, which are weaker than covalent or ionic bonds:
    - Water of hydration can be removed by heating the hydrate to form a partially dehydrated compound or the anhydrous compound.

KEY TERMS

hydrate
water of hydration
water of crystallization

13.7 WATER, A UNIQUE LIQUID

- Water is our most common resource, covering 75% of the Earth's surface.
- Water is a colorless, odorless, tasteless liquid with a melting point of 0°C and a boiling point of 100°C at 1 atm.
- The water molecule consists of 2 H atoms and 1 O atom bonded together at a 105° bond angle, making the molecule polar.
- Water can be formed in a variety of ways including:
    - Reclamation of wastewater
    - Reverse osmosis
    - Desalination
    - Low-temperature distillation
    - Hydrogen combustion

KEY TERM

semipermeable membrane

REVIEW QUESTIONS

1. In what state (solid, liquid, or gas) would H_2S, H_2Se, and H_2Te be at 0°C? (Table 13.2)
2. What property or properties of liquids are similar to solids?
3. What property or properties of liquids are similar to gases?
4. If water were placed in the containers in Figure 13.5, would they all have the same vapor pressure at the same temperature? Explain.
5. Why doesn't the vapor pressure of a liquid depend on the amount of liquid and vapor present?
6. In Figure 13.5, in which case, (a), (b) or (c) will the atmosphere above the liquid reach a point of saturation?
7. Suppose a solution of ethyl ether and ethyl alcohol is placed in the closed bottle in Figure 13.5. (Use Figure 13.7 for information on the substances.)
   (a) Are both substances present in the vapor?
   (b) If the answer to part (a) is yes, which has more molecules in the vapor?
8. Explain why rubbing alcohol warmed to body temperature still feels cold when applied to your skin.
9. The vapor pressure at 20°C for the following substances is

methyl alcohol	96 torr
acetic acid	11.7 torr
benzene	74.7 torr
bromine	173 torr
water	17.5 torr
carbon tetrachloride	91 torr
mercury	0.0012 torr
toluene	23 torr

   (a) Arrange these substances in order of increasing rate of evaporation.
   (b) Which substance listed has the highest boiling point? the lowest?
10. On the basis of the kinetic-molecular theory, explain why vapor pressure increases with temperature.

11. The temperature of the water in the pan on the burner (Figure 13.11) reads 100°C. What is the pressure of the atmosphere?
12. If ethyl alcohol is boiling in a flask and the atmospheric pressure is 543 torr, what is the temperature of the boiling liquid? (Use Figure 13.7.)
13. At approximately what temperature would each of the substances shown in Figure 13.7 boil when the pressure is 30 torr?
14. Use the graph in Figure 13.7 to find the following:
    (a) boiling point of water at 500 torr
    (b) normal boiling point of ethyl alcohol
    (c) boiling point of ethyl ether at 0.50 atm
15. Suggest a method whereby water could be made to boil at 50°C.
16. Explain why a higher temperature is obtained in a pressure cooker than in an ordinary cooking pot.
17. What is the relationship between vapor pressure and boiling point?
18. The boiling point of ammonia, NH_3, is −33.4°C and that of sulfur dioxide, SO_2, is −10.0°C. Which has the higher vapor pressure at −40°C?
19. Explain what is occurring physically when a substance is boiling.
20. At what specific temperature will ethyl ether have a vapor pressure of 760 torr?
21. Compare the potential energy of the three states of water shown in Figure 13.11.
22. Consider Figure 13.8.
    (a) Why is line *BC* horizontal? What is happening in this interval?
    (b) What phases are present in the interval *BC*?
    (c) When heating is continued after point *C*, another horizontal line, *DE*, is reached at a higher temperature. What does this line represent?
23. Account for the fact that an ice–water mixture remains at 0°C until all the ice is melted, even though heat is applied to it.
24. Which contains less energy: ice at 0°C or water at 0°C? Explain.
25. Why does a boiling liquid maintain a constant temperature when heat is continuously being added?

26. Define the terms *intermolecular forces* and *intramolecular forces*.
27. What kind of covalent bond is formed when atoms do not share electrons equally?
28. What property of a molecule most affects its ability to form instantaneous dipoles?
29. If water molecules were linear instead of bent, would the heat of vaporization be higher or lower? Explain.
30. The heat of vaporization for ethyl ether is 351 J/g and that for ethyl alcohol is 855 J/g. From these data, which of these compounds has hydrogen bonding? Explain.
31. Would there be more or less H bonding if water molecules were linear instead of bent? Explain.
32. In which condition are there fewer hydrogen bonds between molecules: water at 40°C or water at 80°C? Explain.
33. Which compound

$$H_2NCH_2CH_2NH_2 \quad or \quad CH_3CH_2CH_2NH_2$$

would you expect to have the higher boiling point? Explain. (Both compounds have similar molar masses.)
34. Why does water have such a relatively high boiling point?

35. Explain why HF (bp = 19.4°C) has a higher boiling point than HCl (bp = −85°C), whereas F_2 (bp = −188°C) has a lower boiling point than Cl_2 (bp = −34°C).
36. How do we specify 1, 2, 3, 4, 5, 6, 7, and 8 molecules of water in the formulas of hydrates? (Table 13.3)
37. Diagram a water molecule and point out the negative and positive ends of the dipole.
38. If the water molecule were linear, with all three atoms in a straight line rather than in the shape of a V, as shown in Figure 13.12, what effect would this have on the physical properties of water?
39. List six physical properties of water.
40. What condition is necessary for water to have its maximum density? What is its maximum density?
41. Why does ice float in water? Would ice float in ethyl alcohol (*d* = 0.789 g/mL)? Explain.
42. What water temperature would you theoretically expect to find at the bottom of a very deep lake? Explain.
43. Is the formation of hydrogen and oxygen from water an exothermic or an endothermic reaction? How do you know?

Most of the exercises in this chapter are available for assignment via the online homework management program, WileyPLUS (www.wileyplus.com) All exercises with blue numbers have answers in Appendix VI.

PAIRED EXERCISES

1. Rank the following molecules in order of increasing polarity: CH_3Cl, CH_3F, and CH_3Br.

3. Rank the following molecules in order of increasing London dispersion forces: CCl_4, Cl_4, and CBr_4.

5. In which of the following substances would you expect to find hydrogen bonding?
   (a) C_3H_7OH (d) PH_3
   (b) H_2O_2 (e) HF
   (c) $CHCl_3$

7. For each of the compounds in Question 5 that form hydrogen bonds, draw a diagram of the two molecules using a dotted line to indicate where the hydrogen bonding will occur.

9. You leave the house wearing a cotton T-shirt and are surprised by a sudden rainstorm. You notice that the water soaks into the T-shirt whereas it just beads up on your raincoat. In this example, are the adhesive forces or cohesive forces stronger between water and your T-shirt? Between water and your raincoat? Explain.

11. Name these hydrates:
    (a) $BaBr_2 \cdot 2\,H_2O$
    (b) $AlCl_3 \cdot 6\,H_2O$
    (c) $FePO_4 \cdot 4\,H_2O$

13. How many moles of compound are in 25.0 g of $Na_2CO_3 \cdot 10\,H_2O$?

15. Upon heating 125 g $MgSO_4 \cdot 7\,H_2O$:
    (a) how many grams of water can be obtained?
    (b) how many grams of anhydrous compound can be obtained?

17. Cobalt(II) chloride hexahydrate, $CoCl_2 \cdot 6\,H_2O$, is often used as a humidity indicator. This is due to the fact that the hydrate is a deep magenta color while the anhydrous form is a pale blue. As the humidity level changes, the color changes as well. What is the mass percent water in cobalt(II) hexahydrate?

19. A hydrated iron chloride compound was found to contain 20.66% Fe, 39.35% Cl, and 39.99% water. Determine the empirical formula of this hydrated compound.

2. Rank the following molecules in order of increasing polarity: H_2S, H_2Se, and H_2O.

4. Rank the following molecules in order of increasing London dispersion forces: CO_2, SO_2, and CS_2.

6. In which of the following substances would you expect to find hydrogen bonding?
   (a) HI   (d) C_2H_5OH
   (b) NH_3 (e) H_2O
   (c) CH_2F_2

8. For each of the compounds in Question 6 that form hydrogen bonds, draw a diagram of the two molecules using a dotted line to indicate where the hydrogen bonding will occur.

10. Rainex is a product that causes water to "bead" instead of spread when sprayed on car windshields. When the water forms these droplets, are adhesive forces or cohesive forces stronger? Explain.

12. Name these hydrates:
    (a) $MgNH_4PO_4 \cdot 6\,H_2O$
    (b) $FeSO_4 \cdot 7\,H_2O$
    (c) $SnCl_4 \cdot 5\,H_2O$

14. How many moles of compound are in 25.0 g of $Na_2B_4O_7 \cdot 10\,H_2O$?

16. Upon heating 125 g $AlCl_3 \cdot 6\,H_2O$:
    (a) how many grams of water can be obtained?
    (b) how many grams of anhydrous compound can be obtained?

18. Gypsum, $CaSO_4 \cdot 2\,H_2O$, is used to manufacture cement. What is the mass percent of water in this compound?

20. A hydrated nickel chloride compound was found to contain 24.69% Ni, 29.83% Cl, and 45.48% water. Determine the empirical formula of this hydrated compound.

21. How many joules of energy are needed to change 275 g of water from 15°C to steam at 100.°C?

22. How many joules of energy must be removed from 325 g water at 35°C to form ice at 0°C?

23. Suppose 100. g of ice at 0°C are added to 300. g of water at 25°C. Is this sufficient ice to lower the temperature of the system to 0°C and still have ice remaining? Show evidence for your answer.

24. Suppose 35.0 g of steam at 100.°C are added to 300. g of water at 25°C. Is this sufficient steam to heat all the water to 100.°C and still have steam remaining? Show evidence for your answer.

25. If 75 g of ice at 0.0°C were added to 1.5 L of water at 75°C, what would be the final temperature of the mixture?

26. If 9560 J of energy were absorbed by 500. g of ice at 0.0°C, what would be the final temperature?

ADDITIONAL EXERCISES

27. You walk out to your car just after a rain shower and notice that small droplets of water are scattered on the hood. How was it possible for the water to form those droplets?

28. Which causes a more severe burn: liquid water at 100°C or steam at 100°C? Why?

29. You have a shallow dish of alcohol set into a tray of water. If you blow across the tray, the alcohol evaporates, while the water cools significantly and eventually freezes. Explain why.

30. Regardless of how warm the outside temperature may be, we always feel cool when stepping out of a swimming pool, the ocean, or a shower. Why is this so?

31. Sketch a heating curve for a substance X whose melting point is 40°C and whose boiling point is 65°C.
   (a) Describe what you will observe as a 60.-g sample of X is warmed from 0°C to 100°C.
   (b) If the heat of fusion of X is 80. J/g, the heat of vaporization is 190. J/g, and if 3.5 J are required to warm 1 g of X each degree, how much energy will be needed to accomplish the change in (a)?

32. For the heating curve of water (see Figure 13.8), why doesn't the temperature increase when ice is melting into liquid water and liquid water is changing into steam?

33. Why does the vapor pressure of a liquid increase as the temperature is increased?

34. Mount McKinley in Alaska is the tallest peak in North America with a summit at 20,320 ft or 6194 m. It is also known as Denali, meaning "the high one" in the native Koyukon Athabaskan language. The average atmospheric pressure at the top of Denali is approximately 330 torr. Use Figure 13.7 to determine the approximate boiling temperature of water on Denali.

35. Explain how anhydrous copper(II) sulfate ($CuSO_4$) can act as an indicator for moisture.

36. Write formulas for magnesium sulfate heptahydrate and disodium hydrogen phosphate dodecahydrate.

37. Analyze the graph below and draw conclusions as appropriate. The boiling point and molar mass are graphed for the noble gases? What generalizations can you draw from these data?

38. You decide to go to the mountains of Santa Fe, New Mexico (7200 ft elevation), for your spring holiday. One of the important tasks for you to complete is to boil the eggs to color and hide for the egg hunt. You are in a hurry and know that eggs boiled for exactly 8 minutes at home at the beach in Santa Barbara, California, come out perfectly cooked. As everyone is cracking their eggs for breakfast after the egg hunt, you discover to your dismay that your eggs are not fully cooked. Explain this observation.

39. Look at the drawing on the right of the water in the glass-graduated cylinder and the water in the plastic graduated cylinder. What differences do you observe in the shape of the meniscus? Plastics are composed of long-chain hydrocarbons and glass is composed of silicon oxides. Based on this information, propose a reason for the differences in these two water samples.

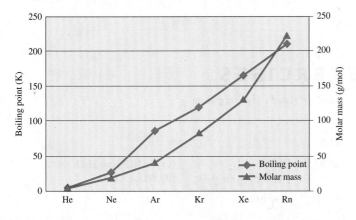

Plastic cylinder   Glass cylinder

40. The flow rate of a liquid is related to the strength of the cohesive forces between molecules. The stronger the cohesive forces, the harder it is for the molecules to slide past one another. Honey is a very thick liquid that pours slowly at ordinary temperatures. A student tried an experiment. She filled five eyedroppers with honey at different temperatures and then measured the rate of dripping for each sample (see table below). Graph the rate of dripping versus temperature for the honey and propose a relationship between temperature and flow rate. Propose an explanation for the trend that you observe.

Temperature (°C)	Number of drops/minute
10	3
20	17
30	52
50	87
70	104

41. You work in a hardware store and notice that whenever you spill water on the waxed floors it tends to bead up and stay fairly well confined. When you spill hexane, however, it spreads over the entire floor and is a big mess to clean up. Based on your knowledge of adhesive and cohesive forces, explain this difference in behavior.

42. A sample of pure acetic acid is also known as glacial acetic acid. This name comes from the fact that flasks of acetic acid would often freeze in the old laboratories of Europe. When acetic acid freezes, it forms many cracks resembling a glacier and thus the name. A heating curve for a sample of glacial acetic acid is pictured below.

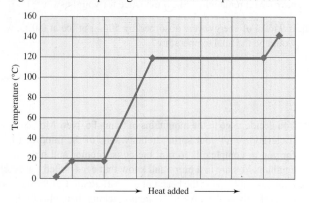

(a) Using this graph, determine the melting point and the boiling point of acetic acid.
(b) Redraw this graph and indicate regions where the solid, liquid, and gas phases will exist.
(c) Indicate regions where a transition of states is occurring and tell what phases are present.

43. How many calories are required to change 225 g of ice at 0°C to steam at 100.°C?

44. The molar heat of vaporization is the number of joules required to change 1 mol of a substance from liquid to vapor at its boiling point. What is the molar heat of vaporization of water?

45. The specific heat of zinc is 0.096 cal/g°C. Determine the energy required to raise the temperature of 250. g of zinc from room temperature (20.0°C) to 150.°C.

46. How many joules of energy would be liberated by condensing 50.0 mol of steam at 100.0°C and allowing the liquid to cool to 30.0°C?

47. How many kilojoules of energy are needed to convert 100. g of ice at −10.0°C to water at 20.0°C? (The specific heat of ice at −10.0°C is 2.01 J/g°C.)

48. What mass of water must be decomposed to produce 25.0 L of oxygen at STP?

49. Suppose 1.00 mol of water evaporates in 1.00 day. How many water molecules, on average, leave the liquid each second?

50. Compare the volume occupied by 1.00 mol of liquid water at 0°C and 1.00 mol of water vapor at STP.

51. A mixture of 80.0 mL of hydrogen and 60.0 mL of oxygen is ignited by a spark to form water.
(a) Does any gas remain unreacted? Which one, H_2 or O_2?
(b) What volume of which gas (if any) remains unreacted? (Assume the same conditions before and after the reaction.)

52. A student (with slow reflexes) puts his hand in a stream of steam at 100.°C until 1.5 g of water have condensed. If the water then cools to room temperature (20.0°C), how many joules have been absorbed by the student's hand?

53. Determine which of the following molecules would hydrogen-bond with other molecules like it. For those that do not hydrogen-bond, explain why. For those that hydrogen-bond, draw a diagram of two molecules using a dotted line to indicate where the hydrogen bonding will occur.
(a) Br_2 (d) H_2O (c) $CH_3—O—CH_3$
(b) $CH_3—O—H$ (e) H_2S

54. The heat of fusion of a substance is given in units of J/g. The specific heat of a substance is given in units of J/g °C. Why is a temperature factor not needed in the units for heat of fusion?

55. How many joules of energy are required to change 50.0 g Cu from 25.0°C to a liquid at its melting point, 1083°C?
Specific heat of Cu = 0.385 J/g°C
Heat of fusion for Cu = 134 J/g

56. You pour a steaming hot bowl of soup. After one taste, you burn your tongue and decide to cool it down with some ice. If you add 75 g of ice at 0°C to your soup and the final temperature of the ice–soup mixture is 87°C, how many joules of energy did the ice remove from the bowl of soup?

57. Write and balance the chemical equation for the decomposition of water into its component elements.

58. Use Figure 13.7 to determine the temperature in degrees Fahrenheit required to purify water by distillation at atmospheric pressures of 500 torr, 300 torr, and 100 torr.

CHALLENGE EXERCISES

59. You buy a box of borax ($Na_2B_4O_7 \cdot 10\ H_2O$) from the corner market in Phoenix, Arizona, in the middle of the summer. You open up the box and pour the borax into a weighed beaker. After all, you do want to be sure that you were not cheated by the manufacturer. You are distracted from your task and do not get back to weigh the filled beaker for several days. Upon weighing the beaker you get the following data from the 5.0-lb box of borax.

Empty beaker   492.5 g
Filled beaker   2467.4 g

Were you cheated by the manufacturer? Why or why not?

60. Why does a lake freeze from the top down? What significance does this have for life on Earth?

61. Suppose 150. g of ice at 0.0°C are added to 0.120 L of water at 45°C. If the mixture is stirred and allowed to cool to 0.0°C, how many grams of ice remain?

ANSWERS TO PRACTICE EXERCISES

13.1 Vapor pressure is the pressure exerted by a vapor in equilibrium with its liquid. The vapor pressure of a liquid increases with increasing temperature.

13.2 They are the same.

13.3 28°C, 73°C, 93°C

13.4 approximately 95°C

13.5 44.8 kJ

13.6 SO_2 and CH_3F are polar.

13.7 (a) yes, (b) yes, (c) no

13.8 I_2

13.9 (a) $BeCO_3 \cdot 4\ H_2O$ (c) $Cr(NO_3)_3 \cdot 9\ H_2O$
      (b) $Cd(MnO_4)_2 \cdot 6\ H_2O$ (d) $PtO_2 \cdot 3\ H_2O$

13.10 51.17% H_2O

Photodisc

Brass, a solid solution of zinc and copper, is used to make musical instruments and many other objects.

CHAPTER **14**

SOLUTIONS

Most substances we encounter in our daily lives are mixtures. Often they are homogeneous mixtures, which are called *solutions*. Some solutions we commonly encounter are shampoo, soft drinks, and wine. Blood plasma is a complex mixture composed of compounds and ions dissolved in water and proteins suspended in the solution. These solutions all have water as a main component, but many common items, such as air, gasoline, and steel, are also solutions that do not contain water. What are the necessary components of a solution? Why do some substances dissolve, while others do not? What effect does a dissolved substance have on the properties of the solution? Answering these questions is the first step in understanding the solutions we encounter in our daily lives.

CHAPTER OUTLINE

14.1 General Properties of Solutions
14.2 Solubility
14.3 Rate of Dissolving Solids
14.4 Concentration of Solutions
14.5 Colligative Properties of Solutions
14.6 Osmosis and Osmotic Pressure

14.1 GENERAL PROPERTIES OF SOLUTIONS

LEARNING OBJECTIVE ● List the properties of a true solution.

KEY TERMS

solution
solute
solvent

The term **solution** is used in chemistry to describe a system in which one or more substances are homogeneously mixed or dissolved in another substance. A simple solution has two components: a solute and a solvent. The **solute** is the component that is dissolved or is the least abundant component in the solution. The **solvent** is the dissolving agent or the most abundant component in the solution. For example, when salt is dissolved in water to form a solution, salt is the solute and water is the solvent. Complex solutions containing more than one solute and/or more than one solvent are common.

The three states of matter—solid, liquid, and gas—give us nine different types of solutions: solid dissolved in solid, solid dissolved in liquid, solid dissolved in gas, liquid dissolved in liquid, and so on. Of these, the most common solutions are solid dissolved in liquid, liquid dissolved in liquid, gas dissolved in liquid, and gas dissolved in gas. Some common types of solutions are listed in Table 14.1.

A true solution is one in which the particles of dissolved solute are molecular or ionic in size, generally in the range of 0.1 to 1 nm (10^{-8} to 10^{-7} cm). The properties of a true solution are as follows:

1. A mixture of two or more components—solute and solvent—is homogeneous and has a variable composition; that is, the ratio of solute to solvent can be varied.
2. The dissolved solute is molecular or ionic in size.
3. It is either colored or colorless and is usually transparent.
4. The solute remains uniformly distributed throughout the solution and will not settle out with time.
5. The solute can generally be separated from the solvent by purely physical means (for example, by evaporation).

Let's illustrate these properties using water solutions of sugar and of potassium permanganate.

EXAMPLE 14.1

Consider two sugar solutions: solution A containing 10 g of sugar added to 100 mL of water and solution B containing 20 g of sugar added to 100 mL of water. Each is stirred until all the sugar dissolves. Explain why these solutions are considered to be true solutions.

SOLUTION

1. All the sugar dissolves in each solution showing we can vary the composition of a sugar solution.
2. Every portion of each solution has the same sweet taste uniformly (although you should not taste solutions in the laboratory) because the sugar molecules are uniformly distributed throughout.
3. If the solution is confined so that no water is lost, the solution will look and taste the same a week or more later.
4. The solutions cannot be separated by filtering.
5. Careful evaporation will separate the solution into its components: sugar and water.

PRACTICE 14.1

Identify the solute and the solvent in each of these solutions:

(a) air   (b) seawater   (c) carbonated water

TABLE 14.1 Common Types of Solutions			
Phase of solution	**Solute**	**Solvent**	**Example**
Gas	gas	gas	air
Liquid	gas	liquid	soft drinks
Liquid	liquid	liquid	antifreeze
Liquid	solid	liquid	salt water
Solid	gas	solid	H_2 in Pt
Solid	solid	solid	brass

Note the beautiful purple trails of $KMnO_4$ as the crystals dissolve.

To observe the dissolving of potassium permanganate ($KMnO_4$), we drop a few crystals of it in a beaker of water. Almost at once, the beautiful purple color of dissolved permanganate ions (MnO_4^-) appears and streams to the bottom of the beaker as the crystals dissolve. After a while, the purple color disperses until it's evenly distributed throughout the solution. This dispersal demonstrates that molecules and ions move about freely and spontaneously (diffuse) in a liquid or solution.

Solution permanency is explained in terms of the kinetic-molecular theory (see Section 12.2). According to this theory, both the solute and solvent particles (molecules or ions) are in constant random motion. This motion is energetic enough to prevent the solute particles from settling out under the influence of gravity.

14.2 SOLUBILITY

Define solubility and the factors that affect it.

The term **solubility** describes the amount of one substance (solute) that will dissolve in a specified amount of another substance (solvent) under stated conditions. For example, 36.0 g of sodium chloride will dissolve in 100 g of water at 20°C. We say then that the solubility of NaCl in water is 36.0 g/100 g H_2O at 20°C. Solubility is often used in a relative way. For instance, we say that a substance is very soluble, moderately soluble, slightly soluble, or insoluble. Although these terms do not accurately indicate how much solute will dissolve, they are frequently used to describe the solubility of a substance qualitatively.

Two other terms often used to describe solubility are *miscible* and *immiscible*. Liquids that are capable of mixing and forming a homogeneous solution are **miscible**; those that do not form solutions or are generally insoluble in each other are **immiscible**. Methyl alcohol and water are miscible in each other in all proportions. Oil and water are immiscible, forming two separate layers when they are mixed, as shown in **Figure 14.1**.

The general guidelines for the solubility of common ionic compounds (salts) are given in **Figure 14.2**. These guidelines have some exceptions, but they provide a solid foundation for the compounds considered in this course. The solubilities of over 200 compounds are given in the Solubility Table in Appendix V. Solubility data for thousands of compounds can be found by consulting standard reference sources.*

The quantitative expression of the amount of dissolved solute in a particular quantity of solvent is known as the **concentration of a solution**. Several methods of expressing concentration are described in Section 14.4.

Predicting solubilities is complex and difficult. Many variables, such as size of ions, charge on ions, interaction between ions, interaction between solute and solvent, and temperature, complicate the problem. Because of the factors involved, the general rules of solubility given in Figure 14.2 have many exceptions. However, these rules are useful because they do apply to many of the more common compounds that we encounter in the study of chemistry. Keep in mind that these are rules, not laws, and are therefore subject to exceptions. Fortunately, the solubility of a solute is relatively easy to determine experimentally. Now let's examine the factors related to solubility.

● **LEARNING OBJECTIVE**

KEY TERMS

solubility
miscible
immiscible
concentration of a solution
saturated solution
unsaturated solution
supersaturated solution

Many chemists use the term *salt* interchangeably with *ionic compound*.

Figure 14.1

An immiscible mixture of oil and water.

---

*Two commonly used handbooks are *Lange's Handbook of Chemistry*, 16th ed. (New York: McGraw-Hill, 2009), and the *CRC Handbook of Chemistry and Physics*, 93rd ed. (Cleveland: Chemical Rubber Co., 2012).

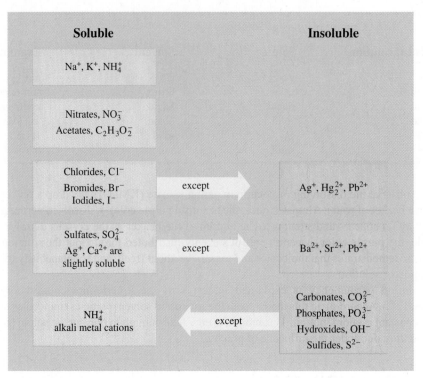

Figure 14.2

The solubility of various common ions. Substances containing the ions on the left are generally soluble in cold water, while those substances containing the ions on the right are insoluble in cold water. The arrows point to the exceptions.

= Water

= Na^+

= Cl^-

Figure 14.3

Dissolution of sodium chloride in water. Polar water molecules are attracted to Na^+ and Cl^- ions in the salt crystal, weakening the attraction between the ions. As the attraction between the ions weakens, the ions move apart and become surrounded by water dipoles. The hydrated ions slowly diffuse away from the crystal to become dissolved in solution.

The Nature of the Solute and Solvent

The old adage "like dissolves like" has merit, in a general way. Polar or ionic substances tend to be more compatible with other polar substances. Nonpolar substances tend to be compatible with other nonpolar substances and less soluble with polar substances. Thus ionic compounds, which are polar, tend to be much more soluble in water, which is polar, than in solvents such as ether, hexane, or benzene, which are essentially nonpolar. Sodium chloride, an ionic substance, is soluble in water, slightly soluble in ethyl alcohol (less polar than water), and insoluble in ether and benzene. Pentane, C_5H_{12}, a nonpolar substance, is only slightly soluble in water but is very soluble in benzene and ether.

At the molecular level the formation of a solution from two nonpolar substances, such as hexane and benzene, can be visualized as a process of simple mixing. These nonpolar molecules, having little tendency to either attract or repel one another, easily intermingle to form a homogeneous solution.

Solution formation between polar substances is much more complex. See, for example, the process by which sodium chloride dissolves in water (**Figure 14.3**). Water molecules are very polar and are attracted to other polar molecules or ions. When salt crystals are put into water, polar water molecules become attracted to the sodium and chloride ions on the crystal surfaces and weaken the attraction between Na^+ and Cl^- ions. The positive end of the water dipole is attracted to the Cl^- ions, and the negative end of the water dipole to the Na^+ ions. The weakened attraction permits the ions to move apart, making room for more water dipoles. Thus the surface ions are surrounded by water molecules, becoming hydrated ions, $Na^+(aq)$ and $Cl^-(aq)$, and slowly diffuse away from the crystals and dissolve in solution:

$$NaCl(crystal) \xrightarrow{H_2O} Na^+(aq) + Cl^-(aq)$$

Examination of the data in Table 14.2 reveals some of the complex questions relating to solubility.

TABLE 14.2 Solubility of Alkali Metal Halides in Water

| Salt | Solubility (g salt/100 g H₂O) | |
	0°C	100°C
LiF	0.12	0.14 (at 35°C)
LiCl	63.7	127.5
LiBr	143	266
LiI	151	481
NaF	4	5
NaCl	35.7	39.8
NaBr	79.5	121
NaI	158.7	302
KF	92.3 (at 18°C)	Very soluble
KCl	27.6	57.6
KBr	53.5	56.7
KI	127.5	102

The Effect of Temperature on Solubility

Temperature affects the solubility of most substances, as shown by the data in Table 14.2. Most solutes have a limited solubility in a specific solvent at a fixed temperature. For most solids dissolved in a liquid, an increase in temperature results in increased solubility (see Figure 14.4). However, no single rule governs the solubility of solids in liquids with change in temperature. Some solids increase in solubility only slightly with increasing temperature (see NaCl in Figure 14.4); other solids decrease in solubility with increasing temperature (see Li₂SO₄ in Figure 14.4).

On the other hand, the solubility of a gas in water usually decreases with increasing temperature (see HCl and SO₂ in Figure 14.4). The tiny bubbles that form when water is heated are due to the decreased solubility of air at higher temperatures. The decreased solubility of gases at higher temperatures is explained in terms of the kinetic-molecular theory (KMT) by assuming that, in order to dissolve, the gas molecules must have attraction of some sort with the molecules of the liquid. An increase in temperature decreases the solubility of the gas because it increases the kinetic energy (speed) of the gas molecules and thereby decreases their ability to interact with the liquid molecules.

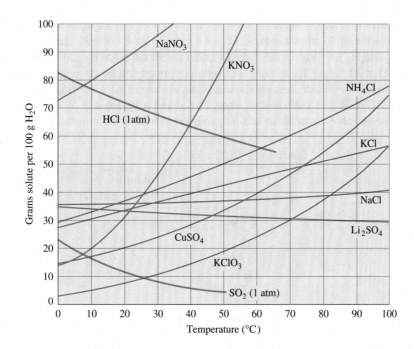

Figure 14.4

Solubility of various compounds in water. Solids are shown in red and gases are shown in blue.

Pouring root beer into a glass illustrates the effect of pressure on solubility. The escaping CO_2 produces the foam.

The Effect of Pressure on Solubility

Small changes in pressure have little effect on the solubility of solids in liquids or liquids in liquids but have a marked effect on the solubility of gases in liquids. The solubility of a gas in a liquid is directly proportional to the pressure of that gas above the solution. Thus the amount of a gas dissolved in solution will double if the pressure of that gas over the solution is doubled. For example, carbonated beverages contain dissolved carbon dioxide under pressures greater than atmospheric pressure. When a can of carbonated soda is opened, the pressure is immediately reduced to the atmospheric pressure, and the excess dissolved carbon dioxide bubbles out of the solution.

Saturated, Unsaturated, and Supersaturated Solutions

At a specific temperature there is a limit to the amount of solute that will dissolve in a given amount of solvent. When this limit is reached, the resulting solution is said to be *saturated*. For example, when we put 40.0 g of KCl into 100 g of H_2O at 20°C, we find that 34.0 g of KCl dissolve and 6.0 g of KCl remain undissolved. The solution formed is a saturated solution of KCl.

Two processes are occurring simultaneously in a saturated solution. The solid is dissolving into solution, and, at the same time, the dissolved solute is crystallizing out of solution. This may be expressed as

$$\text{solute (undissolved)} \rightleftharpoons \text{solute (dissolved)}$$

When these two opposing processes are occurring at the same rate, the amount of solute in solution is constant, and a condition of equilibrium is established between dissolved and undissolved solute. Therefore, a **saturated solution** contains dissolved solute in equilibrium with undissolved solute. Thus, any point on any solubility curve (Figure 14.4) represents a saturated solution of that solute. For example, a solution containing 60 g NH_4Cl per 100 g H_2O is saturated at 70°C.

It's important to state the temperature of a saturated solution, because a solution that is saturated at one temperature may not be saturated at another. If the temperature of a saturated solution is changed, the equilibrium is disturbed, and the amount of dissolved solute will change to reestablish equilibrium.

A saturated solution may be either dilute or concentrated, depending on the solubility of the solute. A saturated solution can be conveniently prepared by dissolving a little more than the saturated amount of solute at a temperature somewhat higher than room temperature. Then the amount of solute in solution will be in excess of its solubility at room temperature, and, when the solution cools, the excess solute will crystallize, leaving the solution saturated. (In this case, the solute must be more soluble at higher temperatures and must not form a supersaturated solution.) Examples expressing the solubility of saturated solutions at two different temperatures are given in Table 14.3.

An **unsaturated solution** contains less solute per unit of volume than does its corresponding saturated solution. In other words, additional solute can be dissolved in an unsaturated solution without altering any other conditions. Consider a solution made by adding 40 g of KCl to 100 g of H_2O at 20°C (see Table 14.3). The solution formed will be saturated and will contain about 6 g of undissolved salt, because the maximum amount of KCl that can dissolve in 100 g of H_2O at 20°C is 34 g. If the solution is now heated and maintained at 50°C, all the salt will dissolve and even more can be dissolved. The solution at 50°C is unsaturated.

TABLE 14.3 Saturated Solutions at 20°C and 50°C

Solute	Solubility (g solute/100 g H_2O)	
	20°C	50°C
NaCl	36.0	37.0
KCl	34.0	42.6
$NaNO_3$	88.0	114.0
$KClO_3$	7.4	19.3
$AgNO_3$	222.0	455.0
$C_{12}H_{22}O_{11}$	203.9	260.4

In some circumstances, solutions can be prepared that contain more solute than needed for a saturated solution at a particular temperature. These solutions are said to be supersaturated. However, we must qualify this definition by noting that a **supersaturated solution** is unstable. Disturbances, such as jarring, stirring, scratching the walls of the container, or dropping in a "seed" crystal, cause the supersaturation to return to saturation, releasing heat. When a supersaturated solution is disturbed, the excess solute crystallizes out rapidly, returning the solution to a saturated state.

Supersaturated solutions are not easy to prepare but may be made from certain substances by dissolving, in warm solvent, an amount of solute greater than that needed for a saturated solution at room temperature. The warm solution is then allowed to cool very slowly. With the proper solute and careful work, a supersaturated solution will result.

The heat released in this hot pack results from the crystallization of a supersaturated solution of sodium acetate.

EXAMPLE 14.2

Will a solution made by adding 2.5 g of $CuSO_4$ to 10 g of H_2O be saturated or unsaturated at 20°C?

SOLUTION

We first need to know the solubility of $CuSO_4$ at 20°C. From Figure 14.4, we see that the solubility of $CuSO_4$ at 20°C is about 21 g per 100 g of H_2O. This amount is equivalent to 2.1 g of $CuSO_4$ per 10 g of H_2O.

Since 2.5 g per 10 g of H_2O is greater than 2.1 g per 10 g of H_2O, the solution will be saturated and 0.4 g of $CuSO_4$ will be undissolved.

PRACTICE 14.2

Will a solution made by adding 9.0 g NH_4Cl to 20 g of H_2O be saturated or unsaturated at 50°C?

14.3 RATE OF DISSOLVING SOLIDS

Describe the factors that affect the rate at which a solid dissolves.

LEARNING OBJECTIVE

The rate at which a solid dissolves is governed by (1) the size of the solute particles, (2) the temperature, (3) the concentration of the solution, and (4) agitation or stirring. Let's look at each of these conditions:

1. *Particle size.* A solid can dissolve only at the surface that is in contact with the solvent. Because the surface-to-volume ratio increases as size decreases, smaller crystals dissolve faster than large ones. For example, if a salt crystal 1 cm on a side (a surface area of 6 cm^2) is divided into 1000 cubes, each 0.1 cm on a side, the total surface of the smaller cubes is 60 cm^2—a 10-fold increase in surface area (see **Figure 14.5**).

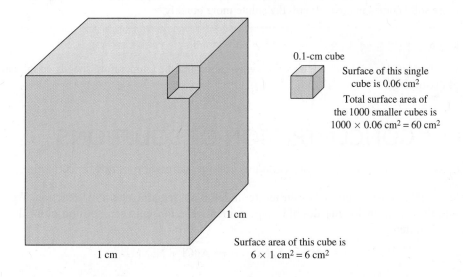

Figure 14.5

Surface area of crystals. A crystal 1 cm on a side has a surface area of 6 cm^2. Subdivided into 1000 smaller crystals, each 0.1 cm on a side, the total surface area is increased to 60 cm^2.

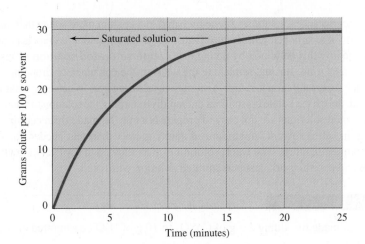

Figure 14.6

Rate of dissolution of a solid solute in a solvent. The rate is maximum at the beginning and decreases as the concentration approaches saturation.

2. *Temperature.* In most cases, the rate of dissolving of a solid increases with temperature. This increase is due to kinetic effects. The solvent molecules move more rapidly at higher temperatures and strike the solid surfaces more often, causing the rate of dissolving to increase.

3. *Concentration of the solution.* When the solute and solvent are first mixed, the rate of dissolving is at its maximum. As the concentration of the solution increases and the solution becomes more nearly saturated with the solute, the rate of dissolving decreases greatly. The rate of dissolving is graphed in **Figure 14.6**. Note that about 17 g dissolve in the first five-minute interval, but only about 1 g dissolves in the fourth five-minute interval. Although different solutes show different rates, the rate of dissolving always becomes very slow as the concentration approaches the saturation point.

4. *Agitation or stirring.* The effect of agitation or stirring is kinetic. When a solid is first put into water, it comes in contact only with solvent in its immediate vicinity. As the solid dissolves, the amount of dissolved solute around the solid becomes more and more concentrated, and the rate of dissolving slows down. If the mixture is not stirred, the dissolved solute diffuses very slowly through the solution; weeks may pass before the solid is entirely dissolved. Stirring distributes the dissolved solute rapidly through the solution, and more solvent is brought into contact with the solid, causing it to dissolve more rapidly.

EXAMPLE 14.3

What does it mean to increase the surface area of a solid? Explain how this change causes an increase in the rate of dissolving.

SOLUTION

Increasing the surface area means breaking up the solid to form smaller particles that have a larger total surface area. Increasing the surface area speeds up dissolving because more solid comes in contact with the solute more quickly.

PRACTICE 14.3

Use a molecular explanation to explain why increasing the temperature speeds up the rate of dissolving a solid in a liquid.

14.4 CONCENTRATION OF SOLUTIONS

LEARNING OBJECTIVE ● Solve problems involving mass percent, volume percent, molarity, and dilution.

KEY TERMS

dilute solution
concentrated solution
parts per million (ppm)
molarity (*M*)

Many solids must be put into solution to undergo appreciable chemical reaction. We can write the equation for the double-displacement reaction between sodium chloride and silver nitrate:

$$NaCl + AgNO_3 \longrightarrow AgCl + NaNO_3$$

But suppose we mix solid NaCl and solid AgNO₃ and look for a chemical change. If any reaction occurs, it is slow and virtually undetectable. In fact, the crystalline structures of NaCl and AgNO₃ are so different that we could separate them by tediously picking out each kind of crystal from the mixture. But if we dissolve the NaCl and AgNO₃ separately in water and mix the two solutions, we observe the immediate formation of a white, curdlike precipitate of silver chloride.

Molecules or ions must collide with one another in order to react. In the foregoing example, the two solids did not react because the ions were securely locked within their crystal structures. But when the NaCl and AgNO₃ are dissolved, their crystal lattices are broken down and the ions become mobile. When the two solutions are mixed, the mobile Ag^+ and Cl^- ions come into contact and react to form insoluble AgCl, which precipitates out of solution. The soluble Na^+ and NO_3^- ions remain mobile in solution but form the crystalline salt NaNO₃ when the water is evaporated:

$$NaCl(aq) + AgNO_3(aq) \longrightarrow AgCl(s) + NaNO_3(aq)$$

$$Na^+(aq) + Cl^-(aq) + Ag^+(aq) + NO_3^-(aq) \longrightarrow AgCl(s) + Na^+(aq) + NO_3^-(aq)$$

| sodium chloride solution | silver nitrate solution | silver chloride | sodium nitrate in solution |

The mixture of the two solutions provides a medium or space in which the Ag^+ and Cl^- ions can react. (See Chapter 15 for further discussion of ionic reactions.)

Solutions also function as diluting agents in reactions in which the undiluted reactants would combine with each other too violently. Moreover, a solution of known concentration provides a convenient method for delivering a specific amount of reactant.

The concentration of a solution expresses the amount of solute dissolved in a given quantity of solvent or solution. Because reactions are often conducted in solution, it's important to understand the methods of expressing concentration and to know how to prepare solutions of particular concentrations. The concentration of a solution may be expressed qualitatively or quantitatively. Let's begin with a look at the qualitative methods of expressing concentration.

Dilute and Concentrated Solutions

When we say that a solution is *dilute* or *concentrated*, we are expressing, in a relative way, the amount of solute present. One gram of a compound and 2 g of a compound in solution are both dilute solutions when compared with the same volume of a solution containing 20 g of a compound. Ordinary concentrated hydrochloric acid contains 12 mol of HCl per liter of solution. In some laboratories, the dilute acid is made by mixing equal volumes of water and the concentrated acid. In other laboratories the concentrated acid is diluted with two or three volumes of water, depending on its use. The term **dilute solution**, then, describes a solution that contains a relatively small amount of dissolved solute. Conversely, a **concentrated solution** contains a relatively large amount of dissolved solute.

Mass Percent Solution

The mass percent method expresses the concentration of the solution as the percent of solute in a given mass of solution. It says that for a given mass of solution, a certain percent of that mass is solute. Suppose we take a bottle from the reagent shelf that reads "sodium hydroxide, NaOH, 10%." This statement means that for every 100 g of this solution, 10 g will be NaOH and 90 g will be water. (Note that this amount of solution is 100 g, not 100 mL.) We could also make this same concentration of solution by dissolving 2.0 g of NaOH in 18 g of water. Mass percent concentrations are most generally used for solids dissolved in liquids:

$$\text{mass percent} = \frac{\text{g solute}}{\text{g solute} + \text{g solvent}} \times 100 = \frac{\text{g solute}}{\text{g solution}} \times 100$$

Note that mass percent is independent of the formula for the solute.

As instrumentation advances are made in chemistry, our ability to measure the concentration of dilute solutions is increasing as well. In addition to mass percent, chemists now commonly use **parts per million (ppm)**:

$$\text{parts per million} = \frac{\text{g solute}}{\text{g solute} + \text{g solvent}} \times 1,000,000$$

Currently, air and water contaminants, drugs in the human body, and pesticide residues are measured in parts per million.

WileyPLUS

ENHANCED EXAMPLE

EXAMPLE 14.4

What is the mass percent of sodium hydroxide in a solution that is made by dissolving 8.00 g NaOH in 50.0 g H_2O?

SOLUTION

Read •   **Knowns:**   8.00 g NaOH
                                50.0 g H_2O

          Solving for: mass percent

Calculate • $\left(\dfrac{8.00 \text{ g NaOH}}{8.00 \text{ g NaOH} + 50.0 \text{ g } H_2O}\right) = 13.8\%$ NaOH solution

---

EXAMPLE 14.5

What masses of potassium chloride and water are needed to make 250. g of 5.00% solution?

SOLUTION

Read •   **Knowns:**   250. g solution
                                5.00% solution

          Solving for: masses KCl and water

Plan •   5.00% of 250. g = (0.0500)(250. g) = 12.5 g KCl (solute)

Calculate • 250. g − 12.5 g = 238 g H_2O
          Dissolving 12.5 g KCl in 238 g H_2O gives a 5.00% solution.

EXAMPLE 14.6

A 34.0% sulfuric acid solution had a density of 1.25 g/mL. How many grams of H_2SO_4 are contained in 1.00 L of this solution?

SOLUTION

Read •   **Knowns:** $d = 1.25$ g/mL
                              $V = 1.00$ L
                              34.0% H_2SO_4 solution

          Solving for: mass H_2SO_4

Plan •   Find the mass of the solution from the density and then use the mass percent to determine the mass of H_2SO_4.

Calculate • $d = $ mass/V

$$\text{mass of solution} = \left(\frac{1.25 \text{ g}}{\text{mL}}\right)(1.00 \times 10^3 \text{ mL}) = 1250 \text{ g (solution)}$$

$$\text{mass percent} = \left(\frac{\text{g solute}}{\text{g solution}}\right)100$$

$$\text{g solute} = \frac{(\text{mass percent})(\text{g solution})}{100}$$

$$\text{g solute} = \frac{(34.0)(1250 \text{ g})}{100} = 425 \text{ g } H_2SO_4$$

Therefore, 1.00 L of 34.0% H_2SO_4 solution contains 425 g H_2SO_4.

PRACTICE 14.4

What is the mass percent of Na_2SO_4 in a solution made by dissolving 25.0 g Na_2SO_4 in 225.0 g H_2O?

Mass/Volume Percent (m/v)

This method expresses concentration as grams of solute per 100 mL of solution. With this system, a 10.0%-m/v-glucose solution is made by dissolving 10.0 g of glucose in water, diluting to 100 mL, and mixing. The 10.0%-m/v solution could also be made by diluting 20.0 g to 200 mL, 50.0 g to 500 mL, and so on. Of course, any other appropriate dilution ratio may be used:

$$\text{mass/volume percent} = \frac{\text{g solute}}{\text{mL solution}} \times 100$$

EXAMPLE 14.7

A 3.0% H_2O_2 solution is commonly used as a topical antiseptic to prevent infection. What volume of this solution will contain 10. g of H_2O_2?

SOLUTION

First solve the mass/volume percent equation for mL of solution:

$$\text{mL solution} = \frac{\text{(g solute)}}{\text{(m/v percent)}} (100)$$

$$\text{mL solution} = \frac{\text{(10. g solute)}}{\text{(3.0 m/v percent)}} (100) = 330 \text{ mL}$$

Volume Percent

Solutions that are formulated from two liquids are often expressed as *volume percent* with respect to the solute. The volume percent is the volume of a liquid in 100 mL of solution. The label on a bottle of ordinary rubbing alcohol reads "isopropyl alcohol, 70% by volume." Such a solution could be made by mixing 70 mL of alcohol with water to make a total volume of 100 mL, but we cannot use 30 mL of water, because the two volumes are not necessarily additive:

$$\text{volume percent} = \frac{\text{volume of liquid in question}}{\text{total volume of solution}} \times 100$$

Volume percent is used to express the concentration of alcohol in beverages. Wines generally contain 12% alcohol by volume. This translates into 12 mL of alcohol in each 100 mL of wine. The beverage industry also uses the concentration unit of *proof* (twice the volume percent). Pure ethyl alcohol is 100% and therefore 200 proof. Scotch whiskey is 86 proof, or 43% alcohol by volume.

Molarity

Mass percent solutions do not equate or express the molar masses of the solute in solution. For example, 1000. g of 10.0% NaOH solution contains 100. g NaOH; 1000. g of 10.0% KOH solution contains 100. g KOH. In terms of moles of NaOH and KOH, these solutions contain

$$\text{mol NaOH} = (100. \text{ g NaOH})\left(\frac{1 \text{ mol NaOH}}{40.00 \text{ g NaOH}}\right) = 2.50 \text{ mol NaOH}$$

$$\text{mol KOH} = (100. \text{ g KOH})\left(\frac{1 \text{ mol KOH}}{56.11 \text{ g KOH}}\right) = 1.78 \text{ mol KOH}$$

From these figures, we see that the two 10.0% solutions do not contain the same number of moles of NaOH and KOH. As a result, we find that a 10.0% NaOH solution contains more reactive base than a 10.0% KOH solution.

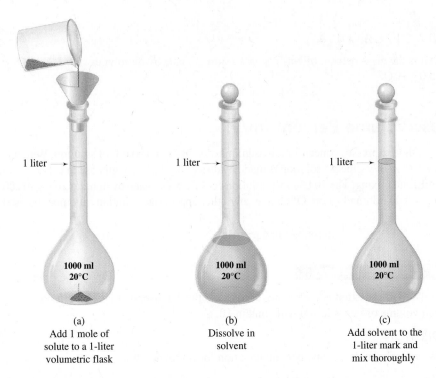

1 liter →

1 liter →

1 liter →

1000 ml
20°C

1000 ml
20°C

1000 ml
20°C

(a)
Add 1 mole of
solute to a 1-liter
volumetric flask

(b)
Dissolve in
solvent

(c)
Add solvent to the
1-liter mark and
mix thoroughly

Figure 14.7

Preparation of a 1 M solution.

We need a method of expressing concentration that will easily indicate how many moles of solute are present per unit volume of solution. For this purpose, the concentration known as molarity is used in calculations involving chemical reactions.

A 1-molar solution contains 1 mol of solute per liter of solution. For example, to make a 1-molar solution of sodium hydroxide (NaOH) we dissolve 40.00 g NaOH (1 mol) in water and dilute the solution with more water to a volume of 1 L. The solution contains 1 mol of the solute in 1 L of solution and is said to be 1 molar in concentration. **Figure 14.7** illustrates the preparation of a 1-molar solution. Note that the volume of the solute and the solvent together is 1 L.

The concentration of a solution can, of course, be varied by using more or less solute or solvent; but in any case the **molarity (M)** of a solution is the number of moles of solute per liter of solution. The abbreviation for molarity is M. The units of molarity are moles per liter. The expression "2.0 M NaOH" means a 2.0-molar solution of NaOH (2.0 mol, or 80. g, of NaOH dissolved in water to make 1.0 L of solution):

$$\text{molarity} = M = \frac{\text{number of moles of solute}}{\text{liter of solution}} = \frac{\text{moles}}{\text{liter}}$$

Flasks that are calibrated to contain specific volumes at a particular temperature are used to prepare solutions of a desired concentration. These *volumetric flasks* have a calibration mark on the neck that accurately indicates the measured volume. Molarity is based on a specific volume of solution and therefore will vary slightly with temperature because volume varies with temperature (1000 mL H_2O at 20°C = 1001 mL at 25°C).

Suppose we want to make 500 mL of 1 M solution. This solution can be prepared by determining the mass of 0.5 mol of the solute and diluting with water in a 500-mL (0.5-L) volumetric flask. The molarity will be

$$M = \frac{0.5 \text{ mol solute}}{0.5 \text{ L solution}} = 1 \text{ molar}$$

You can see that it isn't necessary to have a liter of solution to express molarity. All we need to know is the number of moles of dissolved solute and the volume of solution. Thus, 0.001 mol NaOH in 10 mL of solution is 0.1 M:

$$\left(\frac{0.001 \text{ mol}}{10 \text{ mL}}\right)\left(\frac{1000 \text{ mL}}{1 \text{ L}}\right) = 0.1 \ M$$

When we stop to think that a balance is not calibrated in moles, but in grams, we can incorporate grams into the molarity formula. We do so by using the relationship

$$\text{moles} = \frac{\text{grams of solute}}{\text{molar mass}}$$

Substituting this relationship into our expression for molarity, we get

$$M = \frac{\text{mol}}{\text{L}} = \frac{\text{g solute}}{\text{molar mass solute} \times \text{L solution}}$$

We can now determine the mass of any amount of a solute that has a known formula, dilute it to any volume, and calculate the molarity of the solution using this formula.

Molarities of concentrated acids commonly used in the laboratory:	
HCl	12 M
$HC_2H_3O_2$	17 M
HNO_3	16 M
H_2SO_4	18 M

WileyPLUS

ENHANCED EXAMPLE

EXAMPLE 14.8

What is the molarity of a solution containing 1.4 mol of acetic acid ($HC_2H_3O_2$) in 250. mL of solution?

SOLUTION

By the unit-conversion method, we note that the concentration given in the problem statement is 1.4 mol per 250. mL (mol/mL). Since molarity = mol/L, the needed conversion is

Solution map: $\dfrac{\text{mol}}{\text{mL}} \longrightarrow \dfrac{\text{mol}}{\text{L}} = M$

Calculate $\left(\dfrac{1.4 \text{ mol}}{250. \text{ mL}}\right)\left(\dfrac{1000 \text{ mL}}{\text{L}}\right) = \dfrac{5.6 \text{ mol}}{\text{L}} = 5.6\ M$

EXAMPLE 14.9

What is the molarity of a solution made by dissolving 2.00 g of potassium chlorate in enough water to make 150. mL of solution?

SOLUTION

Knowns: mass $KClO_3$ = 2.00 g

volume = 150. mL

molar mass $KClO_3$ = 122.6 g/mol

Solution map: $\dfrac{\text{g } KClO_3}{\text{mL}} \longrightarrow \dfrac{\text{g } KClO_3}{\text{L}} \longrightarrow \dfrac{\text{mol } KClO_3}{\text{L}} = M$

Calculate $\left(\dfrac{2.00 \text{ g } KClO_3}{150. \text{ mL}}\right)\left(\dfrac{1000 \text{ mL}}{\text{L}}\right)\left(\dfrac{1 \text{ mol } KClO_3}{122.6 \text{ g } KClO_3}\right) = \dfrac{0.109 \text{ mol}}{\text{L}}$

$= 0.109\ M\ KClO_3$

EXAMPLE 14.10

How many grams of potassium hydroxide are required to prepare 600. mL of 0.450 M KOH solution?

SOLUTION

Plan

Solution map: milliliters $\longrightarrow$ liters $\longrightarrow$ moles $\longrightarrow$ grams

Knowns: volume = 600. mL $M = \dfrac{0.450 \text{ mol}}{\text{L}}$

molar mass KOH $= \dfrac{56.11 \text{ g}}{\text{mol}}$

Calculate $(600. \text{ mL})\left(\dfrac{1 \text{ L}}{1000 \text{ mL}}\right)\left(\dfrac{0.450 \text{ mol}}{\text{L}}\right)\left(\dfrac{56.11 \text{ g KOH}}{\text{mol}}\right) = 15.1 \text{ g KOH}$

PRACTICE 14.5

What is the molarity of a solution made by dissolving 7.50 g of magnesium nitrate [$Mg(NO_3)_2$] in enough water to make 25.0 mL of solution?

PRACTICE 14.6

How many grams of sodium chloride are needed to prepare 125 mL of a 0.037 M NaCl solution?

EXAMPLE 14.11

Calculate the number of moles of nitric acid in 325 mL of 16 M HNO_3 solution.

SOLUTION

Read •   **Knowns:** $V = 325$ mL

                        16 M HNO_3 solution

Plan •   Use the equation   moles = liters $\times$ M

Calculate •   moles = $(0.325 \,\cancel{L})\left(\dfrac{16 \text{ mol HNO}_3}{1 \,\cancel{L}}\right) = 5.2$ mol HNO_3

EXAMPLE 14.12

What volume of 0.250 M solution can be prepared from 16.0 g of potassium carbonate?

SOLUTION

Read •   **Knowns:**   mass = 16.0 g K_2CO_3

                        $M = 0.250$ mol/L

                        Molar mass K_2CO_3 = 138.2 g/mol

Plan •   Find the volume that can be prepared from 16.0 g K_2CO_3.

          Solution map: g $K_2CO_3 \rightarrow$ mol $K_2CO_3 \rightarrow$ L solution

Calculate • $(16.0 \text{ g }\cancel{K_2CO_3})\left(\dfrac{1 \text{ mol }\cancel{K_2CO_3}}{138.2 \text{ g }\cancel{K_2CO_3}}\right)\left(\dfrac{1 \text{ L}}{0.250 \text{ mol }\cancel{K_2CO_3}}\right) = 0.463$ L

                                                 = 463 mL

EXAMPLE 14.13

How many milliliters of 2.00 M HCl will react with 28.0 g NaOH?

SOLUTION

Read •   **Knowns:**   mass = 28.0 g NaOH

                            $M = 2.00$ mol/L HCl

Plan •   Write a balanced equation for the reaction

              $HCl(aq) + NaOH(aq) \longrightarrow NaCl(aq) + H_2O(l)$

          Find the number of moles NaOH in 28.0 g NaOH.

          Solution map: grams NaOH $\rightarrow$ moles NaOH

              $(28.0 \text{ g }\cancel{NaOH})\left(\dfrac{1 \text{ mole}}{40.00 \text{ g }\cancel{NaOH}}\right) = 0.700$ mol NaOH

Setup •   **Solution map:**   mol NaOH $\rightarrow$ mol HCl $\rightarrow$ L HCl $\rightarrow$ mL HCl

Calculate • $(0.700 \text{ mol }\cancel{NaOH})\left(\dfrac{1 \text{ mol }\cancel{HCl}}{1 \text{ mol }\cancel{NaOH}}\right)\left(\dfrac{1 \text{ L HCl}}{2.00 \text{ mol }\cancel{HCl}}\right) = 0.350$ L HCl

                                                   = 350. mL HCl

PRACTICE 14.7

What volume of 0.035 M AgNO$_3$ can be made from 5.0 g of AgNO$_3$?

PRACTICE 14.8

How many milliliters of 0.50 M NaOH are required to react with 25.00 mL of 1.5 M HCl?

We've now examined several ways to measure concentration of solutions quantitatively. A summary of these concentration units is found in Table 14.4.

TABLE 14.4   Concentration Units for Solutions

Units	Symbol	Definition
Mass percent	% m/m	$\dfrac{\text{mass solute}}{\text{mass solution}} \times 100$
Parts per million	ppm	$\dfrac{\text{mass solute}}{\text{mass solution}} \times 1{,}000{,}000$
Mass/volume percent	% m/v	$\dfrac{\text{mass solute}}{\text{mL solution}} \times 100$
Volume percent	% v/v	$\dfrac{\text{mL solute}}{\text{mL solution}} \times 100$
Molarity	M	$\dfrac{\text{moles solute}}{\text{L solution}}$
Molality	m	$\dfrac{\text{moles solute}}{\text{kg solvent}}$

Molality is covered in Section 14.5.

Dilution Problems

Chemists often find it necessary to dilute solutions from one concentration to another by adding more solvent to the solution. If a solution is diluted by adding pure solvent, the volume of the solution increases, but the number of moles of solute in the solution remains the same. Thus the moles/liter (molarity) of the solution decreases. Always read a problem carefully to distinguish between (1) how much solvent must be added to dilute a solution to a particular concentration and (2) to what volume a solution must be diluted to prepare a solution of a particular concentration.

EXAMPLE 14.14

Calculate the molarity of a sodium hydroxide solution that is prepared by mixing 100. mL of 0.20 M NaOH with 150. mL of water. Assume that the volumes are additive.

SOLUTION

Read •   **Knowns:** $V_1 = 100.$ mL NaOH

$V_2 = 150.$ mL water

$V_{\text{solution}} = 250.$ mL

$M_{\text{initial}} = 0.20\ M$

Plan •   In the dilution, the moles NaOH remain the same; the molarity and volume change.

1. Calculate the moles of NaOH in the original solution.

$$\text{mol} = (0.100\ \cancel{L})\left(\frac{0.20\ \text{mol NaOH}}{1\ \cancel{L}}\right) = 0.020\ \text{mol NaOH}$$

2. Divide the mol NaOH by the final volume of the solution to find the new molarity.

WileyPLUS

ENHANCED EXAMPLE

Calculate • $M = \dfrac{0.020 \text{ mol NaOH}}{0.250 \text{ L}} = 0.080 \, M \text{ NaOH}$

Check • If we double the volume of the solution, the concentration is half. Therefore the concentration of the new solution here should be less than 0.10 M.

PRACTICE 14.9

Calculate the molarity of a solution prepared by diluting 125 mL of 0.400 M $K_2Cr_2O_7$ with 875 mL of water.

EXAMPLE 14.15

How many grams of silver chloride will be precipitated by adding sufficient silver nitrate to react with 1500. mL of 0.400 M barium chloride solution?

$$2 \, AgNO_3(aq) + BaCl_2(aq) \longrightarrow 2 \, AgCl(s) + Ba(NO_3)_2(aq)$$

SOLUTION

Read • **Knowns:** $V = 1500.$ mL $BaCl_2$

$M = 0.400 \, M \, BaCl_2$

Plan • Solve as a stoichiometry problem.

Determine the mol $BaCl_2$ in 1500. mL of 0.400 M solution.

$$\text{mol} = M \times V = (1.500 \, \cancel{L})\left(\dfrac{0.400 \text{ mol } BaCl_2}{\cancel{L}}\right) = 0.600 \text{ mol } BaCl_2$$

Solution map: moles $BaCl_2 \rightarrow$ moles $AgCl \rightarrow$ grams $AgCl$

Calculate • $(0.600 \text{ mol } \cancel{BaCl_2})\left(\dfrac{2 \text{ mol } AgCl}{1 \text{ mol } \cancel{BaCl_2}}\right)\left(\dfrac{143.4 \text{ g } AgCl}{\cancel{\text{mol } AgCl}}\right) = 172 \text{ g } AgCl$

PRACTICE 14.10

How many grams of lead(II) iodide will be precipitated by adding sufficient $Pb(NO_3)_2$ to react with 750 mL of 0.250 M KI solution?

$$2 \, KI(aq) + Pb(NO_3)_2(aq) \longrightarrow PbI_2(s) + 2 \, KNO_3(aq)$$

14.5 COLLIGATIVE PROPERTIES OF SOLUTIONS

LEARNING OBJECTIVE ● Use the concept of colligative properties to calculate molality, freezing point, boiling point, freezing point depression, and boiling point elevation of various solutions.

KEY TERMS

colligative properties
molality (m)

Two solutions—one containing 1 mol (60.06 g) of urea (NH_2CONH_2) and the other containing 1 mol (342.3 g) of sucrose ($C_{12}H_{22}O_{11}$) each in 1 kg of water—both have a freezing point of $-1.86°C$, not 0°C as for pure water. Urea and sucrose are distinct substances, yet they lower the freezing point of the water by the same amount. The only thing apparently common to these two solutions is that each contains 1 mol (6.022×10^{23} molecules) of solute and 1 kg of solvent. In fact, when we dissolve 1 mol of any nonionizable solute in 1 kg of water, the freezing point of the resulting solution is $-1.86°C$.

These results lead us to conclude that the freezing point depression for a solution containing 6.022×10^{23} solute molecules (particles) and 1 kg of water is a constant, namely, 1.86°C. Freezing point depression is a general property of solutions. Furthermore, the amount by which the freezing point is depressed is the same for all solutions

TABLE 14.5 Freezing Point Depression and Boiling Point Elevation Constants of Selected Solvents

Solvent	Freezing point of pure solvent (°C)	Freezing point depression constant, K_f $\left(\dfrac{°C\ kg\ solvent}{mol\ solute}\right)$	Boiling point of pure solvent (°C)	Boiling point elevation constant, K_b $\left(\dfrac{°C\ kg\ solvent}{mol\ solute}\right)$
Water	0.00	1.86	100.0	0.512
Acetic acid	16.6	3.90	118.5	3.07
Benzene	5.5	5.1	80.1	2.53
Camphor	178	40	208.2	5.95

made with a given solvent; that is, each solvent shows a characteristic *freezing point depression constant*. Freezing point depression constants for several solvents are given in Table 14.5.

The solution formed by the addition of a nonvolatile solute to a solvent has a lower freezing point, a higher boiling point, and a lower vapor pressure than that of the pure solvent. These effects are related and are known as colligative properties. The **colligative properties** are properties that depend only on the number of solute particles in a solution, not on the nature of those particles. Freezing point depression, boiling point elevation, and vapor pressure lowering are colligative properties of solutions.

The colligative properties of a solution can be considered in terms of vapor pressure. The vapor pressure of a pure liquid depends on the tendency of molecules to escape from its surface. If 10% of the molecules in a solution are nonvolatile solute molecules, the vapor pressure of the solution is 10% lower than that of the pure solvent. The vapor pressure is lower because the surface of the solution contains 10% nonvolatile molecules and 90% of the volatile solvent molecules. A liquid boils when its vapor pressure equals the pressure of the atmosphere. We can thus see that the solution just described as having a lower vapor pressure will have a higher boiling point than the pure solvent. The solution with a lowered vapor pressure doesn't boil until it has been heated above the boiling point of the solvent (see Figure 14.8a). Each solvent has its own characteristic boiling point elevation constant (Table 14.5). The boiling point elevation constant is based on a solution that contains 1 mol of solute particles per kilogram of solvent. For example, the boiling point elevation constant for a solution containing 1 mol of solute particles per kilogram of water is 0.512°C, which means that this water solution will boil at 100.512°C.

The freezing behavior of a solution can also be considered in terms of lowered vapor pressure. Figure 14.8b shows the vapor pressure relationships of ice, water, and a solution containing 1 mol of solute per kilogram of water. The freezing point of water is at the intersection of the liquid and solid vapor pressure curves (i.e., at the point where water and ice have the same

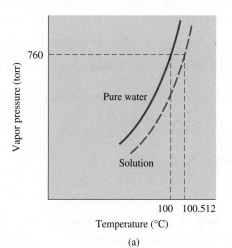

(a)

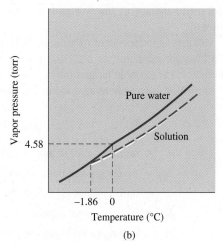

(b)

Figure 14.8

Vapor pressure curves of pure water and water solutions, showing (a) boiling point elevation and (b) freezing point depression effects (concentration: 1 mol solute/1 kg water).

Engine coolant is one application of colligative properties. The addition of coolant to the water in a radiator raises its boiling point and lowers its freezing point.

vapor pressure). Because the vapor pressure of the liquid is lowered by the solute, the vapor pressure curve of the solution does not intersect the vapor pressure curve of the solid until the solution has been cooled below the freezing point of pure water. So the solution must be cooled below 0°C in order for it to freeze.

The foregoing discussion dealing with freezing point depressions is restricted to *un-ionized* substances. The discussion of boiling point elevations is restricted to *nonvolatile* and un-ionized substances. The colligative properties of ionized substances are not under consideration at this point; we will discuss them in Chapter 15.

Some practical applications involving colligative properties are (1) use of salt–ice mixtures to provide low freezing temperatures for homemade ice cream, (2) use of sodium chloride or calcium chloride to melt ice from streets, and (3) use of ethylene glycol–water mixtures as antifreeze in automobile radiators (ethylene glycol also raises the boiling point of radiator fluid, thus allowing the engine to operate at a higher temperature).

Sodium chloride or calcium chloride is used to melt ice on snowy streets and highways.

Both the freezing point depression and the boiling point elevation are directly proportional to the number of moles of solute per kilogram of solvent. When we deal with the colligative properties of solutions, another concentration expression, *molality*, is used. The **molality (m)** of a solution is the number of moles of solute per kilogram of solvent:

$$m = \frac{\text{mol solute}}{\text{kg solvent}}$$

Note that a lowercase *m* is used for molality concentrations and a capital *M* for molarity. The difference between molality and molarity is that molality refers to moles of solute *per kilogram of solvent*, whereas molarity refers to moles of solute *per liter of solution*. For un-ionized substances, the colligative properties of a solution are directly proportional to its molality.

Molality is independent of volume. It is a mass-to-mass relationship of solute to solvent and allows for experiments, such as freezing point depression and boiling point elevation, to be conducted at variable temperatures.

The following equations are used in calculations involving colligative properties and molality:

$$\Delta t_f = mK_f \qquad \Delta t_b = mK_b \qquad m = \frac{\text{mol solute}}{\text{kg solvent}}$$

m = molality; mol solute/kg solvent

Δt_f = freezing point depression; °C

Δt_b = boiling point elevation; °C

K_f = freezing point depression constant; °C kg solvent/mol solute

K_b = boiling point elevation constant; °C kg solvent/mol solute

EXAMPLE 14.16

What is the molality (m) of a solution prepared by dissolving 2.70 g CH_3OH in 25.0 g H_2O?

SOLUTION

Since $m = \dfrac{\text{mol solute}}{\text{kg solvent}}$, the conversion is

$$\dfrac{2.70 \text{ g } CH_3OH}{25.0 \text{ g } H_2O} \longrightarrow \dfrac{\text{mol } CH_3OH}{25.0 \text{ g } H_2O} \longrightarrow \dfrac{\text{mol } CH_3OH}{1 \text{ kg } H_2O}$$

The molar mass of CH_3OH is (12.01 + 4.032 + 16.00), or 32.04 g/mol:

$$\left(\dfrac{2.70 \text{ g } CH_3OH}{25.0 \text{ g } H_2O}\right)\left(\dfrac{1 \text{ mol } CH_3OH}{32.04 \text{ g } CH_3OH}\right)\left(\dfrac{1000 \text{ g } H_2O}{1 \text{ kg } H_2O}\right) = \dfrac{3.37 \text{ mol } CH_3OH}{1 \text{ kg } H_2O}$$

The molality is 3.37 m.

PRACTICE 14.11

What is the molality of a solution prepared by dissolving 150.0 g $C_6H_{12}O_6$ in 600.0 g H_2O?

EXAMPLE 14.17

A solution is made by dissolving 100. g of ethylene glycol ($C_2H_6O_2$) in 200. g of water. What is the freezing point of this solution?

SOLUTION

To calculate the freezing point of the solution, we first need to calculate Δt_f, the change in freezing point. Use the equation

$$\Delta t_f = mK_f = \dfrac{\text{mol solute}}{\text{kg solvent}} \times K_f$$

K_f (for water): $\dfrac{1.86°C \text{ kg solvent}}{\text{mol solute}}$ (from Table 14.5)

mol solute: $(100. \text{ g } C_2H_6O_2)\left(\dfrac{1 \text{ mol } C_2H_6O_2}{62.07 \text{ g } C_2H_6O_2}\right) = 1.61 \text{ mol } C_2H_6O_2$

kg solvent: $(200. \text{ g } H_2O)\left(\dfrac{1 \text{ kg}}{1000 \text{ g}}\right) = 0.200 \text{ kg } H_2O$

$$\Delta t_f = \left(\dfrac{1.61 \text{ mol } C_2H_6O_2}{0.200 \text{ kg } H_2O}\right)\left(\dfrac{1.86°C \text{ kg } H_2O}{1 \text{ mol } C_2H_6O_2}\right) = 15.0°C$$

The freezing point depression, 15.0°C, must be subtracted from 0°C, the freezing point of the pure solvent (water):

freezing point of solution = freezing point of solvent − Δt_f

= 0.0°C − 15.0°C = −15.0°C

Therefore, the freezing point of the solution is −15.0°C. This solution will protect an automobile radiator down to −15.0°C (5°F).

EXAMPLE 14.18

A solution made by dissolving 4.71 g of a compound of unknown molar mass in 100.0 g of water has a freezing point of −1.46°C. What is the molar mass of the compound?

SOLUTION

First substitute the data in $\Delta t_f = m K_f$ and solve for m:

$\Delta t_f = +1.46$ (since the solvent, water, freezes at 0°C)

$$K_f = \frac{1.86°C \text{ kg H}_2\text{O}}{\text{mol solute}}$$

$$1.46°C = m K_f = m \times \frac{1.86°C \text{ kg H}_2\text{O}}{\text{mol solute}}$$

$$m = \frac{1.46°\cancel{C} \times \text{mol solute}}{1.86°\cancel{C} \times \text{kg H}_2\text{O}} = \frac{0.785 \text{ mol solute}}{\text{kg H}_2\text{O}}$$

Now convert the data, 4.71 g solute/100.0 g H₂O, to g/mol:

$$\left(\frac{4.71 \text{ g solute}}{100.0 \text{ g H}_2\text{O}}\right)\left(\frac{1000 \text{ g H}_2\text{O}}{1 \text{ kg H}_2\text{O}}\right)\left(\frac{1 \text{ kg H}_2\text{O}}{0.785 \text{ mol solute}}\right) = 60.0 \text{ g/mol}$$

The molar mass of the compound is 60.0 g/mol.

PRACTICE 14.12

What is the freezing point of the solution in Practice Exercise 14.11? What is the boiling point?

>CHEMISTRY *IN ACTION*

The Scoop on Ice Cream

Ice cream is mainly composed of water (from milk and cream), milk solids, milk fats, and, frequently, various sweeteners (corn syrup or sugar), flavorings, emulsifiers, and stabilizers. But that smooth, creamy, rich flavor and texture are the result of the chemistry of the mixing and freezing process. The rich, smooth texture of great ice cream results from the milk fat. By law, if the carton is labeled "ice cream," it must contain a minimum of 10% milk fat. That carton of ice cream also contains 20%–50% air whipped into the ingredients during the initial mixing process.

H. Douglas Goff, a professor of food science and an ice cream expert from Ontario, Canada, says, "There are no real chemical reactions that take place when you make ice cream, but that doesn't mean that there isn't plenty of chemistry." The structure of ice cream contributes greatly to its taste. Tiny air bubbles are formed in the initial whipping process. These bubbles are distributed through a network of fat globules and liquid water. The milk fat has surface proteins on the globules in the milk to keep the fat dissolved in solution. Ice cream manufacturers destabilize these globules using emulsifiers (such as egg yolks, mono- or diglycerides) and let them come together in larger networks sort of like grape clusters.

Once the mixture is fully whipped, it is cooled to begin the freezing process. But ice cream does not freeze at 0°C even though it is 55%–64% water. The freezing point is depressed as a colligative property of the ice cream solution. Once the freezing of the water begins, the concentration of the solution increases, which continues to depress the freezing point. Goff tells us that even at a typical serving temperature (−16°C for most ice cream), only about 72% of the water in the ice cream is frozen. The unfrozen solution keeps the ice cream "scoopable."

One last factor that affects the quality of ice cream is the size of the ice crystals. For very smooth ice cream, tiny crystals are needed. To produce these, the ice cream must freeze very slowly. Large crystals give a coarse, grainy texture. Now, as you savor that premium ice cream cone, you'll know just how colligative properties and the chemistry of freezing helped make it so delicious!

14.6 OSMOSIS AND OSMOTIC PRESSURE

Discuss osmosis and osmotic pressure and their importance in living systems.

● **LEARNING OBJECTIVE**

KEY TERMS

semipermeable membrane
osmosis
osmotic pressure

When red blood cells are put into distilled water, they gradually swell and in time may burst. If red blood cells are put in a 5% urea (or a 5% salt) solution, they gradually shrink and take on a wrinkled appearance. The cells behave in this fashion because they are enclosed in semipermeable membranes. A **semipermeable membrane** allows the passage of water (solvent) molecules through it in either direction but prevents the passage of larger solute molecules or ions. When two solutions of different concentrations (or water and a water solution) are separated by a semipermeable membrane, water diffuses through the membrane from the solution of lower concentration into the solution of higher concentration. The diffusion of water, either from a dilute solution or from pure water, through a semipermeable membrane into a solution of higher concentration is called **osmosis**.

A 0.90% (0.15 *M*) sodium chloride solution is known as a *physiological saline solution* because it is *isotonic* with blood plasma; that is, it has the same concentration of NaCl as blood plasma. Because each mole of NaCl yields about 2 mol of ions when in solution, the solute particle concentration in physiological saline solution is nearly 0.30 *M*. A 5% glucose solution (0.28 *M*) is also approximately isotonic with blood plasma. Blood cells neither swell nor shrink in an isotonic solution. The cells described in the preceding paragraph swell in water because water is *hypotonic* to cell plasma. The cells shrink in 5% urea solution because the urea solution is *hypertonic* to the cell plasma. To prevent possible injury to blood cells by osmosis, fluids for intravenous use are usually made up at approximately isotonic concentration.

All solutions exhibit *osmotic pressure*, which is another colligative property. Osmotic pressure is a pressure difference between the system and atmospheric pressure. The osmotic pressure of a system can be measured by applying enough pressure to stop the flow of water due to osmosis in the system. The difference between the applied pressure and atmospheric pressure is the **osmotic pressure**. When pressure greater than the osmotic pressure is applied to a system, the flow of water can be reversed from that of osmosis. This process can be used to obtain useful drinking water from seawater and is known as *reverse osmosis*. Osmotic pressure is dependent only on the concentration of the solute particles and is independent of

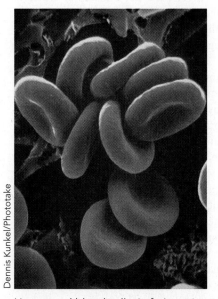

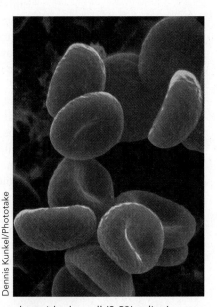

Dennis Kunkel/Phototake

Human red blood cells. Left: In an isotonic solution, the concentration is the same inside and outside the cell (0.9% saline). Center: In a hypertonic solution (1.6% saline), water leaves the cells, causing them to crenate (shrink). Right: In a hypotonic solution (0.2% saline), the cells swell as water moves into the cell center. Magnification is 260,000×.

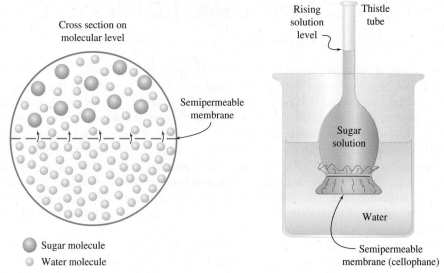

Figure 14.9

Laboratory demonstration of osmosis: As a result of osmosis, water passes through the membrane, causing the solution to rise in the thistle tube.

their nature. The osmotic pressure of a solution can be measured by determining the amount of counterpressure needed to prevent osmosis; this pressure can be very large. The osmotic pressure of a solution containing 1 mol of solute particles in 1 kg of water is about 22.4 atm, which is about the same as the pressure exerted by 1 mol of a gas confined in a volume of 1 L at 0°C.

Osmosis has a role in many biological processes, and semipermeable membranes occur commonly in living organisms. An example is the roots of plants, which are covered with tiny structures called root hairs; soil water enters the plant by osmosis, passing through the semipermeable membranes covering the root hairs. Artificial or synthetic membranes can also be made.

Osmosis can be demonstrated with the simple laboratory setup shown in **Figure 14.9**. As a result of osmotic pressure, water passes through the cellophane membrane into the thistle tube, causing the solution level to rise. In osmosis, the net transfer of water is always from a less concentrated to a more concentrated solution; that is, the effect is toward equalization of the concentration on both sides of the membrane. Note that the effective movement of water in osmosis is always from the region of *higher water concentration* to the region of *lower water concentration.*

Osmosis can be explained by assuming that a semipermeable membrane has passages that permit water molecules and other small molecules to pass in either direction. Both sides of the membrane are constantly being struck by water molecules in random motion. The number of water molecules crossing the membrane is proportional to the number of water molecule-to-membrane impacts per unit of time. Because the solute molecules or ions reduce the concentration of water, there are more water molecules and thus more water molecule impacts on the side with the lower solute concentration (more dilute solution). The greater number of water molecule-to-membrane impacts on the dilute side thus causes a net transfer of water to the more concentrated solution. Again, note that the overall process involves the net transfer, by diffusion through the membrane, of water molecules from a region of higher water concentration (dilute solution) to one of lower water concentration (more concentrated solution).

This is a simplified picture of osmosis. No one has ever seen the hypothetical passages that allow water molecules and other small molecules or ions to pass through them. Alternative explanations have been proposed, but our discussion has been confined to water solutions. Osmotic pressure is a general colligative property, however, and is known to occur in nonaqueous systems.

CHAPTER **14 REVIEW**

14.1 GENERAL PROPERTIES OF SOLUTIONS

- A solution is a homogeneous mixture of two or more substances:
  - Consists of solvent—the dissolving agent—and solute—the component(s) dissolved in the solvent
  - Is a homogeneous mixture
  - Contains molecular or ionic particles
  - Can be colored or colorless
  - Can be separated into solute and solvent by a physical separation process

KEY TERMS

solution
solute
solvent

14.2 SOLUBILITY

- Solubility describes the amount of solute that will dissolve in a specified amount of solvent.
- Solubility can also be qualitative.
- General guidelines for ionic solubility are:

KEY TERMS

solubility
miscible
immiscible
concentration of a solution
saturated solution
unsaturated solution
supersaturated solution

Soluble		Insoluble
Na^+, K^+, NH_4^+		
Nitrates, NO_3^- Acetates, $C_2H_3O_2^-$		
Chlorides, Cl^- Bromides, Br^- Iodides, I^-	except →	Ag^+, Hg_2^{2+}, Pb^{2+}
Sulfates, SO_4^{2-} Ag^+, Ca^{2+} are slightly soluble	except →	Ba^{2+}, Sr^{2+}, Pb^{2+}
NH_4^+ alkali metal cations	← except	Carbonates, CO_3^{2-} Phosphates, PO_4^{3-} Hydroxides, OH^- Sulfides, S^{2-}

- Liquids can also be classified as miscible (soluble in each other) or immiscible (not soluble in each other).
- The concentration of a solution is the quantitative measurement of the amount of solute that is dissolved in a solution.
- Like tends to dissolve like is a general rule for solvents and solutes.
- As temperature increases:
  - Solubility of a solid in a liquid tends to increase.
  - Solubility of a gas in a liquid tends to decrease.
- As pressure increases:
  - Solubility of a solid in a liquid remains constant.
  - Solubility of a gas in a liquid tends to increase.
- At a specific temperature, the amount of solute that can dissolve in a solvent has a limit:
  - Unsaturated solutions contain less solute than the limit.
  - Saturated solutions contain dissolved solute at the limit.
  - Supersaturated solutions contain more solute than the limit and are therefore unstable:
    - If disturbed, the excess solute will precipitate out of solution.

14.3 RATE OF DISSOLVING SOLIDS

- The rate at which a solute dissolves is determined by these factors:
  - Particle size
  - Temperature
  - Concentration of solution
  - Agitation

14.4 CONCENTRATION OF SOLUTIONS

KEY TERMS

dilute solution
concentrated solution
parts per million (ppm)
molarity (M)

- Molecules or ions must collide in order to react.
- Solutions provide a medium for the molecules or ions to collide.
- Concentrations can be measured in many ways:

Mass percent	% m/m	$\dfrac{\text{mass solute}}{\text{mass solution}} \times 100$
Parts per million	ppm	$\dfrac{\text{mass solute}}{\text{mass solution}} \times 1{,}000{,}000$
Mass/volume percent	% m/v	$\dfrac{\text{mass solute}}{\text{mL solution}} \times 100$
Volume percent	% v/v	$\dfrac{\text{mL solute}}{\text{mL solution}} \times 100$
Molarity	M	$\dfrac{\text{moles solute}}{\text{L solution}}$
Molality	m	$\dfrac{\text{moles solute}}{\text{kg solvent}}$

- Dilution of solutions requires the addition of more solvent to an existing solution:
  - The number of moles in the diluted solution is the same as that in the original solution.
  - $M_1V_1 = M_2V_2$.

14.5 COLLIGATIVE PROPERTIES OF SOLUTIONS

KEY TERMS

colligative properties
molality (m)

- Properties of a solution that depend only on the number of solute particles in solution are called colligative properties:
  - Freezing point depression

    $\Delta t_f = mK_f$

  - Boiling point elevation

    $\Delta t_b = mK_b$

  - Osmotic pressure
- Molality is used in working with colligative properties.

14.6 OSMOSIS AND OSMOTIC PRESSURE

KEY TERMS

semipermeable membrane
osmosis
osmotic pressure

- Osmosis is the diffusion of water through a semipermeable membrane:
  - Occurs from dilute solution to a solution of higher concentration
- Osmosis results in osmotic pressure, which is a colligative property of a solution.

REVIEW QUESTIONS

1. What is a true solution?
2. Name and distinguish between the two components of a solution.
3. Is it always apparent in a solution which component is the solute, for example, in a solution of a liquid in a liquid?
4. Explain why the solute does not settle out of a solution.
5. Is it possible to have one solid dissolved in another? Explain.
6. An aqueous solution of KCl is colorless, $KMnO_4$ is purple, and $K_2Cr_2O_7$ is orange. What color would you expect of an aqueous solution of $Na_2Cr_2O_7$? Explain.

7. Why is air considered to be a solution?
8. Sketch the orientation of water molecules (a) about a single sodium ion and (b) about a single chloride ion in solution.
9. Refer to Table 14.2 and estimate the number of grams of potassium bromide that would dissolve in 50 g water at 0°C.
10. Refer to Figure 14.4 to determine the solubility of each of these substances at 25°C:
    (a) ammonium chloride
    (b) copper(II) sulfate
    (c) sodium nitrate

11. As you go down Group 1A from lithium to potassium, what is the trend in solubilities of the chlorides and bromides of these metals? (Table 14.2)

12. State the solubility, in grams of solute per 100 g of H_2O, of: (Figure 14.4)
    (a) NH_4Cl at 35°C (c) SO_2 gas at 30°C
    (b) $CuSO_4$ at 60°C (d) $NaNO_3$ at 15°C

13. Which solid substance in Figure 14.4 shows an overall decrease in solubility with an increase in temperature?

14. At a temperature of 45°C, will a solution made with 6.5 g KNO_3 in 15.0 g water be saturated or unsaturated? What will be the mass percent of solute in the solution? (Figure 14.4)

15. If 40. g Li_2SO_4 are added to 75.0 g water at 40°C, will the solution be saturated or unsaturated? What will be the mass percent of solute in the solution? (Figure 14.4)

16. Explain how a supersaturated solution of $NaC_2H_3O_2$ can be prepared and proven to be supersaturated.

17. Explain why hexane will dissolve benzene but will not dissolve sodium chloride.

18. Some drinks like tea are consumed either hot or cold, whereas others like Coca-Cola are drunk only cold. Why?

19. What is the effect of pressure on the solubility of gases in liquids? of solids in liquids?

20. In a saturated solution containing undissolved solute, solute is continuously dissolving, but the concentration of the solution remains unchanged. Explain.

21. Champagne is usually cooled in a refrigerator prior to opening. It's also opened very carefully. What would happen if a warm bottle of champagne were shaken and opened quickly and forcefully?

22. What would be the total surface area if the 1-cm cube in Figure 14.5 were cut into cubes 0.01 cm on a side?

23. When a solid solute is put into a solvent, does the rate of dissolving increase or decrease as the dissolution proceeds? Explain. (Figure 14.6)

24. In which will a teaspoonful of sugar dissolve more rapidly, 200 mL of iced tea or 200 mL of hot coffee? Explain in terms of the KMT.

25. Why do smaller particles dissolve faster than large ones?

26. Explain why there is no apparent reaction when crystals of $AgNO_3$ and NaCl are mixed, but a reaction is apparent immediately when solutions of $AgNO_3$ and NaCl are mixed.

27. What do we mean when we say that concentrated nitric acid (HNO_3) is 16 molar?

28. Will 1 L of 1 M NaCl contain more chloride ions than 0.5 L of 1 M $MgCl_2$? Explain.

29. Describe how you would prepare 750 mL of 5.0 M NaCl solution.

30. Arrange the following bases (in descending order) according to the volume of each that will react with 1 L of 1 M HCl:
    (a) 1 M NaOH
    (b) 1.5 M Ca(OH)$_2$
    (c) 2 M KOH
    (d) 0.6 M Ba(OH)$_2$

31. Explain in terms of vapor pressure why the boiling point of a solution containing a nonvolatile solute is higher than that of the pure solvent.

32. Explain why the freezing point of a solution is lower than the freezing point of the pure solvent.

33. When water and ice are mixed, the temperature of the mixture is 0°C. But if methyl alcohol and ice are mixed, a temperature of −10°C is readily attained. Explain why the two mixtures show such different temperature behavior.

34. Which would be more effective in lowering the freezing point of 500. g of water?
    (a) 100. g of sucrose, $C_{12}H_{22}O_{11}$, or 100. g of ethyl alcohol, C_2H_5OH
    (b) 100. g of sucrose or 20.0 g of ethyl alcohol
    (c) 20.0 g of ethyl alcohol or 20.0 g of methyl alcohol, CH_3OH.

35. What is the difference between molarity and molality?

36. Is the molarity of a 5 m aqueous solution of NaCl greater or less than 5 M? Explain.

37. Why do salt trucks distribute salt over icy roads in the winter?

38. Assume that the thistle tube in Figure 14.9 contains 1.0 M sugar solution and that the water in the beaker has just been replaced by a 2.0 M solution of urea. Would the solution level in the thistle tube continue to rise, remain constant, or fall? Explain.

39. Explain in terms of the KMT how a semipermeable membrane functions when placed between pure water and a 10% sugar solution.

40. Which has the higher osmotic pressure, a solution containing 100 g of urea (NH_2CONH_2) in 1 kg H_2O or a solution containing 150 g of glucose, $C_6H_{12}O_6$, in 1 kg H_2O?

41. Explain why a lettuce leaf in contact with salad dressing containing salt and vinegar soon becomes wilted and limp whereas another lettuce leaf in contact with plain water remains crisp.

42. A group of shipwreck survivors floated for several days on a life raft before being rescued. Those who had drunk some seawater were found to be suffering the most from dehydration. Explain.

Most of the exercises in this chapter are available for assignment via the online homework management program, WileyPLUS (www.wileyplus.com). All exercises with blue numbers have answers in Appendix VI.

PAIRED EXERCISES

1. Of the following substances, which ones are generally soluble in water? (See Figure 14.2 or Appendix V.)
    (a) AgCl   (d) NaOH
    (b) K_2SO_4 (e) PbI_2
    (c) Na_3PO_4 (f) $SnCO_3$

2. Of the following substances, which ones are generally soluble in water? (See Figure 14.2 or Appendix V.)
    (a) $Ba_3(PO_4)_2$ (d) $NH_4C_2H_3O_2$
    (b) $Cu(NO_3)_2$ (e) MgO
    (c) $Fe(OH)_3$ (f) $AgNO_3$

3. Calculate the mass percent of the following solutions:
    (a) 15.0 g KCl + 100.0 g H_2O
    (b) 2.50 g Na_3PO_4 + 10.0 g H_2O
    (c) 0.20 mol $NH_4C_2H_3O_2$ + 125 g H_2O
    (d) 1.50 mol NaOH in 33.0 mol H_2O

4. Calculate the mass percent of the following solutions:
    (a) 25.0 g $NaNO_3$ in 125.0 g H_2O
    (b) 1.25 g $CaCl_2$ in 35.0 g H_2O
    (c) 0.75 mol K_2CrO_4 in 225 g H_2O
    (d) 1.20 mol H_2SO_4 in 72.5 mol H_2O

5. A bleaching solution requires 5.23 g of sodium hypochlorite. How many grams of a 21.5% by mass solution of sodium hypochorite should be used?

6. A paint requires 42.8 g of iron(III) oxide to give it the right yellow tint. How many grams of a 30.0% by mass solution of iron(III) oxide should be used?

7. In 25 g of a 7.5% by mass solution of $CaSO_4$
   (a) how many grams of solute are present?
   (b) how many grams of solvent are present?

8. In 75 g of a 12.0% by mass solution of $BaCl_2$
   (a) how many grams of solute are present?
   (b) how many grams of solvent are present?

9. Determine the mass/volume percent of a solution made by dissolving:
   (a) 15.0 g of C_2H_5OH (ethanol) in water to make 150.0 mL of solution
   (b) 25.2 g of NaCl in water to make 125.5 mL of solution

10. Determine the mass/volume percent of a solution made by dissolving:
   (a) 175.2 g of table sugar, $C_{12}H_{22}O_{11}$, in water to make 275.5 mL of solution
   (b) 35.5 g of CH_3OH (methanol) in water to make 75.0 mL of solution

11. Determine the volume percent of a solution made by dissolving:
   (a) 50.0 mL of hexanol in enough ethanol to make 125 mL of solution
   (b) 2.0 mL of ethanol in enough methanol to make 15.0 mL of solution
   (c) 15.0 mL of acetone in enough hexane to make 325 mL of solution

12. Determine the volume percent of a solution made by dissolving:
   (a) 37.5 mL of butanol in enough ethanol to make 275 mL of solution
   (b) 4.0 mL of methanol in enough water to make 25.0 mL of solution
   (c) 45.0 mL of isoamyl alcohol in enough acetone to make 750. mL of solution

13. Calculate the molarity of the following solutions:
   (a) 0.25 mol of solute in 75.0 mL of solution
   (b) 1.75 mol of KBr in 0.75 L of solution
   (c) 35.0 g of $NaC_2H_3O_2$ in 1.25 L of solution
   (d) 75 g of $CuSO_4 \cdot 5\,H_2O$ in 1.0 L of solution

14. Calculate the molarity of the following solutions:
   (a) 0.50 mol of solute in 125 mL of solution
   (b) 2.25 mol of $CaCl_2$ in 1.50 L of solution
   (c) 275 g $C_6H_{12}O_6$ in 775 mL of solution
   (d) 125 g $MgSO_4 \cdot 7\,H_2O$ in 2.50 L of solution

15. Calculate the number of moles of solute in each of the following solutions:
   (a) 1.5 L of 1.20 M H_2SO_4
   (b) 25.0 mL of 0.0015 M $BaCl_2$
   (c) 125 mL of 0.35 M K_3PO_4

16. Calculate the number of moles of solute in each of the following solutions:
   (a) 0.75 L of 1.50 M HNO_3
   (b) 10.0 mL of 0.75 M $NaClO_3$
   (c) 175 mL of 0.50 M LiBr

17. Calculate the grams of solute in each of the following solutions:
   (a) 2.5 L of 0.75 M K_2CrO_4
   (b) 75.2 mL of 0.050 M $HC_2H_3O_2$
   (c) 250 mL of 16 M HNO_3

18. Calculate the grams of solute in each of the following solutions:
   (a) 1.20 L of 18 M H_2SO_4
   (b) 27.5 mL of 1.50 M $KMnO_4$
   (c) 120 mL of 0.025 M $Fe_2(SO_4)_3$

19. How many milliliters of 0.750 M H_3PO_4 will contain the following?
   (a) 0.15 mol H_3PO_4
   (b) 35.5 g H_3PO_4
   (c) 7.34×10^{22} molecules of H_3PO_4

20. How many milliliters of 0.250 M NH_4Cl will contain the following?
   (a) 0.85 mol NH_4Cl
   (b) 25.2 g NH_4Cl
   (c) 2.06×10^{20} formula units of NH_4Cl

21. What will be the molarity of the resulting solutions made by mixing the following? Assume that volumes are additive.
   (a) 125 mL of 5.0 M H_3PO_4 with 775 mL of H_2O
   (b) 250 mL of 0.25 M Na_2SO_4 with 750 mL of H_2O
   (c) 75 mL of 0.50 M HNO_3 with 75 mL of 1.5 M HNO_3

22. What will be the molarity of the resulting solutions made by mixing the following? Assume that volumes are additive.
   (a) 175 mL of 3.0 M H_2SO_4 with 275 mL of H_2O
   (b) 350 mL of 0.10 M $CuSO_4$ with 150 mL of H_2O
   (c) 50.0 mL of 0.250 M HCl with 25.0 mL of 0.500 M HCl

23. Calculate the volume of concentrated reagent required to prepare the diluted solutions indicated:
   (a) 15 M H_3PO_4 to prepare 750 mL of 3.0 M H_3PO_4
   (b) 16 M HNO_3 to prepare 250 mL of 0.50 M HNO_3

24. Calculate the volume of concentrated reagent required to prepare the diluted solutions indicated:
   (a) 18 M H_2SO_4 to prepare 225 mL of 2.0 M H_2SO_4
   (b) 15 M NH_3 to prepare 75 mL of 1.0 M NH_3

25. Calculate the molarity of the solutions made by mixing 125 mL of 6.0 M $HC_2H_3O_2$ with the following:
   (a) 525 mL of H_2O
   (b) 175 mL of 1.5 M $HC_2H_3O_2$

26. Calculate the molarity of the solutions made by mixing 175 mL of 3.0 M HCl with the following:
   (a) 250 mL of H_2O
   (b) 115 mL of 6.0 M HCl

27. Use the equation to calculate the following:

$$3\,Ca(NO_3)_2(aq) + 2\,Na_3PO_4(aq) \longrightarrow$$
$$Ca_3(PO_4)_2(s) + 6\,NaNO_3(aq)$$

   (a) the moles of $Ca_3(PO_4)_2$ produced from 2.7 mol Na_3PO_4
   (b) the moles of $NaNO_3$ produced from 0.75 mol $Ca(NO_3)_2$
   (c) the moles of Na_3PO_4 required to react with 1.45 L of 0.225 M $Ca(NO_3)_2$
   (d) the grams of $Ca_3(PO_4)_2$ that can be obtained from 125 mL of 0.500 M $Ca(NO_3)_2$
   (e) the volume of 0.25 M Na_3PO_4 needed to react with 15.0 mL of 0.50 M $Ca(NO_3)_2$
   (f) the molarity (M) of the $Ca(NO_3)_2$ solution when 50.0 mL react with 50.0 mL of 2.0 M Na_3PO_4

28. Use the equation to calculate the following:

$$2\,NaOH(aq) + H_2SO_4(aq) \longrightarrow Na_2SO_4(aq) + 2\,H_2O(l)$$

   (a) the moles of Na_2SO_4 produced from 3.6 mol H_2SO_4
   (b) the moles of H_2O produced from 0.025 mol NaOH
   (c) the moles of NaOH required to react with 2.50 L of 0.125 M H_2SO_4
   (d) the grams of Na_2SO_4 that can be obtained from 25 mL of 0.050 M NaOH
   (e) the volume of 0.250 M H_2SO_4 needed to react with 25.5 mL of 0.750 M NaOH
   (f) the molarity (M) of the NaOH solution when 48.20 mL react with 35.72 mL of 0.125 M H_2SO_4

29. Use the equation to calculate the following:

$$2\ KMnO_4(aq) + 16\ HCl(aq) \longrightarrow$$
$$2\ MnCl_2(aq) + 5\ Cl_2(g) + 8\ H_2O(l) + 2\ KCl(aq)$$

(a) the moles of H_2O that can be obtained from 15.0 mL of 0.250 M HCl

(b) the volume of 0.150 M $KMnO_4$ needed to produce 1.85 mol $MnCl_2$

(c) the volume of 2.50 M HCl needed to produce 125 mL of 0.525 M KCl

(d) the molarity (M) of the HCl solution when 22.20 mL react with 15.60 mL of 0.250 M $KMnO_4$

(e) the liters of Cl_2 gas at STP produced by the reaction of 125 mL of 2.5 M HCl

(f) the liters of Cl_2 gas at STP produced by the reaction of 15.0 mL of 0.750 M HCl and 12.0 mL of 0.550 M $KMnO_4$

30. Use the equation to calculate the following:

$$K_2CO_3(aq) + 2\ HC_2H_3O_2(aq) \longrightarrow$$
$$2\ KC_2H_3O_2(aq) + H_2O(l) + CO_2(g)$$

(a) the moles of H_2O that can be obtained from 25.0 mL of 0.150 M $HC_2H_3O_2$

(b) the volume of 0.210 M K_2CO_3 needed to produce 17.5 mol $KC_2H_3O_2$

(c) the volume of 1.25 M $HC_2H_3O_2$ needed to react with 75.2 mL 0.750 M K_2CO_3

(d) the molarity (M) of the $HC_2H_3O_2$ solution when 10.15 mL react with 18.50 mL of 0.250 M K_2CO_3

(e) the liters of CO_2 gas at STP produced by the reaction of 105 mL of 1.5 M $HC_2H_3O_2$

(f) the liters of CO_2 gas at STP produced by the reaction of 25.0 mL of 0.350 M K_2CO_3 and 25.0 mL of 0.250 M $HC_2H_3O_2$

31. Calculate the molality of each of the following solutions:
(a) 2.0 mol HCl in 175 g water
(b) 14.5 g $C_{12}H_{22}O_{11}$ in 550.0 g water
(c) 25.2 mL methanol, CH_3OH ($d = 0.791$ g/mL) in 595 g ethanol, CH_3CH_2OH

32. Calculate the molality of each of the following solutions:
(a) 1.25 mol $CaCl_2$ in 750.0 g water
(b) 2.5 g $C_6H_{12}O_6$ in 525 g water
(c) 17.5 mL 2-propanol, $(CH_3)_2CHOH$ ($d = 0.785$ g/mL) in 35.5 mL H_2O ($d = 1.00$ g/mL)

33. What is the (a) molality, (b) freezing point, and (c) boiling point of a solution containing 2.68 g of naphthalene ($C_{10}H_8$) in 38.4 g of benzene (C_6H_6)?

34. What is the (a) molality, (b) freezing point, and (c) boiling point of a solution containing 100.0 g of ethylene glycol ($C_2H_6O_2$) in 150.0 g of water?

35. The freezing point of a solution of 8.00 g of an unknown compound dissolved in 60.0 g of acetic acid is 13.2°C. Calculate the molar mass of the compound.

36. What is the molar mass of a compound if 4.80 g of the compound dissolved in 22.0 g of H_2O given a solution that freezes at -2.50°C?

37. Identify which of the following substances are examples of true solutions.
(a) jasmine tea
(b) chromium metal
(c) muddy water
(d) gasoline

38. Identify which of the following substances are examples of true solutions.
(a) red paint
(b) Concord grape juice
(c) oil and vinegar salad dressing
(d) stainless steel

39. For those substances in Exercise 37 that are not true solutions, explain why.

40. For those substances in Exercise 38 that are not true solutions, explain why.

41. For each pairing below, predict which will dissolve faster and explain why.
(a) a teaspoon of sugar or a large crystal of sugar dissolving in hot coffee
(b) 15.0 grams of copper(II) sulfate in 100 mL of water or in 100 mL of a 15% copper(II) sulfate solution
(c) a packet of artificial sweetener dissolving in a cup of iced tea or a cup of hot tea
(d) 20 grams of silver nitrate in 1.5 L of water sitting on a table or in 1.5 L of water sloshing around in a speeding car

42. For each pairing below, predict which will dissolve faster and explain why.
(a) 0.424 g of the amino acid phenylalanine in 50.0 g of isopropyl alcohol at 25°C or in 50.0 g of isopropyl alcohol at 75°C
(b) a 1.42-g crystal of sodium acetate in 300 g of 37°C water or 1.42 g of powdered sodium acetate in 300 g of 37°C water
(c) a 500-g carton of table salt (NaCl) in 5.0 L of water at 35°C or a 500-g salt lick (NaCl) for livestock in 5.0 L of water at 21°C
(d) 25 mg of acetaminophen in 100 mL of infant pain medication containing 160 mg of acetaminophen per 15 mL or in 100 mL of adult pain medication containing 320 mg of acetaminophen per 15 mL

43. Law enforcement uses a quick and easy test for the presence of the illicit drug PCP, reacting it with potassium iodide. The PCP will form a crystalline solid with a long branching needle-like structure with KI. What is the molarity of KI in a stock solution prepared by dissolving 396.1 g of KI to a total volume of 750.0 mL?

44. Drug enforcement agents have a variety of chemical tests they can use to identify crystalline substances they may find at a crime scene. Heroin can be detected by treating a sample with a solution of mercury(II) chloride. The heroin will form a rosette of needle-shaped crystals. What is the molarity of $HgCl_2$ in a stock solution prepared by dissolving 74.15 g of $HgCl_2$ to a final volume of 250.0 mL?

ADDITIONAL EXERCISES

45. What happens to salt (NaCl) crystals when they are dissolved in water?

46. What happens to sugar molecules ($C_{12}H_{22}O_{11}$) when they are dissolved in water?

47. Why do sugar and salt behave differently when dissolved in water?

48. Why don't blood cells shrink or swell in an isotonic sodium chloride solution (0.9% saline)?

49. In the picture of dissolving $KMnO_4$ found in Section 14.1, the compound is forming purple streaks as it dissolves. Why?

50. In Figure 14.4, observe the line for KNO_3. Explain why it slopes up from left to right. How does the slope compare to the slopes of the other substances? What does this mean?

51. An IV bag contains 9.0 g of sodium chloride per liter of solution. What is the molarity of sodium chloride in this solution?

52. How many grams of solution, 10.0% NaOH by mass, are required to neutralize 250.0 g of a 1.0 m solution of HCl?

53. A sugar syrup solution contains 15.0% sugar, $C_{12}H_{22}O_{11}$, by mass and has a density of 1.06 g/mL.
   (a) How many grams of sugar are in 1.0 L of this syrup?
   (b) What is the molarity of this solution?
   (c) What is the molality of this solution?

54. A solution of 3.84 g C_4H_2N (empirical formula) in 250.0 g of benzene depresses the freezing point of benzene 0.614°C. What is the molecular formula for the compound?

55. Hydrochloric acid (HCl) is sold as a concentrated aqueous solution (12.0 mol/L). If the density of the solution is 1.18 g/mL, determine the molality of the solution.

56. How many grams of KNO_3 are needed to make 450 mL of a solution that is to contain 5.5 mg/mL of potassium ion? Calculate the molarity of the solution.

57. Witch hazel solution, an astringent for skin, contains 14% ethyl alcohol, C_2H_5OH, by volume. How many mL of ethyl alcohol are contained in a 16 fluid ounce bottle of witch hazel?

58. Given a solution containing 16.10 g $C_2H_6O_2$ in 82.0 g H_2O that has a boiling point of 101.62°C, verify that the boiling point elevation constant K_f for water is 0.512°C kg H_2O/mole solute.

59. Physiological saline (NaCl) solutions used in intravenous injections have a concentration of 0.90% NaCl (mass/volume).
   (a) How many grams of NaCl are needed to prepare 500.0 mL of this solution?
   (b) How much water must evaporate from this solution to give a solution that is 9.0% NaCl (mass/volume)?

60. A solution is made from 50.0 g KNO_3 and 175 g H_2O. How many grams of water must evaporate to give a saturated solution of KNO_3 in water at 20°C? (See Figure 14.4.)

61. How many liters of a 0.25% (v/v) oil of wintergreen solution can you prepare if you have only 7.35 mL of oil of wintergreen on hand?

62. At 20°C, an aqueous solution of HNO_3 that is 35.0% HNO_3 by mass has a density of 1.21 g/mL.
   (a) How many grams of HNO_3 are present in 1.00 L of this solution?
   (b) What volume of this solution will contain 500. g HNO_3?

63. What is the molarity of a phosphoric acid solution if the solution is 85% by mass H_3PO_4 and has a density of 1.7 g/mL?

64. To what volume must a solution of 80.0 g H_2SO_4 in 500.0 mL of solution be diluted to give a 0.10 M solution?

65. How many pounds of glycerol, $C_3H_8O_3$, are present in 30.0 gallons of a 4.28 M solution?

66. (a) How many moles of hydrogen will be liberated from 200.0 mL of 3.00 M HCl reacting with an excess of magnesium? The equation is

$$Mg(s) + 2\,HCl(aq) \longrightarrow MgCl_2(aq) + H_2(g)$$

   (b) How many liters of hydrogen gas (H_2) measured at 27°C and 720 torr will be obtained?
   (*Hint*: Use the ideal gas law.)

67. Which will be more effective in neutralizing stomach acid, HCl: a tablet containing 1.20 g $Mg(OH)_2$ or a tablet containing 1.00 g $Al(OH)_3$? Show evidence for your answer.

68. Which would be more effective as an antifreeze in an automobile radiator? A solution containing
   (a) 10 kg of methyl alcohol (CH_3OH) or 10 kg of ethyl alcohol (C_2H_5OH)?
   (b) 10 m solution of methyl alcohol or 10 m solution of ethyl alcohol?

69. Automobile battery acid is 38% H_2SO_4 and has a density of 1.29 g/mL. Calculate the molality and the molarity of this solution.

70. What is the (a) molality and (b) boiling point of an aqueous sugar, $C_{12}H_{22}O_{11}$, solution that freezes at −5.4°C?

71. A solution of 6.20 g $C_2H_6O_2$ in water has a freezing point of −0.372°C. How many grams of H_2O are in the solution?

72. What (a) mass and (b) volume of ethylene glycol ($C_2H_6O_2$, density = 1.11 g/mL) should be added to 12.0 L of water in an automobile radiator to protect it from freezing at −20°C? (c) To what temperature Fahrenheit will the radiator be protected?

73. If 150 mL of 0.055 M HNO_3 are needed to completely neutralize 1.48 g of an *impure* sample of sodium hydrogen carbonate (baking soda), what percent of the sample is baking soda?

74. (a) How much water must be added to concentrated sulfuric acid (H_2SO_4) (17.8 M) to prepare 8.4 L of 1.5 M sulfuric acid solution?
   (b) How many moles of H_2SO_4 are in each milliliter of the original concentrate?
   (c) How many moles are in each milliliter of the diluted solution?

75. How would you prepare a 6.00 M HNO_3 solution if only 3.00 M and 12.0 M solutions of the acid are available for mixing?

76. A 20.0-mL portion of an HBr solution of unknown strength is diluted to exactly 240 mL. If 100.0 mL of this diluted solution requires 88.4 mL of 0.37 M NaOH to achieve complete neutralization, what was the strength of the original HBr solution?

77. When 80.5 mL of 0.642 M $Ba(NO_3)_2$ are mixed with 44.5 mL of 0.743 M KOH, a precipitate of $Ba(OH)_2$ forms. How many grams of $Ba(OH)_2$ do you expect?

78. A 0.25 M solution of lithium carbonate (Li_2CO_3), a drug used to treat manic depression, is prepared.
   (a) How many moles of Li_2CO_3 are present in 45.8 mL of the solution?
   (b) How many grams of Li_2CO_3 are in 750 mL of the same solution?
   (c) How many milliliters of the solution would be needed to supply 6.0 g of the solute?
   (d) If the solution has a density of 1.22 g/mL, what is its mass percent?

79. Forensic chemists use Super Glue® to help them see fingerprints left behind at crime scenes. If cyanoacrylate, the main component of Super Glue®, is vaporized, it will be attracted to the oils left behind after someone touches an object. Forensic investigators will then rinse the print with a solution of europium chloride hexahydrate, $EuCl_3 \cdot 6\,H_2O$, which fluoresces when exposed to ultraviolet light. If 0.625 g of europium chloride hexahydrate is dissolved in enough water to make 500.0 mL of solution, what is the molarity of the resulting solution?

80. The pictures below represent solutions of sodium phosphate, potassium chloride, and sucrose dissolved in water. Identify each representation below:

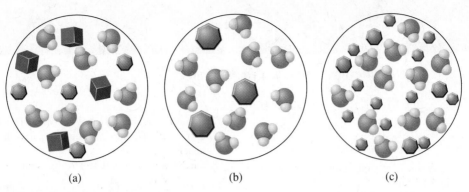

(a)   (b)   (c)

 Represents H_2O molecules

81. Dark streaks have been observed on the surface of Mars. These streaks sometimes seem to run or flow, and other times they appear static. It has been hypothesized that these streaks are salt (sodium chloride) water rivers that are sometimes melted, allowing them to flow, and at other times these rivers are frozen. If the salt water mixtures melt when the temperature rises to $-12.7°C$, what is the molality of salt in the rivers? (*Hint:* The freezing point is determined by the molality of ions. How will the molality of sodium chloride be related to the molality of the ions?)

82. Snails are very sensitive to the concentration of salt in their environment. Because their "skin" is a semipermeable membrane, water can easily flow through this membrane. If you spill salt onto the "skin" of a snail, water from the inside of the body will flow out very quickly. Explain this phenomenon in terms of the chemistry described in this chapter.

83. Eggplant is often considered to be very bitter. One of the secrets chefs use to remove the bitterness is to coat the surface of the eggplant with salt. As the eggplant sits, it becomes wet and when the wetness is rinsed or wiped away the bitterness is gone. Explain what is happening to improve the flavor of the eggplant.

84. To make taffy a sugar water solution is heated until the boiling point is $127°C$. What is the concentration (m) of sugar in taffy?

CHALLENGE EXERCISES

85. When solutions of hydrochloric acid and sodium sulfite react, a salt, water, and sulfur dioxide gas are produced. How many liters of sulfur dioxide gas at 775 torr and 22°C can be produced when 125 mL of 2.50 M hydrochloric acid react with 75.0 mL of 1.75 M sodium sulfite?

86. Consider a saturated solution at 20°C made from 5.549 moles of water and an unknown solute. You determine the mass of the container containing the solution to be 563 g. The mass of the empty container is 375 g. Identify the solute.

ANSWERS TO PRACTICE EXERCISES

14.1 (a) solute, oxygen, CO_2 other gases; solvent, nitrogen (b) solute, salt; solvent, water (c) solute CO_2; solvent, water

14.2 unsaturated

14.3 The rate of dissolving increases as the temperature is increased because the molecules of solvent are moving faster and strike the surfaces of the solute more frequently.

14.4 10.0% Na_2SO_4 solution

14.5 2.02 M

14.6 0.27 g NaCl

14.7 0.84 L (840 mL)

14.8 75 mL NaOH solution

14.9 $5.00 \times 10^{-2} M$

14.10 43 g

14.11 1.387 m

14.12 freezing point $= -2.58°C$, boiling point $= 100.71°C$

Answers for Putting It Together Reviews are found in Appendix VII.

Multiple Choice

Choose the correct answer to each of the following.

1. Which of these statements is *not* one of the principal assumptions of the kinetic-molecular theory for an ideal gas?
   (a) All collisions of gaseous molecules are perfectly elastic.
   (b) A mole of any gas occupies 22.4 L at STP.
   (c) Gas molecules have no attraction for one another.
   (d) The average kinetic energy for molecules is the same for all gases at the same temperature.

2. Which of the following is not equal to 1.00 atm?
   (a) 760. cm Hg   (c) 760. mm Hg
   (b) 29.9 in. Hg   (d) 760. torr

3. If the pressure on 45 mL of gas is changed from 600. torr to 800. torr, the new volume will be
   (a) 60 mL   (b) 34 mL   (c) 0.045 L   (d) 22.4 L

4. The volume of a gas is 300. mL at 740. torr and 25°C. If the pressure remains constant and the temperature is raised to 100.°C, the new volume will be
   (a) 240. mL   (b) 1.20 L   (c) 376 mL   (d) 75.0 mL

5. The volume of a dry gas is 4.00 L at 15.0°C and 745 torr. What volume will the gas occupy at 40.0°C and 700. torr?
   (a) 4.63 L   (b) 3.46 L   (c) 3.92 L   (d) 4.08 L

6. A sample of Cl_2 occupies 8.50 L at 80.0°C and 740. mm Hg. What volume will the Cl_2 occupy at STP?
   (a) 10.7 L   (b) 6.75 L   (c) 11.3 L   (d) 6.40 L

7. What volume will 8.00 g O_2 occupy at 45°C and 2.00 atm?
   (a) 0.462 L   (b) 104 L   (c) 9.62 L   (d) 3.26 L

8. The density of NH_3 gas at STP is
   (a) 0.760 g/mL   (c) 1.32 g/mL
   (b) 0.760 g/L   (d) 1.32 g/L

9. Which temperature is equivalent to 22°C?
   (a) 295 K   (b) 251 K   (c) 191 K   (d) 234°F

10. Measured at 65°C and 500. torr, the mass of 3.21 L of a gas is 3.5 g. The molar mass of this gas is
    (a) 21 g/mole   (c) 24 g/mole
    (b) 46 g/mole   (d) 130 g/mole

11. Box A contains O_2 (molar mass = 32.0) at a pressure of 200 torr. Box B, which is identical to box A in volume, contains twice as many molecules of CH_4 (molar mass = 16.0) as the molecules of O_2 in box A. The temperatures of the gases are identical. The pressure in box B is
    (a) 100 torr   (c) 400 torr
    (b) 200 torr   (d) 800 torr

12. A 300.-mL sample of oxygen (O_2) is collected over water at 23°C and 725 torr. If the vapor pressure of water at 23°C is 21.0 torr, the volume of dry O_2 at STP is
    (a) 256 mL   (b) 351 mL   (c) 341 mL   (d) 264 mL

13. A tank containing 0.01 mol of neon and 0.04 mol of helium shows a pressure of 1 atm. What is the partial pressure of neon in the tank?
    (a) 0.8 atm   (b) 0.01 atm   (c) 0.2 atm   (d) 0.5 atm

14. How many liters of NO_2 (at STP) can be produced from 25.0 g Cu reacting with concentrated nitric acid?

    $Cu(s) + 4\,HNO_3(aq) \longrightarrow Cu(NO_3)_2(aq) + 2\,H_2O(l) + 2\,NO_2(g)$

    (a) 4.41 L   (b) 8.82 L   (c) 17.6 L   (d) 44.8 L

15. How many liters of butane vapor are required to produce 2.0 L CO_2 at STP?

    $$2\,C_4H_{10}(g) + 13\,O_2(g) \longrightarrow 8\,CO_2(g) + 10\,H_2O(g)$$
    butane

    (a) 2.0 L   (b) 4.0 L   (c) 0.80 L   (d) 0.50 L

16. What volume of CO_2 (at STP) can be produced when 15.0 g C_2H_6 and 50.0 g O_2 are reacted?

    $$2\,C_2H_6(g) + 7\,O_2(g) \longrightarrow 4\,CO_2(g) + 6\,H_2O(g)$$

    (a) 20.0 L   (b) 22.4 L   (c) 35.0 L   (d) 5.6 L

17. Which of these gases has the highest density at STP?
    (a) N_2O (b) NO_2 (c) Cl_2 (d) SO_2

18. What is the density of CO_2 at 25°C and 0.954 atm?
    (a) 1.72 g/L   (c) 0.985 g/L
    (b) 2.04 g/L   (d) 1.52 g/L

19. How many molecules are present in 0.025 mol of H_2 gas?
    (a) 1.5×10^{22} molecules   (c) 2.40×10^{25} molecules
    (b) 3.37×10^{23} molecules   (d) 1.50×10^{22} molecules

20. 5.60 L of a gas at STP have a mass of 13.0 g. What is the molar mass of the gas?
    (a) 33.2 g/mol   (c) 66.4 g/mol
    (b) 26.0 g/mol   (d) 52.0 g/mol

21. The heat of fusion of ice at 0°C is
    (a) 4.184 J/g   (c) 2.26 kJ/g
    (b) 335 J/g   (d) 2.26 kJ/mol

22. The heat of vaporization of water is
    (a) 4.184 J/g   (c) 2.26 kJ/g
    (b) 335 J/g   (d) 2.26 kJ/mol

23. The specific heat of water is
    (a) 4.184 J/g°C   (c) 2.26 kJ/g°C
    (b) 335 J/g°C   (d) 18 J/g°C

24. The density of water at 4°C is
    (a) 1.0 g/mL   (c) 18.0 g/mL
    (b) 80 g/mL   (d) 14.7 lb/in.^{3}

25. SO_2 can be properly classified as a(n)
    (a) basic anhydride   (c) anhydrous salt
    (b) hydrate   (d) acid anhydride

26. When compared to H_2S, H_2Se, and H_2Te, water is found to have the highest boiling point because it
(a) has the lowest molar mass
(b) is the smallest molecule
(c) has the highest bonding
(d) forms hydrogen bonds better than the others

27. In which of the following molecules will hydrogen bonding be important?

 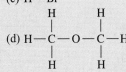

(a) H—F

(c) H—Br

(b) $\underset{\displaystyle H}{S}\!-\!H$

(d) $H\!-\!\underset{\displaystyle H}{\overset{\displaystyle H}{C}}\!-\!O\!-\!\underset{\displaystyle H}{\overset{\displaystyle H}{C}}\!-\!H$

28. Which of the following is an incorrect equation?
(a) $H_2SO_4 + 2\,NaOH \longrightarrow Na_2SO_4 + 2\,H_2O$
(b) $C_2H_6 + O_2 \longrightarrow 2\,CO_2 + 3\,H_2$
(c) $2\,H_2O \xrightarrow[H_2SO_4]{electrolysis} 2\,H_2 + O_2$
(d) $Ca + 2\,H_2O \longrightarrow H_2 + Ca(OH)_2$

29. Which of the following is not an example of an intermolecular force?
(a) London dispersion forces
(b) hydrogen bonds
(c) covalent bonds
(d) dipole–dipole attractions

30. How many kilojoules are required to change 85 g of water at 25°C to steam at 100.°C?
(a) 219 kJ   (b) 27 kJ   (c) 590 kJ   (d) 192 kJ

31. A chunk of 0°C ice, mass 145 g, is dropped into 75 g of water at 62°C. The heat of fusion of ice is 335 J/g. The result, after thermal equilibrium is attained, will be
(a) 87 g ice and 133 g liquid water, all at 0°C
(b) 58 g ice and 162 g liquid water, all at 0°C
(c) 220 g water at 7°C
(d) 220 g water at 17°C

32. The formula for iron(II) sulfate heptahydrate is
(a) $Fe_2SO_4 \cdot 7\,H_2O$ (c) $FeSO_4 \cdot 7\,H_2O$
(b) $Fe(SO_4)_2 \cdot 6\,H_2O$ (d) $Fe_2(SO_4)_3 \cdot 7\,H_2O$

33. The process by which a solid changes directly to a vapor is called
(a) vaporization   (c) sublimation
(b) evaporation   (d) condensation

34. Hydrogen bonding
(a) occurs only between water molecules
(b) is stronger than covalent bonding
(c) can occur between NH_3 and H_2O
(d) results from strong attractive forces in ionic compounds

35. A liquid boils when
(a) the vapor pressure of the liquid equals the external pressure above the liquid
(b) the heat of vaporization exceeds the vapor pressure
(c) the vapor pressure equals 1 atm
(d) the normal boiling temperature is reached

36. Consider two beakers, one containing 50 mL of liquid A and the other 50 mL of liquid B. The boiling point of A is 90°C and that of B is 72°C. Which of these statements is correct?
(a) A will evaporate faster than B.
(b) B will evaporate faster than A.
(c) Both A and B evaporate at the same rate.
(d) Insufficient data to answer the question.

37. 95.0 g of 0.0°C ice are added to exactly 100. g of water at 60.0°C. When the temperature of the mixture first reaches 0.0°C, the mass of ice still present is
(a) 0.0 g   (b) 20.0 g   (c) 10.0 g   (d) 75.0 g

38. Which of the following is not a general property of solutions?
(a) a homogeneous mixture of two or more substances
(b) variable composition
(c) dissolved solute breaks down to individual molecules
(d) the same chemical composition, the same chemical properties, and the same physical properties in every part

39. If NaCl is soluble in water to the extent of 36.0 g NaCl/100 g H_2O at 20°C, then a solution at 20°C containing 45 g NaCl/150 g H_2O would be
(a) dilute   (c) supersaturated
(b) saturated   (d) unsaturated

40. If 5.00 g NaCl are dissolved in 25.0 g of water, the percent of NaCl by mass is
(a) 16.7%   (c) 0.20%
(b) 20.0%   (d) no correct answer given

41. How many grams of 9.0% $AgNO_3$ solution will contain 5.3 g $AgNO_3$?
(a) 47.7 g   (c) 59 g
(b) 0.58 g   (d) no correct answer given

42. The molarity of a solution containing 2.5 mol of acetic acid ($HC_2H_3O_2$) in 400. mL of solution is
(a) 0.063 M (b) 1.0 M (c) 0.103 M (d) 6.3 M

43. What volume of 0.300 M KCl will contain 15.3 g KCl?
(a) 1.46 L   (b) 683 mL   (c) 61.5 mL   (d) 4.60 L

44. What mass of $BaCl_2$ will be required to prepare 200. mL of 0.150 M solution?
(a) 0.750 g   (b) 156 g   (c) 6.25 g   (d) 31.2 g

Problems 45–47 relate to the reaction

$$CaCO_3 + 2\,HCl \longrightarrow CaCl_2 + H_2O + CO_2$$

45. What volume of 6.0 M HCl will be needed to react with 0.350 mol of $CaCO_3$?
(a) 42.0 mL   (b) 1.17 L   (c) 117 mL   (d) 583 mL

46. If 400. mL of 2.0 M HCl react with excess $CaCO_3$, the volume of CO_2 produced, measured at STP, is
(a) 18 L   (b) 5.6 L   (c) 9.0 L   (d) 56 L

47. If 5.3 g $CaCl_2$ are produced in the reaction, what is the molarity of the HCl used if 25 mL of it reacted with excess $CaCO_3$?
(a) 3.8 M (b) 0.19 M (c) 0.38 M (d) 0.42 M

48. If 20.0 g of the nonelectrolyte urea ($CO(NH_2)_2$) is dissolved in 25.0 g of water, the freezing point of the solution will be
(a) −2.47°C   (b) −1.40°C   (c) −24.7°C   (d) −3.72°C

49. When 256 g of a nonvolatile, nonelectrolyte unknown were dissolved in 500. g H_2O, the freezing point was found to be −2.79°C. The molar mass of the unknown solute is
(a) 357   (b) 62.0   (c) 768   (d) 341

50. How many milliliters of 6.0 M H_2SO_4 must you use to prepare 500. mL of 0.20 M sulfuric acid solution?
(a) 30   (b) 17   (c) 12   (d) 100

51. How many milliliters of water must be added to 200. mL of 1.40 M HCl to make a solution that is 0.500 M HCl?
(a) 360. mL   (b) 560. mL   (c) 140. mL   (d) 280. mL

52. Which procedure is most likely to increase the solubility of most solids in liquids?
(a) stirring
(b) pulverizing the solid
(c) heating the solution
(d) increasing the pressure

53. The addition of a crystal of $NaClO_3$ to a solution of $NaClO_3$ causes additional crystals to precipitate. The original solution was
(a) unsaturated
(b) dilute
(c) saturated
(d) supersaturated

54. Which of these anions will not form a precipitate with silver ions, Ag^+?
(a) Cl^-
(b) NO_3^-
(c) Br^-
(d) CO_3^{2-}

55. Which of these salts are considered to be soluble in water?
(a) $BaSO_4$
(b) NH_4Cl
(c) AgI
(d) PbS

56. A solution of ethyl alcohol and benzene is 40% alcohol by volume. Which statement is correct?
(a) The solution contains 40 mL of alcohol in 100 mL of solution.
(b) The solution contains 60 mL of benzene in 100 mL of solution.
(c) The solution contains 40 mL of alcohol in 100 g of solution.
(d) The solution is made by dissolving 40 mL of alcohol in 60 mL of benzene.

57. Which of the following is not a colligative property?
(a) boiling point elevation
(b) freezing point depression
(c) osmotic pressure
(d) surface tension

58. When a solute is dissolved in a solvent
(a) the freezing point of the solution increases
(b) the vapor pressure of the solution increases
(c) the boiling point of the solution increases
(d) the concentration of the solvent increases

59. Which of the following solutions will have the lowest freezing point where X is any element or nonelectrolytic compound?
(a) 1.0 mol X in 1 kg H_2O
(b) 2.0 mol X in 1 kg H_2O
(c) 1.2 mol X in 1 kg H_2O
(d) 0.80 mol X in 1 kg H_2O

60. In the process of osmosis, water passes through a semipermeable membrane
(a) from a more concentrated solution to a dilute solution
(b) from a dilute solution to a more concentrated solution
(c) in order to remove a solute from a solution
(d) so that a sugar solution can become sweeter

Free Response Questions

Answer each of the following. Be sure to include your work and explanations in a clear, logical form.

1. Which solution should have a higher boiling point: 215 mL of a 10.0% (m/v) aqueous KCl solution or 224 mL of a 1.10 *M* aqueous NaCl solution?

2. A glass containing 345 mL of a soft drink (a carbonated beverage) was left sitting out on a kitchen counter. If the CO_2 released at room temperature (25°C) and pressure (1 atm) occupies 1.40 L, at a minimum, what is the concentration (in ppm) of the CO_2 in the original soft drink? (Assume the density of the original soft drink is 0.965 g/mL.)

3. Dina and Murphy were trying to react 100. mL of a 0.10 *M* HCl solution with KOH. The procedure called for a 10% KOH solution. Dina made a 10% mass/volume solution, while Murphy made a 10% by mass solution. (Assume there is no volume change upon dissolving KOH.) Which solution required less volume to fully react with 100. mL of the HCl solution?

4. A flask containing 825 mL of a solution containing 0.355 mol of $CuSO_4$ was left open overnight. The next morning the flask only contained 755 mL of solution.
(a) What is the concentration (molarity) of the $CuSO_4$ solution remaining in the flask?
(b) Which of the pathways shown below best represents the evaporation of water, and why are the others wrong?

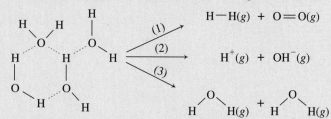

5. Three students at Jamston High—Zack, Gaye, and Lamont—each had the opportunity to travel over spring break. As part of a project, each of them measured the boiling point of pure water at their vacation spot. Zack found a boiling point of 93.9°C, Gaye measured 101.1°C, and Lamont read 100.°C. Which student most likely went snow skiing near Ely, Nevada, and which student most likely went water skiing in Honolulu? From the boiling point information, what can you surmise about the Dead Sea region, the location of the third student's vacation? Explain.

6. Why does a change in pressure of a gas significantly affect its volume, whereas a change in pressure on a solid or liquid has negligible effect on their respective volumes? If the accompanying picture represents a liquid at the molecular level, draw what you might expect a solid and a gas to look like.

7. (a) If you filled up three balloons with equal volumes of hydrogen, argon, and carbon dioxide gas, all at the same temperature and pressure, which balloon would weigh the most? the least? Explain.
(b) If you filled up three balloons with equal masses of nitrogen, oxygen, and neon, all to the same volume at the same temperature, which would have the lowest pressure?

8. Ray ran a double-displacement reaction using 0.050 mol $CuCl_2$ and 0.10 mol $AgNO_3$. The resulting white precipitate was removed by filtration. The filtrate was accidentally left open on the lab bench for over a week, and when Ray returned the flask contained solid, blue crystals. Ray weighed the crystals and found they had a mass of 14.775 g. Was Ray expecting this number? If not, what did he expect?

9. Why is it often advantageous or even necessary to run reactions in solution rather than mixing two solids? Would you expect reactions run in the gas phase to be more similar to solutions or to solids? Why?

10. A solution of 5.36 g of a molecular compound dissolved in 76.8 g benzene (C_6H_6) has a boiling point of 81.48°C. What is the molar mass of the compound?

Boiling point for benzene = 80.1°C

K_b for benzene = 2.53°C kg solvent/mol solute

Lemons and limes are examples of food that contains acidic solutions.

Jan Rihak/iStockphoto

CHAPTER **15**

ACIDS, BASES, AND SALTS

Acids are important chemicals. They are used in cooking to produce the surprise of tartness (from lemons) and to release CO_2 bubbles from leavening agents in baking. Vitamin C is an acid that is an essential nutrient in our diet. Our stomachs release acid to aid in digestion. Excess stomach acid can produce heartburn and indigestion. Bacteria in our mouths produce acids that can dissolve tooth enamel to form cavities. In our recreational activities we are concerned about acidity levels in swimming pools and spas. Acids are essential in the manufacture of detergents, plastics, and storage batteries. The acid–base properties of substances are found in all areas of our lives. In this chapter we consider the properties of acids, bases, and salts.

CHAPTER OUTLINE

15.1 Acids and Bases

15.2 Reactions of Acids and Bases

15.3 Salts

15.4 Electrolytes and Nonelectrolytes

15.5 Introduction to pH

15.6 Neutralization

15.7 Writing Net Ionic Equations

15.8 Acid Rain

15.1 ACIDS AND BASES

Compare the definitions of acids and bases, including Arrhenius, Brønsted–Lowry, and Lewis acids and bases.

KEY TERM
hydronium ion

A characteristic of acids is sour taste, such as when sucking a lemon.

The word *acid* is derived from the Latin *acidus*, meaning "sour" or "tart," and is also related to the Latin word *acetum*, meaning "vinegar." Vinegar has been around since antiquity as a product of the fermentation of wine and apple cider. The sour constituent of vinegar is acetic acid ($HC_2H_3O_2$). Characteristic properties commonly associated with acids include the following:

1. sour taste
2. the ability to change the color of litmus, a vegetable dye, from blue to red
3. the ability to react with

   • metals such as zinc and magnesium to produce hydrogen gas
   • hydroxide bases to produce water and an ionic compound (salt)
   • carbonates to produce carbon dioxide

These properties are due to the hydrogen ions, H^+, released by acids in a water solution.

Classically, a *base* is a substance capable of liberating hydroxide ions, OH^-, in water solution. Hydroxides of the alkali metals (Group 1A) and alkaline earth metals (Group 2A), such as LiOH, NaOH, KOH, $Mg(OH)_2$, $Ca(OH)_2$, and $Ba(OH)_2$, are the most common inorganic bases. Water solutions of bases are called *alkaline solutions* or *basic solutions.* Some of the characteristic properties commonly associated with bases include the following:

1. bitter or caustic taste
2. a slippery, soapy feeling
3. the ability to change litmus from red to blue
4. the ability to interact with acids

Acids change litmus paper from blue to pink.

Several theories have been proposed to answer the question "What is an acid and what is a base?" One of the earliest, most significant of these theories was advanced in 1884 by Svante Arrhenius (1859–1927), a Swedish scientist, who stated that "an acid is a hydrogen-containing substance that dissociates to produce hydrogen ions, and a base is a hydroxide-containing substance that dissociates to produce hydroxide ions in aqueous solutions." Arrhenius postulated that the hydrogen ions are produced by the dissociation of acids in water and that the hydroxide ions are produced by the dissociation of bases in water:

$$\underset{\text{acid}}{HA} \xrightarrow{H_2O} H^+(aq) + A^-(aq)$$

$$\underset{\text{base}}{MOH} \xrightarrow{H_2O} M^+(aq) + OH^-(aq)$$

An Arrhenius acid solution contains an excess of H^+ ions.
An Arrhenius base solution contains an excess of OH^- ions.

Bases change litmus paper from red to blue.

In 1923, the Brønsted–Lowry proton transfer theory was introduced by J. N. Brønsted (1897–1947), a Danish chemist, and T. M. Lowry (1847–1936), an English chemist. This theory states that an acid is a proton donor and a base is a proton acceptor.

A Brønsted–Lowry acid is a proton (H^+) donor.
A Brønsted–Lowry base is a proton (H^+) acceptor.

Consider the reaction of hydrogen chloride gas with water to form hydrochloric acid:

$$HCl(g) + H_2O(l) \longrightarrow H_3O^+(aq) + Cl^-(aq) \tag{1}$$

In the course of the reaction, HCl donates, or gives up, a proton to form a Cl^- ion, and H_2O accepts a proton to form the H_3O^+ ion. Thus, HCl is an acid and H_2O is a base, according to the Brønsted–Lowry theory.

A hydrogen ion (H^+) is nothing more than a bare proton and does not exist by itself in an aqueous solution. In water H^+ combines with a polar water molecule to form a hydrated hydrogen ion (H_3O^+) commonly called a **hydronium ion**. The H^+ is attracted to a polar water molecule, forming a bond with one of the two pairs of unshared electrons:

$$H^+ + H\overset{\cdot\cdot}{\underset{H}{\overset{\cdot\cdot}{O}}}: \longrightarrow \left[H\overset{\cdot\cdot}{\underset{H}{\overset{\cdot\cdot}{O}}}:H \right]^+$$

hydronium ion

Note the electron structure of the hydronium ion. For simplicity we often use H^+ instead of H_3O^+ in equations, with the explicit understanding that H^+ is always hydrated in solution.

When a Brønsted–Lowry acid donates a proton, as illustrated in the equation below, it forms the conjugate base of that acid. When a base accepts a proton, it forms the conjugate acid of that base. A conjugate acid and base are produced as products. The formulas of a conjugate acid–base pair differ by one proton (H^+). Consider what happens when HCl(g) is bubbled through water, as shown by this equation:

conjugate acid–base pair

$$\underset{\text{acid}}{HCl(g)} + \underset{\text{base}}{H_2O(l)} \longrightarrow \underset{\text{base}}{Cl^-(aq)} + \underset{\text{acid}}{H_3O^+(aq)}$$

conjugate acid–base pair

The conjugate acid–base pairs are HCl — Cl^- and H_3O^+ — H_2O. The conjugate base of HCl is Cl^-, and the conjugate acid of Cl^- is HCl. The conjugate base of H_3O^+ is H_2O, and the conjugate acid of H_2O is H_3O^+.

Another example of conjugate acid–base pairs can be seen in this equation:

$$\underset{\text{acid}}{NH_4^+} + \underset{\text{base}}{H_2O} \longrightarrow \underset{\text{acid}}{H_3O^+} + \underset{\text{base}}{NH_3}$$

Here the conjugate acid–base pairs are NH_4^+ — NH_3 and H_3O^+ — H_2O.

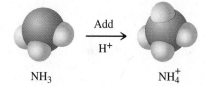

NH_3 Add H^+ NH_4^+

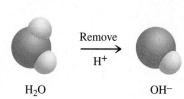

H_2O Remove H^+ OH^-

Remember: The difference between an acid or a base and its conjugate is one proton, H^+.

EXAMPLE 15.1

Write the formula for (a) the conjugate base of H_2O and of HNO_3, and (b) the conjugate acid of SO_4^{2-} and of $C_2H_3O_2^-$.

SOLUTION

(a) To write the conjugate base of an acid, remove one proton from the acid formula:

$$H_2O \xrightarrow{-H^+} OH^- \quad \text{(conjugate base)}$$

$$HNO_3 \xrightarrow{-H^+} NO_3^- \quad \text{(conjugate base)}$$

Note that, by removing an H^+, the conjugate base becomes more negative than the acid by one minus charge.

(b) To write the conjugate acid of a base, add one proton to the formula of the base:

$$SO_4^{2-} \xrightarrow{+H^+} HSO_4^- \quad \text{(conjugate acid)}$$

$$C_2H_3O_2^- \xrightarrow{+H^+} HC_2H_3O_2 \quad \text{(conjugate acid)}$$

In each case the conjugate acid becomes more positive than the base by a $+1$ charge due to the addition of H^+.

PRACTICE 15.1

Write formulas for the conjugate base for these acids:

(a) H_2CO_3 (b) HNO_2 (c) $HC_2H_3O_2$

PRACTICE 15.2

Write formulas for the conjugate acid for these bases:

(a) HSO_4^- (b) NH_3 (c) OH^-

A more general concept of acids and bases was introduced by Gilbert N. Lewis. The Lewis theory deals with the way in which a substance with an unshared pair of electrons reacts in an acid–base type of reaction. According to this theory, a base is any substance that has an unshared pair of electrons (electron pair donor), and an acid is any substance that will attach itself to or accept a pair of electrons.

A Lewis acid is an electron pair acceptor.
A Lewis base is an electron pair donor.

Consider the following reaction:

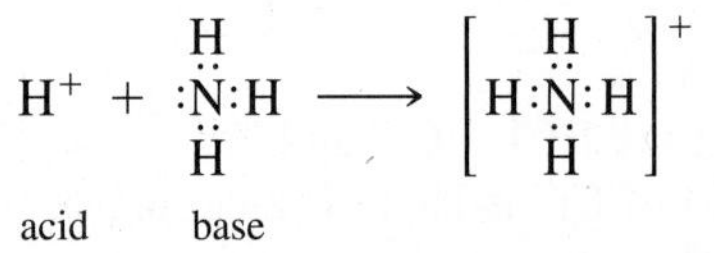

$$H^+ + \overset{\displaystyle H}{\underset{\displaystyle H}{:\!N\!:\!H}} \longrightarrow \left[\overset{\displaystyle H}{\underset{\displaystyle H}{H\!:\!N\!:\!H}} \right]^+$$

acid   base

The H^+ is a Lewis acid, and the $:NH_3$ is a Lewis base. According to the Lewis theory, substances other than proton donors (e.g., BF_3) behave as acids:

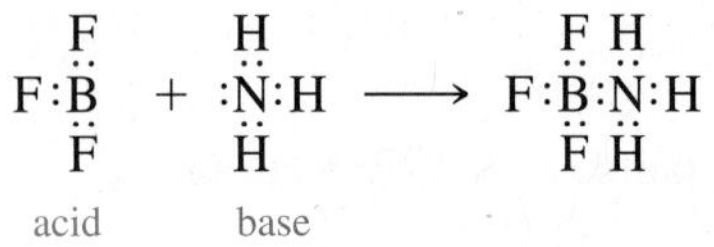

$$\overset{\displaystyle F}{\underset{\displaystyle F}{F\!:\!B}} + \overset{\displaystyle H}{\underset{\displaystyle H}{:\!N\!:\!H}} \longrightarrow \overset{\displaystyle F\ H}{\underset{\displaystyle F\ H}{F\!:\!B\!:\!N\!:\!H}}$$

acid   base

These three theories, which explain how acid–base reactions occur, are summarized in Table 15.1. We will generally use the theory that best explains the reaction under consideration. Most of our examples will refer to aqueous solutions. Note that in an aqueous acidic solution the H^+ ion concentration is always greater than the OH^- ion concentration. And vice versa—in an aqueous basic solution the OH^- ion concentration is always greater than the H^+ ion concentration. When the H^+ and OH^- ion concentrations in a solution are equal, the solution is *neutral*; that is, it is neither acidic nor basic.

TABLE 15.1 Summary of Acid–Base Definitions

Theory	Acid	Base
Arrhenius	A hydrogen-containing substance that produces hydrogen ions in aqueous solution	A hydroxide-containing substance that produces hydroxide ions in aqueous solution
Brønsted–Lowry	A proton (H^+) donor	A proton (H^+) acceptor
Lewis	Any species that will bond to an unshared pair of electrons (electron pair acceptor)	Any species that has an unshared pair of electrons (electron pair donor)

>CHEMISTRY *IN ACTION*

Drug Delivery: An Acid–Base Problem

Have you ever looked at the package insert for some of the injections you have received at the doctor's office? You may have been given lidocaine · HCl while receiving a filling at the dentist, proparacaine · HCl drops during an eye exam, or been pre-scribed pseudoephedrine · HCl for congestion. Why do all of these drugs have HCl in them? Are we taking acid? The answer to that question is a qualified yes. It is not like the acid we use in the laboratory, but it is indeed a salt according to the rules of chemistry. Many pharmaceuticals on the market are molecules with a structure very similar to ammonia. They are called amines, and they are synthesized by replacing one or more of the hydrogen atoms in the ammonia molecule with an organic or carbon-based group. Ammonia and several common amines are pictured below:

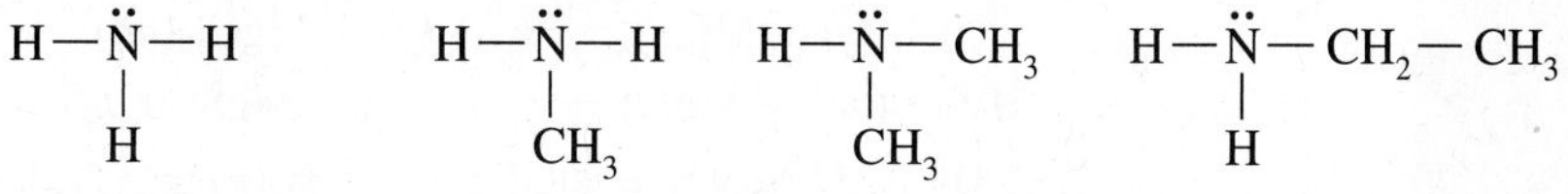

Ammonia   Amines

Most pharmaceuticals are dissolved in water to be administered to patients. For amine-based drugs this can be a problem because many of the pharmacologically active amines are not soluble in water. These amines are generally fairly volatile, meaning they evaporate easily and thus could be vaporized and inhaled, but this is an inconvenient mode of drug delivery. Pharmaceutical chemists discovered a better method of delivering these drugs by taking advantage of one of the chemical properties of ammonia and the amines. These molecules are all Lewis bases and will react with acids to form a salt as shown in the following reaction:

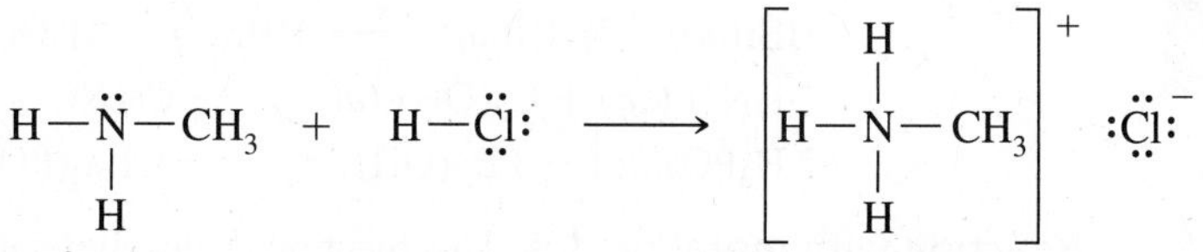

These salts are much more soluble in water than the original amine, which makes it much easier to administer the medi-cation. It can be dissolved in water to be injected into the bloodstream, be dropped into the aqueous environment of the eye or other mucous membranes, or dissolved into an aqueous medium to be consumed orally. So, yes, the hydrochlo-ride salts of many amine-based drugs are very useful. Check the package insert next time you receive an injection or are given a prescription.

Some common pharmacological drugs are shown here:

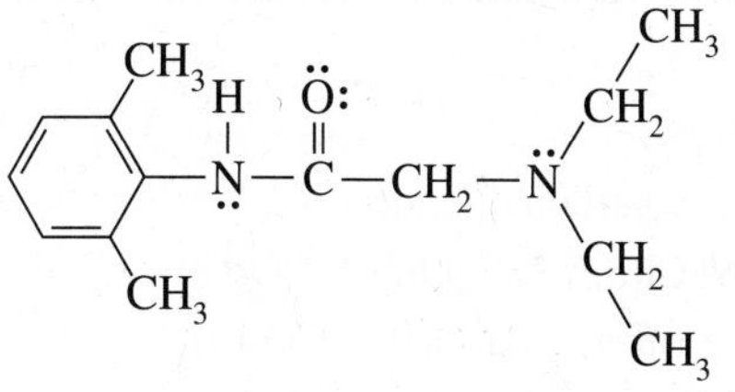

Lidocaine 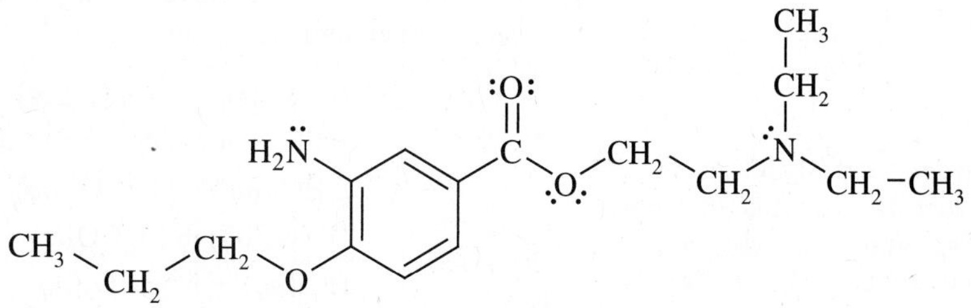   Proparacaine

Amines used as local anesthetic

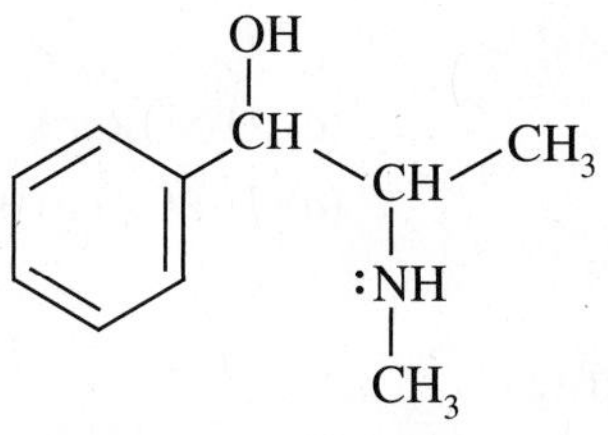

Pseudoephedrine 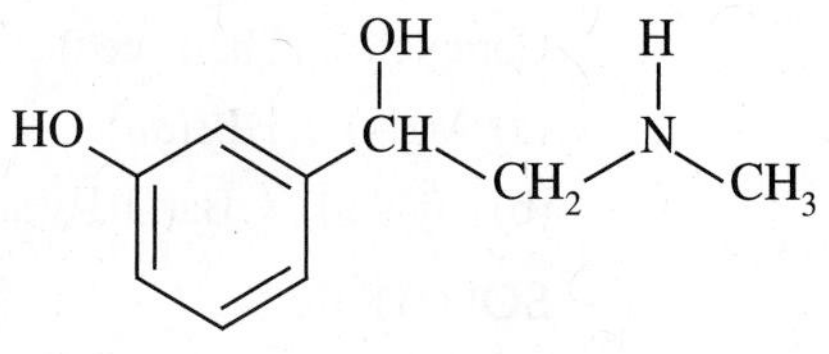   Phenylephrine

Amines used as decongestants

15.2 REACTIONS OF ACIDS AND BASES

LEARNING OBJECTIVE ● Describe the general reactions of acids and bases.

KEY TERM
amphoteric

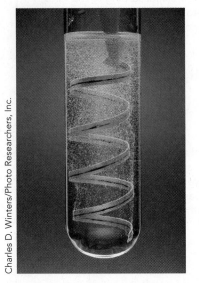

Charles D. Winters/Photo Researchers, Inc.

Magnesium ribbon reacts in HCl solution to produce hydrogen bubbles.

Acid Reactions

In aqueous solutions, the H^+ or H_3O^+ ions are responsible for the characteristic reactions of acids. The following reactions are in an aqueous medium:

Reaction with metals Acids react with metals that lie above hydrogen in the activity series of elements to produce hydrogen and an ionic compound (salt) (see Section 17.5):

$$acid + metal \longrightarrow hydrogen + ionic\ compound$$
$$2\ HCl(aq) + Mg(s) \longrightarrow H_2(g) + MgCl_2(aq)$$
$$H_2SO_4(aq) + Mg(s) \longrightarrow H_2(g) + MgSO_4(aq)$$
$$6\ HC_2H_3O_2(aq) + 2\ Al(s) \longrightarrow 3\ H_2(g) + 2\ Al(C_2H_3O_2)_3(aq)$$

Acids such as nitric acid (HNO_3) are oxidizing substances (see Chapter 17) and react with metals to produce water instead of hydrogen. For example,

$$3\ Zn(s) + 8\ HNO_3(dilute) \longrightarrow 3\ Zn(NO_3)_2(aq) + 2\ NO(g) + 4\ H_2O(l)$$

Reaction with bases The interaction of an acid and a base is called a *neutralization reaction*. In aqueous solutions, the products of this reaction are a salt and water:

$$acid + base \longrightarrow salt + water$$
$$HBr(aq) + KOH(aq) \longrightarrow KBr(aq) + H_2O(l)$$
$$2\ HNO_3(aq) + Ca(OH)_2(aq) \longrightarrow Ca(NO_3)_2(aq) + 2\ H_2O(l)$$
$$2\ H_3PO_4(aq) + 3\ Ba(OH)_2(aq) \longrightarrow Ba_3(PO_4)_2(s) + 6\ H_2O(l)$$

Reaction with metal oxides This reaction is closely related to that of an acid with a base. With an aqueous acid solution, the products are a salt and water:

$$acid + metal\ oxide \longrightarrow salt + water$$
$$2\ HCl(aq) + Na_2O(s) \longrightarrow 2\ NaCl(aq) + H_2O(l)$$
$$H_2SO_4(aq) + MgO(s) \longrightarrow MgSO_4(aq) + H_2O(l)$$
$$6\ HCl(aq) + Fe_2O_3(s) \longrightarrow 2\ FeCl_3(aq) + 3\ H_2O(l)$$

Reaction with carbonates Many acids react with carbonates to produce carbon dioxide, water, and an ionic compound:

$$H_2CO_3(aq) \longrightarrow CO_2(g) + H_2O(l)$$

Carbonic acid (H_2CO_3) is not the product because it is unstable and spontaneously decomposes into water and carbon dioxide.

$$acid + carbonate \longrightarrow salt + water + carbon\ dioxide$$
$$2\ HCl(aq) + Na_2CO_3(aq) \longrightarrow 2\ NaCl(aq) + H_2O(l) + CO_2(g)$$
$$H_2SO_4(aq) + MgCO_3(s) \longrightarrow MgSO_4(aq) + H_2O(l) + CO_2(g)$$
$$HCl(aq) + NaHCO_3(aq) \longrightarrow NaCl(aq) + H_2O(l) + CO_2(g)$$

EXAMPLE 15.2

Complete and balance these equations.

(a) $Mg(s) + HBr(aq) \rightarrow$

(b) $HCl(aq) + Ba(OH)_2(aq) \rightarrow$

(c) $H_2SO_4(aq) + K_2CO_3(aq) \rightarrow$

(d) $H_3PO_4(aq) + NaOH(aq) \rightarrow$

SOLUTION

(a) $Mg + 2\ HBr \rightarrow MgBr_2(aq) + H_2(g)$

(b) $2\ HCl(aq) + Ba(OH)_2(aq) \rightarrow BaCl_2(aq) + 2\ H_2O(l)$

(c) $H_2SO_4(aq) + K_2CO_3c \rightarrow K_2SO_4(aq) + CO_2(g) + H_2O(l)$

(d) $2\ H_3PO_4(aq) + 6\ NaOH(aq) \rightarrow 2\ Na_3PO_4(aq) + 6\ H_2O(l)$

PRACTICE 15.3

Review the four characteristic reactions of acids described in this section and determine which types of reactions are being described.

Base Reactions

The OH^- ions are responsible for the characteristic reactions of bases. The following reactions are in an aqueous medium:

Reaction with acids Bases react with acids to produce a salt and water. See reaction of acids with bases earlier in this section.

Amphoteric hydroxides Hydroxides of certain metals, such as zinc, aluminum, and chromium, are **amphoteric**; that is, they are capable of reacting as either an acid or a base. When treated with a strong acid, they behave like bases; when reacted with a strong base, they behave like acids:

$$Zn(OH)_2(s) + 2\,HCl(aq) \longrightarrow ZnCl_2(aq) + 2\,H_2O(l)$$
$$Zn(OH)_2(s) + 2\,NaOH(aq) \longrightarrow Na_2Zn(OH)_4(aq)$$

Strong acids and bases are discussed in Section 15.4.

Reaction of NaOH and KOH with certain metals Some amphoteric metals react directly with the strong bases sodium hydroxide and potassium hydroxide to produce hydrogen:

$$\text{base} + \text{metal} + \text{water} \longrightarrow \text{salt} + \text{hydrogen}$$
$$2\,NaOH(aq) + Zn(s) + 2\,H_2O(l) \longrightarrow Na_2Zn(OH)_4(aq) + H_2(g)$$
$$2\,KOH(aq) + 2\,Al(s) + 6\,H_2O(l) \longrightarrow 2\,KAl(OH)_4(aq) + 3\,H_2(g)$$

PRACTICE 15.4

Write the formulas and names of the acid and base from which these salts are formed.

(a) K_3PO_4 (b) $MgBr_2$ (c) $LiCl$ (d) $FeCO_3$

15.3 SALTS

LEARNING OBJECTIVE

Explain how a salt is formed and predict the formula of a salt given an acid and a base precursor.

Salts are very abundant in nature. Most of the rocks and minerals of Earth's mantle are salts of one kind or another. Huge quantities of dissolved salts also exist in the oceans. Salts can be considered compounds derived from acids and bases. They consist of positive metal or ammonium ions combined with negative nonmetal ions (OH^- and O^{2-} excluded). The positive ion is the base counterpart, and the nonmetal ion is the acid counterpart:

Chemists use the terms *ionic compound* and *salt* interchangeably.

base NaOH + acid HCl → NaCl salt

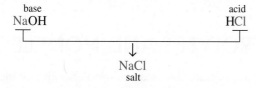

Salts are usually crystalline and have high melting and boiling points.

From a single acid such as hydrochloric acid (HCl), we can produce many chloride compounds by replacing the hydrogen with metal ions (e.g., NaCl, KCl, RbCl, $CaCl_2$, $NiCl_2$). Hence, the number of known salts greatly exceeds the number of known acids and bases. If the hydrogen atoms of a binary acid are replaced by a nonmetal, the resulting compound has covalent bonding and is therefore not considered to be ionic (e.g., PCl_3, S_2Cl_2, Cl_2O, NCl_3, ICl).

You may want to review Chapter 6 for nomenclature of acids, bases, and salts.

>CHEMISTRY *IN ACTION*

A Cool Fizz

Try this experiment with your friends: Chill some soft drinks well. Next, pour the soft drink into a glass. Stick your tongue into the liquid and time how long you can keep it there.

What causes the tingling we feel in our mouth (or on our tongue)? Many people believe it is the bubbles of carbon dioxide, but scientists have found that is not the answer. The tingling is caused by "chemisthesis." Bruce Bryant at Monell Chemical Senses Center in Philadelphia says that chemisthesis is a chemically induced sensation that does not involve taste or odor receptors. The tongue-tingling response to the soft drink is caused by production of protons (H^+ ions) released when an enzyme (carbonic anhydrase) acts on CO_2. The H^+ ions acidify nerve endings, producing the sensation of tingling.

Carbon dioxide also stimulates other neurons when the drink is cold. This means that at a constant pressure of CO_2, a cold drink will produce more tingling than a room-temperature drink. If the drink is at room temperature, CO_2 will increase the cool feeling. Chilling a soda increases the effect of the protons on the nerve endings. At the same time, the high concentration of CO_2 in a freshly opened soda makes the cold soda feel even colder. Together these chemical effects make it really painful to keep your tongue in the chilled soda.

Chemisthesis is important for survival, as well as for having fun with friends.

Bryant says, "It tells you that something is chemically impinging on your body, that tissue is in imminent danger. Burrowing animals can sense toxic levels of carbon dioxide and even feel a sting when exposed to those levels."

Prof. P. Motta/Dept. of Anatomy/University "La Sapienza," Rome/Photo Researchers

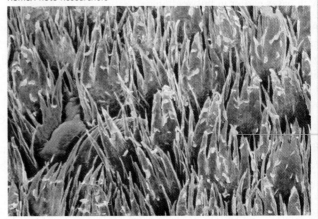

Tongue papillae contain nerve endings that send temperature and tactile information to the brain.

EXAMPLE 15.3

Why are there many more salts than there are acids? Write the formulas for five salts derived from sulfuric acid.

SOLUTION

For any particular acid you can write formulas for many salts by replacing the hydrogen(s) of the acid with a cation: Na_2SO_4, $MgSO_4$, $Fe_2(SO_4)_3$, $KHSO_4$, $Sn(SO_4)_2$, and many more.

PRACTICE 15.5

Given the formulas for three acids and three bases, write the formulas for all the salts that can be formed from the following: HCl, H_2SO_4, H_3PO_4, NaOH, $Mg(OH)_2$, and $Fe(OH)_3$.

15.4 ELECTROLYTES AND NONELECTROLYTES

LEARNING OBJECTIVE   ●   Describe properties, ionization, dissociation, and strength of electrolytes and compare them to nonelectrolytes.

KEY TERMS

electrolyte
nonelectrolyte
dissociation
ionization
strong electrolyte
weak electrolyte

We can show that solutions of certain substances are conductors of electricity with a simple conductivity apparatus, which consists of a pair of electrodes connected to a voltage source through a light bulb and switch (see **Figure 15.1**). If the medium between the electrodes is a conductor of electricity, the light bulb will glow when the switch is closed. When chemically pure water is placed in the beaker and the switch is closed, the light does not glow, indicating that water is a virtual nonconductor. When we dissolve a small amount of sugar in the water and test the solution, the light still does not glow, showing that a sugar solution is also a nonconductor. But when a small amount of salt, NaCl, is dissolved in water and this solution is tested, the light glows brightly. Thus the salt solution conducts electricity. A fundamental difference exists between the chemical bonding in sugar and that in salt. Sugar is a covalently bonded (molecular) substance; common salt is a substance with ionic bonds.

Substances whose aqueous solutions are conductors of electricity are called **electrolytes**. Substances whose solutions are nonconductors are known as **nonelectrolytes**. The major classes of compounds that are electrolytes are acids, bases, and other ionic compounds (salts). Solutions of certain oxides also are conductors because the oxides form an acid or a base when dissolved in water. One major difference between electrolytes and nonelectrolytes is that electrolytes are capable of producing ions in solution, whereas nonelectrolytes do not have this property. Solutions that contain a sufficient number of ions will conduct an electric current. Although pure water is essentially a nonconductor, many city water supplies contain enough dissolved ionic matter to cause the light to glow dimly when the water is tested in a conductivity apparatus. Table 15.2 lists some common electrolytes and nonelectrolytes.

(a) Pure water

TABLE 15.2   Representative Electrolytes and Nonelectrolytes

Electrolytes		Nonelectrolytes	
H_2SO_4	$HC_2H_3O_2$	$C_{12}H_{22}O_{11}$ (sugar)	CH_3OH (methyl alcohol)
HCl	NH_3	C_2H_5OH (ethyl alcohol)	$CO(NH_2)_2$ (urea)
HNO_3	K_2SO_4	$C_2H_4(OH)_2$ (ethylene glycol)	O_2
NaOH	$NaNO_3$	$C_3H_5(OH)_3$ (glycerol)	H_2O

(b) Nonelectrolyte

Dissociation and Ionization of Electrolytes

Arrhenius received the 1903 Nobel Prize in chemistry for his work on electrolytes. He found that a solution conducts electricity because the solute dissociates immediately upon dissolving into electrically charged particles (ions). The movement of these ions toward oppositely charged electrodes causes the solution to be a conductor. According to Arrhenius's theory, solutions that are relatively poor conductors contain electrolytes that are only partly dissociated. Arrhenius also believed that ions exist in solution whether or not an electric current is present. In other words, the electric current does not cause the formation of ions. Remember that positive ions are cations; negative ions are anions.

(c) Strong electrolyte

Figure 15.1

A conductivity apparatus shows the difference in conductivity of solutions. (a) Distilled water does not conduct electricity. (b) Sugar water is a nonelectrolyte. (c) Salt water is a strong electrolyte and conducts electricity.

Acids, bases, and salts are electrolytes.

EXAMPLE 15.4

What is an electrolyte?

SOLUTION

An electrolyte is a substance that forms ions in solution and makes the solution a conductor of electricity.

We have seen that sodium chloride crystals consist of sodium and chloride ions held together by ionic bonds. **Dissociation** is the process by which the ions of a salt separate as the salt dissolves. When placed in water, the sodium and chloride ions are attracted by the polar water molecules, which surround each ion as it dissolves. In water, the salt dissociates,

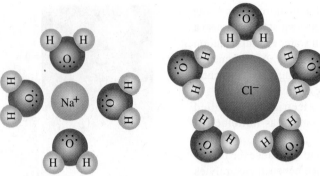

Figure 15.2

Hydrated sodium and chloride ions. When sodium chloride dissolves in water, each Na^+ and Cl^- ion becomes surrounded by water molecules. The negative end of the water dipole is attracted to the Na^+ ion, and the positive end is attracted to the Cl^- ion.

forming hydrated sodium and chloride ions (see **Figure 15.2**). The sodium and chloride ions in solution are surrounded by a specific number of water molecules and have less attraction for each other than they had in the crystalline state. The equation representing this dissociation is

$$NaCl(s) + (x + y)H_2O \longrightarrow Na^+(H_2O)_x + Cl^-(H_2O)_y$$

A simplified dissociation equation in which the water is omitted but understood to be present is

$$NaCl(s) \longrightarrow Na^+(aq) + Cl^-(aq)$$

Remember that sodium chloride exists in an aqueous solution as hydrated ions, not as NaCl units, even though the formula NaCl (or $Na^+ + Cl^-$) is often used in equations.

The chemical reactions of salts in solution are the reactions of their ions. For example, when sodium chloride and silver nitrate react and form a precipitate of silver chloride, only the Ag^+ and Cl^- ions participate in the reaction. The Na^+ and NO_3^- remain as ions in solution:

$$Ag^+(aq) + Cl^-(aq) \longrightarrow AgCl(s)$$

Ionization is the formation of ions; it occurs as a result of a chemical reaction of certain substances with water. Glacial acetic acid (100% $HC_2H_3O_2$) is a liquid that behaves as a nonelectrolyte when tested by the method described earlier. But a water solution of acetic acid conducts an electric current (as indicated by the dull-glowing light of the conductivity apparatus). The equation for the reaction with water, which forms hydronium and acetate ions, is

$$\underset{\text{acid}}{HC_2H_3O_2} + \underset{\text{base}}{H_2O} \rightleftharpoons \underset{\text{acid}}{H_3O^+} + \underset{\text{base}}{C_2H_3O_2^-}$$

or, in the simplified equation,

$$HC_2H_3O_2 \rightleftharpoons H^+ + C_2H_3O_2^-$$

In this ionization reaction, water serves not only as a solvent but also as a base according to the Brønsted–Lowry theory.

Hydrogen chloride is predominantly covalently bonded, but when dissolved in water, it reacts to form hydronium and chloride ions:

$$HCl(g) + H_2O(l) \longrightarrow H_3O^+(aq) + Cl^-(aq)$$

When a hydrogen chloride solution is tested for conductivity, the light glows brilliantly, indicating many ions in the solution.

Ionization occurs in each of the preceding two reactions with water, producing ions in solution. The necessity for water in the ionization process can be demonstrated by dissolving hydrogen chloride in a nonpolar solvent such as hexane and testing the solution for conductivity. The solution fails to conduct electricity, indicating that no ions are produced.

The terms *dissociation* and *ionization* are often used interchangeably to describe processes taking place in water. But, strictly speaking, the two are different. In the dissociation of a salt, the salt already exists as ions; when it dissolves in water, the ions separate, or dissociate, and increase in mobility. In the ionization process, ions are produced by the reaction of a compound with water.

Strong and Weak Electrolytes

Electrolytes are classified as strong or weak depending on the degree, or extent, of dissociation or ionization. **Strong electrolytes** are essentially 100% ionized in solution; **weak electrolytes** are much less ionized (based on comparing 0.1 *M* solutions). Most electrolytes are either strong or weak, with a few classified as moderately strong or weak. Most salts are strong electrolytes. Acids and bases that are strong electrolytes (highly ionized) are called *strong acids* and *strong bases*. Acids and bases that are weak electrolytes (slightly ionized) are called *weak acids* and *weak bases*.

For equivalent concentrations, solutions of strong electrolytes contain many more ions than do solutions of weak electrolytes. As a result, solutions of strong electrolytes are better conductors of electricity. Consider two solutions, 1 M HCl and 1 M HC$_2$H$_3$O$_2$. Hydrochloric acid is almost 100% ionized; acetic acid is about 1% ionized. (See **Figure 15.3**.) Thus, HCl is a strong acid and HC$_2$H$_3$O$_2$ is a weak acid. Hydrochloric acid has about 100 times as many hydronium ions in solution as acetic acid, making the HCl solution much more acidic.

We can distinguish between strong and weak electrolytes experimentally using the apparatus described earlier. A 1 M HCl solution causes the light to glow brilliantly, but a 1 M HC$_2$H$_3$O$_2$ solution causes only a dim glow. The strong base sodium hydroxide (NaOH) can be distinguished in a similar fashion from the weak base ammonia (NH$_3$). The ionization of a weak electrolyte in water is represented by an equilibrium equation showing that both the un-ionized and ionized forms are present in solution. In the equilibrium equation of HC$_2$H$_3$O$_2$ and its ions, we say that the equilibrium lies "far to the left" because relatively few hydrogen and acetate ions are present in solution:

$$HC_2H_3O_2(aq) \rightleftharpoons H^+(aq) + C_2H_3O_2^-(aq)$$

HC$_2$H$_3$O$_2$ C$_2$H$_3$O$_2^-$

We have previously used a double arrow in an equation to represent reversible processes in the equilibrium between dissolved and undissolved solute in a saturated solution. A double arrow ($\rightleftharpoons$) is also used in the ionization equation of soluble weak electrolytes to indicate that the solution contains a considerable amount of the un-ionized compound in equilibrium with its ions in solution. (See Section 16.2 for a discussion of reversible reactions.) A single arrow is used to indicate that the electrolyte is essentially all in the ionic form in the solution. For example, nitric acid is a strong acid; nitrous acid is a weak acid. Their ionization equations in water may be indicated as

$$HNO_3(aq) \xrightarrow{H_2O} H^+(aq) + NO_3^-(aq)$$
$$HNO_2(aq) \xrightleftharpoons{H_2O} H^+(aq) + NO_2^-(aq)$$

Practically all soluble salts, acids (such as sulfuric, nitric, and hydrochloric acids), and bases (such as sodium, potassium, calcium, and barium hydroxides) are strong electrolytes. Weak electrolytes include numerous other acids and bases such as acetic acid, nitrous acid, carbonic acid, and ammonia. The terms *strong acid, strong base, weak acid*, and *weak base* refer to whether an acid or a base is a strong or weak electrolyte. A brief list of strong and weak electrolytes is given in Table 15.3.

Electrolytes yield two or more ions per formula unit upon dissociation—the actual number being dependent on the compound. Dissociation is complete or nearly complete for nearly all soluble ionic compounds and for certain other strong electrolytes, such as those given in Table 15.3. The following are dissociation equations for several strong electrolytes. In all cases, the ions are actually hydrated:

$$NaOH \xrightarrow{H_2O} Na^+(aq) + OH^-(aq) \qquad \text{2 ions in solution per formula unit}$$
$$Na_2SO_4 \xrightarrow{H_2O} 2\,Na^+(aq) + SO_4^{2-}(aq) \qquad \text{3 ions in solution per formula unit}$$
$$Fe_2(SO_4)_3 \xrightarrow{H_2O} 2\,Fe^{3+}(aq) + 3\,SO_4^{2-}(aq) \qquad \text{5 ions in solution per formula unit}$$

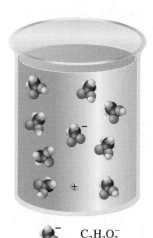

Cl$^-$

H$^+$

C$_2$H$_3$O$_2^-$

H$^+$

HC$_2$H$_3$O$_2$

Figure 15.3

HCl solution (top) is 100% ionized, while in HC$_2$H$_3$O$_2$ solution (bottom) almost all of the solute is in molecular form. HCl is a strong acid, while HC$_2$H$_3$O$_2$ is a weak acid. *Note:* The water molecules in the solution are not shown in this figure.

TABLE 15.3 Strong and Weak Electrolytes

Strong electrolytes		Weak electrolytes	
Most soluble salts	HClO$_4$	HC$_2$H$_3$O$_2$	H$_2$C$_2$O$_4$
H$_2$SO$_4$	NaOH	H$_2$CO$_3$	H$_3$BO$_3$
HNO$_3$	KOH	HNO$_2$	HClO
HCl	Ca(OH)$_2$	H$_2$SO$_3$	NH$_3$
HBr	Ba(OH)$_2$	H$_2$S	HF

One mole of NaCl will give 1 mol of Na^+ ions and 1 mol of Cl^- ions in solution, assuming complete dissociation of the salt. One mole of $CaCl_2$ will give 1 mol of Ca^{2+} ions and 2 mol of Cl^- ions in solution:

$$NaCl \xrightarrow{H_2O} Na^+(aq) + Cl^-(aq)$$
$$\text{1 mol} \qquad \text{1 mol} \qquad \text{1 mol}$$

$$CaCl_2 \xrightarrow{H_2O} Ca^{2+}(aq) + 2\,Cl^-(aq)$$
$$\text{1 mol} \qquad \text{1 mol} \qquad \text{2 mol}$$

WileyPLUS

ENHANCED EXAMPLE

EXAMPLE 15.5

What is the molarity of each ion in a solution of (a) 2.0 M NaCl and (b) 0.40 M K_2SO_4? Assume complete dissociation.

SOLUTION

(a) According to the dissociation equation,

$$NaCl \xrightarrow{H_2O} Na^+(aq) + Cl^-(aq)$$
$$\text{1 mol} \qquad \text{1 mol} \qquad \text{1 mol}$$

the concentration of Na^+ is equal to that of NaCl:

1 mol NaCl $\longrightarrow$ 1 mol Na^+ and the concentration of Cl^- is also equal to that of NaCl. Therefore, the concentrations of the ions in 2.0 M NaCl are 2.0 M Na^+ and 2.0 M Cl^-.

(b) According to the dissociation equation,

$$K_2SO_4 \xrightarrow{H_2O} 2\,K^+(aq) + SO_4^{2-}(aq)$$
$$\text{1 mol} \qquad \text{2 mol} \qquad \text{1 mol}$$

the concentration of K^+ is twice that of K_2SO_4 and the concentration of SO_4^{2-} is equal to that of K_2SO_4. Therefore, the concentrations of the ions in 0.40 M K_2SO_4 are 0.80 M K^+ and 0.40 M SO_4^{2-}.

PRACTICE 15.6

What is the molarity of each ion in a solution of (a) 0.050 M $MgCl_2$ and (b) 0.070 M $AlCl_3$?

Colligative Properties of Electrolyte Solutions

We have learned that when 1 mol of sucrose, a nonelectrolyte, is dissolved in 1000 g of water, the solution freezes at $-1.86°C$. When 1 mol NaCl is dissolved in 1000 g of water, the freezing point of the solution is not $-1.86°C$, as might be expected, but is closer to $-3.72°C$ (-1.86×2). The reason for the lower freezing point is that 1 mol NaCl in solution produces 2 mol of particles ($2 \times 6.022 \times 10^{23}$ ions) in solution. Thus the freezing point depression produced by 1 mol NaCl is essentially equivalent to that produced by 2 mol of a nonelectrolyte. An electrolyte such as $CaCl_2$, which yields three ions in water, gives a freezing point depression of about three times that of a nonelectrolyte. These freezing point data provide additional evidence that electrolytes dissociate when dissolved in water. The other colligative properties are similarly affected by substances that yield ions in aqueous solutions.

PRACTICE 15.7

What is the boiling point of a 1.5 m solution of KCl(aq)?

Ionization of Water

Pure water is a *very* weak electrolyte, but it does ionize slightly. Two equations commonly used to show how water ionizes are

$$H_2O + H_2O \rightleftharpoons H_3O^+ + OH^-$$
$$\text{acid} \quad \text{base} \qquad \text{acid} \quad \text{base}$$

and

$$H_2O \rightleftharpoons H^+ + OH^-$$

The first equation represents the Brønsted–Lowry concept, with water reacting as both an acid and a base, forming a hydronium ion and a hydroxide ion. The second equation is a simplified version, indicating that water ionizes to give a hydrogen and a hydroxide ion. Actually, the proton (H^+) is hydrated and exists as a hydronium ion. In either case equal molar amounts of acid and base are produced so that water is neutral, having neither H^+ nor OH^- ions in excess. The ionization of water at 25°C produces an H^+ ion concentration of 1.0×10^{-7} mol/L and an OH^- ion concentration of 1.0×10^{-7} mol/L. Square brackets, [ ], are used to indicate that the concentration is in moles per liter. Thus $[H^+]$ means the concentration of H^+ is in moles per liter. These concentrations are usually expressed as

$$[H^+] \text{ or } [H_3O^+] = 1.0 \times 10^{-7} \text{ mol/L}$$
$$[OH^-] = 1.0 \times 10^{-7} \text{ mol/L}$$

These figures mean that about two out of every billion water molecules are ionized. This amount of ionization, small as it is, is a significant factor in the behavior of water in many chemical reactions.

15.5 INTRODUCTION TO pH

Calculate the pH of a solution from the hydrogen ion concentration.

● **LEARNING OBJECTIVE**

KEY TERMS

pH

logarithm (log)

The acidity of an aqueous solution depends on the concentration of hydrogen or hydronium ions. The pH scale of acidity provides a simple, convenient, numerical way to state the acidity of a solution. Values on the pH scale are obtained by mathematical conversion of H^+ ion concentrations to pH by the expression

$$pH = -\log[H^+]$$

where $[H^+] = H^+$ or H_3O^+ ion concentration in moles per liter. The **pH** is defined as the *negative* logarithm of the H^+ or H_3O^+ concentration in moles per liter:

$$pH = -\log[H^+] = -\log(1 \times 10^{-7}) = -(-7) = 7$$

For example, the pH of pure water at 25°C is 7 and is said to be neutral; that is, it is neither acidic nor basic, because the concentrations of H^+ and OH^- are equal. Solutions that contain more H^+ ions than OH^- ions have pH values less than 7, and solutions that contain fewer H^+ ions than OH^- ions have pH values greater than 7.

pH < 7.00 is an acidic solution
pH = 7.00 is a neutral solution
pH > 7.00 is a basic solution

$$\text{When } [H^+] = 1 \times 10^{-5} \text{ mol/L, pH} = 5 \text{ (acidic)}$$
$$\text{When } [H^+] = 1 \times 10^{-9} \text{ mol/L, pH} = 9 \text{ (basic)}$$

Instead of saying that the hydrogen ion concentration in the solution is 1×10^{-5} mol/L, it's customary to say that the pH of the solution is 5. The smaller the pH value, the more acidic the solution (see **Figure 15.4**).

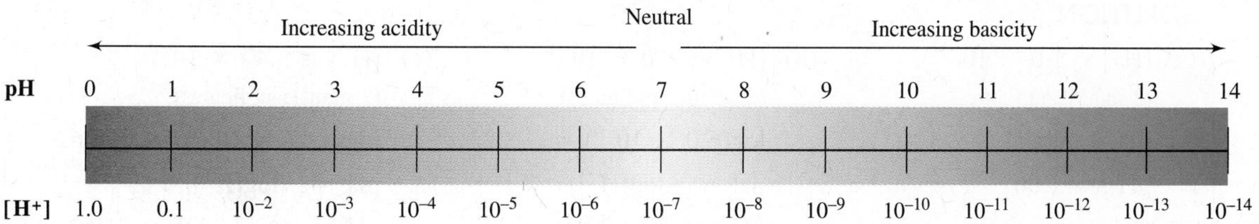

		Increasing acidity					Neutral		Increasing basicity				
pH	0	1	2	3	4	5	6	7	8	9	10	11	12	13	14
[H+]	1.0	0.1	10^{-2}	10^{-3}	10^{-4}	10^{-5}	10^{-6}	10^{-7}	10^{-8}	10^{-9}	10^{-10}	10^{-11}	10^{-12}	10^{-13}	10^{-14}

Figure 15.4

The pH scale of acidity and basicity.

TABLE 15.4	pH Scale for Expressing Acidity	
$[H^+]$ (mol/L)	pH	
1×10^{-14}	14	↑
1×10^{-13}	13	
1×10^{-12}	12	Increasing
1×10^{-11}	11	basicity
1×10^{-10}	10	
1×10^{-9}	9	
1×10^{-8}	8	
1×10^{-7}	7	Neutral
1×10^{-6}	6	
1×10^{-5}	5	
1×10^{-4}	4	
1×10^{-3}	3	Increasing
1×10^{-2}	2	acidity
1×10^{-1}	1	
1×10^{0}	0	↓

TABLE 15.5	The pH of Common Solutions
Solution	pH
Gastric juice	1.0
0.1 M HCl	1.0
Lemon juice	2.3
Vinegar	2.8
0.1 M HC$_2$H$_3$O$_2$	2.9
Orange juice	3.7
Tomato juice	4.1
Coffee, black	5.0
Urine	6.0
Milk	6.6
Pure water (25°C)	7.0
Blood	7.4
Household ammonia	11.0
1 M NaOH	14.0

The pH scale, along with its interpretation, is given in Table 15.4, and Table 15.5 lists the pH of some common solutions. Note that a change of only one pH unit means a 10-fold increase or decrease in H^+ ion concentration. For example, a solution with a pH of 3.0 is 10 times more acidic than a solution with a pH of 4.0. A simplified method of determining pH from $[H^+]$ follows:

$[H^+] = 1 \times 10^{-5}$ ← pH = this number (5)
pH = 5

when this number
is exactly 1

$[H^+] = 2 \times 10^{-5}$ ← pH is between this number and next lower number (4 and 5)
pH = 4.7

when this number
is between 1 and 10

Help on using calculators is found in Appendix II.

Calculating the pH value for H^+ ion concentrations requires the use of logarithms, which are exponents. The **logarithm (log)** of a number is simply the power to which 10 must be raised to give that number. Thus the log of 100 is 2 ($100 = 10^2$), and the log of 1000 is 3 ($1000 = 10^3$). The log of 500 is 2.70, but you can't determine this value easily without a scientific calculator.

Let's determine the pH of a solution with $[H^+] = 2 \times 10^{-5}$ using a calculator. The number $-4.69 \ldots$ will be displayed. The pH is then

Remember: Change the sign on your calculator since pH = −log[H⁺].

$$pH = -\log[H^+] = -(-4.69 \ldots) = 4.7$$

Next we must determine the correct number of significant figures in the logarithm. The rules for logs are different from those we use in other math operations. The number of decimal places for a log must equal the number of significant figures in the original number. Since 2×10^{-5} has one significant figure, we should round the log to one decimal place ($4.69 \ldots$) = 4.7.

WileyPLUS

ENHANCED EXAMPLE

EXAMPLE 15.6

What is the pH of a solution with an $[H^+]$ of (a) 1.0×10^{-11}, (b) 6.0×10^{-4}, and (c) 5.47×10^{-8}?

SOLUTION

(a) $[H^+] = 1.0 \times 10^{-11}$
(2 significant figures)

$pH = -\log(1.0 \times 10^{-11})$

$pH = 11.00$

(2 decimal places)

(b) $[H^+] = 6.0 \times 10^{-4}$
(2 significant figures)

$\log(6.0 \times 10^{-4}) = -3.22$

$pH = -\log[H^+]$

$pH = -(-3.22) = 3.22$
(2 decimal places)

(c) $[H^+] = 5.47 \times 10^{-8}$
(3 significant figures)

$\log(5.47 \times 10^{-8}) = -7.262$

$pH = -\log[H^+]$

$pH = -(-7.262) = 7.262$
(3 decimal places)

PRACTICE 15.8

What is the pH of a solution with $[H^+]$ of (a) $3.9 \times 10^{-12}\, M$, (b) $1.3 \times 10^{-3}\, M$, and (c) $3.72 \times 10^{-6}\, M$?

The measurement and control of pH is extremely important in many fields. Proper soil pH is necessary to grow certain types of plants successfully. The pH of certain foods is too acidic for some diets. Many biological processes are delicately controlled pH systems. The pH of human blood is regulated to very close tolerances through the uptake or release of H^+ by mineral ions, such as HCO_3^-, HPO_4^{2-}, and $H_2PO_4^-$. Changes in the pH of the blood by as little as 0.4 pH unit result in death.

Compounds with colors that change at particular pH values are used as indicators in acid–base reactions. For example, phenolphthalein, an organic compound, is colorless in acid solution and changes to pink at a pH of 8.3. When a solution of sodium hydroxide is added to a hydrochloric acid solution containing phenolphthalein, the change in color (from colorless to pink) indicates that all the acid is neutralized. Commercially available pH test paper contains chemical indicators. The indicator in the paper takes on different colors when wetted with solutions of different pH. Thus the pH of a solution can be estimated by placing a drop on the test paper and comparing the color of the test paper with a color chart calibrated at different pH values. Common applications of pH test indicators are the kits used to measure and adjust the pH of swimming pools, hot tubs, and saltwater aquariums. Electronic pH meters are used for making rapid and precise pH determinations (see **Figure 15.5**).

Figure 15.5

The pH of a substance (in this case garden soil) can be measured by using a pH meter.

>CHEMISTRY *IN ACTION*

Ocean Corals Threatened by Increasing Atmospheric CO_2 Levels

As we continue to burn fossil fuels, we are increasing the atmospheric CO_2 levels beyond anything seen in recent history. Some of this CO_2 is removed from the atmosphere by dissolving in the water of the oceans. This at first seems like an excellent way to remove excess atmospheric CO_2 until we look at the effect it has on the world's oceans. The pH of seawater has dropped from about 8.2 to between 8.05 and 8.1 since the beginning of the Industrial Revolution. Although this may seem like a

small change, it has had dramatic effects on the corals living in the oceans. In a study published in the November 8, 2010, issue of *PNAS*, Rebecca Albright of the University of Miami reported a significant decrease in the number of elkhorn coral eggs fertilized as the pH of seawater drops. Additionally, the ability of fertilized eggs to attach and begin growing has also been affected. Albright suggests that there may be as much as a 73% decrease in the number of new corals able to establish themselves by the year 2100 if acidification of the oceans continues at the current rate. The CO_2 in the atmosphere obviously has significant effects on the chemistry of the oceans.

Healthy coral

Unhealthy coral

15.6 NEUTRALIZATION

Describe a neutralization reaction and do calculations involving titrations.

KEY TERMS

neutralization
spectator ion
titration

The reaction of an acid and a base to form a salt and water is known as **neutralization**. We've seen this reaction before, but now with our knowledge about ions and ionization, let's reexamine the process of neutralization.

Consider the reaction that occurs when solutions of sodium hydroxide and hydrochloric acid are mixed. The ions present initially are Na^+ and OH^- from the base and H^+ and Cl^- from the acid. The products, sodium chloride and water, exist as Na^+ and Cl^- ions and H_2O molecules. A chemical equation representing this reaction is

$$HCl(aq) + NaOH(aq) \longrightarrow NaCl(aq) + H_2O(l)$$

This equation, however, does not show that HCl, NaOH, and NaCl exist as ions in solution. The following total ionic equation gives a better representation of the reaction:

$$(H^+ + Cl^-) + (Na^+ + OH^-) \longrightarrow Na^+ + Cl^- + H_2O(l)$$

This equation shows that the Na^+ and Cl^- ions did not react. These ions are called **spectator ions** because they were present but did not take part in the reaction. The only reaction that occurred was that between the H^+ and OH^- ions. Therefore the equation for the neutralization can be written as this net ionic equation:

$$\underset{\text{acid}}{H^+(aq)} + \underset{\text{base}}{OH^-(aq)} \longrightarrow \underset{\text{water}}{H_2O(l)}$$

This simple net ionic equation represents not only the reaction of sodium hydroxide and hydrochloric acid but also the reaction of any strong acid with any water-soluble hydroxide base in an aqueous solution. The driving force of a neutralization reaction is the ability of an H^+ ion and an OH^- ion to react and form a molecule of water.

The amount of acid, base, or other species in a sample can be determined by **titration**, which measures the volume of one reagent required to react with a measured mass or volume of another reagent. Consider the titration of an acid with a base. A measured volume of acid of unknown concentration is placed in a flask, and a few drops of an indicator solution are added. Base solution of known concentration is slowly added from a buret to the acid until the indicator changes color. The indicator selected is one that changes color when the stoichiometric quantity (according to the equation) of base has been added to the acid. At this point, known as the *end point of the titration*, the titration is complete, and the volume of base used to neutralize the acid is read from the buret. The concentration or amount of acid in solution can be calculated from the titration data and the chemical equation for the reaction. Let's look at some examples.

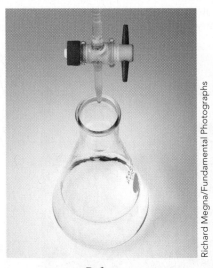

Before
colorless and clear

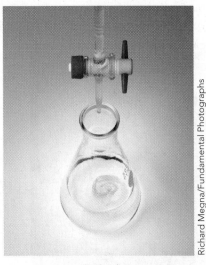

End point
pale pink persists

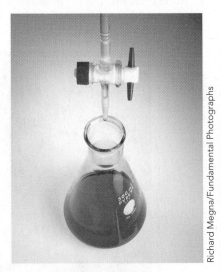

After end point
bright pink (excess base)

The titration process with phenolphthalein indicator.

EXAMPLE 15.7

Suppose that 42.00 mL of 0.150 M NaOH solution are required to neutralize 50.00 mL of hydrochloric acid solution. What is the molarity of the acid solution?

SOLUTION

Read • **Knowns:** 42.00 mL 0.150 M NaOH solution

50.00 mL acid solution

Solving for: acid molarity

Plan • The equation for the reaction is

$$NaOH(aq) + HCl(aq) \longrightarrow NaCl(aq) + H_2O(l)$$

Determine the moles NaOH in the solution.

Setup • $(0.04200 \; \cancel{L})\left(\dfrac{0.150 \; mol \; NaOH}{1 \; \cancel{L}}\right) = 0.00630 \; mol \; NaOH$

Since the acid and base react in a 1 : 1 ratio (from the equation for the reaction), the mol base = mol acid.

Calculate • $M = \dfrac{mol}{L} = \dfrac{0.00630 \; mol \; HCl}{0.05000 \; L} = 0.126 \; M \; HCl$

EXAMPLE 15.8

Suppose that 42.00 mL of 0.150 M NaOH solution are required to neutralize 50.00 mL of H_2SO_4 solution. What is the molarity of the acid solution?

SOLUTION

Read • **Knowns:** 42.00 mL 0.150 M NaOH solution

50.00 mL H_2SO_4 solution

Solving for: acid molarity

Plan • The equation for the reaction is

$$2 \; NaOH(aq) + H_2SO_4(aq) \longrightarrow Na_2SO_4(aq) + 2 \; H_2O(l)$$

Determine the moles NaOH in the solution.

Setup • $(0.04200 \; \cancel{L})\left(\dfrac{0.150 \; mol \; NaOH}{1 \; \cancel{L}}\right) = 0.00630 \; mol \; NaOH$

Since the acid and base react in a 1 : 2 ratio (from the equation for the reaction) 2 mol base = 1 mol acid.

$$(0.00630 \; \cancel{mol \; NaOH})\left(\dfrac{1 \; mol \; H_2SO_4}{2 \; \cancel{mol \; NaOH}}\right) = 0.00315 \; mol \; H_2SO_4$$

Calculate • $M = \dfrac{mol}{L} = \dfrac{0.00315 \; mol \; H_2SO_4}{0.05000 \; L} = 0.0630 \; M \; H_2SO_4$

EXAMPLE 15.9

A 25.00-mL sample of H_2SO_4 solution required 14.26 mL of 0.2240 M NaOH solution for complete neutralization. What is the molarity of the sulfuric acid?

SOLUTION

Read • **Knowns:** 14.26 mL 0.2240 M NaOH solution

25.00 mL H_2SO_4 solution

Solving for: sulfuric acid molarity

Plan •   The equation for the reaction is

$$2\,NaOH(aq) + H_2SO_4(aq) \longrightarrow Na_2SO_4(aq) + 2\,H_2O(l)$$

Determine the moles NaOH in solution.

Setup • $(0.01426\,\cancel{L})\left(\dfrac{0.2240\,mol\,NaOH}{1\,\cancel{L}}\right) = 0.003194\,mol\,NaOH$

Since the mole ratio of acid to base is $\dfrac{1\,mol\,H_2SO_4}{2\,mol\,NaOH}$,

$(0.0031940\,mol\,NaOH)\left(\dfrac{1\,mol\,H_2SO_4}{2\,mol\,\,NaOH}\right) = 0.001597\,mol\,H_2SO_4$

Calculate • $M = \dfrac{mol}{L} = \dfrac{0.001597\,mol\,H_2SO_4}{0.02500\,L} = 0.06388\,M\,H_2SO_4$

PRACTICE 15.9

A 50.0-mL sample of HCl required 24.81 mL of 0.1250 M NaOH for neutralization. What is the molarity of the acid?

15.7 WRITING NET IONIC EQUATIONS

LEARNING OBJECTIVE ● Write net ionic equations using the stated rules.

KEY TERMS

formula equation
total ionic equation
net ionic equation

In Section 15.6, we wrote the reaction of hydrochloric acid and sodium hydroxide in three different equations:

1. $HCl(aq) + NaOH(aq) \longrightarrow NaCl(aq) + H_2O(l)$

2. $(H^+ + Cl^-) + (Na^+ + OH^-) \longrightarrow Na^+ + Cl^- + H_2O(l)$

3. $H^+ + OH^- \longrightarrow H_2O$

In the **formula equation** (1), compounds are written in their molecular, or formula, expressions. In the **total ionic equation** (2), compounds are written to show the form in which they are predominantly present: strong electrolytes as ions in solution and nonelectrolytes, weak electrolytes, precipitates, and gases in their molecular forms. In the **net ionic equation** (3), only those molecules or ions that have reacted are included in the equation; ions or molecules that do not react (the spectators) are omitted.

When balancing equations thus far, we've been concerned only with the atoms of the individual elements. Because ions are electrically charged, ionic equations often end up with a net electrical charge. A balanced equation must have the same net charge on each side, whether that charge is positive, negative, or zero. Therefore, when balancing ionic equations, we must make sure that both the same number of each kind of atom and the same net electrical charge are present on each side.

RULES FOR WRITING IONIC EQUATIONS

1. Strong electrolytes in solution are written in their ionic form.

2. Weak electrolytes are written in their molecular form.

3. Nonelectrolytes are written in their molecular form.

4. Insoluble substances, precipitates, and gases are written in their molecular forms.

5. The net ionic equation should include only substances that have undergone a chemical change. Spectator ions are omitted from the net ionic equation.

6. Equations must be balanced, both in atoms and in electrical charge.

Study the following examples. In each one, the formula equation is given. Write the total ionic equation and the net ionic equation for each.

EXAMPLE 15.10

$$HNO_3(aq) + KOH(aq) \longrightarrow KNO_3(aq) + H_2O(l)$$
formula equation

SOLUTION

$$(H^+ + NO_3^-) + (K^+ + OH^-) \longrightarrow K^+ + NO_3^- + H_2O$$
total ionic equation

$$H^+ + OH^- \longrightarrow H_2O$$
net ionic equation

The HNO_3, KOH, and KNO_3 are soluble, strong electrolytes. The K^+ and NO_3^- ions are spectator ions, have not changed, and are not included in the net ionic equation. Water is a nonelectrolyte and is written in the molecular form.

EXAMPLE 15.11

$$2\ AgNO_3(aq) + BaCl_2(aq) \longrightarrow 2\ AgCl(s) + Ba(NO_3)_2(aq)$$
formula equation

SOLUTION

$$(2\ Ag^+ + 2\ NO_3^-) + (Ba^{2+} + 2\ Cl^-) \longrightarrow 2\ AgCl(s) + Ba^{2+} + 2\ NO_3^-$$
total ionic equation

$$Ag^+ + Cl^- \longrightarrow AgCl(s)$$
net ionic equation

Although AgCl is an ionic compound, it is written in the un-ionized form on the right side of the ionic equations because most of the Ag^+ and Cl^- ions are no longer in solution but have formed a precipitate of AgCl. The Ba^{2+} and NO_3^- ions are spectator ions.

EXAMPLE 15.12

$$Na_2CO_3(aq) + H_2SO_4(aq) \longrightarrow Na_2SO_4(aq) + H_2O(l) + CO_2(g)$$
formula equation

SOLUTION

$$(2\ Na^+ + CO_3^{2-}) + (2\ H^+ + SO_4^{2-}) \longrightarrow 2\ Na^+ + SO_4^{2-} + H_2O(l) + CO_2(g)$$
total ionic equation

$$CO_3^{2-} + 2\ H^+ \longrightarrow H_2O(l) + CO_2(g)$$
net ionic equation

Carbon dioxide (CO_2) is a gas and evolves from the solution; Na^+ and SO_4^{2-} are spectator ions.

EXAMPLE 15.13

$$HC_2H_3O_2(aq) + NaOH(aq) \longrightarrow NaC_2H_3O_2(aq) + H_2O(l)$$
formula equation

SOLUTION

$$HC_2H_3O_2 + (Na^+ + OH^-) \longrightarrow Na^+ + C_2H_3O_2^- + H_2O$$
total ionic equation

$$HC_2H_3O_2 + OH^- \longrightarrow C_2H_3O_2^- + H_2O$$
net ionic equation

Acetic acid, $HC_2H_3O_2$, a weak acid, is written in the molecular form, but sodium acetate, $NaC_2H_3O_2$, a soluble salt, is written in the ionic form. The Na^+ ion is the only spectator ion in this reaction. Both sides of the net ionic equation have a -1 electrical charge.

EXAMPLE 15.14

$$Mg(s) + 2\,HCl(aq) \longrightarrow MgCl_2(aq) + H_2(g)$$
formula equation

SOLUTION

$$Mg + (2\,H^+ + 2\,Cl^-) \longrightarrow Mg^{2+} + 2\,Cl^- + H_2(g)$$
total ionic equation

$$Mg + 2\,H^+ \longrightarrow Mg^{2+} + H_2(g)$$
net ionic equation

The net electrical charge on both sides of the equation is $+2$.

EXAMPLE 15.15

$$H_2SO_4(aq) + Ba(OH)_2(aq) \longrightarrow BaSO_4(s) + 2\,H_2O(l)$$
formula equation

SOLUTION

$$(2\,H^+ + SO_4^{2-}) + (Ba^{2+} + 2\,OH^-) \longrightarrow BaSO_4(s) + 2\,H_2O(l)$$
total ionic equation

$$2\,H^+ + SO_4^{2-} + Ba^{2+} + 2\,OH^- \longrightarrow BaSO_4(s) + 2\,H_2O(l)$$
net ionic equation

Barium sulfate ($BaSO_4$) is a highly insoluble salt. If we conduct this reaction using the conductivity apparatus described in Section 15.4, the light glows brightly at first but goes out when the reaction is complete because almost no ions are left in solution. The $BaSO_4$ precipitates out of solution, and water is a nonconductor of electricity.

PRACTICE 15.10

Write the net ionic equation for

$$3\,H_2S(aq) + 2\,Bi(NO_3)_3(aq) \longrightarrow Bi_2S_3(s) + 6\,HNO_3(aq)$$

15.8 ACID RAIN

LEARNING OBJECTIVE ● Describe how acid rain forms and its effects on society.

Acid rain is defined as any atmospheric precipitation that is more acidic than usual. The increase in acidity might be from natural or industrial sources. Rain acidity varies throughout the world and across the United States. The pH of rain is generally lower in the eastern United States and higher in the West. Unpolluted rain has a pH of 5.6, and so is slightly acidic. This acidity results from the dissolution of carbon dioxide in the water producing carbonic acid:

$$CO_2(g) + H_2O(l) \rightleftharpoons H_2CO_3(aq) \rightleftharpoons H^+(aq) + HCO_3^-(aq)$$

The general process involves the following steps:

1. Emission of nitrogen and sulfur oxides into the air
2. Transportation of these oxides throughout the atmosphere
3. Chemical reactions between the oxides and water, forming sulfuric acid (H_2SO_4) and nitric acid (HNO_3)
4. Rain or snow, which carries the acids to the ground

The oxides may also be deposited directly on a dry surface and become acidic when normal rain falls on them.

Acid rain is not a new phenomenon. Rain was probably acidic in the early days of our planet as volcanic eruptions, fires, and decomposition of organic matter released large volumes of nitrogen and sulfur oxides into the atmosphere. Use of fossil fuels, especially since the Industrial Revolution, has made significant changes in the amounts of pollutants being released into the atmosphere. As increasing amounts of fossil fuels have been burned, more and more sulfur and nitrogen oxides have poured into the atmosphere, thus increasing the acidity of rain.

Acid rain affects a variety of factors in our environment. For example, freshwater plants and animals decline significantly when rain is acidic; large numbers of fish and plants die when acidic water from spring thaws enters the lakes. Acidic rainwater leaches aluminum from the soil into lakes, where the aluminum compounds adversely affect the gills of fish. In addition to leaching aluminum from the soil, acid rain also causes other valuable minerals, such as magnesium and calcium, to dissolve and run into lakes and streams. It can also dissolve the waxy protective coat on plant leaves, making them vulnerable to attack by bacteria and fungi.

In our cities, acid rain is responsible for extensive and continuing damage to buildings, monuments, and statues. It reduces the durability of paint and promotes the deterioration of paper, leather, and cloth. In short, we are just beginning to explore the effects of acid rain on human beings and on our food chain.

This ancient stone carving on the Burgos Cathedral in northern Spain shows the destructive power of acid rain over time.

Mark Boulton/Photo Researchers, Inc.

CHAPTER 15 REVIEW

15.1 ACIDS AND BASES

- Characteristic properties of acids include:
  - They taste sour.
  - They change litmus from blue to red.
  - They react with:
    - Metals to form hydrogen gas
    - Hydroxide bases to form water and a salt
    - Carbonates to produce CO_2
- Characteristic properties of bases include:
  - They taste bitter or caustic.
  - They have a slippery feeling.
  - They change litmus from red to blue.
  - They interact with acids to form water and a salt.
- Arrhenius definition of acids and bases:
  - Acids contain excess H^+ ions in aqueous solutions.
  - Bases contain excess OH^- ions in aqueous solutions.
- Brønsted–Lowry definition of acids and bases:
  - Acids are proton donors.
  - Bases are proton acceptors.
- Lewis definition of acids and bases:
  - Acids are electron pair acceptors.
  - Bases are electron pair donors.

KEY TERM
hydronium ion

15.2 REACTIONS OF ACIDS AND BASES

KEY TERM

amphoteric

- Acids react with metals above hydrogen in the activity series to form hydrogen gas and a salt.
- Acids react with bases to form water and a salt (neutralization reaction).
- Acids react with metal oxides to form water and a salt.
- Acids react with carbonates to form water, a salt, and carbon dioxide.
- Bases react with acids to form water and a salt (neutralization reaction).
- Some amphoteric metals react with NaOH or KOH to form hydrogen and a salt.

15.3 SALTS

- Salts can be considered to be derived from the reaction of an acid and a base.
- Salts are crystalline and have high melting and boiling points.

15.4 ELECTROLYTES AND NONELECTROLYTES

KEY TERMS

electrolyte
nonelectrolyte
dissociation
ionization
strong electrolyte
weak electrolyte

- A substance whose aqueous solution conducts electricity is called an electrolyte:
  - Aqueous solution contains ions.
- A substance whose aqueous solution does not conduct electricity is called a nonelectrolyte:
  - Aqueous solution contains molecules, not ions.
- Dissociation is the process by which the ions of a salt separate as the salt dissolves.
- Ionization is the formation of ions that occurs as a result of a chemical reaction with water.
- Strong electrolytes are 100% ionized in solution:
  - Strong acids and strong bases
- Weak electrolytes are much less ionized in solution:
  - Weak acids and bases
- Colligative properties of electrolyte solutions depend on the number of particles produced during the ionization or dissociation of the electrolyte.
- Water can self-ionize to form H^+ and OH^- ions.
- Concentrations of ions in water at 25°C:
  - $[H^+] = 1.0 \times 10^{-7}$
  - $[OH^-] = 1.0 \times 10^{-7}$

15.5 INTRODUCTION TO pH

KEY TERMS

pH
logarithm (log)

- $pH = -\log[H^+]$:
  - $pH < 7$ for acidic solution
  - $pH = 7$ in neutral solution
  - $pH > 7$ for basic solution
- The number of decimal places in a logarithm equals the number of significant figures in the original number.

15.6 NEUTRALIZATION

KEY TERMS

neutralization
spectator ion
titration

- The reaction of an acid and a base to form water and a salt is called neutralization:
  - The general equation for a neutralization reaction is

$$H^+(aq) + OH^-(aq) \longrightarrow H_2O(l)$$

- The quantitative study of a neutralization reaction is called a titration.

15.7 WRITING NET IONIC EQUATIONS

KEY TERMS

formula equation
total ionic equation
net ionic equation

- In a formula equation the compounds are written in their molecular, or formula, expressions.
- In the total ionic equation compounds are written as ions if they are strong electrolytes in solution and as molecules if they are precipitates, nonelectrolytes, or weak electrolytes in solution.
- The net ionic equation shows only the molecules or ions that have changed:
  - The spectators (nonreactors) are omitted.

15.8 ACID RAIN

- Acid rain is any atmospheric precipitation that is more acidic than usual.
- Acid rain causes significant damage and destruction to our environment.

REVIEW QUESTIONS

1. Since a hydrogen ion and a proton are identical, what differences exist between the Arrhenius and Brønsted–Lowry definitions of an acid? (Table 15.1)
2. Use the three acid–base theories (Arrhenius, Brønsted–Lowry, and Lewis) to define an acid and a base.
3. For each acid–base theory referred to in Question 2, write an equation illustrating the neutralization of an acid with a base.
4. Write the Lewis structure for the (a) bromide ion,(b) hydroxide ion, and (c) cyanide ion. Why are these ions considered to be bases according to the Brønsted–Lowry and Lewis acid–base theories?
5. What kinds of substances will form hydrogen gas when they react with an acid?
6. What kinds of substances will form carbon dioxide gas when they react with an acid?
7. Write the formulas and names of three salts that can be formed from nitric acid.
8. Write the formulas and names of three salts that can be formed from the base lithium hydroxide.
9. According to Figure 15.1, what type of substance must be in solution for the bulb to light?
10. Which of the following classes of compounds are electrolytes: acids, alcohols, bases, salts? (Table 15.2)
11. What two differences are apparent in the arrangement of water molecules about the hydrated ions as depicted in Figure 15.2?
12. Into what three classes of compounds do electrolytes generally fall?
13. Name each compound listed in Table 15.3.
14. A solution of HCl in water conducts an electric current, but a solution of HCl in hexane does not. Explain this behavior in terms of ionization and chemical bonding.

15. How do ionic compounds exist in their crystalline structure? What occurs when they are dissolved in water?
16. An aqueous methyl alcohol, CH_3OH, solution does not conduct an electric current, but a solution of sodium hydroxide, $NaOH$, does. What does this information tell us about the OH group in the alcohol?
17. Why does molten NaCl conduct electricity?
18. Explain the difference between dissociation of ionic compounds and ionization of molecular compounds.
19. Distinguish between strong and weak electrolytes.
20. Explain why ions are hydrated in aqueous solutions.
21. What is the main distinction between water solutions of strong and weak electrolytes?
22. The solubility of HCl gas in water, a polar solvent, is much greater than its solubility in hexane, a nonpolar solvent. How can you account for this difference?
23. The pH of a solution with a hydrogen ion concentration of 0.003 M is between what two whole numbers? (Table 15.4)
24. Which is more acidic, tomato juice or blood? (Table 15.5)
25. What are the relative concentrations of $H^+(aq)$ and $OH^-(aq)$ in (a) a neutral solution, (b) an acid solution, and (c) a basic solution?
26. Pure water, containing equal concentrations of both acid and base ions, is neutral. Why?
27. A solution with a pH of 7 is neutral. A solution with a pH less than 7 is acidic. A solution with a pH greater than 7 is basic. What do these statements mean?
28. Explain the purpose of a titration.
29. Write the net ionic equation for the reaction of a strong acid with a water-soluble hydroxide base in an aqueous solution.
30. How does acid rain form?

Most of the exercises in this chapter are available for assignment via the online homework management program, WileyPLUS (www.wileyplus.com) All exercises with blue numbers have answers in Appendix VI.

PAIRED EXERCISES

1. Identify the conjugate acid–base pairs in each of the following equations:
   (a) $NH_3 + H_2O \rightleftharpoons NH_4^+ + OH^-$
   (b) $HC_2H_3O_2 + H_2O \rightleftharpoons C_2H_3O_2^- + H_3O^+$
   (c) $H_2PO_4^- + OH^- \rightleftharpoons HPO_4^{2-} + H_2O$
   (d) $HCl + H_2O \rightarrow Cl^- + H_3O^+$

2. Identify the conjugate acid–base pairs in each of the following equations:
   (a) $H_2S + NH_3 \rightleftharpoons NH_4^+ + HS^-$
   (b) $HSO_4^- + NH_3 \rightleftharpoons SO_4^{2-} + NH_4^+$
   (c) $HBr + CH_3O^- \rightleftharpoons Br^- + CH_3OH$
   (d) $HNO_3 + H_2O \rightarrow NO_3^- + H_3O^+$

3. Complete and balance these equations:
   (a) $Zn(s) + HCl(aq) \longrightarrow$
   (b) $Al(OH)_3(s) + H_2SO_4(aq) \longrightarrow$
   (c) $Na_2CO_3(aq) + HC_2H_3O_2(aq) \longrightarrow$
   (d) $MgO(s) + HI(aq) \longrightarrow$
   (e) $Ca(HCO_3)_2(s) + HBr(aq) \longrightarrow$
   (f) $KOH(aq) + H_3PO_4(aq) \longrightarrow$

4. Complete and balance these equations:
   (a) $Fe_2O_3(s) + HBr(aq) \longrightarrow$
   (b) $Al(s) + H_2SO_4(aq) \longrightarrow$
   (c) $NaOH(aq) + H_2CO_3(aq) \longrightarrow$
   (d) $Ba(OH)_2(s) + HClO_4(aq) \longrightarrow$
   (e) $Mg(s) + HClO_4(aq) \longrightarrow$
   (f) $K_2O(s) + HI(aq) \longrightarrow$

5. For each of the formula equations in Question 3, write total and net ionic equations.

6. For each of the formula equations in Question 4, write total and net ionic equations.

7. Write the equation for the reaction that will occur when each pair of substances is allowed to react.
   (a) HNO_3 and NaOH
   (b) $HC_2H_3O_2$ and $Ba(OH)_2$
   (c) $HClO_4$ and NH_4OH

8. Write the equation for the reaction that will occur when each pair of substances is allowed to react.
   (a) HBr and $Mg(OH)_2$
   (b) H_3PO_4 and KOH
   (c) H_2SO_4 and NH_4OH

9. Predict the acid and base that must be reacted together to make the salt Li_2S. Write the equation for the reaction to form this salt.

10. Predict the acid and base that must be reacted together to make the salt $CaCO_3$. Write the equation for the reaction to form this salt.

11. Determine whether each of the following substances is an electrolyte or a nonelectrolyte. All are mixed with water.
(a) SO_3
(b) K_2CO_3
(c) CH_3OH (methyl alcohol)
(d) O_2 (insoluble)
(e) $CuBr_2$
(f) HI

13. Determine the molarity of each of the ions present in the following aqueous salt solutions: (assume 100% ionization)
(a) 1.25 M $CuBr_2$
(b) 0.75 M $NaHCO_3$
(c) 3.50 M K_3AsO_4
(d) 0.65 M $(NH_4)_2SO_4$

15. In Question 13, how many grams of each ion would be present in 100. mL of each solution?

17. Calculate the $[H^+]$ for
(a) black coffee, with a pH of 5.0
(b) black tea with a pH of 6.8
(c) green tea with a pH of 7.9
(d) laundry detergent with a pH of 9.7

19. Determine the molar concentration of each ion present in the solutions that result from each of the following mixtures: (Disregard the concentration of H^+ and OH^- from water and assume that volumes are additive.)
(a) 55.5 mL of 0.50 M HCl and 75.0 mL of 1.25 M HCl
(b) 125 mL of 0.75 M $CaCl_2$ and 125 mL of 0.25 M $CaCl_2$
(c) 35.0 mL of 0.333 M NaOH and 22.5 mL of 0.250 M HCl
(d) 12.5 mL of 0.500 M H_2SO_4 and 23.5 mL of 0.175 M NaOH

21. Write an equation showing how nitric acid ionizes in water. Would this be a strong electrolyte? A strong acid?

23. Given the data for the following separate titrations, calculate the molarity of the HCl:

	mL HCl	Molarity HCl	mL NaOH	Molarity NaOH
(a)	40.13	M	37.70	0.728
(b)	19.00	M	33.66	0.306
(c)	27.25	M	18.00	0.555

25. Balance each of the following equations; then change them into balanced net ionic equations:
(a) $K_3PO_4(aq) + Ca(NO_3)_2(aq) \rightarrow Ca_3(PO_4)_2(s) + KNO_3(aq)$
(b) $Al(s) + H_2SO_4(aq) \rightarrow H_2(g) + Al_2(SO_4)_3(aq)$
(c) $Na_2CO_3(aq) + HCl(aq) \rightarrow NaCl(aq) + H_2O(l) + CO_2(g)$

27. For each of the given pairs, determine which solution is more acidic. All are water solutions. Explain your answer.
(a) 1 M HCl or 1 M H_2SO_4?
(b) 1 M HCl or 1 M $HC_2H_3O_2$?

29. Determine how many milliliters of 0.525 M $HClO_4$ will be required to neutralize 50.25 g $Ca(OH)_2$ according to the reaction:

$$2\,HClO_4(aq) + Ca(OH)_2(s) \rightarrow Ca(ClO_4)_2(aq) + 2\,H_2O(l)$$

31. A 0.200-g sample of impure NaOH requires 18.25 mL of 0.2406 M HCl for neutralization. What is the percent of NaOH in the sample?

33. What volume of H_2 gas, measured at 27°C and 700. torr, can be obtained by reacting 5.00 g of zinc metal with 100. mL of 0.350 M HCl? The equation is

$$Zn(s) + 2\,HCl(aq) \longrightarrow ZnCl_2(aq) + H_2(g)$$

12. Determine whether each of the following substances is an electrolyte or a nonelectrolyte. All are mixed with water.
(a) $C_6H_{12}O_6$ (glucose)
(b) P_2O_5
(c) NaClO
(d) LiOH
(e) C_2H_5OH (ethyl alcohol)
(f) $KMnO_4$

14. Determine the molarity of each of the ions present in the following aqueous salt solutions: (assume 100% ionization)
(a) 2.25 M $FeCl_3$
(b) 1.20 M $MgSO_4$
(c) 0.75 M NaH_2PO_4
(d) 0.35 M $Ca(ClO_3)_2$

16. In Question 14, how many grams of each ion would be present in 100. mL of each solution?

18. Calculate the $[H^+]$ for
(a) pure water
(b) lemon lime soda with a pH of 4.3
(c) root beer with a pH of 5.8
(d) window cleaner with a pH of 10.3

20. Determine the molar concentration of each ion present in the solutions that result from each of the following mixtures: (Disregard the concentration of H^+ and OH^- from water and assume that volumes are additive.)
(a) 45.5 mL of 0.10 M NaCl and 60.5 mL of 0.35 M NaCl
(b) 95.5 mL of 1.25 M HCl and 125.5 mL of 2.50 M HCl
(c) 15.5 mL of 0.10 M $Ba(NO_3)_2$ and 10.5 mL of 0.20 M $AgNO_3$
(d) 25.5 mL of 0.25 M NaCl and 15.5 mL of 0.15 M $Ca(C_2H_3O_2)_2$

22. Write an equation showing how hydrocyanic acid ionizes in water. How is this ionization different than the ionization of nitric acid? Would this be a strong electrolyte? A strong acid?

24. Given the data for the following separate titrations, calculate the molarity of the NaOH:

	mL HCl	Molarity HCl	mL NaOH	Molarity NaOH
(a)	37.19	0.126	31.91	M
(b)	48.04	0.482	24.02	M
(c)	13.13	1.425	39.39	M

26. Balance each of the following equations; then change them into balanced net ionic equations:
(a) $Mg(s) + Cu(NO_3)_2(aq) \rightarrow Cu(s) + Mg(NO_3)_2(aq)$
(b) $H_2SO_4(aq) + NaOH(aq) \rightarrow Na_2SO_4(aq) + H_2O(l)$
(c) $K_2SO_3(aq) + HClO_4(aq) \rightarrow KClO_4(aq) + H_2O(l) + SO_2(g)$

28. For each of the given pairs, determine which solution is more acidic. All are water solutions. Explain your answer.
(a) 1 M HCl or 2 M HCl?
(b) 1 M HNO_3 or 1 M H_2SO_4?

30. Determine how many grams of $Al(OH)_3$ will be required to neutralize 275 mL of 0.125 M HCl according to the reaction:

$$3\,HCl(aq) + Al(OH)_3(s) \rightarrow AlCl_3(aq) + 3\,H_2O(l)$$

32. A batch of sodium hydroxide was found to contain sodium chloride as an impurity. To determine the amount of impurity, a 1.00-g sample was analyzed and found to require 49.90 mL of 0.466 M HCl for neutralization. What is the percent of NaCl in the sample?

34. What volume of H_2 gas, measured at 27°C and 700. torr, can be obtained by reacting 5.00 g of zinc metal with 200. mL of 0.350 M HCl? The equation is

$$Zn(s) + 2\,HCl(aq) \longrightarrow ZnCl_2(aq) + H_2(g)$$

35. Calculate the pH of solutions having the following $[H^+]$:
    (a) 0.35 M (b) 1.75 M (c) $2.0 \times 10^{-5} M$

36. Calculate the pH of solutions having the following $[H^+]$:
    (a) 0.0020 M (b) $7.0 \times 10^{-8} M$ (c) 3.0 M

37. Calculate the pH of
    (a) orange juice, $3.7 \times 10^{-4} M H^+$
    (b) vinegar, $2.8 \times 10^{-3} M H^+$
    (c) shampoo, $2.4 \times 10^{-6} M H^+$
    (d) dishwashing detergent, $3.6 \times 10^{-8} M H^+$

38. Calculate the pH of
    (a) black coffee, $5.0 \times 10^{-5} M H^+$
    (b) limewater, $3.4 \times 10^{-11} M H^+$
    (c) fruit punch, $2.1 \times 10^{-4} M H^+$
    (d) cranberry apple drink, $1.3 \times 10^{-3} M H^+$

39. Determine whether each of the following is a strong acid, weak acid, strong base, or weak base. Then write an equation describing the process that occurs when the substance is dissolved in water.
    (a) NH_3 (b) HCl (c) KOH (d) $HC_2H_3O_2$

40. Determine whether each of the following is a strong acid, weak acid, strong base, or weak base. Then write an equation describing the process that occurs when the substance is dissolved in water.
    (a) $H_2C_2O_4$ (b) $Ba(OH)_2$ (c) $HClO_4$ (d) HBr

ADDITIONAL EXERCISES

41. Write the formula for the conjugate acid of the following bases:
    (a) NH_2CH_3 (b) HS^-

42. Write the formula for the conjugate base of the following acids:
    (a) $HBrO_3$ (b) NH_4^+ (c) $H_2PO_4^-$

43. Write the equation for the reaction between the acid $HC_2H_3O_2$ and the base NH_3. Identify the conjugate acid–base pairs in the reaction.

44. Write a reaction between the sulfide ion (S^{2-}) and water showing how it acts as a Brønsted–Lowry base.

45. Write the reaction of hydrochloric acid with the active metal magnesium.

46. Write the reaction of sulfuric acid with calcium carbonate.

47. Write an equation showing how sodium sulfate dissociates in water.

48. Determine whether each of the following describes a substance that is acidic, basic, or neutral:
    (a) Red litmus paper turns blue   (d) $[OH^-] = 1 \times 10^{-8}$
    (b) pH = 6   (e) Blue litmus paper turns red
    (c) $[H^+] = 1 \times 10^{-7}$ (f) pH = 12

49. Draw pictures similar to those found in Figure 15.2 to show what happens to the following compounds when each is dissolved in water.
    (a) $CaCl_2$ (b) KF (c) $AlBr_3$

50. What is the concentration of Al^{3+} ions in a solution of $AlBr_3$ having a Br^- ion concentration of 0.142 M?

51. In an acid–base titration, 25.22 mL of H_2SO_4 were used to neutralize 35.22 mL of 0.313 M NaOH. What was the molarity of the H_2SO_4 solution?

$$H_2SO_4 + 2\,NaOH \rightarrow Na_2SO_4 + 2\,H_2O$$

52. A 1 m solution of acetic acid, $HC_2H_3O_2$, in water freezes at a lower temperature than a 1 m solution of ethyl alcohol, C_2H_5OH, in water. Explain.

53. At the same cost per pound, which alcohol, CH_3OH or C_2H_5OH, would be more economical to purchase as an antifreeze for your car? Why?

54. How does a hydronium ion differ from a hydrogen ion?

55. Arrange, in decreasing order of freezing points, 1 m aqueous solutions of HCl, $HC_2H_3O_2$, $C_{12}H_{22}O_{11}$ (sucrose), and $CaCl_2$. (List the one with the highest freezing point first.)

56. At 100°C the H^+ concentration in water is about 1×10^{-6} mol/L, about 10 times that of water at 25°C. At which of these temperatures is
    (a) the pH of water the greater?

(b) the hydrogen ion (hydronium ion) concentration the higher?
(c) the water neutral?

57. What is the relative difference in H^+ concentration in solutions that differ by one pH unit?

58. What is the mole percent of a 1.00 m aqueous solution?

59. A sample of pure sodium carbonate with a mass of 0.452 g was dissolved in water and neutralized with 42.4 mL of hydrochloric acid. Calculate the molarity of the acid:

$$Na_2CO_3(aq) + 2\,HCl(aq) \longrightarrow 2\,NaCl(aq) + CO_2(g) + H_2O(l)$$

60. What volume (mL) of 0.4233 M H_2SO_4 is needed to neutralize 6.38 g KOH?

61. How many grams of KOH are required to neutralize 50.00 mL of 0.240 M HNO_3?

62. Two drops (0.1 mL) of 1.0 M HCl are added to water to make 1.0 L of solution. What is the pH of this solution if the HCl is 100% ionized?

63. When 250.0 mL of water are added to 10.0 mL of 12 M HCl, what will be the new molarity?

64. If 3.0 g NaOH are added to 500. mL of 0.10 M HCl, will the resulting solution be acidic or basic? Show evidence for your answer.

65. If 380 mL of 0.35 M $Ba(OH)_2$ are added to 500. mL of 0.65 M HCl, will the mixture be acidic or basic? Find the pH of the resulting solution.

66. If 50.00 mL of 0.2000 M HCl are titrated with 0.2000 M NaOH, find the pH of the solution after the following amounts of base have been added:
    (a) 0.000 mL   (d) 49.00 mL   (f) 49.99 mL
    (b) 10.00 mL   (e) 49.90 mL   (g) 50.00 mL
    (c) 25.00 mL

    Plot your answers on a graph with pH on the y-axis and mL NaOH on the x-axis.

67. Sulfuric acid reacts with NaOH:
    (a) Write a balanced equation for the reaction producing Na_2SO_4.
    (b) How many milliliters of 0.10 M NaOH are needed to react with 0.0050 mol H_2SO_4?
    (c) How many grams of Na_2SO_4 will also form?

68. A 10.0-mL sample of HNO_3 was diluted to a volume of 100.00 mL. Then 25 mL of that diluted solution were needed to neutralize 50.0 mL of 0.60 M KOH. What was the concentration of the original nitric acid?

69. The pH of a solution of a strong acid was determined to be 3. If water is then added to dilute this solution, would the pH change? Why or why not? Could enough water ever be added to raise the pH of an acid solution above 7?

70. If 425 mL of 0.94 M H_2SO_4 are added to 750. mL of 0.83 M NaOH, will the mixture be acidic or basic? Calculate the pH of the resulting solution.

71. For each of the following substances determine the concentration of each ion/molecule in solution.
    (a) 1 M $CuSO_4$
    (b) 1 M HNO_3
    (c) 1 M H_2SO_4
    (d) $CaS(s)$
    (e) 1 M $HC_2H_3O_2$

72. Write the formula, total ionic, and net ionic equations for the following reactions:
    (a) HNO_3 + LiOH
    (b) HBr + $Ba(OH)_2$
    (c) HF + NaOH

73. A 25-L sample of rain from Pennsylvania was collected and titrated with 7.2 mL of a 0.125 M solution of sodium hydroxide. What was the pH of the rain?

74. Sodium rhodizonate is often used in forensic investigations to detect the presence of any residue from shooting a firearm. Burned and unburned particles from the gunpowder will be ejected with the projectile. The residue is then stained red using sodium rhodizonate at a pH of 2.8. How many mL of 0.60 M hydrochloric acid need to be added to a neutral solution of sodium rhodizonate to make 500.0 mL of solution with a pH of 2.800?

75. One of the tests used to detect the presence of LSD or some of the other psilocybin-like alkaloids often found in psychedelic mushrooms is the Ehrlich test. For this test a solution of p-dimethylaminobenzaldehyde, $C_9H_{11}NO$, in 3.25 M HCl is used to detect the presence of any of these compounds. How many mL of concentrated HCl (12.1 M) are required to make 3.00 L of 3.25 M HCl for preparing Ehrlich reagent?

76. One of the tests used by geologists to identify rocks containing calcium carbonate, such as limestone and marble, is to drop dilute hydrochloric acid onto the rock. Fizzing or bubbling results if there are carbonates in the rocks. Write the balanced chemical reaction between calcium carbonate and aqueous hydrochloric acid.

CHALLENGE EXERCISES

77. An HCl solution has a pH of 0.300. What volume of water must be added to 200 mL of this solution to change the pH to 0.150?

78. Lactic acid (found in sour milk) has an empirical formula of $HC_3H_5O_3$. A 1.0-g sample of lactic acid required 17.0 mL of 0.65 M NaOH to reach the end point of a titration. What is the molecular formula for lactic acid?

ANSWERS TO PRACTICE EXERCISES

15.1 (a) HCO_3^-, (b) NO_2^-, (c) $C_2H_3O_2^-$

15.2 (a) H_2SO_4, (b) NH_4^+, (c) H_2O

15.3 (a) Reaction with metals: Single displacement

(b) Reaction with bases: Double displacement

(c) Reaction with metal oxides: Double displacement

(d) Reaction with carbonates: Double displacement followed by decomposition

15.4 (a) KOH, potassium hydroxide; H_3PO_4, phosphoric acid

(b) $Mg(OH)_2$, magnesium hydroxide; HBr, hydrobromic acid

(c) LiOH, lithium hydroxide; HCl, hydrochloric acid

(d) $Fe(OH)_2$, iron(II) hydroxide; H_2CO_3, carbonic acid

15.5 The cation of a salt comes from the cation of the base, and the anion of a salt comes from the anion of the acid.

NaCl, Na_2SO_4, Na_3PO_4,

$MgCl_2$, $MgSO_4$, $Mg_3(PO_4)_2$,

$FeCl_3$, $Fe_2(SO_4)_3$, $FePO_4$.

15.6 (a) 0.050 M Mg^{2+}, 0.10 M Cl^-,

(b) 0.070 M Al^{3+}, 0.21 M Cl^-

15.7 approximately 101.54°C

15.8 (a) 11.41, (b) 2.89, (c) 5.429

15.9 0.0620 M HCl

15.10 $3\,H_2S(aq) + 2\,Bi^{3+}(aq) \longrightarrow Bi_2S_3(s) + 6\,H^+(aq)$

Keeping fish in an aquarium requires maintaining an equilibrium among the living organisms and the water.

CHAPTER **16**

CHEMICAL EQUILIBRIUM

Thus far, we've considered chemical change as proceeding from reactants to products. Does that mean that the change then stops? No, but often it appears to be the case at the macroscopic level. A solute dissolves until the solution becomes saturated. Once a solid remains undissolved in a container, the system appears to be at rest. The human body is a marvelous chemical factory, yet from day to day it appears to be quite the same. For example, the blood remains at a constant pH, even though all sorts of chemical reactions are taking place. Another example is a terrarium, which can be watered and sealed for long periods of time with no ill effects. Or an antacid, which neutralizes excess stomach acid. In all of these cases, reactions are proceeding, even though visible signs of chemical change are absent. Similarly, when a system is at equilibrium, chemical reactions are dynamic at the molecular level. In this chapter, we will consider chemical systems as they approach equilibrium conditions.

CHAPTER OUTLINE

16.1 Rates of Reaction
16.2 Chemical Equilibrium
16.3 Le Châtelier's Principle
16.4 Equilibrium Constants
16.5 Ion Product Constant for Water
16.6 Ionization Constants
16.7 Solubility Product Constant
16.8 Buffer Solutions: The Control of pH

16.1 RATES OF REACTION

LEARNING OBJECTIVE   ● List the factors that affect the rate of a chemical reaction.

KEY TERM

chemical kinetics

Every reaction has a rate, or speed, at which it proceeds. Some reactions are fast; some are extremely slow. The study of reaction rates and reaction mechanisms is known as **chemical kinetics**.

The rate of a reaction is variable and depends on the concentration of the reacting species, the temperature, the presence of catalysts, and the nature of the reactants. Consider the hypothetical reaction

$$A + B \longrightarrow C + D \quad \text{(forward reaction)}$$

$$C + D \longrightarrow A + B \quad \text{(reverse reaction)}$$

in which a collision between A and B is necessary for a reaction to occur. The rate at which A and B react depends on the concentration, or the number, of A and B molecules present; it will be fastest, for a fixed set of conditions, when they are first mixed (as shown by the height of the red line in **Figure 16.1**). As the reaction proceeds, the number of A and B molecules available for reaction decreases, and the rate of reaction slows down (seen as the red line flattens in Figure 16.1). If the reaction is reversible, the speed of the reverse reaction is zero at first (blue line in Figure 16.1) and gradually increases as the concentrations of C and D increase. As the number of A and B molecules decreases, the forward rate slows down because A and B cannot find one another as often in order to accomplish a reaction. To counteract this diminishing rate of reaction, an excess of one reagent is often used to keep the reaction from becoming impractically slow. Collisions between molecules may be compared to video games. When many objects are on the screen, collisions occur frequently; but if only a few objects are present, collisions can usually be avoided.

EXAMPLE 16.1

When two reactants are mixed, what happens to the forward and the reverse reactions as the reaction proceeds to equal rates of reaction?

SOLUTION

The rate of the forward reaction slows down and the reverse reaction speeds up until equilibrium is reached in which both reactions are proceeding at the same rate.

PRACTICE 16.1

When two reactants are mixed, why does the rate of the forward reaction slow down?

Figure 16.1

The rates of the forward and reverse reactions become equal at some point in time. The forward reaction rate (red) decreases as a result of decreasing amounts of reactants. The reverse reaction rate (blue) starts at zero and increases as the amount of product increases. When the two rates become equal (purple), a state of chemical equilibrium has been reached.

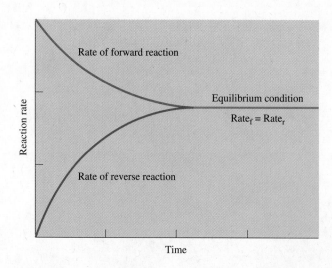

16.2 CHEMICAL EQUILIBRIUM

Define a reversible chemical reaction and explain what is occurring in a chemical system of equilibrium.

● **LEARNING OBJECTIVE**

Reversible Reactions

In the preceding chapters, we treated chemical reactions mainly as reactants changing to products. However, many reactions do not go to completion. Some reactions do not go to completion because they are reversible; that is, when the products are formed, they react to produce the starting reactants.

We've encountered reversible systems before. One is the vaporization of a liquid by heating and its subsequent condensation by cooling:

$$\text{liquid} + \text{heat} \longrightarrow \text{vapor}$$

$$\text{vapor} + \text{cooling} \longrightarrow \text{liquid}$$

The conversion between nitrogen dioxide, NO_2, and dinitrogen tetroxide, N_2O_4, shows us visible evidence of the reversibility of a reaction. The NO_2 is a reddish-brown gas that changes with cooling to N_2O_4, a colorless gas. The reaction is reversible by heating N_2O_4:

$$2\,NO_2(g) \xrightarrow{\text{cooling}} N_2O_4(g)$$

$$N_2O_4(g) \xrightarrow{\text{heating}} 2\,NO_2(g)$$

These two reactions may be represented by a single equation with a double arrow, $\rightleftharpoons$, to indicate that the reactions are taking place in both directions at the same time:

$$2\,NO_2(g) \rightleftharpoons N_2O_4(g)$$

This reversible reaction can be demonstrated by sealing samples of NO_2 in two tubes and placing one tube in warm water and the other in ice water (see **Figure 16.2**).

A **reversible chemical reaction** is one in which the products formed react to produce the original reactants. Both the forward and reverse reactions occur simultaneously. The forward reaction is also called *the reaction to the right*, and the reverse reaction is also called *the reaction to the left*. A double arrow is used in the equation to indicate that the reaction is reversible.

Any system at **equilibrium** represents a dynamic state in which two or more opposing processes are taking place at the same time and at the same rate. A chemical equilibrium is a dynamic system in which two or more opposing chemical reactions are going on at the same time and at the same rate. When the rate of the forward reaction is exactly equal to the rate of the reverse reaction, a condition of **chemical equilibrium** exists (see purple line in

KEY TERMS

reversible chemical reaction
equilibrium
chemical equilibrium

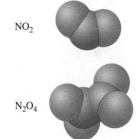

NO_2

N_2O_4

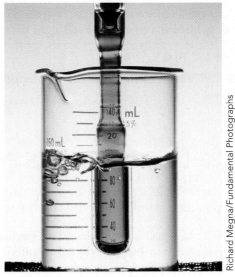

Figure 16.2

Reversible reaction of NO_2 and N_2O_4. More of the dark brown molecules are visible in the heated container on the right than in the room-temperature tube on the left.

Richard Megna/Fundamental Photographs

Richard Megna/Fundamental Photographs

Figure 16.1). The concentrations of the products and the reactants are not changing, and the system appears to be at a standstill because the products are reacting at the same rate at which they are being formed.

> Chemical equilibrium:
> rate of forward reaction = rate of reverse reaction

A saturated salt solution is in a condition of equilibrium:

$$NaCl(s) \rightleftharpoons Na^+(aq) + Cl^-(aq)$$

At equilibrium, salt crystals are continuously dissolving, and Na^+ and Cl^- ions are continuously crystallizing. Both processes are occurring at the same rate.

The ionization of weak electrolytes is another chemical equilibrium system:

$$HC_2H_3O_2(aq) + H_2O(l) \rightleftharpoons H_3O^+(aq) + C_2H_3O_2^-(aq)$$

In this reaction, the equilibrium is established in a 1 M solution when the forward reaction has gone about 1%—that is, when only 1% of the acetic acid molecules in solution have ionized. Therefore, only a relatively few ions are present, and the acid behaves as a weak electrolyte.

The reaction represented by

$$H_2(g) + I_2(g) \xrightleftharpoons{700\ K} 2\ HI(g)$$

provides another example of chemical equilibrium. Theoretically, 1.00 mol of hydrogen should react with 1.00 mol of iodine to yield 2.00 mol of hydrogen iodide. Actually, when 1.00 mol H_2 and 1.00 mol I_2 are reacted at 700 K, only 1.58 mol HI are present when equilibrium is attained. Since 1.58 is 79% of the theoretical yield of 2.00 mol HI, the forward reaction is only 79% complete at equilibrium. The equilibrium mixture will also contain 0.21 mol each of unreacted H_2 and I_2 (1.00 mol − 0.79 mol = 0.21 mol):

$$H_2(g) + I_2(g) \xrightarrow{700\ K} 2\ HI(g)$$

This equation represents the condition if the reaction were 100% complete; 2.00 mol HI would be formed, and no H_2 and I_2 would be left unreacted.

$$H_2(g) + I_2(g) \xrightleftharpoons{700\ K} 2\ HI(g)$$
$$\;0.21\qquad 0.21\qquad\qquad 1.58$$
$$\;\text{mol}\qquad \text{mol}\qquad\qquad \text{mol}$$

This equation represents the actual equilibrium attained starting with 1.00 mol each of H_2 and I_2. It shows that the forward reaction is only 79% complete.

EXAMPLE 16.2

Is a reversible chemical reaction at equilibrium a static or a dynamic system? Explain.

SOLUTION

A reversible chemical reaction is a dynamic system in which two opposing reactions are taking place at the same time and at the same rate of reaction.

PRACTICE 16.2

What symbolism is used in a chemical equation to indicate that a chemical reaction is reversible?

16.3 LE CHÂTELIER'S PRINCIPLE

LEARNING OBJECTIVE ● Use Le Châtelier's principle to predict the changes that occur when concentration, temperature, or volume is changed in a system at equilibrium.

KEY TERMS

Le Châtelier's principle
catalyst
activation energy

In 1888, the French chemist Henri Le Châtelier (1850–1936) set forth a simple, far-reaching generalization on the behavior of equilibrium systems. This generalization, known as **Le Châtelier's principle**, states:

>CHEMISTRY *IN ACTION*

New Ways in Fighting Cavities and Avoiding the Drill

Dentists have understood for more than 20 years what causes cavities, but, until now, there have been only a limited number of over-the-counter products to help us avoid our dates with the drill. Bacteria in the mouth break down sugars remaining in the mouth after eating. Acids produced during this process slip through tooth enamel, dissolving minerals below the surface in a process called demineralization. Saliva works to rebuild teeth by adding calcium and phosphate back in a process called remineralization. Under ideal conditions (assuming that you brush after eating), these two processes form an equilibrium.

Unfortunately, bacteria in plaque (resulting from not brushing) shift the equilibrium toward demineralization (shown in the figure), and a cavity can begin to form. Scientists realized that fluoride encourages remineralization in teeth by replacing hydroxyl ions in nature's calcium phosphate (hydroxyapatite). The substitution changes the hydroxyapatite to fluorapatite, which is more acid resistant.

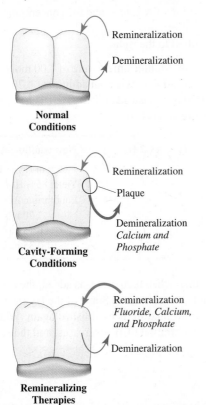

The good news is that scientists have now figured out how to shift the equilibrium between demineralization and remineralization toward remineralization. This new understanding has resulted in products that actually stop cavities and prevent the need for a filling. Scientists now know that adding calcium and phosphate to the mouth regularly speeds up remineralization of the teeth. The largest problem in doing it this way is to figure out how to dispense the calcium and phosphate through the surface enamel gradually. If a calcium ion finds a phosphate ion before this happens, a precipitate forms that can't penetrate the enamel. Researchers have developed new products such as chewing gum and sticky substances (see photo) that can be applied between the teeth. These products work better than brushing or rinsing because they last longer—a brushing or rinsing lasts only a minute.

Several types of remineralizing chewing gums are currently available, including Trident Advantage and Trident for Kids. The calcium and phosphate ions are stablilized by using a system that mimics nature. Researchers looked at how newborns get their large supply of calcium and phosphate. Eric Reynolds of the University of Melbourne in Australia says, "Nature's evolved this system in milk to carry very high levels of calcium and phosphate in a highly bioavailable form." They used casein (a milk protein) to produce microscopic particles of stabilized calcium phosphate that diffuse through the surface of teeth. When these particles were placed into the chewing gum and chewed, the results were better than hoped. In Japan, researchers at Tokyo's FAP Dental Institute have designed a toothpaste that repairs cavities smaller than 50 μm. The toothpaste grows nanocrystals of hydroxyapatite, which is treated with fluoride right on the cavity. The synthetic enamel not only repairs the tiny cavity as you brush it but also strengthens the natural enamel. Brushing can now repair cavities and prevent new ones at the same time.

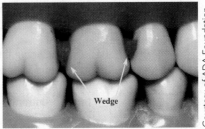

These sticky wedges dissolve slowly to release fluoride, calcium, and phosphate to teeth.

> If a stress is applied to a system in equilibrium, the system will respond in such a way as to relieve that stress and restore equilibrium under a new set of conditions.

The application of Le Châtelier's principle helps us predict the effect of changing conditions in chemical reactions. We will examine the effect of changes in concentration, temperature, and volume.

Effect of Concentration on Equilibrium

The manner in which the rate of a chemical reaction depends on the concentration of the reactants must be determined experimentally. Many simple, one-step reactions result from a collision between two molecules or ions. The rate of such one-step reactions can be altered by changing the concentration of the reactants or products. An increase in concentration of the reactants provides more individual reacting species for collisions and results in an increase in the rate of reaction.

An equilibrium is disturbed when the concentration of one or more of its components is changed. As a result, the concentration of all species will change, and a new equilibrium mixture will be established. Consider the hypothetical equilibrium represented by the equation

$$A + B \rightleftharpoons C + D$$

where A and B react in one step to form C and D. When the concentration of B is increased, the following results occur:

1. The rate of the reaction to the right (forward) increases. This rate is proportional to the concentration of A times the concentration of B.
2. The rate to the right becomes greater than the rate to the left.
3. Reactants A and B are used faster than they are produced; C and D are produced faster than they are used.
4. After a period of time, rates to the right and left become equal, and the system is again in equilibrium.
5. In the new equilibrium the concentration of A is less, and the concentrations of B, C, and D are greater than in the original equilibrium.
   Conclusion: The equilibrium has shifted to the right.

Concentration	Change
[H_2]	?
[I_2]	increase
[HI]	?

Applying this change in concentration to the equilibrium mixture of 1.00 mol of hydrogen and 1.00 mol of iodine from Section 16.2, we find that, when an additional 0.20 mol I_2 is added, the yield of HI (based on H_2) is 85% (1.70 mol) instead of 79%. Here is how the two systems compare after the new equilibrium mixture is reached:

Original equilibrium	$H_2 + I_2 \rightleftharpoons$ 2 HI	New equilibrium
1.00 mol H_2 + 1.00 mol I_2		1.00 mol H_2 + 1.20 mol I_2
Yield: 79% HI		Yield: 85% HI (based on H_2)
Equilibrium mixture contains:		Equilibrium mixture contains:
1.58 mol HI		1.70 mol HI
0.21 mol H_2		0.15 mol H_2
0.21 mol I_2		0.35 mol I_2

Concentration	Change
[H_2]	decrease
[I_2]	increase
[HI]	increase

Analyzing this new system, we see that, when 0.20 mol I_2 is added, the equilibrium shifts to the right to counteract the increase in I_2 concentration. Some of the H_2 reacts with added I_2 and produces more HI, until an equilibrium mixture is established again. When I_2 is added, the concentration of I_2 increases, the concentration of H_2 decreases, and the concentration of HI increases.

PRACTICE 16.3

Use a chart like the one in the margin to show what would happen to the concentrations of each substance in the system

$$H_2(g) + I_2(g) \rightleftharpoons 2 \text{ HI}(g)$$

upon adding (a) more H_2 and (b) more HI

The equation

$$Fe^{3+}(aq) + SCN^-(aq) \rightleftharpoons Fe(SCN)^{2+}(aq)$$

pale yellow   colorless   red

represents an equilibrium that is used in certain analytical procedures as an indicator because of the readily visible, intense red color of the complex $Fe(SCN)^{2+}$ ion. A very dilute solution of iron(III) (Fe^{3+}) and thiocyanate (SCN^-) is light red. When the concentration of either Fe^{3+} or SCN^- is increased, the equilibrium shift to the right is observed by an increase in the intensity of the red color, resulting from the formation of additional $Fe(SCN)^{2+}$.

If either Fe^{3+} or SCN^- is removed from solution, the equilibrium will shift to the left, and the solution will become lighter in color. When Ag^+ is added to the solution, a white precipitate of silver thiocyanate (AgSCN) is formed, thus removing SCN^- ion from the equilibrium:

$$Ag^+(aq) + SCN^-(aq) \rightleftharpoons AgSCN(s)$$

The system accordingly responds to counteract the change in SCN^- concentration by shifting the equilibrium to the left. This shift is evident by a decrease in the intensity of the red color due to a decreased concentration of $Fe(SCN)^{2+}$.

Now consider the effect of changing the concentrations in the equilibrium mixture of chlorine water. The equilibrium equation is

$$Cl_2(aq) + 2 H_2O(l) \rightleftharpoons HOCl(aq) + H_3O^+(aq) + Cl^-(aq)$$

The variation in concentrations and the equilibrium shifts are tabulated in the following table. An X in the second or third column indicates that the reagent is increased or decreased. The fourth column indicates the direction of the equilibrium shift.

| Reagent | Concentration | | Equilibrium shift |
	Increase	Decrease	
Cl_2	—	X	Left
H_2O	X	—	Right
HOCl	X	—	Left
H_3O^+	—	X	Right
Cl^-	X	—	Left

Consider the equilibrium in a 0.100 M acetic acid solution:

$$HC_2H_3O_2(aq) + H_2O(l) \rightleftharpoons H_3O^+(aq) + C_2H_3O_2^-(aq)$$

In this solution, the concentration of the hydronium ion (H_3O^+), which is a measure of the acidity, is 1.34×10^{-3} mol/L, corresponding to a pH of 2.87. What will happen to the acidity when 0.100 mol of sodium acetate ($NaC_2H_3O_2$) is added to 1 L of 0.100 M acetic acid ($HC_2H_3O_2$)? When $NaC_2H_3O_2$ dissolves, it dissociates into sodium ions (Na^+) and acetate ions ($C_2H_3O_2^-$). The acetate ion from the salt is a common ion to the acetic acid equilibrium system and increases the total acetate ion concentration in the solution. As a result, the equilibrium shifts to the left, decreasing the hydronium ion concentration and lowering the acidity of the solution. Evidence of this decrease in acidity is shown by the fact that the pH of a solution that is 0.100 M in $HC_2H_3O_2$ and 0.100 M in $NaC_2H_3O_2$ is 4.74. The pH of several different solutions of $HC_2H_3O_2$ and $NaC_2H_3O_2$ is shown in the table that follows. Each time the acetate ion is increased, the pH increases, indicating a further shift in the equilibrium toward un-ionized acetic acid.

Concentration	Change
$[HC_2H_3O_2]$	increase
$[H_3O^+]$	decrease
$[C_2H_3O_2^-]$	increase

Solution	pH
1 L 0.100 M $HC_2H_3O_2$	2.87
1 L 0.100 M $HC_2H_3O_2$ + 0.100 mol $NaC_2H_3O_2$	4.74
1 L 0.100 M $HC_2H_3O_2$ + 0.200 mol $NaC_2H_3O_2$	5.05
1 L 0.100 M $HC_2H_3O_2$ + 0.300 mol $NaC_2H_3O_2$	5.23

In summary, we can say that when the concentration of a reagent on the left side of an equation is increased, the equilibrium shifts to the right. When the concentration of a reagent on the right side of an equation is increased, the equilibrium shifts to the left. In accordance with Le Châtelier's principle, the equilibrium always shifts in the direction that tends to reduce the concentration of the added reactant.

PRACTICE 16.4

Aqueous chromate ion, CrO_4^{2-}, exists in equilibrium with aqueous dichromate ion, $Cr_2O_7^{2-}$, in an acidic solution. What effect will (a) increasing the dichromate ion and (b) adding HCl have on the equilibrium?

$$2\, CrO_4^{2-}(aq) + 2\, H^+(aq) \rightleftharpoons Cr_2O_7^{2-}(aq) + H_2O(l)$$

Effect of Volume on Equilibrium

Changes in volume significantly affect the reaction rate only when one or more of the reactants or products is a gas and the reaction is run in a closed container. In these cases the effect of decreasing the volume of the reacting gases is equivalent to increasing their concentrations. In the reaction

$$CaCO_3(s) \overset{\Delta}{\rightleftharpoons} CaO(s) + CO_2(g)$$

calcium carbonate decomposes into calcium oxide and carbon dioxide when heated about 825°C. Decreasing the volume of the container speeds up the reverse reaction and causes the equilibrium to shift to the left. Decreasing the volume increases the concentration of CO_2, the only gaseous substance in the reaction.

If the volume of the container is decreased, the pressure of the gas will increase. In a system composed entirely of gases, this decrease in the volume of the container will cause the reaction and the equilibrium to shift to the side that contains the smaller number of molecules. To clarify what's happening: When the container volume is decreased, the pressure in the container is increased. The system tries to lower this pressure by reducing the number of molecules. Let's consider an example that shows these effects.

Prior to World War I, Fritz Haber (1868–1934) invented the first major process for the fixation of nitrogen. In this process, nitrogen and hydrogen are reacted together in the presence of a catalyst at moderately high temperature and pressure to produce ammonia:

$$N_2(g) \;+\; 3\,H_2(g) \rightleftharpoons 2\,NH_3(g) \;+\; 92.5\ kJ$$

| 1 mol | 3 mol | 2 mol |
| 1 volume | 3 volumes | 2 volumes |

The left side of the equation in the Haber process represents 4 mol of gas combining to give 2 mol of gas on the right side of the equation. A decrease in the volume of the container shifts the equilibrium to the right. This decrease in volume results in a higher concentration of both reactants and products. The equilibrium shifts to the right toward fewer molecules.

Haber received the Nobel Prize in chemistry for this process in 1918.

Gaseous ammonia is often used to add nitrogen to the fields before planting and during early growth.

When the total number of gaseous molecules on both sides of an equation is the same, a change in volume does not cause an equilibrium shift. The following reaction is an example:

$$N_2(g) \quad\quad + \quad O_2(g) \quad\quad\quad \rightleftharpoons \quad 2\,NO(g)$$

1 mol	1 mol	2 mol
1 volume	1 volume	2 volumes
6.022×10^{23} molecules	6.022×10^{23} molecules	$2(6.022 \times 10^{23})$ molecules

When the volume of the container is decreased, the rate of both the forward and the reverse reactions will increase because of the higher concentrations of N_2, O_2, and NO. But the equilibrium will not shift because the number of molecules is the same on both sides of the equation and the effects on concentration are the same on both forward and reverse rates.

WileyPLUS

ENHANCED EXAMPLE

EXAMPLE 16.3

What effect would a decrease in volume of the container have on the position of equilibrium in these reactions?

(a) $2\,SO_2(g) + O_2(g) \rightleftharpoons 2\,SO_3(g)$

(b) $H_2(g) + Cl_2(g) \rightleftharpoons 2\,HCl(g)$

(c) $N_2O_4(g) \rightleftharpoons 2\,NO_2(g)$

SOLUTION

(a) The equilibrium will shift to the right because the substance on the right has a smaller number of moles than the substances on the left.

(b) The equilibrium position will be unaffected because the moles of gases on both sides of the equation are the same.

(c) The equilibrium will shift to the left because $N_2O_4(g)$ represents the smaller number of moles.

PRACTICE 16.5

What effect would a decrease in the container's volume have on the position of the equilibrium in these reactions?

(a) $2\,NO(g) + Cl_2(g) \rightleftharpoons 2\,NOCl(g)$

(b) $COBr_2(g) \rightleftharpoons CO(g) + Br_2(g)$

Effect of Temperature on Equilibrium

When the temperature of a system is raised, the rate of reaction increases because of increased kinetic energy and more frequent collisions of the reacting species. In a reversible reaction, the rate of both the forward and the reverse reactions is increased by an increase in temperature; however, the reaction that absorbs heat increases to a greater extent, and the equilibrium shifts to favor that reaction.

High temperatures can also cause the destruction or decomposition of the reactants or products.

An increase in temperature generally increases the rate of reaction. Molecules at elevated temperatures have more kinetic energy; their collisions are thus more likely to result in a reaction.

When heat is applied to a system in equilibrium, the reaction that absorbs heat is favored. When the process, as written, is endothermic, the forward reaction is increased. When the reaction is exothermic, the reverse reaction is favored. In this sense heat may be treated as a reactant in endothermic reactions or as a product in exothermic reactions. Therefore, temperature is analogous to concentration when applying Le Châtelier's principle to heat effects on a chemical reaction.

Hot coke (C) is a very reactive element. In the reaction

$$C(s) + CO_2(g) + heat \rightleftharpoons 2\,CO(g)$$

very little if any CO is formed at room temperature. At 1000°C, the equilibrium mixture contains about an equal number of moles of CO and CO_2. Since the reaction is endothermic, the equilibrium is shifted to the right at higher temperatures.

Light sticks. The chemical reaction that produces light in these light sticks is endothermic. Placing the light stick in hot water (right) favors this reaction, producing a brighter light than when the light stick is in ice water (left).

When phosphorus trichloride reacts with dry chlorine gas to form phosphorus pentachloride, the reaction is exothermic:

$$PCl_3(l) + Cl_2(g) \rightleftharpoons PCl_5(s) + 88 \text{ kJ}$$

Heat must continuously be removed during the reaction to obtain a good yield of the product. According to Le Châtelier's principle, heat will cause the product, PCl_5, to decompose, re-forming PCl_3 and Cl_2. The equilibrium mixture at 200°C contains 52% PCl_5, and at 300°C it contains 3% PCl_5, verifying that heat causes the equilibrium to shift to the left.

EXAMPLE 16.4

What effect would an increase in temperature have on the position of the equilibrium in these reactions?

$4 HCl(g) + O_2(g) \rightleftharpoons 2 H_2O(g) + 2 Cl_2(g) + 95.4 \text{ kJ}$	(1)
$H_2(g) + Cl_2(g) \rightleftharpoons 2 HCl(g) + 185 \text{ kJ}$	(2)
$CH_4(g) + 2 O_2(g) \rightleftharpoons CO_2(g) + 2 H_2O(g) + 890 \text{ kJ}$	(3)
$N_2O_4(g) + 58.6 \text{ kJ} \rightleftharpoons 2 NO_2(g)$	(4)
$2 CO_2(g) + 566 \text{ kJ} \rightleftharpoons 2 CO(g) + O_2(g)$	(5)
$H_2(g) + I_2(g) + 51.9 \text{ kJ} \rightleftharpoons 2 HI(g)$	(6)

SOLUTION

Reactions 1, 2, and 3 are exothermic; an increase in temperature will cause the equilibrium to shift to the left. Reactions 4, 5, and 6 are endothermic; an increase in temperature will cause the equilibrium to shift to the right.

PRACTICE 16.6

What effect would an increase in temperature have on the position of the equilibrium in these reactions?

(a) $2 SO_2(g) + O_2(g) \rightleftharpoons 2 SO_3(g) + 198 \text{ kJ}$

(b) $H_2(g) + CO_2(g) + 41 \text{ kJ} \rightleftharpoons H_2O(g) + CO(g)$

Effect of Catalysts on Equilibrium

A **catalyst** is a substance that influences the rate of a chemical reaction and can be recovered essentially unchanged at the end of the reaction. A catalyst does not shift the equilibrium of a reaction; it affects only the speed at which the equilibrium is reached. It does this by lowering the activation energy for the reaction (see **Figure 16.3**). **Activation energy** is the minimum energy required for the reaction to occur. A catalyst speeds up a reaction by lowering the activation energy while not changing the energies of reactants or products. If a catalyst does not

Figure 16.3

Energy diagram for an exothermic reaction. Energy is put into the reaction (activation energy) to initiate the process. In the reaction shown, all of the activation energy and the net energy are released as the reaction proceeds to products. Note that the presence of a catalyst lowers the activation energy but does not change the energies of the reactants or the products.

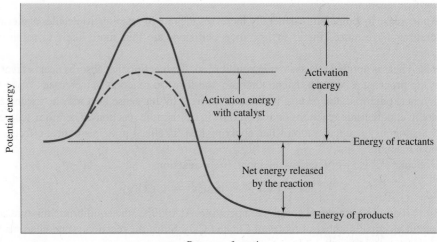

Progress of reaction

affect the equilibrium, then it follows that it must affect the rate of both the forward and the reverse reactions equally.

The reaction between phosphorus trichloride and sulfur is highly exothermic, but it's so slow that very little product, thiophosphoryl chloride, is obtained, even after prolonged heating. When a catalyst such as aluminum chloride is added, the reaction is complete in a few seconds:

$$PCl_3(l) + S(s) \xrightarrow{AlCl_3} PSCl_3(l)$$

The lab preparation of oxygen uses manganese dioxide as a catalyst to increase the rates of decomposition of both potassium chlorate and hydrogen peroxide:

$$2\ KClO_3(s) \xrightarrow[\Delta]{MnO_2} 2\ KCl(s) + 3\ O_2(g)$$

$$2\ H_2O_2(aq) \xrightarrow{MnO_2} 2\ H_2O(l) + O_2(g)$$

Catalysts are extremely important to industrial chemistry. Hundreds of chemical reactions that are otherwise too slow to be of practical value have been put to commercial use once a suitable catalyst was found. And in the area of biochemistry, catalysts are of supreme importance because nearly all chemical reactions in all forms of life are completely dependent on biochemical catalysts known as *enzymes*.

16.4 EQUILIBRIUM CONSTANTS

LEARNING OBJECTIVE

Write the general expression for the equilibrium constant and calculate equilibrium constants.

KEY TERM

equilibrium constant, K_{eq}

In a reversible chemical reaction at equilibrium, the concentrations of the reactants and products are constant. At equilibrium, the rates of the forward and reverse reactions are equal, and an equilibrium constant expression can be written relating the products to the reactants. For the general reaction

$$aA + bB \rightleftharpoons cC + dD$$

at a given temperature, the equilibrium constant expression can be written as

$$K_{eq} = \frac{[C]^c[D]^d}{[A]^a[B]^b}$$

where K_{eq}, the **equilibrium constant**, is constant at a particular temperature. The quantities in brackets are the concentrations of each substance in moles per liter. The superscript letters a, b, c, and d are the coefficients of the substances in the balanced equation. According to convention, we place the concentrations of the products (the substances on the right side of the equation as written) in the numerator and the concentrations of the reactants in the denominator.

Note: The exponents (a, b, c and d) are the same as the coefficients in the balanced equation.

EXAMPLE 16.5

Write equilibrium constant expressions for
(a) $3\ H_2(g) + N_2(g) \rightleftharpoons 2\ NH_3(g)$
(b) $CO(g) + 2\ H_2(g) \rightleftharpoons CH_3OH(g)$

SOLUTION

(a) The only product, NH_3, has a coefficient of 2. Therefore, the numerator will be $[NH_3]^2$. Two reactants are present: H_2, with a coefficient of 3, and N_2, with a coefficient of 1. The denominator will thus be $[H_2]^3[N_2]$. The equilibrium constant expression is

$$K_{eq} = \frac{[NH_3]^2}{[H_2]^3[N_2]}$$

(b) For this equation, the numerator is $[CH_3OH]$ and the denominator is $[CO][H_2]^2$. The equilibrium constant expression is

$$K_{eq} = \frac{[CH_3OH]}{[CO][H_2]^2}$$

WileyPLUS

ENHANCED EXAMPLE

PRACTICE 16.7

Write equilibrium constant expressions for

(a) $2 N_2O_5(g) \rightleftharpoons 4 NO_2(g) + O_2(g)$

(b) $4 NH_3(g) + 3 O_2(g) \rightleftharpoons 2 N_2(g) + 6 H_2O(g)$

The magnitude of an equilibrium constant indicates the extent to which the forward and reverse reactions take place. When K_{eq} is greater than 1, the amount of products at equilibrium is greater than the amount of reactants. When K_{eq} is less than 1, the amount of reactants at equilibrium is greater than the amount of products. A very large value for K_{eq} indicates that the forward reaction goes essentially to completion. A very small K_{eq} means that the reverse reaction goes nearly to completion and that the equilibrium is far to the left (toward the reactants). Consider the following two examples:

$$H_2(g) + I_2(g) \rightleftharpoons 2 HI(g) \qquad K_{eq} = 54.8 \text{ at } 425°C$$

$$COCl_2(g) \rightleftharpoons CO(g) + Cl_2(g) \qquad K_{eq} = 7.6 \times 10^{-4} \text{ at } 400°C$$

In the first example, K_{eq} indicates that more product than reactant exists at equilibrium.

In the second equation, K_{eq} indicates that $COCl_2$ is stable and that very little decomposition to CO and Cl_2 occurs at 400°C. The equilibrium is far to the left.

When the molar concentrations of all species in an equilibrium reaction are known, the K_{eq} can be calculated by substituting the concentrations into the equilibrium constant expression.

> Units are generally not included in values of K_{eq} for reasons beyond the scope of this book.

EXAMPLE 16.6

Calculate K_{eq} for the following reaction based on concentrations of $PCl_5 = 0.030 \text{ mol/L}$, $PCl_3 = 0.97 \text{ mol/L}$, and $Cl_2 = 0.97 \text{ mol/L}$ at 300°C.

$$PCl_5(g) \rightleftharpoons PCl_3(g) + Cl_2(g)$$

SOLUTION

First write the K_{eq} expression; then substitute the respective concentrations into this equation and solve:

$$K_{eq} = \frac{[PCl_3][Cl_2]}{[PCl_5]} = \frac{(0.97)(0.97)}{(0.030)} = 31$$

This K_{eq} is considered to be a fairly large value, indicating that at 300°C the decomposition of PCl_5 proceeds far to the right.

> Remember: Units are not included for K_{eq}.

PRACTICE 16.8

Calculate the K_{eq} for this reaction. Is the forward or the reverse reaction favored?

$$2 NO(g) + O_2(g) \rightleftharpoons 2 NO_2(g)$$

when $[NO] = 0.050 \, M$, $[O_2] = 0.75 \, M$, and $[NO_2] = 0.25 \, M$.

16.5 ION PRODUCT CONSTANT FOR WATER

LEARNING OBJECTIVE ● Calculate the concentrations of H^+, OH^-, pH and pOH in a solution using the ion product constant for water.

KEY TERM

ion product constant for water, K_w

We've seen that water ionizes to a slight degree. This ionization is represented by these equilibrium equations:

$$H_2O + H_2O \rightleftharpoons H_3O^+ + OH^- \qquad (1)$$

$$H_2O \rightleftharpoons H^+ + OH^- \qquad (2)$$

TABLE 16.1	Relationship of H^+ and OH^- Concentrations in Water Solutions			
$[H^+]$	$[OH^-]$	K_w	pH	pOH
1.00×10^{-2}	1.00×10^{-12}	1.00×10^{-14}	2.00	12.00
1.00×10^{-4}	1.00×10^{-10}	1.00×10^{-14}	4.00	10.00
2.00×10^{-6}	5.00×10^{-9}	1.00×10^{-14}	5.70	8.30
1.00×10^{-7}	1.00×10^{-7}	1.00×10^{-14}	7.00	7.00
1.00×10^{-9}	1.00×10^{-5}	1.00×10^{-14}	9.00	5.00

Equation 1 is the more accurate representation of the equilibrium because free protons (H^+) do not exist in water. Equation 2 is a simplified and often-used representation of the water equilibrium. The actual concentration of H^+ produced in pure water is minute and amounts to only 1.00×10^{-7} mol/L at 25°C. In pure water,

$$[H^+] = [OH^-] = 1.00 \times 10^{-7} \text{ mol/L}$$

since both ions are produced in equal molar amounts, as shown in equation 2.

The $H_2O \rightleftharpoons H^+ + OH^-$ equilibrium exists in water and in all water solutions. A special equilibrium constant called the **ion product constant for water, K_w**, applies to this equilibrium. The constant K_w is defined as the product of the H^+ ion concentration and the OH^- ion concentration, each in moles per liter:

$$K_w = [H^+][OH^-]$$

The numerical value of K_w is 1.00×10^{-14} for pure water at 25°C,

$$K_w = [H^+][OH^-] = (1.00 \times 10^{-7})(1.00 \times 10^{-7}) = 1.00 \times 10^{-14}$$

The value of K_w for all water solutions at 25°C is the constant 1.00×10^{-14}. It is important to realize that as the concentration of one of these ions, H^+ or OH^-, increases, the other decreases. However, the product of $[H^+]$ and $[OH^-]$ always equals 1.00×10^{-14}. This relationship can be seen in the examples shown in Table 16.1. If the concentration of one ion is known, the concentration of the other can be calculated from the K_w expression.

$$K_w = [H^+][OH^-] \qquad [H^+] = \frac{K_w}{[OH^-]} \qquad [OH^-] = \frac{K_w}{[H^+]}$$

EXAMPLE 16.7

WileyPLUS

ENHANCED EXAMPLE

What is the concentration of (a) H^+ and (b) OH^- in a 0.001 M HCl solution? Remember that HCl is 100% ionized.

SOLUTION

(a) Since all the HCl is ionized, $H^+ = 0.001$ mol/L, or 1×10^{-3} mol/L:

$$HCl \longrightarrow H^+ + Cl^-$$
$$\quad\quad 0.001\,M \quad 0.001\,M$$

$$[H^+] = 1 \times 10^{-3} \text{ mol/L}$$

(b) To calculate the $[OH^-]$ in this solution, use the following equation and substitute the values for K_w and $[H^+]$:

$$[OH^-] = \frac{K_w}{[H^+]}$$

$$[OH^-] = \frac{1.00 \times 10^{-14}}{1 \times 10^{-3}} = 1 \times 10^{-11} \text{ mol/L}$$

Determine the $[H^+]$ and $[OH^-]$ in

(a) $5.0 \times 10^{-5} M$ HNO$_3$ (b) $2.0 \times 10^{-6} M$ KOH

EXAMPLE 16.8

What is the pH of a 0.010 M NaOH solution? Assume that NaOH is 100% ionized.

SOLUTION

Since all the NaOH is ionized, $[OH^-] = 0.010$ mol/L or 1.0×10^{-2} mol/L.

$$NaOH \longrightarrow Na^+ + OH^-$$
$$\quad\quad 0.010\,M \quad 0.010\,M$$

To find the pH of the solution, we first calculate the H^+ ion concentration. Use the following equation and substitute the values for K_w and $[OH^-]$:

$$[H^+] = \frac{K_w}{[OH^-]} = \frac{1.00 \times 10^{-14}}{1.0 \times 10^{-2}} = 1.0 \times 10^{-12} \text{ mol/L}$$

$$pH = -\log[H^+] = -\log(1.0 \times 10^{-12}) = 12.00$$

PRACTICE 16.10

Determine the pH for the following solutions:

(a) $5.0 \times 10^{-5} M$ HNO$_3$ (b) $2.0 \times 10^{-6} M$ KOH

Just as pH is used to express the acidity of a solution, pOH is used to express the basicity of an aqueous solution. The pOH is related to the OH^- ion concentration in the same way that the pH is related to the H^+ ion concentration:

$$pOH = -\log[OH^-]$$

Thus a solution in which $[OH^-] = 1.0 \times 10^{-2}$, as in Example 16.8, will have pOH = 2.00.

In pure water, where $[H^+] = 1.00 \times 10^{-7}$ and $[OH^-] = 1.00 \times 10^{-7}$, the pH is 7.0 and the pOH is 7.0. The sum of the pH and pOH is always 14.00:

$$pH + pOH = 14.00$$

In Example 16.8, the pH can also be found by first calculating the pOH from the OH^- ion concentration and then subtracting from 14.00.

$$pH = 14.00 - pOH = 14.00 - 2.00 = 12.00$$

Table 16.1 summarizes the relationship between $[H^+]$ and $[OH^-]$ in water solutions.

16.6 IONIZATION CONSTANTS

LEARNING OBJECTIVE   ● Use the ionization constant of a reactant in an equilibrium expression to find the percent ionization of a substance in solution and to find the pH of a weak acid.

KEY TERM

acid ionization constant, K_a

In addition to K_w, several other equilibrium constants are commonly used. Strong acids are essentially 100% ionized. Weak acids are only slightly ionized. Let's consider the equilibrium constant for acetic acid in solution. Because it is a weak acid, an equilibrium is established between molecular HC$_2$H$_3$O$_2$ and its ions in solution:

$$HC_2H_3O_2(aq) \rightleftharpoons H^+(aq) + C_2H_3O_2^-(aq)$$

The ionization constant expression is the concentration of the products divided by the concentration of the reactants:

$$K_a = \frac{[H^+][C_2H_3O_2^-]}{[HC_2H_3O_2]}$$

The constant is called the **acid ionization constant, K_a**, a special type of equilibrium constant. The concentration of water in the solution is large compared to other concentrations and does not change appreciably. It is therefore part of the constant K_{eq}.

At 25°C, a 0.100 M $HC_2H_3O_2$ solution is 1.34% ionized and has an $[H^+]$ of 1.34×10^{-3} mol/L. From this information we can calculate the ionization constant for acetic acid.

A 0.100 M solution initially contains 0.100 mol of acetic acid per liter. Of this 0.100 mol, only 1.34%, or 1.34×10^{-3} mol, is ionized, which gives an $[H^+] = 1.34 \times 10^{-3}$ mol/L. Because each molecule of acid that ionizes yields one H^+ and one $C_2H_3O_2^-$, the concentration of $C_2H_3O_2^-$ ions is also 1.34×10^{-3} mol/L. This ionization leaves $0.100 - 0.00134 = 0.099$ mol/L of un-ionized acetic acid.

Substituting these concentrations in the equilibrium expression, we obtain the value for K_a:

$$K_a = \frac{[H^+][C_2H_3O_2^-]}{[HC_2H_3O_2]} = \frac{(1.34 \times 10^{-3})(1.34 \times 10^{-3})}{(0.099)} = 1.8 \times 10^{-5}$$

Acid	Initial concentration (mol/L)	Equilibrium concentration (mol/L)
$[HC_2H_3O_2]$	0.100	0.099
$[H^+]$	0	0.00134
$[C_2H_3O_2^-]$	0	0.00134

The K_a for acetic acid, 1.8×10^{-5}, is small and indicates that the position of the equilibrium is far toward the un-ionized acetic acid. In fact, a 0.100 M acetic acid solution is 99% un-ionized.

Once the K_a for acetic acid is established, it can be used to describe other systems containing H^+, $C_2H_3O_2^-$, and $HC_2H_3O_2$ in equilibrium at 25°C. The ionization constants for several other weak acids are listed in Table 16.2.

TABLE 16.2   Ionization Constants (K_a) of Weak Acids at 25°C

Acid	Formula	K_a	Acid	Formula	K_a
Acetic	$HC_2H_3O_2$	1.8×10^{-5}	Hydrocyanic	HCN	4.0×10^{-10}
Benzoic	$HC_7H_5O_2$	6.3×10^{-5}	Hypochlorous	$HClO$	3.5×10^{-8}
Carbolic (phenol)	HC_6H_5O	1.3×10^{-10}	Nitrous	HNO_2	4.5×10^{-4}
Cyanic	$HCNO$	2.0×10^{-4}	Hydrofluoric	HF	6.5×10^{-4}
Formic	$HCHO_2$	1.8×10^{-4}			

EXAMPLE 16.9

What is the $[H^+]$ in a 0.50 M $HC_2H_3O_2$ solution? The ionization constant, K_a, for $HC_2H_3O_2$ is 1.8×10^{-5}.

WileyPLUS

ENHANCED EXAMPLE

SOLUTION

To solve this problem, first write the equilibrium equation and the K_a expression:

$$HC_2H_3O_2 \rightleftharpoons H^+ + C_2H_3O_2^- \qquad K_a = \frac{[H^+][C_2H_3O_2^-]}{[HC_2H_3O_2]} = 1.8 \times 10^{-5}$$

We know that the initial concentration of $HC_2H_3O_2$ is 0.50 M. We also know from the ionization equation that one $C_2H_3O_2^-$ is produced for every H^+ produced; that is, the $[H^+]$ and the $[C_2H_3O_2^-]$ are equal. To solve, let $Y = [H^+]$, which also equals the $[C_2H_3O_2^-]$. The un-ionized $[HC_2H_3O_2]$ remaining will then be $0.50 - Y$, the starting concentration minus the amount that ionized:

$$[H^+] = [C_2H_3O_2^-] = Y \qquad [HC_2H_3O_2] = 0.50 - Y$$

	Initial	Equilibrium
$[H^+]$	0	Y
$[C_2H_3O_2^-]$	0	Y
$[HC_2H_3O_2]$	0.5	$0.5 - Y$

Substituting these values into the K_a expression, we obtain

$$K_a = \frac{(Y)(Y)}{0.50 - Y} = \frac{Y^2}{0.50 - Y} = 1.8 \times 10^{-5}$$

An exact solution of this equation for Y requires the use of a mathematical equation known as the *quadratic equation*. However, an approximate solution is obtained if we assume that Y is small and can be neglected compared with 0.50. Then $0.50 - Y$ will be equal to approximately 0.50. The equation now becomes

The quadratic equation is
$$y = \frac{-b \pm \sqrt{b^2 - 4ac}}{2a}$$ for the equation $ay^2 + by + c = 0$.

$$\frac{Y^2}{0.50} = 1.8 \times 10^{-5}$$

$$Y^2 = 0.50 \times 1.8 \times 10^{-5} = 0.90 \times 10^{-5} = 9.0 \times 10^{-6}$$

Taking the square root of both sides of the equation, we obtain

$$Y = \sqrt{9.0 \times 10^{-6}} = 3.0 \times 10^{-3} \text{ mol/L}$$

Thus, the $[H^+]$ is approximately 3.0×10^{-3} mol/L in a 0.50 M $HC_2H_3O_2$ solution. The exact solution to this problem, using the quadratic equation, gives a value of 2.99×10^{-3} mol/L for $[H^+]$, showing that we were justified in neglecting Y compared with 0.50.

PRACTICE 16.11

Calculate the hydrogen ion concentration in (a) 0.100 M hydrocyanic acid (HCN) solution and (b) 0.0250 M carbolic acid (HC_6H_5O) solution.

EXAMPLE 16.10

Calculate the percent ionization in a 0.50 M $HC_2H_3O_2$ solution.

SOLUTION

The percent ionization of a weak acid, $HA(aq) \rightleftharpoons H^+(aq) + A^-(aq)$, is found by dividing the concentration of the H^+ or A^- ions at equilibrium by the initial concentration of HA and multiplying by 100. For acetic acid,

$$\frac{\text{concentration of } [H^+] \text{ or } [C_2H_3O_2^-]}{\text{initial concentration of } [HC_2H_3O_2]} \times 100 = \text{percent ionized}$$

To solve this problem, we first need to calculate the $[H^+]$. This calculation has already been done for a 0.50 M $HC_2H_3O_2$ solution:

$$[H^+] = 3.0 \times 10^{-3} \text{ mol/L in a 0.50 } M \text{ solution} \quad \text{(from Example 16.9)}$$

This $[H^+]$ represents a fractional amount of the initial 0.50 M $HC_2H_3O_2$. Therefore,

$$\frac{3.0 \times 10^{-3} \text{ mol/L}}{0.50 \text{ mol/L}} \times 100 = 0.60\% \text{ ionized}$$

A 0.50 M $HC_2H_3O_2$ solution is 0.60% ionized.

PRACTICE 16.12

Calculate the percent ionization for

(a) 0.100 M hydrocyanic acid (HCN)   (b) 0.0250 M carbolic acid (HC_6H_5O)

16.7 SOLUBILITY PRODUCT CONSTANT

LEARNING OBJECTIVE ● Use the solubility product constant to calculate the solubility of a slightly soluble salt and to determine whether a precipitate will form in a solution.

KEY TERMS

solubility product constant, K_{sp}
common ion effect

The **solubility product constant, K_{sp}**, is the equilibrium constant of a slightly soluble salt. To evaluate K_{sp}, consider this example. The solubility of AgCl in water is 1.3×10^{-5} mol/L at 25°C. The equation for the equilibrium between AgCl and its ions in solution is

$$AgCl(s) \rightleftharpoons Ag^+(aq) + Cl^-(aq)$$

The equilibrium constant expression is

$$K_{eq} = \frac{[Ag^+][Cl^-]}{[AgCl(s)]}$$

The amount of solid AgCl does not affect the equilibrium system, provided that some is present. In other words, the concentration of solid AgCl is constant whether 1 mg or 10 g of the salt are present. Therefore, the product obtained by multiplying the two constants K_{eq} and $[AgCl(s)]$ is also a constant. This is the solubility product constant, K_{sp}:

$$K_{eq} \times [AgCl(s)] = [Ag^+][Cl^-] = K_{sp}$$

$$K_{sp} = [Ag^+][Cl^-]$$

The K_{sp} is equal to the product of the $[Ag^+]$ and the $[Cl^-]$, each in moles per liter. When 1.3×10^{-5} mol/L of AgCl dissolves, it produces 1.3×10^{-5} mol/L each of Ag^+ and Cl^-. From these concentrations, the K_{sp} can be calculated:

$$[Ag^+] = 1.3 \times 10^{-5}\,\text{mol/L} \qquad [Cl^-] = 1.3 \times 10^{-5}\,\text{mol/L}$$
$$K_{sp} = [Ag^+][Cl^-] = (1.3 \times 10^{-5})(1.3 \times 10^{-5}) = 1.7 \times 10^{-10}$$

Once the K_{sp} value for AgCl is established, it can be used to describe other systems containing Ag^+ and Cl^-.

The K_{sp} expression does not have a denominator. It consists only of the concentrations (mol/L) of the ions in solution. As in other equilibrium expressions, each of these concentrations is raised to a power that is the same number as its coefficient in the balanced equation. Here are equilibrium equations and the K_{sp} expressions for several other substances:

$$AgBr(s) \rightleftharpoons Ag^+(aq) + Br^-(aq) \qquad K_{sp} = [Ag^+][Br^-]$$
$$BaSO_4(s) \rightleftharpoons Ba^{2+}(aq) + SO_4^{2-}(aq) \qquad K_{sp} = [Ba^{2+}][SO_4^{2-}]$$
$$Ag_2CrO_4(s) \rightleftharpoons 2Ag^+(aq) + CrO_4^{2-}(aq) \qquad K_{sp} = [Ag^+]^2[CrO_4^{2-}]$$
$$CuS(s) \rightleftharpoons Cu^{2+}(aq) + S^{2-}(aq) \qquad K_{sp} = [Cu^{2+}][S^{2-}]$$
$$Mn(OH)_2(s) \rightleftharpoons Mn^{2+}(aq) + 2OH^-(aq) \qquad K_{sp} = [Mn^{2+}][OH^-]^2$$
$$Fe(OH)_3(s) \rightleftharpoons Fe^{3+}(aq) + 3OH^-(aq) \qquad K_{sp} = [Fe^{3+}][OH^-]^3$$

Table 16.3 lists K_{sp} values for these and several other substances.

When the product of the molar concentration of the ions in solution (each raised to its proper power) is greater than the K_{sp} for that substance, precipitation will occur. If the ion product is less than the K_{sp} value, no precipitation will occur.

TABLE 16.3 Solubility Product Constants (K_{sp}) at 25°C

Compound	K_{sp}	Compound	K_{sp}
AgCl	1.7×10^{-10}	CaF$_2$	3.9×10^{-11}
AgBr	5.2×10^{-13}	CuS	8.5×10^{-45}
AgI	8.5×10^{-17}	Fe(OH)$_3$	6.1×10^{-38}
AgC$_2$H$_3$O$_2$	2.1×10^{-3}	PbS	3.4×10^{-28}
Ag$_2$CrO$_4$	1.9×10^{-12}	PbSO$_4$	1.3×10^{-8}
BaCrO$_4$	8.5×10^{-11}	Mn(OH)$_2$	2.0×10^{-13}
BaSO$_4$	1.5×10^{-9}		

EXAMPLE 16.11

Write K_{sp} expressions for AgI and PbI$_2$, both of which are slightly soluble salts.

SOLUTION

First write the equilibrium equations:

$$AgI(s) \rightleftharpoons Ag^+(aq) + I^-(aq)$$
$$PbI_2(s) \rightleftharpoons Pb^{2+}(aq) + 2I^-(aq)$$

Since the concentration of the solid crystals is constant, the K_{sp} equals the product of the molar concentrations of the ions in solution. In the case of PbI$_2$, the $[I^-]$ must be squared:

$$K_{sp} = [Ag^+][I^-] \qquad K_{sp} = [Pb^{2+}][I^-]^2$$

EXAMPLE 16.12

The K_{sp} value for lead sulfate is 1.3×10^{-8}. Calculate the solubility of $PbSO_4$ in grams per liter.

SOLUTION

First write the equilibrium equation and the K_{sp} expression:

$$PbSO_4 \rightleftharpoons Pb^{2+}(aq) + SO_4^{2-}(aq)$$

$$K_{sp} = [Pb^{2+}][SO_4^{2-}] = 1.3 \times 10^{-8}$$

Since the lead sulfate that is in solution is completely dissociated, the $[Pb^{2+}]$ or $[SO_4^{2-}]$ is equal to the solubility of $PbSO_4$ in moles per liter. Let

$$Y = [Pb^{2+}] = [SO_4^{2-}]$$

Substitute Y into the K_{sp} equation and solve:

$$[Pb^{2+}][SO_4^{2-}] = (Y)(Y) = 1.3 \times 10^{-8}$$

$$Y^2 = 1.3 \times 10^{-8}$$

$$Y = 1.1 \times 10^{-4} \, mol/L$$

The solubility of $PbSO_4$, therefore, is 1.1×10^{-4} mol/L. Now convert mol/L to g/L:

1 mol of $PbSO_4$ has a mass of (207.2 g + 32.07 g + 64.00 g) 303.3 g

$$\left(\frac{1.1 \times 10^{-4} \, mol}{L}\right)\left(\frac{303.3 \, g}{mol}\right) = 3.3 \times 10^{-2} \, g/L$$

The solubility of $PbSO_4$ is 3.3×10^{-2} g/L.

PRACTICE 16.13

Write the K_{sp} expression for

(a) $Cr(OH)_3$ (b) $Cu_3(PO_4)_2$

PRACTICE 16.14

The K_{sp} value for CuS is 8.5×10^{-45}. Calculate the solubility of CuS in grams per liter.

An ion added to a solution already containing that ion is called a *common ion*. When a common ion is added to an equilibrium solution of a weak electrolyte or a slightly soluble salt, the equilibrium shifts according to Le Châtelier's principle. For example, when silver nitrate ($AgNO_3$) is added to a saturated solution of AgCl,

$$AgCl(s) \rightleftharpoons Ag^+ + Cl^-$$

the equilibrium shifts to the left due to the increase in the $[Ag^+]$. As a result, the $[Cl^-]$ and the solubility of AgCl decrease. The AgCl and $AgNO_3$ have the common ion Ag^+. A shift in the equilibrium position upon addition of an ion already contained in the solution is known as the **common ion effect**.

EXAMPLE 16.13

Silver nitrate is added to a saturated AgCl solution until the $[Ag^+]$ is 0.10 *M*. What will be the $[Cl^-]$ remaining in solution?

SOLUTION

This is an example of the common ion effect. The addition of $AgNO_3$ puts more Ag^+ in solution; the Ag^+ combines with Cl^- and causes the equilibrium to shift to the left, reducing the $[Cl^-]$ in solution. After the addition of Ag^+ to the mixture, the $[Ag^+]$ and $[Cl^-]$ in solution are no longer equal.

We use the K_{sp} to calculate the $[Cl^-]$ remaining in solution. The K_{sp} is constant at a particular temperature and remains the same no matter how we change the concentration of the species involved:

$$K_{sp} = [Ag^+][Cl^-] = 1.7 \times 10^{-10} \qquad [Ag^+] = 0.10 \text{ mol/L}$$

We then substitute the $[Ag^+]$ into the K_{sp} expression and calculate the $[Cl^-]$:

$$[0.10][Cl^-] = 1.7 \times 10^{-10}$$

$$[Cl^-] = \frac{1.7 \times 10^{-10}}{0.10} = 1.7 \times 10^{-9} \text{ mol/L}$$

This calculation shows a 10,000-fold reduction of Cl^- ions in solution. It illustrates that Cl^- ions may be quantitatively removed from solution with an excess of Ag^+ ions.

PRACTICE 16.15

Sodium sulfate (Na_2SO_4) is added to a saturated solution of $BaSO_4$ until the concentration of the sulfate ion is 2.0×10^{-2} M. What will be the concentration of the Ba^{2+} ions remaining in solution?

16.8 BUFFER SOLUTIONS: THE CONTROL OF pH

Describe the function of a buffer solution.

● **LEARNING OBJECTIVE**

KEY TERM
buffer solution

The control of pH within narrow limits is critically important in many chemical applications and vitally important in many biological systems. For example, human blood must be maintained between pH 7.35 and 7.45 for the efficient transport of oxygen from the lungs to the cells. This narrow pH range is maintained by buffer systems in the blood.

A **buffer solution** resists changes in pH when diluted or when small amounts of acid or base are added. Two common types of buffer solutions are (1) a weak acid mixed with a salt of its conjugate base and (2) a weak base mixed with a salt of its conjugate acid.

The action of a buffer system can be understood by considering a solution of acetic acid and sodium acetate. The weak acid ($HC_2H_3O_2$) is mostly un-ionized and is in equilibrium with its ions in solution. The sodium acetate is completely ionized:

$$HC_2H_3O_2(aq) \rightleftharpoons H^+(aq) + C_2H_3O_2^-(aq)$$
$$NaC_2H_3O_2(aq) \longrightarrow Na^+(aq) + C_2H_3O_2^-(aq)$$

Because the sodium acetate is completely ionized, the solution contains a much higher concentration of acetate ions than would be present if only acetic acid were in solution. The acetate ion represses the ionization of acetic acid and also reacts with water, causing the solution to have a higher pH (be more basic) than an acetic acid solution (see Section 16.3). Thus, a 0.1 M acetic acid solution has a pH of 2.87, but a solution that is 0.1 M in acetic acid and 0.1 M in sodium acetate has a pH of 4.74. This difference in pH is the result of the common ion effect.

A buffer solution has a built-in mechanism that counteracts the effect of adding acid or base. Consider the effect of adding HCl or NaOH to an acetic acid–sodium acetate buffer. When a small amount of HCl is added, the acetate ions of the buffer combine with the H^+ ions from HCl to form un-ionized acetic acid, thus neutralizing the added acid and maintaining the approximate pH of the solution. When NaOH is added, the OH^- ions react with acetic acid to neutralize the added base and thus maintain the approximate pH. The equations for these reactions are

$$H^+(aq) + C_2H_3O_2^-(aq) \rightleftharpoons HC_2H_3O_2(aq)$$
$$OH^-(aq) + HC_2H_3O_2(aq) \rightleftharpoons H_2O(l) + C_2H_3O_2^-(aq)$$

Mark E. Gibson/Corbis

The aquariums at the MGM Grand Hotel are a buffer system.

>CHEMISTRY *IN ACTION*

Exchange of Oxygen and Carbon Dioxide in the Blood

The transport of oxygen and carbon dioxide between the lungs and tissues is a complex process that involves several reversible reactions, each of which behaves in accordance with Le Châtelier's principle.

The binding of oxygen to hemoglobin is a reversible reaction. The oxygen molecule must attach to the hemoglobin (Hb) and then later detach. The equilibrium equation for this reaction can be written:

$$Hb + O_2 \rightleftharpoons HbO_2$$

In the lungs the concentration of oxygen is high and favors the forward reaction. Oxygen quickly binds to the hemoglobin until it is saturated with oxygen.

In the tissues the concentration of oxygen is lower and in accordance with Le Châtelier's principle: The equilibrium position shifts to the left and the hemoglobin releases oxygen to the tissues. Approximately 45% of the oxygen diffuses out of the capillaries into the tissues, where it may be picked up by *myoglobin*, another carrier molecule.

Myoglobin functions as an oxygen-storage molecule, holding the oxygen until it is required in the energy-producing portions of the cell. The reaction between myoglobin (Mb) and oxygen can be written as an equilibrium reaction:

$$Mb + O_2 \rightleftharpoons MbO_2$$

The hemoglobin and myoglobin equations are very similar, so what accounts for the transfer of the oxygen from the hemoglobin to the myoglobin? Although both equilibria involve similar interactions, the affinity between oxygen and hemoglobin is different from the affinity between myoglobin and oxygen. In the tissues the position of the hemoglobin equilibrium is such that it is 55% saturated with oxygen, whereas the myoglobin is at 90% oxygen saturation. Under these conditions hemoglobin will release oxygen, while myoglobin will bind oxygen. Thus oxygen is loaded onto hemoglobin in the lungs and unloaded in the tissue's cells.

Carbon dioxide produced in the cells must be removed from the tissues. Oxygen-depleted hemoglobin molecules accomplish this by becoming carriers of carbon dioxide. The carbon dioxide does not bind at the heme site as the oxygen does, but rather at one end of the protein chain.

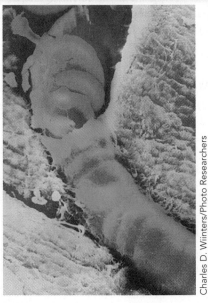

Oxygen and carbon dioxide are exchanged in the red blood cells when they are in capillaries.

Charles D. Winters/Photo Researchers

When carbon dioxide dissolves in water, some of the CO_2 reacts to release hydrogen ions:

$$CO_2 + H_2O \rightleftharpoons HCO_3^- + H^+$$

To facilitate the removal of CO_2 from the tissues, this equilibrium needs to be moved toward the right. This shift is accomplished by the removal of H^+ from the tissues by the hemoglobin molecule. The deoxygenated hemoglobin molecule can bind H^+ ions as well as CO_2. In the lungs this whole process is reversed so the CO_2 is removed from the hemoglobin and exhaled.

Molecules that are similar in structure to the oxygen molecule can become involved in competing equilibria. Hemoglobin is capable of binding with carbon monoxide (CO), nitrogen monoxide (NO), and cyanide (CN⁻). The extent of the competition depends on the affinity. Since these molecules have a greater affinity for hemoglobin than oxygen, they will effectively displace oxygen from hemoglobin. For example,

$$HbO_2 + CO \rightleftharpoons HbCO + O_2$$

Since the affinity of hemoglobin for CO is 150 times stronger than its affinity for oxygen, the equilibrium position lies far to the right. This explains why CO is a poisonous substance and why oxygen is administered to victims of CO poisoning. The hemoglobin molecules can only transport oxygen if the CO is released and the oxygen shifts the equilibrium toward the left.

Data comparing the changes in pH caused by adding HCl and NaOH to pure water and to an acetic acid–sodium acetate buffer solution are shown in Table 16.4. Notice that adding 0.01 mol HCl or 0.01 mol NaOH to 100 mL water changes the pH of the solution by 5 units. But adding the same amount of HCl or NaOH to a buffer solution causes a change of only 0.08 or 0.09 pH units. So buffers really protect the pH of a solution.

Buffers cannot withstand the addition of large amounts of acids or bases. They overpower the capacity of the buffer to absorb acid or base. The maximum buffering effect occurs when the weak acid or base and its conjugate are of equal molar concentrations.

TABLE 16.4 Changes in pH Caused by the Addition of HCl and NaOH

Solution	pH	Change in pH
H_2O (1000 mL)	7	—
H_2O + 0.010 mol HCl	2	5
H_2O + 0.010 mol NaOH	12	5
Buffer solution (1000 mL)		
0.10 M $HC_2H_3O_2$ + 0.10 M $NaC_2H_3O_2$	4.74	—
Buffer + 0.010 mol HCl	4.66	0.08
Buffer + 0.010 mol NaOH	4.83	0.09

The human body has a number of buffer systems. One of these, the hydrogen carbonate–carbonic acid buffer, HCO_3^- — H_2CO_3, maintains the blood plasma at a pH of 7.4. The phosphate system, HPO_4^{2-} — $H_2PO_4^-$, is an important buffer in the red blood cells, as well as in other places in the body.

EXAMPLE 16.14

What is the purpose of a buffer solution?

SOLUTION

A buffer solution resists change in pH when small amounts of acid or base are added to it.

PRACTICE 16.16

Describe how a buffer solution made from HF and NaF behaves when a small amount of a strong acid is added to it: $HF \rightleftharpoons H^+ + F^-$.

CHAPTER **16 REVIEW**

16.1 RATES OF REACTION

- The rate of a reaction is variable and depends on:
  - Concentration of reacting species
  - Temperature
  - Presence of catalysts
  - Nature of the reactants

KEY TERM

chemical kinetics

16.2 CHEMICAL EQUILIBRIUM

- In a reversible chemical reaction, the products formed react to produce the original reaction mixture.
- The forward reaction is called the reaction to the right.
- The reverse reaction is called the reaction to the left.
- A system at equilibrium is dynamic.
- At chemical equilibrium, the rate of the forward reaction equals the rate of the reverse reaction.

KEY TERMS

reversible chemical reaction
equilibrium
chemical equilibrium

16.3 LE CHÂTELIER'S PRINCIPLE

- If stress is applied to a system at equilibrium, the system will respond in such a way as to relieve the stress and restore equilibrium under a new set of conditions.
- When the concentration of a reactant is increased, the equilibrium shifts toward the right.
- When the concentration of a product on the right side of the equation is increased, the equilibrium shifts to the left.
- If the concentration of a reactant is decreased, the equilibrium shifts toward the side where the reactant is decreased.
- In gaseous reactions:
  - A decrease in volume causes the equilibrium position to shift toward the side of the reaction with the fewest molecules.

KEY TERMS

Le Châtelier's principle
catalyst
activation energy

- If the number of molecules is the same on both sides of a reaction, a decrease in volume has no effect on the equilibrium position.
- When heat is added to a reaction, the side of the equation that absorbs heat is favored:
  - If the reaction is endothermic, the forward reaction increases.
  - If the reaction is exothermic, the reverse reaction increases.
- A catalyst does not shift the equilibrium of a reaction; it only affects the speed at which equilibrium is reached.
- The activation energy for a reaction is the minimum energy required for the reaction to occur:
  - A catalyst lowers the activation energy for a reaction by providing a different pathway for the reaction.

16.4 EQUILIBRIUM CONSTANTS

KEY TERM

equilibrium constant, K_{eq}

- For the general reaction $aA + bB \rightleftharpoons cC + dD$

$$K_{eq} = \frac{[C]^c[D]^d}{[A]^a[B]^b}$$

- The magnitude of K_{eq} indicates the extent of the reaction:
  - If $K_{eq} > 1$: Amount of products is greater than reactants at equilibrium.
  - If $K_{eq} < 1$: Amount of reactants is less than products at equilibrium.

16.5 ION PRODUCT CONSTANT FOR WATER

KEY TERM

ion product constant for water, K_w

- $K_w = [H^+][OH^-] = 1 \times 10^{-14}$

- $[H^+] = \dfrac{K_w}{[OH^-]}$

- $[OH^-] = \dfrac{K_w}{[H^+]}$

16.6 IONIZATION CONSTANTS

KEY TERM

acid ionization constant, K_a

- The ionization constant can be used in the equilibrium expression to calculate the concentration of a reactant, the percent ionization of a substance, or the pH of a weak acid.

16.7 SOLUBILITY PRODUCT CONSTANT

KEY TERMS

solubility product constant, K_{sp}
common ion effect

- The solubility product constant is used to calculate the solubility of a slightly soluble salt in water.
- The solubility product constant can also be used to determine whether or not precipitation will occur in a solution:
  - If the product of the molar concentrations of the ions in solution is greater than the K_{sp}, precipitation will occur.
- A shift in the equilibrium position upon addition of an ion already contained in the solution is known as the common ion effect.

16.8 BUFFER SOLUTIONS: THE CONTROL OF pH

KEY TERM

buffer solution

- A buffer solution could contain:
  - A weak acid and a salt of its conjugate base
  - A weak base and a salt of its conjugate acid
- A buffer solution resists a change in pH by neutralizing small amounts of acid or base added to it.

REVIEW QUESTIONS

1. At equilibrium how do the forward and reverse reaction rates compare? (Figure 16.1)
2. Why does the rate of a reaction usually increase when the concentration of one of the reactants is increased?
3. Why does an increase in temperature cause the rate of reaction to increase?
4. How would you expect the two tubes in Figure 16.2 to appear if both are at 25°C?

5. Is the reaction $N_2O_4 \rightleftharpoons 2 NO_2$ exothermic or endothermic? (Figure 16.2)
6. If pure hydrogen iodide, HI, is placed in a vessel at 700 K, will it decompose? Explain.
7. Of the acids listed in Table 16.2, which ones are stronger than acetic acid and which are weaker?
8. Why is heat treated as a reactant in an endothermic process and as a product in an exothermic process?

9. What does a catalyst do?

10. Is it important to specify the temperature when reporting an equilibrium constant? Why or why not?

11. If the value of an equilibrium constant is very large, will the equilibrium lie far to the right or far to the left?

12. If the value of an equilibrium constant is very small, will the equilibrium lie far to the right or far to the left?

13. Why don't free protons (H^+) exist in water?

14. For each solution in Table 16.1, what is the sum of the pH plus the pOH? What would be the pOH of a solution whose pH is -1?

15. What would cause two separate samples of pure water to have slightly different pH values?

16. Why are the pH and pOH equal in pure water?

17. With dilution, aqueous solutions of acetic acid ($HC_2H_3O_2$) show increased ionization. For example, a 1.0 M solution of acetic acid is 0.42% ionized, whereas a 0.10 M solution is 1.34% ionized. Explain the behavior using the ionization equation and equilibrium principles.

18. A 1.0 M solution of acetic acid ionizes less and has a higher concentration of H^+ ions than a 0.10 M acetic acid solution. Explain this behavior. (See Question 17 for data.)

19. Explain why silver acetate is more soluble in nitric acid than in water. (*Hint*: Write the equilibrium equation first and then consider the

effect of the acid on the acetate ion.) What would happen if hydrochloric acid were used in place of nitric acid?

20. Dissolution of sodium acetate ($NaC_2H_3O_2$) in pure water gives a basic solution. Why?
(*Hint*: A small amount of $HC_2H_3O_2$ is formed.)

21. Tabulate the relative order of molar solubilities of AgCl, AgBr, AgI, $AgC_2H_3O_2$, $PbSO_4$, $BaSO_4$, $BaCrO_4$, and PbS. List the most soluble first. (Table 16.3)

22. Which compound in the following pairs has the greater molar solubility? (Table 16.3)
(a) $Mn(OH)_2$ or Ag_2CrO_4
(b) $BaCrO_4$ or Ag_2CrO_4

23. Explain why a precipitate of NaCl forms when HCl gas is passed into a saturated aqueous solution of NaCl.

24. Describe the similarities and differences between K_a, K_b, K_w, and K_{sp}.

25. Explain how the acetic acid–sodium acetate buffer system maintains its pH when 0.010 mol of HCl is added to 1 L of the buffer solution. (Table 16.4)

26. Describe why the pH of a buffer solution remains almost constant when a small amount of acid or base is added to it.

27. Describe how equilibrium is reached when the substances A and B are first mixed and react as

$$A + B \rightleftharpoons C + D$$

Most of the exercises in this chapter are available for assignment via the online homework management program, WileyPLUS (www.wileyplus.com). All exercises with blue numbers have answers in Appendix VI.

PAIRED EXERCISES

1. Write equations for the following reversible systems:
(a) solid $KMnO_4$ in a saturated aqueous solution of $KMnO_4$
(b) a mixture of solid and gaseous forms of CO_2 in a closed container

3. Consider the following system at equilibrium:

$$SiF_4(g) + 2 H_2O(g) + 103.8 \text{ kJ} \rightleftharpoons SiO_2(g) + 4 HF(g)$$

(a) Is the reaction endothermic or exothermic?
(b) If HF is added, in which direction will the reaction shift in order to reestablish equilibrium? After the new equilibrium has been established, will the final molar concentrations of SiF_4, H_2O, SiO_2, and HF increase, decrease, or remain the same?
(c) If heat is added, in which direction will the equilibrium shift?

5. Consider the following system at equilibrium:

$$N_2(g) + 3 H_2(g) \rightleftharpoons 2 NH_3(g) + 92.5 \text{ kJ}$$

Complete the table that follows. Indicate changes in moles by entering I, D, N, or ? in the table. (I = increase, D = decrease, N = no change, ? = insufficient information to determine.)

Change of stress imposed on the system at equilibrium	Direction of reaction, left or right, to reestablish equilibrium	Change in number of moles		
		N_2	H_2	NH_3
(a) Add N_2				
(b) Remove H_2				
(c) Decrease volume of reaction vessel				
(d) Increase temperature				

2. Write equations for the following reversible systems:
(a) a closed container of solid and gaseous forms of I_2
(b) solid $NaNO_3$ in a saturated aqueous solution of $NaNO_3$

4. Consider the following system at equilibrium:

$$4 HCl(g) + O_2(g) \rightleftharpoons 2 H_2O(g) + 2 Cl_2(g) + 114.4 \text{ kJ}$$

(a) Is the reaction endothermic or exothermic?
(b) If O_2 is added, in which direction will the reaction shift in order to reestablish equilibrium? After the new equilibrium has been established, will the final molar concentrations of HCl, O_2, H_2O, and Cl_2 increase or decrease?
(c) If heat is added, in which direction will the equilibrium shift?

6. Consider the following system at equilibrium:

$$N_2(g) + 3 H_2(g) \rightleftharpoons 2 NH_3(g) + 92.5 \text{ kJ}$$

Complete the table that follows. Indicate changes in moles by entering I, D, N, or ? in the table. (I = increase, D = decrease, N = no change, ? = insufficient information to determine.)

Change of stress imposed on the system at equilibrium	Direction of reaction, left or right, to reestablish equilibrium	Change in number of moles		
		N_2	H_2	NH_3
(a) Add NH_3				
(b) Increase volume of reaction vessel				
(c) Add catalyst				
(d) Add both H_2 and NH_3				

7. For the following equations, tell in which direction, left or right, the equilibrium will shift when these changes are made: The temperature is increased, the pressure is increased by decreasing the volume of the reaction vessel, and a catalyst is added.
   (a) $3 O_2(g) + 271 \text{ kJ} \rightleftharpoons 2 O_3(g)$
   (b) $CH_4(g) + Cl_2(g) \rightleftharpoons CH_3Cl(g) + HCl(g) + 110 \text{ kJ}$
   (c) $2 NO(g) + 2 H_2(g) \rightleftharpoons N_2(g) + 2 H_2O(g) + 665 \text{ kJ}$

8. For the following equations, tell in which direction, left or right, the equilibrium will shift when these changes are made: The temperature is increased, the pressure is increased by decreasing the volume of the reaction vessel, and a catalyst is added.
   (a) $2 SO_3(g) + 197 \text{ kJ} \rightleftharpoons 2 SO_2(g) + O_2(g)$
   (b) $4 NH_3(g) + 3 O_2(g) \rightleftharpoons 2 N_2(g) + 6 H_2O(g) + 1531 \text{ kJ}$
   (c) $OF_2(g) + H_2O(g) \rightleftharpoons O_2(g) + 2 HF(g) + 318 \text{ kJ}$

9. Applying Le Châtelier's principle, in which direction will the equilibrium shift (if at all)?

   $$CH_4(g) + 2 O_2(g) \rightleftharpoons CO_2(g) + 2 H_2O(g) + 802.3 \text{ kJ}$$

   (a) if the temperature is increased
   (b) if a catalyst is added
   (c) if CH_4 is added
   (d) if the volume of the reaction vessel is decreased

10. Applying Le Châtelier's principle, in which direction will the equilibrium shift (if at all)?

   $$2 CO_2(g) + N_2(g) + 1095.9 \text{ kJ} \rightleftharpoons C_2N_2(g) + 2 O_2(g)$$

   (a) if the volume of the reaction vessel is increased
   (b) if O_2 is added
   (c) if the temperature is increased
   (d) if the concentration of N_2 is increased

11. Write the equilibrium constant expression for these reactions:
    (a) $2 NO_2(g) + 7 H_2(g) \rightleftharpoons 2 NH_3(g) + 4 H_2O(g)$
    (b) $H_2CO_3(aq) \rightleftharpoons H^+(aq) + HCO_3^-(aq)$
    (c) $2 COF_2(g) \rightleftharpoons CO_2(g) + CF_4(g)$

12. Write the equilibrium constant expression for these reactions:
    (a) $H_3PO_4(aq) \rightleftharpoons H^+(aq) + H_2PO_4^-(aq)$
    (b) $CS_2(g) + 4 H_2(g) \rightleftharpoons CH_4(g) + 2 H_2S(g)$
    (c) $4 NO_2(g) + O_2(g) \rightleftharpoons 2 N_2O_5(g)$

13. Write the solubility product expression, K_{sp}, for these substances:
    (a) AgCl (c) $Zn(OH)_2$
    (b) $PbCrO_4$ (d) $Ca_3(PO_4)_2$

14. Write the solubility product expression, K_{sp}, for these substances:
    (a) $MgCO_3$ (c) $Tl(OH)_3$
    (b) CaC_2O_4 (d) $Pb_3(AsO_4)_2$

15. What effect will decreasing the $[H^+]$ of a solution have on (a) pH, (b) pOH, (c) $[OH^-]$, and (d) K_w?

16. What effect will increasing the $[H^+]$ of a solution have on (a) pH, (b) pOH, (c) $[OH^-]$, and (d) K_w?

17. One of the important pH-regulating systems in the blood consists of a carbonic acid–sodium hydrogen carbonate buffer:

    $$H_2CO_3(aq) \rightleftharpoons H^+(aq) + HCO_3^-(aq)$$
    $$NaHCO_3(aq) \longrightarrow Na^+(aq) + HCO_3^-(aq)$$

    Explain how this buffer resists changes in pH when excess acid, H^+, gets into the bloodstream.

18. One of the important pH-regulating systems in the blood consists of a carbonic acid–sodium hydrogen carbonate buffer:

    $$H_2CO_3(aq) \rightleftharpoons H^+(aq) + HCO_3^-(aq)$$
    $$NaHCO_3(aq) \longrightarrow Na^+(aq) + HCO_3^-(aq)$$

    Explain how this buffer resists changes in pH when excess base, OH^-, gets into the bloodstream.

19. For a solution of $1.2 M H_2CO_3$ ($K_a = 4.4 \times 10^{-7}$), calculate:
    (a) $[H^+]$
    (b) pH
    (c) percent ionization

20. For a solution of $0.025 M$ lactic acid, $HC_3H_5O_2$ ($K_a = 8.4 \times 10^{-4}$), calculate:
    (a) $[H^+]$
    (b) pH
    (c) percent ionization

21. A $0.025 M$ solution of a weak acid, HA, is 0.45% ionized. Calculate the ionization constant, K_a, for the acid.

22. A $0.500 M$ solution of a weak acid, HA, is 0.68% ionized. Calculate the ionization constant, K_a, for the acid.

23. Calculate the percent ionization and the pH of each of the following solutions of phenol, HC_6H_5O ($K_a = 1.3 \times 10^{-10}$):
    (a) $1.0 M$
    (b) $0.10 M$
    (c) $0.010 M$

24. Calculate the percent ionization and the pH of each of the following solutions of benzoic acid, $HC_7H_5O_2$ ($K_a = 6.3 \times 10^{-5}$):
    (a) $1.0 M$
    (b) $0.10 M$
    (c) $0.010 M$

25. A $0.37 M$ solution of a weak acid (HA) has a pH of 3.7. What is the K_a for this acid?

26. A $0.23 M$ solution of a weak acid (HA) has a pH of 2.89. What is the K_a for this acid?

27. A student needs a sample of $1.0 M$ NaOH for a laboratory experiment. Calculate the $[H^+]$, $[OH^-]$, pH, and pOH of this solution.

28. A laboratory cabinet contains a stock solution of $3.0 M HNO_3$. Calculate the $[H^+]$, $[OH^-]$, pH, and pOH of this solution.

29. Calculate the pH and the pOH of these solutions:
    (a) $0.250 M$ HBr
    (b) $0.333 M$ KOH
    (c) $0.895 M HC_2H_3O_2$ ($K_a = 1.8 \times 10^{-5}$)

30. Calculate the pH and the pOH of these solutions:
    (a) $0.0010 M$ NaOH
    (b) $0.125 M$ HCl
    (c) $0.0250 M HC_6H_5O$ ($K_a = 1.3 \times 10^{-10}$)

31. Calculate the $[OH^-]$ in each of the following solutions:
    (a) $[H^+] = 1.0 \times 10^{-2}$ (c) $1.25 M$ KOH
    (b) $[H^+] = 3.2 \times 10^{-7}$ (d) $0.75 M HC_2H_3O_2$

32. Calculate the $[OH^-]$ in each of the following solutions:
    (a) $[H^+] = 4.0 \times 10^{-9}$ (c) $1.25 M$ HCN
    (b) $[H^+] = 1.2 \times 10^{-5}$ (d) $0.333 M$ NaOH

33. Calculate the $[H^+]$ in each of the following solutions:
    (a) $[OH^-] = 1.0 \times 10^{-8}$
    (b) $[OH^-] = 2.0 \times 10^{-4}$

34. Calculate the $[H^+]$ in each of the following solutions:
    (a) $[OH^-] = 4.5 \times 10^{-2}$
    (b) $[OH^-] = 5.2 \times 10^{-9}$

35. Given the following solubility data, calculate the solubility product constant for each substance:
(a) $BaSO_4$, 3.9×10^{-5} mol/L
(b) Ag_2CrO_4, 7.8×10^{-5} mol/L
(c) $CaSO_4$, 0.67 g/L
(d) AgCl, 0.0019 g/L

36. Given the following solubility data, calculate the solubility product constant for each substance:
(a) ZnS, 3.5×10^{-12} mol/L
(b) $Pb(IO_3)_2$, 4.0×10^{-5} mol/L
(c) Ag_3PO_4, 6.73×10^{-3} g/L
(d) $Zn(OH)_2$, 2.33×10^{-4} g/L

37. Calculate the molar solubility for these substances:
(a) CaF_2, $K_{sp} = 3.9 \times 10^{-11}$
(b) $Fe(OH)_3$, $K_{sp} = 6.1 \times 10^{-38}$

38. Calculate the molar solubility for these substances:
(a) $PbSO_4$, $K_{sp} = 1.3 \times 10^{-8}$
(b) $BaCrO_4$, $K_{sp} = 8.5 \times 10^{-11}$

39. For each substance in Question 37, calculate the solubility in grams per 100. mL of solution.

40. For each substance in Question 38, calculate the solubility in grams per 100. mL of solution.

41. Solutions containing 100. mL of 0.010 M Na_2SO_4 and 100. mL of 0.001 M $Pb(NO_3)_2$ are mixed. Show by calculation whether or not a precipitate will form. Assume that the volumes are additive. (K_{sp} for $PbSO_4 = 1.3 \times 10^{-8}$)

42. Solutions containing 50.0 mL of 1.0×10^{-4} M $AgNO_3$ and 100. mL of 1.0×10^{-4} M NaCl are mixed. Show by calculation whether or not a precipitate will form. Assume the volumes are additive. (K_{sp} for AgCl $= 1.7 \times 10^{-10}$)

43. How many moles of AgBr will dissolve in 1.0 L of 0.10 M NaBr? ($K_{sp} = 5.2 \times 10^{-13}$ for AgBr)

44. How many moles of AgBr will dissolve in 1.0 L of 0.10 M $MgBr_2$? ($K_{sp} = 5.2 \times 10^{-13}$ for AgBr)

45. Calculate the [H^+] and the pH of a buffer solution that is 0.20 M in $HC_2H_3O_2$ and contains sufficient sodium acetate to make the [$C_2H_3O_2^-$] equal to 0.10 M. (K_a for $HC_2H_3O_2 = 1.8 \times 10^{-5}$)

46. Calculate the [H^+] and the pH of a buffer solution that is 0.20 M in $HC_2H_3O_2$ and contains sufficient sodium acetate to make the [$C_2H_3O_2^-$] equal to 0.20 M. (K_a for $HC_2H_3O_2 = 1.8 \times 10^{-5}$)

47. When 1.0 mL of 1.0 M HCl is added to 50. mL of 1.0 M NaCl, the [H^+] changes from 1×10^{-7} M to 2.0×10^{-2} M. Calculate the initial pH and the pH change in the solution.

48. When 1.0 mL of 1.0 M HCl is added to 50. mL of a buffer solution that is 1.0 M in $HC_2H_3O_2$ and 1.0 M in $NaC_2H_3O_2$, the [H^+] changes from 1.8×10^{-5} M to 1.9×10^{-5} M. Calculate the initial pH and the pH change in the solution.

ADDITIONAL EXERCISES

49. In a K_{sp} expression, the concentration of solid salt is not included. Why?

50. The energy diagram in Figure 16.3 is for an exothermic reaction. How can you tell?

51. Sketch the energy diagram for an endothermic reaction using a solid line. Sketch what would happen to the energy profile if a catalyst is added to the reaction using a dotted line.

52. What is the maximum number of moles of HI that can be obtained from a reaction mixture containing 2.30 mol I_2 and 2.10 mol H_2?

53. (a) How many moles of HI are produced when 2.00 mol H_2 and 2.00 mol I_2 are reacted at 700 K? (Reaction is 79% complete.)
(b) Addition of 0.27 mol I_2 to the system increases the yield of HI to 85%. How many moles of H_2, I_2, and HI are now present?
(c) From the data in part (a), calculate K_{eq} for the reaction at 700 K.

54. After equilibrium is reached in the reaction of 6.00 g H_2 with 200. g I_2 at 500. K, analysis shows that the flask contains 64.0 g of HI. How many moles of H_2, I_2, and HI are present in this equilibrium mixture?

55. What is the equilibrium constant of the reaction

$$PCl_3(g) + Cl_2(g) \rightleftharpoons PCl_5(g)$$

if a 20.-L flask contains 0.10 mol PCl_3, 1.50 mol Cl_2, and 0.22 mol PCl_5?

56. If the rate of a reaction doubles for every 10°C rise in temperature, how much faster will the reaction go at 100°C than at 30°C?

57. Would you believe that an NH_4Cl solution is acidic? Calculate the pH of a 0.30 M NH_4Cl solution.

$$(NH_4^+ \rightleftharpoons NH_3 + H^+, K_{eq} = 5.6 \times 10^{-10})$$

58. Calculate the pH of an acetic acid buffer composed of $HC_2H_3O_2$ (0.30 M) and $C_2H_3O_2^-$ (0.20 M).

59. Solutions of 50.0 mL of 0.10 M $BaCl_2$ and 50.0 mL of 0.15 M Na_2CrO_4 are mixed, forming a precipitate of $BaCrO_4$. Calculate the concentration of Ba^{2+} that remains in solution.

60. Calculate the ionization constant for the given acids. Each acid ionizes as follows: HA $\rightleftharpoons$ H^+ + A^-.

Acid	Acid concentration	[H^+]
Hypochlorous, HOCl	0.10 M	5.9×10^{-5} mol/L
Propanoic, $HC_3H_5O_2$	0.15 M	1.4×10^{-3} mol/L
Hydrocyanic, HCN	0.20 M	8.9×10^{-6} mol/L

61. The K_{sp} of CaF_2 is 3.9×10^{-11}. Calculate (a) the molar concentrations of Ca^{2+} and F^- in a saturated solution and (b) the grams of CaF_2 that will dissolve in 500. mL of water.

62. Calculate whether or not a precipitate will form in the following mixed solutions:
(a) 100 mL of 0.010 M Na_2SO_4 and 100 mL of 0.001 M $Pb(NO_3)_2$
(b) 50.0 mL of 1.0×10^{-4} M $AgNO_3$ and 100. mL of 1.0×10^{-4} M NaCl
(c) 1.0 g $Ca(NO_3)_2$ in 150 mL H_2O and 250 mL of 0.01 M NaOH

$$K_{sp}PbSO_4 = 1.3 \times 10^{-8}$$
$$K_{sp}AgCl = 1.7 \times 10^{-10}$$
$$K_{sp}Ca(OH)_2 = 1.3 \times 10^{-6}$$

63. If $BaCl_2$ is added to a saturated $BaSO_4$ solution until the [Ba^{2+}] is 0.050 M,
(a) what concentration of SO_4^{2-} remains in solution?
(b) how much $BaSO_4$ remains dissolved in 100. mL of the solution? ($K_{sp} = 1.5 \times 10^{-9}$ for $BaSO_4$)

64. The K_{sp} for $PbCl_2$ is 2.0×10^{-5}. Will a precipitate form when 0.050 mol $Pb(NO_3)_2$ and 0.010 mol $NaCl$ are in 1.0 L solution? Show evidence for your answer.

65. Suppose the concentration of a solution is 0.10 M Ba^{2+} and 0.10 M Sr^{2+}. Which sulfate, $BaSO_4$ or $SrSO_4$, will precipitate first when a dilute solution of H_2SO_4 is added dropwise to the solution? Show evidence for your answer. ($K_{sp} = 1.5 \times 10^{-9}$ for $BaSO_4$ and $K_{sp} = 3.5 \times 10^{-7}$ for $SrSO_4$)

66. Calculate the K_{eq} for the reaction

$$SO_2(g) + O_2(g) \rightleftharpoons SO_3(g)$$

when the equilibrium concentrations of the gases at 530°C are $[SO_3] = 11.0\ M$, $[SO_2] = 4.20\ M$, and $[O_2] = 0.60 \times 10^{-3}\ M$.

67. If it takes 0.048 g BaF_2 to saturate 15.0 mL of water, what is the K_{sp} of BaF_2?

68. The K_{eq} for the formation of ammonia gas from its elements is 4.0. If the equilibrium concentrations of nitrogen gas and hydrogen gas are both 2.0 M, what is the equilibrium concentration of the ammonia gas?

69. The K_{sp} of $SrSO_4$ is 7.6×10^{-7}. Should precipitation occur when 25.0 mL of $1.0 \times 10^{-3}\ M$ $SrCl_2$ solution are mixed with 15.0 mL of $2.0 \times 10^{-3}\ M$ Na_2SO_4? Show proof.

70. The solubility of Hg_2I_2 in H_2O is 3.04×10^{-7} g/L. The reaction $Hg_2I_2 \rightleftharpoons Hg_2^{2+} + 2I^-$ represents the equilibrium. Calculate the K_{sp}.

71. Under certain circumstances, when oxygen gas is heated, it can be converted into ozone according to the following reaction equation:

$$3\ O_2(g) + \text{heat} \rightleftharpoons 2\ O_3(g)$$

Name three different ways that you could increase the production of the ozone.

72. One day in a laboratory, some water spilled on a table. In just a few minutes the water had evaporated. Some days later, a similar amount of water spilled again. This time, the water remained on the table after 7 or 8 hours. Name three conditions that could have changed in the lab to cause this difference.

73. For the reaction $CO(g) + H_2O(g) \rightleftharpoons CO_2(g) + H_2(g)$ at a certain temperature, K_{eq} is 1. At equilibrium which of the following would you expect to find?
(a) only CO and H_2
(b) mostly CO_2 and H_2
(c) about equal concentrations of CO and H_2O, compared to CO_2 and H_2
(d) mostly CO and H_2O
(e) only CO and H_2O
Explain your answer briefly.

74. Write the equilibrium constant expressions for these reactions:
(a) $3\ O_2(g) \rightleftharpoons 2\ O_3(g)$
(b) $H_2O(g) \rightleftharpoons H_2O(l)$
(c) $MgCO_3(s) \rightleftharpoons MgO(s) + CO_2(g)$
(d) $2\ Bi^{3+}(aq) + 3\ H_2S(aq) \rightleftharpoons Bi_2S_3(s) + 6\ H^+(aq)$

75. Reactants A and B are mixed, each initially at a concentration of 1.0 M. They react to produce C according to this equation:

$$2\ A + B \rightleftharpoons C$$

When equilibrium is established, the concentration of C is found to be 0.30 M. Calculate the value of K_{eq}.

76. At a certain temperature, K_{eq} is 2.2×10^{-3} for the reaction

$$2\ ICl(g) \rightleftharpoons I_2(g) + Cl_2(g)$$

Now calculate the K_{eq} value for the reaction

$$I_2(g) + Cl_2(g) \rightleftharpoons 2\ ICl(g)$$

77. One drop of 1 M OH^- ion is added to a 1 M solution of HNO_2. What will be the effect of this addition on the equilibrium concentration of the following?

$$HNO_2(aq) \rightleftharpoons H^+(aq) + NO_2^-(aq)$$

(a) $[OH^-]$
(b) $[H^+]$
(c) $[NO_2^-]$
(d) $[HNO_2]$

78. How many grams of $CaSO_4$ will dissolve in 600. mL of water? ($K_{sp} = 2.0 \times 10^{-4}$ for $CaSO_4$)

79. A student found that 0.098 g of PbF_2 was dissolved in 400. mL of saturated PbF_2. What is the K_{sp} for the lead(II) fluoride?

80. Humidity indicators are available that use the equilibrium reaction between the cobalt(II) hexahydrate ion (pink) and chloride ions, which forms the cobalt(II) tetrachloride ion (blue) and water. The equilibrium reaction used by these indicators is

$$Co(H_2O)_6^{2+} + 4\ Cl^- \rightleftharpoons [CoCl_4]^{2-} + 6\ H_2O$$

If you started with a pink indicator, what would you expect to happen if you
(a) blew a hair dryer on it?
(b) put it over a steam vent?
(c) left it in Houston before a rainstorm?
(d) left it in Phoenix in the middle of the summer?

81. Using Le Châtelier's principle, predict how the following reaction will respond to the specified changes.

$$HCHO_2(g) \rightleftharpoons CO(g) + H_2O(g)$$

(a) increase the concentration of $HCHO_2(g)$
(b) increase the concentration of $CO(g)$
(c) decrease the concentration of $H_2O(g)$
(d) decrease the volume of the reaction vessel

82. Aspirin is synthesized by reacting salicylic acid with acetic anhydride, as shown in the following reaction:

$C_7H_6O_3$
Salicylic acid

$C_4H_6O_3$
Acetic anhydride

$C_9H_8O_4$
Aspirin

$C_2H_4O_2$
Acetic acid

The cost of salicylic acid is $58.90 per 500.0 g and acetic anhydride costs $53.50 for 1.00 kg. Given these data, which is the more cost-effective reagent to use in excess to shift the equilibrium far to the right and get the maximum possible yield?

83. Maintaining the proper ratio of carbonic acid and bicarbonate ion in the blood is important to keeping the pH of the blood in the correct range.
(a) Write the reaction of the bicarbonate ion (HCO_3^-) with an acid (H^+).
(b) Write the reaction of the bicarbonate ion with a base (OH^-).

CHALLENGE EXERCISES

84. All a snowbound skier had to eat were walnuts! He was carrying a bag holding 12 dozen nuts. With his mittened hands, he cracked open the shells. Each nut that was opened resulted in one kernel and two shell halves. When he tired of the cracking and got ready to do some eating, he discovered he had 194 total pieces (whole nuts, shell halves, and kernels). What is the K_{eq} for this reaction?

85. At 500°C, the reaction

$$SO_2(g) + NO_2(g) \rightleftharpoons NO(g) + SO_3(g)$$

has $K_{eq} = 81$. What will be the equilibrium concentrations of the four gases if the two reactants begin with equal concentrations of 0.50 M?

ANSWERS TO PRACTICE EXERCISES

16.1 The reactants must come into contact to produce the product(s). When the reactants are first mixed, the rate of the reaction is at its peak. As the reaction proceeds, the amount of the reactants available for reaction becomes smaller and the number of contacts between reactants becomes smaller so the rate of the reaction slows down.

16.2 A double arrow between reactants and products. $\rightleftharpoons$

16.3 (a)

Concentration	Change
$[H_2]$	increase
$[I_2]$	decrease
$[HI]$	increase

(b)

Concentration	Change
$[H_2]$	increase
$[I_2]$	increase
$[HI]$	increase

16.4 (a) equilibrium shifts to the left;
(b) equilibrium shifts to the right

16.5 (a) equilibrium shifts to the right;
(b) equilibrium shifts to the left

16.6 (a) equilibrium shifts to the left;
(b) equilibrium shifts to the right

16.7 (a) $K_{eq} = \dfrac{[NO_2]^4[O_2]}{[N_2O_5]^2}$; (b) $K_{eq} = \dfrac{[N_2]^2[H_2O]^6}{[NH_3]^4[O_2]^3}$

16.8 $K_{eq} = 33$; the forward reaction is favored

16.9 (a) $[H^+] = 5.0 \times 10^{-5}$ $[OH^-] = 2.0 \times 10^{-10}$
(b) $[H^+] = 5.0 \times 10^{-9}$ $[OH^-] = 2.0 \times 10^{-6}$

16.10 (a) 4.30   (b) 8.30

16.11 (a) 6.3×10^{-6} (b) 1.8×10^{-6}

16.12 (a) 6.3×10^{-3} % ionized
(b) 7.2×10^{-3} % ionized

16.13 (a) $K_{sp} = [Cr^{3+}][OH^-]^3$
(b) $K_{sp} = [Cu^{2+}]^3[PO_4^{3-}]^2$

16.14 8.8×10^{-21} g/L

16.15 7.5×10^{-8} mol/L

16.16 Addition of an acid (H^+) increases the acidity of the buffer solution. The increased acid will combine with F^- in the solution to form un-ionized HF, neutralizing the H^+ added. Thus, the pH of the buffer solution will remain about the same as in the original solution.

The car is powered by an electric motor and lithium-ion batteries. The batteries are recharged by plugging the car into a household circuit. Oxidation–reduction reactions provide the energy in the batteries to drive the electric motor.

Icon SMI/NewsCom

CHAPTER **17**

OXIDATION–REDUCTION

The variety of oxidation–reduction reactions that affect us every day is amazing. Our society runs on batteries—in our calculators, laptop computers, cars, toys, radios, televisions, and more. We paint iron railings and galvanize nails to combat corrosion. We electroplate jewelry and computer chips with very thin coatings of gold or silver. We bleach our clothes using chemical reactions that involve electron transfer. We test for glucose in urine or alcohol in the breath with reactions that show vivid color changes. Plants turn energy into chemical compounds through a series of reactions called photosynthesis. These reactions all involve the transfer of electrons between substances in a chemical process called *oxidation–reduction*.

CHAPTER OUTLINE

17.1 Oxidation Number
17.2 Balancing Oxidation–Reduction Equations
17.3 Balancing Ionic Redox Equations
17.4 Activity Series of Metals
17.5 Electrolytic and Voltaic Cells

17.1 OXIDATION NUMBER

Assign oxidation numbers to the atoms in a compound.

● LEARNING OBJECTIVE

KEY TERMS

oxidation number
oxidation state
oxidation–reduction
redox
oxidation
reduction
oxidizing agent
reducing agent

The oxidation number of an atom (sometimes called its *oxidation state*) represents the number of electrons lost, gained, or unequally shared by an atom. Oxidation numbers can be zero, positive, or negative. An oxidation number of zero means the atom has the same number of electrons assigned to it as there are in the free neutral atom. A positive oxidation number means the atom has fewer electrons assigned to it than in the neutral atom, and a negative oxidation number means the atom has more electrons assigned to it than in the neutral atom.

The oxidation number of an atom that has lost or gained electrons to form an ion is the same as the positive or negative charge of the ion. (See Table 17.1.) In the ionic compound NaCl, the oxidation numbers are clearly $+1$ for the Na^+ ion and -1 for the Cl^- ion. The Na^+ ion has one less electron than the neutral Na atom, and the Cl^- ion has one more electron than the neutral Cl atom. In $MgCl_2$, two electrons have transferred from the Mg atom to the two Cl atoms; the oxidation number of Mg is $+2$.

In covalently bonded substances, where electrons are shared between two atoms, oxidation numbers are assigned by an arbitrary system based on relative electronegativities. For symmetrical covalent molecules such as H_2 and Cl_2, each atom is assigned an oxidation number of zero because the bonding pair of electrons is shared equally between two like atoms, neither of which is more electronegative than the other:

$$H{:}H \qquad {:}\ddot{C}l{:}\ddot{C}l{:}$$

When the covalent bond is between two unlike atoms, the bonding electrons are shared unequally because the more electronegative element has a greater attraction for them. In this case the oxidation numbers are determined by assigning both electrons to the more electronegative element.

Thus in compounds with covalent bonds such as NH_3 and H_2O,

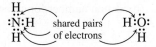

shared pairs of electrons

the pairs of electrons are unequally shared between the atoms and are attracted toward the more electronegative elements, N and O. This causes the N and O atoms to be relatively negative with respect to the H atoms. At the same time, it causes the H atoms to be relatively positive with respect to the N and O atoms. In H_2O, both pairs of shared electrons are assigned to the O atom, giving it two electrons more than the neutral O atom, and each H atom is assigned one electron less than the neutral H atom. Therefore, the oxidation number of the O atom is -2, and the oxidation number of each H atom is $+1$. In NH_3, the three pairs of shared electrons are assigned to the N atom, giving it three electrons more than the neutral N atom, and each H atom has one electron less than the neutral atom. Therefore, the oxidation number of the N atom is -3, and the oxidation number of each H atom is $+1$.

Assigning correct oxidation numbers to elements is essential for balancing oxidation–reduction equations. The **oxidation number** or **oxidation state** of an element is an integer value assigned to each element in a compound or ion that allows us to keep track of electrons associated with each atom. Oxidation numbers have a variety of uses in chemistry—from writing formulas to predicting properties of compounds and assisting in the balancing of oxidation–reduction reactions in which electrons are transferred.

As a starting point, the oxidation number of an uncombined element, regardless of whether it is monatomic or diatomic, is zero. Rules for assigning oxidation numbers are summarized below.

TABLE 17.1 Oxidation Numbers for Common Ions

Ion	Oxidation number
H^+	$+1$
Na^+	$+1$
K^+	$+1$
Li^+	$+1$
Ag^+	$+1$
Cu^{2+}	$+2$
Ca^{2+}	$+2$
Ba^{2+}	$+2$
Fe^{2+}	$+2$
Mg^{2+}	$+2$
Zn^{2+}	$+2$
Al^{3+}	$+3$
Fe^{3+}	$+3$
Cl^-	-1
Br^-	-1
F^-	-1
I^-	-1
S^{2-}	-2
O^{2-}	-2

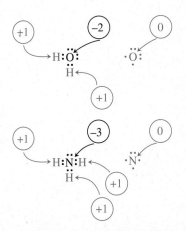

RULES FOR ASSIGNING OXIDATION NUMBERS

1. All elements in their free state (uncombined with other elements) have an oxidation number of zero (e.g., Na, Cu, Mg, H_2, O_2, Cl_2, N_2).

2. H is $+1$, except in metal hydrides, where it is -1 (e.g., NaH, CaH_2).

3. O is -2, except in peroxides, where it is -1, and in OF_2, where it is $+2$.

4. The metallic element in an ionic compound has a positive oxidation number.

5. In covalent compounds the negative oxidation number is assigned to the most electronegative atom.

6. The algebraic sum of the oxidation numbers of the elements in a compound is zero.

7. The algebraic sum of the oxidation numbers of the elements in a polyatomic ion is equal to the charge of the ion.

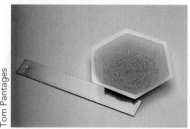

The oxidation number for Cu metal is 0 while the oxidation number for Cu^{2+} ions in the crystal is +2.

WileyPLUS

ENHANCED EXAMPLE

Many elements have multiple oxidation numbers; for example, nitrogen:

	N_2	N_2O	NO	N_2O_3	NO_2	N_2O_5	NO_3^-
Oxidation number	0	+1	+2	+3	+4	+5	+5

PROBLEM-SOLVING STRATEGY: Finding the Oxidation Number of an Element in a Compound

1. Write the oxidation number of each known atom below the atom in the formula.

2. Multiply each oxidation number by the number of atoms of that element in the compound.

3. Write an expression indicating the sum of all the oxidation numbers in the compound.
Remember: The sum of the oxidation numbers in a compound must equal zero.

EXAMPLE 17.1

Determine the oxidation number for carbon in carbon dioxide:

SOLUTION CO_2

1. -2

2. $(-2)2$

3. $C + (-4) = 0$

                $C = +4$ (oxidation number for carbon)

EXAMPLE 17.2

Determine the oxidation number for sulfur in sulfuric acid:

SOLUTION H_2SO_4

1. $+1$ -2

2. $2(+1) = +2$ $4(-2) = -8$

3. $+2 + S + (-8) = 0$

              $S = +6$ (oxidation number for sulfur)

PRACTICE 17.1

Determine the oxidation number of (a) S in Na_2SO_4, (b) As in K_3AsO_4, and (c) C in $CaCO_3$.

Oxidation numbers in a polyatomic ion (ions containing more than one atom) are determined in a similar fashion, except that in a polyatomic ion the sum of the oxidation numbers must equal the charge on the ion instead of zero.

EXAMPLE 17.3

Determine the oxidation number for manganese in the permanganate ion MnO_4^-:

SOLUTION MnO_4^-

1. -2

2. $(-2)4$

3. $Mn + (-8) = -1$ (the charge on the ion)

              $Mn = +7$ (oxidation number for manganese)

EXAMPLE 17.4

Determine the oxidation number for carbon in the oxalate ion $C_2O_4^{2-}$:

SOLUTION $C_2O_4^{2-}$

$$-2$$

1. $(-2)4$
2. $2C + (-8) = -2$ (the charge on the ion)
3. $2C = +6$

$C = +3$ (oxidation number for C)

PRACTICE 17.2

Determine the oxidation numbers of (a) N in NH_4^+, (b) Cr in $Cr_2O_7^{2-}$, and (c) P in PO_4^{3-}.

EXAMPLE 17.5

Determine the oxidation number of each element in (a) KNO_3 and (b) SO_4^{2-}.

SOLUTION

(a) Potassium is a Group 1A metal; therefore, it has an oxidation number of $+1$. The oxidation number of each O atom is -2 (see Rules for Assigning Oxidation Numbers, Rule 3). Using these values and the fact that the sum of the oxidation numbers of all the atoms in a compound is zero, we can determine the oxidation number of N:

$$KNO_3$$
$$+1 + N + 3(-2) = 0$$
$$N = +6 - 1 = +5$$

The oxidation numbers are K, $+1$; N, $+5$; O, -2.

(b) Because SO_4^{2-} is an ion, the sum of oxidation numbers of the S and the O atoms must be -2, the charge of the ion. The oxidation number of each O atom is -2 (Rule 3). Then

$$SO_4^{2-}$$
$$S + 4(-2) = -2, \quad S - 8 = -2$$
$$S = -2 + 8 = +6$$

The oxidation numbers are S, $+6$; O, -2.

PRACTICE 17.3

Determine the oxidation number of each element in these species:

(a) $BeCl_2$ (b) HClO   (c) H_2O_2 (d) NH_4^+ (e) BrO_3^-

Oxidation–Reduction

Oxidation–reduction, also known as **redox**, is a chemical process in which the oxidation number of an element is changed. The process may involve the complete transfer of electrons to form ionic bonds or only a partial transfer or shift of electrons to form covalent bonds.

Oxidation occurs whenever the oxidation number of an element increases as a result of losing electrons. Conversely, **reduction** occurs whenever the oxidation number of an element decreases as a result of gaining electrons. See **Figure 17.1**, for example, a change in oxidation number from $+2$ to $+3$ or from -1 to 0 is oxidation; a change from $+5$ to $+2$ or from -2 to -4 is reduction (see **Figure 17.2**). Oxidation and reduction occur simultaneously in a chemical reaction; one cannot take place without the other.

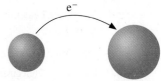

Figure 17.1

In an oxidation–reduction reaction, the element losing an electron is oxidized, while the element gaining an electron is reduced.

Figure 17.2

Oxidation and reduction. Oxidation results in an increase in the oxidation number, and reduction results in a decrease in the oxidation number.

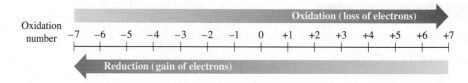

In the reaction between Zn and H_2SO_4, the Zn is oxidized while hydrogen is reduced.

Tom Pantages

Many combination, decomposition, and single-displacement reactions involve oxidation–reduction. Let's examine the combustion of hydrogen and oxygen from this point of view:

$$2\,H_2 + O_2 \longrightarrow 2\,H_2O$$

Both reactants, hydrogen and oxygen, are elements in the free state and have an oxidation number of zero. In the product (water), hydrogen has been oxidized to $+1$ and oxygen reduced to -2. The substance that causes an increase in the oxidation state of another substance is called an **oxidizing agent**. The substance that causes a decrease in the oxidation state of another substance is called a **reducing agent**. In this reaction the oxidizing agent is free oxygen, and the reducing agent is free hydrogen. In the reaction

$$Zn(s) + H_2SO_4(aq) \longrightarrow ZnSO_4(aq) + H_2(g)$$

metallic zinc is oxidized, and hydrogen ions are reduced. Zinc is the reducing agent, and hydrogen ions, the oxidizing agent. Electrons are transferred from the zinc metal to the hydrogen ions. The electron transfer is more clearly expressed as

$$Zn^0 + 2\,H^+ + SO_4^{2-} \longrightarrow Zn^{2+} + SO_4^{2-} + H_2^0$$

Oxidation:	Increase in oxidation number
	Loss of electrons
Reduction:	Decrease in oxidation number
	Gain of electrons

The oxidizing agent is reduced and gains electrons. The reducing agent is oxidized and loses electrons. The transfer of electrons is characteristic of all redox reactions.

EXAMPLE 17.6

Analyze the following balanced redox reactions and identify the element oxidized, the element reduced, the oxidizing agent, and the reducing agent.

(a) $Mg(s) + CuSO_4(aq) \longrightarrow Cu(s) + MgSO_4(aq)$

(b) $C_5H_{12}(l) + 8\,O_2(g) \longrightarrow 5\,CO_2(g) + 6\,H_2O(g)$

SOLUTION

(a)	**(b)**
1. Assign oxidation numbers to each element.	
$Mg(s) + CuSO_4(aq) \longrightarrow Cu(s) + MgSO_4(aq)$ $0 \quad\ \ +2\ +6\ -2 \qquad\quad 0 \quad\ +2\ +6\ -2$	$C_5H_{12}(l) + 8\,O_2(g) \longrightarrow 5\,CO_2(g) + 6\,H_2O(g)$ $-2\ \ +1 \qquad\ 0 \qquad\quad +4\ \ -2 \qquad +1\ -2$
2. Determine which element loses electrons. This is the element oxidized.	
Mg^0 loses 2 electrons to become Mg^{2+}. Magnesium is oxidized.	C^{2-} loses 6 electrons to become C^{4+}. Carbon is oxidized.
3. Determine which element gains electrons. This is the element reduced.	
Cu^{2+} gains 2 electrons to become Cu^0. Copper is reduced.	O^0 gains 2 electrons to become O^{2-}. Oxygen is reduced.
4. The substance that is oxidized is the reducing agent. (Remember the reducing agent must be a substance, not just an atom.)	
Mg is oxidized so it is the reducing agent.	C is oxidized so C_5H_{10} is the reducing agent.
5. The substance that is reduced is the oxidizing agent. (Remember the oxidizing agent must be a substance, not just an atom.)	
Cu is reduced so $CuSO_4$ is the oxidizing agent.	O is reduced so O_2 is the oxidizing agent.

PRACTICE 17.4

Analyze the following balanced redox reactions and identify the element oxidized, the element reduced, the oxidizing agent, and the reducing agent.

(a) $2\,Al(s) + 6\,HNO_3(aq) \longrightarrow 2\,Al(NO_3)_3(aq) + 3\,H_2(g)$

(b) $Ni(s) + CuSO_4(aq) \longrightarrow Cu(s) + NiSO_4(aq)$

(c) $2\,C_2H_6(s) + 7\,O_2(g) \longrightarrow 4\,CO_2(g) + 6\,H_2O(g)$

17.2 BALANCING OXIDATION–REDUCTION EQUATIONS

Balance equations for oxidation–reduction reactions.

● **LEARNING OBJECTIVE**

Many simple redox equations can be balanced readily by inspection, or by trial and error:

$$Na + Cl_2 \longrightarrow NaCl \quad \text{(unbalanced)}$$
$$2\,Na + Cl_2 \longrightarrow 2\,NaCl \quad \text{(balanced)}$$

Balancing this equation is certainly not complicated. But as we study more complex reactions and equations such as

$$P + HNO_3 + H_2O \longrightarrow NO + H_3PO_4 \quad \text{(unbalanced)}$$
$$3\,P + 5\,HNO_3 + 2\,H_2O \longrightarrow 5\,NO + 3\,H_3PO_4 \quad \text{(balanced)}$$

the trial-and-error method of balancing equations takes an unnecessarily long time.

One systematic method for balancing oxidation–reduction equations is based on the transfer of electrons between the oxidizing and reducing agents. Consider the first equation again:

$$Na^0 + Cl_2^0 \longrightarrow Na^+Cl^- \quad \text{(unbalanced)}$$

The superscript 0 shows that the oxidation number is 0 for elements in their uncombined state.

In this reaction, sodium metal loses one electron per atom when it changes to a sodium ion. At the same time chlorine gains one electron per atom. Because chlorine is diatomic, two electrons per molecule are needed to form a chloride ion from each atom. These electrons are furnished by two sodium atoms. Stepwise, the reaction may be written as two half-reactions, the oxidation half-reaction and the reduction half-reaction:

$$2\,Na^0 \longrightarrow 2\,Na^+ + 2\,e^- \quad \text{oxidation half-reaction}$$
$$\underline{Cl_2^0 + 2\,e^- \longrightarrow \qquad\qquad 2\,Cl^-} \quad \text{reduction half-reaction}$$
$$Cl_2^0 + 2\,Na^0 \longrightarrow 2\,Na^+Cl^-$$

When the two half-reactions, each containing the same number of electrons, are added together algebraically, the electrons cancel out. In this reaction there are no excess electrons; the two electrons lost by the two sodium atoms are utilized by chlorine. In all redox reactions, the loss of electrons by the reducing agent must equal the gain of electrons by the oxidizing agent. Here sodium is oxidized and chlorine is reduced. Chlorine is the oxidizing agent; sodium is the reducing agent.

In the following examples, we use the change-in-oxidation-number method, a system for balancing more complicated redox equations.

PROBLEM-SOLVING STRATEGY: Using Change in Oxidation Numbers to Balance Oxidation–Reduction Reactions

1. Assign oxidation numbers for each element to identify the elements being oxidized and reduced. Look for those elements that have changed oxidation number.
2. Write two half-reactions using only the elements that have changed oxidation numbers. One half-reaction must produce electrons, and the other must use electrons.
3. Multiply the half-reactions by the smallest whole numbers that will make the electrons lost by oxidation equal the electrons gained by reduction.
4. Transfer the coefficient in front of each substance in the balanced half-reactions to the corresponding substance in the original equation.
5. Balance the remaining elements that are not oxidized or reduced to give the final balanced equation.
6. Check to make sure both sides of the equation have the same number of atoms of each element.

EXAMPLE 17.7

Balance the equation

$$Sn + HNO_3 \longrightarrow SnO_2 + NO_2 + H_2O \quad \text{(unbalanced)}$$

SOLUTION

Use the Change in Oxidation Number Problem-Solving Strategy

1. Assign oxidation numbers to each element to identify the elements being oxidized and those being reduced. Write the oxidation numbers below each element to avoid confusing them with ionic charge:

$$\underset{0}{Sn} + \underset{+1\ +5\ -2}{HNO_3} \longrightarrow \underset{+4\ -2}{SnO_2} + \underset{+4\ -2}{NO_2} + \underset{+1\ -2}{H_2O}$$

Note that the oxidation numbers of Sn and N have changed.

2. Now write two new equations (half-reactions), using only the elements that change in oxidation number. Then add electrons to bring the equations into electrical balance. One equation represents the oxidation step; the other represents the reduction step. *Remember:* Oxidation produces electrons; reduction uses electrons.

$$Sn^0 \longrightarrow Sn^{4+} + 4\,e^- \quad \text{(oxidation)}$$
$$\text{Sn}^0 \text{ loses 4 electrons}$$

$$N^{5+} + 1\,e^- \longrightarrow N^{4+} \quad \text{(reduction)}$$
$$\text{N}^{5+} \text{ gains 1 electron}$$

3. Multiply the two equations by the smallest whole numbers that will make the electrons lost by oxidation equal to the number of electrons gained by reduction. In this reaction the oxidation step is multiplied by 1 and the reduction step by 4. The equations become

$$Sn^0 \longrightarrow Sn^{4+} + 4\,e^- \quad \text{(oxidation)}$$
$$\text{Sn}^0 \text{ loses 4 electrons}$$

$$4\,N^{5+} + 4\,e^- \longrightarrow 4\,N^{4+} \quad \text{(reduction)}$$
$$4\,\text{N}^{5+} \text{ gains 4 electrons}$$

We have now established the ratio of the oxidizing to the reducing agent as being four atoms of N to one atom of Sn.

4. Transfer the coefficient in front of each substance in the balanced oxidation–reduction equations to the corresponding substance in the original equation. We need to use 1 Sn, 1 SnO_2, 4 HNO_3, and 4 NO_2:

$$Sn + 4\,HNO_3 \longrightarrow SnO_2 + 4\,NO_2 + H_2O \quad \text{(unbalanced)}$$

5. In the usual manner, balance the remaining elements that are not oxidized or reduced to give the final balanced equation:

$$Sn + 4\,HNO_3 \longrightarrow SnO_2 + 4\,NO_2 + 2\,H_2O \quad \text{(balanced)}$$

6. Finally, check to ensure that both sides of the equation have the same number of atoms of each element. The final balanced equation contains 1 atom of Sn, 4 atoms of N, 4 atoms of H, and 12 atoms of O on each side.

In balancing the final elements, we must not change the ratio of the elements that were oxidized and reduced.

Because each new equation presents a slightly different problem and because proficiency in balancing equations requires practice, let's work through two more examples.

EXAMPLE 17.8

Balance the equation

$$I_2 + Cl_2 + H_2O \longrightarrow HIO_3 + HCl \quad \text{(unbalanced)}$$

SOLUTION

Use the Problem-Solving Strategy: Using Change in Oxidation Numbers to Balance Oxidation–Reduction Reactions.

1. Assign oxidation numbers:

$$I_2 + Cl_2 + H_2O \longrightarrow HIO_3 + HCl$$

$$\begin{array}{ccccccc} 0 & 0 & +1\ -2 & & +1\ +5\ -2 & & +1\ -1 \end{array}$$

The oxidation numbers of I_2 and Cl_2 have changed, I_2 from 0 to +5, and Cl_2 from 0 to −1.

2. Write the oxidation and reduction steps. Balance the number of atoms and then balance the electrical charge using electrons:

$$I_2 \longrightarrow 2\ I^{5+} + 10\ e^- \quad \text{(oxidation)} \quad \text{(10 } e^- \text{ are needed to balance the +10 charge)}$$

I_2 loses 10 electrons

$$Cl_2 + 2\ e^- \longrightarrow 2\ Cl^- \quad \text{(reduction)} \quad \text{(2 } e^- \text{ are needed to balance the −2 charge)}$$

Cl_2 gains 2 electrons

3. Adjust loss and gain of electrons so that they are equal. Multiply the oxidation step by 1 and the reduction step by 5:

$$I_2 \longrightarrow 2\ I^{5+} + 10\ e^- \quad \text{(oxidation)}$$

I_2 loses 10 electrons

$$5\ Cl_2 + 10\ e^- \longrightarrow 10\ Cl^- \quad \text{(reduction)}$$

5 Cl_2 gain 10 electrons

4. Transfer the coefficients from the balanced redox equations into the original equation. We need to use 1 I_2, 2 HIO_3, 5 Cl_2, and 10 HCl:

$$I_2 + 5\ Cl_2 + H_2O \longrightarrow 2\ HIO_3 + 10\ HCl \quad \text{(unbalanced)}$$

5. Balance the remaining elements, H and O:

$$I_2 + 5\ Cl_2 + 6\ H_2O \longrightarrow 2\ HIO_3 + 10\ HCl \quad \text{(balanced)}$$

6. *Check:* The final balanced equation contains 2 atoms of I, 10 atoms of Cl, 12 atoms of H, and 6 atoms of O on each side.

EXAMPLE 17.9

Balance the equation

$$K_2Cr_2O_7 + FeCl_2 + HCl \longrightarrow CrCl_3 + KCl + FeCl_3 + H_2O \quad \text{(unbalanced)}$$

SOLUTION

Use the Problem-Solving Strategy: Using the Change in Oxidation Numbers to Balance Oxidation–Reduction Reactions.

1. Assign oxidation numbers (Cr and Fe have changed):

$$K_2Cr_2O_7 + FeCl_2 + HCl \longrightarrow CrCl_3 + KCl + FeCl_3 + H_2O$$

$$\begin{array}{cccccccc} +1\ +6\ -2 & & +2\ -1 & & +1\ -1 & +3\ -1 & +1\ -1 & +3\ -1 & +1\ -2 \end{array}$$

2. Write the oxidation and reduction steps. Balance the number of atoms and then balance the electrical charge using electrons:

$$Fe^{2+} \longrightarrow Fe^{3+} + 1\ e^- \quad \text{(oxidation)}$$

Fe^{2+} loses 1 electron

$$2\ Cr^{6+} + 6\ e^- \longrightarrow 2\ Cr^{3+} \quad \text{(reduction)}$$

2 Cr^{6+} gain 6 electrons

3. Balance the loss and gain of electrons. Multiply the oxidation step by 6 and the reduction step by 1 to equalize the transfer of electrons.

$$6\ Fe^{2+} \longrightarrow 6\ Fe^{3+} + 6\ e^- \quad \text{(oxidation)}$$

6 Fe^{2+} lose 6 electrons

$$2\ Cr^{6+} + 6\ e^- \longrightarrow 2\ Cr^{3+} \quad \text{(reduction)}$$

2 Cr^{6+} gain 6 electrons

4. Transfer the coefficients from the balanced redox equations into the original equation. (Note that one formula unit of $K_2Cr_2O_7$ contains two Cr atoms.) We need to use 1 $K_2Cr_2O_7$, 2 $CrCl_3$, 6 $FeCl_2$, and 6 $FeCl_3$:

$$K_2Cr_2O_7 + 6\,FeCl_2 + HCl \longrightarrow 2\,CrCl_3 + KCl + 6\,FeCl_3 + H_2O$$
(unbalanced)

5. Balance the remaining elements in the order K, Cl, H, O:

$$K_2Cr_2O_7 + 6\,FeCl_2 + 14\,HCl \longrightarrow 2\,CrCl_3 + 2\,KCl + 6\,FeCl_3 + 7\,H_2O$$
(balanced)

6. *Check:* The final balanced equation contains 2 K atoms, 2 Cr atoms, 7 O atoms, 6 Fe atoms, 26 Cl atoms, and 14 H atoms on each side.

PRACTICE 17.5

Balance these equations using the change-in-oxidation-number method:

(a) $HNO_3 + S \longrightarrow NO_2 + H_2SO_4 + H_2O$

(b) $CrCl_3 + MnO_2 + H_2O \longrightarrow MnCl_2 + H_2CrO_4$

(c) $KMnO_4 + HCl + H_2S \longrightarrow KCl + MnCl_2 + S + H_2O$

17.3 BALANCING IONIC REDOX EQUATIONS

LEARNING OBJECTIVE ● Balance equations for ionic oxidation–reduction reactions.

The main difference between balancing ionic redox equations and molecular redox equations is in how we handle the ions. In the ionic redox equations, besides having the same number of atoms of each element on both sides of the final equation, we must also have equal net charges. In assigning oxidation numbers, we must therefore remember to consider the ionic charge.

Several methods are used to balance ionic redox equations, including, with slight modification, the oxidation-number method just shown for molecular equations. But the most popular method is probably the ion–electron method.

The ion–electron method uses ionic charges and electrons to balance ionic redox equations. Oxidation numbers are not formally used, but it is necessary to determine what is being oxidized and what is being reduced.

PROBLEM-SOLVING STRATEGY: Ion–Electron Strategy for Balancing Oxidation–Reduction Reactions

1. Write the two half-reactions that contain the elements being oxidized and reduced using the entire formula of the ion or molecule.

2. Balance the elements other than oxygen and hydrogen.

3. Balance oxygen and hydrogen.

   Acidic solution: For reactions in acidic solution, use H^+ and H_2O to balance oxygen and hydrogen. For each oxygen needed, use one H_2O. Then add H^+ as needed to balance the hydrogen atoms.

   Basic solution: For reactions in alkaline solutions, first balance as though the reaction were in an acidic solution, using Steps 1–3. Then add as many OH^- ions to each side of the equation as there are H^+ ions in the equation. Now combine the H^+ and OH^- ions into water (for example, 4 H^+ and 4 OH^- give 4 H_2O). Rewrite the equation, canceling equal numbers of water molecules that appear on opposite sides of the equation.

4. Add electrons (e^-) to each half-reaction to bring them into electrical balance.

5. Since the loss and gain of electrons must be equal, multiply each half-reaction by the appropriate number to make the number of electrons the same in each half-reaction.

6. Add the two half-reactions together, canceling electrons and any other identical substances that appear on opposite sides of the equation.

EXAMPLE 17.10

Balance this equation using the ion–electron method:

$MnO_4^- + S^{2-} \longrightarrow Mn^{2+} + S^0$ (acidic solution)

SOLUTION

1. Write two half-reactions, one containing the element being oxidized and the other the element being reduced (use the entire molecule or ion):

   $S^{2-} \longrightarrow S^0$ (oxidation)

   $MnO_4^- \longrightarrow Mn^{2+}$ (reduction)

2. Balance elements other than oxygen and hydrogen (accomplished in Step 1: 1 S and 1 Mn on each side).

3. Balance O and H. Remember the solution is acidic. The oxidation requires neither O nor H, but the reduction equation needs 4 H_2O on the right and 8 H^+ on the left:

   $S^{2-} \longrightarrow S^0$

   $8\,H^+ + MnO_4^- \longrightarrow Mn^{2+} + 4\,H_2O$

4. Balance each half-reaction electrically with electrons:

   $S^{2-} \longrightarrow S^0 + 2\,e^-$

   net charge $= -2$ on each side

   $5\,e^- + 8\,H^+ + MnO_4^- \longrightarrow Mn^{2+} + 4\,H_2O$

   net charge $= +2$ on each side

5. Equalize loss and gain of electrons. In this case, multiply the oxidation equation by 5 and the reduction equation by 2:

   $5\,S^{2-} \longrightarrow 5\,S^0 + 10\,e^-$

   $10\,e^- + 16\,H^+ + 2\,MnO_4^- \longrightarrow 2\,Mn^{2+} + 8\,H_2O$

6. Add the two half-reactions together, canceling the 10 e^- from each side, to obtain the balanced equation:

   $5\,S^{2-} \longrightarrow 5\,S^0 + \cancel{10\,e^-}$

   $\underline{\cancel{10\,e^-} + 16\,H^+ + 2\,MnO_4^- \longrightarrow 2\,Mn^{2+} + 8\,H_2O}$ (balanced)

   $16\,H^+ + 2\,MnO_4^- + 5\,S^{2-} \longrightarrow 2\,Mn^{2+} + 5\,S^0 + 8\,H_2O$

   Check: Both sides of the equation have a charge of +4 and contain the same number of atoms of each element.

EXAMPLE 17.11

Balance this equation:

$CrO_4^{2-} + Fe(OH)_2 \longrightarrow Cr(OH)_3 + Fe(OH)_3$ (basic solution)

SOLUTION

Use the Ion–Electron Problem-Solving Strategy.

1. Write the two half-reactions:

   $Fe(OH)_2 \longrightarrow Fe(OH)_3$ (oxidation)

   $CrO_4^{2-} \longrightarrow Cr(OH)_3$ (reduction)

2. Balance elements other than H and O (accomplished in Step 1).

3. Remember the solution is basic. Balance O and H as though the solution were acidic. Use H_2O and H^+. To balance O and H in the oxidation equation, add 1 H_2O on the left and 1 H^+ on the right side:

   $Fe(OH)_2 + H_2O \longrightarrow Fe(OH)_3 + H^+$

   Add 1 OH^- to each side:

   $Fe(OH)_2 + H_2O + OH^- \longrightarrow Fe(OH)_3 + H^+ + OH^-$

   Combine H^+ and OH^- as H_2O and rewrite, canceling H_2O on each side:

   $Fe(OH)_2 + \cancel{H_2O} + OH^- \longrightarrow Fe(OH)_3 + \cancel{H_2O}$

   $\boxed{Fe(OH)_2 + OH^- \longrightarrow Fe(OH)_3}$ (oxidation)

To balance O and H in the reduction equation, add $1\ H_2O$ on the right and $5\ H^+$ on the left:

$$CrO_4^{2-} + 5\ H^+ \longrightarrow Cr(OH)_3 + H_2O$$

Add $5\ OH^-$ to each side:

$$CrO_4^{2-} + 5\ H^+ + 5\ OH^- \longrightarrow Cr(OH)_3 + H_2O + 5\ OH^-$$

Combine $5\ H^+ + 5\ OH^- \longrightarrow 5\ H_2O$:

$$CrO_4^{2-} + 5\ H_2O \longrightarrow Cr(OH)_3 + H_2O + 5\ OH^-$$

Rewrite, canceling $1\ H_2O$ from each side:

$$\boxed{CrO_4^{2-} + 4\ H_2O \longrightarrow Cr(OH)_3 + 5\ OH^-}\quad \text{(reduction)}$$

4. Balance each half-reaction electrically with electrons:

$$Fe(OH)_2 + OH^- \longrightarrow Fe(OH)_3 + e^-\quad \text{(balanced oxidation equation)}$$
$$CrO_4^{2-} + 4\ H_2O + 3\ e^- \longrightarrow Cr(OH)_3 + 5\ OH^-\quad \text{(balanced reduction equation)}$$

5. Equalize the loss and gain of electrons. Multiply the oxidation reaction by 3:

$$3\ Fe(OH)_2 + 3\ OH^- \longrightarrow 3\ Fe(OH)_3 + 3\ e^-$$
$$CrO_4^{2-} + 4\ H_2O + 3\ e^- \longrightarrow Cr(OH)_3 + 5\ OH^-$$

6. Add the two half-reactions together, canceling the $3\ e^-$ and $3\ OH^-$ from each side of the equation:

$$3\ Fe(OH)_2 + 3\ OH^- \longrightarrow 3\ Fe(OH)_3 + \cancel{3\ e^-}$$
$$\underline{CrO_4^{2-} + 4\ H_2O + \cancel{3\ e^-} \longrightarrow Cr(OH)_3 + 5\ OH^-}$$
$$CrO_4^{2-} + 3\ Fe(OH)_2 + 4\ H_2O \longrightarrow Cr(OH)_3 + 3\ Fe(OH)_3 + 2\ OH^-\quad \text{(balanced)}$$

Check: Each side of the equation has a charge of -2 and contains the same number of atoms of each element.

PRACTICE 17.6

Balance these equations using the ion–electron method:

(a) $I^- + NO_2^- \longrightarrow I_2 + NO$ (acidic solution)

(b) $Cl_2 + IO_3^- \longrightarrow IO_4^- + Cl^-$ (basic solution)

(c) $AuCl_4^- + Sn^{2+} \longrightarrow Sn^{4+} + AuCl + Cl^-$

>CHEMISTRY *IN ACTION*

Sensitive Sunglasses

Oxidation–reduction reactions are the basis for many interesting applications. Consider photochromic glass, which is used for lenses in light-sensitive glasses. These lenses, manufactured by the Corning Glass Company, can change from transmitting 85% of light to transmitting only 22% of light when exposed to bright sunlight.

Photochromic glass is composed of linked tetrahedrons of silicon and oxygen atoms jumbled in a disorderly array, with crystals of silver chloride caught between the silica tetrahedrons. When the glass is clear, the visible light passes right through the molecules. The glass absorbs ultraviolet light, however, and this energy triggers an oxidation–reduction reaction between Ag^+ and Cl^-:

$$Ag^+ + Cl^- \xrightarrow{\text{UV light}} Ag^0 + Cl^0$$

To prevent the reaction from reversing itself immediately, a few ions of Cu^+ are incorporated into the silver chloride crystal. These Cu^+ ions react with the newly formed chlorine atoms:

$$Cu^+ + Cl^0 \longrightarrow Cu^{2+} + Cl^-$$

The silver atoms move to the surface of the crystal and form small colloidal clusters of silver metal. This metallic silver absorbs visible light, making the lens appear dark (colored).

As the glass is removed from the light, the Cu^{2+} ions slowly move to the surface of the crystal, where they interact with the silver metal:

$$Cu^{2+} + Ag^0 \longrightarrow Cu^+ + Ag^+$$

The glass clears as the silver ions rejoin chloride ions in the crystals.

Mary Ann Price/John Wiley & Sons

Mary Ann Price/John Wiley & Sons

An oxidation–reduction reaction causes these photochromic glasses to change from light to dark in bright sunlight.

Ionic equations can also be balanced using the change-in-oxidation-number method shown in Example 17.7. To illustrate this method, let's use the equation from Example 17.11.

EXAMPLE 17.12

Balance this equation using the change-in-oxidation-number method:

$$CrO_4^{2-} + Fe(OH)_2 \longrightarrow Cr(OH)_3 + Fe(OH)_3 \quad \text{(basic solution)}$$

SOLUTION

1. and 2. Assign oxidation numbers and balance the charges with electrons:

$$Cr^{6+} + 3\,e^- \longrightarrow Cr^{3+} \quad \text{(reduction)}$$

$$\text{Cr}^{6+} \text{ gains 3 } e^-$$

$$Fe^{2+} \longrightarrow Fe^{3+} + e^- \quad \text{(oxidation)}$$

$$\text{Fe}^{2+} \text{ loses 1 } e^-$$

3. Equalize the loss and gain of electrons, by multiplying the oxidation step by 3:

$$Cr^{6+} + 3\,e^- \longrightarrow Cr^{3+} \quad \text{(reduction)}$$

$$\text{Cr}^{6+} \text{ gains 3 } e^-$$

$$3\,Fe^{2+} \longrightarrow 3\,Fe^{3+} + 3\,e^- \quad \text{(oxidation)}$$

$$3\,\text{Fe}^{2+} \text{ loses 3 } e^-$$

4. Transfer coefficients back to the original equation:

$$CrO_4^{2-} + 3\,Fe(OH)_2 \longrightarrow Cr(OH)_3 + 3\,Fe(OH)_3$$

5. Balance electrically. Because the solution is basic, use OH^- to balance charges. The charge on the left side is -2 and on the right side is 0. Add $2\,OH^-$ ions to the right side of the equation:

$$CrO_4^{2-} + 3\,Fe(OH)_2 \longrightarrow Cr(OH)_3 + 3\,Fe(OH)_3 + 2\,OH^-$$

Adding $4\,H_2O$ to the left side balances the equation:

$$CrO_4^{2-} + 3\,Fe(OH)_2 + 4\,H_2O \longrightarrow Cr(OH)_3 + 3\,Fe(OH)_3 + 2\,OH^- \quad \text{(balanced)}$$

Check: Each side of the equation has a charge of -2 and contains the same number of atoms of each element.

PRACTICE 17.7

Balance these equations using the change-in-oxidation-number method:

(a) $Zn \longrightarrow Zn(OH)_4^{2-} + H_2$ (basic solution)

(b) $H_2O_2 + Sn^{2+} \longrightarrow Sn^{4+}$ (acidic solution)

(c) $Cu + Cu^{2+} \longrightarrow Cu_2O$ (basic solution)

17.4 ACTIVITY SERIES OF METALS

Use the activity series of metals to predict whether a reaction will occur.

● **LEARNING OBJECTIVE**

KEY TERM

activity series of metals

Knowledge of the relative chemical reactivities of the elements helps us predict the course of many chemical reactions. For example, calcium reacts with cold water to produce hydrogen, and magnesium reacts with steam to produce hydrogen. Therefore, calcium is considered a more reactive metal than magnesium:

$$Ca(s) + 2\,H_2O(l) \longrightarrow Ca(OH)_2(aq) + H_2(g)$$

$$Mg(s) + H_2O(g) \longrightarrow MgO(s) + H_2(g)$$

$$\text{steam}$$

The difference in their activity is attributed to the fact that calcium loses its two valence electrons more easily than magnesium and is therefore more reactive and/or more readily oxidized than magnesium.

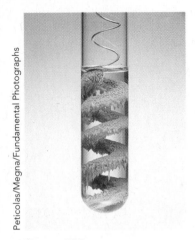

Figure 17.3

A coil of copper placed in a silver nitrate solution forms silver crystals on the wire. The pale blue of the solution indicates the presence of copper ions.

When a coil of copper is placed in a solution of silver nitrate ($AgNO_3$), free silver begins to plate out on the copper. (See **Figure 17.3**.) After the reaction has continued for some time, we can observe a blue color in the solution, indicating the presence of copper(II) ions. The equations are

$$Cu^0(s) + 2\,AgNO_3(aq) \longrightarrow 2\,Ag^0(s) + Cu(NO_3)_2(aq)$$
$$Cu^0(s) + 2\,Ag^+(aq) \longrightarrow 2\,Ag^0(s) + Cu^{2+}(aq) \qquad \text{(net ionic equation)}$$
$$Cu^0(s) \longrightarrow Cu^{2+}(aq) + 2\,e^- \qquad \text{(oxidation of } Cu^0\text{)}$$
$$Ag^+(aq) + e^- \longrightarrow Ag^0(s) \qquad \text{(reduction of } Ag^+\text{)}$$

If a coil of silver is placed in a solution of copper(II) nitrate, $Cu(NO_3)_2$, no reaction is visible.

$$Ag^0(s) + Cu(NO_3)_2(aq) \longrightarrow \text{no reaction}$$

In the reaction between Cu and $AgNO_3$, electrons are transferred from Cu^0 atoms to Ag^+ ions in solution. Copper has a greater tendency than silver to lose electrons, so an electrochemical force is exerted upon silver ions to accept electrons from copper atoms. When an Ag^+ ion accepts an electron, it is reduced to an Ag^0 atom and is no longer soluble in solution. At the same time, Cu^0 is oxidized and goes into solution as Cu^{2+} ions. From this reaction, we can conclude that copper is more reactive than silver.

Metals such as sodium, magnesium, zinc, and iron that react with solutions of acids to liberate hydrogen are more reactive than hydrogen. Metals such as copper, silver, and mercury that do not react with solutions of acids to liberate hydrogen are less reactive than hydrogen. By studying a series of reactions such as these, we can list metals according to their chemical activity, placing the most active at the top and the least active at the bottom. This list is called the **activity series of metals**. Table 17.2 lists some of the common metals in the series. The arrangement corresponds to the ease with which the elements are oxidized or lose electrons, with the most easily oxidizable element listed first. More extensive tables are available in chemistry reference books.

TABLE 17.2 Activity Series of Metals

Ease of oxidation ↑

$$K \longrightarrow K^+ + e^-$$
$$Ba \longrightarrow Ba^{2+} + 2\,e^-$$
$$Ca \longrightarrow Ca^{2+} + 2\,e^-$$
$$Na \longrightarrow Na^+ + e^-$$
$$Mg \longrightarrow Mg^{2+} + 2\,e^-$$
$$Al \longrightarrow Al^{3+} + 3\,e^-$$
$$Zn \longrightarrow Zn^{2+} + 2\,e^-$$
$$Cr \longrightarrow Cr^{3+} + 3\,e^-$$
$$Fe \longrightarrow Fe^{2+} + 2\,e^-$$
$$Ni \longrightarrow Ni^{2+} + 2\,e^-$$
$$Sn \longrightarrow Sn^{2+} + 2\,e^-$$
$$Pb \longrightarrow Pb^{2+} + 2\,e^-$$
$$\mathbf{H_2} \longrightarrow \mathbf{2\,H^+ + 2\,e^-}$$
$$Cu \longrightarrow Cu^{2+} + 2\,e^-$$
$$As \longrightarrow As^{3+} + 3\,e^-$$
$$Ag \longrightarrow Ag^+ + e^-$$
$$Hg \longrightarrow Hg^{2+} + 2\,e^-$$
$$Au \longrightarrow Au^{3+} + 3\,e^-$$

PROBLEM-SOLVING STRATEGY: Using the Activity Series to Predict Reactions

1. The reactivity of the metals listed decreases from top to bottom.

2. A free metal can displace the ion of a second metal from solution, provided that the free metal is above the second metal in the activity series.

3. Free metals above hydrogen react with nonoxidizing acids in solution to liberate hydrogen gas.

4. Free metals below hydrogen do not liberate hydrogen from acids.

5. Conditions such as temperature and concentration may affect the relative position of some of these elements.

Here are two examples using the activity series of metals.

WileyPLUS
ENHANCED EXAMPLE

EXAMPLE 17.13

Will zinc metal react with dilute sulfuric acid?

SOLUTION

From Table 17.2, we see that zinc is above hydrogen; therefore, zinc atoms will lose electrons more readily than hydrogen atoms. Hence, zinc atoms will reduce hydrogen ions from the acid to form hydrogen gas and zinc ions. In fact, these reagents are commonly used for the laboratory preparation of hydrogen. The equation is

$$Zn(s) + H_2SO_4(aq) \longrightarrow ZnSO_4(aq) + H_2(g)$$
$$Zn(s) + 2\,H^+(aq) \longrightarrow Zn^{2+}(aq) + H_2(g) \quad \text{(net ionic equation)}$$

EXAMPLE 17.14

Will a reaction occur when copper metal is placed in an iron(II) sulfate solution?

SOLUTION

No, copper lies below iron in the series, loses electrons less easily than iron, and therefore will not displace iron(II) ions from solution. In fact, the reverse is true. When an iron nail is dipped into a copper(II) sulfate solution, it becomes coated with free copper. The equations are

$$Cu(s) + FeSO_4(aq) \longrightarrow \text{no reaction}$$
$$Fe(s) + CuSO_4(aq) \longrightarrow FeSO_4(aq) + Cu(s)$$

From Table 17.2, we may abstract the following pair in their relative position to each other:

$$Fe \longrightarrow Fe^{2+} + 2\,e^-$$
$$Cu \longrightarrow Cu^{2+} + 2\,e^-$$

According to Step 2 in Problem-Solving Strategy: Using the Activity Series, we can predict that free iron will react with copper(II) ions in solution to form free copper metal and iron(II) ions in solution:

$$Fe(s) + Cu^{2+}(aq) \longrightarrow Fe^{2+}(aq) + Cu(s) \quad \text{(net ionic equation)}$$

PRACTICE 17.8

Indicate whether these reactions will occur:
(a) Sodium metal is placed in dilute hydrochloric acid.
(b) A piece of lead is placed in magnesium nitrate solution.
(c) Mercury is placed in a solution of silver nitrate.

17.5 ELECTROLYTIC AND VOLTAIC CELLS

Compare the reactions and functions of electrolytic and voltaic cells.

● LEARNING OBJECTIVE

KEY TERMS
electrolysis
electrolytic cell
cathode
anode
voltaic cell

The process in which electrical energy is used to bring about chemical change is known as **electrolysis**. An **electrolytic cell** uses electrical energy to produce a chemical reaction. The use of electrical energy has many applications in industry—for example, in the production of sodium, sodium hydroxide, chlorine, fluorine, magnesium, aluminum, and pure hydrogen and oxygen, and in the purification and electroplating of metals.

What happens when an electric current is passed through a solution? Let's consider a hydrochloric acid solution in a simple electrolytic cell, as shown in **Figure 17.4**. The cell consists of a source of direct current (a battery) connected to two electrodes that are immersed in a solution of hydrochloric acid. The negative electrode is called the **cathode** because cations are attracted to it. The positive electrode is called the **anode** because anions are attracted to it. The cathode is attached to the negative pole and the anode to the positive pole of the battery. The battery supplies electrons to the cathode.

When the electric circuit is completed, positive hydronium ions (H_3O^+) migrate to the cathode, where they pick up electrons and evolve as hydrogen gas. At the same time the negative chloride ions (Cl^-) migrate to the anode, where they lose electrons and evolve as chlorine gas.

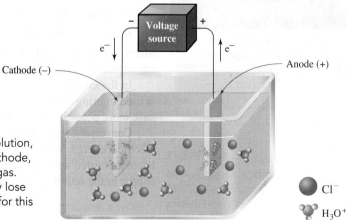

Figure 17.4

During the electrolysis of a hydrochloric acid solution, positive hydronium ions are attracted to the cathode, where they gain electrons and form hydrogen gas. Chloride ions migrate to the anode, where they lose electrons and form chlorine gas. The equation for this process is $2\,HCl(aq) \longrightarrow H_2(g) + Cl_2(g)$.

radio, and so on. When the engine is running, a generator or alternator produces and forces an electric current through the battery and, by electrolytic chemical action, restores it to the charged condition.

The cell unit consists of a lead plate filled with spongy lead and a lead(IV) oxide plate, both immersed in dilute sulfuric acid solution, which serves as the electrolyte (see **Figure 17.7**). When the cell is discharging, or acting as a voltaic cell, these reactions occur:

Pb plate (anode): $Pb^0 \longrightarrow Pb^{2+} + 2\,e^-$ (oxidation)

PbO$_2$ plate (cathode): $PbO_2 + 4\,H^+ + 2\,e^- \longrightarrow Pb^{2+} + 2\,H_2O$ (reduction)

Net ionic redox reaction: $Pb^0 + PbO_2 + 4\,H^+ \longrightarrow 2\,Pb^{2+} + 2\,H_2O$

Precipitation reaction on plates: $Pb^{2+}(aq) + SO_4^{2-}(aq) \longrightarrow PbSO_4(s)$

Because lead(II) sulfate is insoluble, the Pb^{2+} ions combine with SO_4^{2-} ions to form a coating of $PbSO_4$ on each plate. The overall chemical reaction of the cell is

$$Pb(s) + PbO_2(s) + 2\,H_2SO_4(aq) \xrightarrow[\text{cycle}]{\text{discharge}} 2\,PbSO_4(s) + 2\,H_2O(l)$$

The cell can be recharged by reversing the chemical reaction. This reversal is accomplished by forcing an electric current through the cell in the opposite direction. Lead sulfate and water are reconverted to lead, lead(IV) oxide, and sulfuric acid:

$$2\,PbSO_4(s) + 2\,H_2O(l) \xrightarrow[\text{cycle}]{\text{charge}} Pb(s) + PbO_2(s) + 2\,H_2SO_4(aq)$$

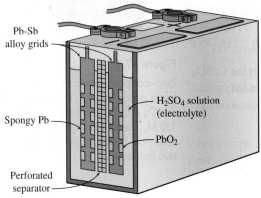

Pb-Sb alloy grids

Spongy Pb

H$_2$SO$_4$ solution (electrolyte)

PbO$_2$

Perforated separator

Figure 17.7

Cross-sectional diagram of a lead storage battery cell.

The electrolyte in a lead storage battery is a 38% by mass sulfuric acid solution having a density of 1.29 g/mL. As the battery is discharged, sulfuric acid is removed, thereby decreasing the density of the electrolyte solution. The state of charge or discharge of the battery can be estimated by measuring the density (or specific gravity) of the electrolyte solution with a hydrometer. When the density has dropped to about 1.05 g/mL, the battery needs recharging.

In a commerical battery, each cell consists of a series of cell units of alternating lead–lead(IV) oxide plates separated and supported by wood, glass wool, or fiberglass. The energy storage capacity of a single cell is limited, and its electrical potential is only about 2 volts. Therefore a bank of six cells is connected in series to provide the 12-volt output of the usual automobile battery.

EXAMPLE 17.15

(a) What is the main difference between a voltaic cell and an electrolytic cell?

(b) In Figure 17.5 what is the purpose of the salt bridge?

SOLUTION

(a) A voltaic cell uses a chemical reaction to produce electrical energy. An electrolytic cell uses electrical energy to produce a chemical reaction.

(b) The salt bridge is used to complete the electrical circuit between the cathode and the anode half-cells.

PRACTICE 17.9

Consider the reaction

$$Sn^{2+}(aq) + Cu(s) \longrightarrow Sn(s) + Cu^{2+}(aq)$$

(a) Which metal is oxidized, and which is reduced?

(b) Write the reaction occurring at the anode and at the cathode.

>CHEMISTRY *IN ACTION*

Superbattery Uses Hungry Iron Ions

Scientists are constantly trying to make longer-lasting environmentally friendly batteries to fuel our many electronic devices. Now a new type of alkaline batteries called "super-iron" batteries have been developed by chemists at Technion-Israel Institute of Technology in Israel. A traditional cell of this type is shown (see Fig. 17.6). In the new battery, the heavy manganese dioxide cathode is replaced with "super iron." (See accompanying diagram.) This special type of iron compound contains iron(VI) in compounds such as K_2FeO_4 or $BaFeO_4$. Iron typically has an oxidation state of +2 or +3. In super iron, each iron atom is missing 6 electrons instead of the usual 2 or 3. This allows the battery to store 1505 J more energy than other alkaline

batteries. When the battery is used (or the cell discharges), the following reaction occurs:

$$2\, MFeO_4 + 3\, Zn \longrightarrow Fe_2O_3 + ZnO + 2\, MZnO_2$$
$$(M = K_2 \text{ or } Ba)$$

The iron compounds used in this battery are much less expensive than the current MnO_2 compounds, and the products are more environmentally friendly (Fe_2O_3 is a form of rust).

Each manganese dioxide molecule in conventional batteries can only accept 1 electron while iron(VI) compounds can absorb 3 electrons each. The super-iron compounds are highly conductive, which means the super-iron battery will work well in our high-drain-rate electronic items. The accompanying graph shows a comparison between a conventional battery and a super-iron battery (both AAA). The conventional AAA battery lasts less than half as long as a super-iron AAA battery.

Super-iron battery

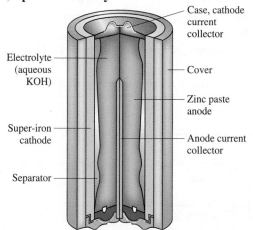

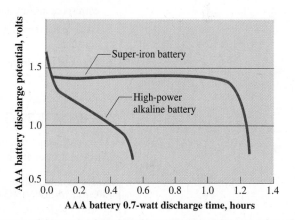

17.1 OXIDATION NUMBER

- To assign an oxidation number:

RULES FOR ASSIGNING OXIDATION NUMBERS

1. All elements in their free state (uncombined with other elements) have an oxidation number of zero (e.g., Na, Cu, Mg, H_2, O_2, Cl_2, N_2).
2. H is +1, except in metal hydrides, where it is −1 (e.g., NaH, CaH_2).
3. O is −2, except in peroxides, where it is −1, and in OF_2, where it is +2.
4. The metallic element in an ionic compound has a positive oxidation number.
5. In covalent compounds the negative oxidation number is assigned to the most electronegative atom.
6. The algebraic sum of the oxidation numbers of the elements in a compound is zero.
7. The algebraic sum of the oxidation numbers of the elements in a polyatomic ion is equal to the charge of the ion.

KEY TERMS

oxidation number
oxidation state
oxidation–reduction
redox
oxidation
reduction
oxidizing agent
reducing agent

- Oxidation–reduction is a chemical process in which electrons are transferred from one atom to another to change the oxidation number of the atom.
- When the oxidation number increases, oxidation occurs, resulting in the loss of electrons.
- When the oxidation number decreases, reduction occurs, resulting in the gain of electrons.

17.2 BALANCING OXIDATION–REDUCTION EQUATIONS

- Trial and error or inspection methods.
- Can be balanced by writing and balancing the half-reactions for the overall reaction.
- Can be balanced by the change-in-oxidation-number method.

> **PROBLEM-SOLVING STRATEGY: Using Change in Oxidation Numbers to Balance Oxidation–Reduction Reactions**
>
> **1.** Assign oxidation numbers for each element to identify the elements being oxidized and reduced. Look for those elements that have changed oxidation number.
>
> **2.** Write two half-reactions using only the elements that have changed oxidation numbers. One half-reaction must produce electrons, and the other must use electrons.
>
> **3.** Multiply the half-reactions by the smallest whole numbers that will make the electrons lost by oxidation equal the electrons gained by reduction.
>
> **4.** Transfer the coefficient in front of each substance in the balanced half-reactions to the corresponding substance in the original equation.
>
> **5.** Balance the remaining elements that are not oxidized or reduced to give the final balanced equation.
>
> **6.** Check to make sure both sides of the equation have the same number of atoms of each element.

17.3 BALANCING IONIC REDOX EQUATIONS

- To balance equations that are ionic, charge must also be balanced (in addition to atoms and ions).
- The ion–electron method for balancing equations is:

> **PROBLEM-SOLVING STRATEGY: Ion–Electron Strategy for Balancing Oxidation–Reduction Reactions**
>
> **1.** Write the two half-reactions that contain the elements being oxidized and reduced using the entire formula of the ion or molecule.
>
> **2.** Balance the elements other than oxygen and hydrogen.
>
> **3.** Balance oxygen and hydrogen.
>
> **Acidic solution:** For reactions in acidic solution, use H^+ and H_2O to balance oxygen and hydrogen. For each oxygen needed, use one H_2O. Then add H^+ as needed to balance the hydrogen atoms.
>
> **Basic solution:** For reactions in alkaline solutions, first balance as though the reaction were in an acidic solution, using Steps 1–3. Then add as many OH^- ions to each side of the equation as there are H^+ ions in the equation. Now combine the H^+ and OH^- ions into water (for example, $4 H^+$ and $4 OH^-$ give $4 H_2O$). Rewrite the equation, canceling equal numbers of water molecules that appear on opposite sides of the equation.
>
> **4.** Add electrons (e^-) to each half-reaction to bring them into electrical balance.
>
> **5.** Since the loss and gain of electrons must be equal, multiply each half-reaction by the appropriate number to make the number of electrons the same in each half-reaction.
>
> **6.** Add the two half-reactions together, canceling electrons and any other identical substances that appear on opposite sides of the equation.

17.4 ACTIVITY SERIES OF METALS

KEY TERM

activity series of metals

- The activity series lists metals from most to least reactive.
- A free metal can displace anything lower on the activity series.

PROBLEM-SOLVING STRATEGY: Using the Activity Series to Predict Reactions

1. The reactivity of the metals listed decreases from top to bottom.

2. A free metal can displace the ion of a second metal from solution, provided that the free metal is above the second metal in the activity series.

3. Free metals above hydrogen react with nonoxidizing acids in solution to liberate hydrogen gas.

4. Free metals below hydrogen do not liberate hydrogen from acids.

5. Conditions such as temperature and concentration may affect the relative position of some of these elements.

17.5 ELECTROLYTIC AND VOLTAIC CELLS

- Electrolysis is the process of using electricity to bring about chemical change.
- Oxidation always occurs at the anode and reduction at the cathode.
- A cell that produces electric current from a spontaneous chemical reaction is a voltaic or galvanic cell.
- A typical electrolytic cell is shown next:

KEY TERMS

electrolysis
electrolytic cell
cathode
anode
voltaic cell

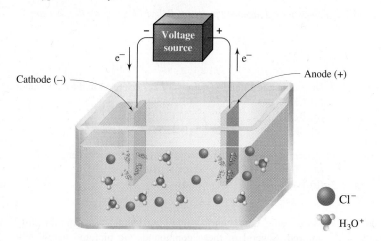

REVIEW QUESTIONS

1. In the equation

$$I_2 + 5\,Cl_2 + 6\,H_2O \longrightarrow 2\,HIO_3 + 10\,HCl$$

(a) has iodine been oxidized, or has it been reduced?
(b) has chlorine been oxidized, or has it been reduced? (Figure 17.2)

2. What is the difference between an oxidation number for an atom in an ionic compound and an oxidation number for an atom in a covalently bonded compound?

3. Why are oxidation and reduction said to be complementary processes?

4. Why are oxidation–reduction equations not usually balanced by inspection?

5. How many electrons are needed to balance the following half-reaction? Should the electrons be placed on the right or left side of the equation?

$$Mn^{2+} \longrightarrow Mn^{+6}$$

6. How many electrons are needed to balance the following half-reaction? Should the electrons be placed on the right or left side of the equation?

$$O^0 \longrightarrow O^{2-}$$

7. What is the important first step in balancing oxidation–reduction reactions by the ion–electron strategy?

8. Describe how you would balance the following balanced half-reaction in a basic solution. Write the equation in basic solution.

$$2\,IO_3^- + 12\,H^+ + 10\,e^- \longrightarrow I_2 + 6\,H_2O$$

9. Why do we say that the more active a metal is, the more easily it will be oxidized?

10. Which element in each pair is more chemically reactive? (Table 17.2)
(a) Mg or Ca   (b) Fe or Ag   (c) Zn or H

11. Complete and balance the equation for all of the reactions that will occur. If the reaction will not occur, explain why. (Table 17.2)
(a) $Al(s) + ZnCl_2(aq) \rightarrow$ (e) $Cr(s) + Ni^{2+}(aq) \rightarrow$
(b) $Sn(s) + HCl(aq) \rightarrow$ (f) $Mg(s) + Ca^{2+}(aq) \rightarrow$
(c) $Ag(s) + H_2SO_4(aq) \rightarrow$ (g) $Cu(s) + H^+(aq) \rightarrow$
(d) $Fe(s) + AgNO_3(aq) \rightarrow$ (h) $Ag(s) + Al^{3+}(aq) \rightarrow$

12. If a copper wire is placed into a solution of lead (II) nitrate, will a reaction occur? Explain.

13. The reaction between powdered aluminum and iron(III) oxide (in the thermite process) producing molten iron is very exothermic.
(a) Write the equation for the chemical reaction that occurs.
(b) Explain in terms of Table 17.2 why a reaction occurs.
(c) Would you expect a reaction between powdered iron and aluminum oxide?
(d) Would you expect a reaction between powdered aluminum and chromium(III) oxide?

14. Write equations for the chemical reaction of aluminum, chromium, gold, iron, copper, magnesium, mercury, and zinc with dilute solutions of (a) hydrochloric acid and (b) sulfuric acid. If a reaction will not occur, write "no reaction" as the product. (Table 17.2)

15. State the charge and purpose of the anode and the cathode in an electrolytic or volatic cell.

16. A $NiCl_2$ solution is placed in the apparatus shown in Figure 17.4, instead of the HCl solution shown. Write equations for the following:
  (a) the anode reaction
  (b) the cathode reaction
  (c) the net electrochemical reaction

17. What is the major distinction between the reactions occurring in Figures 17.4 and 17.5?

18. In the cell shown in Figure 17.5,
  (a) what would be the effect of removing the voltmeter and connecting the wires shown coming to the voltmeter?
  (b) what would be the effect of removing the salt bridge?

19. When molten $CaBr_2$ is electrolyzed, calcium metal and bromine are produced. Write equations for the two half-reactions that occur at the electrodes. Label the anode half-reaction and the cathode half-reaction.

20. Why is direct current used instead of alternating current in the electroplating of metals?

21. What property of lead(IV) oxide and lead(II) sulfate makes it unnecessary to have salt bridges in the cells of a lead storage battery?

22. Explain why the density of the electrolyte in a lead storage battery decreases during the discharge cycle.

23. In one type of alkaline cell used to power devices such as portable radios, Hg^{2+} ions are reduced to metallic mercury when the cell is being discharged. Does this reduction occur at the anode or the cathode? Explain.

24. Differentiate between an electrolytic cell and a voltaic cell.

25. Why is a porous barrier or a salt bridge necessary in some voltaic cells?

Most of the exercises in this chapter are available for assignment via the online homework management program, WileyPLUS (www.wileyplus.com). All exercises with blue numbers have answers in Appendix VI.

PAIRED EXERCISES

1. Determine the oxidation number of each element in the compound:
  (a) $CuCO_3$
  (b) CH_4
  (c) IF
  (d) CH_2Cl_2
  (e) SO_2
  (f) $(NH_4)_2CrO_4$

2. Determine the oxidation number of each element in the compound:
  (a) CHF_3
  (b) P_2O_5
  (c) SF_6
  (d) $SnSO_4$
  (e) CH_3OH
  (f) H_3PO_4

3. Determine the oxidation number for each of the underlined elements:
  (a) $\underline{P}O_3^{3-}$
  (b) $Ca\underline{S}O_4$
  (c) $NaH\underline{C}O_3$
  (d) $\underline{Br}O_4^-$

4. Determine the oxidation number for each of the elements in blue:
  (a) CO_3^{2-}
  (b) H_2SO_4
  (c) NaH_2PO_4
  (d) $Cr_2O_7^{2-}$

5. Many ionic compounds containing metal atoms are brightly colored. Several of these compounds are pictured below. For each compound determine the oxidation number of the specified element(s).

6. Many ionic compounds containing metal atoms are brightly colored. Several of these compounds are pictured below. For each compound determine the oxidation number of the specified element(s).

(a) Sodium chromate (Na_2CrO_4). Determine the oxidation state of chromium.	Courtesy Ondřej Mangl
(b) Potassium ferricyanide $(K_3[Fe(CN)_6])$. Determine the oxidation state of iron.	W. Oelen/http:// woelen. homescience.net/ science/index.html
(c) Cobalt chloride $(CoCl_2)$. Determine the oxidation state of cobalt and chlorine.	W. Oelen/http:// woelen. homescience.net/ science/index.html
(d) Nickel chloride hexahydrate $(NiCl_2 \cdot 6\ H_2O)$. Determine the oxidation state of nickel.	Courtesy Benjah-bmm27 via Wikimedia

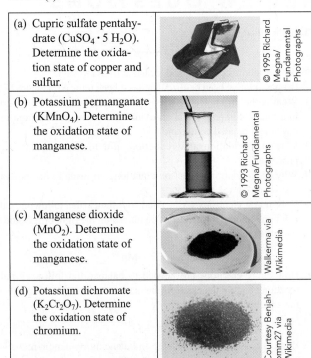

(a) Cupric sulfate pentahydrate $(CuSO_4 \cdot 5\ H_2O)$. Determine the oxidation state of copper and sulfur.	© 1995 Richard Megna/ Fundamental Photographs
(b) Potassium permanganate $(KMnO_4)$. Determine the oxidation state of manganese.	© 1993 Richard Megna/Fundamental Photographs
(c) Manganese dioxide (MnO_2). Determine the oxidation state of manganese.	Walkerma via Wikimedia
(d) Potassium dichromate $(K_2Cr_2O_7)$. Determine the oxidation state of chromium.	Courtesy Benjah-bmm27 via Wikimedia

7. Determine whether each of the following half-reactions represents an oxidation or a reduction. Supply the correct number of electrons to the appropriate side to balance the equation.
   (a) $Na \longrightarrow Na^+$
   (b) $C_2O_4^{2-} \longrightarrow 2\, CO_2$
   (c) $2\, I^- \longrightarrow I_2$
   (d) $Cr_2O_7^{2-} + 14\, H^+ \longrightarrow 2\, Cr^{3+} + 7\, H_2O$

8. Determine whether each of the following half-reactions represents an oxidation or a reduction. Supply the correct number of electrons to the appropriate side to balance the equation.
   (a) $Cu^{2+} \longrightarrow Cu^{1+}$
   (b) $F_2 \longrightarrow 2\, F^-$
   (c) $2\, IO_4^- + 16\, H^+ \longrightarrow I_2 + 8\, H_2O$
   (d) $Mn \longrightarrow Mn^{2+}$

9. In each of the following unbalanced equations, identify the oxidized element and the reduced element. Then, identify the oxidizing agent and the reducing agent.
   (a) $Cu + AgNO_3 \longrightarrow Ag + Cu(NO_3)_2$
   (b) $Zn + HCl \longrightarrow H_2 + ZnCl_2$

10. In each of the following unbalanced equations, identify the oxidized element and the reduced element. Then identify the oxidizing agent and the reducing agent.
    (a) $CH_4 + O_2 \longrightarrow CO_2 + H_2O$
    (b) $Mg + FeCl_3 \longrightarrow Fe + MgCl_2$

11. Determine whether the following oxidation–reduction reactions are balanced correctly. If they are not, provide the correct balanced reaction.
    (a) unbalanced:
    $$CH_4(g) + O_2(g) \longrightarrow CO_2(g) + H_2O(g)$$
    balanced:
    $$CH_4(g) + 2\, O_2(g) \longrightarrow CO_2(g) + 2\, H_2O(g)$$
    (b) unbalanced:
    $$NO^{2-}(aq) + Al(s) \longrightarrow NH_3(g) + AlO_2^-(aq)$$
    $$\text{(basic solution)}$$
    balanced:
    $$2\, H_2O(l) + Al(s) + NO^{2-}(aq) \longrightarrow$$
    $$AlO_2^-(aq) + NH_3(g) + OH^-(aq)$$
    (c) unbalanced:
    $$Mg(s) + HCl(aq) \longrightarrow Mg^{2+}(aq) + Cl^-(aq) + H_2(g)$$
    balanced:
    $$Mg(s) + 2\, HCl(aq) \longrightarrow Mg^{2+}(aq) + Cl^-(aq) + H_2(g)$$
    (d) unbalanced:
    $$CH_3OH(aq) + Cr_2O_7^{2-}(aq) \longrightarrow CH_2O(aq) + Cr^{3+}(aq)$$
    $$\text{(acidic solution)}$$
    balanced:
    $$3\, CH_3OH(aq) + 14\, H^+(aq) + Cr_2O_7^{2-}(aq) \longrightarrow$$
    $$2\, Cr^{3+}(aq) + 3\, CH_2O(aq) + 7\, H_2O(l) + 6\, H^+(aq)$$

12. Determine whether the following oxidation–reduction reactions are balanced correctly. If they are not, provide the correct balanced reaction.
    (a) unbalanced:
    $$MnO_2(s) + Al(s) \longrightarrow Mn(s) + Al_2O_3(s)$$
    balanced:
    $$MnO_2(s) + 2\, Al(s) \longrightarrow Mn(s) + Al_2O_3(s)$$
    (b) unbalanced:
    $$Cu(s) + Ag^+(aq) \longrightarrow Cu^{2+}(aq) + Ag(s)$$
    balanced:
    $$Cu(s) + 2\, Ag^+(aq) \longrightarrow Cu^{2+}(aq) + 2\, Ag(s)$$
    (c) unbalanced:
    $$Br^-(aq) + MnO_4^-(aq) \longrightarrow Br_2(l) + Mn^{2+}(aq)$$
    $$\text{(acidic solution)}$$
    balanced:
    $$16\, H^+(aq) + 10\, Br^-(aq) + 2\, MnO_4^-(aq) \longrightarrow$$
    $$5\, Br_2(l) + 2\, Mn^{2+}(aq) + 8\, H_2O(l)$$
    (d) unbalanced:
    $$MnO_4^-(aq) + S^{2-}(aq) \longrightarrow MnS(s) + S(s)\quad \text{(basic solution)}$$
    balanced:
    $$8\, H^+(aq) + MnO_4^-(aq) + S^{2-}(aq) \longrightarrow$$
    $$S(s) + MnS(s) + 4\, H_2O(l)$$

13. Balance each of the following equations using the change-in-oxidation-number method:
    (a) $Cu + O_2 \longrightarrow CuO$
    (b) $KClO_3 \longrightarrow KCl + O_2$
    (c) $Ca + H_2O \longrightarrow Ca(OH)_2 + H_2$
    (d) $PbS + H_2O_2 \longrightarrow PbSO_4 + H_2O$
    (e) $CH_4 + NO_2 \longrightarrow N_2 + CO_2 + H_2O$

14. Balance each of the following equations using the change-in-oxidation-number method:
    (a) $Cu + AgNO_3 \longrightarrow Ag + Cu(NO_3)_2$
    (b) $MnO_2 + HCl \longrightarrow MnCl_2 + Cl_2 + H_2O$
    (c) $HCl + O_2 \longrightarrow Cl_2 + H_2O$
    (d) $Ag + H_2S + O_2 \longrightarrow Ag_2S + H_2O$
    (e) $KMnO_4 + CaC_2O_4 + H_2SO_4 \longrightarrow$
    $$K_2SO_4 + MnSO_4 + CaSO_4 + CO_2 + H_2O$$

15. Balance these ionic redox equations using the ion–electron method. These reactions occur in acidic solution.
    (a) $Zn + NO_3^- \longrightarrow Zn^{2+} + NH_4^+$
    (b) $NO_3^- + S \longrightarrow NO_2 + SO_4^{2-}$
    (c) $PH_3 + I_2 \longrightarrow H_3PO_2 + I^-$
    (d) $Cu + NO_3^- \longrightarrow Cu^{2+} + NO$
    (e) $ClO_3^- + Cl^- \longrightarrow Cl_2$

16. Balance these ionic redox equations using the ion–electron method. These reactions occur in acidic solution.
    (a) $ClO_3^- + I^- \longrightarrow I_2 + Cl^-$
    (b) $Cr_2O_7^{2-} + Fe^{2+} \longrightarrow Cr^{3+} + Fe^{3+}$
    (c) $MnO_4^- + SO_2 \longrightarrow Mn^{2+} + SO_4^{2-}$
    (d) $H_3AsO_3 + MnO_4^- \longrightarrow H_3AsO_4 + Mn^{2+}$
    (e) $Cr_2O_7^{2-} + H_3AsO_3 \longrightarrow Cr^{3+} + H_3AsO_4$

17. Balance these ionic redox equations using the ion–electron method. These reactions occur in basic solution.
    (a) $Cl_2 + IO_3^- \longrightarrow Cl^- + IO_4^-$
    (b) $MnO_4^- + ClO_2^- \longrightarrow MnO_2 + ClO_4^-$
    (c) $Se \longrightarrow Se^{2-} + SeO_3^{2-}$
    (d) $Fe_3O_4 + MnO_4^- \longrightarrow Fe_2O_3 + MnO_2$
    (e) $BrO^- + Cr(OH)_4^- \longrightarrow Br^- + CrO_4^{2-}$

18. Balance these ionic redox equations using the ion–electron method. These reactions occur in basic solution.
    (a) $MnO_4^- + SO_3^{2-} \longrightarrow MnO_2 + SO_4^{2-}$
    (b) $ClO_2 + SbO_2^- \longrightarrow ClO_2^- + Sb(OH)_6^-$
    (c) $Al + NO_3^- \longrightarrow NH_3 + Al(OH)_4^-$
    (d) $P_4 \longrightarrow HPO_3^{2-} + PH_3$
    (e) $Al + OH^- \longrightarrow Al(OH)_4^- + H_2$

19. Balance these reactions:

(a) $IO_3^-(aq) + I^-(aq) \longrightarrow I_2(aq)$ (acid solution)

(b) $Mn^{2+}(aq) + S_2O_8^{2-}(aq) \longrightarrow MnO_4^-(aq) + SO_4^{2-}(aq)$
     (acid solution)

(c) $Co(NO_2)_6^{3-}(aq) + MnO_4^-(aq) \longrightarrow$
     $Co^{2+}(aq) + Mn^{2+}(aq) + NO_3^-(aq)$ (acid solution)

20. Balance these reactions:

(a) $Mo_2O_3(s) + MnO_4^-(aq) \longrightarrow MoO_3(s) + Mn^{2+}(aq)$
     (acid solution)

(b) $BrO^-(aq) + Cr(OH)_4^-(aq) \longrightarrow$
     $Br^-(aq) + CrO_4^{2-}(aq)$ (basic solution)

(c) $S_2O_3^{2-}(aq) + MnO_4^-(aq) \longrightarrow$
     $SO_4^{2-}(aq) + MnO_2(s)$ (basic solution)

ADDITIONAL EXERCISES

21. Vitamin C (ascorbic acid) is often touted as a good nutritional supplement because of its antioxidant properties. Vitamin C is very easily oxidized and, therefore, will be preferentially oxidized in the presence of an oxidizing agent. The hope is that vitamin C will coexist in cellular solutions and thus prevent the oxidation of other important biomolecules. Write and balance the equation for the oxidation of vitamin C ($C_6H_8O_6$) by oxygen gas. The products of this reaction are dehydroascorbic acid ($C_6H_6O_6$) and water.

22. Most explosive reactions are complex redox reactions with multiple oxidations and reductions. The reaction of gunpowder is shown below. Determine the element(s) oxidized and reduced.

$4 KNO_3(s) + 7C(s) + S(s) \longrightarrow$
$3 CO_2(g) + 3 CO(g) + 2 N_2(g) + K_2CO_3(s) + K_2S(s)$

23. Draw a picture of an electrolytic cell made from an aqueous HBr solution.

24. The chemical reactions taking place during discharge in a lead storage battery are

$Pb + SO_4^{2-} \longrightarrow PbSO_4$
$PbO_2 + SO_4^{2-} + 4 H^+ \longrightarrow PbSO_4 + 2 H_2O$

(a) Complete each half-reaction by supplying electrons.
(b) Which reaction is oxidation, and which is reduction?
(c) Which reaction occurs at the anode of the battery?

25. Use this unbalanced redox equation

$KMnO_4 + HCl \longrightarrow KCl + MnCl_2 + H_2 + Cl_2$

to indicate
(a) the oxidizing agent
(b) the reducing agent
(c) the number of electrons that are transferred per mole of oxidizing agent

26. Brass is an alloy of zinc and copper. When brass is in contact with salt water, it corrodes as the zinc dissolves from the alloy, leaving almost pure copper. Explain why the zinc is preferentially dissolved.

27. How many liters of NO_2 gas at STP will be formed by the reaction of 75.5 g of copper with 55.0 g of nitric acid? Which substance is oxidized, and which one is reduced?

$Cu + HNO_3 \longrightarrow Cu(NO_3)_2 + NO_2 + H_2O$ (acid solution)

28. How many liters of CO_2 gas at STP will be formed by the reaction of 25.5 g of $K_2S_2O_8$ and 35.5 g of $H_2C_2O_4$? Which substance is oxidized, and which one is reduced?

$K_2S_2O_8 + H_2C_2O_4 \longrightarrow K_2SO_4 + H_2SO_4 + CO_2$

29. The amount of vitamin C can be determined experimentally by titration. The balanced chemical reaction between vitamin C or ascorbic acid and iodine is shown below. If the vitamin C is extracted from 100.0 grams of green peppers and reacted with 0.03741 M potassium triiodide, it requires 32.61 mL of the solution to react with all of the vitamin C. Calculate the number of milligrams of vitamin C per gram of green pepper.

$C_6H_8O_6(aq) + KI_3(aq) \longrightarrow C_6H_6O_6(aq) + KI(aq) + 2 HI(aq)$

30. Alcohols can be oxidized very easily by potassium dichromate. This fact formed the basis for many of the early breathalyzers used by law enforcement officers to determine the approximate blood alcohol levels of suspected drunk drivers. The unbalanced equation for the reaction is

$CH_3CH_2OH + Cr_2O_7^{2-} \longrightarrow CH_3COOH + Cr^{3+}$

In this reaction, the potassium dichromate is orange and the chromium(III) ions are blue. The bluer a solution of potassium dichromate becomes, the greater the concentration of alcohol in the breath. Write a balanced ionic equation for this reaction in an acidic environment.

31. What mass of $KMnO_4$ is needed to react with 100. mL H_2O_2 solution? ($d = 1.031$ g/mL, 9.0% H_2O_2 by mass)

$H_2O_2 + KMnO_4 + H_2SO_4 \longrightarrow$
$O_2 + MnSO_4 + K_2SO_4 + H_2O$ (acid solution)

32. How many grams of zinc are required to reduce Fe^{3+} when 25.0 mL of 1.2 M $FeCl_3$ are reacted?

$Zn + Fe^{3+} \longrightarrow Zn^{2+} + Fe$

33. What volume of 0.200 M $K_2Cr_2O_7$ will be required to oxidize the Fe^{2+} ion in 60.0 mL of 0.200 M $FeSO_4$ solution?

$Cr_2O_7^{2-} + Fe^{2+} \longrightarrow Cr^{3+} + Fe^{3+}$ (acid solution)

34. How many moles of H_2 can be produced from 100.0 g Al according to this reaction?

$Al + OH^- \longrightarrow Al(OH)_4^- + H_2$ (basic solution)

35. There is something incorrect about these half-reactions:
(a) $Cu^+ + e^- \longrightarrow Cu^{2+}$ (b) $Pb^{2+} + e^{2-} \longrightarrow Pb$
Identify what is wrong and correct it.

36. Why can oxidation *never* occur without reduction?

37. The following observations were made concerning metals A, B, C, and D.
(a) When a strip of metal A is placed in a solution of B^{2+} ions, no reaction is observed.
(b) Similarly, A in a solution containing C^+ ions produces no reaction.
(c) When a strip of metal D is placed in a solution of C^+ ions, black metallic C deposits on the surface of D, and the solution tests positively for D^{2+} ions.
(d) When a piece of metallic B is placed in a solution of D^{2+} ions, metallic D appears on the surface of B and B^{2+} ions are found in the solution.
Arrange the ions—A^+, B^{2+}, C^+, and D^{2+}—in order of their ability to attract electrons. List them in order of increasing ability.

38. Tin normally has oxidation numbers of 0, +2, and +4. Which of these species can be an oxidizing agent, which can be a reducing agent, and which can be both? In each case, what product would you expect as the tin reacts?

39. Manganese is an element that can exist in numerous oxidation states. In each of these compounds, identify the oxidation number of the manganese. Which compound would you expect to be the best oxidizing agent and why?
(a) $Mn(OH)_2$ (c) MnO_2 (e) $KMnO_4$
(b) MnF_3 (d) K_2MnO_4

40. Which equations represent oxidations?
(a) $Mg \longrightarrow Mg^{2+}$ (c) $KMnO_4 \longrightarrow MnO_2$
(b) $SO_2 \longrightarrow SO_3$ (d) $Cl_2O_3 \longrightarrow Cl^-$

41. In the following equation, note the reaction between manganese(IV) oxide and bromide ions:

$$MnO_2 + Br^- \longrightarrow Br_2 + Mn^{2+}$$

(a) Balance this redox reaction in acidic solution.
(b) How many grams of MnO_2 would be needed to produce 100.0 mL of 0.05 M Mn^{2+}?
(c) How many liters of bromine vapor at 50°C and 1.4 atm would also result?

42. Use the table shown to complete the following reactions. If no reaction occurs, write NR:

(a) $F_2 + Cl^- \longrightarrow$
(b) $Br_2 + Cl^- \longrightarrow$
(c) $I_2 + Cl^- \longrightarrow$
(d) $Br_2 + I^- \longrightarrow$

Activity
↑ F_2
Cl_2
Br_2
I_2

ease of reduction

43. Manganese metal reacts with HCl to give hydrogen gas and the Mn^{2+} ion in solution. Write a balanced equation for the reaction.

44. If zinc is allowed to react with dilute nitric acid, zinc is oxidized to the 2+ ion, while the nitrate ion can be reduced to ammonium, NH_4^+. Write a balanced equation for the reaction in acidic solution.

45. In the following equations, identify the
(a) atom or ion oxidized
(b) atom or ion reduced
(c) oxidizing agent
(d) reducing agent
(e) change in oxidation number associated with each oxidizing process
(f) change in oxidation number associated with each reducing process

1. $C_3H_8 + O_2 \longrightarrow CO_2 + H_2O$
2. $HNO_3 + H_2S \longrightarrow NO + S + H_2O$
3. $CuO + NH_3 \longrightarrow N_2 + H_2O + Cu$
4. $H_2O_2 + Na_2SO_3 \longrightarrow Na_2SO_4 + H_2O$
5. $H_2O_2 \longrightarrow H_2O + O_2$

46. In the galvanic cell shown in the diagram, a strip of silver is placed in a solution of silver nitrate, and a strip of lead is placed in a solution of lead(II) nitrate. The two beakers are connected with a salt bridge. Determine
(a) the anode
(b) the cathode
(c) where oxidation occurs
(d) where reduction occurs
(e) in which direction electrons flow through the wire
(f) in which direction ions flow through the solution

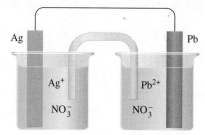

CHALLENGE EXERCISES

47. A sample of crude potassium iodide was analyzed using this reaction (not balanced):

$$I^- + SO_4^{2-} \longrightarrow I_2 + H_2S \quad \text{(acid solution)}$$

If a 4.00-g sample of crude KI produced 2.79 g of iodine, what is the percent purity of the KI?

48. What volume of NO gas, measured at 28°C and 744 torr, will be formed by the reaction of 0.500 mol Ag reacting with excess nitric acid?

$$Ag + HNO_3 \longrightarrow AgNO_3 + NO + H_2O \quad \text{(acid solution)}$$

ANSWERS TO PRACTICE EXERCISES

17.1 (a) S = +6, (b) As = +5, (c) C = +4

17.2 (*Note:* H = +1 even though it comes second in the formula: N is a nonmetal.)
(a) N = −3, (b) Cr = +6, (c) P = +5

17.3 (a) Be = +2; Cl = −1,
(b) H = +1; Cl = +1; O = −2,
(c) H = +1; O = −1,
(d) N = −3; H = +1,
(e) Br = +5; O = −2

17.4 (a) Al is oxidized and loses 3 electrons. H is reduced and gains 1 electron. Al is the reducing agent and HNO_3 is the oxidizing agent.
(b) Ni is oxidized and loses 2 electrons. Cu is reduced and gains 2 electrons. Ni is the reducing agent and $CuSO_4$ is the oxidizing agent.
(c) C is oxidized and loses 1 electron. O is reduced and gains 2 electrons. C_2H_6 is the reducing agent and O_2 is the oxidizing agent.

17.5 (a) $6 HNO_3 + S \longrightarrow 6 NO_2 + H_2SO_4 + 2 H_2O$
(b) $2 CrCl_3 + 3 MnO_2 + 2 H_2O \longrightarrow 3 MnCl_2 + 2 H_2CrO_4$
(c) $2 KMnO_4 + 6 HCl + 5 H_2S \longrightarrow$
$\qquad 2 KCl + 2 MnCl_2 + 5 S + 8 H_2O$

17.6 (a) $4 H^+ + 2 I^- + 2 NO_2^- \longrightarrow I_2 + 2 NO + 2 H_2O$
(b) $2 OH^- + Cl_2 + IO_3^- \longrightarrow IO_4^- + H_2O + 2 Cl^-$
(c) $AuCl_4^- + Sn^{2+} \longrightarrow Sn^{4+} + AuCl + 3 Cl^-$

17.7 (a) $Zn + 2 H_2O + 2 OH^- \longrightarrow Zn(OH)_4^{2-} + H_2$
(b) $H_2O_2 + Sn^{2+} + 2 H^+ \longrightarrow Sn^{4+} + 2 H_2O$
(c) $Cu + Cu^{2+} + 2 OH^- \longrightarrow Cu_2O + H_2O$

17.8 (a) yes, (b) no, (c) no

17.9 (a) Cu is oxidized; Sn is reduced
(b) anode: $Cu(s) \longrightarrow Cu^{2+}(aq) + 2e^-$
cathode: $Sn^{2+}(aq) + 2e^- \longrightarrow Sn(s)$

PUTTING IT TOGETHER:
Review for Chapters 15–17

Answers for Putting It Together Reviews are found in Appendix VII.

Multiple Choice

Choose the correct answer to each of the following.

1. When the reaction

$$Al + HCl \longrightarrow$$

   is completed and balanced, this term appears in the balanced equation:
   - (a) 3 HCl
   - (b) $AlCl_2$
   - (c) $3 H_2$
   - (d) 4 Al

2. When the reaction

$$CaO + HNO_3 \longrightarrow$$

   is completed and balanced, this term appears in the balanced equation:
   - (a) H_2
   - (b) $2 H_2$
   - (c) $2 CaNO_3$
   - (d) H_2O

3. When the reaction

$$H_3PO_4 + KOH \longrightarrow$$

   is completed and balanced, this term appears in the balanced equation:
   - (a) H_3PO_4
   - (b) $6 H_2O$
   - (c) KPO_4
   - (d) 3 KOH

4. When the reaction

$$HCl + Cr_2(CO_3)_3 \longrightarrow$$

   is completed and balanced, this term appears in the balanced equation:
   - (a) Cr_2Cl
   - (b) 3 HCl
   - (c) $3 CO_2$
   - (d) H_2O

5. Which of these is not a salt?
   - (a) $K_2Cr_2O_7$
   - (b) $NaHCO_3$
   - (c) $Ca(OH)_2$
   - (d) $Na_2C_2O_4$

6. Which of these is not an acid?
   - (a) H_3PO_4
   - (b) H_2S
   - (c) H_2SO_4
   - (d) NH_3

7. Which of these is a weak electrolyte?
   - (a) NH_4OH
   - (b) $Ni(NO_3)_2$
   - (c) K_3PO_4
   - (d) NaBr

8. Which of these is a nonelectrolyte?
   - (a) $HC_2H_3O_2$
   - (b) $MgSO_4$
   - (c) $KMnO_4$
   - (d) CCl_4

9. Which of these is a strong electrolyte?
   - (a) H_2CO_3
   - (b) HNO_3
   - (c) NH_4OH
   - (d) H_3BO_3

10. Which of these is a weak electrolyte?
    - (a) NaOH
    - (b) NaCl
    - (c) $HC_2H_3O_2$
    - (d) H_2SO_4

11. A solution has an H^+ concentration of $3.4 \times 10^{-5} M$. The pH is
    - (a) 4.47
    - (b) 5.53
    - (c) 3.53
    - (d) 5.47

12. A solution with a pH of 5.85 has an H^+ concentration of
    - (a) $7.1 \times 10^{-5} M$
    - (b) $7.1 \times 10^{-6} M$
    - (c) $3.8 \times 10^{-4} M$
    - (d) $1.4 \times 10^{-6} M$

13. If 16.55 mL of 0.844 M NaOH are required to titrate 10.00 mL of a hydrochloric acid solution, the molarity of the acid solution is
    - (a) 0.700 M
    - (b) 0.510 M
    - (c) 1.40 M
    - (d) 0.255 M

14. What volume of 0.462 M NaOH is required to neutralize 20.00 mL of 0.391 M HNO_3?
    - (a) 23.6 mL
    - (b) 16.9 mL
    - (c) 9.03 mL
    - (d) 11.8 mL

15. 25.00 mL of H_2SO_4 solution requires 18.92 mL of 0.1024 M NaOH for complete neutralization. The molarity of the acid is
    - (a) 0.1550 M
    - (b) 0.03875 M
    - (c) 0.07750 M
    - (d) 0.06765 M

16. Dilute hydrochloric acid is a typical acid, as shown by its
    - (a) color
    - (b) odor
    - (c) solubility
    - (d) taste

17. What is the pH of a 0.00015 M HCl solution?
    - (a) 4.0
    - (b) 2.82
    - (c) between 3 and 4
    - (d) no correct answer given

18. The chloride ion concentration in 300. mL of 0.10 M $AlCl_3$ is
    - (a) 0.30 M
    - (b) 0.10 M
    - (c) 0.030 M
    - (d) 0.90 M

19. The amount of $BaSO_4$ that will precipitate when 100. mL of 0.10 M $BaCl_2$ and 100. mL of 0.10 M Na_2SO_4 are mixed is
    - (a) 0.010 mol
    - (b) 0.10 mol
    - (c) 23 g
    - (d) no correct answer given

20. The freezing point of a 0.50 m NaCl aqueous solution will be about what?
    - (a) $-1.86°C$
    - (b) $-0.93°C$
    - (c) $-2.79°C$
    - (d) no correct answer given

21. The equation

$$HC_2H_3O_2 + H_2O \rightleftharpoons H_3O^+ + C_2H_3O_2^-$$

    implies that
    - (a) If you start with 1.0 mol $HC_2H_3O_2$, 1.0 mol H_3O^+ and 1.0 mol $C_2H_3O_2^-$ will be produced.
    - (b) An equilibrium exists between the forward reaction and the reverse reaction.
    - (c) At equilibrium, equal molar amounts of all four will exist.
    - (d) The reaction proceeds all the way to the products and then reverses, going all the way back to the reactants.

22. If the reaction $A + B \rightleftharpoons C + D$ is initially at equilibrium and then more A is added, which of the following is not true?
    - (a) More collisions of A and B will occur; the rate of the forward reaction will thus be increased.
    - (b) The equilibrium will shift toward the right.
    - (c) The moles of B will be increased.
    - (d) The moles of D will be increased.

23. What will be the H^+ concentration in a 1.0 M HCN solution? ($K_a = 4.0 \times 10^{-10}$)
    - (a) $2.0 \times 10^{-5} M$
    - (b) 1.0 M
    - (c) $4.0 \times 10^{-10} M$
    - (d) $2.0 \times 10^{-10} M$

24. What is the percent ionization of HCN in Question 23?
    - (a) 100%
    - (b) 2.0×10^{-8} %
    - (c) 2.0×10^{-3} %
    - (d) 4.0×10^{-8} %

25. If $[H^+] = 1 \times 10^{-5} M$, which of the following is not true?
    - (a) pH = 5
    - (b) pOH = 9
    - (c) $[OH^-] = 1 \times 10^{-5} M$
    - (d) The solution is acidic.

26. If $[H^+] = 2.0 \times 10^{-4} M$, then $[OH^-]$ will be
    - (a) $5.0 \times 10^{-9} M$
    - (b) 3.70 M
    - (c) $2.0 \times 10^{-4} M$
    - (d) $5.0 \times 10^{-11} M$

27. The solubility product of $PbCrO_4$ is 2.8×10^{-13}. The solubility of $PbCrO_4$ is
    (a) $5.3 \times 10^{-7} M$ (c) $7.8 \times 10^{-14} M$
    (b) $2.8 \times 10^{-13} M$ (d) $1.0 M$

28. The solubility of AgBr is $7.2 \times 10^{-7} M$. The value of the solubility product is
    (a) 7.2×10^{-7} (c) 5.2×10^{-48}
    (b) 5.2×10^{-13} (d) 5.2×10^{-15}

29. Which of these solutions would be the best buffer solution?
    (a) $0.10\ M\ HC_2H_3O_2 + 0.10\ M\ NaC_2H_3O_2$
    (b) $0.10\ M\ HCl$
    (c) $0.10\ M\ HCl + 0.10\ M\ NaCl$
    (d) pure water

30. For the reaction $H_2(g) + I_2(g) \rightleftharpoons 2\ HI(g)$, at 700 K, $K_{eq} = 56.6$. If an equilibrium mixture at 700 K was found to contain $0.55\ M$ HI and $0.21\ M\ H_2$, the I_2 concentration must be
    (a) $0.046\ M$ (b) $0.025\ M$ (c) $22\ M$ (d) $0.21\ M$

31. The equilibrium constant for the reaction $2\ A + B \rightleftharpoons 3\ C + D$ is
    (a) $\dfrac{[C]^3[D]}{[A]^2[B]}$ (b) $\dfrac{[2A][B]}{[3C][D]}$ (c) $\dfrac{[3C][D]}{[2A][B]}$ (d) $\dfrac{[A]^2[B]}{[C]^3[D]}$

32. In the equilibrium represented by
    $$N_2(g) + O_2(g) \rightleftharpoons 2\ NO_2(g)$$
    as the pressure is increased, the amount of NO_2 formed
    (a) increases   (c) remains the same
    (b) decreases   (d) increases and decreases irregularly

33. Which factor will not increase the concentration of ammonia as represented by this equation?
    $$3\ H_2(g) + N_2(g) \rightleftharpoons 2\ NH_3(g) + 92.5\ kJ$$
    (a) increasing the temperature
    (b) increasing the concentration of N_2
    (c) increasing the concentration of H_2
    (d) increasing the pressure

34. If $HCl(g)$ is added to a saturated solution of AgCl, the concentration of Ag^+ in solution
    (a) increases   (c) remains the same
    (b) decreases   (d) increases and decreases irregularly

35. The solubility of $CaCO_3$ at 20°C is 0.013 g/L. What is the K_{sp} for $CaCO_3$?
    (a) 1.3×10^{-8} (c) 1.7×10^{-8}
    (b) 1.3×10^{-4} (d) 1.7×10^{-4}

36. The K_{sp} for $BaCrO_4$ is 8.5×10^{-11}. What is the solubility of $BaCrO_4$ in grams per liter?
    (a) 9.2×10^{-6} (c) 2.3×10^{-3}
    (b) 0.073 (d) 8.5×10^{-11}

37. What will be the $[Ba^{2+}]$ when 0.010 mol Na_2CrO_4 is added to 1.0 L of saturated $BaCrO_4$ solution? See Question 36 for K_{sp}.
    (a) $8.5 \times 10^{-11}\ M$ (c) $9.2 \times 10^{-6}\ M$
    (b) $8.5 \times 10^{-9}\ M$ (d) $9.2 \times 10^{-4}\ M$

38. Which would occur if a small amount of sodium acetate crystals, $NaC_2H_3O_2$, were added to 100 mL of 0.1 $M\ HC_2H_3O_2$ at constant temperature?
    (a) The number of acetate ions in the solution would decrease.
    (b) The number of acetic acid molecules would decrease.
    (c) The number of sodium ions in solution would decrease.
    (d) The H^+ concentration in the solution would decrease.

39. If the temperature is decreased for the endothermic reaction
    $$A + B \rightleftharpoons C + D$$

which of the following is true?
    (a) The concentration of A will increase.
    (b) No change will occur.
    (c) The concentration of B will decrease.
    (d) The concentration of D will increase.

40. In K_2SO_4, the oxidation number of sulfur is
    (a) $+2$ (c) $+6$
    (b) $+4$ (d) -2

41. In $Ba(NO_3)_2$, the oxidation number of N is
    (a) $+5$ (c) $+4$
    (b) -3 (d) -1

42. In the reaction
    $$H_2S + 4\ Br_2 + 4\ H_2O \longrightarrow H_2SO_4 + 8\ HBr$$
    the oxidizing agent is
    (a) H_2S (b) Br_2 (c) H_2O (d) H_2SO_4

43. In the reaction
    $$VO_3^- + Fe^{2+} + 4\ H^+ \longrightarrow VO^{2+} + Fe^{3+} + 2\ H_2O$$
    the element reduced is
    (a) V   (b) Fe   (c) O   (d) H

Questions 44–46 pertain to the activity series below.

$$K\quad Ca\quad Mg\quad Al\quad Zn\quad Fe\quad H\quad Cu\quad Ag$$

44. Which of these pairs will not react in water solution?
    (a) $Zn, CuSO_4$ (c) $Fe, AgNO_3$
    (b) $Cu, Al_2(SO_4)_3$ (d) $Mg, Al_2(SO_4)_3$

45. Which element is the most easily oxidized?
    (a) K   (b) Mg   (c) Zn   (d) Cu

46. Which element will reduce Cu^{2+} to Cu but will not reduce Zn^{2+} to Zn?
    (a) Fe   (b) Ca   (c) Ag   (d) Mg

47. In the electrolysis of fused (molten) $CaCl_2$, the product at the negative electrode is
    (a) Ca^{2+} (b) Cl^- (c) Cl_2 (d) Ca

48. In its reactions, a free element from Group 2A in the periodic table is most likely to
    (a) be oxidized   (c) be unreactive
    (b) be reduced   (d) gain electrons

49. In the partially balanced redox equation
    $$3\ Cu + HNO_3 \longrightarrow 3\ Cu(NO_3)_2 + 2\ NO + H_2O$$
    the coefficient needed to balance H_2O is
    (a) 8   (b) 6   (c) 4   (d) 3

50. Which reaction does not involve oxidation–reduction?
    (a) burning sodium in chlorine
    (b) chemical union of Fe and S
    (c) decomposition of $KClO_3$
    (d) neutralization of NaOH with H_2SO_4

51. How many moles of Fe^{2+} can be oxidized to Fe^{3+} by 2.50 mol Cl_2 according to this equation?
    $$Fe^{2+} + Cl_2 \longrightarrow Fe^{3+} + Cl^-$$
    (a) 2.50 mol   (c) 1.00 mol
    (b) 5.00 mol   (d) 22.4 mol

52. How many grams of sulfur can be produced in this reaction from 100. mL of 6.00 $M\ HNO_3$?
    $$HNO_3 + H_2S \longrightarrow S + NO + H_2O$$
    (a) 28.9 g   (b) 19.3 g   (c) 32.1 g   (d) 289 g

53. Which of these ions can be reduced by H_2?
   (a) Hg^{2+} (b) Sn^{2+} (c) Zn^{2+} (d) K^+

54. Which of the following is *not* true of a zinc–mercury cell?
   (a) It provides current at a steady potential.
   (b) It has a short service life.
   (c) It is self-contained.
   (d) It can be stored for long periods of time.

Balancing Oxidation–Reduction Equations

Balance each equation.

55. $P + HNO_3 \longrightarrow HPO_3 + NO + H_2O$

56. $MnSO_4 + PbO_2 + H_2SO_4 \longrightarrow HMnO_4 + PbSO_4 + H_2O$

57. $Cr_2O_7^{2-} + Cl^- \longrightarrow Cr^{3+} + Cl_2$ (acidic solution)

58. $MnO_4^- + AsO_3^{3-} \longrightarrow Mn^{2+} + AsO_4^{3-}$ (acidic solution)

59. $S^{2-} + Cl_2 \longrightarrow SO_4^{2-} + Cl^-$ (basic solution)

60. $Zn + NO_3^- \longrightarrow Zn(OH)_4^{2-} + NH_3$ (basic solution)

61. $KOH + Cl_2 \longrightarrow KCl + KClO + H_2O$ (basic solution)

62. $As + ClO_3^- \longrightarrow H_3AsO_3 + HClO$ (acidic solution)

63. $MnO_4^- + Cl^- \longrightarrow Mn^{2+} + Cl_2$ (acidic solution)

64. $H_2O_2 + Cl_2O_7 \longrightarrow ClO_2^- + O_2$ (basic solution)

Free Response Questions

Answer each of the following. Be sure to include your work and explanations in a clear, logical form.

1. You are investigating the properties of two new metallic elements found on Pluto, Bz and Yz. Bz reacts with aqueous HCl. However, the formula for the compound it forms with chlorine is $YzCl_2$. Write the balanced reaction that should occur if a galvanic cell is set up with Bz and Yz electrodes in solutions containing the metallic ions.

2. Suppose that 25 mL of an iron(II) nitrate solution are added to a beaker containing aluminum metal. Assume the reaction went to completion with no excess reagents. The solid iron produced was removed by filtration.
   (a) Write a balanced redox equation for the reaction.
   (b) For which solution, the initial iron(II) nitrate or the solution after the solid iron was filtered out, would the freezing point be lower? Explain your answer.

3. 50. mL of 0.10 *M* HCl solution are poured equally into two flasks.
   (a) What is the pH of the HCl solution?
   (b) Next, 0.050 mol Zn is added to Flask A and 0.050 mol Cu is added to Flask B. Determine the pH of each solution after approximately 20 minutes.

4. (a) Write a balanced acid–base reaction that produces Na_2S.
   (b) If Na_2S is added to an aqueous solution of H_2S ($K_a = 9.1 \times 10^{-8}$), will the pH of the solution rise or fall? Explain.

5. (a) Would you expect a reaction to take place between HCN (*aq*) and $AgNO_3$ (*aq*)

$$(K_{sp} \text{ for AgCN} = 5.97 \times 10^{-17})?$$

   Explain, and if a reaction occurs, write the net ionic equation.
   (b) If NaCN is added to distilled water, would you expect the solution to be acidic, basic, or neutral? Explain using any chemical equations that may be appropriate.

6. For each set of beakers below, draw a picture you might expect to see when the contents of the beakers are mixed together and allowed to react.

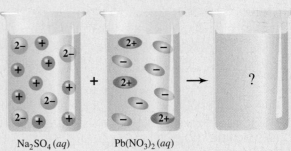

(a)

Na_2SO_4 (*aq*) $Pb(NO_3)_2$ (*aq*)

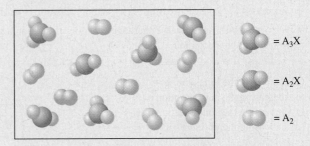

(b)

K_2SO_4 (*aq*) $CuCl_2$ (*aq*)

7. The picture below represents the equilibrium condition of $2 A_3X \rightleftharpoons 2 A_2X + A_2$.

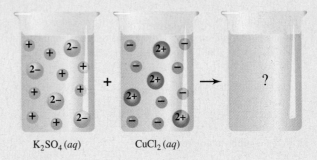

= A_3X

= A_2X

= A_2

   (a) What is the equilibrium constant?
   (b) Does the equilibrium lie to the left or to the right?
   (c) Do you think the reaction is a redox reaction? Explain your answer.

8. The picture below represents the equilibrium condition of the reaction $X_2 + 2G \rightleftharpoons X_2G_2$.

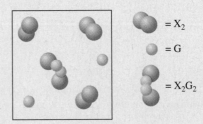

= X_2

= G

= X_2G_2

   (a) What is the equilibrium constant?
   (b) If the ratio of reactants to products increased when the temperature was raised, was the reaction exothermic or endothermic?
   (c) Provide a logical explanation for why the equilibrium shifts to the right when the pressure is increased at constant temperature.

9. The hydroxide ion concentration of a solution is 3.4×10^{-10} *M*. Balance the following equation:

$$Fe^{2+}(aq) + MnO_4^-(aq) \longrightarrow Fe^{3+}(aq) + Mn^{2+}(aq)$$

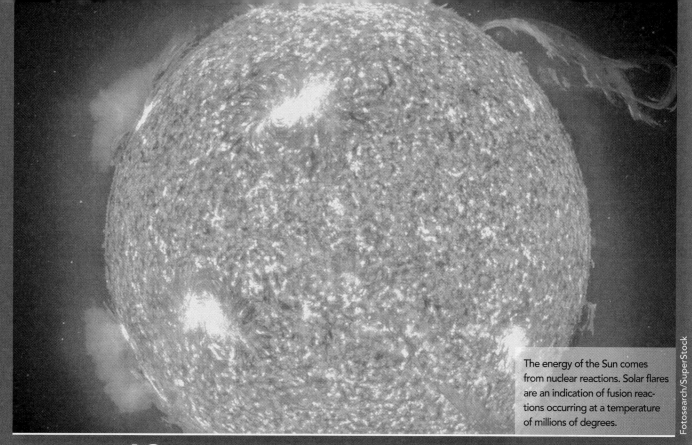

The energy of the Sun comes from nuclear reactions. Solar flares are an indication of fusion reactions occurring at a temperature of millions of degrees.

Fotosearch/SuperStock

CHAPTER **18**

NUCLEAR CHEMISTRY

The nucleus of the atom is a source of tremendous energy. Harnessing this energy has enabled us to fuel power stations, treat cancer, and preserve food. We use isotopes in medicine to diagnose illness and to detect minute quantities of drugs or hormones. Researchers use radioactive tracers to sequence the human genome. We also use nuclear processes to detect explosives in luggage and to establish the age of objects such as human artifacts and rocks. In this chapter we'll consider properties of atomic nuclei and their applications in our lives.

CHAPTER OUTLINE

18.1 Discovery of Radioactivity

18.2 Alpha Particles, Beta Particles, and Gamma Rays

18.3 Radioactive Disintegration Series

18.4 Measurement of Radioactivity

18.5 Nuclear Energy

18.6 Mass–Energy Relationship in Nuclear Reactions

18.7 Biological Effects of Radiation

18.1 DISCOVERY OF RADIOACTIVITY

Describe the particles associated with nuclear chemistry and radioactivity and use half-life to calculate the age of an object.

KEY TERMS

radioactivity
nucleon
nuclide
radioactive decay
half-life ($t_{1/2}$)

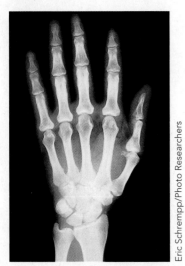

X-ray technology opened the door for a new world of diagnosis and treatment.

In 1895, Wilhelm Conrad Roentgen (1845–1923) made an important breakthrough that eventually led to the discovery of radioactivity. Roentgen discovered X-rays when he observed that a vacuum discharge tube enclosed in a thin, black cardboard box caused a nearby piece of paper coated with the salt barium platinocyanide to glow with a brilliant phosphorescence. From this and other experiments, he concluded that certain rays, which he called X-rays, were emitted from the discharge tube, penetrated the box, and caused the salt to glow. Roentgen's observations that X-rays could penetrate other bodies and affect photographic plates led to the development of X-ray photography.

Shortly after this discovery, Antoine Henri Becquerel (1852–1908) attempted to show a relationship between X-rays and the phosphorescence of uranium salts. In one experiment he wrapped a photographic plate in black paper, placed a sample of uranium salt on it, and exposed it to sunlight. The developed photographic plate showed that rays emitted from the salt had penetrated the paper. When Becquerel attempted to repeat the experiment, the sunlight was intermittent, so he placed the entire setup in a drawer. Several days later he developed the photographic plate, expecting to find it only slightly affected. To his amazement, he found an intense image on the plate. He repeated the experiment in total darkness and obtained the same results, proving that the uranium salt emitted rays that affected the photographic plate without its being exposed to sunlight. Thus did the discovery of radioactivity come about, a combination of numerous experiments by the finest minds of the day—and serendipity. Becquerel later showed that the rays coming from uranium are able to ionize air and are also capable of penetrating thin sheets of metal.

The name *radioactivity* was coined two years later (in 1898) by Marie Curie. **Radioactivity** is the spontaneous emission of particles and/or rays from the nucleus of an atom. Elements having this property are said to be radioactive.

In 1898, Marie Sklodowska Curie (1867–1934) and her husband Pierre Curie (1859–1906) turned their research interests to radioactivity. In a short time, the Curies discovered two new elements, polonium and radium, both of which are radioactive. To confirm their work on radium, they processed 1 ton of pitchblende residue ore to obtain 0.1 g of pure radium chloride, which they used to make further studies on the properties of radium and to determine its atomic mass.

In 1899, Ernest Rutherford began to investigate the nature of the rays emitted from uranium. He found two particles, which he called *alpha* and *beta particles*. Soon he realized that uranium, while emitting these particles, was changing into another element. By 1912, over 30 radioactive isotopes were known, and many more are known today. The *gamma ray*, a third type of emission from radioactive materials similar to an X-ray, was discovered by Paul Villard (1860–1934) in 1900. Rutherford's description of the nuclear atom led scientists to attribute the phenomenon of radioactivity to reactions taking place in the nuclei of atoms.

The symbolism and notation we described for isotopes in Chapter 5 are also very useful in nuclear chemistry:

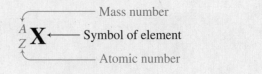

For example, $^{238}_{92}\text{U}$ represents a uranium isotope with an atomic number of 92 and a mass number of 238. This isotope is also designated as U-238, or uranium-238, and contains 92 protons

TABLE 18.1 Isotopic Notation for Several Particles (and Small Isotopes) Associated with Nuclear Chemistry

Particle	Symbol	Z Atomic number	A Mass number
Neutron	^1_0n	0	1
Proton	^1_1H	1	1
Beta particle (electron)	$^0_{-1}\text{e}$	−1	0
Positron (positive electron)	$^0_{+1}\text{e}$	1	0
Alpha particle (helium nucleus)	^4_2He	2	4
Deuteron (heavy hydrogen nucleus)	^2_1H	1	2

and 146 neutrons. The protons and neutrons collectively are known as **nucleons**. The mass number is the total number of nucleons in the nucleus. Table 18.1 shows the isotopic notations for several particles associated with nuclear chemistry.

When we speak of isotopes, we mean atoms of the same element with different masses, such as $^{16}_8\text{O}$, $^{17}_8\text{O}$, $^{18}_8\text{O}$. In nuclear chemistry we use the term **nuclide** to mean any isotope of any atom. Thus, $^{16}_8\text{O}$ and $^{235}_{92}\text{U}$ are referred to as nuclides. Nuclides that spontaneously emit radiation are referred to as *radionuclides*.

Natural Radioactivity

Radioactive elements continuously undergo **radioactive decay**, or disintegration, to form different elements. The chemical properties of an element are associated with its electronic structure, but radioactivity is a property of the nucleus. Therefore, neither ordinary changes of temperature and pressure nor the chemical or physical state of an element has any effect on its radioactivity.

The principal emissions from the nuclei of radionuclides are known as alpha particles, beta particles, and gamma rays. Upon losing an alpha or beta particle, the radioactive element changes into a different element. We will explain this process in detail later.

Each radioactive nuclide disintegrates at a specific and constant rate, which is expressed in units of half-life. The **half-life ($t_{1/2}$)** is the time required for one-half of a specific amount of a radioactive nuclide to disintegrate. The half-lives of the elements range from a fraction of a second to billions of years. To illustrate, suppose we start with 1.0 g of $^{226}_{88}\text{Ra}$ ($t_{1/2} = 1620$ years):

$$1.0 \text{ g } ^{226}_{88}\text{Ra} \xrightarrow[\text{1620 years}]{t_{1/2}} 0.50 \text{ g } ^{226}_{88}\text{Ra} \xrightarrow[\text{1620 years}]{t_{1/2}} 0.25 \text{ g } ^{226}_{88}\text{Ra}$$

Element	$t_{1/2}$
$^{238}_{92}\text{U}$	4.5×10^9 years
$^{226}_{88}\text{Ra}$	1620 years
$^{15}_6\text{C}$	2.4 seconds

The half-lives of the various radioisotopes of the same element differ dramatically. Half-lives for certain isotopes of radium, carbon, and uranium are listed in Table 18.2.

TABLE 18.2 Half-Lives for Radium, Carbon, and Uranium Isotopes

Isotope	Half-life	Isotope	Half-life
Ra-223	11.7 days	C-14	5730 years
Ra-224	3.64 days	C-15	2.4 seconds
Ra-225	14.8 days	U-235	7.1×10^8 years
Ra-226	1620 years	U-238	4.5×10^9 years
Ra-228	6.7 years		

WileyPLUS

ENHANCED EXAMPLE

EXAMPLE 18.1

The half-life of $^{131}_{53}I$ is 8 days. How much $^{131}_{53}I$ from a 32-g sample remains after five half-lives?

SOLUTION

Using the following graph, we can find the number of grams of ^{131}I remaining after one half-life:

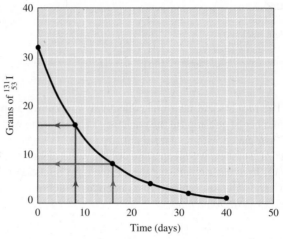

Trace a perpendicular line from 8 days on the x-axis to the line on the graph. Now trace a horizontal line from this point on the plotted line to the y-axis and read the corresponding grams of ^{131}I. Continue this process for each half-life, adding 8 days to the previous value on the x-axis:

Half-lives	0	1	2	3	4	5
Number of days		8	16	24	32	40
Amount remaining	32 g	16 g	8 g	4 g	2 g	1 g

Starting with 32 g, one gram of ^{131}I remains after five half-lives (40 days).

EXAMPLE 18.2

In how many half-lives will 10.0 g of a radioactive nuclide decay to less than 10% of its original value?

SOLUTION

We know that 10% of the original amount is 1.0 g. After the first half-life, half the original material remains and half has decayed (5.00 g). After the second half-life, one-fourth of the original material remains (i.e., one-half of 5.00 g). This progression continues, reducing the quantity remaining by half for each half-life that passes.

Half-lives	0	1	2	3	4
Percent remaining	100%	50%	25%	12.5%	6.25%
Amount remaining	10.0 g	5.00 g	2.50 g	1.25 g	0.625 g

Therefore, the amount remaining will be less than 10% sometime between the third and the fourth half-lives.

PRACTICE 18.1

The half-life of $^{14}_{6}C$ is 5730 years. How much $^{14}_{6}C$ will remain after six half-lives in a sample that initially contains 25.0 g?

Nuclides are said to be either *stable* (nonradioactive) or *unstable* (radioactive). Elements that have atomic numbers greater than 83 (bismuth) are naturally radioactive, although some of the nuclides have extremely long half-lives. Some of the naturally occurring nuclides of elements 81, 82, and 83 are radioactive, and some are stable. Only a few naturally occurring elements that have atomic numbers less than 81 are radioactive. However, no stable isotopes of element 43 (technetium) or of element 61 (promethium) are known.

Radioactivity is believed to be a result of an unstable ratio of neutrons to protons in the nucleus. Stable nuclides of elements up to about atomic number 20 generally have about a 1 : 1 neutron-to-proton ratio. In elements above number 20, the neutron-to-proton ratio in the stable nuclides gradually increases to about 1.5 : 1 in element number 83 (bismuth). When the neutron-to-proton ratio is too high or too low, alpha, beta, or other particles are emitted to achieve a more stable nucleus.

18.2 ALPHA PARTICLES, BETA PARTICLES, AND GAMMA RAYS

Describe the change in atomic number or atomic mass associated with emission of an alpha particle, beta particle, and gamma ray.

● **LEARNING OBJECTIVE**

KEY TERMS

alpha particle (α)
beta particle (β)
gamma ray (γ)

The classical experiment proving that alpha and beta particles are oppositely charged was performed by Marie Curie (see **Figure 18.1**). She placed a radioactive source in a hole in a lead block and positioned two poles of a strong electromagnet so that the radiations that were given off passed between them. The paths of three different kinds of radiation were detected by means of a photographic plate placed some distance beyond the electromagnet. The lighter beta particles were strongly deflected toward the positive pole of the electromagnet; the heavier alpha particles were less strongly deflected and in the opposite direction. The uncharged gamma rays were not affected by the electromagnet and struck the photographic plates after traveling along a path straight out of the lead block.

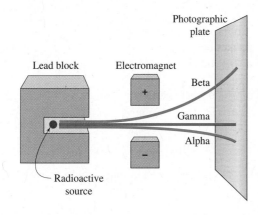

Figure 18.1

The effect of an electromagnetic field on alpha particles, beta particles, and gamma rays. Lighter beta particles are deflected considerably more than alpha particles. Alpha and beta particles are deflected in opposite directions. Gamma radiation is not affected by the electromagnetic field.

Alpha Particles

An **alpha particle (α)** consists of two protons and two neutrons, has a mass of about 4 amu, and has a charge of $+2$. It is a helium nucleus that is usually given one of the following symbols: α or ^4_2He. When an alpha particle is emitted from the nucleus, a different element is formed. The atomic number of the new element is 2 less, and the mass is 4 amu less, than that of the starting element.

Loss of an alpha particle from the nucleus results in
  loss of 4 in the mass number (A)
  loss of 2 in the atomic number (Z)

For example, when $^{238}_{92}\text{U}$ loses an alpha particle, $^{234}_{90}\text{Th}$ is formed, because two neutrons and two protons are lost from the uranium nucleus. This disintegration may be written as a nuclear equation:

$$^{238}_{92}\text{U} \longrightarrow {}^{234}_{90}\text{Th} + \alpha \quad \text{or} \quad {}^{238}_{92}\text{U} \longrightarrow {}^{234}_{90}\text{Th} + {}^4_2\text{He}$$

For the loss of an alpha particle from $^{226}_{88}\text{Ra}$, the equation is

$$^{226}_{88}\text{Ra} \longrightarrow {}^{222}_{86}\text{Rn} + {}^4_2\text{He} \quad \text{or} \quad {}^{226}_{88}\text{Ra} \longrightarrow {}^{222}_{86}\text{Rn} + \alpha$$

A nuclear equation, like a chemical equation, consists of reactants and products and must be balanced. To have a balanced nuclear equation, the sum of the mass numbers (superscripts) on both sides of the equation must be equal and the sum of the atomic numbers (subscripts) on both sides of the equation must be equal:

sum of mass numbers equals 226   sum of mass numbers is 230

$$^{226}_{88}\text{Ra} \longrightarrow \ ^{222}_{86}\text{Rn} + \ ^{4}_{2}\text{He} \qquad\qquad ^{230}_{90}\text{Th} \longrightarrow \ ^{226}_{88}? + \ ^{4}_{2}\text{He}$$

sum of atomic numbers equals 88   sum of atomic numbers is 90

What new nuclide will be formed when $^{230}_{90}\text{Th}$ loses an alpha particle? The new nuclide will have a mass of 226 amu and will contain 88 protons, so its atomic number is 88. Locate the corresponding element on the periodic table—in this case, $^{226}_{88}\text{Ra}$ or radium-226.

Beta Particles

The **beta particle (β)** is identical in mass and charge to an electron; its charge is -1. Both a beta particle and a proton are produced by the decomposition of a neutron:

$$^{1}_{0}\text{n} \longrightarrow \ ^{1}_{1}\text{p} + \ ^{0}_{-1}\text{e}$$

The beta particle leaves, and the proton remains in the nucleus. When an atom loses a beta particle from its nucleus, a different element is formed that has essentially the same mass but an atomic number that is 1 greater than that of the starting element. The beta particle is written as β or $^{0}_{-1}\text{e}$.

Loss of a beta particle from the nucleus results in
no change in the mass number (A)
increase of 1 in the atomic number (Z)

Examples of equations in which a beta particle is lost are

$$^{234}_{90}\text{Th} \longrightarrow \ ^{234}_{91}\text{Pa} + \beta$$

$$^{234}_{91}\text{Pa} \longrightarrow \ ^{234}_{92}\text{U} + \ ^{0}_{-1}\text{e}$$

$$^{210}_{82}\text{Pb} \longrightarrow \ ^{210}_{83}\text{Bi} + \beta$$

Gamma Rays

Gamma rays (γ) are photons of energy. A gamma ray is similar to an X-ray but is more energetic. It has no electrical charge and no measurable mass. Gamma rays are released from the nucleus in many radioactive changes along with either alpha or beta particles. Gamma radiation does not result in a change of atomic number or the mass of an element.

Loss of a gamma ray from the nucleus results in
no change in mass number (A) or atomic number (Z)

WileyPLUS
ENHANCED EXAMPLE

EXAMPLE 18.3

(a) Write an equation for the loss of an alpha particle from the nuclide $^{194}_{78}\text{Pt}$.

(b) What nuclide is formed when $^{228}_{88}\text{Ra}$ loses a beta particle from its nucleus?

SOLUTION

(a) Loss of an alpha particle, $^{4}_{2}\text{He}$, results in a decrease of 4 in the mass number and a decrease of 2 in the atomic number:

Mass of new nuclide: $A - 4$ or $194 - 4 = 190$
Atomic number of new nuclide: $Z - 2$ or $78 - 2 = 76$

Looking up element number 76 on the periodic table, we find it to be osmium, Os. The equation then is

$$^{194}_{78}\text{Pt} \longrightarrow ^{190}_{76}\text{Os} + ^{4}_{2}\text{He}$$

(b) The loss of a beta particle from a $^{228}_{88}\text{Ra}$ nucleus means a gain of 1 in the atomic number with no essential change in mass. The new nuclide will have an atomic number of $(Z + 1)$, or 89, which is actinium, Ac:

$$^{228}_{88}\text{Ra} \longrightarrow ^{228}_{89}\text{Ac} + ^{0}_{-1}\text{e}$$

EXAMPLE 18.4

What nuclide will be formed when $^{214}_{82}\text{Pb}$ successively emits two beta particles, then one alpha particle from its nucleus? Write successive equations showing these changes.

SOLUTION

The changes brought about in the three steps outlined are as follows:

β loss: Increase of 1 in the atomic number; no change in mass

β loss: Increase of 1 in the atomic number; no change in mass

α loss: Decrease of 2 in the atomic number; decrease of 4 in the mass

The equations are

$$^{214}_{82}\text{Pb} \longrightarrow ^{214}_{83}\text{X} + \beta \longrightarrow ^{214}_{84}\text{X} + \beta \longrightarrow ^{210}_{82}\text{X} + \alpha$$

where X stands for the new nuclide formed. Looking up each of these elements by their atomic numbers, we rewrite the equations

$$^{214}_{82}\text{Pb} \xrightarrow[\beta]{} ^{214}_{83}\text{Bi} \xrightarrow[\beta]{} ^{214}_{84}\text{Po} \xrightarrow[\alpha]{} ^{210}_{82}\text{Pb}$$

PRACTICE 18.2

What nuclide will be formed when $^{222}_{86}\text{Rn}$ emits an alpha particle?

The ability of radioactive rays to pass through various objects is in proportion to the speed at which they leave the nucleus. Gamma rays travel at the velocity of light (186,000 miles per second) and are capable of penetrating several inches of lead. The velocities of beta particles are variable, the fastest being about nine-tenths the velocity of light. Alpha particles have velocities less than one-tenth the velocity of light. **Figure 18.2** illustrates the relative penetrating power of these rays. A few sheets of paper will stop alpha particles; a thin sheet

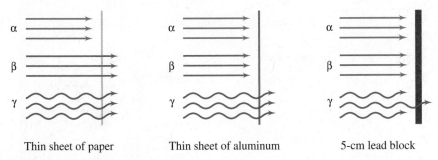

Thin sheet of paper   Thin sheet of aluminum   5-cm lead block

Figure 18.2
Relative penetrating ability of alpha, beta, and gamma radiation.

TABLE 18.3 Characteristics of Nuclear Radiation

Radiation	Symbol	Mass (amu)	Electrical charge	Velocity	Composition	Ionizing power
Alpha	α, $_2^4\text{He}$	4	+2	Variable, less than 10% the speed of light	He nucleus	High
Beta	β, $_{-1}^{0}\text{e}$	$\dfrac{1}{1837}$	-1	Variable, up to 90% the speed of light	Identical to an electron	Moderate
Gamma	γ	0	0	Speed of light	Photons or electromagnetic waves of energy	Almost none

of aluminum will stop both alpha and beta particles; and a 5-cm block of lead will reduce, but not completely stop, gamma radiation. In fact, it is difficult to stop all gamma radiation. Table 18.3 summarizes the properties of alpha, beta, and gamma radiation.

18.3 RADIOACTIVE DISINTEGRATION SERIES

LEARNING OBJECTIVE ● Discuss the transmutation of elements and both natural and artificial disintegration series.

KEY TERMS

transmutation
artificial radioactivity
induced radioactivity
transuranium elements

These three series begin with the elements uranium, thorium, and actinium.

The naturally occurring radioactive elements with a higher atomic number than lead (Pb) fall into three orderly disintegration series. Each series proceeds from one element to the next by the loss of either an alpha or a beta particle, finally ending in a nonradioactive nuclide. The uranium series starts with $_{92}^{238}\text{U}$ and ends with $_{82}^{206}\text{Pb}$. The thorium series starts with $_{90}^{232}\text{Th}$ and ends with $_{82}^{208}\text{Pb}$. The actinium series starts with $_{92}^{235}\text{U}$ and ends with $_{82}^{207}\text{Pb}$. A fourth series, the neptunium series, starts with the synthetic element $_{94}^{241}\text{Pu}$ and ends with the stable bismuth nuclide $_{83}^{209}\text{Bi}$. The uranium series is shown in **Figure 18.3**. Gamma radiation, which accompanies alpha and beta radiation, is not shown.

By using these series and the half-lives of their members, scientists have been able to approximate the age of certain geologic deposits. This approximation is done by comparing the amount of $_{92}^{238}\text{U}$ with the amount of $_{82}^{206}\text{Pb}$ and other nuclides in the series that are present in a particular geologic formation. Rocks found in Canada and Finland have been calculated to be about 3.0×10^9 (3 billion) years old. Some meteorites have been determined to be 4.5×10^9 years old.

Figure 18.3

The uranium disintegration series. $_{92}^{238}\text{U}$ decays by a series of alpha (α) and beta (β) emissions to the stable nuclide $_{82}^{206}\text{Pb}$.

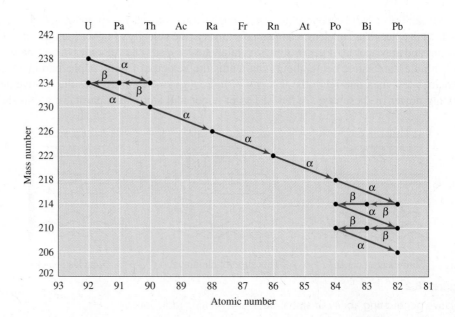

EXAMPLE 18.5

In Figure 18.3 identify the last five nuclides in the disintegration series of $^{381}_{92}U \longrightarrow {}^{206}_{82}Pb$. Write the symbol, the mass number, and the atomic number of each nuclide.

SOLUTION

The five nuclides are $^{214}_{84}Po \longrightarrow {}^{210}_{82}Pb \longrightarrow {}^{210}_{83}Bi \longrightarrow {}^{210}_{84}Po \longrightarrow {}^{206}_{82}Pb$.

PRACTICE 18.3

What nuclides are formed when $^{238}_{92}U$ undergoes the following decays?

(a) loss of an alpha particle and a beta particle

(b) loss of three alpha particles and two beta particles

Transmutation of Elements

Transmutation is the conversion of one element into another by either natural or artificial means. Transmutation occurs spontaneously in natural radioactive disintegrations. Alchemists tried for centuries to convert lead and mercury into gold by artificial means, but transmutation by artificial means was not achieved until 1919, when Ernest Rutherford succeeded in bombarding the nuclei of nitrogen atoms with alpha particles and produced oxygen nuclides and protons. The nuclear equation for this transmutation can be written as

$$^{14}_{7}N + \alpha \longrightarrow {}^{17}_{8}O + {}^{1}_{1}H \quad or \quad {}^{14}_{7}N + {}^{4}_{2}He \longrightarrow {}^{17}_{8}O + {}^{1}_{1}H$$

It is believed that the alpha particle enters the nitrogen nucleus, forming $^{18}_{9}F$ as an intermediate, which then decomposes into the products.

Rutherford's experiments opened the door to nuclear transmutations of all kinds. Atoms were bombarded by alpha particles, neutrons, protons, deuterons ($^{2}_{1}H$), electrons, and so forth. Massive instruments were developed for accelerating these particles to very high speeds and energies to aid their penetration of the nucleus. The famous cyclotron was developed by E. O. Lawrence (1901–1958) at the University of California; later instruments include the Van de Graaf electrostatic generator, the betatron, and the electron and proton synchrotrons. With these instruments many nuclear transmutations became possible. Equations for a few of these are as follows:

$$^{7}_{3}Li + {}^{1}_{1}H \longrightarrow 2\,{}^{4}_{2}He$$
$$^{40}_{18}Ar + {}^{1}_{1}H \longrightarrow {}^{40}_{19}K + {}^{1}_{0}n$$
$$^{23}_{11}Na + {}^{1}_{1}H \longrightarrow {}^{23}_{12}Mg + {}^{1}_{0}n$$
$$^{209}_{83}Bi + {}^{2}_{1}H \longrightarrow {}^{210}_{84}Po + {}^{1}_{0}n$$
$$^{16}_{8}O + {}^{1}_{0}n \longrightarrow {}^{13}_{6}C + {}^{4}_{2}He$$
$$^{238}_{92}U + {}^{12}_{6}C \longrightarrow {}^{244}_{98}Cf + 6\,{}^{1}_{0}n$$

On November 1, 2004, the IUPAC officially approved the name roentgenium, symbol Rg, for the element of atomic number 111. Roentgenium was named for Wilhelm Conrad Roentgen, who discovered X-rays in 1895. The nuclear equation for its formation is

$$^{209}_{83}Bi + {}^{64}_{28}Ni \longrightarrow {}^{272}_{111}Rg + {}^{1}_{0}n$$

The IUPAC has recently named elements 112, 114, and 116, as copernicium (Cn), flerovium (Fl) and livermorium (Lv) respectively.

Artificial Radioactivity

Irene Joliot-Curie (daughter of Pierre and Marie Curie) and her husband Frederic Joliot-Curie observed that when aluminum-27 is bombarded with alpha particles, neutrons and positrons (positive electrons) are emitted as part of the products. When the source of alpha

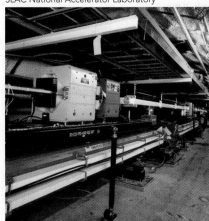

SLAC National Accelerator Laboratory

Electrons and positrons are accelerated along the tunnels at the SLAC National Accelerator Laboratory.

particles is removed, neutrons cease to be produced, but positrons continue to be emitted. This observation suggested that the neutrons and positrons come from two separate reactions. It also indicated that a product of the first reaction is radioactive. After further investigation, they discovered that, when aluminum-27 is bombarded with alpha particles, phosphorus-30 and neutrons are produced. Phosphorus-30 is radioactive, has a half-life of 2.5 minutes, and decays to silicon-30 with the emission of a positron. The equations for these reactions are

$$^{27}_{13}\text{Al} + {}^{4}_{2}\text{He} \longrightarrow {}^{30}_{15}\text{P} + {}^{1}_{0}\text{n}$$

$$^{30}_{15}\text{P} \longrightarrow {}^{30}_{14}\text{Si} + {}^{0}_{+1}\text{e}$$

The radioactivity of nuclides produced in this manner is known as **artificial radioactivity** or **induced radioactivity**. Artificial radionuclides behave like natural radioactive elements in two ways: They disintegrate in a definite fashion and they have a specific half-life. The Joliot-Curies received the Nobel Prize in chemistry in 1935 for the discovery of artificial, or induced, radioactivity.

Transuranium Elements

The elements that follow uranium on the periodic table and that have atomic numbers greater than 92 are known as the **transuranium elements**. They are synthetic radioactive elements; none of them occur naturally.

The first transuranium element, number 93, was discovered in 1939 by Edwin M. McMillan (1907–1991) at the University of California while he was investigating the fission of uranium. He named it neptunium for the planet Neptune. In 1941, element 94, plutonium, was identified as a beta-decay product of neptunium:

$$^{238}_{93}\text{Np} \longrightarrow {}^{238}_{94}\text{Pu} + {}^{0}_{-1}\text{e}$$

$$^{239}_{93}\text{Np} \longrightarrow {}^{239}_{94}\text{Pu} + {}^{0}_{-1}\text{e}$$

Plutonium is one of the most important fissionable elements known today.

Since 1964, the discoveries of 15 new transuranium elements, numbers 104–118, have been announced. These elements have been produced in minute quantities by high-energy particle accelerators.

18.4 MEASUREMENT OF RADIOACTIVITY

LEARNING OBJECTIVE ● Describe the units and instruments used in the measurement of radioactivity.

KEY TERMS

curie (Ci)
roentgen (R)
rad (radiation absorbed *dose*)
rem (roentgen equivalent to *man*)

Radiation from radioactive sources is so energetic that it is called *ionizing radiation.* When it strikes an atom or a molecule, one or more electrons are knocked off, and an ion is created. The Geiger counter, an instrument commonly used to detect and measure radioactivity, depends on this property. The instrument consists of a Geiger–Müller detecting tube and a counting device. The detector tube is a pair of oppositely charged electrodes in an argon gas–filled chamber fitted with a thin window. When radiation, such as a beta particle, passes through the window into the tube, some argon is ionized, and a momentary pulse of current (discharge) flows between the electrodes. These current pulses are electronically amplified in the counter and appear as signals in the form of audible clicks, flashing lights, meter deflections, or numerical readouts (**Figure 18.4**).

The amount of radiation that an individual encounters can be measured by a film badge. This badge contains a piece of photographic film in a lightproof holder and is worn in areas where radiation might be encountered. The silver grains in the film will darken when exposed to radiation. The badges are processed after a predetermined time interval to determine the amount of radiation the wearer has been exposed to.

A scintillation counter is used to measure radioactivity for biomedical applications. A scintillator contains molecules that emit light when exposed to ionizing radiation. A light-sensitive detector counts the flashes and converts them to a numerical readout.

Figure 18.4
Geiger–Müller survey meter.

EXAMPLE 18.6

What three types of radiation detectors are commonly used to detect and measure radioactivity?

SOLUTION

Three types of radiation detectors are a Geiger–Müller detecting tube and counter, a radiation film badge, and a scintillation counter.

PRACTICE 18.4

How does a radiation film badge detect radiation?

The *curie* is the unit used to express the amount of radioactivity produced by an element. One **curie (Ci)** is defined as the quantity of radioactive material giving 3.7×10^{10} disintegrations per second. The basis for this figure is pure radium, which has an activity of 1 Ci/g. Because the curie is such a large quantity, the millicurie and microcurie, representing one-thousandth and one-millionth of a curie, respectively, are more practical and more commonly used.

The curie only measures radioactivity emitted by a radionuclide. Different units are required to measure exposure to radiation. The **roentgen (R)** quantifies exposure to gamma or X-rays; 1 roentgen is defined as the amount of radiation required to produce 2.1×10^9 ions/cm^3 of dry air. The **rad (radiation absorbed dose)** is defined as the amount of radiation that provides 0.01 J of energy per kilogram of matter. The amount of radiation absorbed will change depending on the type of matter. The roentgen and the rad are numerically similar; 1 roentgen of gamma radiation provides 0.92 rad in bone tissue.

Neither the rad nor the roentgen indicates the biological damage caused by radiation. One rad of alpha particles has the ability to cause 10 times more damage than 1 rad of gamma rays or beta particles. Another unit, **rem (roentgen equivalent to man)**, takes into account the degree of biological effect caused by the type of radiation exposure; 1 rem is equal to the dose in rads multiplied by a factor specific to the form of radiation. The factor is 10 for alpha particles and 1 for both beta particles and gamma rays. Units of radiation are summarized in Table 18.4.

TABLE 18.4 Radiation Units

Unit	Measure	Equivalent
curie (Ci)	rate of decay of a radioactive substance	$1\ Ci = 3.7 \times 10^{10}$ disintegrations/sec
roentgen (R)	exposure based on the quantity of ionization produced in air	$1\ R = 2.1 \times 10^9$ ions/cm^3
rad	absorbed dose of radiation	$1\ rad = 0.01$ J/kg matter
rem	roentgen equivalent in man	$1\ rem = 1\ rad \times factor$
gray (Gy) (SI unit)	energy absorbed by tissue	$1\ Gy = 1$ J/kg tissue ($1\ Gy = 100\ rad$)

18.5 NUCLEAR ENERGY

Describe the characteristics of a chain reaction and compare nuclear fission and nuclear fusion.

● **LEARNING OBJECTIVE**

Nuclear Fission

In **nuclear fission** a heavy nuclide splits into two or more intermediate-sized fragments when struck in a particular way by a neutron. The fragments are called *fission products*. As the atom splits, it releases energy and two or three neutrons, each of which can cause another nuclear fission. The first instance of nuclear fission was reported in January 1939 by the German scientists Otto Hahn (1879–1968) and Fritz Strassmann (1902–1980). Detecting isotopes of barium, krypton, cerium, and lanthanum after bombarding uranium with neutrons led scientists to believe that the uranium nucleus had been split.

KEY TERMS

nuclear fission
chain reaction
critical mass
nuclear fusion

>CHEMISTRY *IN ACTION*

Does Your Food Glow in the Dark?

The Food and Drug Administration (FDA) has approved irradiation to delay ripening or kill microbes and insects in wheat, potatoes, fresh fruits, and poultry as well as in spices. But the only regular use of irradiation on food in the United States has been on spices.

Now the FDA has approved irradiation of red meat, and new legislation allows the labels about irradiated food to be in much smaller type. The general public has not yet accepted the use of radiation to reduce microbes and help preserve food. Why? Many people believe that irradiating food makes it radioactive. This is not true; in fact, the most that irradiation of food does is to produce compounds similar to those created by cooking and also to reduce the vitamin content of some food.

How does irradiation of food work? The radioactive source currently used is cobalt-60. It is contained in a concrete cell (see diagram) with 6-foot-thick walls. Inside the cell is a pool of water with racks of thin Co-60 rods suspended above. When the rods are not being used, they are submerged in the water, which absorbs the gamma radiation. When food to be irradiated moves into the cells, the Co-60 rods are lifted from the water, and the boxes of food move among them on a conveyor being irradiated from all sides.

Scientists who have investigated the process say that food irradiation is safe. The Centers for Disease Control and Prevention in Atlanta estimates that food-borne illness causes as many as 9000 deaths per year. Irradiation provides one way to reduce microbial contamination in food. It does not solve the problem of careless handling of food by processors, and long-term studies on humans have not yet been concluded regarding irradiated food supplies. Once again, we face the issue of balancing benefit and risk.

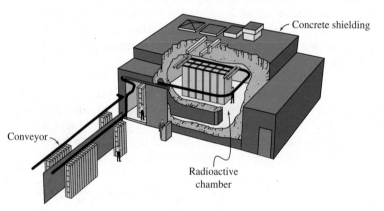

Concrete shielding

Conveyor

Radioactive chamber

Characteristics of nuclear fission are as follows:

1. Upon absorption of a neutron, a heavy nuclide splits into two or more smaller nuclides (fission products).

2. The mass of the nuclides formed ranges from about 70 to 160 amu.

3. Two or more neutrons are produced from the fission of each atom.

4. Large quantities of energy are produced as a result of the conversion of a small amount of mass into energy.

5. Most nuclides produced are radioactive and continue to decay until they reach a stable nucleus.

One process by which this fission takes place is illustrated in **Figure 18.5**. When a heavy nucleus captures a neutron, the energy increase may be sufficient to cause deformation of the nucleus until the mass finally splits into two fragments, releasing energy and usually two or more neutrons.

In a typical fission reaction, a $^{235}_{92}U$ nucleus captures a neutron and forms unstable $^{236}_{92}U$. This $^{236}_{92}U$ nucleus undergoes fission, quickly disintegrating into two fragments, such as $^{139}_{56}Ba$ and $^{94}_{36}Kr$, and three neutrons. The three neutrons in turn may be captured by three other $^{235}_{92}U$ atoms, each of which undergoes fission, producing nine neutrons, and so on. A reaction of this

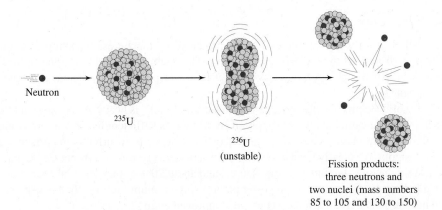

Figure 18.5

The fission process. When a neutron is captured by a heavy nucleus, the nucleus becomes more unstable. The more energetic nucleus begins to deform, resulting in fission. Two nuclear fragments and three neutrons are produced by this fission process.

Neutron

^{235}U

^{236}U
(unstable)

Fission products:
three neutrons and
two nuclei (mass numbers
85 to 105 and 130 to 150)

kind, in which the products cause the reaction to continue or magnify, is known as a **chain reaction**. For a chain reaction to continue, enough fissionable material must be present so that each atomic fission causes, on average, at least one additional fission. The minimum quantity of an element needed to support a self-sustaining chain reaction is called the **critical mass**. Since energy is released in each atomic fission, chain reactions provide a steady supply of energy. (A chain reaction is illustrated in **Figure 18.6**.) Two of the many possible ways in which $^{235}_{92}$U may fission are shown by these nuclear equations:

$$^{235}_{92}\text{U} + ^{1}_{0}\text{n} \longrightarrow ^{139}_{56}\text{Ba} + ^{94}_{36}\text{Kr} + 3\,^{1}_{0}\text{n}$$

$$^{235}_{92}\text{U} + ^{1}_{0}\text{n} \longrightarrow ^{144}_{54}\text{Xe} + ^{90}_{38}\text{Sr} + 2\,^{1}_{0}\text{n}$$

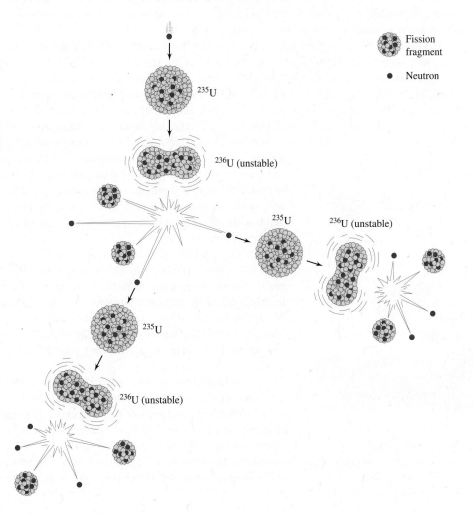

Fission fragment

Neutron

^{235}U

^{236}U (unstable)

^{235}U ^{236}U (unstable)

^{235}U

^{236}U (unstable)

Figure 18.6

Fission and chain reaction of $^{235}_{92}$U. Each fission produces two major fission fragments and three neutrons, which may be captured by other $^{235}_{92}$U nuclei, continuing the chain reaction.

Nuclear Power

Nearly all electricity for commercial use is produced by machines consisting of a turbine linked by a driveshaft to an electrical generator. The energy required to run the turbine can be supplied by falling water, as in hydroelectric power plants, or by steam, as in thermal power plants.

The world's demand for energy, largely from fossil fuels, is heavy. At the present rates of consumption, the estimated world supply of fossil fuels is sufficient for only a few centuries. Although the United States has large coal and oil shale deposits, it currently imports over 40% of its oil supply. We clearly need to develop alternative energy sources. At present, uranium is the most productive alternative energy source, and about 20% of the electrical energy used in the United States is generated from power plants using uranium fuel. The first new permits to build nuclear plants in the United States were approved in 2012.

A nuclear power plant is a thermal power plant in which heat is produced by a nuclear reactor instead of by combustion of fossil fuel. The major components of a nuclear reactor are

1. an arrangement of nuclear fuel, called the reactor core
2. a control system, which regulates the rate of fission and thereby the rate of heat generation
3. a cooling system, which removes the heat from the reactor and also keeps the core at the proper temperature

One type of reactor uses metal slugs containing uranium enriched from the normal 0.7% U-235 to about 3% U-235. The self-sustaining fission reaction is moderated, or controlled, by adjustable control rods containing substances that slow down and capture some of the neutrons produced. Ordinary water, heavy water, and molten sodium are typical coolants used. Energy obtained from nuclear reactions in the form of heat is used in the production of steam to drive turbines for generating electricity. (See **Figure 18.7**.)

The potential dangers of nuclear power were tragically demonstrated by the accidents at Three Mile Island, Pennsylvania (1979), Chernobyl in the former Soviet Union (1986), and the earthquake and tsunami in Japan in 2011. These crises resulted from the loss of coolant to the reactor core. The reactors at Three Mile Island were covered by concrete containment buildings and, therefore, released a relatively small amount of radioactive material into the atmosphere. But because the Soviet Union did not require containment structures on nuclear power plants, the Chernobyl accident resulted in 31 deaths and the resettlement of 135,000 people. The release of large quantities of I-131, Cs-134, and Cs-137 appears to be causing long-term health problems in that exposed population. Japan is currently dealing with radiation contamination in areas around its damaged plants and estimates it could take 40 years to clean up the reactors completely.

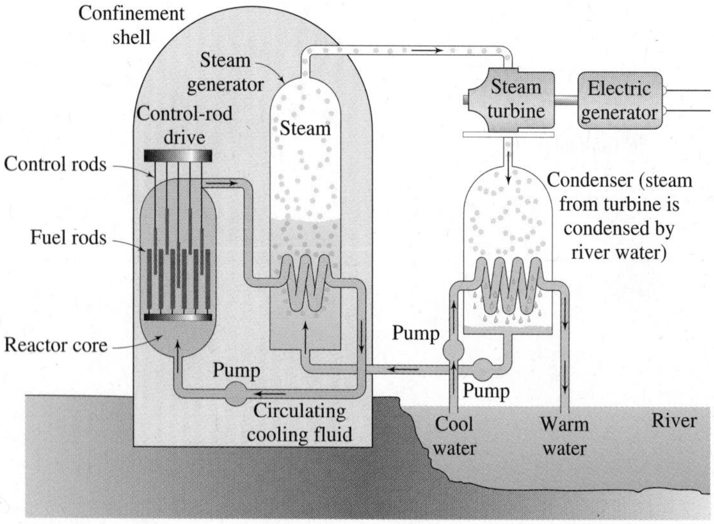

Another problem with nuclear power plants is their radioactive waste production. The United States has debated where and how to permanently store nuclear wastes. In July 2002, an underground site was authorized at Yucca Mountain, Nevada, about 90 miles north of Las Vegas. The Yucca Mountain facility was closed in 2009. A permanent site for nuclear waste storage has not yet been decided upon.

In the United States, reactors designed for commercial power production use uranium oxide, U_3O_8, that is enriched with the relatively scarce fissionable U-235 isotope. Because the supply of U-235 is limited, a new type of reactor known as the *breeder reactor* has been developed. Breeder reactors produce additional fissionable material at the same time that the fission reaction is occurring. In a

Figure 18.7

Diagram of a nuclear power plant. Heat produced by the reactor core is carried by a cooling fluid to a steam generator, which turns a turbine that produces electricity.

breeder reactor, excess neutrons convert nonfissionable isotopes, such as U-238 or Th-232, to fissionable isotopes, Pu-239 or U-233, as shown below:

$$\,^{238}_{92}\text{U} + \,^{1}_{0}\text{n} \longrightarrow \,^{239}_{92}\text{U} \underset{\beta}{\searrow} \,^{239}_{93}\text{Np} \underset{\beta}{\searrow} \,^{239}_{94}\text{Pu}$$

$$\,^{232}_{90}\text{Th} + \,^{1}_{0}\text{n} \longrightarrow \,^{233}_{90}\text{Th} \underset{\beta}{\searrow} \,^{233}_{91}\text{Pa} \underset{\beta}{\searrow} \,^{233}_{92}\text{U}$$

These transmutations make it possible to greatly extend the supply of fuel for nuclear reactors. No breeder reactors are presently in commercial operation in the United States, but a number of them are being operated in Europe and Great Britain.

EXAMPLE 18.7

What are the functions of the component parts of a nuclear power plant?

SOLUTION

A nuclear power plant consists of a nuclear reactor core that generates heat energy to change water into steam. The steam drives a turbine that produces electrical energy. Surplus steam is condensed back to water and recycled to the nuclear reactor.

The Atomic Bomb

The atomic bomb is a fission bomb; it operates on the principle of a very fast chain reaction that releases a tremendous amount of energy. An atomic bomb and a nuclear reactor both depend on self-sustaining nuclear fission chain reactions. The essential difference is that in a bomb the fission is "wild," or uncontrolled, whereas in a nuclear reactor the fission is moderated and carefully controlled. A minimum critical mass of fissionable material is needed for a bomb; otherwise a major explosion will not occur. When a quantity smaller than the critical mass is used, too many neutrons formed in the fission step escape without combining with another nucleus, and a chain reaction does not occur. Therefore, the fissionable material of an atomic bomb must be stored as two or more subcritical masses and brought together to form the critical mass at the desired time of explosion. The temperature developed in an atomic bomb is believed to be about 10 million degrees Celsius.

The nuclides used in atomic bombs are U-235 and Pu-239. Uranium deposits contain about 0.7% of the U-235 isotope, the remainder being U-238. Uranium-238 does not undergo fission except with very high-energy neutrons. It was discovered, however, that U-238 captures a low-energy neutron without undergoing fission and that the product, U-239, changes to Pu-239 (plutonium) by a beta-decay process. Plutonium-239 readily undergoes fission upon capture of a neutron and is therefore useful for nuclear weapons. The equations for the nuclear transformations are

$$\,^{238}_{92}\text{U} + \,^{1}_{0}\text{n} \longrightarrow \,^{239}_{92}\text{U} \underset{\beta}{\searrow} \,^{239}_{93}\text{Np} \underset{\beta}{\searrow} \,^{239}_{94}\text{Pu}$$

The hazards of an atomic bomb explosion include not only shock waves from the explosive pressure and tremendous heat but also intense radiation in the form of alpha particles, beta particles, gamma rays, and ultraviolet rays. Gamma rays and X-rays can penetrate deeply into the body, causing burns, sterilization, and gene mutation, which can adversely affect future generations. Both radioactive fission products and unfissioned material are present after the explosion. If the bomb explodes near the ground, many tons of dust are lifted into the air. Radioactive material adhering to this dust, known as *fallout*, is spread by air currents over wide areas of the land and constitutes a lingering source of radiation hazard.

Today nuclear war is probably the most awesome threat facing civilization. Only two rather primitive fission-type atom bombs were used to destroy the Japanese cities of

The mushroom cloud is a signature of uncontrolled fission in an atomic bomb.

NASA

GSO Images/Getty Images, Inc.

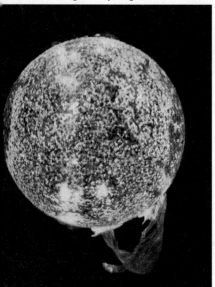

Solar flares such as these are indications of fusion reactions occurring at temperatures of millions of degrees.

Hiroshima and Nagasaki and bring World War II to an early end. The threat of nuclear war is increased by the fact that the number of nations possessing nuclear weapons is steadily increasing.

Nuclear Fusion

The process of uniting the nuclei of two light elements to form one heavier nucleus is known as **nuclear fusion**. Such reactions can be used for producing energy, because the masses of the two nuclei that fuse into a single nucleus are greater than the mass of the nucleus formed by their fusion. The mass differential is liberated in the form of energy. Fusion reactions are responsible for the tremendous energy output of the Sun. Thus, aside from relatively small amounts from nuclear fission and radioactivity, fusion reactions are the ultimate source of our energy, even the energy from fossil fuels. They are also responsible for the devastating power of the thermonuclear, or hydrogen, bomb.

Fusion reactions require temperatures on the order of tens of millions of degrees for initiation. Such temperatures are present in the Sun but have been produced only momentarily on Earth. For example, the hydrogen, or fusion, bomb is triggered by the temperature of an exploding fission bomb. Two typical fusion reactions are

$$\underset{\text{tritium}}{{}^{3}_{1}\text{H}} \quad + \quad \underset{\text{deuterium}}{{}^{2}_{1}\text{H}} \quad \longrightarrow \quad {}^{4}_{2}\text{He} + {}^{1}_{0}\text{n} + \text{energy}$$

$$\underset{\substack{3.0150 \\ \text{amu}}}{{}^{3}_{1}\text{H}} \quad + \quad \underset{\substack{1.0079 \\ \text{amu}}}{{}^{1}_{1}\text{H}} \quad \longrightarrow \quad \underset{\substack{4.0026 \\ \text{amu}}}{{}^{4}_{2}\text{He}} + \text{energy}$$

The total mass of the reactants in the second equation is 4.0229 amu, which is 0.0203 amu greater than the mass of the product. This difference in mass is manifested in the great amount of energy liberated.

A great deal of research in the United States and in other countries, especially the former Soviet Union, has focused on controlled nuclear fusion reactions. The goal of controlled nuclear fusion has not yet been attained, although the required ignition temperature has been reached in several devices. Evidence to date leads us to believe that we can develop a practical fusion power reactor. Fusion power, if we can develop it, will be far superior to fission power for the following reasons:

1. Virtually infinite amounts of energy are possible from fusion. Uranium supplies for fission power are limited, but heavy hydrogen, or deuterium (the most likely fusion fuel), is abundant. It is estimated that the deuterium in a cubic mile of seawater used as fusion fuel could provide more energy than the petroleum reserves of the entire world.

2. From an environmental viewpoint, fusion power is much "cleaner" than fission power because fusion reactions (in contrast to uranium and plutonium fission reactions) do not produce large amounts of long-lived and dangerously radioactive isotopes.

NASA/Science Source/Photo Researchers, Inc.

Remnants of a supernova show temperatures of about 10 million degrees Celsius created when the star exploded.

PRACTICE 18.5

Briefly distinguish between nuclear fission and nuclear fusion.

18.6 MASS–ENERGY RELATIONSHIP IN NUCLEAR REACTIONS

Explain mass defect and binding energy.

KEY TERMS

mass defect
nuclear binding energy

Large amounts of energy are released in nuclear reactions; thus, significant amounts of mass are converted to energy. We stated earlier that the amount of mass converted to energy in chemical changes is insignificant compared to the amount of energy released in a nuclear reaction. In fission reactions, about 0.1% of the mass is converted into energy. In fusion reactions as much as 0.5% of the mass may be changed into energy. The Einstein equation, $E = mc^2$, can be used to calculate the energy liberated, or available, when the mass loss is known. For example, in the fusion reaction

$$\underset{7.016\,g}{^{7}_{3}\text{Li}} \quad + \quad \underset{1.008\,g}{^{1}_{1}\text{H}} \quad \longrightarrow \quad \underset{4.003\,g}{^{4}_{2}\text{He}} \quad + \quad \underset{4.003\,g}{^{4}_{2}\text{He}} \quad + \quad \text{energy}$$

the mass difference between the reactants and products ($8.024\text{ g} - 8.006\text{ g}$) is 0.018 g. The energy equivalent to this amount of mass is 1.62×10^{12} J. By comparison, this is more than 4 million times greater than the 3.9×10^5 J of energy obtained from the complete combustion of 12.01 g (1 mol) of carbon.

The mass of a nucleus is actually less than the sum of the masses of the protons and neutrons that make up that nucleus. The difference between the mass of the protons and the neutrons in a nucleus and the mass of the nucleus is known as the **mass defect**. The energy equivalent to this difference in mass is known as the **nuclear binding energy**. This energy is the amount that would be required to break a nucleus into its individual protons and neutrons. The higher the binding energy, the more stable the nucleus. Elements of intermediate atomic masses have high binding energies. For example, iron (element number 26) has a very high binding energy and therefore a very stable nucleus. Just as electrons attain less energetic and more stable arrangements through ordinary chemical reactions, neutrons and protons attain less energetic and more stable arrangements through nuclear fission or fusion reactions. Thus, when uranium undergoes fission, the products have less mass (and greater binding energy) than the original uranium. In like manner, when hydrogen and lithium fuse to form helium, the helium has less mass (and greater binding energy) than the hydrogen and lithium. It is this conversion of mass to energy that accounts for the very large amounts of energy associated with both fission and fusion reactions.

EXAMPLE 18.8

WileyPLUS
ENHANCED EXAMPLE

Calculate the mass defect and the nuclear binding energy for an α particle (helium nucleus).

SOLUTION

Knowns: proton mass = 1.0073 g/mol
neutron mass = 1.0087 g/mol
α mass = 4.0015 g/mol
$1.0\text{ g} = 9.0 \times 10^{13}$ J

Plan •   First calculate the sum of the individual parts of an α particle (2 protons +2 neutrons).

2 protons: 2×1.0073 g/mol = 2.0146 g/mol
2 neutrons: 2×1.0087 g/mol = $\underline{2.0174 \text{ g/mol}}$
4.0320 g/mol

The mass defect is the difference between the mass of the α-particle and its component parts:

4.0320 g/mol
$\underline{4.0015 \text{ g/mol}}$
0.0305 g/mol   (mass defect)

Calculate • To find the nuclear binding energy, convert the mass defect to its energy equivalent:

$$(0.0305 \text{ g/mol})(9.0 \times 10^{13} \text{ J/g})$$

$$= 2.7 \times 10^{12} \text{ J/mol} \, (6.5 \times 10^{11} \text{ cal})$$

PRACTICE 18.6

Why is the mass of an atomic nucleus less than the sum of the masses of the protons and neutrons that are in that nucleus?

18.7 BIOLOGICAL EFFECTS OF RADIATION

LEARNING OBJECTIVE ● Describe the effects of ionizing radiation on living organisms.

KEY TERM
ionizing radiation

Radiation with energy to dislocate bonding electrons and create ions when passing through matter is classified as **ionizing radiation**. Alpha particles, beta particles, gamma rays, and X-rays fall into this classification. Ionizing radiation can damage or kill living cells and can be particularly devastating when it strikes the cell nuclei and affects molecules involved in cell reproduction. The effects of radiation on living organisms fall into these general categories: (1) acute or short-term effects, (2) long-term effects, and (3) genetic effects.

Acute Radiation Damage

High levels of radiation, especially from gamma rays or X-rays, produce nausea, vomiting, and diarrhea. The effect has been likened to a sunburn throughout the body. If the dosage is high enough, death will occur in a few days. The damaging effects of radiation appear to be centered in the nuclei of the cells, and cells that are undergoing rapid cell division are most susceptible to damage. It is for this reason that cancers are often treated with gamma radiation from a Co-60 source. Cancerous cells multiply rapidly and are destroyed by a level of radiation that does not seriously damage normal cells.

Long-Term Radiation Damage

Protracted exposure to low levels of any form of ionizing radiation can weaken an organism and lead to the onset of malignant tumors, even after fairly long time delays. The largest exposure to synthetic sources of radiation is from X-rays. Evidence suggests that the lives of early workers in radioactivity and X-ray technology may have been shortened by long-term radiation damage.

Strontium-90 isotopes are present in the fallout from atmospheric testing of nuclear weapons. Strontium is in the same periodic-table group as calcium, and its chemical behavior is similar to that of calcium. Hence, when foods contaminated with Sr-90 are eaten, Sr-90 ions are laid down in the bone tissue along with ordinary calcium ions. Strontium-90 is a beta emitter with a half-life of 28 years. Blood cells manufactured in bone marrow are affected by the radiation from Sr-90. Hence, there is concern that Sr-90 accumulation in the environment may cause an increase in the incidence of leukemia and bone cancers.

Genetic Effects

The information needed to create an individual of a particular species, be it a bacterial cell or a human being, is contained within the nucleus of a cell. This genetic information is encoded in the structure of DNA (deoxyribonucleic acid) molecules, which make up genes. The DNA molecules form precise duplicates of themselves when cells divide, thus passing genetic information from one generation to the next. Radiation can damage DNA molecules. If the damage is not severe enough to prevent the individual from reproducing, a mutation may result. Most mutation-induced traits are undesirable. Unfortunately, if the bearer of the altered genes

>CHEMISTRY *IN ACTION*

A Window into Living Organisms

Imagine viewing a living process as it is occurring. This was the dream of many scientists in the past as they tried to extract this knowledge from dead tissue. Today, because of innovations in nuclear chemistry, this dream is a common, everyday occurrence.

Compounds containing a radionuclide are described as being *labeled*, or *tagged*. These compounds undergo their normal chemical reactions, but their location can be detected because of their radioactivity. When such compounds are given to a plant or an animal, the movement of the nuclide can be traced through the organism by the use of a Geiger counter or other detecting device.

In an early use of the tracer technique, the pathway by which CO_2 becomes fixed into carbohydrate ($C_6H_{12}O_6$) during photosynthesis was determined. The net equation for photosynthesis is

$$6\,CO_2 + 6\,H_2O \longrightarrow C_6H_{12}O_6 + 6\,O_2$$

Radioactive $^{14}CO_2$ was injected into a colony of green algae, and the algae were then placed in the dark and killed at selected time intervals. When the radioactive compounds were separated by paper chromatography and analyzed, the results elucidated a series of light-independent photosynthetic reactions.

Biological research using tracer techniques have determined

1. the rate of phosphate uptake by plants, using radio-phosphorus

2. the flow of nutrients in the digestive tract using radioactive barium compounds

3. the accumulation of iodine in the thyroid gland, using radioactive iodine

4. the absorption of iron by the hemoglobin of the blood, using radioactive iron

In chemistry, uses for tracers are unlimited. The study of reaction mechanisms, the measurement of the rates of chemical reactions, and the determination of physical constants are just a few of the areas of application.

Radioactive tracers are commonly used in medical diagnosis. The radionuclide must be effective at a low concentration and have a short half-life to reduce the possibility of damage to the patient.

Radioactive iodine (I-131) is used to determine the function of the thyroid, where the body concentrates iodine. In this process a small amount of radioactive potassium or sodium iodide is ingested. A detector is focused on the

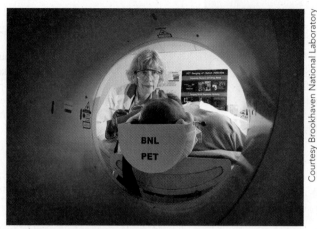

A radioactive tracer is injected into this patient and absorbed by the brain. The PET scanner detects photons emitted by the tracer and produces an image used in medical diagnosis and research.

thyroid gland and measures the amount of iodine in the gland. This picture is then compared to that of a normal thyroid to detect any differences.

Doctors examine the heart's pumping performance and check for evidence of obstruction in coronary arteries by *nuclear scanning*. The radionuclide Tl-201, when injected into the bloodstream, lodges in healthy heart muscle. Thallium-201 emits gamma radiation, which is detected by a special imaging device called a *scintillation camera*. The data obtained are simultaneously translated into pictures by a computer. With this technique doctors can observe whether heart tissue has died after a heart attack and whether blood is flowing freely through the coronary passages.

One of the most recent applications of nuclear chemistry is the use of positron emission tomography (PET) in the measurement of dynamic processes in the body, such as oxygen use or blood flow. In this application, a compound is made that contains a positron-emitting nuclide such as C-11, O-15, or N-13. The compound is injected into the body, and the patient is placed in an instrument that detects the positron emission. A computer produces a three-dimensional image of the area.

PET scans have been used to locate the areas of the brain involved in epileptic seizures. Glucose tagged with C-11 is injected, and an image of the brain is produced. Since the brain uses glucose almost exclusively for energy, diseased areas that use glucose at a rate different than normal tissue can then be identified.

survives to reproduce, these traits are passed along to succeeding generations. In other words, the genetic effects of increased radiation exposure are found in future generations, not only in the present generation.

Because radioactive rays are hazardous to health and living tissue, special precautions must be taken in designing laboratories and nuclear reactors, in disposing of waste materials, and in monitoring the radiation exposure of people working in this field.

CHAPTER 18 REVIEW

18.1 DISCOVERY OF RADIOACTIVITY

KEY TERMS

radioactivity
nucleon
nuclide
radioactivite decay
half-life (t₁/₂)

- Radioactivity is the spontaneous emission of particles and energy from the nucleus of an atom.
- Protons and neutrons are known as nucleons.
- In nuclear chemistry isotopes are also known as nuclides.
- Radioactive elements undergo radioactive decay to form different elements.
- Radioactivity is not affected by changes in temperature, pressure, or the state of the element.
- Principal emissions:
    - Alpha particles
    - Beta particles
    - Gamma rays
- The half-life of a nuclide is the time required for one-half of a specific amount of the nuclide to disintegrate.

18.2 ALPHA PARTICLES, BETA PARTICLES, AND GAMMA RAYS

KEY TERMS

alpha particle (α)
beta particle (β)
gamma ray (γ)

- Alpha particles:
    - Consist of 2 protons and 2 neutrons with mass of 4 amu and charge of +2
    - Loss of an alpha particle from the nucleus results in
        - Loss of 4 in mass number
        - Loss of 2 in atomic number
    - High ionizing power, low penetration
- Beta particles:
    - Same as an electron
    - Loss of a beta particle from the nucleus results in
        - No change in mass number
        - Increase of 1 in atomic number
    - Moderate ionizing power and penetration
- Gamma rays:
    - Photons of energy
    - Loss of a gamma ray from the nucleus results in
        - No change in mass number (A)
        - No change in atomic number (Z)
    - Almost no ionizing power, high penetration

18.3 RADIOACTIVE DISINTEGRATION SERIES

KEY TERMS

transmutation
artificial radioactivity
induced radioactivity
transuranium elements

- As elements undergo disintegration, they eventually become stable after the loss of a series of particles and energy.
- The disintegration series can be used to determine the age of geologic deposits.
- Transmutation of an element can occur spontaneously or artificially.
- Artificial radionuclides behave like natural radioactive elements in two ways:
    - They disintegrate in a definite fashion.
    - They have a specific half-life.
- Transuranium elements are the elements that follow uranium on the periodic table.

18.4 MEASUREMENT OF RADIOACTIVITY

KEY TERMS

curie (Ci)
roentgen (R)
rad (radiation absorbed dose)
rem (roentgen equivalent
    to man)

- Radioactivity can be measured in several ways:
    - Geiger–Müller counter
    - Film badge
    - Scintillation counter
- Units for measuring radiation:
    - Gray
    - Curie
    - Roentgen
    - Rad
    - Rem

18.5 NUCLEAR ENERGY

- Characteristics of fission:
  - A heavy nuclide splits into two or more smaller nuclides.
  - The mass of nuclides formed is 70 to 160 amu.
  - Two or more neutrons are produced per fission.
  - Large quantities of energy are released.
  - Most nuclides produced are radioactive and continue to decay until they reach stability.
- A nuclear power plant is a thermal plant with a nuclear fission source of energy.
- Major components of a nuclear reactor:
  - Reactor core
  - Control system
  - Cooling system
- Out-of-control fission reaction results in a bomb
- Nuclear fusion is the process of uniting 2 light nuclei to form 1 heavier nucleus.
- Nuclear fusion requires extreme temperatures for ignition.

KEY TERMS

nuclear fission
chain reaction
critical mass
nuclear fusion

18.6 MASS–ENERGY RELATIONSHIP IN NUCLEAR REACTIONS

- The mass defect is the difference between the actual mass of an atom and the calculated mass of the protons and neutrons in the nucleus of that atom.
- The nuclear binding energy is the energy equivalent to the mass defect.
- $E = mc^2$ is used to determine the energy liberated in a nuclear reaction.

KEY TERMS

mass defect
nuclear binding energy

18.7 BIOLOGICAL EFFECTS OF RADIATION

- Effects of radiation fall into three general categories:
  - Acute or short-term
  - Long-term
  - Genetic

KEY TERM

ionizing radiation

REVIEW QUESTIONS

1. Identify these people and their associations with the early history of radioactivity:
   (a) Antoine Henri Becquerel
   (b) Marie and Pierre Curie
   (c) Wilhelm Roentgen
   (d) Ernest Rutherford
   (e) Otto Hahn and Fritz Strassmann
2. Why is the radioactivity of an element unaffected by the usual factors that affect the rate of chemical reactions, such as ordinary changes of temperature and concentration?
3. Distinguish between the terms *isotope* and *nuclide.*
4. The half-life of Pu-244 is 76 million years. If Earth's age is about 5 billion years, discuss the feasibility of finding this nuclide as a naturally occurring nuclide.
5. Describe the radiocarbon method for dating archaeological artifacts.
6. Anthropologists have found bones whose age suggests that the human line may have emerged in Africa as much as 4 million years ago. If wood or charcoal were found with such bones, would C-14 dating be useful in dating the bones? Explain.
7. To afford protection from radiation injury, which kind of radiation requires (a) the most shielding? (b) the least shielding?
8. Why is an alpha particle deflected less than a beta particle in passing through an electromagnetic field?

9. Tell how types of alpha, beta, and gamma radiation are distinguished from the standpoint of
   (a) charge   (c) nature of particle or ray
   (b) relative mass   (d) relative penetrating power
10. Bismuth-211 decays by alpha emission to give a nuclide that in turn decays by beta emission to yield a stable nuclide. Show these two steps with nuclear equations.
11. Name three pairs of nuclides that might be obtained by fissioning U-235 atoms.
12. Distinguish between natural and artificial radioactivity.
13. What is a radioactive disintegration series?
14. Briefly discuss the transmutation of elements.
15. Stable Pb-208 is formed from Th-232 in the thorium disintegration series by successive α, β, β, α, α, α, α, β, β, α particle emissions. Write the symbol (including mass and atomic number) for each nuclide formed in this series.
16. The nuclide Np-237 loses a total of seven alpha particles and four beta particles. What nuclide remains after these losses?
17. How might radioactivity be used to locate a leak in an underground pipe?
18. What is a scintillation counter and how does it work?
19. What unit is used to describe the amount of radioactivity produced by an element?
20. What does the unit **rem** stand for and what does it measure?

21. What were Otto Hahn and Fritz Strassmann's contribution to nuclear physics?
22. What is a breeder reactor? Explain how it accomplishes the "breeding."
23. What is the essential difference between the nuclear reactions in a nuclear reactor and those in an atomic bomb?
24. Why must a certain minimum amount of fissionable material be present before a self-supporting chain reaction can occur?
25. What are the disadvantages of nuclear power?
26. What are the hazards associated with an atomic bomb explosion?

27. What types of elements undergo fission? How about fusion?
28. Using Figure 18.7, describe how nuclear power is generated.
29. What are mass defect and nuclear binding energy?
30. Explain why radioactive rays are classified as ionizing radiation.
31. Give a brief description of the biological hazards associated with radioactivity.
32. Strontium-90 has been found to occur in radioactive fallout. Why is there so much concern about this radionuclide being found in cow's milk? (Half-life of Sr-90 is 28 years.)
33. What is a radioactive tracer? How is it used?

Most of the exercises in this chapter are available for assignment via the online homework management program, WileyPLUS (www.wileyplus.com). All exercises with blue numbers have answers in Appendix VI.

PAIRED EXERCISES

1. Indicate the number of protons, neutrons, and nucleons in these nuclei:
   (a) $^{207}_{82}Pb$ (b) $^{70}_{31}Ga$

2. Indicate the number of protons, neutrons, and nucleons in these nuclei:
   (a) $^{128}_{52}Te$ (b) $^{32}_{16}S$

3. How are the mass and the atomic number of a nucleus affected by the loss of an alpha particle?

4. How are the mass and the atomic number of a nucleus affected by the loss of a beta particle?

5. Write nuclear equations for the alpha decay of
   (a) $^{210}_{83}Bi$ (b) $^{238}_{92}U$

6. Write nuclear equations for the alpha decay of
   (a) $^{238}_{90}Th$ (b) $^{239}_{94}Pu$

7. Write nuclear equations for the beta decay of
   (a) $^{13}_{7}N$ (b) $^{234}_{90}Th$

8. Write nuclear equations for the beta decay of
   (a) $^{28}_{13}Al$ (b) $^{239}_{93}Np$

9. Determine the type of emission or emissions (alpha, beta, or gamma) that occurred in the following transitions:
   (a) $^{226}_{88}Ra$ to $^{222}_{86}Rn$
   (b) $^{222}_{88}Ra$ to $^{222}_{87}Fr$ to $^{222}_{87}Fr$
   (c) $^{238}_{92}U$ to $^{238}_{93}Np$

10. Determine the type of emission or emissions (alpha, beta, or gamma) that occurred in the following transitions:
    (a) $^{210}_{82}Pb$ to $^{210}_{82}Pb$
    (b) $^{234}_{91}Pa$ to $^{230}_{89}Ac$ to $^{230}_{90}Th$
    (c) $^{234}_{90}Th$ to $^{230}_{88}Ra$ to $^{230}_{88}Ra$

11. Write a nuclear equation for the conversion of $^{26}_{13}Al$ to $^{26}_{12}Mg$.

12. Write a nuclear equation for the conversion of $^{32}_{15}P$ to $^{32}_{16}S$.

13. Complete and balance these nuclear equations by supplying the missing particles:
    (a) $^{66}_{29}Cu \longrightarrow ^{66}_{30}Zn +$ _____
    (b) $^{0}_{-1}e +$ _____ $\longrightarrow ^{7}_{3}Li$
    (c) $^{27}_{13}Al + ^{4}_{2}He \longrightarrow ^{30}_{14}Si +$ _____
    (d) $^{85}_{37}Rb +$ _____ $\longrightarrow ^{82}_{35}Br + ^{4}_{2}He$

14. Complete and balance these nuclear equations by supplying the missing particles:
    (a) $^{27}_{13}Al + ^{4}_{2}He \longrightarrow ^{30}_{15}P +$ _____
    (b) $^{27}_{14}Si \longrightarrow ^{0}_{+1}e +$ _____
    (c) _____ $+ ^{2}_{1}H \longrightarrow ^{13}_{7}N + ^{1}_{0}n$
    (d) _____ $\longrightarrow ^{82}_{36}Kr + ^{0}_{-1}e$

15. Strontium-90 has a half-life of 28 years. If a 1.00-mg sample was stored for 112 years, what mass of Sr-90 would remain?

16. Strontium-90 has a half-life of 28 years. If a sample was tested in 1980 and found to be emitting 240 counts/ min, in what year would the same sample be found to be emitting 30 counts/min? How much of the original Sr-90 would be left?

17. By what series of emissions does $^{233}_{91}Pa$ disintegrate to $^{225}_{89}Ac$?

18. By what series of emissions does $^{238}_{90}Th$ disintegrate to $^{212}_{82}Pb$?

19. Consider the fission reaction
   $$^{235}_{92}U + ^{1}_{0}n \longrightarrow ^{94}_{38}Sr + ^{139}_{54}Xe + 3\,^{1}_{0}n + energy$$
   Calculate the following using these mass data (1.0 g is equivalent to 9.0×10^{13} J):

U-235 = 235.0439 amu	Sr-94 = 93.9154 amu
Xe-139 = 138.9179 amu	n = 1.0087 amu

   (a) the energy released in joules for a single event (one uranium atom splitting)
   (b) the energy released in joules per mole of uranium splitting
   (c) the percentage of mass lost in the reaction

20. Consider the fusion reaction
   $$^{1}_{1}H + ^{2}_{1}H \longrightarrow ^{3}_{2}He + energy$$
   Calculate the following using these mass data (1.0 g is equivalent to 9.0×10^{13} J):

   $$^{1}_{1}H = 1.00794 \text{ amu}$$
   $$^{2}_{1}H = 2.01410 \text{ amu}$$
   $$^{3}_{2}H = 3.01603 \text{ amu}$$

   (a) the energy released in joules per mole of He-3 formed
   (b) the percentage of mass lost in the reaction

ADDITIONAL EXERCISES

21. Many radioactive isotopes are used in nuclear medicine. Some of the useful isotopes are listed below. For each of these isotopes, give the number of protons, neutrons, and electrons.
    (a) Chromium-51—this isotope is used to label red blood cells and quantify gastrointestinal protein loss.
    (b) Holmium-166—this isotope is being tested as a diagnostic and treatment tool for liver tumors.
    (c) Palladium-103—this isotope is used for treatment of early stage prostate cancer.
    (d) Strontium-89—this isotope is used to treat the pain of bone cancer

22. The Th-232 disintegration series starts with $^{232}_{90}$Th and emits the following rays successively: α, β, β, α, α, α, β, α, β, α. The series ends with the stable $^{208}_{82}$Pb. Write the formula for each nuclide in the series.

23. Potassium-40 has a half-life of 1.25×10^9 years. How many months will it take for one-half of a 25.0-g sample to disappear?

24. When the nuclide $^{249}_{98}$Cf was bombarded with $^{15}_{7}$N, four neutrons and a new transuranium element were formed. Write the nuclear equation for this transmutation.

25. What percent of the mass of the nuclide $^{226}_{88}$Ra is neutrons? electrons?

26. If radium costs $90,000 a gram, how much will 0.0100 g of ^{226}RaCl$_2$ cost if the price is based only on the radium content?

27. An archaeological specimen was analyzed and found to be emitting only 25% as much C-14 radiation per gram of carbon as newly cut wood. How old is this specimen?

28. Barium-141 is a beta emitter. What is the half-life if a 16.0-g sample of the nuclide decays to 0.500 g in 90 minutes?

29. Calculate (a) the mass defect and (b) the binding energy of $^7_3$Li using the mass data:

$$^7_3\text{Li} = 7.0160 \text{ g} \qquad n = 1.0087 \text{ g}$$
$$p = 1.0073 \text{ g} \qquad e^- = 0.00055 \text{ g}$$
$$1.0 \text{ g} \equiv 9.0 \times 10^{13} \text{ J (from } E = mc^2)$$

30. In the disintegration series $^{235}_{92}$U $\longrightarrow$ $^{207}_{82}$Pb, how many alpha and beta particles are emitted?

31. List three devices used for radiation detection and explain their operation.

32. The half-life of I-123 is 13 hours. If 10 mg of I-123 are administered to a patient, how much I-123 remains after 3 days and 6 hours?

33. Iodine-131 is often used to treat Graves' disease or hyperthyroidism. In Graves' disease the thyroid begins to overproduce thyroid hormone. If a small amount of iodine-131 is consumed, it will concentrate in the thyroid gland and destroy its ability to function. If a tablet containing 15.0 μg of ^{131}I is ingested by a patient, how many days will it take for the amount of iodine to decay to less than 1 μg? (Iodine-131 has a half-life of 8 days.)

34. One of the ways to determine the extent of termite infestation in a wooden structure is to feed the termites a wood substitute containing a radioactive element. The extent of the infestation can then be determined by monitoring the spread of the radioactivity. If you were given the task to prepare a wood substitute, would you use bismuth-213 (half-life 46 minutes), rhenium-186 (half-life 4 days), or cobalt-60 (half-life 5 years). Explain your reasoning.

35. Many home smoke detectors use americunium-241 as a radiation source. In the smoke detector, ^{241}Am emits alpha particles. These alpha particles are detected by the detector unless smoke particles block their passage. Thus, as long as there are no smoke particles, the detector gets a continuous stream of alpha particles hitting it; smoke causes a disruption in this stream and activates the alarm. Write out the nuclear equation for the decay of ^{241}Am by alpha emission.

36. Clearly distinguish between fission and fusion. Give an example of each.

37. Starting with 1 g of a radioactive isotope whose half-life is 10 days, sketch a graph showing the pattern of decay for that material. On the x-axis, plot time (you may want to simply show multiples of the half-life), and on the y-axis, plot mass of material remaining. Then after completing the graph, explain why a sample never really gets to the point where *all* of its radioactivity is considered to be gone.

38. Identify each missing product (name the element and give its atomic number and mass number) by balancing the following nuclear equations:
    (a) ^{235}U + $^1_0$n $\longrightarrow$ ^{143}Xe + $3\,^1_0$n + _____
    (b) ^{235}U + $^1_0$n $\longrightarrow$ ^{102}Y + $3\,^1_0$n + _____
    (c) ^{14}N + $^1_0$n $\longrightarrow$ ^{1}H + _____

39. Consider these reactions:
    (a) $H_2O(l) \longrightarrow H_2O(g)$
    (b) $2\,H_2(g) + O_2(g) \longrightarrow 2\,H_2O(g)$
    (c) $^2_1$H + $^2_1$H $\longrightarrow$ $^3_1$H + $^1_1$H
    The following energy values belong to one of these equations:

    energy$_1$ 115.6 kcal released
    energy$_2$ 10.5 kcal absorbed
    energy$_3$ 7.5×10^7 kcal released

    Match the equation to the energy value and briefly explain your choices.

40. When $^{235}_{92}$U is struck by a neutron, the unstable isotope $^{236}_{92}$U results. When that daughter isotope undergoes fission, there are numerous possible products. If strontium-90 and three neutrons are the results of one such fission, what is the other product?

41. Write balanced nuclear equations for
    (a) beta emission by $^{29}_{12}$Mg
    (b) alpha emission by $^{150}_{60}$Nd
    (c) positron emission by $^{72}_{33}$As

42. How much of a sample of cesium-137 ($t_{1/2} = 30$ years) must have been present originally if, after 270 years, 15.0 g remain?

43. The curie is equal to 3.7×10^{10} disintegration/sec, and the becquerel is equivalent to just 1 disintegration/sec. Suppose a hospital has a 150-g radioactive source with an activity of 1.24 Ci. What is its activity in becquerels?

44. Cobalt-60 has a half-life of 5.26 years. If 1.00 g of ^{60}Co were allowed to decay, how many grams would be left after
    (a) one half-life?   (c) four half-lives?
    (b) two half-lives?   (d) ten half-lives?

45. Write balanced equations to show these changes:
    (a) alpha emission by boron-11
    (b) beta emission by strontium-88
    (c) neutron absorption by silver-107
    (d) proton emission by potassium-41
    (e) electron absorption by antimony-116

46. The $^{14}_6$C content of an ancient piece of wood was found to be one-sixteenth of that in living trees. How many years old is this piece of wood if the half-life of carbon-14 is 5730 years?

47. How might human exposure to ionizing radiation today affect future generations?

48. What are some of the effects of long-term exposure to low-level ionizing radiation?

49. Explain how scientists can measure the rate of phosphate uptake by plants using radiophosphorus.

CHALLENGE EXERCISES

50. Rubidium-87, a beta emitter, is the product of positron emission. Identify
   (a) the product of rubidium-87 decay
   (b) the precursor of rubidium-87

51. Potassium-42 is used to locate brain tumors. Its half-life is 12.5 hours. Starting with 15.4 mg, what fraction will remain after 100 hours? If it was necessary to have at least 1 μg for a particular procedure, could you hold the original sample for 200 hours before using it?

ANSWERS TO PRACTICE EXERCISES

18.1   0.391 g

18.2 $^{218}_{84}\text{Po}$

18.3   (a) $^{234}_{91}\text{Pa}$ (b) $^{226}_{88}\text{Ra}$

18.4   The silver grains in the photographic film badge darken when exposed to ionizing radiation. Badges worn by workers around radiation sources are processed periodically to measure the amount of radiation to which workers have been exposed.

18.5   In nuclear fission a heavy nuclide is bombarded by and absorbs a neutron, which causes the nuclide to become unstable and to break (fission) into two or more lighter nuclear fragments and two or three neutrons plus energy. In a fusion reaction two light nuclides are combined (fused) with a release of energy.

18.6   The mass difference is the nuclear binding energy, which is the energy equivalent needed to hold the nucleus together.

The parachutes used by people parasailing are composed of nylon, an organic compound. All living organisms, plant and animal, are composed of organic compounds, too.

Mauritius/SuperStock

CHAPTER **19**

INTRODUCTION TO ORGANIC CHEMISTRY

Numerous substances throughout nature incorporate silicon or carbon in their structures. Silicon is the staple of the geologist—it combines with oxygen in various ways to produce silica and a family of compounds known as the silicates. These compounds form the chemical foundation of most sand, rocks, and soil. In the living world, carbon combines with hydrogen, oxygen, nitrogen, and sulfur to form millions of compounds.

The petroleum and the polymer products we find indispensable are two of the many industries that depend on carbon chemistry. Synthetic fibers (clothing and carpeting) and plastics (containers, compact discs, computer terminals, and pens) are made from carbon compounds. Cold remedies, cleaning products, nutrients for space travel, convenience foods, and countless drugs, both legal and illegal, have all come about from our understanding of the chemistry of carbon.

CHAPTER OUTLINE

19.1 The Beginnings of Organic Chemistry

19.2 Why Carbon?

19.3 Alkanes

19.4 Alkenes and Alkynes

19.5 Aromatic Hydrocarbons

19.6 Hydrocarbon Derivatives

19.7 Alcohols

19.8 Ethers

19.9 Aldehydes and Ketones

19.10 Carboxylic Acids

19.11 Esters

19.12 Polymers—Macromolecules

19.1 THE BEGINNINGS OF ORGANIC CHEMISTRY

LEARNING OBJECTIVE ● Define organic chemistry.

KEY TERMS

vital-force theory
organic chemistry

During the late eighteenth and early nineteenth centuries, the fact that compounds obtained from animal and vegetable sources defied the established rules for inorganic compounds baffled chemists. They knew that compound formation in inorganic compounds is due to a simple attraction between positively and negatively charged elements, and usually only one, or at most a few, compounds could be made from a given group of two or three elements. But one group of four elements—carbon, hydrogen, oxygen, and nitrogen—gave rise to a large number of remarkably stable compounds.

No organic compounds had ever been synthesized from inorganic substances and chemists had no other explanation for the complexities of organic compounds, so they believed these compounds were formed by a "vital force." The **vital-force theory** held that organic substances could originate only from living material. In 1828, the results of a simple experiment by German chemist Friedrich Wöhler (1800–1882) proved to be the end of this theory. As he attempted to prepare ammonium cyanate (NH_4CNO) by heating cyanic acid (HCNO), and ammonia (NH_3), Wöhler obtained a white crystalline substance he identified as urea, $H_2N-CO-NH_2$. Wöhler knew urea to be an authentic organic substance because it had been isolated from urine. The implications of Wöhler's results were not immediately recognized, but the fact that one organic compound had been isolated from inorganic compounds changed the face of chemistry forever.

With Wöhler's work, it was apparent that no vital force other than skill and knowledge was needed to make organic chemicals in the laboratory. Today the branch of chemistry that deals with carbon compounds, **organic chemistry**, does not imply that these compounds must originate from living matter. A few special kinds of carbon compounds (e.g., carbon oxides, metal carbides, and metal carbonates) are excluded from the organic classification because their chemistry is more closely related to that of inorganic substances.

The field of organic chemistry is vast; it includes not only all living organisms but also a great many other materials that we use daily. Foodstuffs (fats, proteins, carbohydrates); fuels; fabrics; wood and paper products; paints and varnishes; plastics; dyes; soaps and detergents; cosmetics; medicinals; and rubber products—all are organic materials.

The sources of organic compounds are carbon-containing raw materials—petroleum and natural gas, coal, carbohydrates, fats, and oils. In the United States we produce more than 250 billion pounds of organic chemicals from these sources, which amounts to about 1100 pounds per year for every man, woman, and child. About 90% of this 250 billion pounds comes from petroleum and natural gas. Because world reserves of petroleum and natural gas are finite, we will sometime have to rely on other sources to make the vast amount of organic substances that we depend on. Fortunately, we know how to synthesize many organic compounds from sources other than petroleum, although at a much greater expense.

19.2 WHY CARBON?

LEARNING OBJECTIVE ● Describe the arrangement of electrons around a carbon atom and how it leads to the chemical composition of hydrocarbons.

KEY TERMS

functional group
hydrocarbons
saturated hydrocarbon
unsaturated hydrocarbon

The carbon atom is central to all organic compounds. The atomic number of carbon is 6, and its electron structure is $1s^2 2s^2 2p^2$. Two stable isotopes of carbon exist: C-12 and C-13. In addition, carbon has several radioactive isotopes, C-14 being the most widely known of these because of its use in radiocarbon dating.

A carbon atom usually forms four covalent bonds. The most common geometric arrangement of these bonds is tetrahedral (see **Figure 19.1**). In this structure, the four covalent bonds are not planar about the carbon atom but are directed toward the corners of a regular tetrahedron. (A tetrahedron is a solid figure with four sides.) The angle between these tetrahedral bonds is 109.5°.

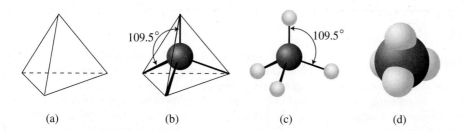

(a) (b) (c) (d)

Figure 19.1

Tetrahedral structure of carbon.
(a) A regular tetrahedron;
(b) a carbon atom within a regular
tetrahedron; (c) a methane
molecule, CH_4; (d) spacefilling
model of methane, CH_4.

With four valence electrons, the carbon atom ($\cdot\overset{\displaystyle\cdot}{\underset{\displaystyle\cdot}{C}}\cdot$) forms four single covalent bonds by sharing electrons with other atoms. The structures of methane and carbon tetrachloride illustrate this point:

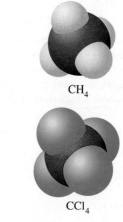

$$\begin{array}{ccc} & H & \\ H\!:\!\overset{\displaystyle H}{\underset{\displaystyle \ddot{H}}{C}}\!:\!H & H\!-\!\overset{\displaystyle |}{\underset{\displaystyle |}{C}}\!-\!H & \\ & H & \end{array}$$

methane

$$\begin{array}{cc} & Cl \\ Cl\!:\!\overset{\displaystyle Cl}{\underset{\displaystyle \ddot{Cl}}{C}}\!:\!Cl & Cl\!-\!\overset{\displaystyle |}{\underset{\displaystyle |}{C}}\!-\!Cl \\ & Cl \end{array}$$

carbon tetrachloride

CH_4

CCl_4

Figure 19.2

Spacefilling models of CH_4 and
CCl_4.

Actually, these compounds have a tetrahedral shape with bond angles of 109.5° (see **Figure 19.2**), but the bonds are often drawn at right angles. In methane each bond is formed by the sharing of electrons between a carbon and a hydrogen atom.

Carbon–carbon bonds are formed because carbon atoms can share electrons with other carbon atoms. One, two, or three pairs of electrons can be shared between two carbon atoms, forming a single, double, or triple bond, respectively:

$$\cdot\dot{C}\!:\!\dot{C}\cdot \qquad \cdot\dot{C}\!::\!\dot{C}\cdot \qquad \cdot C\!:::\!C\cdot$$

$$\cdot\dot{C}\!-\!\dot{C}\cdot \qquad \cdot\dot{C}\!=\!\dot{C}\cdot \qquad \cdot C\!\equiv\!C\cdot$$

single bond   double bond   triple bond

Each dash represents a covalent bond. Carbon, more than any other element, has the ability to form chains of covalently bonded atoms. This bonding ability is the main reason for the large number of organic compounds. Three examples are shown here. It's easy to see how, through this bonding ability, long chains of carbon atoms form by linking one carbon atom to another through covalent bonds:

$$\cdot\dot{C}\!:\!\dot{C}\!:\!\dot{C}\cdot \qquad \cdot\dot{C}\!-\!\dot{C}\!-\!\dot{C}\cdot \qquad$$

three carbon atoms bonded by single bonds

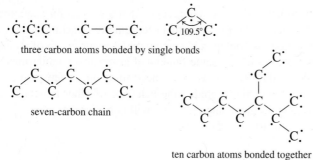

seven-carbon chain

ten carbon atoms bonded together

EXAMPLE 19.1

How many bonds can a carbon atom have to other atoms? Draw the various types of bonds using carbon-to-carbon bonds. Are these bonds ionic or covalent?

SOLUTION

Carbon can make four bonds to other atoms.

$$-\overset{\displaystyle |}{\underset{\displaystyle |}{C}}\!-\!\overset{\displaystyle |}{\underset{\displaystyle |}{C}}\!- \qquad \overset{\diagdown}{\diagup}C\!=\!C\overset{\diagup}{\diagdown} \qquad -C\!\equiv\!C\!-$$

Each bond is a covalent bond.

> CHEMISTRY *IN ACTION*

Biodiesel: Today's Alternative Fuel

Rudolf Diesel chose peanut oil for his fuel when he first invented the diesel engine in 1892. While diesel engines now power many of our vehicles (farm equipment, trucks, barges), the diesel fuel currently in use is a mixture of thousands of hydrocarbons obtained from crude oil. Diesel fuel is known for its crude and dirty smoke-containing noxious gases and particles of sulfur, organic compounds, and soot.

Today scientists and environmentalists are beginning to return to a vegetable-based renewable fuel. It turns out that vegetable oils can be converted to esters, which are less viscous and make good fuels. These new fuels are called biodiesel, and they can be made by using virtually any vegetable oil including soybean oil or used restaurant oil.

Most biodiesel used today is a blend with regular diesel fuel. Diesel engines can run on biodiesel with little or no engine modification. Biodiesel has been in common use in Europe for years. All diesel in France is 5% biodiesel. European governments have maintained policies to reduce their dependence on foreign oil. In the United States, some city and government fleets use biodiesel. A number of states, including Maine, Nevada, California, and Hawaii, even have biodiesel available to the public at pumps. The popularity of biodiesel is increasing significantly, as shown in the accompanying graph.

The Veggie Van, designed by Joshua Tickell, has run more than 25,000 miles on biodiesel. Tickell even collected his fuel on the road, stopping at fast-food places for gallons of used oil for a fill-up. Tickell says, "It's instantly available renewable energy, one we can begin using today and use for the next 10 to 15 years until higher technology comes to fruition." So hang onto that used French fry oil—you may soon use it to run your car!

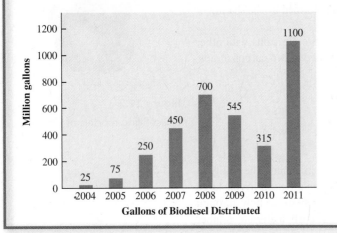

Gallons of Biodiesel Distributed

Courtesy of Joshua Tickell/veggievan.org

Carbon forms so many different compounds that a system for grouping the molecules is necessary. Organic molecules are classified according to structural features. The members of each class of compounds contain a characteristic atom or group of atoms called a **functional group**. Molecules with the same functional group share similarities in structure that result in similar chemical properties. Thus, we need study only a few members of a particular class of compounds to be able to predict the behavior of other molecules in that class. In this brief introduction to organic chemistry we will consider two categories of compounds: hydrocarbons and hydrocarbon derivatives.

Hydrocarbons

Hydrocarbons are compounds composed entirely of carbon and hydrogen atoms bonded to each other by covalent bonds. These molecules are further classified as saturated or unsaturated. **Saturated hydrocarbons** have only single bonds between carbon atoms. These hydrocarbons are classified as *alkanes*. **Unsaturated hydrocarbons** contain a double or triple bond between two carbon atoms and include *alkenes, alkynes,* and *aromatic compounds*. These classifications are summarized in **Figure 19.3**.

Fossil fuels—natural gas, petroleum, and coal—are the principal sources of hydrocarbons. Natural gas is primarily methane with small amounts of ethane, propane, and butane. Petroleum is a mixture of hydrocarbons from which gasoline, kerosene, fuel oil, lubricating oil, paraffin wax, and petrolatum (themselves mixtures of hydrocarbons) are separated. Coal tar, a volatile by-product of the steel industry's process of making coke from coal, is the source of many valuable chemicals, including the aromatic hydrocarbons benzene, toluene, and naphthalene.

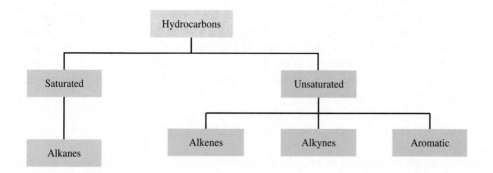

Figure 19.3
General classification of hydrocarbons.

19.3 ALKANES

Define an alkane, write structural formulas for alkanes, and name alkanes.

The **alkanes**, also known as *paraffins* or *saturated hydrocarbons,* are straight- or branched-chain hydrocarbons with only single covalent bonds between the carbon atoms. We will study the alkanes in some detail because many other classes of organic compounds are derivatives of these substances. Be sure to learn the names of the first 10 members of the alkane series, because they are the basis for naming other classes of compounds.

Methane, CH_4, is the first member of the alkane series. Members having two, three, and four carbon atoms are ethane, propane, and butane, respectively. The first four alkanes have common names and must be memorized, but the next six names are derived from Greek numbers. The names and formulas of the first 10 alkanes are given in Table 19.1.

Successive compounds in the alkane series differ from each other in composition by one carbon and two hydrogen atoms (CH_2). When each member of a series differs from the next member by a CH_2 group, the series is called a **homologous series**. The members of a homologous series are similar in structure but differ in formula. All common classes of organic compounds exist in homologous series, which can be represented by a general formula. For open-chain alkanes the general formula is C_nH_{2n+2}, where n corresponds to the number of carbon atoms in the molecule. The formulas of specific alkanes are easily determined from this general formula. Thus for pentane, $n = 5$ and $2n + 2 = 12$, so its formula is C_5H_{12}. For hexadecane, a 16-carbon alkane, the formula is $C_{16}H_{34}$.

LEARNING OBJECTIVE

KEY TERMS
alkane
homologous series
isomerism
isomers
alkyl group

Learn the names of the alkanes in Table 19.1; they form the stem of many other names.

Oil contains alkanes as well as other hydrocarbons.

TABLE 19.1   Names, Formulas, and Physical Properties of Straight-Chain Alkanes

Name	Molecular formula C_nH_{2n+2}	Condensed structural formula	Boiling point (°C)	Melting point (°C)
Methane	CH_4	CH_4	−161	−183
Ethane	C_2H_6	CH_3CH_3	−88	−172
Propane	C_3H_8	$CH_3CH_2CH_3$	−45	−187
Butane	C_4H_{10}	$CH_3CH_2CH_2CH_3$	−0.5	−138
Pentane	C_5H_{12}	$CH_3CH_2CH_2CH_2CH_3$	36	−130
Hexane	C_6H_{14}	$CH_3CH_2CH_2CH_2CH_2CH_3$	69	−95
Heptane	C_7H_{16}	$CH_3CH_2CH_2CH_2CH_2CH_2CH_3$	98	−90
Octane	C_8H_{18}	$CH_3CH_2CH_2CH_2CH_2CH_2CH_2CH_3$	125	−57
Nonane	C_9H_{20}	$CH_3CH_2CH_2CH_2CH_2CH_2CH_2CH_2CH_3$	151	−54
Decane	$C_{10}H_{22}$	$CH_3CH_2CH_2CH_2CH_2CH_2CH_2CH_2CH_2CH_3$	174	−30

One single reaction of alkanes has inspired people to explore equatorial jungles, to endure the heat and sandstorms of the deserts of Africa and the Middle East, to mush across the frozen Arctic, and to drill holes in Earth more than 30,000 feet deep! The substance is oil, and the reaction is combustion with oxygen to produce heat energy. Combustion reactions overshadow all other reactions of alkanes in economic importance. For example, note the heat generated when methane reacts with oxygen:

$$CH_4(g) + 2\,O_2(g) \longrightarrow CO_2(g) + 2\,H_2O(g) + 802.5\ kJ\ (191.8\ kcal)$$

Thermal energy is converted to mechanical and electrical energy all over the world. But combustion reactions are not usually of great interest to organic chemists, because carbon dioxide and water are the only chemical products of complete combustion. Aside from their combustibility, alkanes are limited in reactivity.

Michael Dalton/Fundamental Photographs

We burn natural gas (a hydrocarbon) in our daily lives. Combustion is the most common reaction for hydrocarbons.

Structural Formulas and Isomerism

The properties of an organic substance are dependent on its molecular structure. By structure, we mean the way in which the atoms bond within the molecule. The majority of organic compounds are made from relatively few elements—carbon, hydrogen, oxygen, nitrogen, and the halogens. In these compounds, carbon has four bonds to each atom, nitrogen three bonds, oxygen two bonds, and hydrogen and the halogens one bond to each atom:

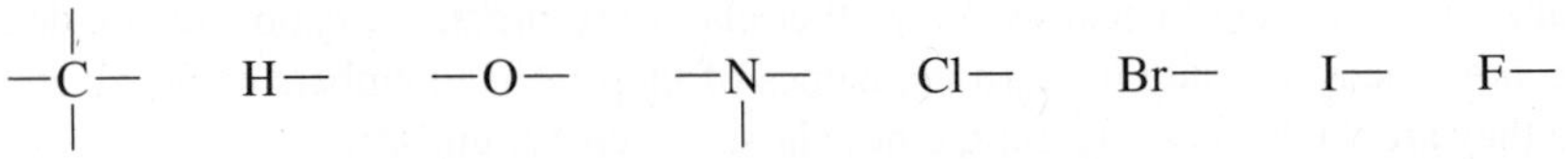

$$-\overset{|}{\underset{|}{C}}-\qquad H-\qquad -O-\qquad -\overset{}{\underset{|}{N}}-\qquad Cl-\qquad Br-\qquad I-\qquad F-$$

Alkane molecules contain only carbon–carbon and carbon–hydrogen bonds. Each carbon atom is joined to four other atoms by four single covalent bonds. These bonds are separated by angles of 109.5° (corresponding to those angles formed by lines drawn from the center of a regular tetrahedron to its corners). Alkane molecules are essentially nonpolar. Because of this low polarity, these molecules have very little intermolecular attraction and therefore relatively low boiling points compared with other organic compounds of similar molar mass.

The three-dimensional character of atoms and molecules is difficult to portray without models or computer-generated drawings. Methane and ethane are shown here in Lewis structure and line structure form:

Lewis diagrams can be drawn for alkanes, but these molecules are usually represented by replacing bonding electron pairs with single lines.

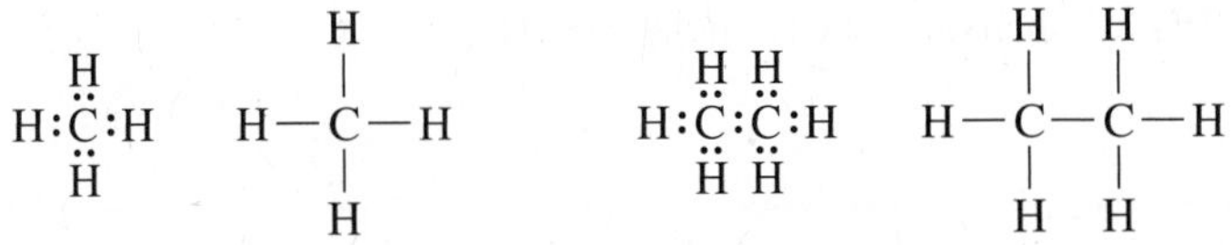

methane   ethane

To write the correct structural formula for propane (C_3H_8), the next member of the alkane series, we need to place each atom in the molecule. An alkane contains only single bonds, so each carbon atom must be bonded to four other atoms by either C—C or C—H bonds. Hydrogen must be bonded to only one carbon atom by a C—H bond, since C—H—C bonds do not occur, and an H—H bond would simply represent a hydrogen molecule. Thus, the only possible structure for propane is

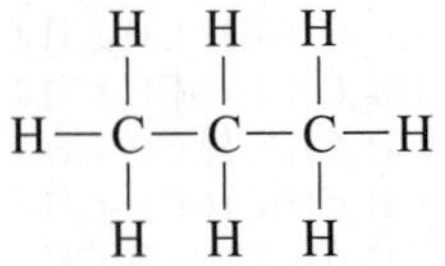

propane

However, it's possible to write two structural formulas corresponding to the molecular formula C_4H_{10} (butane). Two C_4H_{10} compounds with these structural formulas actually exist:

$$
\begin{array}{cccc}
& \text{H} & \text{H} & \text{H} & \text{H} \\
& | & | & | & | \\
\text{H} - & \text{C} - & \text{C} - & \text{C} - & \text{C} - \text{H} \\
& | & | & | & | \\
& \text{H} & \text{H} & \text{H} & \text{H}
\end{array}
\qquad \text{and}
$$

normal butane   2-methylpropane

The butane with the unbranched carbon chain is called *normal butane* (abbreviated *n*-butane); it boils at 0.5°C and melts at −138.3°C. The branched-chain butane is called 2-methylpropane; it boils at −11.7°C and melts at −159.5°C. These differences in physical properties are sufficient to establish that the two compounds, though they have the same molecular formula, are different substances. The structural arrangements of the atoms in methane, ethane, propane, butane, and 2-methylpropane are shown in **Figure 19.4**.

This phenomenon of two or more compounds having the same molecular formula, but different structural arrangements, is called **isomerism**. The individual compounds are called **isomers**. Butane, C_4H_{10}, shown above, has two isomers. Isomerism is common among organic compounds and is another reason for the large number of known compounds. There are 3 isomers of pentane, 5 isomers of hexane, 9 isomers of heptane, 18 isomers of octane, 35 isomers of nonane, and 75 isomers of decane. The phenomenon of isomerism is a compelling reason for using structural formulas.

Isomers are compounds with the same molecular formula but different structural formulas.

To save time and space, condensed structural formulas, in which the atoms and groups attached to a carbon atom are written to the right of that carbon atom, are often used. For example, the condensed structural formula for pentane is $CH_3CH_2CH_2CH_2CH_3$ or $CH_3(CH_2)_3CH_3$. Some condensed structural formulas are shown in Figure 19.4.

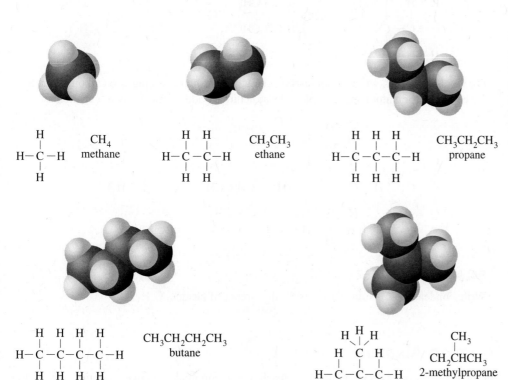

Figure 19.4

Spacefilling models illustrating structural formulas of methane, ethane, propane, butane, and 2-methylpropane. Condensed structural formulas are shown above the names, line structures to the left of the names.

Let's interpret the condensed structural formula for propane:

$$\overset{1}{C}H_3\overset{2}{C}H_2\overset{3}{C}H_3$$

Carbon-1 has three hydrogen atoms attached to it and is bonded to C-2, which has two hydrogen atoms on it and is bonded to C-3. Carbon-3 has three hydrogen atoms bonded to it.

EXAMPLE 19.2

Pentane (C_5H_{12}) has three isomers. Write their structural formulas and their condensed structural formulas.

SOLUTION

In a problem of this kind, it's best to first write the carbon skeleton with the longest continuous carbon chain—in this case, five carbon atoms. We complete the structure by attaching hydrogen atoms around each carbon atom so that each carbon atom has four bonds. The carbon atoms at the ends of the chain need three hydrogen atoms. The three inner carbon atoms each need two hydrogen atoms to give them four bonds:

$$C-C-C-C-C \qquad H-\underset{\underset{H}{|}}{\overset{\overset{H}{|}}{C}}-\underset{\underset{H}{|}}{\overset{\overset{H}{|}}{C}}-\underset{\underset{H}{|}}{\overset{\overset{H}{|}}{C}}-\underset{\underset{H}{|}}{\overset{\overset{H}{|}}{C}}-\underset{\underset{H}{|}}{\overset{\overset{H}{|}}{C}}-H \qquad CH_3CH_2CH_2CH_2CH_3$$

For the next isomer, we write a four-carbon chain and attach the fifth carbon atom to either of the middle carbon atoms (don't use the end ones):

$$C-C-\overset{\overset{C}{|}}{C}-C \qquad C-\overset{\overset{C}{|}}{C}-C-C \qquad \text{These structures represent the same compound.}$$

Now add the 12 hydrogen atoms to complete the structure:

$$H-\underset{\underset{H}{|}}{\overset{\overset{H}{|}}{C}}-\underset{\underset{H}{|}}{\overset{\overset{H}{|}}{C}}-\overset{\overset{\overset{H}{\underset{|}{C}}\overset{H}{\diagdown}}{\diagup}}{\underset{\underset{H}{|}}{C}}-\underset{\underset{H}{|}}{\overset{\overset{H}{|}}{C}}-H \qquad CH_3CH_2CHCH_3 \quad \text{or} \quad CH_3CH_2CH(CH_3)_2$$

For the third isomer, write a three-carbon chain, attach the remaining two carbon atoms to the central carbon atom, and complete the structure by adding the 12 hydrogen atoms:

$$C-\overset{\overset{C}{|}}{\underset{\underset{C}{|}}{C}}-C \qquad H-\overset{\overset{\overset{H\diagdown\overset{H}{|}\diagup H}{C}}{\underset{|}{}}}{\underset{\underset{H\diagup\underset{\underset{H}{|}}{C}\diagdown H}{}}{C}}-\underset{\underset{H}{|}}{\overset{\overset{H}{|}}{C}}-\overset{\overset{H}{|}}{\underset{\underset{H}{|}}{C}}-H \qquad CH_3\overset{\overset{CH_3}{|}}{\underset{\underset{CH_3}{|}}{C}}CH_3 \quad \text{or} \quad C(CH_3)_4$$

PRACTICE 19.2

Write condensed structural formulas for the isomer of hexane, C_6H_{14}.

Naming Alkanes

In the early years of organic chemistry, each new compound was given a name, usually by the person who had isolated or synthesized it. Names were not systematic but did carry some

TABLE 19.2   Names and Formulas of Selected Alkyl Groups

Formula	Name	Formula	Name
CH_3-	methyl	CH_3CH- (with CH_3 branch)	isopropyl
CH_3CH_2-	ethyl		
$CH_3CH_2CH_2-$	propyl	CH_3CHCH_2- (with CH_3 branch)	isobutyl
$CH_3CH_2CH_2CH_2-$	butyl	CH_3CH_2CH- (with CH_3 branch)	sec-butyl (secondary butyl)
$CH_3(CH_2)_3CH_2-$	pentyl		
$CH_3(CH_2)_4CH_2-$	hexyl	CH_3C- (with two CH_3 branches)	tert-butyl (t-butyl) (tertiary butyl)
$CH_3(CH_2)_5CH_2-$	heptyl		
$CH_3(CH_2)_6CH_2-$	octyl		
$CH_3(CH_2)_7CH_2-$	nonyl		
$CH_3(CH_2)_8CH_2-$	decyl		

information—often about the origin of the substance. Wood alcohol (methanol), for example, was so named because it was obtained by destructive distillation or pyrolysis of wood.

It soon became apparent that a naming system was needed, and, in 1892, such a system was proposed and adopted. In its present form, the International Union of Pure and Applied Chemistry (IUPAC) system is generally unambiguous and internationally accepted. However, a great many well-established common names and abbreviations (such as TNT and DDT) have continued to be used because of their brevity and/or convenience. So you need a knowledge of both the IUPAC system and the common names.

To name organic compounds using the IUPAC system, you must recognize certain common alkyl groups. **Alkyl groups** have the general formula C_nH_{2n+1} (one less hydrogen atom than the corresponding alkane). The name of the group is formed from the name of the corresponding alkane by simply dropping -ane and substituting a -yl ending. The names and formulas of selected alkyl groups are given in Table 19.2. The letter "R" is often used in formulas to represent any of the possible alkyl groups:

$$R = C_nH_{2n+1} \text{ (any alkyl group)}$$

The following IUPAC rules are all that are needed to name a great many alkanes. In later sections these rules will be extended to cover other classes of compounds, but advanced texts or references must be consulted for the complete system.

RULES FOR NAMING ALKANES

1. Select the longest continuous chain of carbon atoms as the parent compound, and consider all alkyl groups attached to it as branch chains or substituents that have replaced hydrogen atoms of the parent hydrocarbon. If two chains of equal length are found, use the chain that has the larger number of substituents attached to it as the parent compound. The alkane's name consists of the parent compound's name prefixed by the names of the alkyl groups attached to it.

2. Number the carbon atoms in the parent carbon chain starting from the end closest to the first carbon atom that has an alkyl or other group substituted for a hydrogen atom. If the first substituent from each end is on the same-numbered carbon, go to the next substituent to determine which end of the chain to start numbering.

3. Name each alkyl group and designate its position on the parent carbon chain by a number (e.g., 2-methyl means a methyl group attached to C-2).

4. When the same alkyl-group branch chain occurs more than once, indicate this repetition by a prefix (*di-*, *tri-*, *tetra-*, and so forth) written in front of the alkyl-group name (e.g., *dimethyl* indicates two methyl groups). The numbers indicating the alkyl-group positions are separated by a comma and followed by a hyphen and are placed in front of the name (e.g., 2,3-dimethyl).

5. When several different alkyl groups are attached to the parent compound, list them in alphabetical order (e.g., ethyl before methyl in 3-ethyl-4-methyloctane). Prefixes are not included in alphabetical ordering (ethyl comes before dimethyl).

Let's use the IUPAC system to name this compound:

$$\overset{4}{CH_3}-\overset{3}{CH_2}-\overset{2}{CH}-\overset{1}{CH_3} \quad \text{or} \quad \overset{1}{CH_3}-\overset{2}{CH}-\overset{3}{CH_2}-\overset{4}{CH_3}$$
$$\underset{CH_3}{|} \qquad\qquad\qquad \underset{CH_3}{|}$$

2-methylbutane

The longest continuous chain contains four carbon atoms. Therefore, we use the parent compound name, butane. The methyl group, CH_3—, attached to C-2 is named as a prefix to butane, the "2-" indicating the point of its attachment on the butane chain.

How would we write the structural formula for 2-methylpentane? Its name tells us how. The parent compound, pentane, contains five carbons. We write and number the five-carbon skeleton of pentane, put a methyl group on C-2 (because of the "2-methyl" in the name), and add hydrogens to give each carbon four bonds:

$$\overset{5}{C}-\overset{4}{C}-\overset{3}{C}-\overset{2}{C}-\overset{1}{C} \quad \overset{5}{C}-\overset{4}{C}-\overset{3}{C}-\overset{2}{C}-\overset{1}{C} \quad CH_3-CH_2-CH_2-CH-CH_3$$
$$\underset{CH_3}{|} \qquad\qquad\qquad \underset{CH_3}{|}$$

2-methylpentane

Could this compound be called 4-methylpentane? No; in the IUPAC system, the parent carbon chain is numbered starting from the end *nearest* the branch chain.

It is very important to understand that the *sequence* of carbon atoms and groups—not the way the sequence is written—determines the name of a compound. These formulas all represent 2-methylpentane; note that carbon numbering does not have to follow a straight line:

$$\overset{1}{CH_3}-\overset{2}{CH}-\overset{3}{CH_2}-\overset{4}{CH_2}-\overset{5}{CH_3} \qquad\qquad \overset{5}{CH_3}-\overset{4}{CH_2}-\overset{3}{CH_2}-\overset{2}{CH}-\overset{1}{CH_3}$$
$$\underset{CH_3}{|} \qquad\qquad\qquad\qquad\qquad \overset{CH_3}{|}$$

$$\overset{1}{CH_3}-\overset{2}{CH}-\overset{3}{CH_2} \qquad\qquad CH_3$$
$$\overset{CH_3}{|} \qquad\qquad\qquad\qquad \overset{2}{CH}$$
$$\underset{4}{|} \qquad\qquad\qquad\qquad\qquad CH_3 \quad CH_2$$
$$\underset{CH_2-CH_3}{4 \quad 5} \qquad\qquad CH_3 \quad CH_2 \quad CH_3$$

The following formulas and names demonstrate other aspects of the IUPAC nomenclature system:

$$\overset{4}{CH_3}-\overset{3}{CH}-\overset{2}{CH}-\overset{1}{CH_3} \qquad\qquad \overset{4}{CH_3}-\overset{3}{CH_2}-\overset{2}{C}-\overset{1}{CH_3}$$
$$\underset{CH_3}{|}\ \underset{CH_3}{|} \qquad\qquad\qquad \underset{CH_3}{|}$$

2,3-dimethylbutane   2,2-dimethylbutane

In 2,3-dimethylbutane, the longest carbon chain is four, indicating a butane; "dimethyl" indicates two methyl groups; "2,3-" means that one CH_3 is on C-2 and one is on C-3.

In 2,2-dimethylbutane, both methyl groups are on the same carbon atom; both numbers are required.

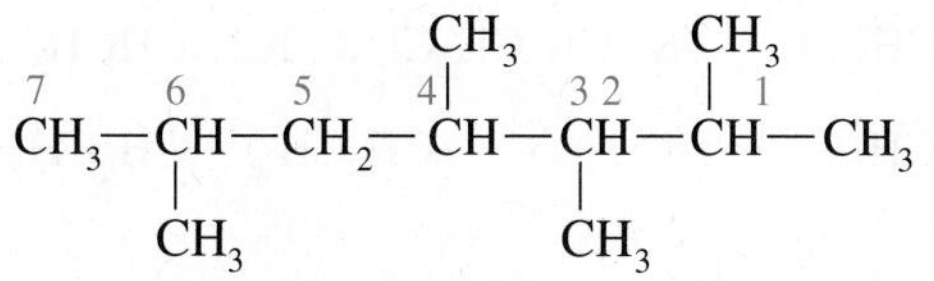

$$\underset{7}{CH_3}-\underset{6}{CH}-\underset{5}{CH_2}-\underset{4}{\overset{CH_3}{CH}}-\underset{3\,2}{CH}-\underset{1}{\overset{CH_3}{CH}}-CH_3$$

2,3,4,6-tetramethylheptane (not 2,4,5,6-)

Note that this molecule is numbered from right to left.

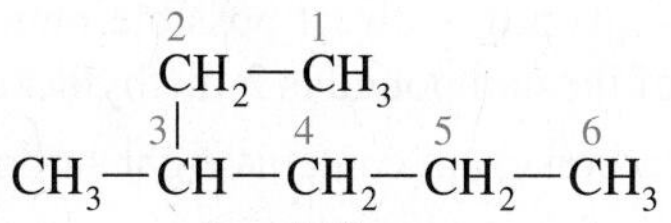

$$\overset{2}{CH_2}-\overset{1}{CH_3}$$
$$\underset{3}{CH_3}-\underset{}{\overset{}{CH}}-\underset{4}{CH_2}-\underset{5}{CH_2}-\underset{6}{CH_3}$$

3-methylhexane

The longest continuous chain in 3-methylhexane has six carbons.

In the next structure, the longest carbon chain is eight. The groups attached or substituted for hydrogen on the octane chain are named in alphabetical order.

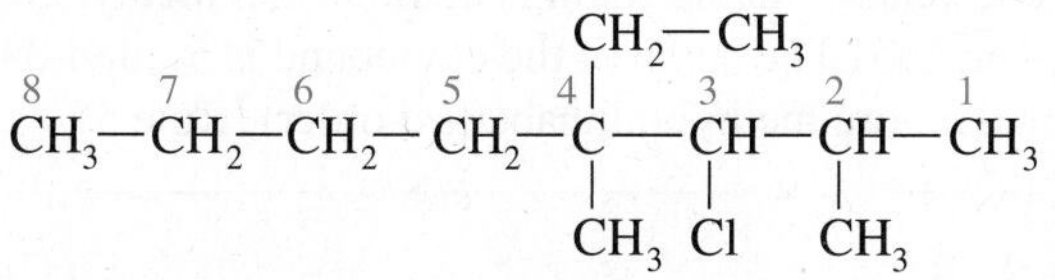

$$\overset{4}{\overset{CH_2-CH_3}{|}}$$
$$\underset{8}{CH_3}-\underset{7}{CH_2}-\underset{6}{CH_2}-\underset{5}{CH_2}-\underset{4}{C}-\underset{3}{CH}-\underset{2}{CH}-\underset{1}{CH_3}$$
$$\underset{CH_3\ \ Cl\ \ \ CH_3}{}$$

3-chloro-4-ethyl-2,4-dimethyloctane

EXAMPLE 19.3

Write the formulas for (a) 3-ethylpentane and (b) 2,2,4-trimethylpentane.

SOLUTION

(a) The name *pentane* indicates a five-carbon chain. Write five connecting carbon atoms and number them. Attach an ethyl group, CH_3CH_2—, to C-3. Now add hydrogen atoms to give each carbon atom four bonds: C-1 and C-5 each need three hydrogen atoms; C-2 and C-4 each need two hydrogen atoms; and C-3 needs one hydrogen atom:

$$\overset{1}{C}-\overset{2}{C}-\overset{3}{C}-\overset{4}{C}-\overset{5}{C}\quad \overset{1}{C}-\overset{2}{C}-\overset{3}{C}-\overset{4}{C}-\overset{5}{C}\quad CH_3CH_2CHCH_2CH_3$$
$$\underset{CH_2CH_3}{}\qquad\qquad \underset{CH_2CH_3}{}$$

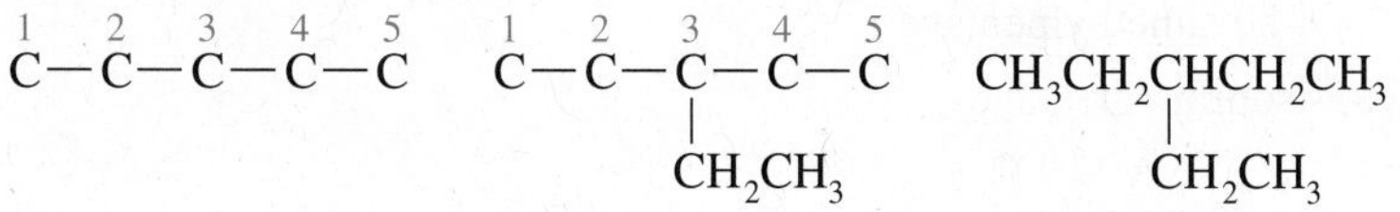

3-ethylpentane

(b) Pentane indicates a five-carbon chain. Write five connecting carbon atoms and number them. There are three methyl groups, CH_3—, in the compound (trimethyl), two attached to C-2 and one attached to C-4. Attach these three methyl groups to their respective carbon atoms. Now add hydrogen atoms to give each carbon atom four bonds. Thus C-1 and C-5 each need three hydrogen atoms; C-2 does not need any hydrogen atoms; C-3 needs two hydrogen atoms; and C-4 needs one hydrogen atom. The formula is complete:

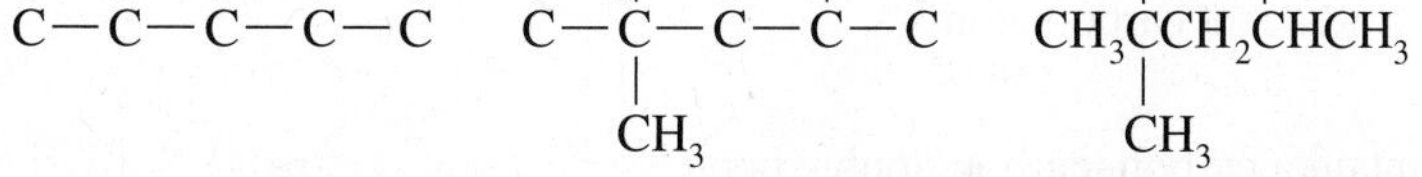

$$\overset{1}{C}-\overset{2}{C}-\overset{3}{C}-\overset{4}{C}-\overset{5}{C}\quad \overset{1}{C}-\overset{2}{C}-\overset{3}{C}-\overset{4}{C}-\overset{5}{C}\quad CH_3CCH_2CHCH_3$$
$$\underset{CH_3}{}\qquad\qquad \underset{CH_3}{}$$

2,2,4-trimethylpentane

WileyPLUS

ENHANCED EXAMPLE

EXAMPLE 19.4

Name these compounds:

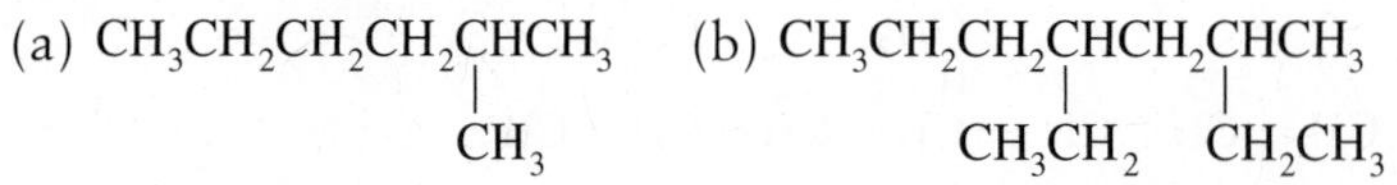

(a) $CH_3CH_2CH_2CH_2CHCH_3$ (b) $CH_3CH_2CH_2CHCH_2CHCH_3$
                  $|$ $|$ $|$
                CH_3 CH_3CH_2 CH_2CH_3

SOLUTION

(a) The longest continuous carbon chain contains six carbon atoms (Rule 1). Thus, the parent name is hexane. Number the carbon chain from right to left so that the methyl group attached to C-2 is given the lowest possible number (Rule 2). With a methyl group on C-2, the name of the compound is 2-methylhexane (Rule 3).

(b) The longest continuous carbon chain contains eight carbon atoms:

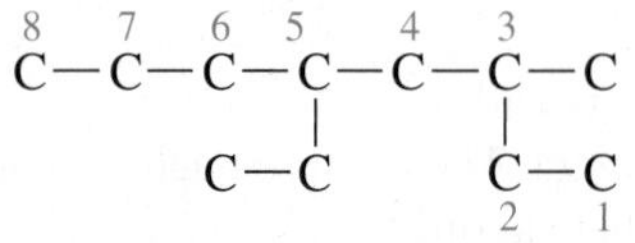

The parent name is octane. As the chain is numbered, a methyl group is on C-3 and an ethyl group is on C-5. The name of the compound is 5-ethyl-3-methyloctane. Note that ethyl is named before methyl (alphabetical order) (Rule 5).

PRACTICE 19.3

Name these alkanes:

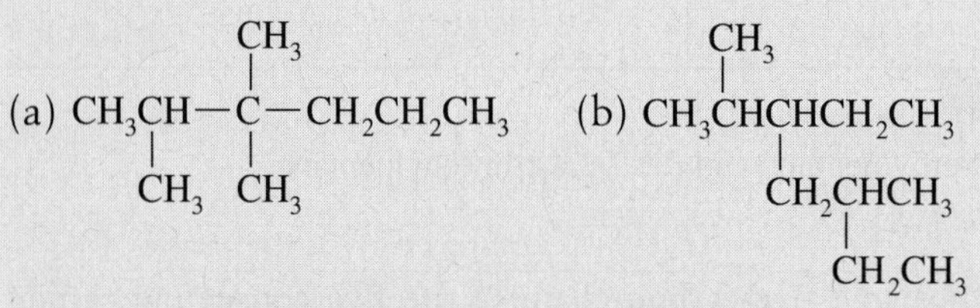

                     CH_3 CH_3
                     $|$ $|$
(a) $CH_3CH-C-CH_2CH_2CH_3$ (b) $CH_3CHCHCH_2CH_3$
          $|$ $|$ $|$
         CH_3 CH_3 CH_2CHCH_3
                                                  $|$
                                              CH_2CH_3

PRACTICE 19.4

Write the structure for the following:

(a) 3,4-dimethylheptane

(b) 2,2,3,3,4,4-hexamethylpentane

(c) 2-ethyl-3-isopropylheptane

19.4 ALKENES AND ALKYNES

LEARNING OBJECTIVE ● Define alkenes and alkynes, write names for alkenes and alkynes, and discuss reactions for alkenes and alkynes.

KEY TERMS

alkene
alkyne
addition reaction
hydrogenation

Alkenes and alkynes are classified as unsaturated hydrocarbons. They are said to be unsaturated because, unlike alkanes, their molecules do not contain the maximum possible number of hydrogen atoms. **Alkenes** have two fewer hydrogen atoms, and **alkynes** have four fewer hydrogen atoms than alkanes with a comparable number of carbon atoms. Alkenes contain at least one double bond between adjacent carbon atoms, while alkynes contain at least one triple bond between adjacent carbon atoms.

Remember: In a homologous series, the formulas of successive members differ by increments of CH_2.

Alkenes contain a carbon–carbon double bond.
Alkynes contain a carbon–carbon triple bond.

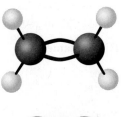

ethylene (common name)
ethene (IUPAC)

acetylene (common name)
ethyne (IUPAC)

Figure 19.5
Ball-and-stick and spacefilling models for ethylene and acetylene.

The simplest alkene is ethylene (or ethene), $CH_2\!=\!CH_2$, and the simplest alkyne is acetylene (or ethyne), $CH\!\equiv\!CH$ (**Figure 19.5**). Ethylene and acetylene are the first members of a homologous series (e.g., $CH_2\!=\!CH_2$, $CH_3CH\!=\!CH_2$, and $CH_3CH_2CH\!=\!CH_2$). Huge quantities of alkenes are made by cracking and dehydrogenating alkanes during the processing of crude oil. These alkenes are used to manufacture motor fuels, polymers, and petrochemicals. Alkene molecules, like those of alkanes, have very little polarity. Hence, the physical properties of alkenes are similar to those of the corresponding saturated hydrocarbons.

General formula for alkenes: C_nH_{2n}
General formula for alkynes: C_nH_{2n-2}

Naming Alkenes and Alkynes

The names of alkenes and alkynes are derived from the corresponding alkanes.

RULES FOR NAMING ALKENES AND ALKYNES (IUPAC)

1. Select the longest continuous carbon–carbon chain that contains a double or triple bond.
2. Name this parent compound as you would an alkane but change the *-ane* ending to *-ene* for an alkene or to *-yne* for an alkyne; thus, propane changes to propene or propyne:

$$CH_3CH_2CH_3 \qquad CH_3CH\!=\!CH_2 \qquad CH_3C\!\equiv\!CH$$
propane   propene   propyne

3. Number the carbon chain of the parent compound starting with the end nearer to the double or triple bond. Use the smaller of the two numbers on the double- or triple-bonded carbon atoms to indicate the position of the double or triple bond. Place this number in front of the alkene or alkyne name; 2-butene means that the carbon–carbon double bond is between C-2 and C-3.
4. Branched chains and other groups are treated as in naming alkanes. Name the substituent group, and designate its position on the parent chain with a number.

Roberto A Sanchez/iStockphoto

Ethylene gas causes bananas to ripen.

Craig Aurness/Corbis Images

Acetylene is used for welding steel girders.

TABLE 19.3   Names and Formulas for Several Alkenes and Alkynes

Formula	IUPAC name
$CH_2{=}CH_2$	ethene
$CH_3CH{=}CH_2$	propene
$CH_3CH_2CH{=}CH_2$	1-butene
$CH_3CH{=}CHCH_3$	2-butene
$CH_3C{=}CH_2$ 　　$\vert$ 　　CH_3	2-methylpropene
$CH{\equiv}CH$	ethyne
$CH_3C{\equiv}CH$	propyne
$CH_3CH_2C{\equiv}CH$	1-butyne
$CH_3C{\equiv}CCH_3$	2-butyne

Table 19.3 gives the names and formulas for several alkenes and alkynes. Study the following examples of named alkenes and alkynes:

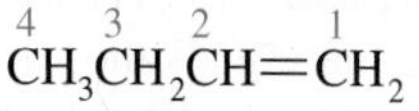

$$\overset{4}{C}H_3\overset{3}{C}H_2\overset{2}{C}H{=}\overset{1}{C}H_2$$
1-butene

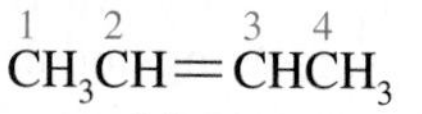

$$\overset{1}{C}H_3\overset{2}{C}H{=}\overset{3}{C}H\overset{4}{C}H_3$$
2-butene

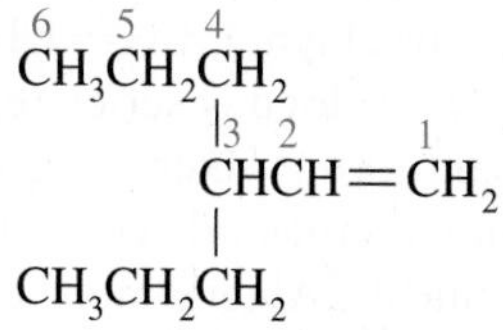

$$\overset{6}{C}H_3\overset{5}{C}H_2\overset{4}{C}H_2$$
$$\overset{\vert 3}{C}H\overset{2}{C}H{=}\overset{1}{C}H_2$$
$$\vert$$
$$CH_3CH_2CH_2$$
3-propyl-1-hexene

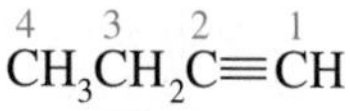

$$\overset{4}{C}H_3\overset{3}{C}H_2\overset{2}{C}{\equiv}\overset{1}{C}H$$
1-butyne

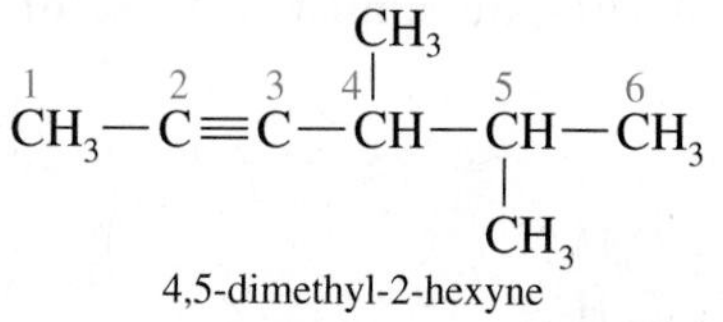

$$\overset{1}{C}H_3{-}\overset{2}{C}{\equiv}\overset{3}{C}{-}\overset{4}{C}H\overset{CH_3}{\vert}{-}\overset{5}{C}H{-}\overset{6}{C}H_3$$
$$\vert$$
$$CH_3$$
4,5-dimethyl-2-hexyne

To write a structural formula from a systematic name, the naming process is reversed. For example, how would we write the structural formula for 4-methyl-2-pentene? The name indicates (1) five carbons in the longest chain, (2) a double bond between C-2 and C-3, and (3) a methyl group on C-4.

Write five carbon atoms in a row. Place a double bond between C-2 and C-3, and place a methyl group on C-4. Now add hydrogen atoms to give each carbon atom four bonds. Carbons 1 and 5 each need three hydrogen atoms; C-2, C-3, and C-4 each need one hydrogen atom.

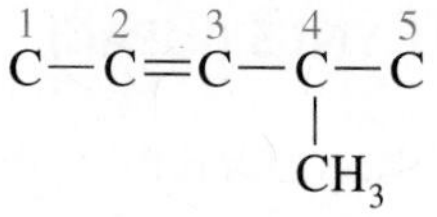

$$\overset{1}{C}{-}\overset{2}{C}{=}\overset{3}{C}{-}\overset{4}{C}{-}\overset{5}{C}$$
$$\vert$$
$$CH_3$$
carbon skeleton

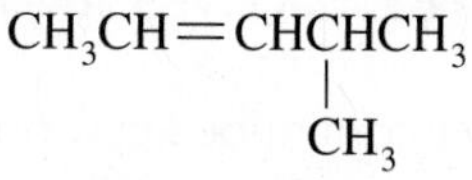

$$CH_3CH{=}CHCHCH_3$$
$$\vert$$
$$CH_3$$
4-methyl-2-pentene

EXAMPLE 19.5

Write structural formulas for (a) 7-methyl-2-octene and (b) 3-hexyne.

SOLUTION

(a) Octene, like octane, indicates an eight-carbon chain. The chain contains a double bond between C-2 and C-3 and a methyl group on C-7. Write eight carbon atoms in a row, place a double bond between C-2 and C-3, and place a methyl group on C-7. Now add hydrogen atoms to give each carbon atom four bonds:

$$\overset{1}{C}{-}\overset{2}{C}{=}\overset{3}{C}{-}\overset{4}{C}{-}\overset{5}{C}{-}\overset{6}{C}{-}\overset{7}{C}{-}\overset{8}{C}$$
$$\vert$$
$$CH_3$$
carbon skeleton

$$CH_3CH{=}CHCH_2CH_2CH_2CHCH_3$$
$$\vert$$
$$CH_3$$
7-methyl-2-octene

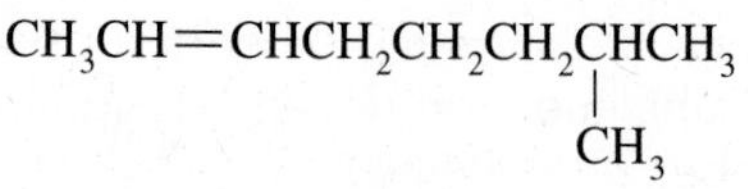

Remember: A double bond counts as two bonds.

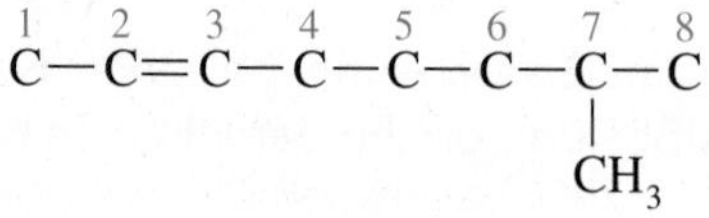

(b) The stem *hex-* indicates a six-carbon chain; the suffix *-yne* indicates a carbon–carbon triple bond; the number 3 locates the triple bond between C-3 and C-4. Write six carbon atoms in a row and place a triple bond between C-3 and C-4. Now add hydrogen atoms to give each carbon atom four bonds; C-3 and C-4 don't need hydrogen atoms:

$$\overset{1}{C}-\overset{2}{C}-\overset{3}{C}\equiv\overset{4}{C}-\overset{5}{C}-\overset{6}{C} \qquad CH_3CH_2C\equiv CCH_2CH_3$$

3-hexyne

PRACTICE 19.5

Name these compounds

(a) CH$_3$CHCH=CHCHCH$_3$
         | |
        CH$_3$ CH$_2$CH$_3$

(b) CH$_3$C≡CCHCH$_2$CH$_2$CH$_2$CH$_3$
                      |
                   CH$_2$CH$_3$

Reactions of Alkenes

The alkenes are much more reactive than the corresponding alkanes. This greater reactivity is due to the carbon–carbon double bonds. In our brief overview, we will limit our discussion to the most common reaction of alkenes.

Addition

In organic chemistry, a reaction in which two substances join to produce one compound is called an **addition reaction**. Addition at the carbon–carbon double bond is the most common reaction of alkenes. Hydrogen, halogens (Cl_2 or Br_2), hydrogen halides, and water are some of the reagents that can be added to unsaturated hydrocarbons. Ethylene, for example, reacts in this fashion:

$$CH_2{=}CH_2 + H_2 \xrightarrow[\text{1 atm}]{\text{Pt \quad 25°C}} CH_3{-}CH_3$$

ethylene   ethane

$$CH_2{=}CH_2 + Br{-}Br \longrightarrow BrCH_2{-}CH_2Br$$

1,2-dibromoethane

Visible evidence of the Br_2 addition is the disappearance of the reddish-brown color of bromine as it reacts.

$$CH_2{=}CH_2 + HCl \longrightarrow CH_3CH_2Cl$$

chloroethane

$$CH_2{=}CH_2 + HOH \xrightarrow{\text{H}^+} CH_3CH_2OH$$

ethanol
(ethyl alcohol)

Note that the double bond is broken and the unsaturated alkene molecules become saturated by an addition reaction. Reactions of this kind can occur with almost any molecule that contains a carbon–carbon double bond.

When hydrogen adds across a double bond, the process is called **hydrogenation**. Many of our processed foods contain partially hydrogenated vegetable oils, which give them longer shelf lives. In these oils some of the double bonds have been converted to single bonds by adding hydrogen. Unfortunately, these partially hydrogenated oils have become known for their health risk. They are frequently called *trans* fats and have been implicated in increasing people's risk of heart disease. Physicians now recommend that we should limit our intake of *trans* fats to improve our health.

$\xrightarrow[\text{1 atm}]{\text{Pt 25°C}}$ indicates the catalyst, Pt, and other necessary conditions for the reaction.

The H$^+$ indicates that the reaction is carried out under acidic conditions.

Nutrition labels now tell us how much *trans* fat is contained in the product.

19.5 AROMATIC HYDROCARBONS

LEARNING OBJECTIVE ● Describe the structure of aromatic hydrocarbons and name them.

KEY TERM
aromatic compound

Benzene and all substances with structures and chemical properties resembling benzene are classified as **aromatic compounds**. The word *aromatic* originally referred to the rather pleasant odor many of these substances possess, but this meaning has been dropped. Benzene, the parent substance of the aromatic hydrocarbons, was first isolated by Michael Faraday in 1825. Its correct molecular formula, C_6H_6, was established a few years later, but finding a reasonable structural formula that would account for the properties of benzene was difficult.

Finally, in 1865, August Kekulé proposed that the carbon atoms in a benzene molecule are arranged in a six-membered ring with one hydrogen atom bonded to each carbon atom and with three carbon–carbon double bonds:

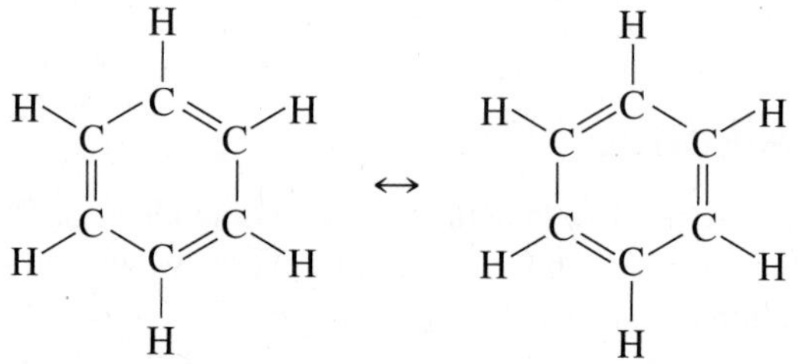

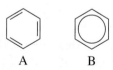

benzene

A turning point in the history of chemistry, Kekulé's structures mark the beginning of our understanding of structure in aromatic compounds. His formulas do have one serious shortcoming, though: They represent benzene and related substances as highly unsaturated compounds. Yet benzene does not readily undergo addition reactions like a typical alkene. For example, benzene does not decolorize bromine solutions rapidly. Instead, the chemical behavior of benzene resembles that of an alkane. Its typical reactions are the substitution type, where a hydrogen atom is replaced by some other group:

$$C_6H_6 + Cl_2 \xrightarrow{Fe} C_6H_5Cl + HCl$$

Modern theory suggests that the benzene molecule is a hybrid of the two Kekulé structures shown earlier.

For convenience, chemists usually write the structure of benzene as one or the other of these abbreviated forms:

benzene

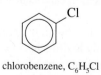

In both representations, it is understood that there is a carbon atom and a hydrogen atom at each corner of the hexagon. The classical Kekulé structure is represented by A; the modern molecular orbital structure is represented by B. These hexagonal structures are also used to represent the structural formulas of benzene derivatives—that is, substances in which one or more hydrogen atoms in the ring have been replaced by other atoms or groups. Chlorobenzene (C_6H_5Cl), for example, is written in this fashion:

chlorobenzene, C_6H_5Cl

This notation indicates that the chlorine atom has replaced a hydrogen atom and is bonded directly to a carbon atom in the ring. Thus, the correct formula for chlorobenzene is C_6H_5Cl, not C_6H_6Cl.

Naming Aromatic Compounds

A substituted benzene is derived by replacing one or more of benzene's hydrogen atoms with another atom or group of atoms. Thus, a monosubstituted benzene has the formula C_6H_5G, where G is the group replacing a hydrogen atom. Because all the hydrogen atoms in benzene are equivalent, it doesn't matter which hydrogen is replaced by the monosubstituted group.

Monosubstituted Benzenes

Some monosubstituted benzenes are named by adding the name of the substituent group as a prefix to the word *benzene*. The name is written as one word. Here are some examples:

nitrobenzene   ethylbenzene   chlorobenzene   bromobenzene

—NO_2 is a nitro group.

Certain monosubstituted benzenes have special names that should be learned. These are parent names for further substituted compounds:

toluene   phenol   benzoic acid   aniline
(methylbenzene) (hydroxybenzene)   (aminobenzene)

—NH_2 is an amino group.

The C_6H_5— group is known as the phenyl group (pronounced *fen-ill*). The name *phenyl* is used for compounds that cannot be easily named as benzene derivatives. For example, the following compounds are named as derivatives of alkanes:

3-chloro-2-phenylpentane   diphenylmethane

Disubstituted Benzenes

When two substituent groups replace two hydrogen atoms in a benzene molecule, three isomers are possible. The prefixes *ortho-*, *meta-*, and *para-* (abbreviated *o-*, *m-*, and *p-*) are used to name these disubstituted benzenes. In the *ortho*-compound the substituents are located on adjacent carbon atoms, in the *meta*-compound they are one carbon apart, and in the *para*-compound they are on opposite sides of the ring. When the two groups are different, name them alphabetically followed by the word *benzene*.

The dichlorobenzenes ($C_6H_4Cl_2$) illustrate this method of naming. The three isomers have different physical properties, indicating that they are different substances. Note that the *para*-isomer is a solid and the other two are liquids at room temperature.

ortho-dichlorobenzene   *meta*-dichlorobenzene   *para*-dichlorobenzene
(1,2-dichlorobenzene)   (1,3-dichlorobenzene)   (1,4-dichlorobenzene)
mp 17.2°C, bp 180.4°C   mp 24.8°C, bp 172°C   mp 53.1°C, bp 174.4°C

Reuters/Corbis Images

TNT was used to demolish the Kingdome in Seattle.

When one substituent corresponds to a monosubstituted benzene with a special name, the monosubstituted compound becomes the parent name for the disubstituted compound. In the following examples, the parent compounds are phenol, aniline, toluene, and benzoic acid:

o-nitrophenol   *p*-bromoaniline   *m*-nitrotoluene   *p*-aminobenzoic acid

Tri- and Polysubstituted Benzenes

When a benzene ring has more than two substituents, the carbon atoms in the ring are numbered. Numbering starts at one of the substituted groups and goes either clockwise or counterclockwise, but it must be done in the direction that gives the lowest possible numbers to the substituent groups. When the compound is named as a derivative of the special parent compound, the substituent of the parent compound is considered to be on C-1 of the ring (the CH_3 group is on C-1 in 2,4,6-trinitrotoluene). The following examples illustrate this system:

1,3,5-trinitrobenzene   1,2,4-tribromobenzene
(not 1,4,6-)

2,4,6-trinitrotoluene   5-bromo-2-chlorophenol
(TNT)

EXAMPLE 19.6

Write formulas and names for all possible isomers of (a) chloronitrobenzene, $C_6H_4Cl(NO_2)$, and (b) tribromobenzene, $C_6H_3Br_3$.

SOLUTION

(a) The name and formula indicate a chloro-group, Cl, and a nitro-group, NO_2, attached to a benzene ring. There are six positions in which to place these two groups. They can be *ortho-*, *meta-*, or *para-* to each other.

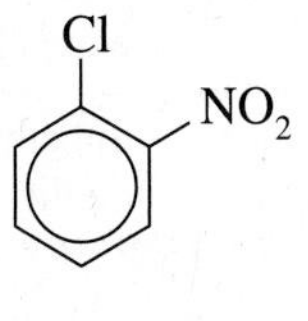

o-chloronitrobenzene 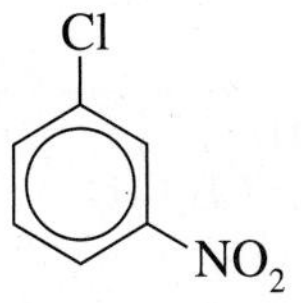   *m*-chloronitrobenzene 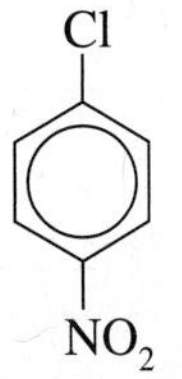   *p*-chloronitrobenzene

(b) For tribromobenzene, start by placing the three bromo-groups in the 1, 2, and 3 positions, then the 1, 2, and 4 positions, and so on, until all the possible isomers are formed. Name each isomer to check that no duplicate formulas have been written:

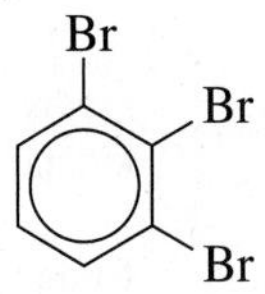

1,2,3-tribromobenzene 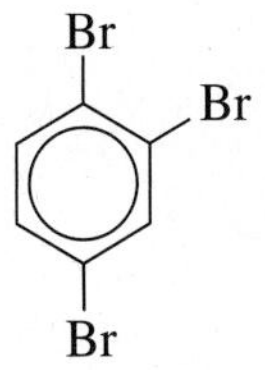   1,2,4-tribromobenzene 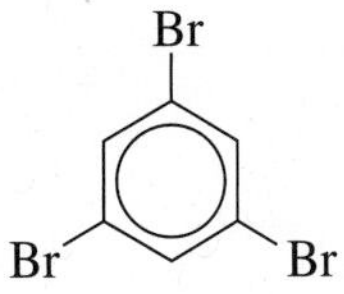   1,3,5-tribromobenzene

Tribromobenzene has only three isomers. However, if one Br is replaced by a Cl, six isomers are possible.

PRACTICE 19.6

Name these compounds:

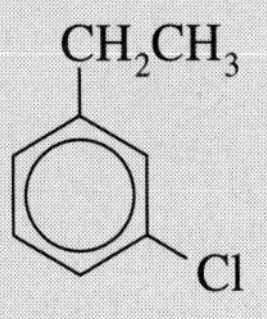

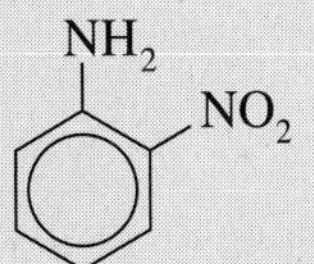

19.6 HYDROCARBON DERIVATIVES

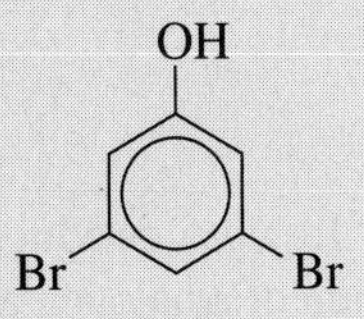

Draw the structure of each of the hydrocarbon derivatives shown in Table 19.4.   ● **LEARNING OBJECTIVE**

Hydrocarbon derivatives are compounds that can be synthesized from a hydrocarbon. These derivatives contain not only carbon and hydrogen but also such additional elements as oxygen, nitrogen, or a halogen. The compounds in each class have similarities in structure and properties. We will consider the classes of hydrocarbon derivatives shown in Table 19.4, which is divided into two sections of different color. The compounds in the first section don't contain a C=O group in their molecules, while those in the second section all contain a C=O group.

TABLE 19.4   Classes of Hydrocarbon Derivatives

Class	General formula	Structure of functional group	Sample structural formula	Name IUPAC	Name Common
Alkyl halide	R*X	—X X = F, Cl, Br, I	CH_3Cl CH_3CH_2Cl	chloromethane chloroethane	methyl chloride ethyl chloride
Alcohol	ROH	—OH	CH_3OH CH_3CH_2OH	methanol ethanol	methyl alcohol ethyl alcohol
Ether	R—O—R	R—O—R	CH_3—O—CH_3 CH_3CH_2—O—CH_2CH_3	methoxymethane ethoxyethane	dimethyl ether diethyl ether
Aldehyde	$R-\overset{\overset{H}{\mid}}{C}=O$	$-\overset{\overset{H}{\mid}}{C}=O$	$H-\overset{\overset{H}{\mid}}{C}=O$ $CH_3-\underset{\underset{H}{\mid}}{C}=O$	methanal ethanal	formaldehyde acetaldehyde
Ketone	$R-\underset{\underset{O}{\parallel}}{C}-R$	$R-\underset{\underset{O}{\parallel}}{C}-R$	$CH_3-\underset{\underset{O}{\parallel}}{C}-CH_3$ $CH_3-\underset{\underset{O}{\parallel}}{C}-CH_2CH_3$	propanone 2-butanone	acetone methyl ethyl ketone
Carboxylic acid	$R-C\overset{\overset{\textstyle O}{\diagup}}{\diagdown}_{OH}$	$-C\overset{\overset{\textstyle O}{\diagup}}{\diagdown}_{OH}$	HCOOH CH_3COOH	methanoic acid ethanoic acid	formic acid acetic acid
Ester	$R-C\overset{\overset{\textstyle O}{\diagup}}{\diagdown}_{OR}$	$-C\overset{\overset{\textstyle O}{\diagup}}{\diagdown}_{OR}$	$HCOOCH_3$ CH_3COOCH_3	methyl methanoate methyl ethanoate	methyl formate methyl acetate

*The letter "R" is used to indicate any of the many possible alkyl groups.

Alkyl Halides

Alkanes react with a halogen in ultraviolet light to produce an alkyl halide, R—X. *Halogenation* is a general term for the substitution of a halogen atom for a hydrogen atom.

$$RH + X_2 \xrightarrow[\text{light}]{\text{UV}} RX + HX \quad (X = Cl \text{ or } Br)$$

When a specific halogen is used, the name reflects this. For example, when chlorine is the halogen, the process is called *chlorination*.

$$CH_3CH_3 + Cl_2 \xrightarrow[\text{light}]{\text{UV}} \underset{\text{chloroethane}}{CH_3CH_2Cl} + HCl$$

The reaction yields alkyl halides, RX, which are useful as intermediates for the manufacture of other substances.

A well-known reaction of methane and chlorine is shown by the equation

$$CH_4 + Cl_2 \xrightarrow[\text{light}]{\text{UV}} \underset{\text{chloromethane}}{CH_3Cl} + HCl$$

According to IUPAC rules, organic halides are named by giving the halogen substituent as a prefix of the parent alkane (e.g., CH_3Cl is chloromethane).

EXAMPLE 19.7

Write two names for each of the following compounds.

(a) CH_3CH_2Cl

(b) $CH_3CH_2CH_2Cl$

(c) CH_3CHCl
　　　　|
　　　　CH_3

(d) $CH_3CHClCH_2Cl$

(e) CBr_4

SOLUTION

(a) ethyl chloride, chloroethane

(b) propyl chloride, 1-chloropropane

(c) isopropyl chloride, 2-chloropropane

(d) 2,3-dichloropropane

(e) carbon tetrabromide, tetrabromomethane

PRACTICE 19.7

Write condensed structural formulas for all the isomers of dichloropropane.

Alkyl halides are primarily used as industrial solvents. Carbon tetrachloride, CCl_4, was once used extensively in the dry-cleaning process, but it has been replaced because of its toxicity and carcinogenic effects. (See the Chemistry in Action on p. 472.)

Organic halides are also used as anesthetics. Chloroform, $CHCl_3$, was once a popular anesthetic. It's no longer used, however, because it is harmful to the respiratory system. Instead, such compounds as halothane ($CF_3CHClBr$) are being used.

Composed of carbon, hydrogen, fluorine, and chlorine, chlorinated fluorocarbons (CFCs) are alkyl halides used in aerosol propellants and refrigerants. CFCs have been replaced by other compounds because they generate chlorine atoms in the upper atmosphere, which deplete the ozone layer. Many aerosols are now propelled by carbon dioxide. Different refrigerant molecules (called hydrofluorocarbons, HFCs) are being used in refrigerators and air conditioners.

19.7 ALCOHOLS

Draw structures for alcohols and name alcohols.

Alcohols are organic compounds whose molecules contain a hydroxyl ($-OH$) group covalently bonded to a saturated carbon atom. Thus if we substitute an $-OH$ for an H in CH_4, we get CH_3OH, methyl alcohol. The functional group of the alcohols is $-OH$. The general formula for alcohols is ROH, where R is an alkyl or a substituted alkyl group.

Alcohols differ from metal hydroxides in that they do not dissociate or ionize in water. The $-OH$ group is attached to the carbon atom by a covalent bond, not by an ionic bond. The alcohols form a homologous series with methanol, CH_3OH, as the first member of the series. Models of the structural arrangements of the atoms in methanol and ethanol are shown in **Figure 19.6**.

● **LEARNING OBJECTIVE**

KEY TERMS

alcohol
primary alcohol
secondary alcohol
tertiary alcohol
polyhydroxy alcohol

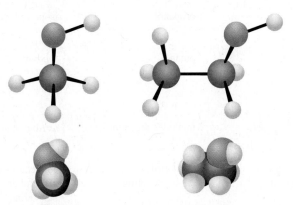

Figure 19.6

Models illustrating structural formulas of methanol and ethanol.

CH_3OH
methanol

CH_3CH_2OH
ethanol

Alcohols are classified as **primary** (1°), **secondary** (2°), or **tertiary** (3°), depending on whether the carbon atom to which the —OH group is attached is bonded to one, two, or three other carbon atoms, respectively. Generalized formulas for 1°, 2°, and 3° alcohols are

$$
\begin{array}{ccc}
\text{H} & \text{R} & \text{R} \\
| & | & | \\
\text{R}-\text{C}-\text{OH} & \text{R}-\text{C}-\text{OH} & \text{R}-\text{C}-\text{OH} \\
| & | & | \\
\text{H} & \text{H} & \text{R}
\end{array}
$$

primary alcohol   secondary alcohol   tertiary alcohol

Examples of these alcohols are listed in Table 19.5. Methanol, CH_3OH, is grouped with the primary alcohols.

Molecular structures with more than one —OH group bonded to a single carbon atom are generally not stable. But an alcohol molecule can contain two or more —OH groups if each —OH is bonded to a different carbon atom. Accordingly, alcohols are also classified as monohydroxy, dihydroxy, trihydroxy, and so on, on the basis of the number of hydroxyl groups per molecule. **Polyhydroxy alcohol** is a general term for an alcohol that has more than one —OH group per molecule.

Methanol

Methanol is a volatile (bp 65°C), highly flammable liquid. It is poisonous and capable of causing blindness or death if taken internally. Exposure to methanol vapors for even short periods of time is dangerous. Despite the hazards, over 8 billion pounds of methanol (3.6×10^9 kg) are manufactured annually and used

- in the conversion of methanol to formaldehyde, which is primarily used to make polymers
- in the manufacture of other chemicals, especially various kinds of esters

TABLE 19.5   Names and Classifications of Alcohols

Class	Formula	IUPAC name	Common name*	Boiling point (°C)
Primary	CH_3OH	methanol	Methyl alcohol	65.0
Primary	CH_3CH_2OH	ethanol	Ethyl alcohol	78.5
Primary	$CH_3CH_2CH_2OH$	1-propanol	*n*-propyl alcohol	97.4
Primary	$CH_3CH_2CH_2CH_2OH$	1-butanol	*n*-butyl alcohol	118
Primary	$CH_3(CH_2)_3CH_2OH$	1-pentanol	*n*-pentyl alcohol	138
Primary	$CH_3(CH_2)_6CH_2OH$	1-octanol	*n*-octyl alcohol	195

Primary	CH_3CHCH_2OH
	     $\mid$
	    CH_3
		2-methyl-1-propanol	isobutyl alcohol	108

Secondary	CH_3CHCH_3
	     $\mid$
	    OH
		2-propanol	isopropyl alcohol	82.5

Secondary	$CH_3CH_2CHCH_3$
	        $\mid$
	       OH
		2-butanol	*sec*-butyl alcohol	91.5

Tertiary	$\quad\quad CH_3$ $\quad\quad\mid$ CH_3-C-OH $\quad\quad\mid$ $\quad\quad CH_3$	2-methyl-2-propanol	*tert*-butyl alcohol	82.9
Dihydroxy	$HOCH_2CH_2OH$	1,2-ethanediol	ethylene glycol	197

Trihydroxy	$HOCH_2CHCH_2OH$
	        $\mid$
	       OH
		1,2,3-propanetriol	glycerol or glycerine	290

*The abbreviations *n*, *sec*, and *tert* stand for normal, secondary, and tertiary, respectively.

- to denature ethyl alcohol (rendering it unfit as a beverage)
- as an industrial solvent
- as an inexpensive and temporary antifreeze for radiators (not a satisfactory permanent antifreeze because its boiling point is lower than that of water)

Ethanol

Ethanol is without doubt the most widely known alcohol. Huge quantities of this substance are prepared by fermentation, using starch and sugar as the raw materials. Conversion of simple sugars to ethanol is accomplished by the yeast enzyme zymase:

$$\underset{\text{glucose}}{C_6H_{12}O_6} \xrightarrow{\text{zymase}} 2\ \underset{\text{ethanol}}{CH_3CH_2OH} + 2\ CO_2$$

Ethanol is economically significant as

- an intermediate in the manufacture of other chemicals such as acetaldehyde, acetic acid, ethyl acetate, and diethyl ether
- a solvent for many organic substances (e.g., pharmaceuticals, perfumes, flavorings)
- an ingredient in alcoholic beverages
- an alternative fuel for automobiles

Ethanol is now an alternative fuel used in flexible-fuel vehicles.

Ethanol acts physiologically as a food, a drug, and a poison. It's a food in the limited sense that the body metabolizes small amounts of it to carbon dioxide and water with the production of energy. In moderate quantities, ethanol causes drowsiness and depresses brain functions so that activities requiring skill and judgment (such as automobile driving) are impaired. In larger quantities, it causes nausea, vomiting, impaired perception, and lack of coordination. If a very large amount is consumed, unconsciousness and ultimately death may occur.

For industrial use, ethanol is often denatured (rendered unfit for drinking) by adding small amounts of methanol and other denaturants that are extremely difficult to remove. Denaturing is required by the federal government to protect the beverage alcohol tax source. Special tax-free use permits are issued to scientific and industrial users who require pure ethanol for nonbeverage uses.

Other widely used alcohols include the following:

- Isopropyl alcohol (2-propanol) is the principal ingredient in rubbing alcohol formulations.
- Ethylene glycol (ethanediol) is the main compound in antifreezes; it is also used in the manufacture of synthetic fibers and in the paint industry.
- Glycerol (1,2,3-propanetriol), also known as glycerine, is a syrupy, sweet-tasting liquid used in the manufacture of polymers and explosives, as an emollient in cosmetics, as a humectant in tobacco, and as a sweetener.

Naming Alcohols

If you know how to name alkanes, it's easy to name alcohols by the IUPAC system. Unfortunately, several alcohols are also known by common names, so it's often necessary to know both names for a given alcohol. The common name is usually formed from the name of the alkyl group that is attached to the —OH group, followed by the word *alcohol*. (See Table 19.5.)

RULES FOR NAMING ALCOHOLS (IUPAC)

1. Select the longest continuous chain of carbon atoms containing the hydroxyl group.
2. Number the carbon atoms in this chain so that the one bonded to the —OH group has the lowest possible number.

Rubbing alcohol is a 70% solution of isopropyl alcohol and is often used for disinfecting the skin.

3. Form the parent alcohol name by replacing the final -e of the corresponding alkane name by -ol. When isomers are possible, locate the position of the —OH by placing the number (hyphenated) of the carbon atom to which the —OH is bonded immediately before the parent alcohol name.

4. Name each alkyl branch chain (or other group) and designate its position by number.

Study the following examples of this naming system and also those shown in Table 19.5.

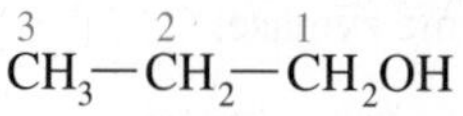

$$\overset{3}{C}H_3—\overset{2}{C}H_2—\overset{1}{C}H_2OH$$

1-propanol

$$\overset{1}{C}H_3—\overset{2}{C}H—\overset{3}{C}H_3$$
$$\underset{OH}{|}$$

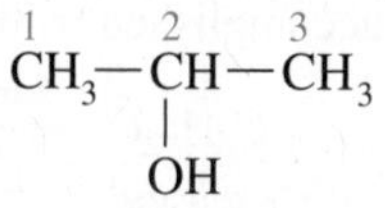

2-propanol
(isopropyl alcohol)

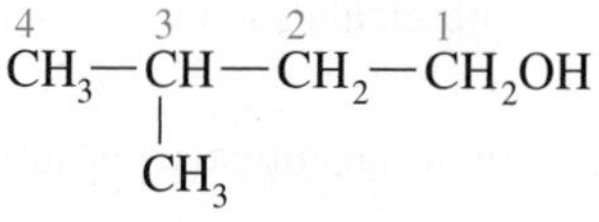

$$\overset{4}{C}H_3—\overset{3}{C}H—\overset{2}{C}H_2—\overset{1}{C}H_2OH$$
$$\underset{CH_3}{|}$$

3-methyl-1-butanol

$$HO\overset{2}{C}H_2—\overset{1}{C}H_2OH$$

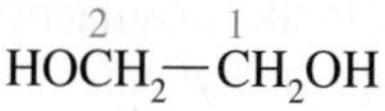

1,2-ethanediol
(ethylene glycol)

EXAMPLE 19.8

Name the following alcohol by the IUPAC system:

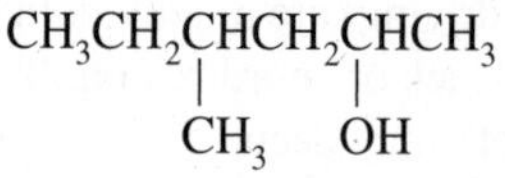

$$CH_3CH_2CHCH_2CHCH_3$$
$$\underset{CH_3}{|} \quad \underset{OH}{|}$$

SOLUTION

1. The longest continuous carbon chain containing the —OH group has six carbon atoms.

2. This carbon chain is numbered from right to left so that the carbon bonded to the —OH group has the smallest possible number. In this case the —OH is on C-2:

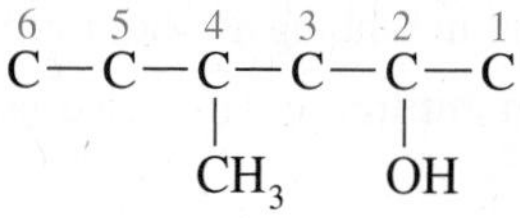

$$\overset{6}{C}—\overset{5}{C}—\overset{4}{C}—\overset{3}{C}—\overset{2}{C}—\overset{1}{C}$$
$$\underset{CH_3}{|} \quad \underset{OH}{|}$$

3. The name of the six-carbon alkane is hexane. Replace the final -e in hexane by -ol, forming the name hexanol. Since the —OH is on C-2, place a 2- before hexanol to give it the parent alcohol name of 2-hexanol.

4. A methyl group, —CH₃, is located on C-4. Therefore, the full name of the compound is 4-methyl-2-hexanol.

EXAMPLE 19.9

Write the structural formula of 3,3-dimethyl-2-hexanol.

SOLUTION

The 2-hexanol refers to a six-carbon chain with an —OH group on C-2. Write the skeleton structure with an —OH on C-2. Now place the two methyl groups (3,3-dimethyl) on C-3:

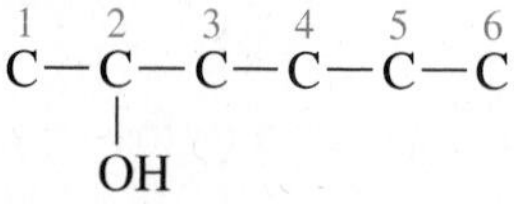

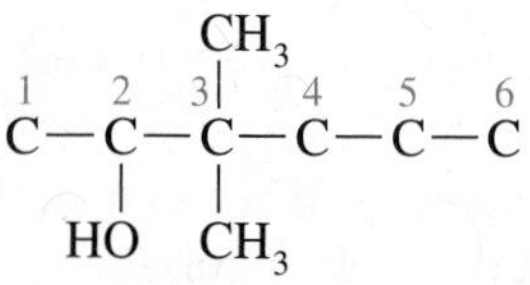

$$\overset{1}{C}—\overset{2}{C}—\overset{3}{C}—\overset{4}{C}—\overset{5}{C}—\overset{6}{C}$$
$$\underset{OH}{|}$$

$$\overset{1}{C}—\overset{2}{C}—\overset{3}{C}—\overset{4}{C}—\overset{5}{C}—\overset{6}{C}$$
$$\underset{HO}{|} \quad \overset{CH_3}{\overset{|}{\underset{CH_3}{|}}}$$

Finally, add hydrogen atoms to give each carbon atom four bonds:

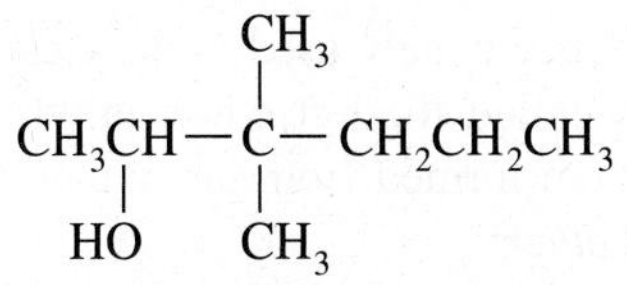

$$CH_3CH - \overset{\overset{\displaystyle CH_3}{|}}{\underset{\underset{\displaystyle CH_3}{|}}{C}} - CH_2CH_2CH_3$$
$$\underset{}{HO}$$

3,3-dimethyl-2-hexanol

PRACTICE 19.8

Name these alcohols and classify each as primary, secondary, or tertiary:

(a) 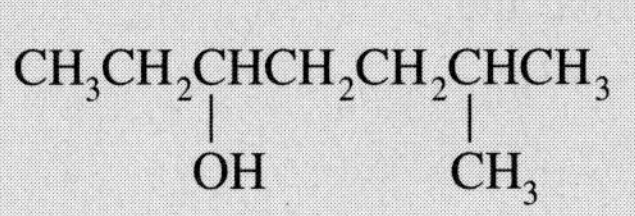 $CH_3CH_2\underset{\underset{\displaystyle OH}{|}}{CH}CH_2CH_2\underset{\underset{\displaystyle CH_3}{|}}{CH}CH_3$ (b) 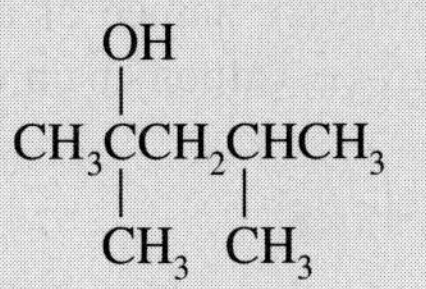 $CH_3\overset{\overset{\displaystyle OH}{|}}{C}CH_2\underset{\underset{\displaystyle CH_3}{|}}{CH}CH_3$ with $\underset{\underset{\displaystyle CH_3}{|}}{}$

19.8 ETHERS

Draw structures for ethers and name ethers.

KEY TERM
ether

Ethers have the general formula ROR′. The two groups, R and R′, may be derived from saturated, unsaturated, or aromatic hydrocarbons and, for a given ether, may be alike or different. Table 19.6 lists structural formulas and names for different ethers.

Saturated ethers have little chemical reactivity, but because they readily dissolve a great many organic substances, ethers are often used as solvents in both laboratory and manufacturing operations.

Alcohols (ROH) and ethers (ROR′) are isomeric, having the same molecular formula but different structural formulas. For example, the molecular formula for ethanol and dimethyl ether is C_2H_6O, but the structural formulas are

$$CH_3CH_2OH \qquad CH_3-O-CH_3$$
ethanol   dimethyl ether

These two molecules are extremely different in both physical and chemical properties. Ethanol boils at 78.3°C, and dimethyl ether boils at −23.7°C. Ethanol is capable of intermolecular hydrogen bonding and therefore has a much higher boiling point. In addition, it also has a greater solubility in water than dimethyl ether.

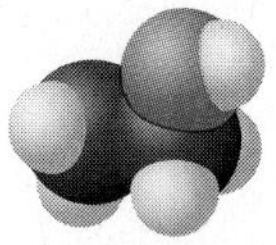

CH_3CH_2OH
ethanol

CH_3CH_3OH
dimethyl ether

TABLE 19.6   Names and Structural Formulas of Ethers

Name*	Formula	Boiling point (°C)
Dimethyl ether (methoxymethane)	CH_3-O-CH_3	−24
Methyl ethyl ether (methoxyethane)	$CH_3CH_2-O-CH_3$	8
Diethyl ether (ethoxyethane)	$CH_3CH_2-O-CH_2CH_3$	35
Ethyl isopropyl ether (2-ethoxypropane)	$CH_3CH_2-O-\underset{\underset{\displaystyle CH_3}{|}}{C}HCH_3$	54
Divinyl ether	$CH_2=CH-O-CH=CH_2$	39
Anisole (methoxybenzene)	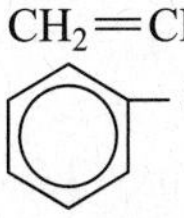 ⬡—OCH_3	154

*The IUPAC name is in parentheses.

Naming Ethers

Like alcohols, individual ethers can have several names. Once widely used as an anesthetic, the ether with the formula $CH_3CH_2-O-CH_2CH_3$ is called diethyl ether, ethyl ether, ethoxyethane, or simply ether. Common names of ethers are formed from the names of the groups attached to the oxygen atom followed by the word *ether*:

$$CH_3 \overset{\uparrow}{-\boxed{O}-} CH_3 \qquad CH_3 \overset{\uparrow}{-\boxed{O}-} CH_2CH_3$$

$$\underset{\text{methyl}}{} \underset{\text{ether}}{} \underset{\text{methyl}}{} \qquad \underset{\text{methyl}}{} \underset{\text{ether}}{} \underset{\text{ethyl}}{}$$

dimethyl ether   methyl ethyl ether

In the IUPAC system, ethers are named as alkoxy, $RO-$, derivatives of the alkane corresponding to the longest carbon–carbon chain in the molecule.

RULES FOR NAMING ETHERS

alkyl + oxy = alkoxyl

1. Select the longest carbon–carbon chain and label it with the name of the corresponding alkane.

2. Change the -*yl* ending of the other hydrocarbon group to -*oxy* to obtain the alkoxy group name. For example, CH_3O- is called *methoxy*.

3. Combine the two names from Steps 1 and 2, giving the alkoxy name first, to form the ether name:

$$CH_3 - O - CH_2CH_3$$

This is the longest carbon–carbon chain, so call it *ethane*.

This is the other group; modify its name to *methoxy* and combine with *ethane* to obtain the IUPAC name of the ether, *methoxyethane*.

Thus,

$CH_3CH_2-O-CH_2CH_3$ is ethoxyethane.

$CH_3CH_2CH_2-O-CH_2CH_2CH_2CH_3$ is propoxybutane.

EXAMPLE 19.10

Name these ethers.

(a) $CH_3-O-CH_2CH_2CH_3$ (2 names)

(b) $CH_3-O-\underset{\underset{CH_3}{|}}{CHCH_3}$ (2 names)

(c) 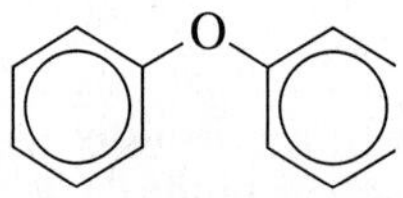   (1 name)

(d) $CH_3CH_2-O-CH_2CH_2CH_2CH_3$ (2 names)

SOLUTION

(a) 1-methoxypropane, methyl propyl ether

(b) 2-methoxypropane, methyl isopropyl ether

(c) diphenyl ether

(d) 1-ethoxybutane, ethyl butyl ether

PRACTICE 19.9

Write formulas for these ethers.

(a) diisopropyl ether   (c) 1-ethoxy-2,2-dimethylbutane

(b) ethyl phenyl ether

19.9 ALDEHYDES AND KETONES

Draw structures for aldehydes and ketones and name aldehydes and ketones.

● LEARNING OBJECTIVE

KEY TERMS

carbonyl group
aldehyde
ketone

The aldehydes and ketones are closely related classes of compounds. Their structures contain the **carbonyl group**, $\diagdown C{=}O$, a carbon double bonded to oxygen. **Aldehydes** have at least one hydrogen atom bonded to the carbonyl group, while **ketones** have two alkyl (R) or aromatic (Ar) groups bonded to the carbonyl group:

$$
\underset{\text{aldehydes}}{\overset{\displaystyle \quad O \qquad\qquad O}{R-\overset{\|}{C}-H \qquad Ar-\overset{\|}{C}-H}} \qquad\qquad \underset{\text{ketones}}{\overset{\displaystyle O \qquad\qquad O \qquad\qquad O}{R-\overset{\|}{C}-R \qquad R-\overset{\|}{C}-Ar \qquad Ar-\overset{\|}{C}-Ar}}
$$

In a linear expression, the aldehyde group is often written as CHO or CH=O. For example,

$$
CH_3CHO \quad \text{is equivalent to} \quad CH_3\overset{\displaystyle O}{\overset{\|}{C}}-H
$$

In the linear expression of a ketone, the carbonyl group is written as CO; thus,

$$
CH_3COCH_3 \quad \text{is equivalent to} \quad CH_3\overset{\displaystyle O}{\overset{\|}{C}}CH_3
$$

Formaldehyde is the most widely used aldehyde. It is a poisonous, irritating gas that is very soluble in water. Marketed as a 40% aqueous solution called *formalin*, the largest use of this chemical is in the manufacture of polymers. About 2.1 billion pounds (9.6×10^8 kg) of formaldehyde are manufactured annually in the United States. Formaldehyde vapors are intensely irritating to the mucous membranes, and ingestion may cause severe abdominal pains, leading to coma and death.

Acetone and methyl ethyl ketone are widely used organic solvents. Acetone, in particular, is used in very large quantities for this purpose. The U.S. production of acetone is about 1.9 billion pounds (8.7×10^8 kg) annually. It is used as a solvent in the manufacture of drugs, chemicals, and explosives; for removal of paints, varnishes, and fingernail polish; and as a solvent in the plastics industry. Methyl ethyl ketone (MEK) is also widely used as a solvent, especially for lacquers.

acetaldehyde

acetone

The solvent acetone is the main ingredient in nail polish remover for natural nails.

Naming Aldehydes

The IUPAC names of aldehydes are obtained by dropping the final *-e* and adding *-al* to the name of the parent hydrocarbon (i.e., the longest carbon–carbon chain carrying the —CHO group). The aldehyde carbon is always at the beginning of the carbon chain, is understood to be carbon number 1, and does not need to be numbered. The first member of the homologous series, $H_2C{=}O$, is methanal. The name *methanal* is derived from the hydrocarbon methane, which contains one carbon atom. The second member of the series is ethanal, the third member of the series is propanal, and so on:

$$
\underset{\substack{\text{methane}}}{CH_4} \qquad \underset{\substack{\text{methanal} \\ \text{(from methan}e + \text{-}al)}}{H-\overset{\displaystyle O}{\overset{\|}{C}}-H} \qquad\qquad \underset{\substack{\text{ethane}}}{CH_3CH_3} \qquad \underset{\substack{\text{ethanal} \\ \text{(from ethan}e + \text{-}al)}}{CH_3\overset{\displaystyle O}{\overset{\|}{C}}-H}
$$

The longest carbon chain containing the aldehyde group is the parent compound. Other groups attached to this chain are numbered and named as we have done previously. For example,

$$
\underset{\text{4-methylhexanal}}{CH_3CH_2\overset{\displaystyle \overset{CH_3}{|}}{C}HCH_2CH_2\overset{\displaystyle O}{\overset{\|}{C}}-H}
$$

Common names for some aldehydes are widely used. The common names for the aldehydes are derived from the common names of the carboxylic acids. The *-ic acid* or *-oic acid* ending of the acid name is dropped and is replaced with the suffix *-aldehyde*. Thus, the name of the one-carbon acid, formic acid, becomes formaldehyde for the one-carbon aldehyde:

$$\underset{\text{formic acid}}{H-\overset{\displaystyle O}{\overset{\|}{C}}-OH} \qquad \underset{\text{formaldehyde}}{H-\overset{\displaystyle O}{\overset{\|}{C}}-H} \qquad \underset{\text{acetic acid}}{CH_3\overset{\displaystyle O}{\overset{\|}{C}}-OH} \qquad \underset{\text{acetaldehyde}}{CH_3\overset{\displaystyle O}{\overset{\|}{C}}-H}$$

Naming Ketones

The IUPAC name of a ketone is derived from the name of the alkane corresponding to the longest carbon chain that contains the ketone–carbonyl group. The parent name is formed by changing the *-e* ending of the alkane to *-one*. If the chain is longer than four carbons, it's numbered so that the carbonyl carbon has the smallest number possible, and this number precedes the name of the ketone. Other groups bonded to the parent chain are named and numbered as previously indicated for hydrocarbons and alcohols. For example,

$$\underset{\text{propanone}}{CH_3-\overset{\displaystyle O}{\overset{\|}{C}}-CH_3} \qquad \underset{\text{2-pentanone}}{\overset{5\ \ 4\ \ 3\ \ 2\quad 1}{CH_3CH_2CH_2-\overset{\displaystyle O}{\overset{\|}{C}}-CH_3}} \qquad \underset{\substack{\displaystyle CH_3 \\ \text{4-methyl-3-hexanone}}}{\overset{1\quad 2\quad 3\ \ 4\ \ 5\ \ 6}{CH_3CH_2-\overset{\displaystyle O}{\overset{\|}{C}}-CHCH_2CH_3}}$$

Note that in 4-methyl-3-hexanone the carbon chain is numbered from left to right to give the ketone group the lowest possible number.

An alternative non-IUPAC method commonly used to name simple ketones is to list the names of the alkyl or aromatic groups attached to the carbonyl carbon together with the word *ketone*.

$$\underset{\substack{\uparrow\qquad\qquad\uparrow \\ \text{methyl}\qquad\text{methyl} \\ \text{(dimethyl ketone, acetone)}}}{CH_3-\overset{\displaystyle O}{\overset{\|}{C}}-CH_3} \qquad \underset{\substack{\uparrow\qquad\uparrow\qquad\uparrow \\ \text{methyl ketone}\quad\text{ethyl} \\ \text{(methyl ethyl ketone, MEK)}}}{CH_3-\overset{\displaystyle O}{\overset{\|}{C}}-CH_2CH_3} \qquad \underset{\substack{\uparrow\quad\uparrow\qquad\uparrow \\ \text{methyl ketone}\quad\text{isopropyl} \\ \text{(methyl isopropyl ketone)}}}{CH_3-\overset{\displaystyle O}{\overset{\|}{C}}-\overset{\displaystyle CH_3}{\overset{|}{C}HCH_3}}$$

Two of the most widely used ketones are commonly known by non-IUPAC names: Propanone is called acetone, and butanone is known as methyl ethyl ketone, or MEK.

WileyPLUS

ENHANCED EXAMPLE

EXAMPLE 19.11

Write the formulas and the names for the open-chain five- and six-carbon aldehydes.

SOLUTION

The IUPAC names are based on the five- and six-carbon alkanes. Drop the *-e* of the alkane name and add the suffix *-al*. Pentane, C_5, becomes pentanal and hexane, C_6, becomes hexanal.

$$\underset{\text{pentanal}}{CH_3CH_2CH_2CH_2\overset{\displaystyle O}{\overset{\|}{C}}-H} \qquad\qquad \underset{\text{hexanal}}{CH_3CH_2CH_2CH_2CH_2\overset{\displaystyle O}{\overset{\|}{C}}-H}$$

EXAMPLE 19.12

Give two names for these ketones:

(a) $\overset{\displaystyle O}{\overset{\displaystyle \|}{CH_3CH_2C}}CH_2\overset{\displaystyle CH_3}{\overset{\displaystyle |}{CH}}CH_3$

(b) $CH_3CH_2CH_2\overset{\displaystyle O}{\overset{\displaystyle \|}{C}}$—⬡

SOLUTION

(a) The parent carbon chain that contains the carbonyl group has six carbons. Number this chain from the end nearest to the carbonyl group. The ketone group is on C-3, and a methyl group is on C-5. The six-carbon alkane is hexane. Drop the -e from hexane and add -*one* to give the parent name hexanone. Prefix the name hexanone with 3- to locate the ketone group and with 5-methyl- to locate the methyl group. The name is 5-methyl-3-hexanone. The common name is ethyl isobutyl ketone since the C=O has an ethyl group and an isobutyl group bonded to it.

(b) The longest chain has four carbons. The parent ketone name is butanone, derived by dropping the -e of butane and adding -*one*. The butanone has a phenyl group attached to C-1. The IUPAC name is therefore 1-phenyl-1-butanone. The common name is phenyl *n*-propyl ketone, since the C=O group has a phenyl and an *n*-propyl group bonded to it.

PRACTICE 19.10

Write the IUPAC names for these molecules:

(a) $H—\overset{\displaystyle O}{\overset{\displaystyle \|}{C}}—\underset{\displaystyle \underset{\displaystyle CH_3}{|}}{CH}\underset{\displaystyle \underset{\displaystyle CH_2CH_3}{|}}{CH_2}CH_2$

(b) $CH_3CH_2CH_2\underset{\displaystyle \underset{\displaystyle CH_3}{|}}{CH}CH_2\underset{\displaystyle \underset{\displaystyle CH_3}{|}}{CH}\overset{\displaystyle O}{\overset{\displaystyle \|}{C}}CH_3$

19.10 CARBOXYLIC ACIDS

Draw structures for carboxylic acids and name carboxylic acids.

● LEARNING OBJECTIVE

KEY TERMS
carboxylic acid
carboxyl group

Organic acids, known as **carboxylic acids**, are characterized by the functional group called a **carboxyl group**. The carboxyl group is represented in the following ways:

$$\overset{\displaystyle O}{\overset{\displaystyle \|}{—C—OH}} \quad or \quad —COOH \quad or \quad —CO_2H$$

Open-chain carboxylic acids form a homologous series. The carboxyl group is always at the beginning of a carbon chain, and the carbon atom in this group is understood to be C-1 in naming the compound.

To name a carboxylic acid by the IUPAC system, first identify the longest carbon chain, including the carboxyl group. Then form the acid name by dropping the -e from the corresponding parent hydrocarbon name and adding -*oic acid*. Thus, the names corresponding to the C-1, C-2, and C-3 acids are methanoic acid, ethanoic acid, and propanoic acid. These names are, of course, derived from methane, ethane, and propane:

CH_4	methane	HCOOH	methanoic acid
CH_3CH_3	ethane	CH_3COOH	ethanoic acid
$CH_3CH_2CH_3$	propane	CH_3CH_2COOH	propanoic acid

The IUPAC system is neither the only nor the most generally used method of naming acids. Organic acids are usually known by common names. Methanoic, ethanoic, and propanoic acids are commonly called formic, acetic, and propionic acids, respectively. These

TABLE 19.7 Formulas and Names of Saturated Carboxylic Acids

Formula	IUPAC name	Common name
HCOOH	methanoic acid	formic acid
CH_3COOH	ethanoic acid	acetic acid
CH_3CH_2COOH	propanoic acid	propionic acid
$CH_3(CH_2)_2COOH$	butanoic acid	butyric acid
$CH_3(CH_2)_3COOH$	pentanoic acid	valeric acid
$CH_3(CH_2)_4COOH$	hexanoic acid	caproic acid
$CH_3(CH_2)_6COOH$	octanoic acid	caprylic acid
$CH_3(CH_2)_8COOH$	decanoic acid	capric acid
$CH_3(CH_2)_{10}COOH$	dodecanoic acid	lauric acid
$CH_3(CH_2)_{12}COOH$	tetradecanoic acid	myristic acid
$CH_3(CH_2)_{14}COOH$	hexadecanoic acid	palmitic acid
$CH_3(CH_2)_{16}COOH$	octadecanoic acid	stearic acid
$CH_3(CH_2)_{18}COOH$	eicosanoic acid	arachidic acid

Corbis/SuperStock

The bark of a willow tree is a natural source of salicylic acid.

names usually refer to a natural source of the acid and are not really systematic. Formic acid was named from the Latin word *formica*, meaning "ant." This acid contributes to the stinging sensation of ant bites. Acetic acid is found in vinegar and is so named from the Latin word for vinegar. The name of butyric acid is derived from the Latin term for butter, since it is a constituent of butterfat. Many of the carboxylic acids, principally those having even numbers of carbon atoms ranging from 4 to about 20, exist in combined form in plant and animal fats. These acids are called *saturated fatty acids*. Table 19.7 lists the IUPAC and common names of the more important saturated acids.

The simplest aromatic acid is benzoic acid. *Ortho*-hydroxybenzoic acid is known as salicylic acid, the basis for many salicylate drugs such as aspirin. There are three methylbenzoic acids, known as *o*-, *m*-, and *p*-toluic acids.

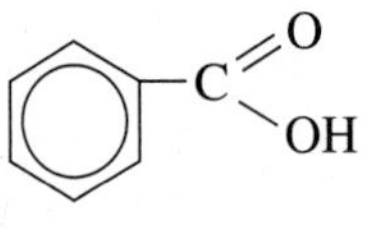

benzoic acid

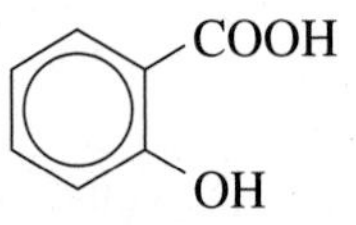

salicylic acid
(*o*-hydroxybenzoic acid)

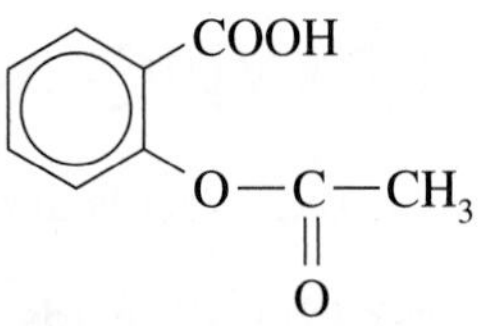

acetylsalicylic acid
(aspirin)

p-toluic acid

EXAMPLE 19.13

The acid hydrogen in carboxylic acids is the hydrogen bonded to the OH in the carboxyl group. It is an acidic hydrogen like any other acid. Complete the following neutralization equation.

$$CH_3CH_2\overset{\overset{\displaystyle O}{\|}}{C}-OH + NaOH \longrightarrow$$

SOLUTION

$$CH_3CH_2\overset{\overset{\displaystyle O}{\|}}{C}-OH + NaOH \longrightarrow CH_3CH_2\overset{\overset{\displaystyle O}{\|}}{C}-O^-Na^+ + H_2O$$

PRACTICE 19.11

Write the name and the formula for the 17- and 19-carbon carboxylic acids.

19.11 ESTERS

Draw structures for esters and name esters.

● LEARNING OBJECTIVE

KEY TERM
ester

Carboxylic acids react with alcohols in an acidic medium to form esters. **Esters** have the general formula RCOOR′ where R′ can be any type of saturated, unsaturated, or aromatic hydrocarbon group. The functional group of the ester is —COOR′:

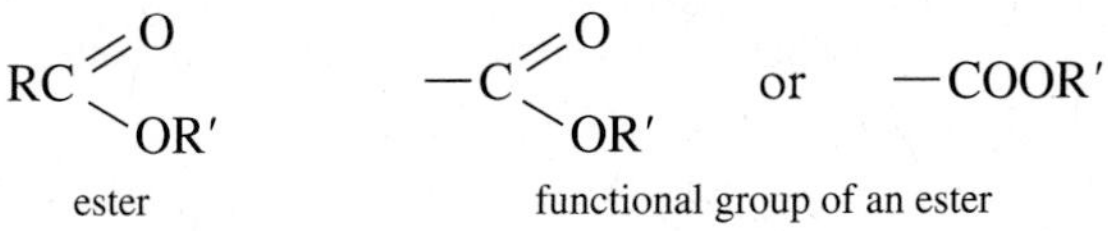

$$RC\overset{O}{\underset{OR'}{<}} \qquad -C\overset{O}{\underset{OR'}{<}} \quad \text{or} \quad -COOR'$$

ester   functional group of an ester

The reaction of acetic acid and ethyl alcohol is shown as an example of the group of reactions called *esterification*. In addition to the ester, a molecule of water is formed as a product.

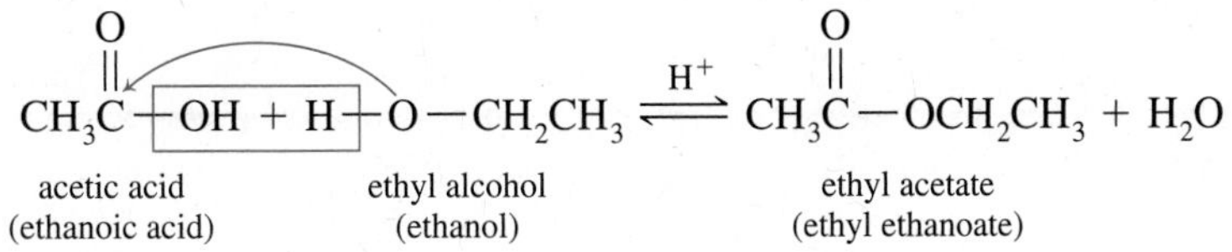

$$\underset{\substack{\text{acetic acid}\\(\text{ethanoic acid})}}{CH_3\overset{O}{\overset{\|}{C}}\boxed{-OH} + H\boxed{-O}} \underset{\substack{\text{ethyl alcohol}\\(\text{ethanol})}}{-CH_2CH_3} \underset{H^+}{\rightleftharpoons} \underset{\substack{\text{ethyl acetate}\\(\text{ethyl ethanoate})}}{CH_3\overset{O}{\overset{\|}{C}}-OCH_2CH_3} + H_2O$$

Esters are alcohol derivatives of carboxylic acids. They are named in much the same way as salts. The alcohol part (R′ in OR′) is named first, followed by the name of the acid modified to end in *-ate*. The *-ic* ending of the organic acid name is replaced by the ending *-ate*. Thus, in the IUPAC system, ethanoic acid becomes ethanoate. In the common names, acetic acid becomes acetate. To name an ester, be sure to recognize the portion of the ester molecule that comes from the acid and the portion that comes from the alcohol. In the general formula for an ester, the RC=O comes from the acid and the R′O comes from the alcohol:

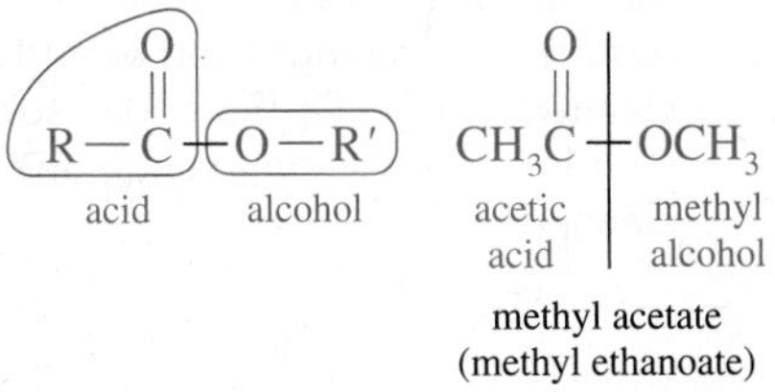

$$\underset{\substack{\text{acid} \quad \text{alcohol}}}{R-\overset{O}{\overset{\|}{C}}-O-R'} \qquad \underset{\substack{\text{acetic} \quad \text{methyl}\\ \text{acid} \quad \text{alcohol}}}{CH_3\overset{O}{\overset{\|}{C}}-OCH_3}$$

methyl acetate
(methyl ethanoate)

Esters occur naturally in many varieties of plant life. They often have pleasant, fragrant, fruity odors and are used as flavoring and scenting agents. Esters are insoluble in water but soluble in alcohol. **Table 19.8** lists selected esters.

The smell of ripe pineapple is a result of the ester ethyl butanoate.

Sarah Bossert/iStockphoto

PRACTICE 19.12

Give two names for each of the following:

(a) $CH_3CH_2\overset{\|}{\underset{O}{C}}-OCH_3$

(b) $CH_3CH_2CH_2\overset{O}{\overset{\|}{C}}-OCH_2CH_3$

(c) $CH_3(CH_2)_{10}\overset{O}{\overset{\|}{C}}-OCH_3$

TABLE 19.8 Odors and Flavors of Selected Esters

Formula	IUPAC name	Common name	Odor or flavor

$CH_3COOCH_2CH_2CHCH_3$
    $\vert$
    CH_3
	isopentyl ethanoate	isoamyl acetate	banana, pear

$CH_3CH_2CH_2COOCH_2CH_3$	ethyl butanoate	ethyl butyrate	pineapple

$HCOOCH_2CHCH_3$
    $\vert$
    CH_3
	isobutyl methanoate	isobutyl formate	raspberry

$CH_3COOCH_2(CH_2)_6CH_3$	octyl ethanoate	n-octyl acetate	orange
⬡—$COOCH_3$ —OH	methyl-2-hydroxy benzoate	methyl salicylate	wintergreen

> CHEMISTRY *IN ACTION*

Getting Clothes CO₂ Clean!

Dry cleaning has been used for many years to clean clothes that can't be placed in water (such as silk, rayon, wool) without severe damage or shrinking. The dry-cleaning process involves no water. Instead, clothes are treated for stains and then washed in perchloro-ethylene (perclene), an organic liquid. Unfortunately, perclene is a health hazard and is classified as both an air pollutant and environmental contaminant. This means big trouble for small dry cleaners who now must dispose of it as hazardous waste and also worry about health risks to employees. Consequently, researchers are looking for a solvent that can replace perclene for dry cleaning.

One interesting alternative is to use carbon dioxide for dry cleaning. You may think of CO_2 as a greenhouse gas, the gas you breathe in and out, or the solid called dry ice. One thing's for sure—we don't often consider CO_2 as a liquid. Scientists have already started to consider liquid CO_2 as a dry-cleaning alternative. Why? There is no shortage of CO_2 in the world. It can be collected from the waste generated in industry. No special disposal procedures are needed since it can be recycled and leaks would be nontoxic to humans and the environment in small quantities.

The largest obstacle for the dry-cleaning process is that CO_2 doesn't dissolve most polymers, oils, waxes, and substances that get caught in clothes. Of course, neither does water; it requires the help of detergents. So scientists have now developed polymers that can be used as detergents in liquid CO_2. Joseph DeSimone (University of North Carolina at Chapel Hill) made detergent polymers called copolymers, which are really just two polymers joined together. One end is soluble in CO_2; the other attracts oils and waxes. The new detergent polymers form micelles that trap the oils and waxes within the micelle and dissolve in CO_2.

Now machines that look like bank vaults can clean 50–70 pounds of clothes per load with CO_2 and about 5 ounces of polymer detergent. One extra bonus—since the machine operates at room temperature, stains don't require pretreating! Leather and suede can also be cleaned using CO_2. Someday soon you may walk into your local dry-cleaning shop and get your clothes CO_2 cleaned!

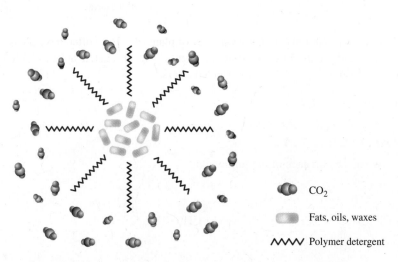

CO₂

Fats, oils, waxes

Polymer detergent

Copolymers act as detergents, dissolving in CO_2 and attracting and trapping organic substances in a micelle.

19.12 POLYMERS—MACROMOLECULES

Discuss the chemical composition of polymers.

● **LEARNING OBJECTIVE**

KEY TERMS
polymerization
polymer
monomer
copolymer

Some very large molecules (macromolecules) exist in nature, containing tens of thousands of atoms. Some of these, such as starch, glycogen, cellulose, proteins, and DNA, have molar masses in the millions and are central to many of our life processes. Synthetic macromolecules touch every phase of modern living. Today it's hard to imagine a world without polymers. Textiles for clothing, carpeting, and draperies; shoes; toys; automobile parts; construction materials; synthetic rubber; chemical equipment; medical supplies; cooking utensils; synthetic leather; recreational equipment—the list goes on. The vast majority of these polymeric materials are based on petroleum. Petroleum is nonreplaceable; therefore, our dependence on polymers is another good reason for not squandering our limited world supply of petroleum.

The process of forming very large, high-molar-mass molecules from smaller units is called **polymerization**. The large molecule, or unit, is called the **polymer** and the small repeating unit, the **monomer**. Polymers containing more than one kind of monomer are called **copolymers**. The term *polymer* is derived from the Greek word *polumerēs*, meaning "having many parts." Ethylene is a monomer and polyethylene is a polymer. Because of their large size, polymers are often called *macromolecules.* Some synthetic polymers are called *plastics.* The word *plastic* means "capable of being molded, or pliable."

Polyethylene is an example of a synthetic polymer. Ethylene, derived from petroleum, is made to react with itself to form polyethylene (or polythene).

Polyethylene is a long-chain hydrocarbon made from many ethylene units:

$$n\,CH_2{=}CH_2 \longrightarrow -CH_2CH_2[CH_2CH_2]_nCH_2CH_2CH_2CH_2-$$
$$\text{ethylene} \qquad\qquad \text{polyethylene}$$

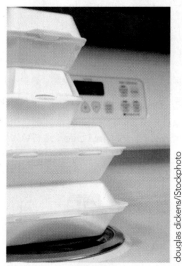

Molded foam take-out containers are made of polystyrene.

A typical polyethylene molecule is made up of about 2500–25,000 ethylene molecules joined in a continuous structure. Over 30 billion pounds of polyethylene are produced annually in the United States. Its uses are as varied as any single substance known and include chemical equipment, packaging material, electrical insulation, films, industrial protective clothing, and toys.

Ethylene derivatives, in which one or more hydrogen atoms have been replaced by other atoms or groups, can also be polymerized. Many of our commercial synthetic polymers are made from such modified ethylene monomers. For example, $CH_2{=}CH-$ is a vinyl group and $CH_2{=}CHCl$ is vinyl chloride. Vinyl chloride can be polymerized to polyvinyl chloride (PVC). The names, structures, and some uses for several of these polymers are given in Table 19.9.

EXAMPLE 19.14

What is the fundamental meaning of a polymer?

SOLUTION

A polymer is a very large molecule (macromolecule) made from repeating small units called monomers. An example of the simplest synthetic polymer is made from ethylene $(CH_2 = CH_2)$, which polymerizes into polyethylene of various densities with molar masses ranging from 10,000 to 100,000.

PRACTICE 19.13

Select two monomers from Table 19.9 and write two units of a copolymer that can be made from these two monomer units.

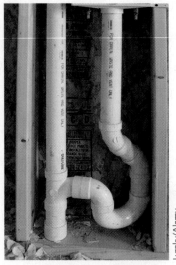

Much of the plumbing in our homes is polyvinyl chloride (PVC).

Polyethylene is used every day in many commercial products such as milk containers.

TABLE 19.9 Polymers Derived from Modified Ethylene Monomers

Monomer	Polymer	Uses
$CH_2{=}CH_2$ ethylene	$-(CH_2{-}CH_2)_n$ polyethylene	Packing material, molded articles, containers, toys
$CH_2{=}CH$ 　$\vert$ 　CH_3 propylene	$\left(CH_2{-}CH\right.$ 　　　$\vert$ 　　$\left.CH_3\right)_n$ polypropylene	Textile fibers, molded articles, lightweight ropes, autoclavable biological equipment
$CH_2{=}CH$ 　$\vert$ 　Cl vinyl chloride	$\left(CH_2{-}CH\right.$ 　　　$\vert$ 　　$\left.Cl\right)_n$ polyvinyl chloride	Garden hoses, pipes, molded articles, floor tile, electrical insulation, vinyl leather
$CH_2{=}CH$ 　$\vert$ 　CN acrylonitrile	$\left(CH_2{-}CH\right.$ 　　　$\vert$ 　　$\left.CN\right)_n$ Orlon, Acrilan	Textile fibers
$CF_2{=}CF_2$ tetrafluoroethylene	$-(CF_2{-}CF_2)_n$ Teflon	Gaskets, valves, insulation, heat-resistant and chemical-resistant coatings, linings for pots and pans
$CH_2{=}CH$ styrene	$\left(CH_2{-}CH\right)_n$ polystyrene	Molded articles, Styrofoam, insulation, toys, disposable food containers
$CH_2{=}C{-}CH_3$ 　　　$\vert$ 　　$C{-}O{-}CH_3$ 　　$\Vert$ 　　O methylmethacrylate	$\left(CH_2{-}C\begin{smallmatrix}CH_3\\\vert\\\\O{=}C\\\vert\\OCH_3\end{smallmatrix}\right)_n$ Lucite, Plexiglas (acrylic resins)	Contact lenses, clear sheets for windows and optical uses, molded articles, automobile finishes

CHAPTER **19 REVIEW**

19.1 THE BEGINNINGS OF ORGANIC CHEMISTRY

KEY TERMS

vital-force theory
organic chemistry

• Originally, organic compounds were thought to be formed by a "vital force."
• Organic chemistry today is the chemistry of carbon compounds:
  • This does not imply that the compounds come from living matter.
  • A few carbon compounds are considered to be inorganic since their chemistry matches the chemistry of inorganic compounds better.

19.2 WHY CARBON?

KEY TERMS

functional group
hydrocarbons
saturated hydrocarbon
unsaturated hydrocarbon

• Carbon is central to all organic compounds:
  • Usually forms four covalent bonds
  • Has tetrahedral structure
  • Can also form double bonds and triple bonds
• Hydrocarbons are composed of carbon and hydrogen:
  • Saturated—contain only single bonds
  • Unsaturated—contain a double or triple bond between carbon atoms

19.3 ALKANES

- Saturated hydrocarbons
- General formula is C_nH_{2n+2}.
- Most important reaction is combustion to form water, carbon dioxide, and heat.
- Two or more compounds that have the same molecular formula but different structural arrangements are isomers.
- Formulas can be written to show all the bonds or condensed, using the groups attached to a carbon atom to save space.
- An alkyl group has the general formula C_nH_{2n+1}.
- To name an alkane:
  - Find the longest continuous carbon chain.
  - Count the carbons in the parent chain starting from the end closest to a branch point.
  - Name each alkyl group and designate its position on the parent chain by a number.
  - If the same alkyl group appears multiple times, use a prefix on the alkyl name to indicate how many are in the compound.
  - If multiple different alkyl groups are on the parent chain, list them alphabetically (prefixes do not count).

KEY TERMS

alkane
homologous series
isomerism
isomers
alkyl group

19.4 ALKENES AND ALKYNES

- Alkenes:
  - Contain a carbon–carbon double bond
  - General formula: C_nH_{2n}
- Alkynes:
  - Contain a carbon–carbon triple bond
  - General formula: C_nH_{2n-2}
- To name an alkene or alkyne:
  - Find the longest continuous carbon chain containing a double or triple carbon–carbon bond.
  - Number the carbon atoms in the parent chain starting with the end closest to the double or triple bond and name the parent chain as you would for an alkane but change the ending to -ene or -yne.
  - Name each alkyl group and designate its position on the parent chain by a number.
  - If the same alkyl group appears multiple times, use a prefix on the alkyl name to indicate how many are in the compound.
  - If multiple different alkyl groups are on the parent chain, list them alphabetically (prefixes do not count).
- The most common reaction for alkenes and alkynes is addition:
  - Occurs at the carbon–carbon double or triple bond
  - Hydrogen, halogens, hydrogen halides, or water can be added across the double bond.
  - Complete addition results in a saturated compound.

KEY TERMS

alkene
alkyne
addition reaction
hydrogenation

19.5 AROMATIC HYDROCARBONS

- Compounds with structures and chemical properties resembling benzene are called aromatic compounds.
- Substituted benzenes have the general formula of C_6H_5G.
- Three major classes:
  - Monosubstituted:
    - Named by adding substituent name as prefixes to *benzene*.
  - Disubstituted:
    - *Ortho-*, *meta-*, *para-* used as prefixes for the substituent groups to show the position of the substituents.
    - Alternatively, numbers can be used to show the positions.
    - They are named by adding substituent names as a prefix to *benzene*.
  - Tri- and polysubstituted:
    - Atoms in the benzene ring are numbered, and the substituents start on the C-1 atom.
    - They are named by adding substituent names as prefixes to *benzene*.

KEY TERM

aromatic compound

19.6 HYDROCARBON DERIVATIVES

- These are compounds that can be synthesized from a hydrocarbon.
- They can contain additional elements such as O, N, or a halogen.
- See Table 19.4 for a list of derivatives.
- $RH + X_2 \xrightarrow[\text{light}]{\text{UV}} RX + HX$ (X = Cl or Br)

19.7 ALCOHOLS

KEY TERMS

alcohol
primary alcohol
secondary alcohol
tertiary alcohol
polyhydroxy alcohol

- Compounds that contain a hydroxyl group bonded to a saturated carbon atom
- Classification:
  - Primary (1°)
  - Secondary (2°)
  - Tertiary (3°)
- To name an alcohol:
  - Select the longest continuous chain of C atoms containing the alcohol group.
  - Number the C atoms in this chain so the —OH group has the lowest number.
  - Form the parent alcohol name by replacing the *-e* in *alkane* with *-ol*.
  - Name any alkyl branches and show their positions by the corresponding numbers.

19.8 ETHERS

KEY TERM

ether

- Ethers are isomers of alcohols.
- Naming ethers:
  - Select the longest continuous chain of C atoms and name it as the corresponding alkane.
  - Change the *-yl* ending on the other hydrocarbon group to *-oxy* to get the alkoxy group name.
  - Combine the two names, giving the alkoxy name first, to form the ether name.

19.9 ALDEHYDES AND KETONES

KEY TERMS

carbonyl group
aldehyde
ketone

- Aldehydes and ketones both contain a carbonyl group:
  - Aldehydes have a hydrogen bonded to the carbonyl carbon atom.
  - Ketones have two alkyl or aromatic groups bonded to the carbonyl carbon atom.
- Naming aldehydes:
  - Drop the final *-e* and add *-al* to the name of the parent hydrocarbon.
- Naming ketones:
  - Name the alkane corresponding to the longest chain containing the ketone group by changing the *-e* to *-one*.
  - If the chain is longer than four C atoms, make the carbonyl the lowest number possible.
  - Name groups attached to the parent chain as for hydrocarbons and alcohols.

19.10 CARBOXYLIC ACIDS

KEY TERMS

carboxyl acid
carboxyl group

- The carboxyl group is represented in the following ways:

$$\begin{matrix} & C \\ & \| \\ -C & -OH \end{matrix} \qquad -COOH \qquad -CO_2H$$

- To name a carboxylic acid:
  - Find the longest carbon chain including the carboxyl group.
  - Drop the *-e* from the corresponding parent hydrocarbon name and add *-oic acid*.
- Other systems are used to name carboxylic acids.

19.11 ESTERS

KEY TERM

ester

- Carboxylic acids react with alcohols to form esters.
- General formula RCOOR′
- Esters are named like salts:
  - Alcohol part is named first.
  - Name of the acid is given second, with the acid ending modified to *-ate*.

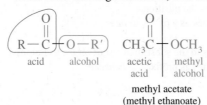

19.12 POLYMERS—MACROMOLECULES

KEY TERMS

polymerization
polymer
monomer
copolymer

- Polymers are called macromolecules.
- Polymers are composed of small repeating units called monomers or copolymers.

REVIEW QUESTIONS

1. What is the vital-force theory and who disproved this theory?
2. What property do all organic compounds have in common?
3. What are some examples of consumer products that are composed of organic compounds?
4. What bonding characteristic of carbon is primarily responsible for the existence of so many organic compounds?
5. What is the most common geometric arrangement of covalent carbon bonds? Illustrate this structure.
6. In addition to single bonds, what other types of bonds can carbon atoms form? Give examples.
7. Draw a Lewis structure for
   (a) a single carbon atom
   (b) molecules of methane, ethene, and ethyne
8. Write the names and draw the structural formulas for the first 10 normal alkanes.
9. Write the names and draw the structural formulas for all possible alkyl groups, C_nH_{2n+1}, containing from one to four carbon atoms.
10. What is the single most important reaction of alkanes?
11. What is wrong with the name 1-methylpentane?
12. Which one of these compounds belongs to a different homologous series than the others?
   (a) C_2H_4 (c) C_6H_{14}
   (b) CH_4 (d) C_5H_{12}
13. Which word does not belong with the others?
   (a) alkane   (d) ethane
   (b) paraffin   (e) ethylene
   (c) saturated   (f) pentane
14. Draw the structure for vinyl acetylene, which has the formula C_4H_4 and contains one double bond and one triple bond.
15. An open-chain hydrocarbon has the formula C_6H_8. What possible combinations of carbon–carbon double bonds and/or carbon–carbon triple bonds can be in this compound?

16. What is the parent compound for most aromatic compounds?
17. What do the prefixes *ortho-*, *meta-*, and *para-* mean when used in naming disubstituted benzene rings?
18. What kinds of organic compounds contain an —OH group?
19. What kinds of organic compounds contain a carbonyl group,

$$-\overset{\overset{\textstyle O}{\|}}{C}-?$$

20. What kinds of organic compounds contain a $-\overset{\overset{\textstyle O}{\|}}{C}-OH$ group?
21. Why is ethylene glycol (1,2-ethanediol) superior to methyl alcohol (methanol) as an antifreeze for automobile radiators?
22. Alcohols are considered to be toxic to the human body—with ethanol, the alcohol in alcoholic beverages, being the least toxic one. What are the hazards of ingesting (a) methanol and (b) ethanol?
23. What is the general formula for an ether?
24. What is the suffix used for naming an RO— group?
25. How do ethers differ from alcohols?
26. What is the functional group found in aldehydes and ketones?
27. How does an aldehyde differ from a ketone?
28. What is the suffix that indicates a compound is an aldehyde?
29. What is the suffix that indicates a compound is a ketone?
30. What is the chemical structure that identifies a carboxylic acid?
31. What is the suffix that indicates a compound is a carboxylic acid?
32. What is the common name for methanoic acid, and what is the origin of this name?
33. How do esters differ in structure from carboxylic acids?
34. What is the suffix that indicates a compound is an ester?
35. What is a common property of esters?
36. What are the small repeating units of polymers?
37. What simple organic molecule is the parent molecule of many polymers?

Most of the exercises in this chapter are available for assignment via the online homework management program, WileyPLUS (www.wileyplus.com). All exercises with blue numbers have answers in Appendix VI.

PAIRED EXERCISES

1. Which of the following are organic compounds?
   (a) CH_3CH_3
   (b) NH_3
   (c) CH_3SCH_3
   (d) CH_3CH_2OH
   (e) $Mg(OH)_2$

2. Which of the following are organic compounds?
   (a) H_2SO_4
   (b) CH_3NH_2
   (c) Na_2CO_3
   (d) $CH_3CH_2CH_2CH_2CH_3$
   (e) $CH_3CH_2OCH_3$

3. For each organic compound in Question 1, explain why it is classified as an organic compound.

4. For each organic compound in Question 2, explain why it is classified as an organic compound.

5. Name these normal alkyl groups:
   (a) $C_5H_{11}-$
   (b) $C_7H_{15}-$

6. Name these normal alkyl groups:
   (a) $C_8H_{17}-$
   (b) $C_{10}H_{21}-$

7. Write condensed structural formulas for the five isomers of hexane.

8. Write condensed structural formulas for the nine isomers of heptane.

9. Give IUPAC names for the following:
   (a) CH_3CH_2Cl
   (b) $CH_3CHClCH_3$
   (c) $(CH_3)_2CHCH_2Cl$

10. Give IUPAC names for the following:
   (a) $CH_3CH_2CH_2Cl$
   (b) $(CH_3)_3CCl$
   (c) $CH_3CHClCH_2CH_3$

11. Give IUPAC names for these compounds:

(a) $CH_3CHCH_2CHCH_2CH_2CH_3$
       | |
      CH_3 CH_2CH_3

(b) $CH_3CHCH_2CHCH_2-CHCH_2CH_3$
       | | |
      CH_3 CH_2CH_3 CH_2CH_3

12. Give IUPAC names for these compounds:

(a) $CH_3CHCH_2CH_2CHCH_3$
       | |
      CH_2 CH_2
       | |
      CH_3 CH_3

(b) $CH_3CH_2CH_2CHCH_2CH_3$
                    |
                 CH_3CHCH_3

13. Draw structural formulas of these compounds:
(a) 2,4-dimethylpentane
(b) 2,2-dimethylpentane
(c) 3-isopropyloctane

14. Draw structural formulas of these compounds:
(a) 4-ethyl-2-methylhexane
(b) 4-*tert*-butylheptane
(c) 4-ethyl-7-isopropyl-2,4,8-trimethyldecane

15. One name in each pair is incorrect. Draw structures corresponding to each name and indicate which name is incorrect.
(a) 2-methylbutane and 3-methylbutane
(b) 2-ethylbutane and 3-methylpentane

16. One name in each pair is incorrect. Draw structures corresponding to each name and indicate which name is incorrect.
(a) 2-dimethylbutane and 2,2-dimethylbutane
(b) 2,4-dimethylhexane and 2-ethyl-4-methyl-pentane

17. Draw structural formulas for all the isomers of
(a) CH_3Br (c) C_2H_5Cl
(b) CH_2Cl_2 (d) C_3H_7Br

18. Draw structural formulas for all the isomers of
(a) C_4H_9I (c) C_3H_6BrCl
(b) $C_3H_6Cl_2$ (d) $C_4H_8Cl_2$

19. Using alkenes and any other necessary inorganic reagents, write reactions showing the formation of
(a) $CH_3CH_2CHClCH_2Cl$
(b) CH_3CH_2Br

20. Using alkenes and any other necessary inorganic reagents, write reactions showing the formation of
(a) $CH_3CHBrCH_2Br$
(b) $CH_3CHBrCH_3$

21. Draw structures containing two carbon atoms for the following classes of compounds:
(a) alkene
(b) alkyne
(c) alkyl halide
(d) alcohol

22. Draw structures containing two carbon atoms for the following classes of compounds:
(a) ether
(b) aldehyde
(c) carboxylic acid
(d) ester

23. Give the IUPAC name for each compound in Question 21.

24. Give the IUPAC name for each compound in Question 22.

25. Draw structural formulas for the following:
(a) chloromethane
(b) vinyl chloride
(c) chloroform
(d) 1,1-dibromoethene

26. Draw structural formulas for the following:
(a) hexachloroethane
(b) iodoethyne
(c) 6-bromo-3-methyl-3-hexene-1-yne
(d) 1,2-dibromoethene

27. Draw structural formulas for the following:
(a) 2,5-dimethyl-3-hexene
(b) 2-ethyl-3-methyl-1-pentene
(c) 4-methyl-2-pentene

28. Draw structural formulas for the following:
(a) 1,2-diphenylethene
(b) 3-pentene-1-yne
(c) 3-phenyl-1-butyne

29. Name these compounds:

(a) $CH_3CH=CCH_2CH_2CH_3$
                |
               CH_3

(b) $CH_3C=C-CH_3$
        | |
       CH_3 CH_3

30. Name these compounds:

(a) $CH_3CH_2CHCH=CH_2$
              |
             CH
          H_3C CH_3

(b) $CH_3CH_2CH=CCH_2CH_3$
                   |
                  CH_3

31. Complete these reactions and name the products:
(a) $CH_2=CHCH_3 + Br_2 \longrightarrow$
(b) $CH_2=CH_2 + HBr \longrightarrow$
(c) $CH_3CH=CHCH_3 + H_2 \xrightarrow[\text{1 atm}]{\text{Pt 25°C}}$

32. Complete these reactions and name the products:
(a) $CH_2=CH_2 + H_2O \xrightarrow{H^+}$
(b) $CH\equiv CH + 2Br_2 \longrightarrow$
(c) $CH_2=CH_2 + H_2 \xrightarrow[\text{1 atm}]{\text{Pt 25°C}}$

33. Name these compounds:

(a) 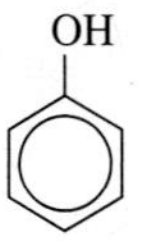 OH

(b) 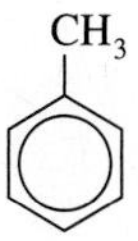 CH$_3$

(c)  COOH

(d) 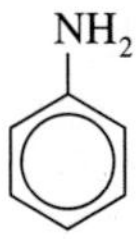 NH$_2$

(e) 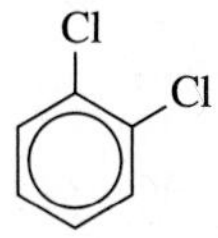 Cl
Cl

(f) 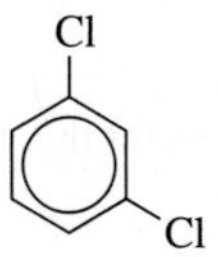 Cl
Cl

34. Name these compounds:

(a)  Cl
Cl

(b) 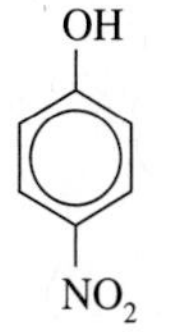 OH
NO$_2$

(c) 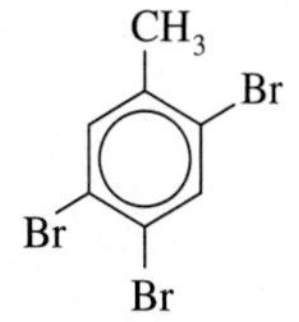 CH$_3$
Br
Br
Br

(d) 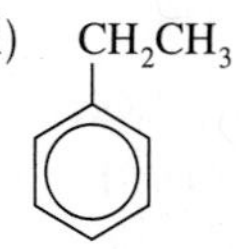 CH$_2$CH$_3$

(e) 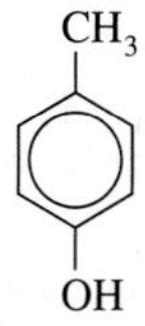 CH$_3$
OH

(f) 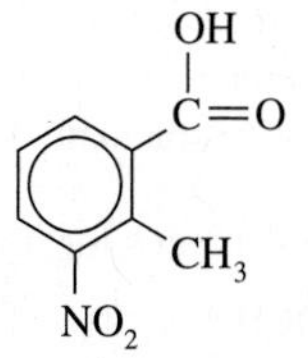 OH
C=O
CH$_3$
NO$_2$

35. Draw structural formulas for
(a) benzene   (c) benzoic acid
(b) toluene   (d) aniline

36. Draw structural formulas for
(a) phenol   (c) 1,3-dichloro-5-nitrobenzene
(b) o-bromochlorobenzene   (d) m-dinitrobenzene

37. Draw structural formulas for
(a) ethylbenzene   (b) 1,3,5-tribromobenzene

38. Draw structural formulas for
(a) tert-butylbenzene   (b) 1,1-diphenylethane

39. Draw structural formulas and write the IUPAC names for all isomers of trichlorobenzene (C$_6$H$_3$Cl$_3$).

40. Draw structural formulas and write the IUPAC names for all isomers of dichlorobromobenzene (C$_6$H$_3$Cl$_2$Br).

41. Write the IUPAC name for these alcohols. Classify each as primary, secondary, or tertiary alcohols.

(a) 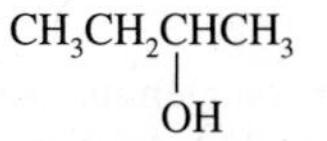 CH$_3$CH$_2$CHCH$_3$
         |
         OH

(b) 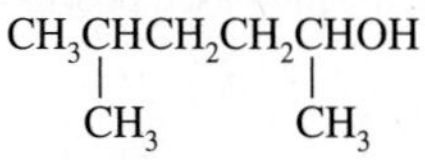 CH$_3$CHCH$_2$CH$_2$CHOH
      | |
      CH$_3$ CH$_3$

42. Write the IUPAC name for these alcohols. Classify each as primary, secondary, or tertiary alcohols.

(a) 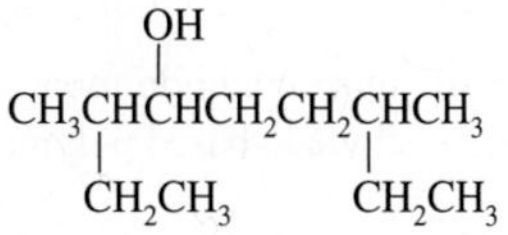 OH
    |
CH$_3$CHCHCH$_2$CH$_2$CHCH$_3$
       | |
       CH$_2$CH$_3$ CH$_2$CH$_3$

(b) 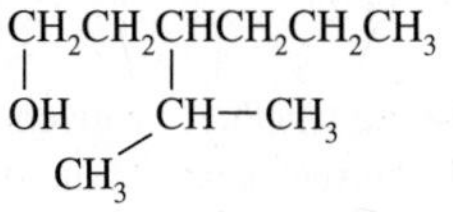 CH$_2$CH$_2$CHCH$_2$CH$_2$CH$_3$
    | |
    OH   CH—CH$_3$
              |
              CH$_3$

43. Draw structural formulas for
(a) 2-pentanol   (b) isopropyl alcohol

44. Draw structural formulas for
(a) 2,2-dimethyl-1-heptanol   (b) 1,3-propanediol

45. Name these aldehydes:

(a) CH$_2$=O

(b) 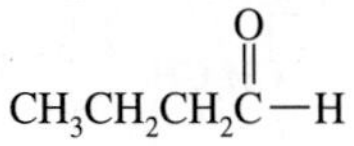 CH$_3$CH$_2$CH$_2$C—H
                   ‖
                   O

(c) CH$_3$CHCH$_2$C—H
       | ‖
       CH$_3$ O

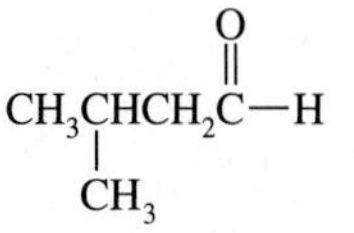

46. Name these aldehydes:

(a) 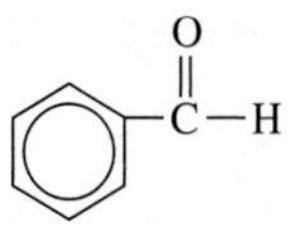
    O
    ‖
    C—H

(b) O=CCH$_2$CH$_2$C—H
       | ‖
       H   O

(c) CH$_3$CHCH$_2$C—H
       | ‖
       OH   O

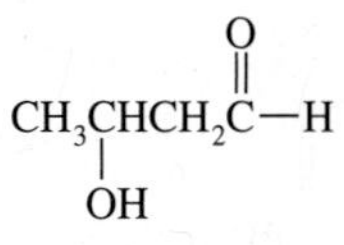

47. Name these ketones:

(a) CH$_3$COCH$_3$

(b) CH$_3$CH$_2$COCH$_3$

(c) 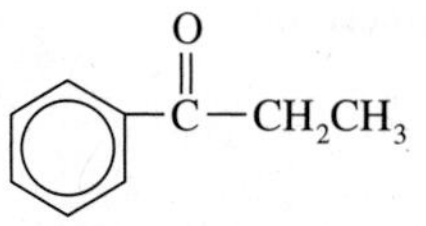
    O
    ‖
    C—CH$_2$CH$_3$

48. Name these ketones:

(a) 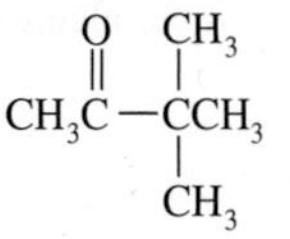 CH$_3$C—CCH$_3$
       ‖ |
       O   CH$_3$
            |
            CH$_3$

(b) CH$_3$CCH$_2$CH$_2$CCH$_3$
       ‖   ‖
       O   O

(c) CH$_3$C—CH$_2$CCH$_3$
       | ‖
       CH$_3$ O
       |
       OH

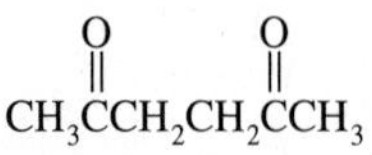

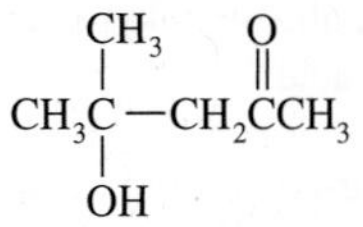

49. Name these acids:

(a) $CH_3CHBrCOOH$

(b) $CH_2{=}CHCH_2COOH$

(c) $CH_3CH_2CH_2COOH$

(d)

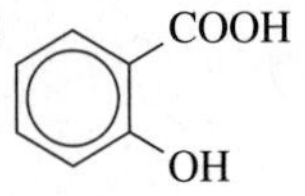

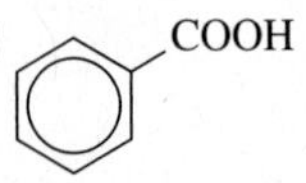

51. Draw structural formulas for these esters:
(a) ethyl formate
(b) methyl ethanoate
(c) isopropyl propanoate

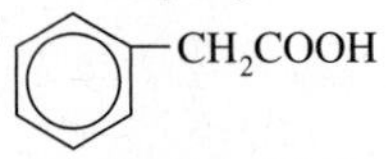

53. Write IUPAC names for these esters:

(a) $CH_3\underset{\underset{O}{\|}}{C}OCH_2CH_3$

(b) 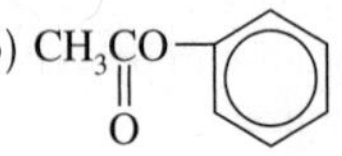 $CH_3\underset{\underset{O}{\|}}{C}O$—

(c) $CH_3CH_2CH_2\overset{\overset{O}{\|}}{C}{-}O{-}\overset{\overset{CH_3}{|}}{C}HCH_3$

55. Complete these equations:

(a) $CH_3COOH + NaOH \longrightarrow$

(b) $CH_3\underset{\underset{OH}{|}}{C}HCOOH + NH_3 \longrightarrow$

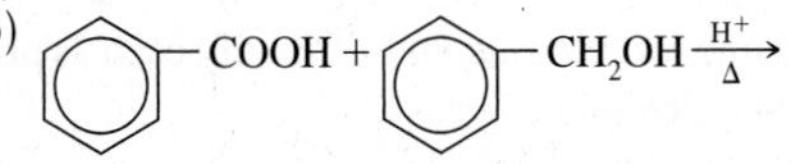

57. Ethylene and its derivatives are the most common monomers for polymers. Write formulas for these ethylene-based polymers:
(a) polyethylene
(b) polyvinyl chloride

59. For the following structures, determine whether the given name is correct. If it is incorrect, provide the correct name for the structure.

(a) 3-methylbutane

$CH_3\underset{\underset{CH_3}{|}}{C}HCH_2CH_3$

(b) *ortho*-dibromobenzene

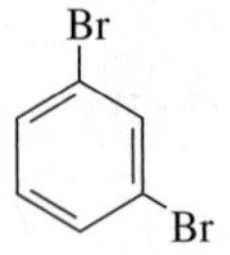

(c) 2-methyl-1-butyne

$HC{\equiv}\underset{\underset{CH_3}{|}}{C}CHCH_3$

(d) pentanol

$CH_3CH_2\underset{\underset{OH}{|}}{C}HCH_2CH_3$

(e) butanoic acid

$CH_3CH_2CH_2\overset{\overset{O}{\|}}{C}{-}OH$

61. Classify the structures in Question 59 as one of the following: hydrocarbon, alkene, alkyne, aromatic, alkyl halide, alcohol, ester, carboxylic acid.

63. Draw a four-carbon saturated hydrocarbon.

50. Name these acids:

(a) $CH_3\underset{\underset{CH_2CH_3}{|}}{C}HCOOH$

(b)
—CH_2COOH

(c) $CH_3CH_2CH_2CH_2COOH$

(d)

52. Draw structural formulas for these esters:
(a) *n*-nonyl acetate
(b) ethyl benzoate
(c) methyl salicylate

54. Write IUPAC names for these esters:

(a) $HC\underset{\underset{O}{\|}}{O}CH(CH_3)_2$

(b)
—$\underset{\underset{O}{\|}}{C}OCH_3$

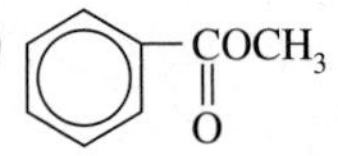

(c) $CH_3\overset{\overset{CH_3}{|}}{C}HCH_2\overset{\overset{O}{\|}}{C}{-}OCH_2CH_3$

56. Complete these equations:

(a) $CH_3COOH + KOH \longrightarrow$

(b)
—$COOH +$ —$CH_2OH \xrightarrow[\Delta]{H^+}$

58. Ethylene and its derivatives are the most common monomers for polymers. Write formulas for these ethylene-based polymers:
(a) polyacrylonitrile
(b) Teflon

60. For the following structures, determine whether the given name is correct. If it is incorrect, provide the correct name for the structure.

(a) 2-chlorobutane

$CH_3CH_2\underset{\underset{Cl}{|}}{C}HCH_3$

(b) 2-butaneketone

$CH_3\overset{\overset{O}{\|}}{C}CH_2CH_3$

(c) 2-methyl-3-pentene

$CH_3CH{=}CHCH\underset{\underset{CH_3}{|}}{}CH_3$

(d) propyl methanoate

$CH_3CH_2\overset{\overset{O}{\|}}{C}OCH_3$

(e) 2-ethylbutane

$CH_3CH_2\underset{\underset{CH_2CH_3}{|}}{C}HCH_3$

62. Classify the structures in Question 60 as one of the following: hydrocarbon, alkene, alkyne, ether, alkyl halide, ester, aldehyde, ketone.

64. Draw a four-carbon unsaturated hydrocarbon.

ADDITIONAL EXERCISES

65. Write the Lewis structure for methane, CH_4, and tell the bond angles.

66. Write the Lewis structure for ethene, C_2H_4, and tell the bond angles.

67. Write the Lewis structure for ethyne, C_2H_2, and tell the bond angles.

68. Why do small alkanes have low boiling points?

69. Draw all possible open-chain structures for
  (a) C_6H_{14}
  (b) C_5H_{10}
  (c) C_7H_{16}

70. Draw the structural formulas and write the IUPAC names for all open-chain isomers of
  (a) pentyne C_5H_8
  (b) hexyne, C_6H_{10}

71. Eight open-chain isomeric alcohols have the formula $C_5H_{11}OH$.
  (a) Draw the structural formula and write the IUPAC name for each of these alcohols.
  (b) Indicate which of these isomers are primary, secondary, and tertiary alcohols.

72. What is the molar mass of an open-chain saturated alcohol containing 30 carbon atoms? This alcohol is present in beeswax as an ester.

73. Write the common name and structure of the ether that is isomeric with
  (a) 1-propanol
  (b) ethanol
  (c) isopropyl alcohol

74. Give the IUPAC and common names and structures of all ethers having the molecular formula $C_5H_{12}O$.

75. Draw structural formulas for propanal and propanone. From these formulas, do you think that aldehydes and ketones are isomeric with each other? Show evidence and substantiate your answer by testing with a four-carbon aldehyde and ketone.

76. Give the common and IUPAC names for the first five straight-chain carboxylic acids.

77. Draw the structural formulas and write the IUPAC names for all the acid isomers of hexanoic acid, $CH_3(CH_2)_4COOH$.

78. Write equations for the preparation of these esters:
  (a) ethyl formate
  (b) methyl propanoate
  (c) *n*-propyl benzoate

79. (a) Draw a structural formula showing the polymer that can be formed from the following monomers (show four units): (1) propylene, (2) 1-butene, (3) 2-butene.
  (b) How many ethylene units are in a polyethylene molecule that has a molar mass of 35,000?

80. How many different dibromobenzenes are there? How many tribromobenzenes? Show structures.

81. A compound has the following composition:
  C = 24.3%
  H = 4.1%
  Cl = 71.7%
  If 140.3 mL of the vapor at 100°C and 740 mm Hg pressure have a mass of 0.442 g, what is the molecular formula of the compound? Draw possible structures of the compound.

82. A compound was found to contain 24 g of carbon, 4 g of hydrogen, and 32 g of oxygen. The molar mass of the compound is 60. Which of these structures is correct for that compound?

  (a) C_2H_5OH
  (b) CH_3CHO
  (c) CH_3OCH_3
  (d) COH
  (e) CH_3COOH

83. Write formulas for
  (a) ethyl alcohol
  (b) iodomethane
  (c) 2-chloropentane
  (d) decane
  (e) *tert*-butyl alcohol

84. Write equations for the reactions in acid solution between
  (a) methanol and formic acid
  (b) 1-butanol and butanoic acid
  (c) 1-hexanol and hexanoic acid
  Name the esters produced in each case.

85. Draw the structural formula for
  (a) a six-carbon alkane with methyl branches on two adjacent carbons
  (b) a six-carbon alcohol with an —OH group not at the end of the chain
  (c) a three-carbon organic (carboxylic) acid
  (d) a four-carbon aldehyde
  (e) the ester formed when a four-carbon alcohol reacts with a three-carbon acid

86. Look at the imaginary molecule pictured below and name the functional groups that are identified.

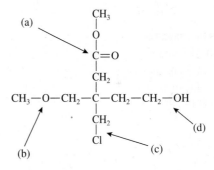

87. Look at the imaginary molecule pictured below and name the functional groups that are identified.

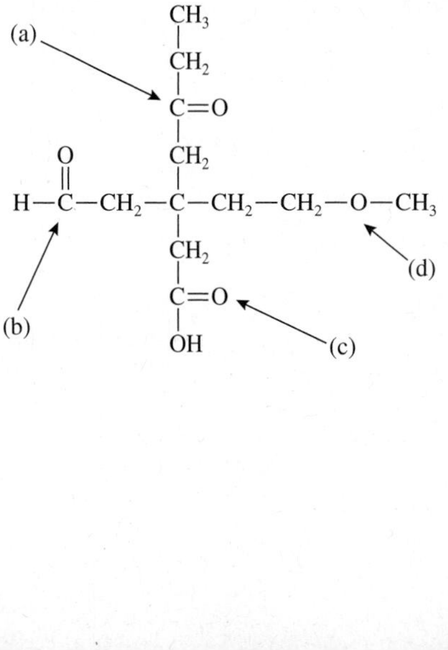

ANSWERS TO PRACTICE EXERCISES

19.1 Alkanes, alkenes, alkynes, and aromatic hydrocarbons

19.2 (a) $CH_3CH_2CH_2CH_2CH_2CH_3$

(b) 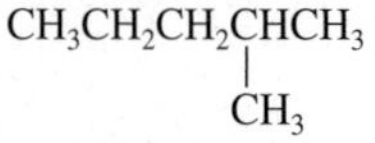 $CH_3CH_2CH_2CHCH_3$
$\qquad\qquad\quad |$
$\qquad\qquad\quad CH_3$

$\qquad\qquad CH_3$
$\qquad\qquad |$
(c) 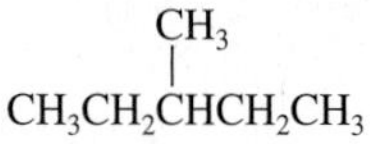 $CH_3CH_2CHCH_2CH_3$

$\qquad\qquad CH_3 \qquad\quad CH_3$
$\qquad\qquad | \qquad\qquad\quad |$
(d) 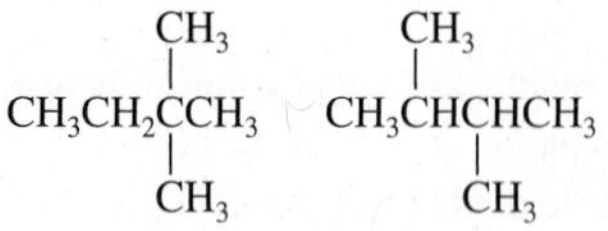 $CH_3CH_2CCH_3 \quad CH_3CHCHCH_3$
$\qquad\qquad | \qquad\qquad\qquad |$
$\qquad\qquad CH_3 \qquad\qquad\quad CH_3$

19.3 (a) 2,3,3-trimethylhexane
(b) 3-ethyl-2,5-dimethylheptane

19.4
$\qquad\qquad\quad CH_3 \quad CH_3$
$\qquad\qquad\quad | \qquad\quad |$
(a) 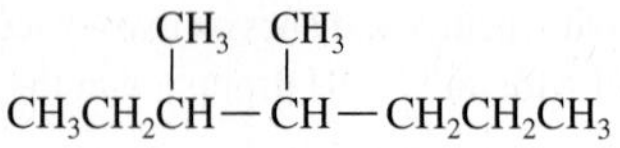 $CH_3CH_2CH — CH — CH_2CH_2CH_3$

$\qquad\quad CH_3 \ CH_3 \ CH_3$
$\qquad\quad | \qquad | \qquad |$
(b) 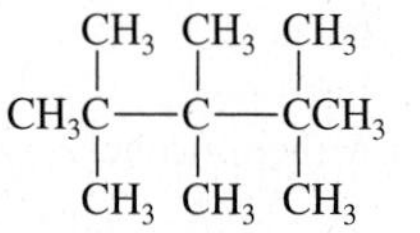 $CH_3C — C — CCH_3$
$\qquad\quad | \qquad | \qquad |$
$\qquad\quad CH_3 \ CH_3 \ CH_3$

$\qquad\qquad\qquad CH_3CHCH_3$
$\qquad\qquad\qquad\qquad |$
(c) $CH_3CH_2 — CH — CH_2CH_2CH_2CH_3$
$\qquad\qquad\qquad\quad |$
$\qquad\qquad\qquad CH_2CH_3$

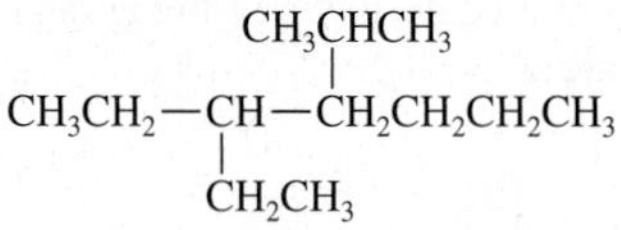

19.5 (a) 2,5-dimethyl-3-heptene
(b) 4-ethyl-2-octyne

19.6 (a) *m*-chloroethylbenzene
(b) 3,5-dibromophenol
(c) *o*-nitroaniline

19.7 $CH_3CH_2CHCl_2, CH_3CHClCH_2Cl, CH_2ClCH_2CH_2Cl, CH_3CCl_2CH_3$

19.8 (a) secondary, 6-methyl-3-heptanol
(b) tertiary, 2,4-dimethyl-2-pentanol

19.9 (a) 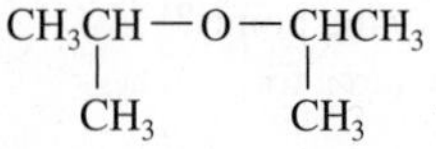 $CH_3CH — O — CHCH_3$
$\qquad\qquad\quad | \qquad\qquad\quad |$
$\qquad\qquad\quad CH_3 \qquad\qquad CH_3$

(b) 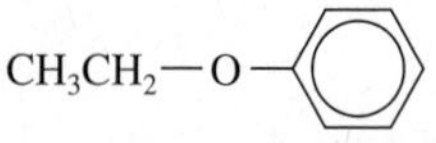 $CH_3CH_2 — O —$

$\qquad\qquad\qquad\qquad\qquad CH_3$
$\qquad\qquad\qquad\qquad\qquad |$
(c) 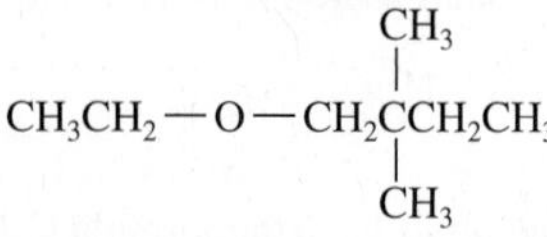 $CH_3CH_2 — O — CH_2CCH_2CH_3$
$\qquad\qquad\qquad\qquad\qquad |$
$\qquad\qquad\qquad\qquad\qquad CH_3$

19.10 (a) 2-methylhexanal
(b) 3,5-dimethyl-2-octanone

19.11 $CH_3(CH_2)_{15}COOH$, heptadecanoic acid
$CH_3(CH_2)_{17}COOH$, nonadecanoic acid

19.12 (a) methyl propanoate
methyl propionate
(b) ethyl butanoate
ethyl butyrate
(c) methyl dodecanoate
methyl laurate

19.13 Answers will vary. Here is an example:

$CH_2{=}CH_2 + F_2C{=}CF_2 \longrightarrow$

$\qquad\quad {-}(CH_2CH_2 — CF_2CF_2 — CH_2CH_2 — CF_2CF_2 {)}_n$

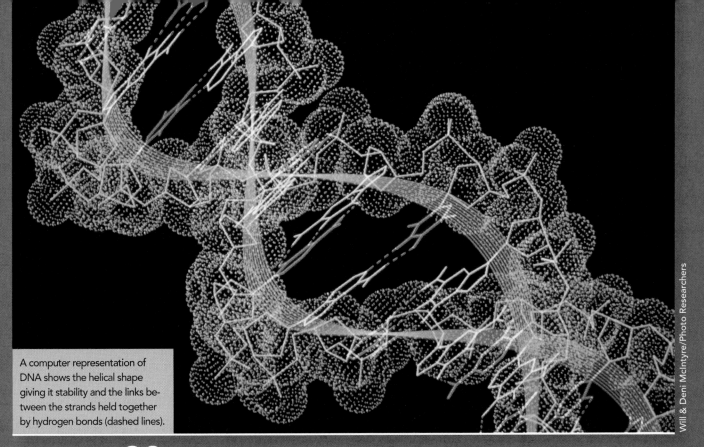

A computer representation of DNA shows the helical shape giving it stability and the links between the strands held together by hydrogen bonds (dashed lines).

Will & Deni McIntyre/Photo Researchers

CHAPTER **20**

INTRODUCTION TO BIOCHEMISTRY

The study of life has long fascinated us—it's probably the most intriguing of all scientific studies, but the answer to our question "What is life?" still eludes us. The chemical bases for certain fundamental biological processes are well understood; we know how sunlight and carbon dioxide are converted into food. We can pinpoint specific genes and identify their function. We even use DNA to "fingerprint" suspects at crime scenes. But still we search for an answer to "What is life?"

Common molecular principles underscore the diversity of life, which enable us to produce essential substances, such as insulin for diabetics, using bacteria as the manufacturing machine. Other substances, such as the human growth hormone, are being genetically engineered as well.

Biochemistry is more important than ever to the practice of medicine and the study of nutrition. Today we routinely measure the severity of a heart attack by the levels of certain enzymes in the blood. The roles of cholesterol, fats, and trace elements in our diet are current topics of great interest.

CHAPTER OUTLINE

20.1 Chemistry in Living Organisms
20.2 Carbohydrates
20.3 Lipids
20.4 Amino Acids and Proteins
20.5 Enzymes
20.6 Nucleic Acids, DNA, and Genetics

20.1 CHEMISTRY IN LIVING ORGANISMS

Name the four major classes of biomolecules.

KEY TERM
biochemistry

Chemical substances present in living organisms—from microbes to humans—range in complexity from water and simple salts to DNA (deoxyribonucleic acid) molecules containing tens of thousands of atoms. Four of the chemical elements—hydrogen, carbon, nitrogen, and oxygen—make up approximately 95% of the mass of living matter. Small amounts of sulfur, phosphorus, calcium, sodium, potassium, chlorine, magnesium, and iron, together with trace amounts of many other elements such as copper, manganese, zinc, cobalt, and iodine, are also found in living organisms. The human body consists of about 60% water, with some tissues having a water content as high as 80%.

Biochemistry is the branch of chemistry concerned with the chemical reactions occurring in living organisms. Its scope includes such processes as growth, digestion, metabolism, and reproduction.

The four major classes of biomolecules upon which all life depends are carbohydrates, lipids, proteins, and nucleic acids. Each kind of living organism has an amazing ability to select and synthesize a large portion of the many complicated molecules needed for its existence. In fact, the processes carried out in a living organism are similar to those of a highly automated, smoothly running chemical factory. But unlike chemical factories, living organisms are able to expand (grow), repair damage (if not too severe), and, finally, reproduce themselves.

20.2 CARBOHYDRATES

State the empirical formula for carbohydrates and the different classes of sugars, giving an example of each.

KEY TERMS
carbohydrate
monosaccharide
disaccharide
polysaccharide

© Rey Rojo/iStockphoto

The carbohydrates in these hay rolls are storage molecules for energy from the Sun.

Chemically, **carbohydrates** are polyhydroxy aldehydes or polyhydroxy ketones or substances that yield these compounds when hydrolyzed. The name *carbohydrate* was given to this class of compounds many years ago by French scientists who called them *hydrates de carbone* because their empirical formulas approximated $(C \cdot H_2O)_n$. However, the hydrogen and oxygen do not actually exist as water or in hydrate form, as we have seen in such compounds as $BaCl_2 \cdot 2 H_2O$. Empirical formulas used to represent carbohydrates are $C_x(H_2O)_y$ and $(CH_2O)_n$.

Carbohydrates, also known as saccharides, occur naturally in plants and are one of the three principal classes of animal food. The other two classes of foods are lipids (fats) and proteins. Plants are able to synthesize carbohydrates by the photosynthetic process. Animals are incapable of this synthesis and are dependent on the plant kingdom for their source of carbohydrates. Animals are, however, capable of converting plant carbohydrate into glycogen (animal carbohydrate) and storing this glycogen throughout the body as a reserve energy source. The amount of energy available from carbohydrates is about 17 kJ/g (4 kcal/g).

Carbohydrates exist as sugars, starches, and cellulose. The simplest of these are the sugars. The common names for sugars end in -*ose* (e.g., glucose, sucrose, maltose). Carbohydrates are classified as monosaccharides, disaccharides, oligosaccharides, and polysaccharides according to the number of monosaccharide units linked in a molecule.

Monosaccharides

Monosaccharides are carbohydrates that cannot be hydrolyzed to simpler carbohydrate units. They are often called simple sugars, the most common of which is glucose. Monosaccharides containing three-, four-, five-, and six-carbon atoms are called *trioses, tetroses, pentoses,* and *hexoses,* respectively. Monosaccharides that contain an aldehyde group on one carbon atom and a hydroxyl group on each of the other carbon atoms are called *aldoses. Ketoses* are monosaccharides that contain a ketone group on one carbon atom and a hydroxyl group on each of the other carbons.

triose	$C_3H_6O_3$
tetrose	$C_4H_8O_4$
pentose	$C_5H_{10}O_5$
hexose	$C_6H_{12}O_6$

$$-\overset{\overset{\displaystyle H}{|}}{C}=O \qquad -OH \qquad \overset{\displaystyle O}{\underset{}{\overset{\|}{C}}}$$

aldehyde   hydroxyl   ketone

We can write structural formulas for many monosaccharides, but only a limited number are of biological importance. Sixteen different isomeric aldohexoses of formula $C_6H_{12}O_6$ are known; glucose and galactose are the most important of these. Most sugars exist predominantly in a cyclic structure, forming a five- or six-membered ring. For example, glucose exists as a ring in which C-1 is bonded through an oxygen atom to C-5:

open-chain form
of glucose

cyclic form of glucose

or

erythrose
(an aldotetrose)

Glucose

Glucose Glucose ($C_6H_{12}O_6$) is the most important of the monosaccharides. An aldohexose found in the free state in plants and animal tissues, glucose is commonly known as *dextrose* or *grape sugar* and is a component of the disaccharides sucrose, maltose, and lactose; it is also the monomer of the polysaccharides starch, cellulose, and glycogen. Among the common sugars, glucose is of intermediate sweetness (see Table 20.1).

Glucose is the key sugar of the body and is carried by the bloodstream to all body parts. The concentration of glucose in the blood is normally 80–100 mg per 100 mL of blood. Because glucose is the most abundant carbohydrate in the blood, it is also sometimes known as *blood sugar*. Glucose requires no digestion and therefore may be given intravenously to patients who cannot take food by mouth. Glucose is found in the urine of those who have diabetes mellitus (sugar diabetes), a condition called *glycosuria*.

Galactose Galactose ($C_6H_{12}O_6$) is also an aldohexose and occurs along with glucose in the disaccharide lactose and in many oligo- and polysaccharides, such as pectin, gums, and mucilages. It is an isomer of glucose, differing only in the spatial arrangement of the —H and —OH groups around C-4. Galactose is synthesized in the mammary glands to make the lactose of milk and is less than half as sweet as glucose.

galactose
(an aldohexose)

cyclic form of galactose

A severe inherited disease, called galactosemia, is the inability of infants to metabolize galactose. The galactose concentration increases markedly in the blood and also appears in the urine. Galactosemia causes vomiting, diarrhea, enlargement of the liver, and often mental retardation. If not recognized within a few days after birth, it can lead to death. But if diagnosis is made early and lactose is excluded from the diet, the symptoms disappear and normal growth is resumed. Sometimes, as people age, their intestines stop producing lactase enzyme. Instead, lactose is metabolized by bacteria in the large intestine, creating gas and intestinal discomfort. This condition is called lactose intolerance and is treated by taking lactase tablets after meals.

Fructose Fructose ($C_6H_{12}O_6$), also known as levulose, is a ketohexose and occurs in fruit juices, honey, and (along with glucose) as a constituent of the disaccharide sucrose. Fructose is the major constituent of the polysaccharide inulin, a starchlike substance present in many plants, such as dahlia tubers, chicory roots, and Jerusalem artichokes. Fructose is the sweetest of all the

TABLE 20.1 Relative Sweetness of Sugars	
Fructose	100
Sucrose	58
Glucose	43
Maltose	19
Galactose	19
Lactose	9.2
Invert sugar	75

sugars, being more than twice as sweet as glucose. This sweetness accounts for the sweet taste of honey; the enzyme invertase, which is present in bees, splits sucrose into glucose and fructose. Fructose is metabolized directly but is also readily converted to glucose in the liver.

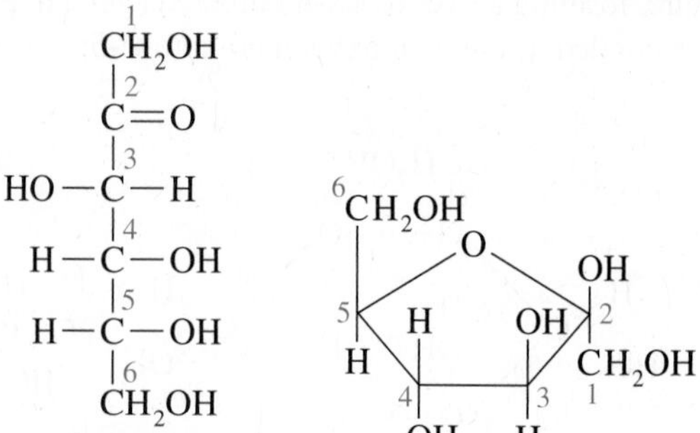

Fructose is the sugar found in honey. It is shown dripping from a honeycomb into the jar.

fructose
(a ketohexose)

cyclic form of fructose

Ribose Ribose ($C_5H_{10}O_5$) is an aldopentose and is present in adenosine triphosphate (ATP), one of the chemical energy carriers in the body. Ribose and one of its derivatives, 2-deoxyribose, are also important components of the nucleic acids RNA and DNA, respectively—the genetic information carriers in the body.

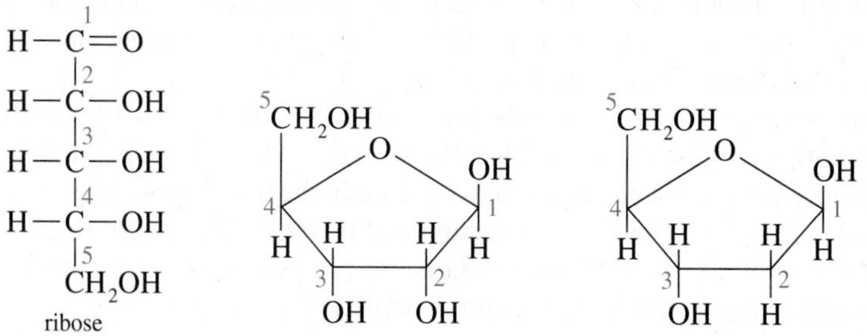

Table sugar is sucrose, a disaccharide, and can be found cubed or granulated.

ribose
(an aldopentose)

cyclic form of ribose

2-deoxyribose

WileyPLUS

ENHANCED EXAMPLE

EXAMPLE 20.1

Classify the structures shown below as aldo-sugars or keto-sugars and indicate which carbon atom is the defining atom for the classification.

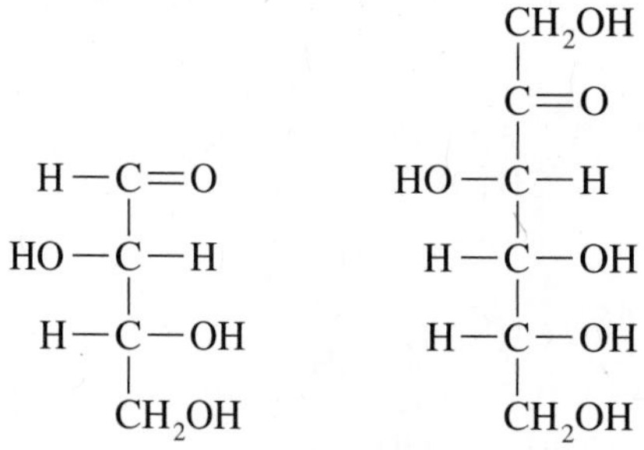

(a)

(b)

SOLUTION

(a) This is an aldotetrose. The aldehyde group is found on C-1.
(b) This is a ketohexose. The ketone group is found on C-2.

PRACTICE 20.1

What happens to the carbonyl oxygen when a sugar is in a cyclic structure?

Disaccharides

Disaccharides are carbohydrates whose molecules yield two molecules of the same or of different monosaccharides when hydrolyzed. The three disaccharides that are especially important from a biological viewpoint are sucrose, lactose, and maltose.

Sucrose, $C_{12}H_{22}O_{11}$, which is commonly known as *table sugar*, is found in the free state throughout the plant kingdom. Sugar cane contains 15–20% sucrose, and sugar beets contain 10–17%. Maple syrup and sorghum are also good sources of sucrose.

Lactose ($C_{12}H_{22}O_{11}$), or *milk sugar*, is found free in nature mainly in the milk of mammals. Human milk contains about 6.7% lactose, and cow's milk contains about 4.5%.

Maltose ($C_{12}H_{22}O_{11}$) is found in sprouting grain but occurs much less commonly (in nature) than either sucrose or lactose. Maltose is prepared commercially by the partial hydrolysis of starch, catalyzed either by enzymes or by dilute acids.

Disaccharides are not used directly in the body but are first hydrolyzed to monosaccharides. Disaccharides yield two monosaccharide molecules when hydrolyzed in the laboratory at elevated temperatures in the presence of hydrogen ions (acids) as catalysts. In biological systems, enzymes (biochemical catalysts) carry out the reaction. A different enzyme is required for the hydrolysis of each of the three disaccharides:

$$\text{sucrose + water} \xrightarrow{\text{H}^+ \text{ or sucrase}} \text{glucose + fructose}$$

$$\text{lactose + water} \xrightarrow{\text{H}^+ \text{ or lactase}} \text{galactose + glucose}$$

$$\text{maltose + water} \xrightarrow{\text{H}^+ \text{ or maltase}} \text{glucose + glucose}$$

The structure of a disaccharide is derived from two monosaccharide molecules by the elimination of a water molecule between them. In maltose, for example, the two monosaccharides are glucose. The water molecule is eliminated between the OH group on C-1 of one glucose unit and the OH group on C-4 of the other glucose unit. Thus, the two glucose units are joined through an oxygen atom at C-1 and C-4.

© SelectStock/iStockphoto

Lactose is a disaccharide found in breast milk.

maltose, a disaccharide

glucose unit   glucose unit

Sucrose consists of a glucose unit and a fructose unit linked through an oxygen atom from C-1 on glucose to C-2 on fructose. In lactose, the linkage is from C-1 of galactose through an oxygen atom to C-4 of glucose.

glucose unit

fructose unit

sucrose, a disaccharide

Image Source/Getty Images, Inc.

Sucrose is the sugar we commonly use in our kitchens for baking.

Cotton is a complex carbohydrate made up of almost pure cellulose.

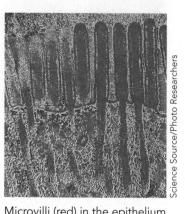

Microvilli (red) in the epithelium of the human duodenum are responsible for the absorption of food. The fingerlike projections maximize absorption capacity.

Polysaccharides

Polysaccharides are also called *complex carbohydrates* and can be hydrolyzed to a large number of monosaccharide units. The molar masses of polysaccharides range up to 1 million or more. Three of the most important polysaccharides are starch, glycogen, and cellulose.

Starch is a polymer of glucose and is found mainly in the seeds, roots, and tubers of plants. Corn, wheat, potatoes, rice, and cassava are the chief sources of starch whose principal use is for food.

Glycogen is the reserve carbohydrate of the animal kingdom and is often called *animal starch*. Glycogen is formed in the body by polymerization of glucose and is stored especially in the liver and in muscle tissue. Glycogen also occurs in some insects and lower plants including fungi and yeasts.

Cellulose, like starch and glycogen, is also a polymer of glucose. It differs from starch and glycogen in the manner in which the cyclic glucose units are linked to form chains. Cellulose is the most abundant organic substance found in nature. It is the chief structural component of plants and wood. Cotton fibers are almost pure cellulose, and wood, after removal of moisture, consists of about 50% cellulose. An important substance in the textile and paper industries, cellulose is also used to make rayon fibers, photographic film, guncotton, and cellophane. Humans cannot utilize cellulose as food because they lack the necessary enzymes to hydrolyze it to glucose. However, cellulose is an important source of fiber in the diet.

The digestion and metabolism of carbohydrates is a complex biochemical process. It starts in the mouth, where the enzyme amylase in the saliva begins the hydrolysis of starch to maltose and temporarily stops in the stomach, where hydrochloric acid deactivates the enzyme. Digestion continues in the intestines, where the hydrochloric acid is neutralized and pancreatic enzymes complete the hydrolysis to maltose. The enzyme maltase then catalyzes the digestion of maltose to glucose:

$$\text{starch} \xrightarrow{\text{amylase}} \text{dextrins} \xrightarrow{\text{amylase}} \text{maltose} \xrightarrow{\text{maltase}} \text{glucose}$$

Other specific enzymes in the intestines convert sucrose and lactose to monosaccharides. Glucose is absorbed through the intestinal walls into the bloodstream, where it is transported to the cells to be used for energy. Excess glucose is rapidly removed by the liver and muscle tissue, where it is polymerized and stored as glycogen. As the body calls for it, glycogen is converted back to glucose, which is ultimately oxidized to carbon dioxide and water with the release of energy. This energy is used by the body for maintenance, growth, and other normal functions.

20.3 LIPIDS

LEARNING OBJECTIVE ● Describe the general characteristics of lipids including fats, oils, and triglycerides.

KEY TERMS

lipids
fats
oils
triacylglycerols (triglycerides)

Lipids are a group of oily, greasy organic substances found in living organisms that are water insoluble but soluble in organic solvents, such as diethyl ether, benzene, and chloroform. Unlike carbohydrates, lipids share no common chemical structure. The most abundant lipids are the fats and oils, which make up one of the three important classes of foods.

Fats and **oils** are esters of glycerol and predominantly long-chain fatty acids (carboxylic acids). Fats and oils are also called **triacylglycerols** or **triglycerides**, since each molecule is derived from one molecule of glycerol and three molecules of fatty acid:

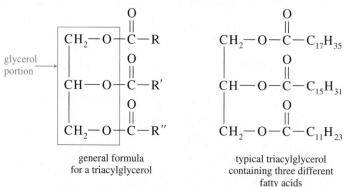

general formula for a triacylglycerol

typical triacylglycerol containing three different fatty acids

The formulas of triacylglycerol molecules vary for the following reasons:

1. The length of the fatty acid chain may vary from 4 to 20 carbons, but the number of carbon atoms in the chain is nearly always even.

2. Each fatty acid may be saturated, or it may be unsaturated and contain one, two, or three carbon–carbon double bonds.

3. An individual triacylglycerol may, and frequently does, contain three different fatty acids.

The most abundant saturated fatty acids in fats and oils are lauric, myristic, palmitic, and stearic acids (see Table 19.7). The most abundant unsaturated acids in fats and oils contain 18 carbon atoms and have one, two, or three carbon–carbon double bonds. Their formulas are

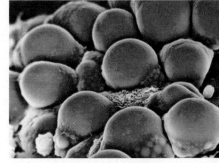

$$CH_3(CH_2)_7CH{=}CH(CH_2)_7COOH$$
oleic acid

$$CH_3(CH_2)_4CH{=}CHCH_2CH{=}CH(CH_2)_7COOH$$
linoleic acid

$$CH_3CH_2CH{=}CHCH_2CH{=}CHCH_2CH{=}CH(CH_2)_7COOH$$
linolenic acid

Fat cells like these can result in clogged arteries, but they also provide a major reserve of potential energy.

Other significant unsaturated fatty acids are palmitoleic acid (with 16 carbons) and arachidonic acid (with 20 carbons).

$$CH_3(CH_2)_5CH{=}CH(CH_2)_7COOH$$
palmitoleic acid

$$CH_3(CH_2)_4(CH{=}CHCH_2)_4CH_2CH_2COOH$$
arachidonic acid

Three unsaturated fatty acids—linoleic, linolenic, and arachidonic—are essential for animal nutrition and must be supplied in the diet. Diets lacking these fatty acids lead to impaired growth and reproduction and such skin disorders as eczema and dermatitis. We don't require fats in our diet except as a source of these three fatty acids.

PRACTICE 20.2

Write the structure of a triacylglycerol that contains one unit of palmitic acid, stearic acid, and oleic acids. How many other structures are possible containing these three acids?

The major physical difference between fats and oils is that fats are solid and oils are liquid at room temperature. Since the glycerol part of the structure is the same for both a fat and an oil, the difference must be due to the fatty acid end of the molecule. Fats contain a higher proportion of saturated fatty acids, whereas oils contain higher amounts of unsaturated fatty acids. The term *polyunsaturated* means that the molecules of a particular product each contain several double bonds.

Fats and oils are obtained from natural sources. In general, fats come from animal sources and oils from vegetable sources. Olive, cottonseed, corn, soybean, linseed, and other oils are obtained from the fruit or seed of their respective vegetable sources. Table 20.2 shows the major constituents of several fats and oils.

Fats are an important energy source for humans and normally account for about 25–50% of caloric intake. When oxidized to carbon dioxide and water, fats supply about 39 kJ/g (9.3 kcal/g), which is more than twice the amount obtained from carbohydrates and proteins.

Fats are digested in the small intestine, where they are first emulsified by the bile salts and then hydrolyzed to di- and monoglycerides, fatty acids, and glycerol. The fatty acids pass through the intestinal wall and are coated with a protein to increase solubility in the blood. They are then transported to various parts of the body where they are broken down in a series of enzyme-catalyzed reactions for the production of energy. Part of the hydrolyzed fat is converted back into fat in the adipose tissue and stored as a reserve source of energy. These fat deposits also function to insulate against loss of heat as well as to protect vital organs against mechanical injury.

Olive oil harvested from these olives is a healthy, monounsaturated oil.

TABLE 20.2 Fatty Acid Composition of Selected Fats and Oils

Fat or oil	Fatty acid (%)				
	Myristic acid	**Palmitic acid**	**Stearic acid**	**Oleic acid**	**Linoleic acid**
Animal fat					
Butter*	7–10	23–26	10–13	30–40	4–5
Lard	1–2	28–30	12–18	41–48	6–7
Vegetable oil					
Olive	0–1	5–15	1–4	49–84	4–12
Peanut	—	6–9	2–6	50–70	13–26
Corn	0–2	7–11	3–4	43–49	34–42
Cottonseed	0–2	19–24	1–2	23–33	40–48
Soybean	0–2	6–10	2–4	21–29	50–59
Linseed**	—	4–7	2–5	9–38	3–43

*Butyric acid, 3–4%
**Linolenic acid, 25–58%

Solid fats are preferable to oils for the manufacture of soaps and for use as certain food products. Hydrogenation of oils to make them solid is carried out on a large commercial scale. In this process hydrogen, bubbled through hot oil containing a finely dispersed nickel catalyst, adds to the carbon–carbon double bonds of the oil to saturate the double bonds and form fats. In practice, only some of the double bonds are allowed to become saturated (partially hydrogenated vegetable oil). The product that is marketed as solid "shortening" is used for cooking and baking. Oils and fats are also partially hydrogenated to improve their shelf life. Rancidity in fats and oils results from air oxidation at points of unsaturation, producing low-molar-mass aldehydes and acids of disagreeable odor and flavor.

Soap is made by reacting fats or oils with aqueous NaOH. This process, called *saponification*, requires 3 moles NaOH per mole of fat:

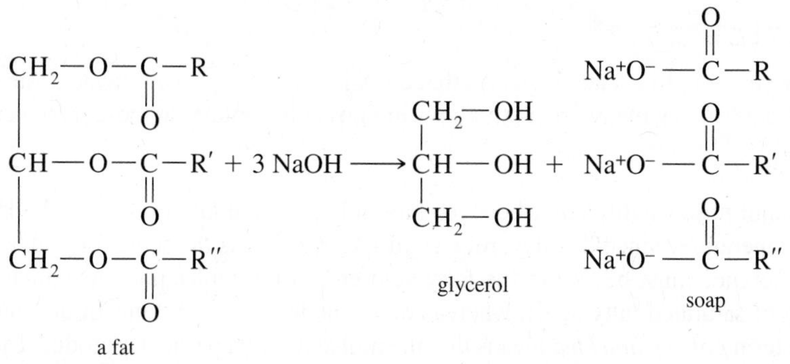

The most common soaps are the sodium salts of long-chain fatty acids, such as sodium stearate ($NaOOCC_{17}H_{35}$), sodium palmitate ($NaOOCC_{15}H_{31}$), and sodium oleate ($NaOOCC_{17}H_{33}$).

EXAMPLE 20.2

Write the reaction for the production of the soap sodium stearate.

SOLUTION

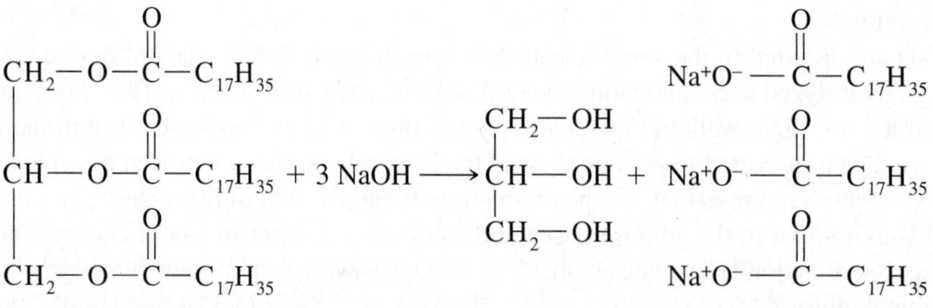

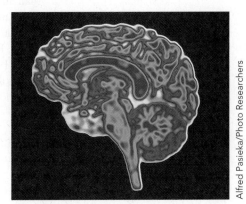

Figure 20.1

Formulas for a phospholipid, a glycolipid, and steroids.

a lecithin (a phospholipid)

a cerebroside (a glycolipid)

cholesterol (a steroid)

Norlutin
(a birth control pill)

Phospholipids and glycolipids are common in the brain.

Other principal classes of lipids, besides fats and oils, are phospholipids, glycolipids, and steroids (see **Figure 20.1**). The phospholipids are found in all animal and vegetable cells and are abundant in the brain, the spinal cord, egg yolk, and liver. Glycolipids (cerebrosides) contain a long-chain alcohol called *sphingosine*. They contain no glycerol but do contain a monosaccharide (usually galactose). Glycolipids are found in many different tissues but, as the name *cerebroside* indicates, occur in large quantities in brain tissue.

Steroids all have a four-fused carbocyclic-ring system (as in cholesterol) with various side groups attached to the rings. Cholesterol is the most abundant steroid in the body. It occurs in the brain, the spinal column, and nervous tissue, and it is the principal constituent of gallstones. The body synthesizes about 1 g of cholesterol per day, while about 0.3 g per day is ingested in the average diet. The major sources of cholesterol in the diet are meat, liver, and egg yolk. The cholesterol level in the blood generally rises with a person's age and body weight. High blood-level cholesterol is associated with atherosclerosis (hardening of the arteries), which results in reduced flow of blood and high blood pressure. Cholesterol is needed by the body to synthesize other steroids, some of which regulate male and female sexual characteristics. Many synthetic birth control pills are modified steroids that interfere with a woman's normal conception cycle.

Ring structure of steroids

20.4 AMINO ACIDS AND PROTEINS

LEARNING OBJECTIVE ● Describe the chemical composition and functions of amino acids and proteins.

KEY TERMS

amino acid
protein
peptide linkage
polypeptide

Proteins are the third important class of foodstuffs. Some common foods with high (over 10%) protein content are gelatin, fish, beans, nuts, cheese, eggs, poultry, and meat. Proteins are present in all body tissue. They can form structural elements, such as hair, fingernails, wool, and silk. Proteins also function as enzymes that regulate the countless chemical reactions taking place in every living organism. About 15% of the human body weight is protein. Chemically, proteins are polymers of amino acids with molar masses ranging up to more than 50 million.

Amino acids are carboxylic acids that contain an amino ($-NH_2$) group attached to C-2 (the alpha carbon) and are thus called *α-amino acids*. They also contain another variable group, R. The R group represents any of the various groups that make up the specific amino acids. For example, when R is $H-$, the amino acid is glycine; when R is CH_3-, the amino acid is alanine; when R is $CH_3SCH_2CH_2-$, the amino acid is methionine.

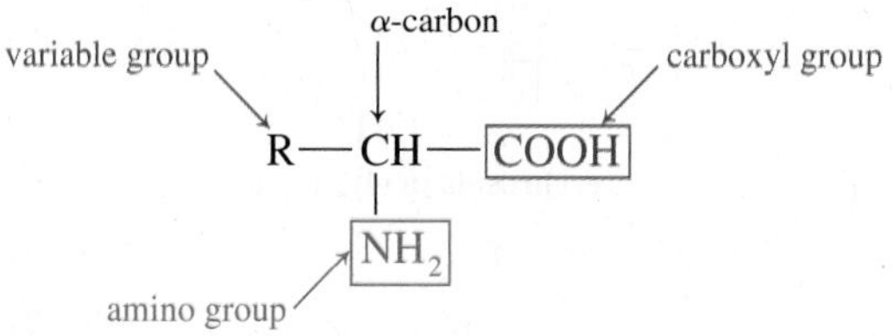

$$\overset{\text{α-carbon}}{\underset{\underset{\text{α-amino acid}}{\underset{\boxed{NH_2}}{|}}}{R-CH-\boxed{COOH}}}$$

variable group   carboxyl group
amino group

Some amino acids have two amino groups and some contain two acid groups.

PRACTICE 20.3

Write formulas for α-aminobutyric acid and β-aminobutyric acid.

There are approximately 200 different known amino acids in nature. Some are found in only one particular species of plant or animal, others in only a few lifeforms. But 20 of these amino acids are found in almost all proteins. Furthermore, these same 20 amino acids are used by all forms of life in the synthesis of proteins; their names, formulas, and abbreviations are given in Table 20.3. Eight are considered essential amino acids, since the human body is *not* capable of synthesizing them. Therefore, they must be supplied in our diets if we are to enjoy normal health.

TABLE 20.3 Common Amino Acids Derived from Proteins

Name	Abbreviation	Formula
Alanine	Ala	$CH_3CHCOOH$ $\quad\ \ |$ $\quad NH_2$
Arginine	Arg	$NH_2-\overset{||}{\underset{NH}{C}}-NH-CH_2CH_2CH_2CHCOOH$ $\qquad\qquad\qquad\qquad\qquad\ |$ $\qquad\qquad\qquad\qquad\quad NH_2$
Asparagine	Asn	$NH_2\overset{||}{\underset{O}{C}}-CH_2CHCOOH$ $\qquad\qquad\qquad |$ $\qquad\qquad\quad NH_2$
Aspartic acid	Asp	$HOOCCH_2CHCOOH$ $\qquad\qquad\quad |$ $\qquad\qquad\ NH_2$

TABLE 20.3 Common Amino Acids Derived from Proteins (Continued)

Name	Abbreviation	Formula
Cysteine	Cys	$\underset{\underset{NH_2}{\mid}}{HSCH_2CHCOOH}$
Glutamic acid	Glu	$\underset{\underset{NH_2}{\mid}}{HOOCCH_2CH_2CHCOOH}$
Glutamine	Gln	$\underset{\underset{O}{\parallel}}{NH_2C}CH_2CH_2\underset{\underset{NH_2}{\mid}}{CHCOOH}$
Glycine	Gly	$\underset{\underset{NH_2}{\mid}}{HCHCOOH}$
Histidine	His	$\begin{array}{c} N = CH \\ \parallel \qquad \parallel \\ HC \qquad C - CH_2\underset{\underset{NH_2}{\mid}}{CHCOOH} \\ \diagdown N \diagup \\ \mid \\ H \end{array}$
Isoleucine*	Ile	$CH_3CH_2\underset{\underset{CH_3}{\mid}}{CH} - \underset{\underset{NH_2}{\mid}}{CHCOOH}$
Leucine*	Leu	$(CH_3)_2CHCH_2\underset{\underset{NH_2}{\mid}}{CHCOOH}$
Lysine*	Lys	$NH_2CH_2CH_2CH_2CH_2\underset{\underset{NH_2}{\mid}}{CHCOOH}$
Methionine*	Met	$CH_3SCH_2CH_2\underset{\underset{NH_2}{\mid}}{CHCOOH}$
Phenylalanine*	Phe	$CH_2\underset{\underset{NH_2}{\mid}}{CHCOOH}$ (phenyl ring)
Proline	Pro	pyrrolidine ring $-COOH$
Serine	Ser	$HOCH_2\underset{\underset{NH_2}{\mid}}{CHCOOH}$
Threonine*	Thr	$CH_3\underset{\underset{OH}{\mid}}{CH} - \underset{\underset{NH_2}{\mid}}{CHCOOH}$
Tryptophan*	Trp	indole ring $C-CH_2\underset{\underset{NH_2}{\mid}}{CHCOOH}$
Tyrosine	Tyr	$HO -$ (phenyl ring) $- CH_2\underset{\underset{NH_2}{\mid}}{CHCOOH}$
Valine*	Val	$(CH_3)_2CH\underset{\underset{NH_2}{\mid}}{CHCOOH}$

*Amino acids essential in human nutrition.

Proteins are polymeric substances that yield primarily amino acids on hydrolysis. The bond connecting the amino acids in a protein is commonly called a **peptide linkage** or peptide bond. If we combine two glycine molecules by eliminating a water molecule between the amino group of one and the carboxyl group of the second glycine, we form a compound containing the amide structure and the peptide linkage. The compound containing the two amino acid groups is called a *dipeptide:*

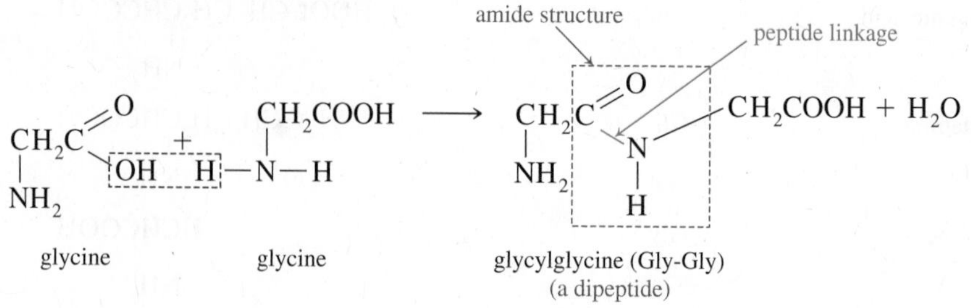

The product formed from two glycine molecules is called *glycylglycine* (abbreviated Gly-Gly). Note that the molecule still has a free amino group at one end and a free carboxyl group at the other end. The formation of Gly-Gly is considered the first step in the synthesis of a protein, since each end of the molecule is capable of joining to another amino acid. We can thus visualize the formation of a protein by joining a great many amino acids in this fashion.

EXAMPLE 20.3

Show how the tripeptide made from tyrosine–alanine–glycine is formed.

SOLUTION

This compound contains two peptide linkages:

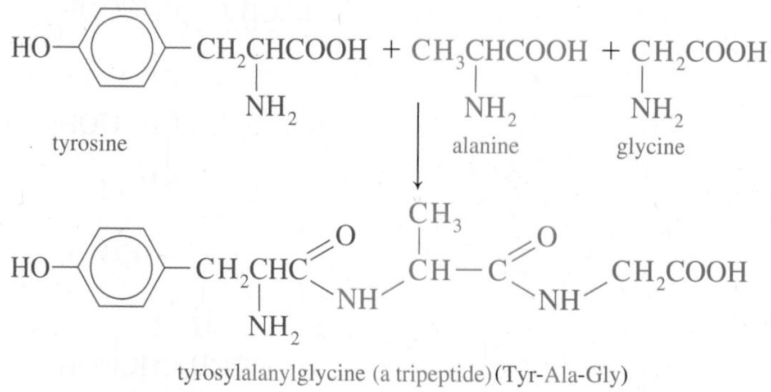

tyrosylalanylglycine (a tripeptide)(Tyr-Ala-Gly)

There are five other tripeptide combinations of these three amino acids using only one unit of each amino acid. Peptides containing up to about 40–50 amino acid units in a chain are **polypeptides**. Still longer chains of amino acids are proteins.

The amino acid units in a peptide are called *amino acid residues* or simply residues. (They no longer are amino acids because they have lost an H atom from their amino groups and an OH from their carboxyl groups.) In linear peptides, one end of the chain has a free amino group and the other end a free carboxyl group. The amino-group end is called the *N-terminal residue* and the other end the *C-terminal residue:*

The sequence of amino acids in a chain is numbered starting with the N-terminal residue, which is usually written to the left with the C-terminal residue at the right.

Peptides are named as acyl derivatives of the C-terminal amino acid, with the C-terminal unit keeping its complete name. The *-ine* ending of all but the C-terminal amino acid is changed to *-yl*, and these are listed in the order in which they appear, starting with the N-terminal amino acid:

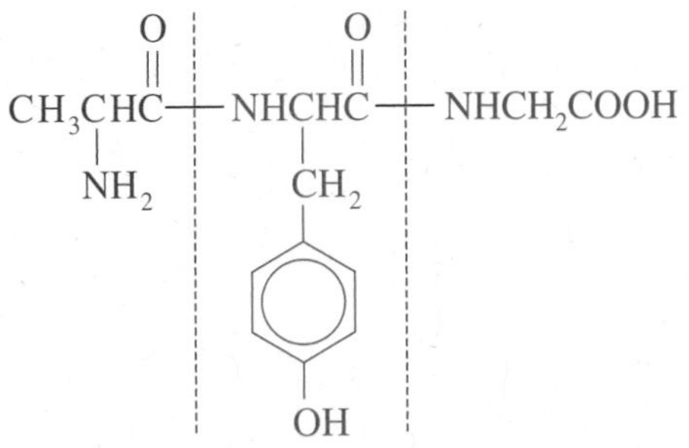

alanyl   tyrosyl   glycine

Ala-Tyr-Gly

Thus, Ala-Tyr-Gly is called *alanyltyrosylglycine*, and Arg-Gln-His-Ala is arginylglutamylhistidylalanine.

PRACTICE 20.4

Write full names for

(a) Pro-Tyr   (b) Gln-His-Leu

PRACTICE 20.5

Draw the complete structure for

(a) Ala-Ser   (b) Gly-Thr-Met

Many small, naturally occurring polypeptides have significant biochemical functions. The amino acid sequences of two of these, oxytocin and vasopressin, are shown in **Figure 20.2**. Oxytocin controls uterine contractions during labor in childbirth and also causes contraction of the smooth muscles of the mammary gland, resulting in milk secretion. Vasopressin in high concentration raises the blood pressure and has been used in surgical shock treatment for this purpose. Vasopressin is also an antidiuretic, regulating the excretion of fluid by the kidneys. The absence of vasopressin leads to diabetes insipidus. This condition, which is characterized by excretion of up to 30 L of urine per day, may be controlled by administering vasopressin or its derivatives. Oxytocin and vasopressin are similar nonapeptides, differing only at positions 3 and 8.

Determining the sequence of the amino acids in even one protein molecule was once a formidable task. The amino acid sequence of beef insulin was announced in 1955 by the British biochemist Frederick Sanger (b. 1918). Finding the sequence of this structure required several

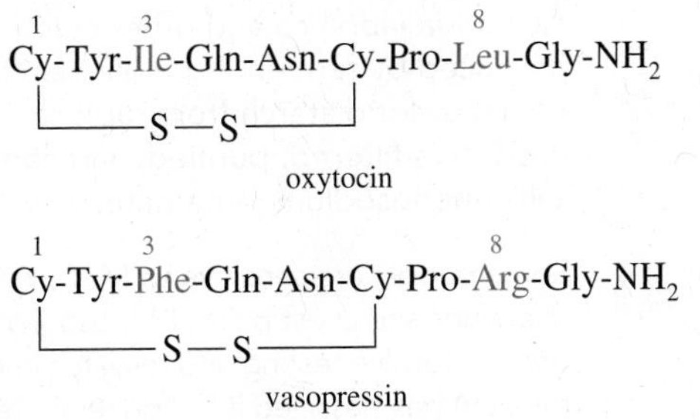

oxytocin

vasopressin

Figure 20.2

Amino acid sequences of oxytocin and vasopressin. The difference in only two amino acids (shown in red) in these two compounds results in very different physiological activity. The C-terminal amino acid has an amide structure (Gly-NH$_2$) instead of —COOH.

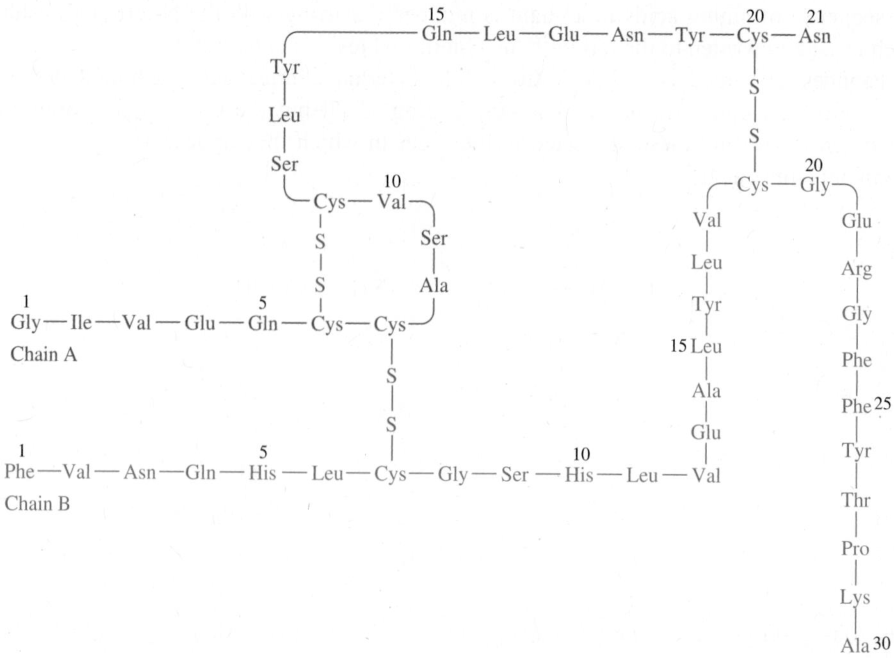

Figure 20.3

Amino acid sequence of beef insulin.

years of effort by Sanger's team. He was awarded the 1958 Nobel Prize in chemistry for this work. As a result of this work, automated amino acid sequencers have been developed that can determine the sequence of an average-sized protein in a few days. Beef insulin consists of 51 amino acid units in two polypeptide chains. The two chains are connected by disulfide linkages ($-S-S-$) of two cysteine residues at two different sites. The structure is shown in **Figure 20.3**. Insulins from other animals, including humans, differ slightly by one, two, or three amino acid residues.

Protein digestion takes place in the stomach and the small intestine. Here, digestive enzymes hydrolyze proteins to smaller peptides and amino acids, which pass through the walls of the intestines, are absorbed by the blood, and are transported to the liver and other tissues of the body. The body does not store free amino acids. They are used to synthesize

- proteins to replace and repair body tissues
- other nitrogen-containing substances, such as certain hormones and heme
- nucleic acids
- enzymes that control the synthesis of other necessary products, such as carbohydrates and fats

---

> CHEMISTRY *IN ACTION*

The Taste of Umami

The four primary tastes were long ago established as sour, sweet, salty, and bitter. A fifth taste has been added to the group. It was discovered by Kikunae Ikeda, a chemistry professor at Imperial University of Tokyo. The taste is called umami, after the Japanese word *umai*, meaning "delicious." Ikeda isolated and identified glutamate as the substance Japanese chefs used to make food taste better. We know glutamate as MSG (monosodium glutamate), sold in North America as Accent. It is the sodium salt of the amino acid glutamic acid (see structure in Table 20.3) and is a fine white crystal that looks similar to salt.

Glutamate is found in lots of natural foods, including meat, poultry, vegetables, fish, cheese, and tomatoes. The flavor-enhancing property of glutamate occurs when it is in its free or uncombined form (not bound into protein). Many Asian seasonings give umami taste to foods. These include soy sauce and fish sauce. The Australian product Vegemite also has a strong umami taste.

Long ago MSG was extracted from a seaweed called *Laminaria japonica* and other plants. Now MSG is produced by fermenting molasses (from sugar cane or beets) or food starch from tapioca. The fermentation product is filtered, purified, and converted to the sodium salt—monosodium glutamate.

Some people are sensitive to MSG, with reactions including migraines and chest pains. Monosodium glutamate has undergone extensive testing and research in the food industry, and the FDA has classified it as "generally recognized as safe." It is widely used to add flavor and umami to many foods.

Proteins are catabolized (degraded) to carbon dioxide, water, and urea. Urea, containing the protein nitrogen, is eliminated from the body in the urine.

Carbohydrates and fats are used primarily to supply heat and energy to the body. Proteins, on the other hand, are used mainly to repair and replace worn-out tissue. Tissue proteins are continuously being broken down and resynthesized. Therefore, protein must be continually supplied to the body in the diet. It is nothing short of amazing how organisms pick out the desired amino acids from the bloodstream and put them together in proper order to synthesize a needed protein.

20.5 ENZYMES

Explain the importance of enzymes in living organisms.

Enzymes are the catalysts of biochemical reactions. Most enzymes are proteins, and they catalyze nearly all of the reactions that occur in living cells. Uncatalyzed reactions that may require hours of boiling in the presence of a strong acid or a strong base can occur in a fraction of a second in the presence of the proper enzyme at room temperature and nearly neutral pH. This process is all the more remarkable when we realize that enzymes do not actually cause chemical reactions. They act as catalysts by greatly lowering the activation energy of specific biochemical reactions. The lowered activation energy permits these reactions to proceed at high speed at body temperature.

● **LEARNING OBJECTIVE**

KEY TERMS
enzyme
substrate

EXAMPLE 20.4

Use a graph to show how an enzyme speeds up a chemical reaction.

SOLUTION

An enzyme lowers the activation energy of a biochemical reaction. This allows more molecules to react forming product more quickly. A graph of an uncatalyzed reaction and an enzyme-catalyzed reaction would look like this:

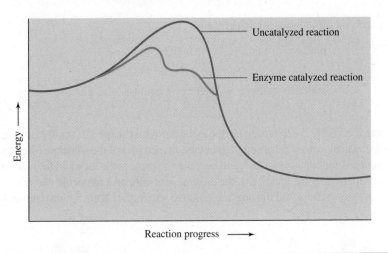

Louis Pasteur (1822–1895) was one of the first scientists to study enzyme-catalyzed reactions. He believed that living yeasts or bacteria were required for these reactions, which he called fermentations—for example, the conversion of glucose to alcohol by yeasts. In 1897, Eduard Büchner (1860–1917) made a cell-free filtrate that contained enzymes prepared by grinding yeast cells with very fine sand. The enzymes in this filtrate converted glucose to alcohol, thus proving that the presence of living cells is not required for enzyme activity. For this work Büchner received the Nobel Prize in chemistry in 1907.

Each organism contains thousands of enzymes. Some enzymes are simple proteins consisting only of amino acid units. Others are conjugated and consist of a protein part, or *apoenzyme*, and a nonprotein part, or *coenzyme*. Both parts are essential, and a functioning enzyme consisting of both the protein and nonprotein parts is called a *holoenzyme*.

$$\text{apoenzyme} + \text{coenzyme} = \text{holoenzyme}$$

Often the coenzyme is a vitamin, and the same coenzyme may be associated with many different enzymes.

For some enzymes, an inorganic component, such as a metal ion (e.g., Ca^{2+}, Mg^{2+}, or Zn^{2+}), is required. This inorganic component is an *activator*. From the standpoint of function, an activator is analogous to a coenzyme, but inorganic components are not called coenzymes.

PRACTICE 20.6

What is the difference between an activator and a coenzyme?

Another remarkable property of enzymes is their specificity of reaction; that is, a certain enzyme will catalyze the reaction of a specific type of substance. For example, the enzyme maltase catalyzes the reaction of maltose and water to form glucose. Maltase has no effect on the other two common disaccharides sucrose and lactose. Each of these sugars requires a specific enzyme—sucrase to break down sucrose, lactase to break down lactose. (See the equations in Section 20.2.)

The substance acted on by an enzyme is called the **substrate**. Sucrose is the substrate of the enzyme sucrase. Common names for enzymes are formed by adding the suffix *-ase* to the root of the substrate name. Note, for example, the derivations of maltase, sucrase, and lactase from maltose, sucrose, and lactose. Many enzymes, especially digestive enzymes, have common names such as pepsin, rennin, trypsin, and so on. These names have no systematic significance.

Enzymes act according to the following general sequence. Enzyme (E) and substrate (S) combine to form an enzyme–substrate intermediate (E–S). This intermediate decomposes to give the product (P) and regenerate the enzyme:

$$\text{E} + \text{S} \longleftarrow \text{E–S} \longrightarrow \text{E} + \text{P}$$

For the hydrolysis of maltose, the sequence is

$$\underset{\text{E}}{\text{maltase}} + \underset{\text{S}}{\text{maltose}} \rightleftharpoons \underset{\text{E–S}}{\text{maltase–maltose}}$$

$$\underset{\text{E–S}}{\text{maltase–maltose}} + \text{H}_2\text{O} \longrightarrow \underset{\text{E}}{\text{maltase}} + \underset{\text{P}}{\text{2 glucose}}$$

Enzyme specificity is believed to be due to the particular shape of a small part of the enzyme, its active site, which exactly fits a complementary-shaped part of the substrate (see **Figure 20.4**). This interaction is analogous to a lock and key; the substrate is the lock and the enzyme, the key. Just as a key opens only the lock it fits, the enzyme acts only on a molecule that fits its particular shape. When the substrate and the enzyme come together, they form a substrate–enzyme com-

Figure 20.4

Enzyme–substrate interaction illustrating specificity of an enzyme by the lock-and-key model.

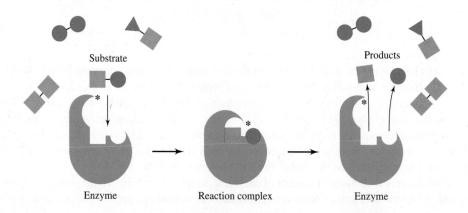

Substrate

Products

Enzyme

Reaction complex

Enzyme

plex unit. The substrate, activated by the enzyme in the complex, reacts to form the products, regenerating the enzyme.

A more recent model of the enzyme–substrate catalytic site, known as the induced-fit model, visualizes a flexible site of enzyme–substrate attachment, with the substrate inducing a change in the enzyme shape to fit the shape of the substrate. This model allows for the possibility that in some cases the enzyme might wrap itself around the substrate and so form the correct shape of lock and key. Thus, the enzyme does not need to have an exact preformed catalytic site to match the substrate.

20.6 NUCLEIC ACIDS, DNA, AND GENETICS

Describe the chemical structure of DNA and explain how it functions in genetics.

● **LEARNING OBJECTIVE**

KEY TERMS
DNA
nucleotide
RNA
transcription
genes
mitosis
meiosis

Explaining how hereditary material duplicates itself was a baffling problem for biologists. This explanation and the answer to the question "Why are the offspring of a given species undeniably of that species?" eluded biologists for many years. Many thought the chemical basis for heredity lay in the structure of proteins. But they couldn't find how protein reproduced itself. The answer to the hereditary problem was finally found in the structure of the nucleic acids.

The unit structure of all living things is the cell. Suspended in the nucleus of cells are chromosomes, which consist largely of proteins and nucleic acids. The nucleic acids and proteins are intimately associated in complexes called *nucleoproteins.* Nucleic acids contain either the sugar deoxyribose or the sugar ribose. Accordingly, they are called deoxyribonucleic acid (DNA) and ribonucleic acid (RNA). DNA was discovered in 1869 by Swiss physiologist Friedrich Miescher (1844–1895), who extracted it from the nuclei of cells.

DNA is a polymeric substance made up of thousands of units called nucleotides. The fundamental components of the nucleotides in DNA are phosphoric acid, deoxyribose (a pentose sugar), and the four nitrogen-containing bases adenine, thymine, guanine, and cytosine (abbreviated as A, T, G, and C). Phosphoric acid is obtained from minerals in the diet; deoxyribose is synthesized in the body from glucose; and the four nitrogen bases are made in the body from amino acids. The formulas for these compounds are given in Figure 20.5.

A **nucleotide** in DNA consists of one of the four bases linked to a deoxyribose sugar, which in turn is linked to a phosphate group. Each nucleotide has the following sequence:

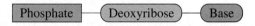

The structures for a single nucleotide and a segment of a polynucleotide (DNA) are shown in Figure 20.6.

In 1953, the American biologist James D. Watson (b. 1928) and the British physicist Francis H. C. Crick (1916–2004) announced their double-stranded helix structure for DNA. This concept was a milestone in the history of biology, and in 1962 Watson and Crick, along with Maurice H. F. Wilkin (1916–2004), who did the brilliant X-ray diffraction studies on DNA, were awarded the Nobel Prize in medicine and physiology.

phosphoric acid

2-deoxyribose

thymine (T)

Figure 20.5

Fundamental components of nucleotides in DNA.

guanine (G)

adenine (A)

cytosine (C)

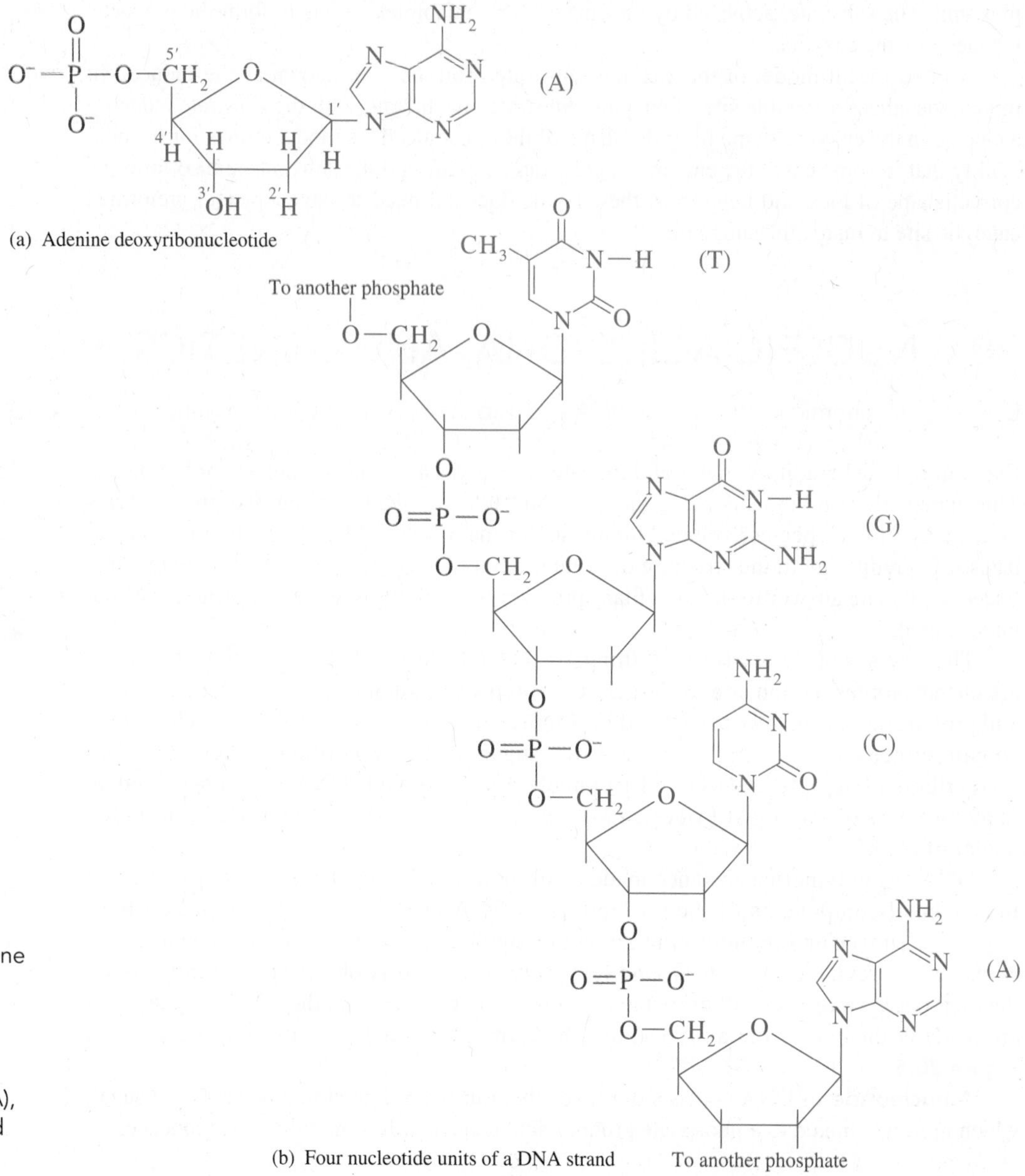

(a) Adenine deoxyribonucleotide

(b) Four nucleotide units of a DNA strand

Figure 20.6

(a) A single nucleotide, adenine deoxyribonucleotide. (b) A segment of one strand of deoxyribonucleic acid (DNA) showing four nucleotides, including those of adenine (A), cytosine (C), guanine (G), and thymine (T).

The structure of DNA, according to Watson and Crick, consists of two polymeric strands of nucleotides in the form of a double helix, with both nucleotide strands coiled around the same axis (see **Figure 20.7**). Along each strand are alternate phosphate and deoxyribose units with one of the four bases adenine, guanine, cytosine, or thymine attached to deoxyribose as a side group. The double helix is held together by hydrogen bonds extending from the base on one strand of the double helix to a complementary base on the other strand. Watson and Crick furthermore ascertained that adenine is always hydrogen-bonded to thymine and guanine is always hydrogen-bonded to cytosine. Previous analytical work by others substantiated this concept of complementary bases by showing that the molar ratio of adenine to thymine in DNA is approximately 1:1 and that of guanine to cytosine is also approximately 1:1.

The structure of DNA has been compared to a ladder twisted into a double helix, with the rungs of the ladder perpendicular to the twisted railings. The phosphate and deoxyribose units alternate along the two railings of the ladder, and two nitrogen bases form each rung of the ladder. The DNA structure is illustrated in Figure 20.7.

For any individual of any species, the sequence of base pairs and the length of the nucleotide chains in DNA molecules contain the coded messages that determine all of the characteristics of that individual. In this sense the DNA molecule is a template that stores information for recall as needed. DNA contains the genetic code of life, which is passed on from one generation to another.

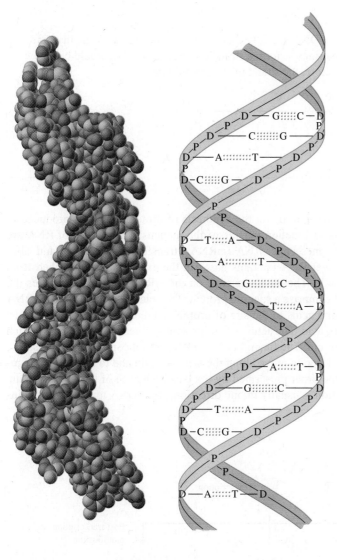

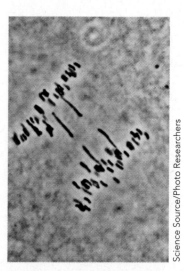

The chromosomes in this mammal cell are in the final stage of cell division. They contain nucleic acids and proteins replicated for the new cell.

In DNA, adenine is always bonded to thymine.

P = Phosphate
D = Deoxyribose
A = Adenine
T = Thymine
C = Cytosine
G = Guanine

Figure 20.7

Double-stranded helix structure of DNA. Adenine is hydrogen-bonded to thymine (A∷T), and cytosine is hydrogen-bonded to guanine (C⋮⋮G).

RNA is a polymer of nucleotides, but it differs from DNA in that (1) it is single-stranded; (2) it contains the pentose sugar ribose instead of deoxyribose; and (3) it contains uracil instead of thymine as one of its four nitrogen bases (see **Figure 20.8**).

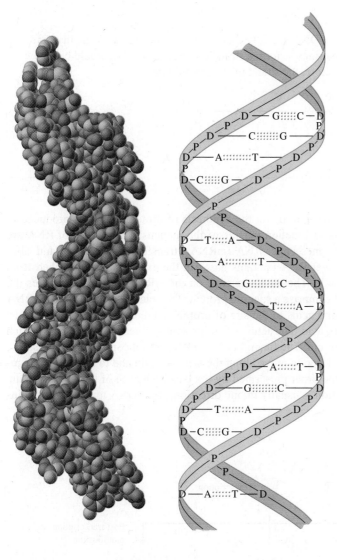

uracil

PRACTICE 20.7

Write the formula for the adenine ribonucleotide.

Transcription is the process by which DNA directs the synthesis of RNA. The nucleotide sequence of only one strand of DNA is transcribed into a single strand of RNA. This transcription occurs in a complementary fashion. Where there is a guanine base in DNA, a cytosine base will occur in RNA. Cytosine is transcribed to guanine, thymine to adenine, and adenine to uracil. Thymine occurs only in DNA and uracil only in RNA.

ribose

Figure 20.8

Components of RNA that differ from DNA.

EXAMPLE 20.5

How would the DNA strand shown below be transcribed to an RNA strand?

G	C	A	T	T	G	A	C	T	G

Guanine – G   Thymine – T
Cytosine – C   Uracil – U
Adenine – A

WileyPLUS

ENHANCED EXAMPLE

SOLUTION

The pairings between bases in DNA are G=C and A=T. Since RNA has no thymine but uses uracil instead, the pairings to form a strand of RNA from DNA would be C=G, T=A, A=U. Therefore, the strand of RNA would be

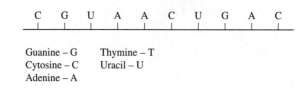

```
C   G   U   A   A   C   U   G   A   C
|   |   |   |   |   |   |   |   |   |
```

Guanine – G   Thymine – T
Cytosine – C   Uracil – U
Adenine – A

The main function of RNA is to direct the synthesis of proteins. RNA is produced in the cell nucleus but performs its function outside of the nucleus. Three kinds of RNA are produced directly from DNA: messenger RNA (*m*RNA), transfer RNA (*t*RNA), and ribosomal RNA (*r*RNA). Messenger RNA contains bases in the exact order transcribed from a strip of the DNA master code. The base sequence on the *m*RNA in turn establishes the sequence of amino acids that comprise a specific protein. The relatively small transfer RNA molecules bring specific amino acids to the site of protein synthesis. There is at least one different *t*RNA for each amino acid. The actual site of protein synthesis is a ribosome, which is composed of *r*RNA and protein. The function of the *r*RNA is not completely understood. However, the ribosome is believed to move along the *m*RNA chain and to aid in the polymerization of amino acids in the order prescribed by the base sequence of the *m*RNA chain. The flow of genetic information is usually in one direction, from DNA to RNA to proteins (**Figure 20.9**).

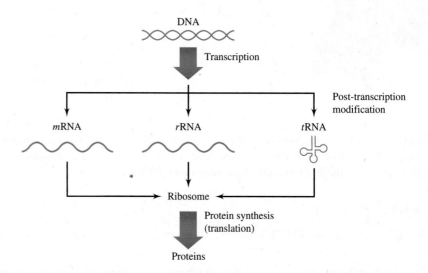

Figure 20.9

The process of cellular genetic information.

DNA and Genetics

Heredity is the process by which the physical and mental characteristics of parents are transferred to their offspring. For this process to occur, the material responsible for genetic transfer must be able to make exact copies of itself. The design for replication is built into Watson and Crick's DNA structure, first by the nature of its double-helical structure and second by the complementary nature of its nitrogen bases, where adenine will bond only to thymine and guanine only to cytosine. The DNA double helix unwinds, or "unzips," into two separate strands at the hydrogen bonds between the bases. Each strand then serves as a template combining only with the proper free nucleotides to produce two identical replicas of itself. This replication of DNA occurs in the cell just before the cell divides, thereby giving each daughter cell the full genetic code of the original cell. This process is illustrated in **Figure 20.10**.

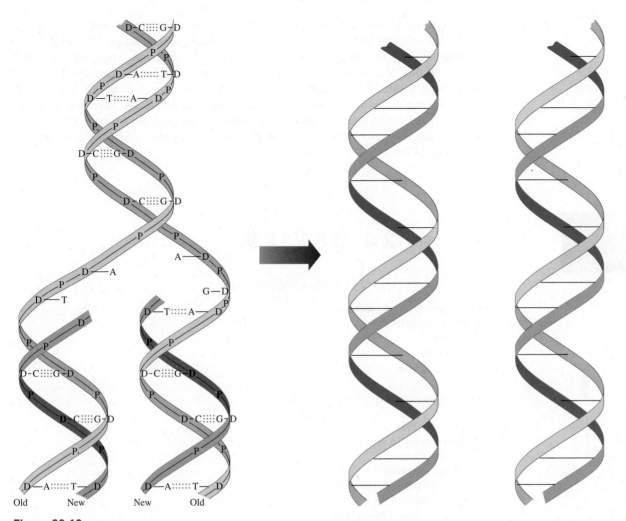

Figure 20.10

Replication of DNA. The two helices unwind, separating at the point of the hydrogen bonds. Each strand then serves as a template, recombining with the proper nucleotides to duplicate itself as a double-stranded helix.

As we have indicated, DNA is an integral part of the chromosomes. Each species carries a specific number of chromosomes in the nucleus of each of its cells. The number of chromosomes varies with different species. Humans have 23 pairs, or 46 chromosomes. Chromosomes are long, threadlike bodies composed of nucleic acids and proteins that contain the basic units of heredity called genes. **Genes** are segments of the DNA chain that contain the codes for the formation of polypeptides and RNAs. Hundreds of genes can exist along a DNA chain.

In ordinary cell division, known as **mitosis**, each DNA molecule forms a duplicate by uncoiling to single strands. Each strand then assembles the complementary portion from available free nucleotides to form a duplicate of the original DNA molecule. After cell division is completed, each daughter gene contains the genetic material that corresponds exactly to that present in the original cell before division.

However, in almost all higher forms of life, reproduction takes place by union of the male sperm with the female egg. Cell splitting to form the sperm cell and the egg cell occurs by a different and more complicated process called *meiosis.* In **meiosis**, the genetic material is divided so that each daughter cell receives one chromosome from each pair. After meiosis, the egg cell and the sperm cell each carry only half of the chromosomes from its original cell. Between them they form a new cell that once again contains the correct number of chromosomes and all the hereditary characteristics of the species. Thus, the offspring derives half of its genetic characteristics from the father and half from the mother.

Nature is not 100% perfect. Occasionally, DNA replication is not perfect, or a section of the DNA molecule is damaged by X-rays, radioactivity, or drugs, and a mutant organism is produced. For example, in the disease sickle-cell anemia, a large proportion of the red blood cells form into sickle shapes instead of the usual globular shape. This irregularity limits the ability of the blood to carry oxygen and causes the person to be weak and unable to fight infection, leading to early death. Sickle-cell anemia is due to one misplaced amino acid in the structure of hemoglobin. The sickle-cell-producing hemoglobin has a valine residue where a glutamic acid residue should be located. Sickle-cell anemia is an inherited disease indicating a fault in the DNA coding transmitted from parent to child. Many biological disorders and ailments have been traced directly to a deficiency in the genetic information of DNA.

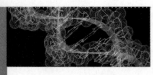

CHAPTER **20 REVIEW**

20.1 CHEMISTRY IN LIVING ORGANISMS

KEY TERM

biochemistry

- Four major classes of biomolecules:
  - Carbohydrates
  - Lipids
  - Proteins
  - Nucleic acids

20.2 CARBOHYDRATES

KEY TERMS

carbohydrate
monosaccharide
disaccharide
polysaccharide

- Empirical formulas: $C_x(H_2O)_y$ or $(CH_2O)_n$.
- Carbohydrates exist as sugars, starches, and cellulose.
- Classifying sugars:
  - Monosaccharide
    - Glucose, galactose, fructose, ribose
  - Disaccharide
    - Sucrose, lactose, maltose
  - Polysaccharide
    - Starch, glycogen, cellulose

20.3 LIPIDS

KEY TERMS

lipids
fats
oils
triacylglycerols (triglycerides)

- Lipids share no common chemical structure.
- Classes of lipids:
  - Fats and oils
  - Phospholipids
  - Glycolipids
  - Steroids

20.4 AMINO ACIDS AND PROTEINS

KEY TERMS

amino acid
protein
peptide linkage
polypeptide

- Form various structural elements and enzymes
- Proteins are polymers that break down mainly into amino acids:
  - Amino acids are connected in a protein by a peptide linkage.
  - Peptides containing up to about 40–50 amino acids are called polypeptides.
  - Longer chains are proteins.

20.5 ENZYMES

KEY TERMS

enzyme
substrate

- Enzymes catalyze biochemical reactions.
- Enzymes are classified as:
  - Simple—only amino acid units
  - Conjugated—protein part and nonprotein part
- Substance acted on by the enzyme is called the substrate.
- Specificity of enzyme action is due to the shape of a small part of the enzyme molecule.

20.6 NUCLEIC ACIDS, DNA, AND GENETICS

- Components of DNA:
  - Phosphoric acid
  - Deoxyribose
  - Four nitrogen-containing bases
    - Adenine
    - Thymine
    - Guanine
    - Cytosine
- A nucleotide is a base linked to deoxyribose (or ribose) sugar linked to a phosphate group.
- The structure of DNA is a double helix composed of two complementary strands joined by hydrogen bonds.
- RNA is a polymer of nucleotides but differs from DNA:
  - Single stranded
  - Contains ribose instead of deoxyribose
  - Contains uracil in place of the base thymine
- Transcription is the process by which DNA directs the synthesis of RNA.

KEY TERMS

DNA
nucleotide
RNA
transcription
genes
mitosis
meiosis

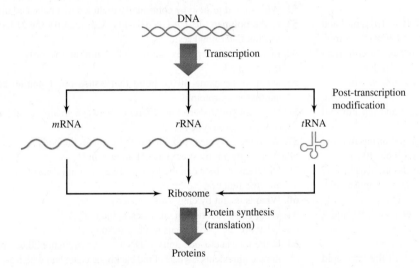

- Genes are segments of DNA that contain codes for the formation of polypeptides and RNAs.
- In mitosis a DNA molecule uncoils and reassembles to form two new duplicate double helixes. Each new cell contains a full complement of DNA.
- In meiosis the union of the male sperm and the female egg forms a new cell that carries half of the chromosomes from the male and half from the female.

REVIEW QUESTIONS AND EXERCISES

1. What are the three principal classes of animal food?
2. Life is dependent on four major classes of biomolecules. What are they?
3. Of the sugars listed in Table 20.1, which is the sweetest disaccharide? Which is the sweetest monosaccharide? (Invert sugar is a mixture of glucose and fructose, so it should not be considered.)
4. Indicate the three types of carbohydrates. Which is the simplest?
5. What is an aldose, an aldotetrose, a ketose, a ketohexose? Give an example of each.
6. Classify the following as a monosaccharide, disaccharide, or polysaccharide: glucose, sucrose, maltose, fructose, cellulose, lactose, glycogen, galactose, starch, and ribose.
7. Draw structural formulas in the open-chain form for ribose, glucose, fructose, and galactose.
8. Draw structural formulas in the cyclic form for ribose, glucose, fructose, and galactose.

9. State the properties and the sources of ribose, glucose, fructose, and galactose.
10. The molecular formula for lactic acid is $C_3H_6O_3$, and its structural formula is $CH_3CH(OH)COOH$. Is this compound a carbohydrate? Explain.
11. What is the monosaccharide composition of
   (a) sucrose
   (b) maltose
   (c) lactose
   (d) starch
   (e) cellulose
   (f) glycogen
12. Draw structural formulas in the cyclic form for sucrose and maltose.
13. If the most common monosaccharides have the formula $C_6H_{12}O_6$, why do the resulting disaccharides have the formula $C_{12}H_{22}O_{11}$, rather than $C_{12}H_{24}O_{12}$?

14. Write equations, using structural formulas in the cyclic form, for the hydrolysis of
    (a) sucrose   (b) maltose
    What enzymes catalyze these reactions?

15. Discuss the similarities and differences between starch and cellulose.

16. In what form is carbohydrate stored in the body?

17. Discuss, in simple terms, the metabolism of carbohydrates in the human body.

18. State the natural sources of sucrose, maltose, lactose, and starch.

19. Invert sugar, obtained by the hydrolysis of sucrose to an equal-molar mixture of fructose and glucose, is commonly used as a sweetener in commercial food preparations. Why is invert sugar sweeter than the original sucrose?

20. Use molecular structure to explain the following:
    (a) Fructose is a ketone.   (b) Glucose is an aldehyde.

21. What is the empirical formula for monosaccharides?

22. What change in oxidation state is experienced by the carbon in the combustion of sugar, $C_{12}H_{22}O_{11}$, to CO_2?

23. The molar mass of cellulose is approximately 6.0×10^5 g/mol and the molar mass of a soluble starch is on the order of 4.0×10^3 g/mol. The monomer unit in both of these molecules has the empirical formula $C_6H_{10}O_5$. The units are about 5.0×10^{-10} m long. About how many units occur in each molecule, and how long are the molecules of cellulose and starch as a result?

24. What are the differences between saturated fats and unsaturated fats?

25. Are the fatty acids in vegetable oils more saturated or unsaturated than those in animal fats? Explain your answer. (Table 20.2)

26. What properties of molecules cause them to be classified as lipids?

27. Write structural formulas for glycerol, stearic acid, palmitic acid, oleic acid, and linoleic acid.

28. Distinguish both chemically and physically between a fat and a vegetable oil.

29. What is a triacylglycerol? Give an example.

30. Write the structure for tristearin, a fat in which all the fatty acid units are stearic acid.

31. Write the structure of a triacylglycerol that contains one unit each of linoleic, stearic, and oleic acids. How many other formulas are possible in which the triacylglycerol contains one unit each of these acids?

32. Write equations for the saponification of
    (a) tripalmitin (a fat in which all the fatty acids are palmitic acid)
    (b) the triacylglycerol of Question 31.
    Which product(s) are soaps?

33. How can vegetable oils be solidified? What is the advantage of solidifying these oils?

34. What functions do fats have in the human body?

35. Which fatty acids are essential to human diets?

36. Draw the structural formula of cholesterol.

37. Draw the ring structure that is common to all steroids.

38. Write the formula for the triacylglycerol formed from
    (a) glycerol and butanoic acid
    (b) glycerol and one molecule each of stearic ($C_{17}H_{35}COOH$), palmitic ($C_{15}H_{31}COOH$), and myristic ($C_{13}H_{27}COOH$) acids

39. In a linear peptide, what are the N-terminal residue and the C-terminal residue?

40. Which of the common amino acids have more than one carboxyl group? Which have more than one amino group? (Table 20.3)

41. How many disulfide linkages are there in each molecule of beef insulin? (Figure 20.3)

42. List six foods that are major sources of proteins.

43. What functional groups are present in amino acids?

44. Why are the amino acids of proteins called α-amino acids?

45. In Table 20.3, which amino acids (a) are aromatic, (b) contain sulfur, (c) contain an —OH group?

46. Write the full structure for the two possible dipeptides containing glycine and phenylalanine.

47. Write structures for
    (a) glycylglycine
    (b) glycylglycylalanine
    (c) leucylmethionylglycylserine

48. Using amino acid abbreviations, write all the possible tripeptides containing one unit each of glycine, phenylalanine, and leucine.

49. The artificial sweetener aspartame is a dipeptide made from two amino acids, aspartic acid and phenylalanine. Write two possible structures for this dipeptide.

50. What are essential amino acids? Write the names of the amino acids that are essential for humans.

51. When proteins are eaten by a human, what are the metabolic fates of the protein material?

52. Why should protein be continually included in a balanced diet?

53. In the polypeptide Tyr-Gly-His-Phe-Val, identify the N-terminal and the C-terminal residues.

54. How are polypeptides numbered? Number the polypeptide in Question 53.

55. What is the name of the bond that bonds one α-amino acid to another in a protein?

56. How can you discern the difference between an amino acid and an ordinary carboxylic acid?

57. What are enzymes, and what is their role in the body?

58. What is meant by specificity of an enzyme?

59. Differentiate between the lock-and-key and induced-fit models for enzyme function.

60. What is meant by enzyme specificity?

61. What is the function of enzymes in the body?

62. What are the components of chromosomes?

63. In the four nucleotide units of DNA shown in Figure 20.6, which of these components are part of the backbone chain and which are off to the side: the nitrogen bases, the deoxyribose, and the phosphoric acid?

64. In the double-stranded helix structure of DNA, which bases are always hydrogen-bonded to the following: cytosine, thymine, adenine, and guanine? (Figure 20.7)

65. Write structural formulas for the compounds that make up DNA.

66. (a) What are the three units that make up a DNA nucleotide?
    (b) List the components of the four types of nucleotides found in DNA.
    (c) Write the structure and name of one of these nucleotides.

67. Briefly describe the structure of DNA as proposed by Watson and Crick.

68. What is the role of hydrogen bonding in the structure of DNA?

69. Explain the concept of complementary bases and how it relates to DNA.

70. A segment of a DNA strand has a base sequence of C-G-A-T-T-G-C-A. What is the base sequence of the other complementary strand of the double helix?

71. Explain the replication process of DNA.

72. Briefly discuss the relationship of DNA to genetics.

73. What are the three differences between DNA and RNA in terms of structure?

74. Distinguish between cell division in mitosis and in meiosis.

75. Specifically, where does protein synthesis occur?

76. What is the relationship between proteins and amino acids?

77. In Figure 20.7, what do the red dotted lines represent?

78. What is galactosemia, and how does it affect humans?

ANSWERS TO PRACTICE EXERCISES

20.1 The carbonyl oxygen becomes part of the ring and bonds to another carbon atom in the sugar.

20.2

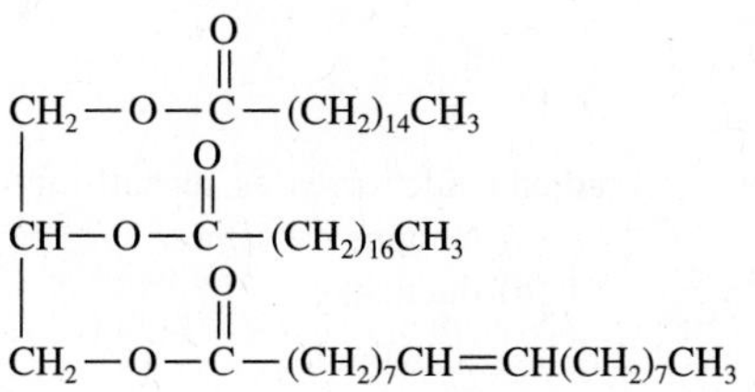

$$CH_2-O-\overset{\displaystyle O}{\overset{\|}{C}}-(CH_2)_{14}CH_3$$
$$CH-O-\overset{\displaystyle O}{\overset{\|}{C}}-(CH_2)_{16}CH_3$$
$$CH_2-O-\overset{\displaystyle O}{\overset{\|}{C}}-(CH_2)_7CH=CH(CH_2)_7CH_3$$

Two other structures are possible.

20.3 α-aminobutyric acid $CH_3CH_2\underset{\underset{\displaystyle NH_2}{|}}{C}HCOOH$

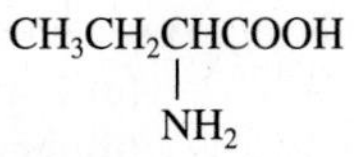

β-aminobutyric acid $CH_3\underset{\underset{\displaystyle NH_2}{|}}{C}HCH_2COOH$

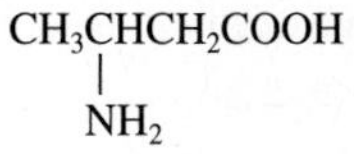

20.4 (a) prolyltyrosine

(b) glycylhistidylleucine

20.5 (a) $CH_3\overset{\displaystyle O}{\overset{\|}{C}}\underset{\underset{\displaystyle NH_2}{|}}{H}C-NH\overset{\displaystyle CH_2OH}{\overset{|}{C}}HCOOH$

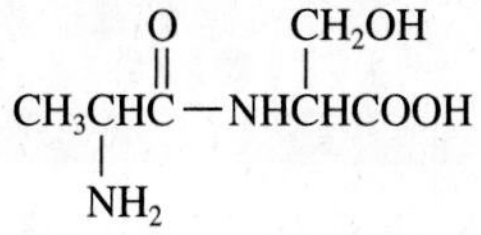

(b) $NH_2CH_2\overset{\displaystyle O}{\overset{\|}{C}}-NH\underset{\underset{\displaystyle O}{\overset{\|}{}}}{C}H\overset{\displaystyle CH_3CHOH}{\overset{|}{C}}NH\overset{\displaystyle CH_2CH_2SCH_3}{\overset{|}{C}}HCOOH$

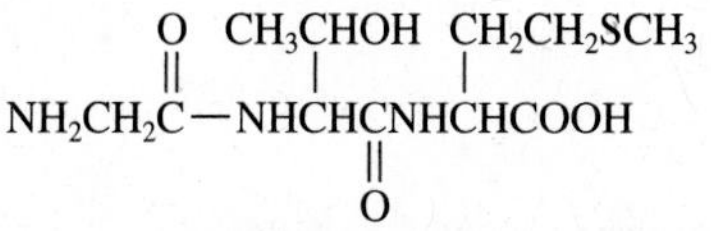

20.6 An activator is an inorganic component of an enzyme, whereas a coenzyme is a nonprotein organic component of an enzyme.

20.7 This nucleotide contains ribose.

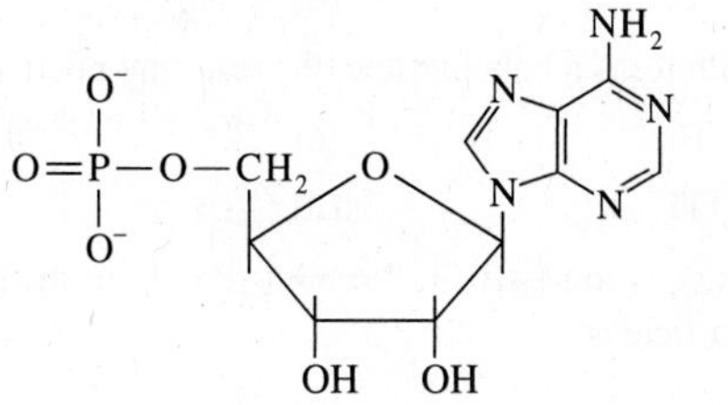

Answers for Putting It Together Reviews are found in Appendix VII.

Multiple Choice

Choose the correct answer to each of the following.

1. If $^{238}_{92}U$ loses an alpha particle, the resulting nuclide is
   (a) $^{237}_{92}U$
   (c) $^{238}_{93}Np$
   (b) $^{234}_{90}Th$
   (d) $^{236}_{90}Th$

2. If $^{210}_{82}Pb$ loses a beta particle, the resulting nuclide is
   (a) $^{209}_{83}Bi$
   (c) $^{206}_{80}Hg$
   (b) $^{210}_{81}Tl$
   (d) $^{210}_{83}Bi$

3. In the equation $^{209}_{83}Bi + ? \longrightarrow ^{210}_{84}Po + ^{1}_{0}n$, the missing bombarding particle is
   (a) $^{2}_{1}H$
   (c) $^{4}_{2}He$
   (b) $^{1}_{0}n$
   (d) $^{0}_{-1}e$

4. Which of the following is not a characteristic of nuclear fission?
   (a) Upon adsorption of a proton, a heavy nucleus splits into two or more smaller nuclei.
   (b) Two or more neutrons are produced from the fission of each atom.
   (c) Large quantities of energy are produced.
   (d) Most nuclei formed are radioactive.

5. The half-life of Sn-121 is 10 days. If you started with 40 g of this isotope, how much would you have left 30 days later?
   (a) 10 g
   (c) 15 g
   (b) none
   (d) 5 g

6. $^{241}_{94}Pu$ successively emits β, α, α, β, α, α. At that point, the nuclide has become
   (a) $^{225}_{94}Pu$
   (c) $^{207}_{84}Po$
   (b) $^{225}_{88}Ra$
   (d) $^{219}_{84}Po$

7. Calculate the nuclear binding energy of $^{56}_{26}Fe$.

   Mass data:

   $^{56}_{26}Fe = 55.9349$ g/mol

   n = 1.0087 g/mol $e^- = 0.00055$ g/mol

   p = 1.0073 g/mol 1.0 g $= 9.0 \times 10^{13}$ J

   (a) 4.8×10^{13} J/mol
   (c) 0.5302 g/mol
   (b) 56.4651 g/mol
   (d) 4.9×10^{15} J/mol

8. The radioactivity ray with the greatest penetrating ability is
   (a) alpha
   (c) gamma
   (b) beta
   (d) proton

9. In a nuclear reaction,
   (a) mass is lost
   (b) mass is gained
   (c) mass is converted into energy
   (d) energy is converted into mass

10. As the temperature of a radionuclide increases, its half-life
    (a) increases
    (c) remains the same
    (b) decreases
    (d) fluctuates

11. The nuclide that has the longest half-life is
    (a) $^{238}_{92}U$
    (c) $^{234}_{90}Th$
    (b) $^{210}_{82}Pb$
    (d) $^{222}_{88}Ra$

12. Which of the following is not a unit of radiation?
    (a) curie
    (c) rod
    (b) roentgen
    (d) rem

13. When $^{235}_{92}U$ is bombarded by a neutron, the atom can fission into
    (a) $^{124}_{53}I + ^{109}_{47}Ag + 2\,^{1}_{0}n$
    (b) $^{124}_{50}Sn + ^{110}_{42}Mo + 2\,^{1}_{0}n$
    (c) $^{134}_{56}Ba + ^{128}_{36}Xe + 2\,^{1}_{0}n$
    (d) $^{90}_{38}Sr + ^{143}_{58}Ce + 2\,^{1}_{0}n$

14. In the nuclear equation

    $$^{45}_{21}Sc + ^{1}_{0}n \longrightarrow X + ^{1}_{1}H$$

    the nuclide X that is formed is
    (a) $^{45}_{22}Ti$
    (c) $^{46}_{22}Ti$
    (b) $^{45}_{20}Ca$
    (d) $^{45}_{20}K$

15. What type of radiation is a very energetic form of photons?
    (a) alpha
    (c) gamma
    (b) beta
    (d) positron

16. When $^{239}_{92}U$ decays to $^{239}_{93}Np$, what particle is emitted?
    (a) positron
    (c) alpha particle
    (b) neutron
    (d) beta particle

17. The roentgen is the unit of radiation that measures
    (a) an absorbed dose of radiation
    (b) exposure to X-rays
    (c) the dose from a different type of radiation
    (d) the rate of decay of a radioactive substance

18. Which of the following is not a correct name for the alkane shown with it?
    (a) C_2H_6, ethane
    (c) C_7H_{16}, heptane
    (b) C_5H_{12}, propane
    (d) $C_{10}H_{22}$, decane

19. The structural formula of *o*-xylene is

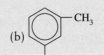

20. The ester

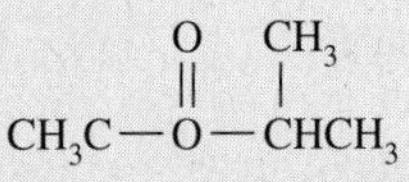

$$CH_3C(=O)-O-CHCH_3 \ (CH_3)$$

can be made from which alcohol and carboxylic acid?

(a) $CH_3C(=O)-OH$, $CH_3CH_2CH_2OH$

(b) $CH_3C(=O)-OH$, $HO-CHCH_2OH$ (CH_3)

(c) $H-C(=O)-OH$, CH_3CHCH_3 (OH)

(d) $CH_3C(=O)-OH$, CH_3CHCH_3 (OH)

21. The product of the reaction

$$CH_3CH{=}CH_2 + H_2O + H^+ \longrightarrow \text{ is a(n)}$$

(a) alcohol
(b) aldehyde
(c) alkyne
(d) carboxylic acid

22. The number of isomers of butyl alcohol, C_4H_9OH, is

(a) 2   (b) 3   (c) 4   (d) 5

23. What is the correct name for this compound?

$$CH_3-CH-CH-CH-CH{=}CH_2$$
with CH_3 on carbon 2, CH_3 on carbon 4, and $CH_3-CH-CH_3$ substituent

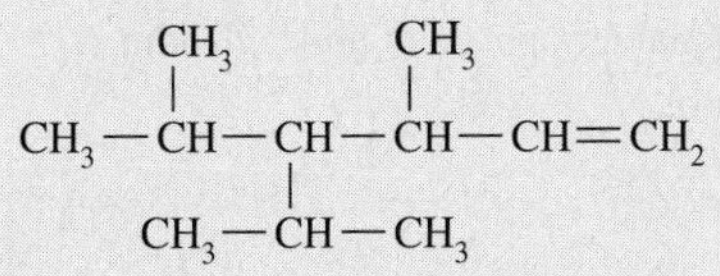

(a) isobutane
(b) 2,4-methyl-3-propyl-5-hexene
(c) 3,5-dimethyl-4-isopropyl-1-hexene
(d) 4,4-diisopropylhexene

24. Which of these acids is named incorrectly?

(a) $CH_3CH_2CH_2COOH$, butyric acid
(d) $HCOOH$, formic acid
(c) CH_3CH_2COOH, propic acid
(d) CH_3COOH, acetic acid

25. With acid as a catalyst, ethanol and formic acid will react to form

(a) $CH_3C(=O)-OH$
(b) $H-C(=O)-O-CH_2CH_3$
(c) $CH_3C(=O)-O-CH_3$
(d) $CH_3C(=O)-O-CCH_3$ (with second $=O$)

26. The number of isomers of C_6H_{14} is

(a) 3   (b) 5   (c) 6   (d) 8

27. An open-chain hydrocarbon of formula C_6H_8 can have in its formula

(a) one carbon–carbon double bond
(b) two carbon–carbon double bonds
(c) one carbon–carbon triple bond
(d) one carbon–carbon double bond and one carbon–carbon triple bond

28. The reaction

$$CH_2{=}CH_2 + Br_2 \longrightarrow CH_2BrCH_2Br$$

represents

(a) dehalogenation   (c) addition
(b) substitution   (d) dehydration

29. The general formula for a ketone is

(a) RCHO   (c) RCOOR
(b) ROR   (d) R_2CO

30. Which of the following cannot be an aromatic compound?

(a) C_6H_5OH (c) C_6H_{14}
(b) C_6H_6 (d) $C_6H_5CH_3$

31. What is the correct name for this compound?

Benzene ring with OH at top, O_2N and NO_2 at bottom positions

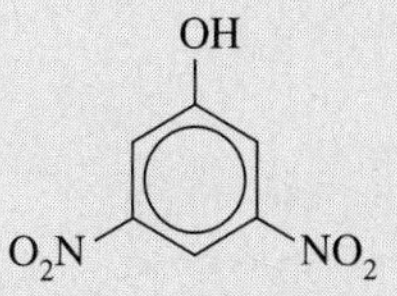

(a) *m*-dinitrophenol   (c) 3,5-dinitrophenol
(b) 2,4-dinitrophenol   (d) 1,3-dinitrophenol

32. Which of these pairs are not isomers?

(a) CH_3OCH_2Cl and CH_2ClCH_2OH
(b) CH_3CH_2CHO and $CH_3OCH_2CH_3$
(c) $CH_3OCH_2OCH_3$ and $CH_3CH(OH)CH_2OH$
(d) $C_6H_4(CH_3)_2$ and $C_6H_5CH_2CH_3$

33. The reaction of $CH_3CH{=}CHCH_3 + HBr$ produces

(a) $CH_3CH_2CH_2CH_3 + Br_2$
(b) $CH_3CHBrCHBrCH_3 + H_2$
(c) $CH_3CHBrCH_2CH_3$
(d) $CH_3CH_2CH_2CH_2Br$

34. Polyvinyl chloride is a polymer of

(a) $CH_2{=}CCl_2$ (c) $CH_2{=}CHCl$
(b) $CF_2{=}CF_2$ (d) $C_6H_5CH{=}CHCl$

35. What is the correct name for this compound?

$$CH_3C(=O)-O-C(CH_3)(H)-CH_3$$

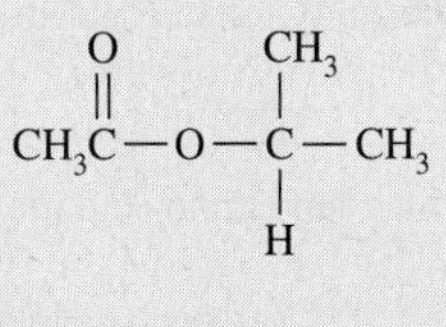

(a) acetyl-2-propanoate   (c) ethyl isopropylate
(b) propyl acetate   (d) isopropyl ethanoate

36. Teflon is a polymer of

(a) $CH_2{=}CCl_2$ (c) $CH_2{=}CHCl$
(b) $CF_2{=}CF_2$ (d) $C_6H_5CH{=}CHCl$

37. Sugars are members of a group of compounds with the general name

(a) carbohydrates   (c) proteins
(b) lipids   (d) steroids

38. The products formed when maltose is hydrolyzed are

(a) glucose and fructose
(b) glucose and galactose
(c) glucose and glucose
(d) galactose and fructose

39. Which is *not* true of starch?

(a) It is a polysaccharide.
(b) It is hydrolyzed to maltose.
(c) It is composed of glucose units.
(d) It is not digestible by humans.

40. Lactose is
    (a) a monosaccharide
    (b) a disaccharide composed of galactose and glucose
    (c) a disaccharide composed of two glucose units
    (d) a decomposition product of starch

41. Which is *not* true of glucose?
    (a) It is a monosaccharide.
    (b) It is a component of sucrose, maltose, lactose, starch, glycogen, and cellulose.
    (c) It is a ketohexose.
    (d) It is the main source of energy for the body.

42. The sweetest of the common sugars is
    (a) fructose
    (b) sucrose
    (c) glucose
    (d) maltose

43. Which of the following is *not* true?
    (a) Lactose is sweeter than sucrose.
    (b) Glycogen is known as animal starch.
    (c) Cellulose is the most abundant organic substance in nature.
    (d) Lactose is a disaccharide known as milk sugar.

44. Which is *not* formed in the saponification of a fat?
    (a) glycerol   (c) soap
    (b) amino acids   (d) a metal salt of a long-chain fatty acid

45. Which of these lipids does *not* contain a glycerol unit as part of its structure?
    (a) a fat   (c) a glycolipid
    (b) a phospholipid   (d) an oil

46. An α-amino acid always contains
    (a) an amino group on the carbon atom adjacent to the carboxyl group
    (b) a carboxyl group at each end of the molecule
    (c) two amino groups
    (d) alternating amino and carboxyl groups

47. Which of these amino acids contains sulfur?
    (a) alanine   (c) cysteine
    (b) histidine   (d) glycine

48. A compound containing ten amino acid molecules linked together is called a
    (a) protein   (c) deca-amino acid
    (b) polypeptide   (d) nucleotide

49. Which of the following is *not* a correct statement about DNA and RNA?
    (a) DNA contains deoxyribose, whereas RNA contains ribose.
    (b) Both DNA and RNA are polymers made up of nucleotides.
    (c) DNA directs the synthesis of proteins and RNA directs the genetic code of life.
    (d) DNA exists as a double helix, whereas RNA exists as a single helix.

50. Which of these bases is found in RNA but not in DNA?
    (a) thymine   (c) guanine
    (b) adenine   (d) uracil

51. Complementary base pairs in DNA are linked through the formation of
    (a) phosphate ester bonds
    (b) peptide linkages
    (c) hydrogen bonds
    (d) ionic bonds

52. In a DNA double helix, hydrogen bonding occurs between
    (a) adenine and thymine
    (b) thymine and guanine
    (c) adenine and uracil
    (d) cytosine and thymine

53. Which of these scientists did *not* receive the Nobel Prize for the structure of DNA?
    (a) Crick
    (b) Watson
    (c) Sanger
    (d) Wilkins

54. The substance acted on by an enzyme is called a(n)
    (a) catalyst   (c) coenzyme
    (b) apoenzyme   (d) substrate

55. The process during which a cell splits to form a sperm or an egg cell is called
    (a) mitosis   (c) translation
    (b) meiosis   (d) transcription

Free Response Questions

Answer each of the following. Be sure to include your work and explanations in a clear, logical form.

1. The nuclide $^{223}_{87}\text{Fr}$ emits three particles losing 8 mass units and 3 atomic number units. Propose a radioactive decay series and write the symbol for the resulting nuclide.

2. Biological molecules such as methionine enkephalin, leucine enkephalin, and β-endorphin begin with a common tetrapeptide tyrosyl-glycyl-glycyl-phenylalanine. Draw the tetrapeptide sequence and write its abbreviated form.

3. What functional groups are in prostaglandin A_2?

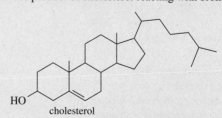

prostaglandin A_2

4. Which of the following molecules would you expect to be significantly soluble in water? Why?

    $CH_3CH_2CH_2OH$, $CH_3CH_2CH_2CHO$

    $CH_3CH_2COCH_3$, $CH_3CH_2CH_2CH_3$

    (*Hint:* Remember that "like dissolves like.")

5. Which of these compounds are structural isomers of each other?

    methyl propyl ether, butanoic acid, acetone, propene, 2-butanol, phenol, methyl propanoate

6. (a) Indicate which two amino acids could form an ester bond using their R group side chains.
   (b) Draw the product of cholesterol reacting with bromine.

HO
cholesterol

    (c) Is cholesterol soluble in water? How do you know?

7. (a) Carboxypeptidase A is a digestive enzyme. A zinc ion is necessary for enzyme activity. What is the zinc ion called?

   (b) Some RNA strands also show enzymatic activity but DNA does not. What are two structural differences between RNA and DNA?

   (c) Do you think we have radioactive isotopes in our bodies? Explain.

8. (a) For each OH group in open-chain glucose, indicate if it is primary, secondary, or tertiary.

   (b) Does tyrosine contain an *ortho-*, *meta-*, or *para*-substituted aromatic ring?

   (c) How many amino acids are in a pentapeptide? How many molecules of water were lost when it was formed?

   (d) Based on what you know about the hydrolysis of disaccharides, if a person has high blood sugar levels, do you think it would be better for this person to cut down on lactose or maltose? Why?

9. How many half-lives are required to change 96 g of a sample of a radioactive isotope to 1.5 g over approximately 24 days? What is the approximate half-life of this isotope?

10. (a) What type of process is $^{6}_{3}\text{Li} + ^{1}_{0}\text{n} \longrightarrow ^{3}_{1}\text{H} + ^{4}_{2}\text{He}$?

    (b) How is this process different from radioactive decay?

    (c) Could nuclear fission be classified as the same type process? Explain.

Mathematical Review

Multiplication

Multiplication is a process of adding any given number or quantity to itself a certain number of times. Thus, 4 times 2 means 4 added two times, or 2 added together four times, to give the product 8. Various ways of expressing multiplication are

$$ab \qquad a \times b \qquad a \cdot b \qquad a(b) \qquad (a)(b)$$

Each of these expressions means a times b, or a multiplied by b, or b times a.

When $a = 16$ and $b = 24$, we have $16 \times 24 = 384$.

The expression $°F = (1.8 \times °C) + 32$ means that we are to multiply 1.8 times the Celsius degrees and add 32 to the product. When $°C$ equals 50,

$$°F = (1.8 \times 50) + 32 = 90 + 32 = 122°F$$

The result of multiplying two or more numbers together is known as the *product.*

Division

The word *division* has several meanings. As a mathematical expression, it is the process of finding how many times one number or quantity is contained in another. Various ways of expressing division are

$$a \div b \qquad \frac{a}{b} \qquad a/b$$

Each of these expressions means a divided by b.

When $a = 15$ and $b = 3$, $\dfrac{15}{3} = 5$.

The number above the line is called the *numerator;* the number below the line is the *denominator.* Both the horizontal and the slanted (/) division signs also mean "per." For example, in the expression for density, we determine the mass per unit volume:

$$\text{density} = \text{mass}/\text{volume} = \frac{\text{mass}}{\text{volume}} = \text{g/mL}$$

The diagonal line still refers to a division of grams by the number of milliliters occupied by that mass. The result of dividing one number into another is called the *quotient.*

Fractions and Decimals

A fraction is an expression of division, showing that the numerator is divided by the denominator. A *proper fraction* is one in which the numerator is smaller than the denominator. In an *improper fraction,* the numerator is the larger number. A decimal or a decimal fraction is a proper fraction in which the denominator is some power of 10. The decimal fraction is determined by carrying out the division of the proper fraction. Examples of proper fractions and their decimal fraction equivalents are shown in the accompanying table.

Proper fraction		Decimal fraction		Proper fraction
$\dfrac{1}{8}$	$=$	0.125	$=$	$\dfrac{125}{1000}$
$\dfrac{1}{10}$	$=$	0.1	$=$	$\dfrac{1}{10}$
$\dfrac{3}{4}$	$=$	0.75	$=$	$\dfrac{75}{100}$
$\dfrac{1}{100}$	$=$	0.01	$=$	$\dfrac{1}{100}$
$\dfrac{1}{4}$	$=$	0.25	$=$	$\dfrac{25}{100}$

Addition of Numbers with Decimals

To add numbers with decimals, we use the same procedure as that used when adding whole numbers, but we always line up the decimal points in the same column. For example, add $8.21 + 143.1 + 0.325$:

$$\begin{array}{r} 8.21 \\ +143.1 \\ +\ \ 0.325 \\ \hline 151.635 \end{array}$$

When adding numbers that express units of measurement, we must be certain that the numbers added together all have the same units. For example, what is the total length of three pieces of glass tubing: 10.0 cm, 125 mm, and 8.4 cm? If we simply add the numbers, we obtain a value of 143.4, but we are not certain what the unit of measurement is. To add these lengths correctly, first change 125 mm to 12.5 cm. Now all the lengths are expressed in the same units and can be added:

$$\begin{array}{r} 10.0 \text{ cm} \\ 12.5 \text{ cm} \\ 8.4 \text{ cm} \\ \hline 30.9 \text{ cm} \end{array}$$

Subtraction of Numbers with Decimals

To subtract numbers containing decimals, we use the same procedure as for subtracting whole numbers, but we always line up the decimal points in the same column. For example, subtract 20.60 from 182.49:

$$\begin{array}{r} 182.49 \\ -\ 20.60 \\ \hline 161.89 \end{array}$$

When subtracting numbers that are measurements, be certain that the measurements are in the same units. For example, subtract 22 cm from 0.62 m. First change m to cm, then do the subtraction.

$$(0.62 \text{ m})\left(\frac{100 \text{ cm}}{\text{m}}\right) = 62 \text{ cm} \qquad \begin{array}{r} 62\ \text{ cm} \\ -22\ \text{ cm} \\ \hline 40.\ \text{cm} \end{array}$$

Multiplication of Numbers with Decimals

To multiply two or more numbers together that contain decimals, we first multiply as if they were whole numbers. Then, to locate the decimal point in the product, we add together the number of digits to the right of the decimal in all the numbers multiplied together. The product should have this same number of digits to the right of the decimal point.

Multiply $2.05 \times 2.05 = 4.2025$ (total of four digits to the right of the decimal).

> If a number is a measurement, the answer must be adjusted to the correct number of significant figures.

Here are more examples:

$14.25 \times 6.01 \times 0.75 = 64.231875$ (six digits to the right of the decimal)
$39.26 \times 60 = 2355.60$ (two digits to the right of the decimal)

Division of Numbers with Decimals

To divide numbers containing decimals, we first relocate the decimal points of the numerator and denominator by moving them to the right as many places as needed to make the denominator a whole number. (Move the decimal of both the numerator and the denominator the same amount and in the same direction.) For example,

$$\frac{136.94}{4.1} = \frac{1369.4}{41}$$

The decimal point adjustment in this example is equivalent to multiplying both numerator and denominator by 10. Now we carry out the division normally, locating the decimal point immediately above its position in the dividend:

$$\frac{33.4}{41)\overline{1269.4}} \qquad \frac{0.441}{26.25} = \frac{44.1}{2625} = \frac{0.0168}{2625)\overline{44.1000}}$$

These examples are guides to the principles used in performing the various mathematical operations illustrated. Every student of chemistry should learn to use a calculator for solving mathematical problems (see Appendix II). The use of a calculator will save many hours of doing tedious calculations. After solving a problem, the student should check for errors and evaluate the answer to see if it is logical and consistent with the data given.

Algebraic Equations

Many mathematical problems that are encountered in chemistry fall into the following algebraic forms. Solutions to these problems are simplified by first isolating the desired term on one side of the equation. This rearrangement is accomplished by treating both sides of the equation in an identical manner until the desired term is isolated.

(a) $a = \dfrac{b}{c}$

To solve for b, multiply both sides of the equation by c:

$$a \times c = \frac{b}{\cancel{c}} \times \cancel{c}$$

$$b = a \times c$$

To solve for c, multiply both sides of the equation by $\dfrac{c}{a}$:

$$\cancel{a} \times \frac{c}{\cancel{a}} = \frac{b}{\cancel{c}} \times \frac{\cancel{c}}{a}$$

$$c = \frac{b}{a}$$

(b) $\dfrac{a}{b} = \dfrac{c}{d}$

To solve for a, multiply both sides of the equation by b:

$$\frac{a}{\cancel{b}} \times \cancel{b} = \frac{c}{d} \times b$$

$$a = \frac{c \times b}{d}$$

To solve for b, multiply both sides of the equation by $\dfrac{b \times d}{c}$:

$$\frac{a}{\cancel{b}} \times \frac{\cancel{b} \times d}{c} = \frac{\cancel{c}}{\cancel{d}} \times \frac{b \times \cancel{d}}{\cancel{c}}$$

$$\frac{a \times d}{c} = b$$

(c) $a \times b = c \times d$

To solve for a, divide both sides of the equation by b:

$$\frac{a \times \cancel{b}}{\cancel{b}} = \frac{c \times d}{b}$$

$$a = \frac{c \times d}{b}$$

(d) $\dfrac{b - c}{a} = d$

To solve for b, first multiply both sides of the equation by a:

$$\frac{\cancel{a}(b - c)}{\cancel{a}} = d \times a$$

$$b - c = d \times a$$

Then add c to both sides of the equation:

$b - \cancel{c} + \cancel{c} = d \times a + c$

$b = (d \times a) + c$

When $a = 1.8, c = 32$, and $d = 35$,

$b = (35 \times 1.8) + 32 = 63 + 32 = 95$

Expression of Large and Small Numbers

scientific, or exponential, notation
exponent

In scientific measurement and calculations, we often encounter very large and very small numbers—for example, 0.00000384 and 602,000,000,000,000,000,000,000. These numbers are troublesome to write and awkward to work with, especially in calculations. A convenient method of expressing these large and small numbers in a simplified form is by means of exponents, or powers, of 10. This method of expressing numbers is known as **scientific, or exponential, notation**.

An **exponent** is a number written as a superscript following another number. Exponents are often called *powers* of numbers. The term *power* indicates how many times the number is used as a factor. In the number 10^2, 2 is the exponent, and the number means 10 squared, or 10 to the second power, or $10 \times 10 = 100$. Three other examples are

$$3^2 = 3 \times 3 = 9$$

$$3^4 = 3 \times 3 \times 3 \times 3 = 81$$

$$10^3 = 10 \times 10 \times 10 = 1000$$

For ease of handling, large and small numbers are expressed in powers of 10. Powers of 10 are used because multiplying or dividing by 10 coincides with moving the decimal point in a number by one place. Thus, a number multiplied by 10^1 would move the decimal point one place to the right; 10^2, two places to the right; 10^{-2}, two places to the left. To express a number in powers of 10, we move the decimal point in the original number to a new position, placing it so that the number is a value between 1 and 10. This new decimal number is multiplied by 10 raised to the proper power. For example, to write the number 42,389 in exponential form, the decimal point is placed between the 4 and the 2 (4.2389), and the number is multiplied by 10^4; thus, the number is 4.2389×10^4:

$$42{,}389 = 4.2389 \times 10^4$$
$$\underset{4\,3\,2\,1}{}$$

The exponent of 10 (4) tells us the number of places that the decimal point has been moved from its original position. If the decimal point is moved to the left, the exponent is a positive number; if it is moved to the right, the exponent is a negative number. To express the number 0.00248 in exponential notation (as a power of 10), the decimal point is moved three places to the right; the exponent of 10 is -3, and the number is 2.48×10^{-3}.

$$0.00248 = 2.48 \times 10^{-3}$$
$$\underset{1\,2\,3}{}$$

Study the following examples of changing a number to scientific notation.

$$1237 = 1.237 \times 10^3$$

$$988 = 9.88 \times 10^2$$

$$147.2 = 1.472 \times 10^2$$

$$2{,}200{,}000 = 2.2 \times 10^6$$

$$0.0123 = 1.23 \times 10^{-2}$$

$$0.00005 = 5 \times 10^{-5}$$

$$0.000368 = 3.68 \times 10^{-4}$$

Exponents in multiplication and division The use of powers of 10 in multiplication and division greatly simplifies locating the decimal point in the answer. In multiplication, first change all numbers to powers of 10, then multiply the numerical portion in the usual manner, and finally add the exponents of 10 algebraically, expressing them as a power of 10 in the product. In multiplication, the exponents (powers of 10) are added algebraically.

$$10^2 \times 10^3 = 10^{(2+3)} = 10^5$$
$$10^2 \times 10^2 \times 10^{-1} = 10^{(2+2-1)} = 10^3$$

Multiply: (40,000)(4200)

Change to powers of 10: $(4 \times 10^4)(4.2 \times 10^3)$

Rearrange: $(4 \times 4.2)(10^4 \times 10^3)$

$16.8 \times 10^{(4+3)}$

16.8×10^7 or 1.68×10^8 (Answer)

Multiply: (380)(0.00020)

$(3.80 \times 10^2)(2.0 \times 10^{-4})$

$(3.80 \times 2.0)(10^2 \times 10^{-4})$

$7.6 \times 10^{(2-4)}$

7.6×10^{-2} or   0.076   (Answer)

Multiply: (125)(284)(0.150)

$(1.25 \times 10^2)(2.84 \times 10^2)(1.50 \times 10^{-1})$

$(1.25)(2.84)(1.50)(10^2 \times 10^2 \times 10^{-1})$

$5.325 \times 10^{(2+2-1)}$

5.33×10^3 (Answer)

In division, after changing the numbers to powers of 10, move the 10 and its exponent from the denominator to the numerator, changing the sign of the exponent. Carry out the division in the usual manner, and evaluate the power of 10. Change the sign(s) of the exponent(s) of 10 in the denominator, and move the 10 and its exponent(s) to the numerator. Then add all the exponents of 10 together. For example,

$$\frac{10^5}{10^3} = 10^5 \times 10^{-3} = 10^{(5-3)} = 10^2$$

$$\frac{10^3 \times 10^4}{10^{-2}} = 10^3 \times 10^4 \times 10^2 = 10^{(3+4+2)} = 10^9$$

Significant Figures in Calculations

The result of a calculation based on experimental measurements cannot be more precise than the measurement that has the greatest uncertainty. (See Section 2.4 for additional discussion.)

Addition and subtraction The result of an addition or subtraction should contain no more digits to the right of the decimal point than are contained in the quantity that has the least number of digits to the right of the decimal point.

Perform the operation indicated and then round off the number to the proper number of significant figures:

(a)   142.8   g

      18.843 g

      36.42   g

    198.063 g

    198.1   g   (Answer)

(b)   93.45 mL

    −18.0   mL

     75.45 mL

     75.5   mL   (Answer)

(a) The answer contains only one digit after the decimal point since 142.8 contains only one digit after the decimal point.

(b) The answer contains only one digit after the decimal point since 18.0 contains one digit after the decimal point.

Multiplication and division In calculations involving multiplication or division, the answer should contain the same number of significant figures as the measurement that has the least number of significant figures. In multiplication or division, the position of the decimal point has nothing to do with the number of significant figures in the answer. Study the following examples:

	Round off to
$(2.05)(2.05) = 4.2025$	4.20
$(18.48)(5.2) = 96.096$	96
$(0.0126)(0.020) = 0.000252$ or	
$(1.26 \times 10^{-2})(2.0 \times 10^{-2}) = 2.520 \times 10^{-4}$	2.5×10^{-4}
$\dfrac{1369.4}{41} = 33.4$	33
$\dfrac{2268}{4.20} = 540.$	540.

Dimensional Analysis

Many problems of chemistry can be readily solved by dimensional analysis using the factor-label or conversion-factor method. Dimensional analysis involves the use of proper units of dimensions for all factors that are multiplied, divided, added, or subtracted in setting up and solving a problem. Dimensions are physical quantities such as length, mass, and time, which are expressed in such units as centimeters, grams, and seconds, respectively. In solving a problem, we treat these units mathematically just as though they were numbers, which gives us an answer that contains the correct dimensional units.

A measurement or quantity given in one kind of unit can be converted to any other kind of unit having the same dimension. To convert from one kind of unit to another, the original quantity or measurement is multiplied or divided by a conversion factor. The key to success lies in choosing the correct conversion factor. This general method of calculation is illustrated in the following examples.

Suppose we want to change 24 ft to inches. We need to multiply 24 ft by a conversion factor containing feet and inches. Two such conversion factors can be written relating inches to feet:

$$\frac{12 \text{ in.}}{1 \text{ ft}} \quad \text{or} \quad \frac{1 \text{ ft}}{12 \text{ in.}}$$

We choose the factor that will mathematically cancel feet and leave the answer in inches. Note that the units are treated in the same way we treat numbers, multiplying or dividing as required. Two possibilities then arise to change 24 ft to inches:

$$(24 \text{ ft})\left(\frac{12 \text{ in.}}{1 \text{ ft}}\right) \quad \text{or} \quad (24 \text{ ft})\left(\frac{1 \text{ ft}}{12 \text{ in.}}\right)$$

In the first case (the correct method), feet in the numerator and the denominator cancel, giving us an answer of 288 in. In the second case, the units of the answer are $\text{ft}^2/\text{in.}$, the answer being $2.0 \text{ ft}^2/\text{in.}$ In the first case, the answer is reasonable because it is expressed in units having the proper dimensions. That is, the dimension of length expressed in feet has been converted to length in inches according to the mathematical expression

$$\text{ft} \times \frac{\text{in.}}{\text{ft}} = \text{in.}$$

In the second case, the answer is not reasonable because the units $(\text{ft}^2/\text{in.})$ do not correspond to units of length. The answer is therefore incorrect. The units are the guiding factor for the proper conversion.

The reason we can multiply 24 ft times 12 in./ft and not change the value of the measurement is that the conversion factor is derived from two equivalent quantities. Therefore, the conversion factor 12 in./ft is equal to unity. When you multiply any factor by 1, it does not change the value:

$$12 \text{ in.} = 1 \text{ ft} \quad \text{and} \quad \frac{12 \text{ in.}}{1 \text{ ft}} = 1$$

Convert 16 kg to milligrams. In this problem it is best to proceed in this fashion:

$$kg \longrightarrow g \longrightarrow mg$$

The possible conversion factors are

$$\frac{1000 \text{ g}}{1 \text{ kg}} \quad \text{or} \quad \frac{1 \text{ kg}}{1000 \text{ g}} \qquad \frac{1000 \text{ mg}}{1 \text{ g}} \quad \text{or} \quad \frac{1 \text{ g}}{1000 \text{ mg}}$$

We use the conversion factor that leaves the proper unit at each step for the next conversion. The calculation is

$$(16 \text{ kg})\left(\frac{1000 \text{ g}}{1 \text{ kg}}\right)\left(\frac{1000 \text{ mg}}{1 \text{ g}}\right) = 1.6 \times 10^7 \text{ mg}$$

Regardless of application, the basis of dimensional analysis is the use of conversion factors to organize a series of steps in the quest for a specific quantity with a specific unit.

Graphical Representation of Data

A graph is often the most convenient way to present or display a set of data. Various kinds of graphs have been devised, but the most common type uses a set of horizontal and vertical coordinates to show the relationship of two variables. It is called an x–y graph because the data of one variable are represented on the horizontal or x-axis (abscissa) and the data of the other variable are represented on the vertical or y-axis (ordinate). See **Figure I.1**.

As a specific example of a simple graph, let us graph the relationship between Celsius and Fahrenheit temperature scales. Assume that initially we have only the information in the table next to **Figure I.2**.

On a set of horizontal and vertical coordinates (graph paper), scale off at least 100 Celsius degrees on the x-axis and at least 212 Fahrenheit degrees on the y-axis. Locate and mark the three points corresponding to the three temperatures given and draw a line connecting these points (see Figure I.2).

Here is how a point is located on the graph: Using the (50°C, 122°F) data, trace a vertical line up from 50°C on the x-axis and a horizontal line across from 122°F on the y-axis and mark

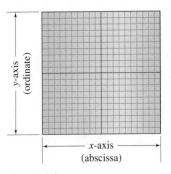

Figure I.1

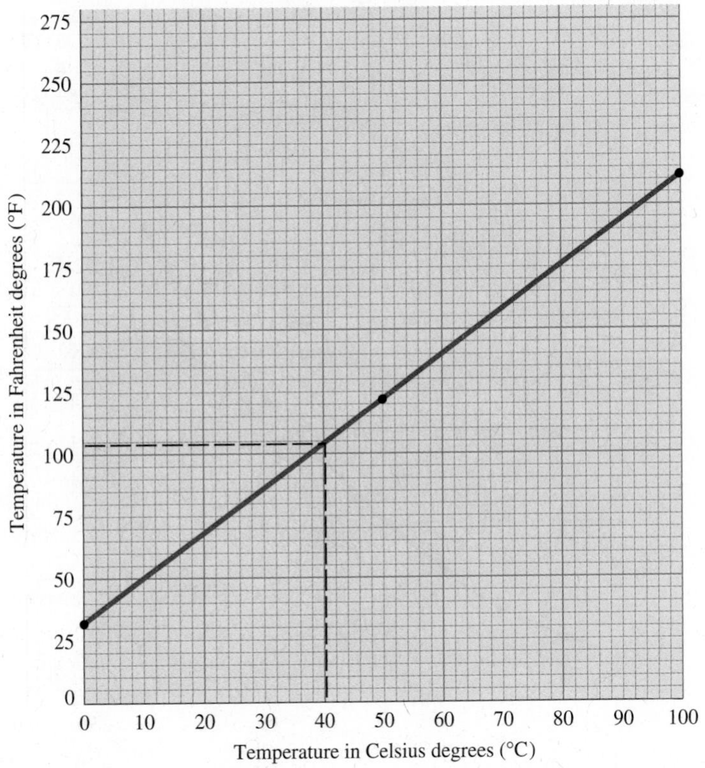

°C	°F
0	32
50	122
100	212

Figure I.2

the point where the two lines intersect. This process is called *plotting*. The other two points are plotted on the graph in the same way. (*Note:* The number of degrees per scale division was chosen to give a graph of convenient size. In this case, there are 5 Fahrenheit degrees per scale division and 2 Celsius degrees per scale division.)

The graph in Figure I.2 shows that the relationship between Celsius and Fahrenheit temperature is that of a straight line. The Fahrenheit temperature corresponding to any given Celsius temperature between 0 and 100° can be determined from the graph. For example, to find the Fahrenheit temperature corresponding to 40°C, trace a perpendicular line from 40°C on the *x*-axis to the line plotted on the graph. Now trace a horizontal line from this point on the plotted line to the *y*-axis and read the corresponding Fahrenheit temperature (104°F). See the dashed lines in Figure I.2. In turn, the Celsius temperature corresponding to any Fahrenheit temperature between 32° and 212° can be determined from the graph by tracing a horizontal line from the Fahrenheit temperature to the plotted line and reading the corresponding temperature on the Celsius scale directly below the point of intersection.

The mathematical relationship of Fahrenheit and Celsius temperatures is expressed by the equation $°F = (1.8 \times °C) + 32$. Figure I.2 is a graph of this equation. Because the graph is a straight line, it can be extended indefinitely at either end. Any desired Celsius temperature can be plotted against the corresponding Fahrenheit temperature by extending the scales along both axes as necessary.

Figure I.3 is a graph showing the solubility of potassium chlorate in water at various temperatures. The solubility curve on this graph was plotted from the data in the table next to the graph.

In contrast to the Celsius–Fahrenheit temperature relationship, there is no simple mathematical equation that describes the exact relationship between temperature and the solubility of potassium chlorate. The graph in Figure I.3 was constructed from experimentally determined solubilities at the six temperatures shown. These experimentally determined solubilities are all located on the smooth curve traced by the unbroken-line portion of the graph. We are therefore confident that the unbroken line represents a very good approximation of the solubility data for potassium chlorate over the temperature range from 10 to 80°C. All points on the plotted

Temperature (°C)	Solubility (g KClO₃/ 100 g water)
10	5.0
20	7.4
30	10.5
50	19.3
60	24.5
80	38.5

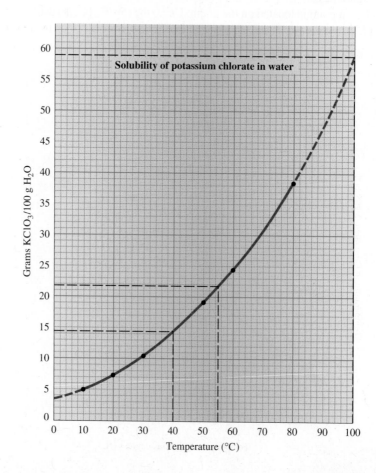

Figure I.3

curve represent the composition of saturated solutions. Any point below the curve represents an unsaturated solution.

The dashed-line portions of the curve are *extrapolations;* that is, they extend the curve above and below the temperature range actually covered by the plotted solubility data. Curves such as this one are often extrapolated a short distance beyond the range of the known data, although the extrapolated portions may not be highly accurate. Extrapolation is justified only in the absence of more reliable information.

The graph in Figure I.3 can be used with confidence to obtain the solubility of $KClO_3$ at any temperature between 10° and 80°C, but the solubilities between 0° and 10°C and between 80° and 100°C are less reliable. For example, what is the solubility of $KClO_3$ at 55°C, at 40°C, and at 100°C?

First draw a perpendicular line from each temperature to the plotted solubility curve. Now trace a horizontal line to the solubility axis from each point on the curve and read the corresponding solubilities. The values that we read from the graph are

40°C	14.2 g $KClO_3$/100 g water
55°C	22.0 g $KClO_3$/100 g water
100°C	59 g $KClO_3$/100 g water

Of these solubilities, the one at 55°C is probably the most reliable because experimental points are plotted at 50° and at 60°C. The 40°C solubility value is a bit less reliable because the nearest plotted points are at 30° and 50°C. The 100°C solubility is the least reliable of the three values because it was taken from the extrapolated part of the curve, and the nearest plotted point is 80°C. Actual handbook solubility values are 14.0 and 57.0 g of $KClO_3$/100 g water at 40°C and 100°C, respectively.

The graph in Figure I.3 can also be used to determine whether a solution is saturated or unsaturated. For example, a solution contains 15 g of $KClO_3$/100 g of water and is at a temperature of 55°C. Is the solution saturated or unsaturated? *Answer:* The solution is unsaturated because the point corresponding to 15 g and 55°C on the graph is below the solubility curve; all points below the curve represent unsaturated solutions.

Using a Scientific Calculator

A calculator is useful for most calculations in this book. You should obtain a scientific calculator, that is, one that has at least the following function keys on its keyboard.

Addition $\boxed{+}$	Second function $\boxed{F}$, $\boxed{INV}$, $\boxed{Shift}$
Subtraction $\boxed{-}$	Change sign $\boxed{+/-}$
Multiplication $\boxed{\times}$	Exponential number $\boxed{Exp}$
Division $\boxed{\div}$	Logarithm $\boxed{Log}$
Equals $\boxed{=}$	Antilogarithm $\boxed{10^x}$

Not all calculators use the same symbolism for these function keys, nor do all calculators work in the same way. The following discussion may not pertain to your particular calculator. Refer to your instruction manual for variations from the function symbols shown above and for the use of other function keys.

Some keys have two functions, upper and lower. In order to use the upper (second) function, the second function key $\boxed{F}$ must be pressed in order to activate the desired upper function after entering the number.

The second function key may have a different designation on your calculator.

The display area of the calculator shows the numbers entered and often shows more digits in the answer than should be used. Therefore, the final answer should be rounded to reflect the proper number of significant figures of the calculations.

Addition and Subtraction

To add numbers using your calculator,

1. Enter the first number to be added followed by the plus key $\boxed{+}$.
2. Enter the second number to be added followed by the plus key $\boxed{+}$.
3. Repeat Step 2 for each additional number to be added, except the last number.
4. After the last number is entered, press the equal key $\boxed{=}$. You should now have the answer in the display area.
5. When a number is to be subtracted, use the minus key $\boxed{-}$ instead of the plus key.

As an example, to add $16.0 + 1.223 + 8.45$, enter 16.0 followed by the $\boxed{+}$ key; then enter 1.223 followed by the $\boxed{+}$ key; then enter 8.45 followed by the $\boxed{=}$ key. The display shows 25.673, which is rounded to the answer 25.7.

Examples of Addition and Subtraction			
Calculation	Enter in sequence	Display	Rounded answer
a. $12.0 + 16.2 + 122.3$	$12.0\boxed{+}16.2\boxed{+}122.3\boxed{=}$	150.5	150.5
b. $132 - 62 + 141$	$132\boxed{-}62\boxed{+}141\boxed{=}$	211	211
c. $46.23 + 13.2$	$46.23\boxed{+}13.2\boxed{=}$	59.43	59.4
d. $129.06 + 49.1 - 18.3$	$129.06\boxed{+}49.1\boxed{-}18.3\boxed{=}$	159.86	159.9

Multiplication

To multiply numbers using your calculator,

1. Enter the first number to be multiplied followed by the multiplication key $\boxed{\times}$.
2. Enter the second number to be multiplied followed by the multiplication key $\boxed{\times}$.
3. Repeat Step 2 for all other numbers to be multiplied except the last number.
4. Enter the last number to be multiplied followed by the equal key $\boxed{=}$. You now have the answer in the display area.

   Round off to the proper number of significant figures.
   As an example, to calculate (3.25)(4.184)(22.2), enter 3.25 followed by the $\boxed{\times}$ key; then enter 4.184 followed by the $\boxed{\times}$ key; then enter 22.2 followed by the $\boxed{=}$ key. The display shows 301.8756, which is rounded to the answer 302.

	Examples of Multiplication		
Calculation	Enter in sequence	Display	Rounded answer
a. $12 \times 14 \times 18$	$12\boxed{\times}14\boxed{\times}18\boxed{=}$	3024	3.0×10^3
b. $122 \times 3.4 \times 60.$	$122\boxed{\times}3.4\boxed{\times}60.\boxed{=}$	24888	2.5×10^4
c. $0.522 \times 49.4 \times 6.33$	$0.522\boxed{\times}49.4\boxed{\times}6.33\boxed{=}$	163.23044	163

Division

To divide numbers using your calculator,

1. Enter the numerator followed by the division key $\boxed{\div}$.
2. Enter the denominator followed by the equal key to give the answer.
3. If there is more than one denominator, enter each denominator followed by the division key except for the last number, which is followed by the equal key.

As an example, to calculate $\dfrac{126}{12}$, enter 126 followed by the $\boxed{\div}$ key; then enter 12 followed by the $\boxed{=}$ key. The display shows 10.5, which is rounded to the answer 11.

	Examples of Division		
Calculation	Enter in sequence	Display	Rounded answer
a. $\dfrac{142}{25}$	$142\boxed{\div}25\boxed{=}$	5.68	5.7
b. $\dfrac{0.422}{5.00}$	$0.422\boxed{\div}5.00\boxed{=}$	0.0844	0.0844
c. $\dfrac{124}{0.022 \times 3.00}$	$124\boxed{\div}0.022\boxed{\div}3.00\boxed{=}$	1878.7878	1.9×10^3

Exponents

In scientific measurements and calculations, we often encounter very large and very small numbers. To express these large and small numbers conveniently, we use exponents, or powers, of 10. A number in exponential form is treated like any other number; that is, it can be added, subtracted, multiplied, or divided.

To enter an exponential number into your calculator, first enter the nonexponential part of the number and then press the exponent key $\boxed{\text{Exp}}$, followed by the exponent. For example, to enter 4.9×10^3, enter 4.94, then press $\boxed{\text{Exp}}$, and then press 3. When the exponent of 10 is a negative number, press the Change of Sign key $\boxed{+/-}$ after entering the exponent. For example, to enter 4.94×10^{-3}, enter in sequence 4.94 [Exp] 3 [+/−]. In most calculators, the exponent will appear in the display a couple of spaces after the nonexponent part of the number—for example, 4.94   03 or 4.94   −03.

Examples Using Exponential Numbers			
Calculation	Enter in sequence	Display	Rounded answer
a. $(4.94 \times 10^3)(21.4)$	4.94 [Exp] 3 [×] 21.4 [=]	105716	1.06×10^5
b. $(1.42 \times 10^4)(2.88 \times 10^{-5})$	1.42 [Exp] 4 [×] 2.88 [Exp] 5 [+/−] [=]	0.40896	0.409
c. $\dfrac{8.22 \times 10^{-5}}{5.00 \times 10^7}$	8.22 [Exp] 5 [+/−] [÷] 5.00 [Exp] 7 [=]	1.644 −12	1.64×10^{-12}

Logarithms

The logarithm of a number is the power (exponent) to which some base number must be raised to give the original number. The most commonly used base number is 10. The base number that we use is 10. For example, the log of 100 is 2.0 (log 100 = $10^{2.0}$). The log of 200 is 2.3 (log 200 = $10^{2.3}$). Logarithms are used in chemistry to calculate the pH of an aqueous acidic solution. The answer (log) should contain the same number of significant figures to the right of the decimal as is in the original number. Thus, log 100 = 2.0, but log 100. is 2.000.

The log key on most calculators is a function key. To determine the log using your calculator, enter the number and then press the log function key. For example, to determine the log of 125, enter 125 and then the [Log] key. The answer is 2.097.

Examples Using Logarithms Determine the log of the following:			
	Enter in sequence	Display	Rounded answer
a. 42	42 [Log]	1.6232492	1.62
b. 1.62×10^5	1.62 [Exp] 5 [Log]	5.209515	5.210
c. 6.4×10^{-6}	6.4 [Exp] 6 [+/−] [Log]	−5.19382	−5.19

Antilogarithms (Inverse Logarithms)

An antilogarithm is the number from which the logarithm has been calculated. It is calculated using the [10ˣ] key on your calculator. For example, to determine the antilogarithm of 2.891, enter 2.891 into your calculator and then press the second function key followed by the [10ˣ] key: 2.891 [F] [10ˣ].

The display shows 778.03655, which rounds to the answer 778.

Examples Using Antilogarithms Determine the antilogarithm of the following:			
	Enter in sequence	Display	Rounded answer
a. 1.628	1.628 [F] [10ˣ]	42.461956	42.5
b. 7.086	7.086 [F] [10ˣ]	12189896	1.22×10^7
c. −6.33	6.33 [+/−] [F] [10ˣ]	4.6773514 −07	4.7×10^{-7}
d. 6.33	6.33 [F] [10ˣ]	2137962	2.1×10^6

Additional Practice Problems*		
Problem	**Display**	**Rounded answer**
1. $143.5 + 14.02 + 1.202$	158.722	158.7
2. $72.06 - 26.92 - 49.66$	-4.52	-4.52
3. $2.168 + 4.288 - 1.62$	4.836	4.84
4. $(12.3)(22.8)(1.235)$	346.3434	346
5. $(2.42 \times 10^6)(6.08 \times 10^{-4})(0.623)$	916.65728	917
6. $\dfrac{(46.0)(82.3)}{19.2}$	197.17708	197
7. $\dfrac{0.0298}{243}$	$1.2263374 \quad -04$	1.23×10^{-4}
8. $\dfrac{(5.4)(298)(760)}{(273)(1042)}$	4.2992554	4.3
9. $(6.22 \times 10^6)(1.45 \times 10^3)(9.00)$	$8.1171 \quad 10$	8.12×10^{10}
10. $\dfrac{(1.49 \times 10^6)(1.88 \times 10^6)}{6.02 \times 10^{23}}$	$4.6531561 \quad -12$	4.65×10^{-12}
11. $\log 245$	2.389166	2.389
12. $\log 6.5 \times 10^{-6}$	-5.1870866	-5.19
13. $\log 24 \times \log 34$	2.1137644	2.11
14. antilog 6.34	2187761.6	2.2×10^6
15. antilog -6.34	$4.5708818 \quad -07$	4.6×10^{-7}

*Only the problem, the display, and the rounded answer are given.

Units of Measurement

Physical Constants		
Constant	**Symbol**	**Value**
Atomic mass unit	amu	1.6606×10^{-27} kg
Avogadro's number	N	6.022×10^{23} particles/mol
Gas constant	R (at STP)	0.08205 L atm/K mol
Mass of an electron	m_e	9.109×10^{-28} g
		5.486×10^{-4} amu
Mass of a neutron	m_n	1.675×10^{-27} kg
		1.00866 amu
Mass of a proton	m_p	1.673×10^{-27} kg
		1.00728 amu
Speed of light	c	2.997925×10^8 m/s

SI Units and Conversion Factors	
Length	**Mass**
SI unit: meter (m)	SI unit: kilogram (kg)
1 meter = 1.0936 yards	1 kilogram = 1000 grams
= 100 centimeters	= 2.20 pounds
= 1000 millimeters	1 gram = 1000 milligrams
1 centimeter = 0.3937 inch	1 pound = 453.59 grams
1 inch = 2.54 centimeters	= 0.45359 kilogram
(exactly)	= 16 ounces
1 kilometer = 0.62137 mile	1 ton = 2000 pounds
1 mile = 5280 feet	= 907.185 kilograms
= 1.609 kilometers	1 ounce = 28.3 grams
1 angstrom = 10^{-10} meter	1 atomic mass unit = 1.6606×10^{-27} kilogram
Volume	**Temperature**
SI unit: cubic meter (m³)	SI unit: kelvin (K)
1 liter = 10^{-3} m³	0 K = -273.15°C
= 1 dm³	= -459.67°F
= 1.0567 quarts	
= 1000 milliliters	K = °C + 273.15
1 gallon = 4 quarts	
= 8 pints	°C = $\dfrac{°F - 32}{1.8}$
= 3.785 liters	
1 quart = 32 fluid ounces	°F = 1.8(°C) + 32
= 0.946 liter	
1 fluid ounce = 29.6 milliliters	°C = $\dfrac{5}{9}$(°F − 32)
Energy	**Pressure**
SI unit: joule (J)	SI unit: pascal (Pa)
1 joule = 1 kg m²/s²	1 pascal = 1 kg/m s²
= 0.23901 calorie	1 atmosphere = 101.325 kilopascals
1 calorie = 4.184 joules	= 760 torr (mm Hg)
	= 14.70 pounds per square inch (psi)

Vapor Pressure of Water at Various Temperatures

Temperature (°C)	Vapor pressure (torr)		Temperature (°C)	Vapor pressure (torr)
0	4.6		26	25.2
5	6.5		27	26.7
10	9.2		28	28.3
15	12.8		29	30.0
16	13.6		30	31.8
17	14.5		40	55.3
18	15.5		50	92.5
19	16.5		60	149.4
20	17.5		70	233.7
21	18.6		80	355.1
22	19.8		90	525.8
23	21.2		100	760.0
24	22.4		110	1074.6
25	23.8			

Solubility Table

	F^-	Cl^-	Br^-	I^-	O^{2-}	S^{2-}	OH^-	NO_3^-	CO_3^{2-}	SO_4^{2-}	$C_2H_3O_2^-$
H^+	aq	aq	aq	aq	aq	sl.aq	aq	aq	sl.aq	aq	aq
Na^+	aq	aq	aq	aq	aq	aq	aq	aq	aq	aq	aq
K^+	aq	aq	aq	aq	aq	aq	aq	aq	aq	aq	aq
NH_4^+	aq	aq	aq	aq	—	aq	aq	aq	aq	aq	aq
Ag^+	aq	I	I	I	I	I	—	aq	I	I	I
Mg^{2+}	I	aq	aq	aq	I	d	I	aq	I	aq	aq
Ca^{2+}	I	aq	aq	aq	I	d	I	aq	I	I	aq
Ba^{2+}	I	aq	aq	aq	sl.aq	d	sl.aq	aq	I	I	aq
Fe^{2+}	sl.aq	aq	aq	aq	I	I	I	aq	sl.aq	aq	aq
Fe^{3+}	I	aq	aq	—	I	I	I	aq	I	aq	I
Co^{2+}	aq	aq	aq	aq	I	I	I	aq	I	aq	aq
Ni^{2+}	sl.aq	aq	aq	aq	I	I	I	aq	I	aq	aq
Cu^{2+}	sl.aq	aq	aq	—	I	I	I	aq	I	aq	aq
Zn^{2+}	sl.aq	aq	aq	aq	I	I	I	aq	I	aq	aq
Hg^{2+}	d	aq	I	I	I	I	I	aq	I	d	aq
Cd^{2+}	sl.aq	aq	aq	aq	I	I	I	aq	I	aq	aq
Sn^{2+}	aq	aq	aq	sl.aq	I	I	I	aq	I	aq	aq
Pb^{2+}	I	I	I	I	I	I	I	aq	I	I	aq
Mn^{2+}	sl.aq	aq	aq	aq	I	I	I	aq	I	aq	aq
Al^{3+}	I	aq	aq	aq	I	d	I	aq	—	aq	aq

Key: aq = soluble in water

sl.aq = slightly soluble in water

I = insoluble in water (less than 1 g/100 g H_2O)

d = decomposes in water

Answers to Selected Exercises

CHAPTER 1

Exercises

2. Two states are present; solid and liquid.

4. The maple leaf represents a heterogeneous mixture.

6. (a) homogeneous

   (b) homogeneous

   (c) heterogeneous

   (d) heterogeneous

9. (a) water, a pure substance

   (c) salt, a pure substance

12. (a) two phases, solid and gas

   (c) two phases, solid and liquid

CHAPTER 2

Exercises

2. (a) 1000 meters = 1 kilometer

   (c) 0.000001 liter = 1 microliter

   (e) 0.001 liter = 1 milliliter

4. (a) mg   (e) Å

   (c) m

6. (a) not significant   (e) significant

   (c) significant

8. (a) 40.0 (3 sig fig)

   (b) 0.081 (2 sig fig)

   (c) 129,042 (6 sig fig)

   (d) 4.090×10^{-3} (4 sig fig)

10. (a) 8.87   (c) 130. (1.30×10^2)

   (b) 21.3   (d) 2.00×10^6

12. (a) 4.56×10^{-2} (c) 4.030×10^1

   (b) 4.0822×10^3 (d) 1.2×10^7

14. (a) 28.1   (e) 2.49×10^{-4}

   (c) 4.0×10^1

16. (a) $\frac{1}{4}$ (c) $1\frac{2}{3}$ or $\frac{5}{3}$

   (b) $\frac{5}{8}$ (d) $\frac{8}{9}$

18. (a) 1.0×10^2 (c) 22

   (b) 4.6 mL

20. (a) 4.5×10^8 Å   (e) 6.5×10^5 mg

   (c) 8.0×10^6 mm   (g) 468 mL

22. (a) 117 ft

   (c) 7.4×10^4 mm^3

   (e) 75.7 L

24. 12 mi/hr

26. 8.33 gr

28. 3.54 days

30. 2.54×10^{-3} lb

32. 5.94×10^3 cm/s

34. (a) 50 ft/min

   (b) 3×10^{-4} km/s

36. 4.03×10^6 tilapia

38. 160 L

40. 4×10^5 m^2

42. 6 gal

44. 113°F Summer!

46. (a) 90.°F   (c) 546 K

   (b) −22.6°C   (d) −300°F

48. −11.4°C = 11.4°F

50. −153°C

52. 0.810 g/mL

54. 7.2 g/mL

56. 63 mL

59. A graduated cylinder would be the best choice for adding 100 mL of solvent to a reaction. While the volumetric flask is also labeled 100 mL, volumetric flasks are typically used for doing dilutions. The other three pieces of glassware could also be used, but they hold smaller volumes, so it would take a longer time to measure out 100 mL. Also, because you would have to repeat the measurement many times using the other glassware, there is a greater chance for error.

60. 14.5 mL

61. 26 mL

63. 3.43 vials potion

64. 10.1 lb sequins

65. Yes, 120 L additional solution

67. (a) −252.88°C

   (b) −423.18°F

68. 97°F − 100°F, similar to mammals

72. 5.1×10^3 L

75. 0.506 km^2

77. 0.965 g/mL

79. ethyl alcohol, because it has the lower density

81. 54.3 mL

82. 18.6 mg NHP

87. 76.9 g

CHAPTER 3

Exercises

2. (d) HI   (f) Cl_2 (g) CO

4. (a) magnesium, bromine

    (c) hydrogen, nitrogen, oxygen

    (e) aluminum, phosphorus, oxygen

6. (a) $AlBr_3$ (c) $PbCrO_4$

8. (a) $C_{55}H_{66}O_{24}$ (b) $Cl_{15}H_{14}O_6$ (c) $Cl_{30}H_{48}O_3$

10. (a) hydrogen, 4 atoms; carbon, 2 atoms; oxygen, 2 atoms

    (c) magnesium, 1 atom; hydrogen, 2 atoms; sulfur 2 atoms; oxygen, 6 atoms

    (e) nickel, 1 atom; carbon, 1 atom; oxygen, 3 atoms

    (g) carbon, 4 atoms; hydrogen, 10 atoms

12. (a) 9 atoms   (c) 12 atoms

    (b) 12 atoms   (d) 12 atoms

14. (a) 6 atoms O   (c) 6 atoms O

    (b) 4 atoms O   (d) 21 atoms O

16. (a) mixture   (e) mixture

    (c) pure substance

18. (c) element   (d) compound

20. (a) compound   (c) mixture

    (b) compound

22. No. The only common liquid elements (at room temperature) are mercury and bromine.

24. 72% solids

26. The formula for hydrogen peroxide is H_2O_2. There are two atoms of oxygen for every two atoms of hydrogen. The molar mass of oxygen is 16.00 g, and the molar mass of hydrogen is 1.008 g. For hydrogen peroxide, the total mass of hydrogen is 2.016 g and the total mass of oxygen is 32.00 g, for a ratio of hydrogen to oxygen of approximately 2:32. or 1:16. Therefore, there is 1 gram of hydrogen for every 16 grams of oxygen.

29. (a) 1 carbon atom and 1 oxygen atom; total number of atoms = 2

    (c) 1 hydrogen atom, 1 nitrogen atom, and 3 oxygen atoms; total number of atoms = 5

    (e) 1 calcium atom, 2 nitrogen atoms, and 6 oxygen atoms; total number of atoms = 9

32. 40 atoms H

34. (a) magnesium, manganese, molybdenum, mendelevium, mercury, meitnerium

    (c) sodium, potassium, iron, silver, tin, antimony

36. 420 atoms

39. (a) NaCl   (g) $C_6H_{12}O_6$

    (c) K_2O (i) $Cr(NO_3)_3$

    (e) K_3PO_4

41. (a) 12C, 22H, 11O   (e) 12C, 19H, 8O, 3Cl

    (c) 14C, 18H, 5O, 2N

CHAPTER 4

Exercises

2. (a) physical   (e) physical

    (c) chemical   (g) chemical

4. The copper wire, like the platinum wire, changed to a glowing red color when heated. Upon cooling, a new substance, black copper(II) oxide, had appeared.

6. Reactant: water

    Products: hydrogen, oxygen

8. (a) physical   (e) chemical

    (c) chemical

10. (a) kinetic   (e) kinetic

    (c) potential

12. the transformation of kinetic energy to thermal energy

14. (a) +   (b) −   (c) +   (d) −   (e) −

16. 6.4×10^2 J

18. 5.58×10^3 J

20. 0.128 J/g°C

22. 31.7°C

27. 0.525 lb gold

30. 1.0×10^2 g Cu

32. 654°C

39. 8 g fat

40. (a) 1.12 kJ

    (b) Glove made of wool felt

    (c) wool felt 23.9°C, paper or cotton 24.8°C, rubber 9.04°C, silicon 22.6°C

43. A chemical change has occurred. Hydrogen molecules and oxygen molecules have combined to form water molecules.

CHAPTER 5

Exercises

2. Mg^{2+}, Cr^{3+}, Ba^{2+}, Ca^{2+}, and Y^{3+}.

4. (a) The nucleus of the atom contains most of the mass.

    (c) The atom is mostly empty space.

6. The nucleus of an atom contains nearly all of its mass.

8. Electrons:

    Dalton—Electrons are not part of his model.

    Thomson—Electrons are scattered throughout the positive mass of matter in the atom.

    Rutherford—Electrons are located out in space away from the central positive mass.

    Positive matter:

    Dalton—No positive matter in his model.

    Thomson—Positive matter is distributed throughout the atom.

    Rutherford—Positive matter is concentrated in a small central nucleus.

10. Yes. The mass of the isotope $^{12}_6C$, 12, is an exact number. The masses of other isotopes are not exact numbers.

12. Three isotopes of hydrogen have the same number of protons and electrons but differ in the number of neutrons.

14. All five isotopes have nuclei that contain 30 protons and 30 electrons. The numbers of neutrons are:

Isotope mass number	Neutrons
64	34
66	36
67	37
68	38
70	40

16. (a) $^{109}_{47}Ag$ (c) $^{57}_{26}Fe$

18. (a) 26 protons, 26 electrons, and 28 neutrons

(c) 35 protons, 35 electrons, and 44 neutrons

20. (a) 56   (c) 79   (e) $^{135}_{56}Ba$

22. 50. amu

24. 35.46 amu; chlorine (Cl)

26. $1.5 \times 10^5 : 1.0$

28. (a) and (c) are isotopes of phosphorus

29. 9.2×10^8 atoms silicon

31. ^{160}Gd, ^{156}Dy, ^{163}Ho, ^{162}Er. This order differs from that on the periodic table.

33. (a) 197 amu   (b) gold (Au)

36.

C^+	6 protons, 5 electrons
O^+	8 protons, 7 electrons
O^{2+}	8 protons, 6 electrons

42. (a) 0.02644% electrons   (c) 0.02356% electrons

44. The electron region is the area around the nucleus where electrons are most likely to be located.

CHAPTER 6

Exercises

2. (a) Al_2S_3 (d) SrO

(c) K_3N (f) $AlCl_3$

4. (a) F^- (i) $Cr_2O_7^{2-}$

(c) I^- (k) PO_4^{3-}

(e) S^{2-} (m) N^{3-}

(g) P^{3-}

6. (a) sodium thiosulfate   (d) sodium chloride

(b) dinitrogen oxide   (e) magnesium hydroxide

(c) aluminum oxide   (f) lead (II) sulfide

8.

Ion	SO_4^{2-}	OH^-	AsO_4^{3-}	$C_2H_3O_2^-$	CrO_4^{2-}
NH_4^+	$(NH_4)_2SO_4$	NH_4OH	$(NH_4)_3AsO_4$	$NH_4C_2H_3O_2$	$(NH_4)_2CrO_4$
Ca^{2+}	$CaSO_4$	$Ca(OH)_2$	$Ca_3(AsO_4)_2$	$Ca(C_2H_3O_2)_2$	$CaCrO_4$
Fe^{3+}	$Fe_2(SO_4)_3$	$Fe(OH)_3$	$FeAsO_4$	$Fe(C_2H_3O_2)_3$	$Fe_2(CrO_4)_3$
Ag^+	Ag_2SO_4	$AgOH$	Ag_3AsO_4	$AgC_2H_3O_2$	Ag_2CrO_4
Cu^{2+}	$CuSO_4$	$Cu(OH)_2$	$Cu_3(AsO_4)_2$	$Cu(C_2H_3O_2)_2$	$CuCrO_4$

10. NH_4^+ compounds: ammonium sulfate, ammonium hydroxide, ammonium arsenate, ammonium acetate, ammonium chromate.

Ca^{2+} compounds: calcium sulfate, calcium hydroxide, calcium arsenate, calcium acetate, calcium chromate.

Fe^{3+} compounds: iron(III) sulfate, iron(III) hydroxide, iron(III) arsenate, iron(III) acetate, iron(III) chromate.

Ag^+ compounds: silver sulfate, silver hydroxide, silver arsenate, silver acetate, silver chromate.

Cu^{2+} compounds: copper(II) sulfate, copper(II) hydroxide, copper(II) arsenate, copper(II) acetate, copper(II) chromate.

12. (a) K_3N (e) Ca_3N_2

(c) FeO (g) MnI_3

14. (a) phosphorus pentabromide

(c) dinitrogen pentasulfide

(e) silicon tetrachloride

(g) tetraphosphorus heptasulfide

16. (a) $SnBr_4$ (d) $Hg(NO_2)_2$

(c) $Ni(BO_3)_2$ (f) $Fe(C_2H_3O_2)_2$

18. (a) $HC_2H_3O_2$ (d) H_3BO_3

(c) H_2S (f) $HClO$

20. (a) phosphoric acid   (f) nitric acid

(c) iodic acid   (g) hydroiodic acid

(d) hydrochloric acid   (i) sulfurous acid

22. (a) Na_2CrO_4 (i) $NaClO$

(c) $Ni(C_2H_3O_2)_2$ (k) $Cr_2(SO_3)_3$

(e) $Mg(BrO_3)_2$ (m) $Na_2C_2O_4$

(g) $Mn(OH)_2$

24. (a) calcium hydrogen sulfate

(c) tin(II) nitrite

(e) potassium hydrogen carbonate

(f) bismuth(III) arsenate

(h) ammonium monohydrogen phosphate

(j) potassium permanganate

26. (a) C_2H_5OH (e) HCl

(c) $CaSO_4 \cdot 2H_2O$ (g) $NaOH$

28. (a) sulfate   (d) chlorate

(c) nitrate   (f) carbonate

30. (a) K_2SO_4 (e) $KClO_3$

(c) KNO_3 (f) K_2CO_3

32. Formula: KCl Name: potassium chloride

34. (a) $CaCO_3$, no concern   (c) H_2O, no concern

(b) $HC_2H_3O_2$, no concern

37. $Li_3Fe(CN)_6$

$AlFe(CN)_6$

$Zn_3[Fe(CN)_6]_2$

CHAPTER 7

Exercises

2. Molar masses

(a) $NaOH$ 40.00

(c) Cr_2O_3 152.0

(e) $Mg(HCO_3)_2$ 146.3

(g) $C_6H_{12}O_6$ 180.2

(i) $BaCl_2 \cdot 2H_2O$ 244.2

4. (a) 0.625 mol $NaOH$

(c) 7.18×10^{-3} mol $MgCl_2$

(e) 2.03×10^{-2} mol Na_2SO_4

6. (a) 0.0417 g H_2SO_4 (c) 0.122 g Ti

(b) 11 g CCl_4 (d) 8.0×10^{-7} g S

8. (a) 5.8×10^{24} molecules C_2H_4

(c) 4.36×10^{23} molecules CH_3OH

10. (a) 16 atoms

(c) 4.5×10^{25} atoms

(e) 1.3×10^{24} atoms

12. (a) 2.180×10^{-22} g Xe

(c) 8.975×10^{-22} g CH_3COOH

14. (a) 4.57×10^{-22} mol W

(c) 1.277×10^{-21} g SO_2

16. (a) 0.492 mol S_2F_{10}

(c) 3.56×10^{24} total atoms

(e) 2.96×10^{24} atoms F

18. (a) 138 atoms H

(b) 1.3×10^{25} atoms H

(c) 2.2×10^{24} atoms H

20. (a) 1.27 g Cl   (c) 0.274 g H

(b) 9.07 g H

22. (a) 47.98% Zn   (e) 23.09% Fe

   52.02% Cl   17.37% N

(c) 12.26% Mg   59.53% O

   31.24% P

   56.48% O

24. (a) KCl   47.55% Cl   (c) $SiCl_4$ 83.46% Cl

(b) $BaCl_2$ 34.05% Cl   (d) LiCl   83.63% Cl

   highest % Cl is LiCl

   lowest % Cl is $BaCl_2$

26. 41.1% S, 6.02% H, 6.82% O, 46.0% C

28. (a) $KClO_3$ lower % Cl

(b) $KHSO_4$ higher % S

(c) Na_2CrO_4 lower % Cr

30. (a) CuCl   (e) $BaCr_2O_7$

(c) Cr_2S_3

32. (a) C_6H_5Br

(c) V_2S_3

34. Ti_2O_3

36. $HgCl_2$

38. Empirical formula is $C_3H_5OS_2$, molecular formula is $C_6H_{10}O_2S_4$.

40. The molecular formula is $C_4H_8O_2$.

42. 40.0% C, 6.73% H, 53.3% O. Empirical formula is CH_2O, molecular formula is $C_6H_{12}O_6$.

44. $Al_2(SiO_3)_3$

45. 8.43×10^{23} atoms P

47. 197 g/mol

49. 5.54×10^{19} m

50. 8.6×10^{13} dollars/person

52. (a) 10.3 cm^3 (b) 2.18 cm

54. 41.58 g Fe_2S_3

56. 1.0 ton of iron ore contains 2×10^4 g Fe

58. (a) 76.98% Hg   (c) 17.27% N

62. 42.10% C

   6.480% H

   51.42% O

64. 4.77 g O in 8.50 g $Al_2(SO_4)_3$

66. The empirical formula is $C_4H_4CaO_6$

68. 1.910×10^{16} years

70. 1.529×10^{-7} g $C_3H_8O_3$

72. $Co_3Mo_2Cl_{11}$

74. The empirical formula is C_5H_5N.

76. (a) CH_2O (c) CH_2O (e) $C_6H_2Cl_2O$

78. 9.0×10^{17} molecules O_2

CHAPTER 8

Exercises

2. (a) endothermic   (e) exothermic

(c) endothermic

4. (a) $H_2 + Br_2 \longrightarrow 2\,HBr$ combination

(c) $Ba(ClO_3)_2 \longrightarrow BaCl_2 + 3\,O_2$ decomposition

(e) $2\,H_2O_2 \longrightarrow 2\,H_2O + O_2$ decomposition

6. A metal and a nonmetal can react to form a salt; also an acid plus a base.

8. (a) $2\,SO_2 + O_2 \longrightarrow 2\,SO_3$

(c) $2\,Na + 2\,H_2O \longrightarrow 2\,NaOH + H_2$

(e) $Bi_2S_3 + 6\,HCl \longrightarrow 2\,BiCl_3 + 3\,H_2S$

(g) $Hg_2(C_2H_3O_2)_2(aq) + 2\,KCl(aq) \longrightarrow$
$$Hg_2Cl_2(s) + 2\,KC_2H_3O_2(aq)$$

(i) $2\,K_3PO_4 + 3\,BaCl_2 \longrightarrow 6\,KCl + Ba_3(PO_4)_2$

10. (a) $MgCO_3(s) \xrightarrow{\Delta} MgO(s) + CO_2(g)$

(c) $Fe_2(SO_4)_3(aq) + 6\,NaOH(aq) \longrightarrow$
$$2\,Fe(OH)_3(s) + 3\,Na_2SO_4(aq)$$

(e) $SO_3(g) + H_2O(l) \longrightarrow H_2SO_4(aq)$

12. (a) $CuSO_4(aq) + 2\,KOH(aq) \longrightarrow Cu(OH)_2(s) + K_2SO_4(aq)$

(c) $3\,NaHCO_3(s) + H_3PO_4(aq) \longrightarrow$
$$Na_3PO_4(aq) + 3\,H_2O(l) + 3\,CO_2(g)$$

14. (a) $Cu(s) + NiCl_2(aq) \longrightarrow$ no reaction

(c) $I_2(s) + CaCl_2(aq) \longrightarrow$ no reaction

16. (a) $Li_2O(s) + H_2O(l) \longrightarrow 2\,LiOH(aq)$

(c) $Zn(s) + CuSO_4(aq) \longrightarrow Cu(s) + ZnSO_4(aq)$

18. (a) $C + O_2 \longrightarrow CO_2$

(c) $CuBr_2 + Cl_2 \longrightarrow CuCl_2 + Br_2$

(e) $2\,NaNO_3 \xrightarrow{\Delta} 2\,NaNO_2 + O_2$

20. (a) 2 mol of Na react with 1 mol of Cl_2 to produce 2 mol of NaCl and release 822 kJ of energy. Exothermic

(b) 1 mol of PCl_5 absorbs 92.9 kJ of energy to produce 1 mol of PCl_3 and 1 mol of Cl_2. Endothermic

22. (a) $Ca(s) + 2\,H_2O(l) \longrightarrow Ca(OH)_2(aq) + H_2(g) + 635.1$ kJ

(b) $2\,BrF_3 + 601.6$ kJ $\longrightarrow Br_2 + 3\,F_2$

24. (a) single displacement,
$$Ni(s) + Pb(NO_3)_2(aq) \longrightarrow Pb(s) + Ni(NO_3)_2(aq)$$

(c) decomposition, $2\,HgO(s) \longrightarrow 2\,Hg(l) + O_2(g)$

25. Combinations that form precipitates

$Ca(NO_3)_2(aq) + (NH_4)_2SO_4(aq) \longrightarrow$
$$CaSO_4(s) + 2\,NH_4NO_3(aq)$$

$Ca(NO_3)_2(aq) + (NH_4)_2CO_3 \longrightarrow CaCO_3(s) + 2\,NH_4NO_3(aq)$

$AgNO_3(aq) + NH_4Cl(aq) \longrightarrow AgCl(s) + NH_4NO_3(aq)$

$2\,AgNO_3(aq) + (NH_4)_2SO_4 \longrightarrow Ag_2SO_4(s) + 2\,NH_4NO_3(aq)$

2 AgNO$_3$(aq) + (NH$_4$)$_2$CO$_3$(aq) $\longrightarrow$
Ag$_2$CO$_3$(s) + 2 NH$_4$NO$_3$(aq)

26. P$_4$O$_{10}$ + 12 HClO$_4$ $\longrightarrow$ 6 Cl$_2$O$_7$ + 4 H$_3$PO$_4$

58 atoms O on each side

34. (a) 4 Cs + O$_2$ $\longrightarrow$ 2 Cs$_2$O

(c) SO$_3$ + H$_2$O $\longrightarrow$ H$_2$SO$_4$

37. (a) 2 ZnO $\xrightarrow{\Delta}$ 2 Zn + O$_2$

(c) Na$_2$CO$_3$ $\xrightarrow{\Delta}$ Na$_2$O + CO$_2$

38. (a) Mg(s) + 2 HCl(aq) $\longrightarrow$ H$_2$(g) + MgCl$_2$(aq)

(c) 3 Zn(s) + 2 Fe(NO$_3$)$_3$(aq) $\longrightarrow$ 2 Fe(s) + 3 Zn(NO$_3$)$_2$(aq)

39. (a) 2 (NH$_4$)$_3$PO$_4$(aq) + 3 Ba(NO$_3$)$_2$(aq) $\longrightarrow$
Ba$_3$(PO$_4$)$_2$(s) + 6 NH$_4$NO$_3$(aq)

(c) CuSO$_4$(aq) + Ca(ClO$_3$)$_2$(aq) $\longrightarrow$
CaSO$_4$(s) + Cu(ClO$_3$)$_2$(aq)

(e) H$_3$PO$_4$(aq) + 3 KOH(aq) $\longrightarrow$ K$_3$PO$_4$(aq) + 3 H$_2$O(l)

42. (a) CH$_4$ + 2 O$_2$ $\longrightarrow$ CO$_2$ + 2 H$_2$O

(b) 2 C$_3$H$_6$ + 9 O$_2$ $\longrightarrow$ 6 CO$_2$ + 6 H$_2$O

(c) C$_6$H$_5$CH$_3$ + 9 O$_2$ $\longrightarrow$ 7 CO$_2$ + 4 H$_2$O

CHAPTER 9

Exercises

2. (a) 25.0 mol NaHCO$_3$ (c) 16 mol CO$_2$

4. (a) 1.31 g NiSO$_4$ (e) 18 g K$_2$CrO$_4$

(c) 373 g Bi$_2$S$_3$

6. HCl

8. (a) $\dfrac{6 \text{ mol O}_2}{1 \text{ mol C}_4\text{H}_9\text{OH}}$ (d) $\dfrac{1 \text{ mol C}_4\text{H}_9\text{OH}}{4 \text{ mol CO}_2}$

(c) $\dfrac{4 \text{ mol CO}_2}{5 \text{ mol H}_2\text{O}}$ (f) $\dfrac{4 \text{ mol CO}_2}{6 \text{ mol O}_2}$

10. (a) 34 mol NaOH   (b) 11 mol Na$_2$SO$_4$

12. (a) 8.33 mol H$_2$O   (b) 0.800 mol Al(OH)$_3$

14. 2 Al + 6 HBr $\longrightarrow$ 2 AlBr$_3$ + 3 H$_2$

249 g AlBr$_3$

16. 117 g H$_2$O, 271 g Fe

18. (a) 1.9 mol CaS   (e) 2.50 × 10^3 g CaS

(c) 0.4179 mol CaO

20. (a)

Li

I

neither is limiting

(b)

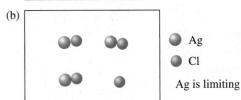

Ag

Cl

Ag is limiting

22. (a) Oxygen is the limiting reactant.

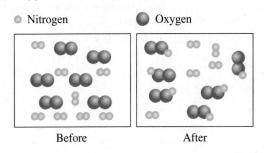

Before   After

(b) Water is the limiting reactant.

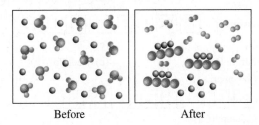

Before   After

24. (a) H$_2$S is the limiting reactant and Bi(NO$_3$)$_3$ is in excess.

(b) H$_2$O is the limiting reactant and Fe is in excess.

26. (a) 0.193 mol PCl$_3$

(b) 71 g PCl$_3$; P$_4$ is in excess

(c) 23 mol PCl$_3$ will be produced; 9 mol P$_4$ unreacted

28. Zinc

30. 89.02%

32. (a) 0.32 mol FeCl$_3$ and 0.48 mol H$_2$O

(b) 9 g Fe$_2$O$_3$ in excess

(c) 48 g FeCl$_3$ actual yield

36. (a) 0.19 mol KO$_2$ (b) 28 g O$_2$/hr

39. 70.7 g O$_2$

41. (a) 0.16 mol MnCl$_2$ (e) 11 g KMnO$_4$ unreacted

(c) 82% yield

44. 46.97 g Ca(OH)$_2$

46. (a) 5.0 mol CO$_2$, 7.5 mol H$_2$O

(b) 430. g CO$_2$, 264 g H$_2$

48. (a) 2.0 mol Cu, 2.0 mol FeSO$_4$, 1 mol CuSO$_4$

(b) 15.9 g Cu, 38.1 g FeSO$_4$, 6.0 g Fe

50. (a) 3.2 × 10^2 g C$_2$H$_5$OH

(b) 1.10 × 10^3 g C$_6$H$_{12}$O$_6$

52. 3.7 × 10^2 kg Li$_2$O

55. 223.0 g H$_2$SO$_4$; 72.90% yield

58. 67.2% KClO$_3$

CHAPTER 10

Exercises

2. (a) Na   11 total electrons; 1 valence electron

(c) P   15 total electrons; 5 valence electrons

4. (a) Mn $1s^2 2s^2 2p^6 3s^2 3p^6 4s^2 3d^5$

(c) Ga $1s^2 2s^2 2p^6 3s^2 3p^6 4s^2 3d^{10} 4p^1$

6. Bohr said that a number of orbits were available for electrons, each corresponding to an energy level. When an electron falls from a higher energy orbit to a lower orbit, energy is given off as a specific wavelength of light. Only those energies in the visible range are seen in the hydrogen spectrum. Each line corresponds to a change from one orbit to another.

8. 18 electrons in third energy level; 2 in s, 6 in p, 10 in d

10. (a) $^{28}_{14}Si$ [↑↓] [↑↓] [↑↓][↑↓][↑↓] [↑↓] [↑] [↑] [ ]

(c) $^{40}_{18}Ar$ [↓↑] [↓↑] [↓↑][↓↑][↓↑] [↓↑] [↓↑][↓↑][↓↑]

(e) $^{31}_{15}P$ [↑↓] [↑↓] [↑↓][↑↓][↑↓] [↑↓] [↑] [↑] [↑]

12. (a) Li $1s^2 2s^1$

(c) Zn $1s^2 2s^2 2p^6 3s^2 3p^6 4s^2 3d^{10}$

(e) K $1s^2 2s^2 2p^6 3s^2 3p^6 4s^1$

14. (a) correct

(c) incorrect [↑↓] [↑↓] [↑↓][↑↓][↑↓] [↑↓] [↑↓][↑↓][↑↓] [↑↓]

16. (a) N (c) Ca

18. (a) boron, B (c) lead, Pb

20. (a) Phosphorus (P)

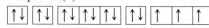

[↑↓] [↑↓] [↑↓][↑↓][↑↓] [↑↓] [↑] [↑] [↑]

(c) Calcium (Ca)

[↑↓] [↑↓] [↑↓][↑↓][↑↓] [↑↓] [↑↓][↑↓][↑↓] [↑↓]

(e) Potassium (K)

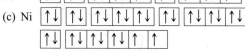

[↑↓] [↑↓] [↑↓][↑↓][↑↓] [↑↓] [↑↓][↑↓][↑↓] [↑]

22. (a) Cl [↑↓] [↑↓] [↑↓][↑↓][↑↓] [↑↓] [↑↓][↑↓][↑]

(c) Ni [↑↓] [↑↓] [↑↓][↑↓][↑↓] [↑↓] [↑↓][↑↓][↑↓]

[↑↓] [↑↓][↑↓][↑↓][↑] [↑]

(e) Ba [↑↓] [↑↓] [↑↓][↑↓][↑↓] [↑↓] [↑↓][↑↓][↑↓]

[↑↓] [↑↓][↑↓][↑↓][↑↓][↑↓] [↑↓][↑↓][↑↓]

[↑↓] [↑↓][↑↓][↑↓][↑↓][↑↓] [↑↓][↑↓][↑↓] [↑↓]

24. (a) (13p 14n) $2e^- 8e^- 3e^-$ $^{27}_{13}Al$

(b) (22p 26n) $2e^- 8e^- 8e^- 4e^-$ $^{48}_{22}Ti$

26. The last electron in potassium is located in the fourth energy level because the 4s orbital is at a lower energy level than the 3d orbital. Also the properties of potassium are similar to the other elements in Group 1A.

28. Noble gases each have filled s and p orbitals in the outermost energy level.

30. The elements in a group have the same number of outer energy level electrons. They are located vertically on the periodic table.

32. Valence shell electrons

(a) 5 (c) 6 (e) 3

34. All of these elements have an $s^2 d^{10}$ electron configuration in their outermost energy levels.

36. (a) and (f) (e) and (h)

38. 7, 33 since they are in the same periodic group.

40. (a) nonmetal, I (c) metal, Mo

42. Period 4, Group 3B

44. Group 3A contains 3 valence electrons. Group 3B contains 2 electrons in the outermost level and 1 electron in an inner d orbital. Group A elements are representative, while Group B elements are transition elements.

46. (a) Pb (b) Sm (c) Ga (d) Ir

48. (a) Mg $1s^2 2s^2 2p^6 \boxed{3s^2}$

(c) K $1s^2 2s^2 2p^6 3s^2 3p^6 \boxed{4s^1}$

(d) F $1s^2 \boxed{2s^2 2p^5}$

(f) N $1s^2 \boxed{2s^2 2p^3}$

50. (a) 7A, Halogens (e) 8A, Noble Gases

(c) 1A, Alkali Metals

56. (a) Any orbital can hold a maximum of two electrons.

(c) The third principal energy level can hold two electrons in 3s, six electrons in 3p, and ten electrons in 3d for a total of eighteen electrons.

(e) An f sublevel can hold a maximum of fourteen electrons.

59. (a) Ne (c) F

(b) Ge (d) N

61. Transition elements are found in Groups 1B–8B, lanthanides and actinides.

63. Elements 7, 15, 33, 51, 83 all have 5 electrons in their valence shell.

65. (a) sublevel p (c) sublevel f

(b) sublevel d

67. If 36 is a noble gas, 35 would be in periodic Group 7A and 37 would be in periodic Group 1A.

CHAPTER 11

Exercises

2. Drawing 1 Cl atom, drawing 2 Cl^- ion

4. (a) Fe^{2+} (e) Rubidium ion.

(c) Chloride ion.

6. + − + −

(a) H Cl (e) Cs I

(c) C Cl

8. (a) ionic (c) covalent

10. (a) $K \longrightarrow K^+ + 1 e^-$

(b) $S + 2 e^- \longrightarrow S^{2-}$

12. (a)

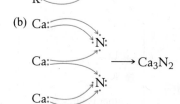

(b) Ca: ... N: ... Ca: ... $\longrightarrow Ca_3N_2$ ... N: ... Ca:

14. (a) Pb (4) (d) Cs (1)

(c) O (6) (f) Ar (8)

16. (a) sulfur atom, gain 2 e^-

(b) calcium atom, lose 2 e^-

(c) nitrogen atom, gain 3 e^-

(d) iodide ion, none

18. (a) Nonpolar covalent compound; O_2.

(c) Polar covalent compound; H_2O.

20. (a) SbH_3, Sb_2O_3 (c) HCl, Cl_2O_7

22. $BeBr_2$, beryllium bromide

$MgBr_2$, magnesium bromide

$SrBr_2$, strontium bromide

$BaBr_2$, barium bromide

$RaBr_2$, radium bromide

24. (a) $\text{Ga}\cdot$ (b) $[Ga]^{3+}$ (c) $[Ca]^{2+}$

26. (a) covalent   (c) covalent

(b) ionic   (d) covalent

28. (a) covalent   (c) covalent

(b) covalent

30. (a) $:\overset{..}{O}::\overset{..}{O}:$ (b) $:\overset{..}{Br}:\overset{..}{Br}:$ (c) $:\overset{..}{I}:\overset{..}{I}:$

32. (a) $:\overset{..}{\underset{H}{S}}:H$ (c) $H:\overset{..}{\underset{H}{N}}:H$

34. (a) $\left[:\overset{..}{\underset{..}{I}}:\right]^{-}$ (e) $\left[:\overset{..}{O}:N::\overset{..}{\underset{:\overset{..}{O}:}{O}}:\right]^{-}$

(c) $\left[:\overset{..}{O}:C::\overset{..}{\underset{:\overset{..}{O}:}{O}}:\right]^{2-}$

36. (a) H_2, nonpolar   (c) CH_3OH, polar

(b) NI_3, polar   (d) CS_2, nonpolar

38. (a) 3 electron pairs, trigonal planar

(b) 4 electron pairs, tetrahedral

(c) 4 electron pairs, tetrahedral

40. (a) tetrahedral   (c) tetrahedral

(b) trigonal pyramidal

42. (a) tetrahedral   (b) bent   (c) bent

44. potassium

47. N_2H_4 $14\ e^-$ $:N-N:$ with H H above and H H below HN_3 $16\ e^-$ $H-\overset{..}{N}=N=\overset{..}{\underset{..}{N}}$

50. (a) Be   (d) F

(b) He   (e) Fr

(c) K   (f) Ne

51. (a) Cl   (c) Ca

55. $SnBr_2$, $GeBr_2$

58. This structure shown is incorrect since the bond is ionic. It should be represented as:

$\left[Na\right]^{+}\left[:\overset{..}{\underset{..}{O}}:\right]^{2-}\left[Na\right]^{+}$

59. Answer: Fluorine, oxygen, nitrogen, chlorine

64. (a) 109.5° (actual angle is closer to 105)   (c) 109.5°

68. Empirical formula is SO_3.

$:\overset{..}{\underset{..}{O}}-\overset{:\overset{..}{O}:}{S}=\overset{..}{O}:$

70. (a) ionic   (c) covalent

(b) both   (d) covalent

CHAPTER 12

Exercises

2.

	mm Hg	lb/in.^{2}	atmospheres
a.	789	15.3	1.04
b.	1700	32	2.2
c.	1100	21	1.4

4. (a) 86.5 kPa   (c) 2770 mm Hg   (e) 82.1 cm Hg

(b) 0.0500 atm   (d) 1.06 atm

6. (a) 2.1 atm   (b) 1600 torr   (c) 210 kPa

8. (a) 1330 torr   (b) 40.0 torr

10. 21.2 L

12. (a) 691 mL   (c) 737 mL

(b) 633 mL

14. 30.8 L

16. 33.4 L CO_2

18. $-150°C$

20. 719 mm Hg

22. 1.450×10^3 torr

24. 1.19 L C_3H_8

26. 132 L

28. (a) 67.0 L SO_3 (c) 7.96 L Cl_2

(b) 170 L C_2H_6

30. 1.78 g C_3H_6

32. 2.82×10^{23} molecules CO_2

34. 2.1 g CH_4

36. 1.8×10^3 L CH_4

38. 0.362 mol CO_2

40. 209 K

42. (a) 9.91 g/L Rn   (c) 3.575 g/L SO_3

(b) 2.054 g/L NO_2 (d) 1.252 g/L C_2H_4

44. (a) 0.85 g/L C_2H_4 (b) 0.154 g/L He

46. (a) 3.95×10^4 mL H_2 (b) 0.0737 mol HCl

48. (a) 36 L O_2 (c) 12 L H_2O vapor

(b) 210 g CO_2

50. 1.12×10^3 L CO_2

54. image (a)

55. (a) the pressure will be cut in half

(c) the pressure will be cut in half

59. 22.4 L

62. SF_6 has the greatest density.

64. (a) CH_3

(b) C_2H_6

(c) $H-\overset{H}{\underset{H}{C}}-\overset{H}{\underset{H}{C}}-H$

67. 2.3×10^7 K

70. 327°C

72. 211 K ($-62°C$)

74. 3.55 atm

78. 1.03×10^4 mm (33.8 ft)

79. 330 mol O_2

81. 43.2 g/mol (molar mass)

83. (a) 8.4 L H_2 (c) 8.48 g/L CO_2

85. 279 L H_2 at STP

86. C_4H_8

CHAPTER 13

Exercises

2. $H_2Se < H_2S < H_2O$

4. $CO_2 < SO_2 < CS_2$

6. (a) HI will not hydrogen bond; hydrogen is not bonded to fluorine, oxygen, or nitrogen.

(b) NH_3 will hydrogen bond; hydrogen is bonded to nitrogen.

(c) CH_2F_2 will not hydrogen bond; hydrogen is not bonded to fluorine, oxygen, or nitrogen.

(d) C_2H_5OH will hydrogen bond; one of the hydrogens is bonded to oxygen.

(e) H_2O will hydrogen bond; hydrogen is bonded to oxygen.

8. (b)

NH_3

(d)

C_2H_5OH

(e)

H_2O

10. Water forming beaded droplets is an example of cohesive forces. The water molecules have stronger attractive forces for other water molecules than they do for the surface.

12. (a) magnesium ammonium phosphate hexahydrate

(b) iron(II) sulfate heptahydrate

(c) tin(IV) chloride pentahydrate

14. 0.0655 mol $Na_2B_4O_7 \cdot 10\ H_2O$

16. (a) 56.0 g H_2O

(b) 69.0 g $AlCl_3$

18. 20.93% H_2O

20. The formula is $NiCl_2 \cdot 6\ H_2O$.

22. 1.57×10^5 J

24. 42 g of steam needed; not enough steam (35 g)

26. The system will be at 0°C. It will be a mixture of ice and water.

28. Steam molecules will cause a more severe burn. Steam molecules contain more energy at 100°C than water molecules at 100°C due to the energy absorbed during the vaporization state (heat of vaporization).

30. When one leaves the swimming pool, water starts to evaporate from the skin of the body. Part of the energy needed for evaporation is absorbed from the skin, resulting in the cool feeling.

36. $MgSO_4 \cdot 7\ H_2O$ $Na_2HPO_4 \cdot 12\ H_2O$

43. 6.78×10^5 J or 1.62×10^5 cal

44. 40.7 kJ/mol

46. 2.30×10^6 J

48. 40.2 g H_2O

50. 1.00 mole of water vapor at STP has a volume of 22.4 L (gas); 1.00 mole of liquid water has a volume of 18.0 mL.

52. 3.9×10^3 J

55. 2.71×10^4 J

56. 5.2×10^4 J

57. $2\ H_2O(l) \longrightarrow 2\ H_2(g) + O_2(g)$

CHAPTER 14

Exercises

2. Generally soluble:

(b) $Cu(NO_3)_2$

(d) $NH_4C_2H_3O_2$

(f) $AgNO_3$

4. (a) 16.7% $NaNO_3$

(c) 39% K_2CrO_4

6. 143 g iron(III) oxide solution

8. (a) 9.0 g $BaCl_2$

(b) 66 g solvent

10. (a) 63.59% $C_{12}H_{22}O_{11}$

(b) 47.3% CH_3OH

12. (a) 13.6% butanol

(b) 16% methanol

(c) 6.00% isoamyl alcohol

14. (a) 4.0 M

(c) 1.97 M $C_6H_{12}O_6$

16. (a) 1.1 mol HNO_3

(b) 7.5×10^{-3} mol $NaClO_3$

(c) 0.088 mol LiBr

18. (a) 2.1×10^3 g H_2SO_4

(b) 6.52 g $KMnO_4$

(c) 1.2 g $Fe_2(SO_4)_3$

20. (a) 3.4×10^3 mL

(b) 1.88×10^3 mL

(c) 1.37 mL

22. (a) 1.2 M

(b) 0.070 M

(c) 0.333 M

24. (a) 25 mL 18 M H_2SO_4

(b) 5.0 mL 15 M NH_3

26. (a) 1.2 M HCl

(b) 4.21 M HCl

28. (a) 3.6 mol Na_2SO_4

(c) 0.625 mol NaOH

(e) 38.25 mL H_2SO_4

30. (a) 1.88×10^{-3} mol H_2O

(c) 90.2 mL $HC_2H_3O_2$

(e) 1.8 L CO_2

32. (a) 1.67 m $CaCl_2$

(b) 0.026 m $C_6H_{12}O_6$

(c) 6.44 m $(CH_3)_2CHOH$

34. (a) 10.74 m

(b) 105.50°C

(c) −20.0°C

36. 163 g/mol

38. Concord grape juice and stainless steel

40. Red paint and oil and vinegar salad dressing are not a true solution because they will separate out if they are not shaken or stirred.

42. (a) The amino acid will dissolve faster in the solvent at 75°C because the rate of dissolving increases with increasing temperature.

    (b) The powdered sodium acetate will dissolve faster because it has a smaller particle size.

    (c) The carton of table salt will dissolve faster because it has a smaller particle size and the temperature is higher.

    (d) The acetaminophen will dissolve faster in the infant pain medication because it has less acetaminophen already dissolved in it.

44. 1.092 M $HgCl_2$

51. 0.15 M NaCl

53. (a) 160 g sugar

    (b) 0.47 M $C_{12}H_{22}O_{11}$

    (c) 0.516 m $C_{12}H_{22}O_{11}$

55. 16.2 m HCl

57. 66 mL C_2H_5OH [ethyl alcohol]

59. (a) 4.5 g NaCl

    (b) 450. mL H_2O must evaporate.

61. 2.94 L solution

63. 15 M H_3PO_4

67. 1.20 g $Mg(OH)_2$

69. 6.2 m H_2SO_4

    5.0 M H_2SO_4

70. (a) 2.9 m

    (b) 101.5°C

72. (a) 8.04×10^3 g $C_2H_6O_2$

    (b) 7.24×10^3 mL $C_2H_6O_2$

    (c) −4.0°F

73. 47% $NaHCO_3$

74. (a) 7.7 L H_2O must be added

    (b) 0.0178 mol H_2SO_4

    (c) 0.0015 mol H_2SO_4 in each mL

77. 2.84 g $Ba(OH)_2$ is formed.

78. (a) 0.011 mol Li_2CO_3

    (c) 3.2×10^2 mL solution

84. 53 m sugar

CHAPTER 15

Exercises

2. Conjugate acid–base pairs:

    (a) $H_2S - HS^-$; $NH_3 - NH_4^+$

    (c) $HBr - Br^-$; $CH_3O^- - CH_3OH$

4. (a) $Fe_2O_3(s) + 6\ HBr(aq) \longrightarrow 2\ FeBr_3(aq) + 3\ H_2O(l)$

    (c) $2\ NaOH(aq) + H_2CO_3(aq) \longrightarrow Na_2CO_3(aq) + 2\ H_2O(l)$

    (e) $Mg(s) + 2\ HClO_4(aq) \longrightarrow Mg(ClO_4)_2(aq) + H_2(g)$

6. (a) $Fe_2O_3 + (6\ H^+ + 6\ Br^-) \longrightarrow (2\ Fe^{3+} + 6\ Br^-) + 3\ H_2O$

    $Fe_2O_3 + 6\ H^+ \longrightarrow 2\ Fe^{3+} + 3\ H_2O$

    (c) $(2\ Na^+ + 2\ OH^-) + H_2CO_3 \longrightarrow (2\ Na^+ + CO_3^{2-}) + 2\ H_2O$

    $2\ OH^- + H_2CO_3 \longrightarrow CO_3^{2-} + 2\ H_2O$

    (e) $Mg + (2\ H^+ + 2\ ClO_4^-) \longrightarrow (Mg^{2+} + 2\ ClO_4^-) + H_2$

    $Mg + 2\ H^+ \longrightarrow Mg^{2+} + H_2$

8. (a) $2\ HBr + Mg(OH)_2 \longrightarrow 2\ H_2O + MgBr_2$

    (b) $H_3PO_4 + 3\ KOH \longrightarrow 3\ H_2O + K_3PO_4$

    (c) $H_2SO_4 + 2\ NH_4OH \longrightarrow 2\ H_2O + (NH_4)_2SO_4$

10. $Ca(OH)_2$ and H_2CO_3; $Ca(OH)_2 + H_2CO_3 \longrightarrow 2\ H_2O + CaCO_3$

12. (a) $C_6H_{12}O_6$ — nonelectrolyte

    (c) NaClO — electrolyte

    (e) C_2H_5OH — nonelectrolyte

14. (a) 2.25 M Fe^{3+}, 6.75 M Cl^-

    (c) 0.75 M Na^+, 0.75 M $H_2PO_4^-$

16. (a) 12.6 g Fe^{3+}, 23.9 g Cl^-

    (c) 1.7 g Na^+, 7.3 g $H_2PO_4^-$

18. (a) $[H^+] = 1 \times 10^{-7}$

    (b) $[H^+] = 5 \times 10^{-5}\ M$

    (c) $[H^+] = 2 \times 10^{-6}\ M$

    (d) $[H^+] = 5 \times 10^{-11}\ M$

20. (a) $[Na^+] = 0.25\ M$, $[Cl^-] = 0.25\ M$

    (c) $[Ba^{2+}] = 0.062\ M$, $[Ag^+] = 0.081\ M$, $[NO_3^-] = 0.20\ M$

22. $HCN + H_2O \rightleftharpoons H_3O^+ + CN^-$, weak electrolyte and weak acid

24. (a) 0.147 M NaOH

    (b) 0.964 M NaOH

    (c) 0.4750 M NaOH

26. (a) $Mg(s) + Cu^{2+}(aq) \longrightarrow Cu(s) + Hg^{2+}(aq)$

    (b) $H^+(aq) + OH^-(aq) \longrightarrow H_2O(l)$

    (c) $SO_3^{2-}(aq) + 2\ H^+(aq) \longrightarrow H_2O(l) + SO_2(g)$

28. (a) 2 M HCl

    (b) 1 M H_2SO_4

30. 0.894 g of $Al(OH)_3$

32. 7.0% NaCl in the sample

34. 0.936 L H_2

36. (a) pH = 2.70

    (b) pH = 7.15

    (c) pH = −0.48

38. (a) pH = 4.30

    (b) pH = 10.47

40. (a) weak acid

    (b) strong base

    (c) strong acid

    (d) strong acid

42. (a) BrO_3^-

    (b) NH_3

    (c) HPO_4^{2-}

44. $S^{2-}(aq) + H_2O(l) \longrightarrow HS^-(aq) + OH^-(aq)$

46. $H_2SO_4(aq) + CaCO_3(s) \longrightarrow CaSO_4(s) + H_2O(l) + CO_2(g)$

48. (a) basic

    (c) neutral

    (e) acidic

51. 0.218 M H_2SO_4

55. Freezing point depression is directly related to the concentration of particles in the solution.

highest freezing point	$C_{12}H_{22}O_{11}$ > $HC_2H_3O_2$ > HCl > $CaCl_2$	lowest freezing point
	1 mol 1 mol + 2 mol 3 mol	
	(particles in solution)	

57. As the pH changes by 1 unit, the concentration of H^+ in solution changes by a factor of 10. For example, the pH of 0.10 M HCl is 1.00, while the pH of 0.0100 M HCl is 2.00.

58. 1.77% solute, 98.23% H_2O

59. 0.201 M HCl

61. 0.673 g KOH

63. 0.462 M HCl

65. pH = 1.1, acidic

67. (a) 2 NaOH(aq) + H_2SO_4(aq) $\longrightarrow$ Na_2SO_4(aq) + 2 H_2O(l)

(b) 1.0×10^2 mL NaOH

(c) 0.71 g Na_2SO_4

70. acidic pH = 0.82

75. 806 mL

CHAPTER 16

Exercises

2. (a) $I_2(s) \Longleftrightarrow I_2(g)$

(b) $NaNO_3(s) \Longleftrightarrow Na^+(aq) + NO_3^-(aq)$

4. (a) Reaction is exothermic.

(b) Reaction shifts to the right. [HCl] will be decreased; $[O_2]$, $[H_2O]$, and $[Cl_2]$ will be increased.

(c) Reaction will shift left.

6. (a) left   I   I   I (add NH_3)

(b) left   I   I   D (increase volume)

(c) no change   N   N   N (add catalyst)

(d) ?   ?   I   I (add H_2 and N_2)

8.

Reaction	Increase temperature	Increase pressure	Add catalyst
(a)	right	left	no change
(b)	left	left	no change
(c)	left	left	no change

10. Equilibrium shift

(a) no change   (b) left   (c) right   (d) right

12. (a) $K_{eq} = \dfrac{[H^+][H_2PO_4^-]}{[H_3PO_4]}$ (c) $K_{eq} = \dfrac{[N_2O_5]^2}{[NO_2]^4[O_2]}$

(b) $K_{eq} = \dfrac{[CH_4][H_2S]^2}{[CS_2][H_2]^4}$

14. (a) $K_{sp} = [Mg^{2+}][CO_3^{2-}]$ (c) $K_{sp} = [Tl^{3+}][OH^-]^3$

(b) $K_{sp} = [Ca^{2+}][C_2O_4^{2-}]$ (d) $K_{sp} = [Pb^{2+}]^3[AsO_4^{3-}]^2$

16. If H^+ is increased,

(a) pH is decreased   (c) OH^- is decreased

(b) pOH is increased   (d) K_W, remains unchanged

18. When excess base gets into the bloodstream, it reacts with H^+ to form water. Then H_2CO_3 ionizes to replace H^+, thus maintaining the approximate pH of the blood.

20. (a) $[H^+] = 4.6 \times 10^{-3}$ M (c) 18% ionized

(b) pH = 2.34

22. $K_a = 2.3 \times 10^{-5}$

24. (a) 0.79% ionized; pH = 2.10

(b) 2.5% ionized; pH = 2.60

(c) 7.9% ionized; pH = 3.10

26. $K_a = 7.3 \times 10^{-6}$

28. $[H^+] = 3.0$ M; pH = -0.48; pOH = 14.48; $[OH^-] = 3.3 \times 10^{-15}$

30. (a) pOH = 3.00; pH = 11.00

(b) pH = 0.903; pOH = 13.10

(c) pH = 5.74; pOH = 8.26

32. (a) $[OH^-] = 2.5 \times 10^{-6}$

(c) $[OH^-] = 4.5 \times 10^{-10}$

34. (a) $[H^+] = 2.2 \times 10^{-13}$

(b) $[H^+] = 1.9 \times 10^{-6}$

36. (a) 1.2×10^{-23}

(c) 1.81×10^{-18}

38. (a) 1.1×10^{-4} M

(b) 9.2×10^{-6} M

40. (a) 3.3×10^{-3} g $PbSO_4$

(b) 2.3×10^{-4} g $BaCrO_4$

42. Precipitation occurs.

44. 2.6×10^{-12} mol AgBr will dissolve.

46. pH = 4.74

48. Change in pH = 4.74 − 4.72 = 0.02 units in the buffered solution. Initial pH = 4.74.

52. 4.20 mol HI

54. 0.500 mol HI, 2.73 mol H_2, 0.538 mol I_2

56. 128 times faster

58. pH = 4.57

60. Hypochlorous acid: $K_a = 3.5 \times 10^{-8}$

Propanoic acid: $K_a = 1.3 \times 10^{-5}$

Hydrocyanic acid: $K_a = 4.0 \times 10^{-10}$

62. (a) Precipitation occurs.

(b) Precipitation occurs.

(c) No precipitation occurs.

64. No precipitate of $PbCl_2$

66. $K_{eq} = 1.1 \times 10^4$

68. 8.0 M NH_3

70. $K_{sp} = 4.00 \times 10^{-28}$

72. (a) The temperature could have been cooler.

(b) The humidity in the air could have been higher.

(c) The air pressure could have been greater.

74. (a) $K_{eq} = \dfrac{[O_3]^2}{[O_2]^3}$ (c) $K_{sp} = [CO_2]$

(b) $K_{eq} = \dfrac{[H_2O(l)]}{[H_2O(g)]}$ (d) $K_{sp} = \dfrac{[H^+]^6}{[Bi^{3+}]^2[H_2S]^3}$

76. $K_{eq} = 450$

78. 1.1 g $CaSO_4$

81. (a) shift right   (b) shift left   (c) shift right   (d) shift left

83. (a) $HCO_3^-(aq) + H^+(aq) \Longleftrightarrow H_2CO_3(aq)$

(b) $HCO_3^-(aq) + OH^-(aq) \Longleftrightarrow H_2O(l) + CO_3^{2-}(aq)$

CHAPTER 17

Exercises

2. (a) CHF_3 C = +2; H = +1; F = −1

(c) SF_6 S = +6; F = −1

(e) CH_3OH C = −2; H = +1; O = −2

4. (a) CO_3^{2-} +4

   (c) NaH_2PO_4 +5

6. (a) Cu +2; S +6   (b) +7   (c) +4   (d) +6

8. (a) $Cu^{2+} + e^- \longrightarrow Cu^{1+}$

   reduction

   (c) $2 IO_4^- + 16 H^+ + 14 e^- \longrightarrow I_2 + 8 H_2O$

   reduction

10. (a) C is oxidized, O is reduced.

     CH_4 is the reducing agent, O_2 is the oxidizing agent.

   (b) Mg is oxidized, Fe is reduced.

     Mg is the reducing agent, $FeCl_3$ is the oxidizing agent.

12. (a) incorrectly balanced

     $3 MnO_2(s) + 4 Al(s) \longrightarrow 3 Mn(s) + 2 Al_2O_3(s)$

   (c) correctly balanced

14. (a) $Cu + 2 AgNO_3 \longrightarrow 2 Ag + Cu(NO_3)_2$

   (c) $4 HCl + O_2 \longrightarrow 2 Cl_2 + 2 H_2O$

   (e) $2 KMnO_4 + 5 CaC_2O_4 + 8 H_2SO_4 \longrightarrow$

             $K_2SO_4 + 2 MnSO_4 + 5 CaSO_4 + 10 CO_2 + 8 H_2O$

16. (a) $6 H^+ + ClO_3^- + 6 I^- \longrightarrow 3 I_2 + Cl^- + 3 H_2O$

   (c) $2 H_2O + 2 MnO_4^- + 5 SO_2 \longrightarrow$

             $4 H^+ + 2 Mn^{2+} + 5 SO_4^{2-}$

   (e) $8 H^+ + Cr_2O_7^{2-} + 3 H_3AsO_3 \longrightarrow$

             $2 Cr^{3+} + 3 H_3AsO_4 + 4 H_2O$

18. (a) $H_2O + 2 MnO_4^- + 3 SO_3^{2-} \longrightarrow$

             $2 MnO_2 + 3 SO_4^{2-} + 2 OH^-$

   (c) $8 Al + 3 NO_3^- + 18 H_2O + 5 OH^- \longrightarrow$

             $3 NH_3 + 8 Al(OH)_4^-$

   (e) $2 Al + 6 H_2O + 2 OH^- \longrightarrow 2 Al(OH)_4^- + 3 H_2$

20. (a) $5 Mo_2O_3 + 6 MnO_4^- + 18 H^+ \longrightarrow$

             $10 MoO_3 + 6 Mn^{2+} + 9 H_2O$

   (b) $3 BrO^- + 2 Cr(OH)_4^- + 2 OH^- \longrightarrow$

             $3 Br^- + 2 CrO_4^{2-} + 5 H_2O$

   (c) $3 S_2O_3^{2-} + 8 MnO_4^- + H_2O \longrightarrow$

             $6 SO_4^{2-} + 8 MnO_2 + 2 OH^-$

24. (a) $Pb + SO_4^{2-} \longrightarrow PbSO_4 + 2 e^-$

     $PbO_2 + SO_4^{2-} + 4 H^+ + 2 e^- \longrightarrow PbSO_4 + 2 H_2O$

   (b) The first reaction is oxidation (Pb^0 is oxidized to Pb^{2+}).

     The second reaction is reduction (Pb^{4+} is reduced to Pb^{2+}).

   (c) The first reaction (oxidation) occurs at the anode of the battery.

26. Zinc is a more reactive metal than copper, so when corrosion occurs, the zinc preferentially reacts. Zinc is above hydrogen in the Activity Series of Metals; copper is below hydrogen.

28. 4.23 L CO_2; $H_2C_2O_4$ oxidized, $K_2S_2O_8$ reduced

32. 2.94 g Zn

34. 5.560 mol H_2

36. The electrons lost by the species undergoing oxidation must be gained (or attracted) by another species, which then undergoes reduction.

38. Sn^{4+} can only be an oxidizing agent.

   Sn^0 can only be a reducing agent.

   Sn^{2+} can be both oxidizing and reducing agents.

40. Equations (a) and (b) represent oxidations.

42. (a) $F_2 + 2 Cl^- \longrightarrow 2 F^- + Cl_2$

   (b) $Br_2 + Cl^- \longrightarrow NR$

   (c) $I_2 + Cl^- \longrightarrow NR$

   (d) $Br_2 + 2 I^- \longrightarrow 2 Br^- + I_2$

44. $4 Zn + NO_3^- + 10 H^+ \longrightarrow 4 Zn^{2+} + NH_4^+ + 3 H_2O$

46. (a) Pb is the anode.

   (c) oxidation occurs at Pb (anode).

   (e) Electrons flow from the lead through the wire to the silver.

CHAPTER 18

Exercises

2.

	Protons	Neutrons	Nucleons
(a) $^{128}_{52}Te$	52	76	128
(b) $^{32}_{16}S$	16	16	32

4. Its atomic number increases by one, and its mass number remains unchanged.

6. Equations for alpha decay:

   (a) $^{238}_{90}Th \longrightarrow ^4_2He + ^{234}_{88}Ra$

   (b) $^{239}_{94}Pu \longrightarrow ^4_2He + ^{235}_{92}U$

8. Equations for beta decay:

   (a) $^{28}_{13}Al \longrightarrow ^0_{-1}e + ^{28}_{14}Si$

   (b) $^{239}_{93}Np \longrightarrow ^0_{-1}e + ^{239}_{94}Pu$

10. (a) gamma-emission

   (b) alpha-emission then beta-emission

   (c) alpha-emission then gamma-emission

12. $^{32}_{15}P \longrightarrow ^0_{-1}e + ^{32}_{16}S$

14. (a) $^{27}_{13}Al + ^4_2He \longrightarrow ^{30}_{15}P + ^1_0n$

   (c) $^{12}_6C + ^2_1H \longrightarrow ^{13}_7N + ^1_0n$

16. The year 2064; 1/8 of Sr-90 remaining

18. $^{228}_{90}Th \xrightarrow{-\alpha} ^{224}_{88}Ra \xrightarrow{-\alpha} ^{220}_{86}Rn \xrightarrow{-\alpha} ^{216}_{84}Po \xrightarrow{-\alpha} ^{212}_{82}Pb$

20. (a) 5.4×10^{11} J/mol

   (b) 0.199% mass loss

21. (a) 24 protons; 27 neutrons; 24 electrons

   (c) 46 protons; 57 neutrons; 46 electrons

23. 1.50×10^{10} months

25. 61.59% neutrons by mass; 0.021% electrons by mass

27. 11,460 years old

29. (a) 0.0424 g/mol (mass defect)

   (b) 3.8×10^{12} J/mol (binding energy)

32. 0.16 mg remaining

33. 112 days

38. (a) $^{235}_{92}U + ^1_0n \longrightarrow ^{143}_{54}Xe + 3 ^1_0n + ^{90}_{38}Sr$

   (b) $^{235}_{92}U + ^1_0n \longrightarrow ^{102}_{39}Y + 3 ^1_0n + ^{131}_{53}I$

   (c) $^{14}_7N + ^1_0n \longrightarrow ^1_1H + ^{14}_6C$

40. $^{236}_{92}U \longrightarrow ^{90}_{38}Sr + 3 ^1_0n + ^{143}_{54}Xe$

42. 7680 g

44. (a) 0.500 g left

   (c) 0.0625 g left

46. (a) 2.29×10^4 years

CHAPTER 19

Exercises

2. (a) inorganic   (e) organic

(c) inorganic

4. Organic compounds are compounds in which carbon combines with hydrogen, oxygen, nitrogen, sulfur and/or other elements. Each compound identified as organic above contains some combination of carbon with hydrogen, oxygen, nitrogen, and/or sulfur.

6. Names of alkyl groups

(a) C_8H_{17} is octyl-

(b) $C_{10}H_{21}$ is decyl-

8. $CH_3CH_2CH_2CH_2CH_2CH_2CH_3$

$CH_3CH_2CH_2CH_2CHCH_3$
$\qquad\qquad\qquad |$
$\qquad\qquad\quad CH_3$

$CH_3CH_2\underset{\underset{CH_3}{|}}{\overset{\overset{CH_3}{|}}{C}}CH_2CH_3 \quad CH_3CH_2CH_2\underset{\underset{CH_3}{|}}{\overset{\overset{CH_3}{|}}{C}}CH_3$

$CH_3CH_2CH_2\underset{\underset{CH_3}{|}}{CH}CH_2CH_3 \quad CH_3\underset{\underset{CH_3}{|}}{CH}CH_2\underset{\underset{CH_3}{|}}{CH}CH_3$

$CH_3CH_2\underset{\underset{CH_3}{|}}{CH}CHCH_3 \quad CH_3\underset{\underset{CH_3}{|}}{CH}-\underset{\underset{CH_3}{|}}{\overset{\overset{CH_3}{|}}{C}}CH_3$

$CH_3CH_2\underset{\underset{CH_2CH_3}{|}}{CH}CH_2CH_3$

10. (a) $CH_3CH_2CH_2Cl$ 1-chloropropane

(b) $(CH_3)_3CCl$ 2-chloro-2-methylpropane

(c) $CH_3CHClCH_2CH_3$ 2-chlorobutane

12. (a) 3,6-dimethyloctane

(b) 3-ethyl-2-methylhexane (or 3-isopropylhexane)

14. (a) $CH_3CH_2CHCH_2CH(CH_3)_2$
$\qquad\qquad\quad |$
$\qquad\qquad CH_2CH_3$

(b) $CH_3CH_2CH_2CHCH_2CH_2CH_3$
$\qquad\qquad\qquad |$
$\qquad\qquad\quad C(CH_3)_3$

(c) $CH_3\underset{}{CH}CH_2\underset{\underset{CH_2CH_3}{|}}{\overset{\overset{CH_3}{|}}{C}}CH_2CH_2\underset{\underset{CH(CH_3)_2}{|}}{\overset{\overset{CH_3}{|}}{CH}}CHCH_2CH_3$

16. (a) $CH_3\underset{\underset{CH_3}{|}}{\overset{\overset{CH_3}{|}}{C}}CH_2CH_3$ Each methyl group requires a number location.

2,2-dimethylbutane

(b) $CH_3\underset{\underset{CH_3}{|}}{CH}CH_2\underset{\underset{CH_2CH_3}{|}}{CH}CH_3$ The longest carbon chain is 6.

2,4-dimethylhexane

18. (a) (4 isomers) $CH_3CH_2CH_2CH_2I$

$CH_3\underset{\underset{CH_2CH_3}{|}}{CH}I \qquad CH_3\underset{\underset{CH_3}{|}}{\overset{\overset{CH_3}{|}}{C}}-I \qquad CH_3\underset{\underset{CH_3}{|}}{CH}CH_2I$

(b) (4 isomers)

$CH_3CH_2CHCl_2$ $\qquad\qquad$ $CH_3CCl_2CH_3$

$CH_3CHClCH_2Cl$ $\qquad\quad$ $CH_2ClCH_2CH_2Cl$

(c) (5 isomers)

$CH_3CH_2CHBrCl$ $\qquad\quad$ $CH_3CHClCH_2Br$

$CH_3CHBrCH_2Cl$ $\qquad\quad$ $CH_2ClCH_2CH_2Br$

$CH_3CClBrCH_3$

(d) (9 isomers)

$CH_3CH_2CH_2CHCl_2$ $\qquad$ $CH_3CH_2CHClCH_2Cl$

$CH_3CHClCH_2CH_2Cl$ $\qquad$ $CH_2ClCH_2CH_2CH_2Cl$

$CH_3CH_2CCl_2CH_3$ $\qquad$ $CH_3CHClCHClCH_3$

$CH_3\underset{\underset{CH_3}{|}}{CH}CHCl_2 \qquad\qquad CH_3\underset{\underset{CH_3}{|}}{\overset{}{C}}ClCH_2Cl$

$CH_3\underset{\underset{CH_2Cl}{|}}{CH}CH_2Cl$

20. (a) $CH_3CH{=}CH_2 + Br_2 \longrightarrow CH_3CHBrCH_2Br$

(b) $CH_3CH{=}CH_2 + HBr \longrightarrow CH_3CHBrCH_3$

22. (a) CH_3OCH_3 ether

(b) CH_3CHO aldehyde

(c) CH_3COOH carboxylic acid

(d) $HCOOCH_3$ ester

24. (a) methoxymethane   (c) ethanoic acid

(b) ethanal   (d) methyl methanoate

26. (a) hexachloroethane CCl_3CCl_3

(b) iodoethyne $CH{\equiv}CI$

(c) 6-bromo-3-methyl-3-hexene-1-yne $BrCH_2CH_2CH{=}\underset{\underset{CH_3}{|}}{C}C{\equiv}CH$

(d) 1,2-dibromoethene $CHBr{=}CHBr$

28. (a) 1,2-diphenylethene

$\underset{}{\overset{H\qquad\quad H}{\underset{}{C{=}C}}}$ (benzene rings attached)

(c) 3-phenyl-1-butyne

$CH{\equiv}CCHCH_3$ (benzene ring attached)

(b) 3-pentene-1-yne

$CH{\equiv}CCH{=}CHCH_3$

30. (a) $CH_3CH_2CHCH{=}CH_2$ 3-isopropyl-1-pentene
$\qquad\qquad\quad |$
$\qquad\qquad CH(CH_3)_2$

(b) $CH_3CH_2CH{=}CCH_2CH_3$ 3-methyl-3-hexene
$\qquad\qquad\qquad\quad |$
$\qquad\qquad\qquad CH_3$

32. (a) $CH_2{=}CH_2 + H_2O \xrightarrow{H^+} CH_2CH_2OH$ (ethanol)

(b) $CH{\equiv}CH + 2\,Br_2 \longrightarrow CHBr_2CHBr_2$
$\qquad\qquad\qquad$ (1,1,2,2-tetrabromoethane)

(c) $CH_2{=}CH_2 + H_2 \xrightarrow[\text{Pt, 25°C}]{\text{1 atm}} CH_3CH_3$ (ethane)

34. (a)

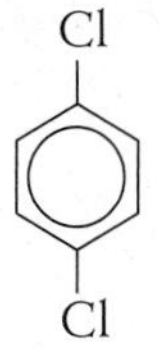

para-dichlorobenzene
(1,4-dichlorobenzene)

(d)

ethylbenzene

(b)

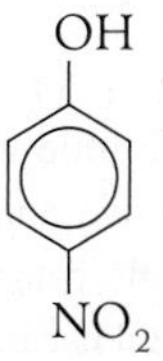

para-nitrophenol

(e)

para-methylphenol

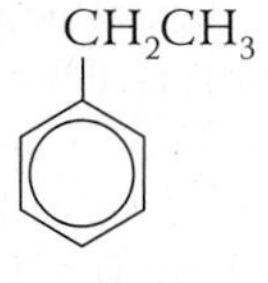

(c)

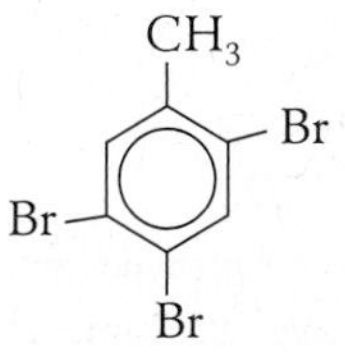

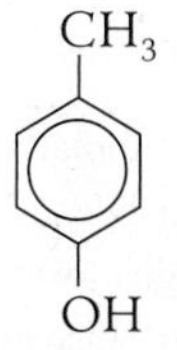

2,4,5-tribromotoluene

(f)

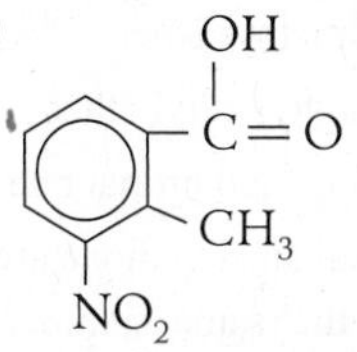

2-methyl-
3-nitrobenzoic acid

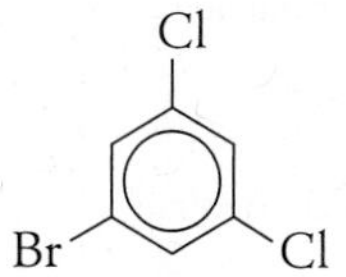

1,3-dichloro-
5-bromobenzene

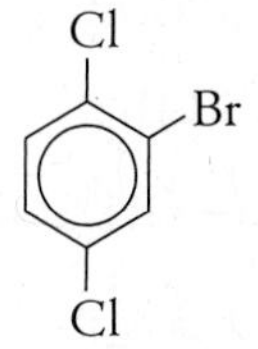

1,4-dichloro-
2-bromobenzene

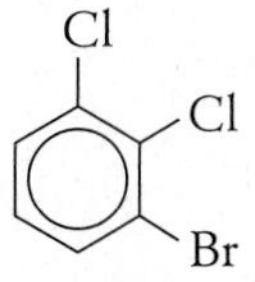

1,2-dichloro-
3-bromobenzene

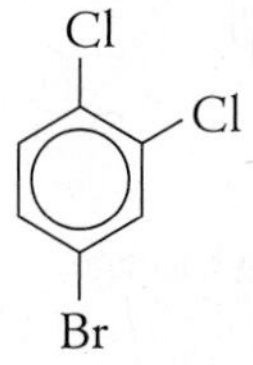

1,2-dichloro-
4-bromobenzene

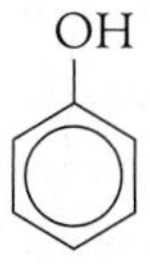

42. (a) $CH_3CH-CHCH_2CH_2CHCH_3$ with OH and two CH_2CH_3 groups

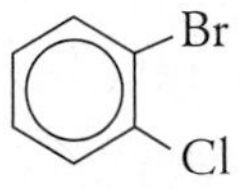

3,7-dimethyl-4-nonanol   secondary

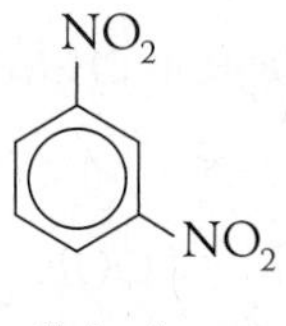

(b) $HO-CH_2CH_2CHCH_2CH_2CH_3$ with CH_3CHCH_3

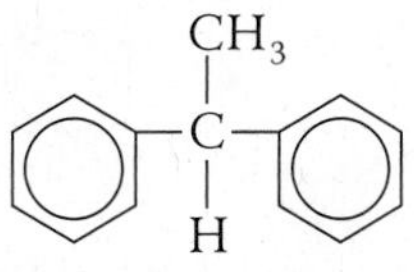

3-isopropyl-1-hexanol   primary

36. (a)

phenol

(c)

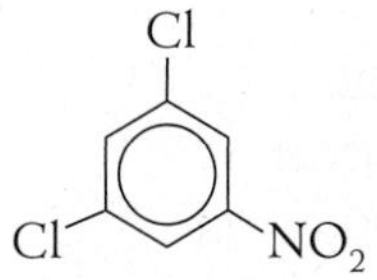

1,3-dichloro-
5-nitrobenzene

44. (a) 2,2-dimethyl-
1-heptanol

$HOCH_2CCH_2CH_2CH_2CH_2CH_3$ with two CH_3 groups

(b) 1,3-propanediol $HOCH_2CH_2CH_2OH$

(b)

o-bromochlorobenzene

(d)

m-dinitrobenzene

46. (a)

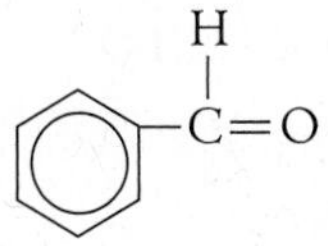

benzaldehyde

(b) $O=CCH_2CH_2C=O$ with H and H

butanedial

(c) CH_3CHCH_2C-H with O and OH

3-hydroxybutanal

38. (a)

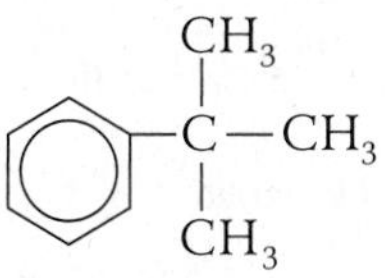

t-butylbenzene

(b)

1,1-diphenylethane

48. (a) CH_3C-CCH_3 with O, CH_3 and CH_3

3,3-dimethyl-2-butanone
(methyl *t*-butyl ketone)

(b) $CH_3CCH_2CH_2CCH_3$ with O and O

2,5-hexanedione

(c) $CH_3CCH_2CCH_3$ with CH_3, O and OH

4-hydroxy-4-methyl-2-pentanone

40.

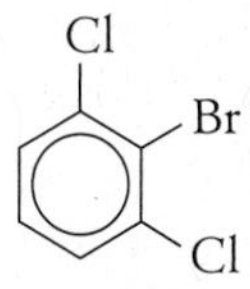

1,3-dichloro-
2-bromobenzene

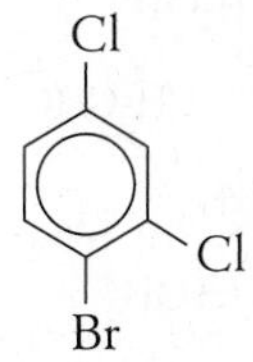

1,3-dichloro-
4-bromobenzene

50. (a) CH₃CHCOOH   2-methylbutanoic acid
       |
       CH₂CH₃

     (b) 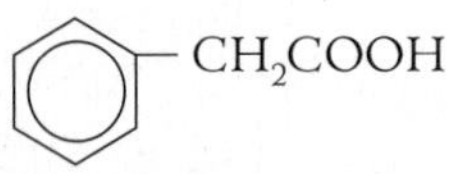 ⬡—CH₂COOH   phenyethanoic acid

     (c) CH₃CH₂CH₂CH₂COOH   pentanoic acid

     (d) 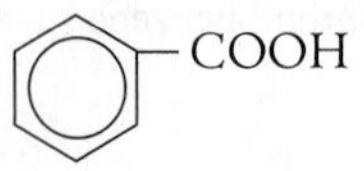 ⬡—COOH   benzoic acid

52. Esters

     (a) *n*-nonyl acetate
             O
             ‖
         CH₃C—O—CH₂(CH₂)₇CH₃

                 O
                 ‖
     (b) ethyl benzoate   ⬡—C—OCH₂CH₃

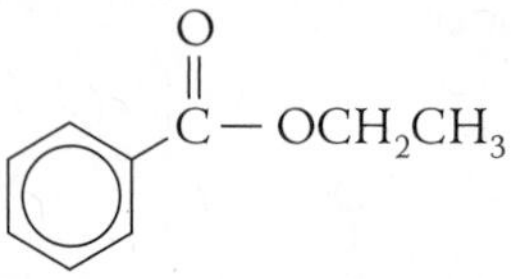

                 O
                 ‖
     (c) methyl salicylate   ⬡—C—O—CH₃
                  |
                  OH

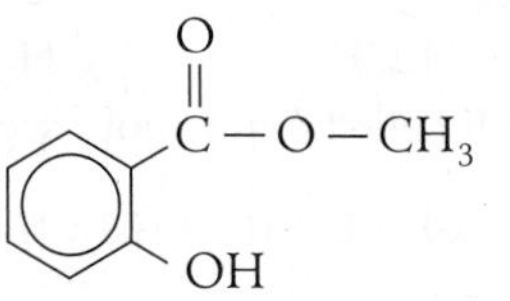

54. (a) H—C—OCH(CH₃)₂   isopropyl methanoate
           ‖
           O

     (b) 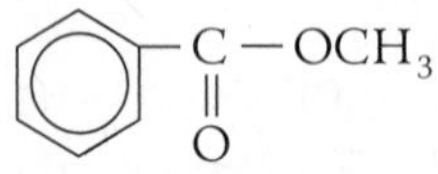 ⬡—C—OCH₃   methyl benzoate
              ‖
              O

         CH₃   O
         | ‖
     (c) CH₃CHCH₂C—OCH₂CH₃   ethyl-3-methylbutanoate

56. (a) CH₃COOH + KOH ⟶ CH₃COO⁻K⁺ + H₂O

     (b) 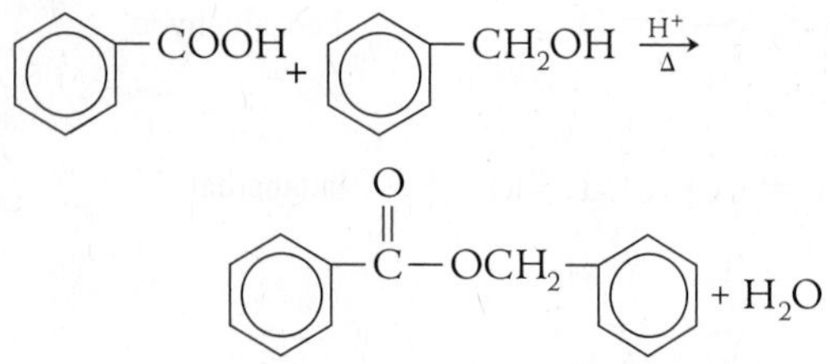 ⬡—COOH + ⬡—CH₂OH $\xrightarrow[\Delta]{H^+}$

                 O
                 ‖
        ⬡—C—OCH₂—⬡ + H₂O

58. (a) ⸺(CH₂—CH)ₙ⸺   polyacrylonitrile
             |
             CN

     (b) ⸺(CF₂—CF₂)ₙ⸺   Teflon

60. (a) correct

     (b) incorrect, should be butanone

     (c) incorrect, should be 4-methyl-2-pentene

     (d) incorrect, should be methyl propanoate

     (e) incorrect, should be 3-methylpentane

62. (a) alkyl halide   (c) alkene   (e) hydrocarbon

     (b) ketone   (d) ester

64. There are several possible 4 carbon unsaturated hydrocarbons. One example is shown below:

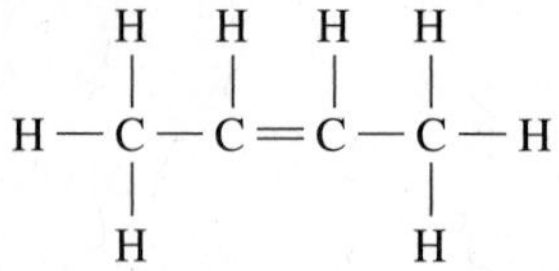

     H   H   H   H
     | | | |
   H—C—C=C—C—H
     | |
     H   H

66. 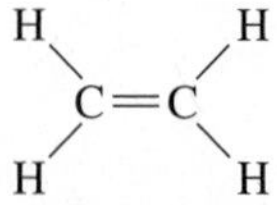   H   H
       \   /
        C=C   120° bond angles
       /   \
     H   H

68. Alkanes are nonpolar molecules. Therefore, there is little attraction between them because there are no partial positive and partial negative charges to attract one another. As a result, it takes very little energy to cause low molar-mass alkanes to boil because there are no intermolecular attractive forces to be overcome.

73. (a) methyl ethyl ether   CH₃CH₂OCH₃

     (b) dimethyl ether   CH₃OCH₃

     (c) methyl ethyl ether , CH₃CH₂OCH₃

75. Propanal and propanone are isomers, C₃H₆O.

     Butanal and butanone are isomers, C₄H₈O. Aldehydes and ketones with the same number of carbons have the same molecular formula.

76. Carboxylic acids, IUPAC name, common name

	IUPAC	**Common**
HCOOH	methanoic acid	formic acid
CH₃COOH	ethanoic acid	acetic acid
CH₃CH₂COOH	propanoic acid	propionic acid
CH₃CH₂CH₂COOH	butanoic acid	butyric acid
CH₃CH₂CH₂CH₂COOH	pentanoic acid	valeric acid

78. (a) HCOOH + CH₃CH₂OH $\xrightarrow[\Delta]{H^+}$ HC—OCH₂CH₃ + H₂O
                                       ‖
                                       O
                                   ethyl formate

     (c) 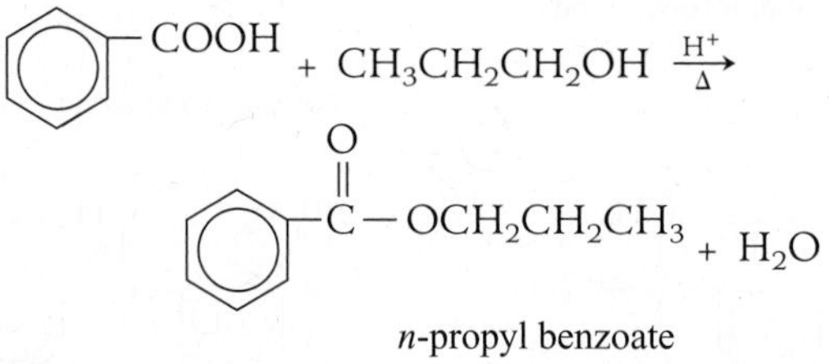 ⬡—COOH + CH₃CH₂CH₂OH $\xrightarrow[\Delta]{H^+}$

                 O
                 ‖
       ⬡—C—OCH₂CH₂CH₃ + H₂O

             n-propyl benzoate

81. The compound is C₂H₄Cl₂

     CH₃CHCl₂ and CH₂ClCH₂Cl

83. (a) CH₃CH₂OH

     (c) CH₃CH₂CH₂CHClCH₃

           CH₃
           |
     (e) CH₃C—OH
           |
           CH₃

85. (a)
$$CH_3CHCHCH_3$$
with CH_3 groups on the two middle carbons

(c) $CH_3CH_2\overset{O}{\overset{\|}{C}}-OH$

(e) $CH_3CH_2\overset{O}{\overset{\|}{C}}-O-CH_2CH_2CH_2CH_3$

CHAPTER 20

Exercises

4. Monosaccharides, disaccharides, and polysaccharides. The simplest type of carbohydrate is the monosaccharide.

6.

Monosaccharides	Disaccharides	Polysaccharides
glucose	sucrose	cellulose
fructose	maltose	glycogen
galactose	lactose	starch
ribose		

10. Lactic acid is not a carbohydrate. It does not have an aldehyde or a ketone group.

11. The monosaccharide composition of:

(a) Sucrose; a disaccharide made from one unit of glucose and one unit of fructose.

(b) Maltose; a disaccharide made from two units of glucose.

(c) Lactose; a disaccharide made from one unit of galactose and one unit of glucose.

(d) Starch; a polysaccharide made from many units of glucose.

(e) Cellulose; a polysaccharide made from many units of glucose.

(f) Glycogen; a polysaccharide made from many units of glucose.

16. Carbohydrates are stored in the body as glycogen.

21. CH_2O

22. The oxidation number of carbon changes from 0 to +4.

23. Cellulose: 3.7×10^3 monomer units

Starch: 25 monomer units

Cellulose: 1.9×10^{-6} m long

Starch: 1.3×10^{-8} m long

25. Fatty acids in vegetable oils are more unsaturated than fatty acids in animal fats.

27. Structural formulas

Glycerol
$$CH_2OH$$
$$CHOH$$
$$CH_2OH$$

Palmitic acid $CH_3(CH_2)_{14}COOH$

Oleic acid $CH_3(CH_2)_7CH=CH(CH_2)_7COOH$

Stearic acid $CH_3(CH_2)_{16}COOH$

Linoleic acid $CH_3(CH_2)_4CH=CHCH_2CH=CH(CH_2)_7COOH$

29. A triacylglycerol (triglyceride) is a triester of glycerol. Most animal fats are triacylglycerols.

35. The three essential fatty acids are linoleic, linolenic, and arachidonic acids. Diets lacking these fatty acids lead to impaired growth and reproduction and skin disorders, such as eczema and dermatitis.

36. The structural formula of cholesterol is

37. The ring structure common to all steroids is

41. There are three disulfide linkages in beef insulin.

43. All amino acids contain a carboxylic acid group ($-COOH$) and an amine group ($-NH_2$).

45. (a) phenylalanine, tryptophan, tyrosine

(b) cysteine, methionine

(c) serine, threonine, tyrosine

47. (a) $H_2N-CH_2-\overset{O}{\overset{\|}{C}}-NH-CH_2COOH$

(b) $H_2N-CH_2-\overset{O}{\overset{\|}{C}}-NH-CH_2-\overset{O}{\overset{\|}{C}}-NH-\underset{CH_3}{\overset{}{CH}}-COOH$

(c) $(CH_3)_2CHCH_2\underset{NH_2}{\overset{}{CH}}C\underset{}{NH}CH\underset{CH_2}{\overset{}{C}}NHCH_2\overset{O}{\overset{\|}{C}}NH\underset{CH_2OH}{\overset{}{CH}}COOH$
with CH_2-S-CH_3 branch

49. Aspartame

$\bigcirc-CH_2\underset{NH_2}{\overset{}{CH}}COOH$ $HOOCCH_2\underset{NH_2}{\overset{}{CH}}COOH$

phenylalanine   aspartic acid

$HOOCCH_2\underset{COOH}{\overset{}{CH}}-NH-\underset{O}{\overset{}{C}}-\underset{NH_2}{\overset{}{CH}}CH_2-\bigcirc$

$\bigcirc-CH_2\underset{COOH}{\overset{}{CH}}-NH-\underset{O}{\overset{}{C}}-\underset{NH_2}{\overset{}{CH}}CH_2COOH$

The methyl ester of this structure is aspartame.

54. Polypeptides are numbered starting with the N-terminal acid.

$$\underset{1}{\text{try}} - \underset{2}{\text{gly}} - \underset{3}{\text{his}} - \underset{4}{\text{phe}} - \underset{5}{\text{val}}$$

61. Enzymes act as catalysts for biochemical reactions. Their function is to lower the activation energy of biochemical reactions.

62. Chromosomes are composed of proteins and nucleic acids.

64. Guanine and cytosine are hydrogen bonded to each other, as are adenine and thymine.

68. The two helices of the double helix are joined together by hydrogen bonds between bases. The structure of the bases is such that one base will hydrogen bond to only one other specific base. That is, adenine is always hydrogen–bonded to thymine, and cytosine is always bonded to guanine. Therefore, the hydrogen bonding requires a specific structure on the adjoining helix.

69. Complementary bases are the pairs that "fit" to each other by hydrogen bonds between the two helices of DNA. For DNA, the complementary pairs are thymine with adenine and cytosine with guanine or T—A and C—G.

73. The structural differences between DNA and RNA are:

(a) RNA exists in the form of a single-stranded helix, whereas DNA is a double helix.

(b) RNA contains the pentose sugar ribose, whereas DNA contains deoxyribose.

(c) RNA contains the base uracil, whereas DNA contains thymine.

75. The location of protein synthesis is in the ribosomes.

77. The red dotted lines in Figure 20.7 represent hydrogen bonds between complementary base pairs.

Answers to Putting It Together Review Exercises

CHAPTERS 1–4

Multiple Choice: **1.** d **2.** a **3.** d **4.** b **5.** d **6.** c **7.** a
8. d **9.** a **10.** b **11.** a **12.** d **13.** b **14.** c **15.** a **16.** d
17. c **18.** b **19.** d **20.** c **21.** c **22.** a **23.** b **24.** c **25.** c
26. a **27.** d **28.** d **29.** c **30.** a **31.** b **32.** a **33.** c **34.** c
35. c **36.** d **37.** d **38.** c **39.** c **40.** b **41.** b **42.** c **43.** b
44. b

Free Response

1. $(1.5 \text{ m})\left(\dfrac{100 \text{ cm}}{1 \text{ m}}\right)\left(\dfrac{1 \text{ in.}}{2.54 \text{ cm}}\right)\left(\dfrac{1 \text{ ft}}{12 \text{ in.}}\right) = 4.9 \text{ ft}$

$(4 \text{ m})\left(\dfrac{100 \text{ cm}}{1 \text{ m}}\right)\left(\dfrac{1 \text{ in.}}{2.54 \text{ cm}}\right)\left(\dfrac{1 \text{ ft}}{12 \text{ in.}}\right) = 13 \text{ ft}$

$(27°C \times 1.8) + 32 = 81°F$

2. Jane needs to time how long it took from starting to heat to when the butter is just melted. From this information, she can determine how much heat the pot and butter absorbed. Jane can look up the specific heat of copper. Jane should weigh the pot and measure the temperature of the pot and the temperature at which the butter just melted. This should allow Jane to calculate how much heat the pot absorbed. Then she simply has to subtract the heat the pot absorbed from the heat the stove put out to find out how much heat the butter absorbed.

3. $CaCO_3 \longrightarrow CaO + CO_2$
   $\underset{75 \text{ g}}{} \qquad \underset{42 \text{ g}}{} \quad \underset{X}{}$

   $X = 75 \text{ g} - 42 \text{ g} = 33 \text{ g}$
   $44 \text{ g } CO_2$ occupies 24 dm^3
   Therefore, $33 \text{ g } CO_2$ occupies

   $(33 \text{ g})\left(\dfrac{24 \text{ dm}^3}{44 \text{ g}}\right)\left(\dfrac{1 \text{ L}}{1 \text{ dm}^3}\right) = 18 \text{ L.}$

4. (a), (b), Picture (2) best represents a homogeneous mixture. Pictures (1) and (c) (3) show heterogeneous mixtures, and picture (4) does not show a mixture, as only one species is present.

   Picture (1) likely shows a compound, as one of the components of the mixture is made up of more than one type of "ball." Picture (2) shows a component with more than one part, but the parts seem identical, and therefore it could be representing a diatomic molecule.

5. (a) Picture (3) because fluorine gas exists as a diatomic molecule.

   (b) Other elements that exist as diatomic molecules are oxygen, nitrogen, chlorine, hydrogen, bromine, and iodine.

   (c) Picture (2) could represent SO_3 gas.

6. (a) Tim's bowl should require less energy. Both bowls hold the same volume, but since snow is less dense than a solid block of ice, the mass of water in Tim's bowl is less than the mass of water in Sue's bowl. (Both bowls contain ice at 12°F.)

   (b) $\dfrac{12°F - 32}{1.8} = -11°C$

   Temperature change: $-11°C$ to $25°C = 36°C$

   (c) temperature change: $-11°C$ to $0°C = 11°C$
   specific heat of ice $= 2.059 \text{ J/g°C}$

   vol. of $H_2O = 1 \text{ qt} = (0.946 \text{ L})\left(\dfrac{1000 \text{ mL}}{\text{L}}\right)$
   $= 946 \text{ mL}$

   mass of ice $=$ mass of water $= (946 \text{ mL})\left(\dfrac{1 \text{ g}}{1 \text{ mL}}\right)$
   $= 946 \text{ g}$

   heat required $= (m)(\text{sp. ht.})(\Delta t)$
   $= (946 \text{ g})\left(\dfrac{2.059 \text{ J}}{\text{g°C}}\right)(11°C)\left(\dfrac{1 \text{ kJ}}{1000 \text{ J}}\right) = 21 \text{ kJ}$

   (d) Physical changes

7. (a) Let $x =$ RDA of iron
   60% of $x = 11 \text{ mg Fe}$
   $x = \dfrac{11 \text{ mg Fe} \times 100\%}{60\%} = 18 \text{ mg Fe}$

   (b) density of iron $= 7.86 \text{ g/mL}$
   $V = \dfrac{m}{d} = (11 \text{ mg Fe})\left(\dfrac{1 \text{ g}}{1000 \text{ mg}}\right)\left(\dfrac{1 \text{ mL}}{7.86 \text{ g}}\right)$
   $= 1.4 \times 10^{-3} \text{ mL Fe}$

8. If Alfred inspects the bottles carefully, he should be able to see whether the contents are solid (silver) or liquid (mercury). Alternatively, since mercury is more dense than silver, the bottle of mercury should weigh more than the bottle of silver (the question indicated that both bottles were of similar size and both were full). Density is mass/volume.

9. (a) Container holds a mixture of sulfur and oxygen.

   (b) No. If the container were sealed, the total mass would remain the same whether a reaction took place or not. The mass of the reactants must equal the mass of the products.

   (c) No. Density is mass/volume. The volume is the container volume, which does not change. Since the total mass remains constant even if a reaction has taken place, the density of the container, including its contents, remains constant. The density of each individual component within the container may have changed, but the total density of the container is constant.

CHAPTERS 5–6

Multiple Choice: **1.** b **2.** d **3.** b **4.** d **5.** b **6.** b **7.** a
8. b **9.** d **10.** c **11.** b **12.** d **13.** a **14.** c **15.** d
Names and Formulas: The following are correct: 1, 2, 4, 5, 6, 7, 9, 11, 12, 15, 16, 17, 18, 19, 21, 22, 25, 28, 30, 32, 33, 34, 36, 37, 38, 40.

Free Response

1. (a) An ion is a charged atom or group of atoms. The charge can be either positive or negative.

   (b) Electrons have negligible mass compared with the mass of protons and neurons. The only difference between Ca and Ca^{2+} is two electrons. The mass of those two electrons is insignificant compared with the mass of the protons and neutrons present (and whose numbers do not change).

2. (a) Let x = abundance of heavier isotope.

   $$303.9303(x) + 300.9326(1 - x) = 303.001$$
   $$303.9303x - 300.9326x = 303.001 - 300.9326$$
   $$2.9977x = 2.068$$
   $$x = 0.6899$$
   $$1 - x = 0.3101$$

   % abundance of heavier isotope = 68.99%

   % abundance of lighter isotope = 31.01%

   (b) $^{304}_{120}Wz$, $^{301}_{120}Wz$

   (c) mass number − atomic number = 303 − 120

   $$= 183 \text{ neutrons}$$

3. Cl_2O_7 Cl: $17p \times 2 = \underline{34 \text{ protons}}$

   O: $8p \times 7 = \underline{56 \text{ protons}}$

   90 protons in Cl_2O_7

   Since the molecule is electrically neutral, the number of electrons is equal to the number of protons, so Cl_2O_7 has 90 electrons. The number of neutrons cannot be precisely determined unless it is known which isotopes of Cl and O are in this particular molecule.

4. Phosphate has a −3 charge; therefore, the formula for the ionic compound is $M_3(PO_4)_2$.

   P has 15 protons; therefore, $M_3(PO_4)_2$ has 30 phosphorus protons.

   $$3 \text{ (number of protons in M)} = \frac{30 \times 6}{5}$$
   $$= 36 \text{ protons in 3 M}$$

   number of protons in M $= \dfrac{36}{3} = 12 \text{ protons}$

   from the periodic table, M is Mg.

5. (a) Iron can form cations with different charges (e.g., Fe^{2+} or Fe^{3+}). The Roman numeral indicating which cation of iron is involved is missing. This name cannot be fixed unless the particular cation of iron is specified.

   (b) $K_2Cr_2O_7$. Potassium is generally involved in ionic compounds. The naming system used was for covalent compounds. The name should be potassium dichromate. (Dichromate is the name of the $Cr_2O_7^{2-}$ anion.)

   (c) Sulfur and oxygen are both nonmetals and form a covalent compound. The number of each atom involved needs to be specified for covalent compounds. There are two oxides of sulfur—SO_2 and SO_3. Both elements are nonmetals, so the names should be sulfur dioxide and sulfur trioxide, respectively.

6. No. Each compound, SO_2 and SO_3, has a definite composition of sulfur and oxygen by mass. The law of multiple proportions says that two elements may combine in different ratios to form more than one compound.

7. (a) Electrons are not in the nucleus.

   (b) When an atom becomes an anion, its size increases.

   (c) An ion of Ca (Ca^{2+}) and an atom of Ar have the same number of electrons.

8. (a) $12 \text{ amu} \times 7.18 = 86.16 \text{ amu}$

   (b) The atom is most likely Rb or Sr. Other remote possibilities are Kr or Y.

   (c) Because of the possible presence of isotopes, the atom cannot be positively identified. The periodic table gives average masses.

   (d) M forms a +1 cation and is most likely in Group 1A. The unknown atom is most likely $^{86}_{37}Rb$.

9. The presence of isotopes contradicts Dalton's theory that all atoms of the same element are identical. Also, the discovery of protons, neutrons, and electrons suggests that there are particles smaller than the atom and that the atom is not indivisible.

   Thomson proposed a model of an atom with no clearly defined nucleus.

   Rutherford passed alpha particles through gold foil and inspected the angles at which the alpha particles were deflected. From his results, he proposed the idea of an atom having a small dense nucleus.

CHAPTERS 7–9

Multiple Choice: 1. a   2. a   3. d   4. d   5. a   6. b   7. a   8. b   9. b   10. d   11. a   12. d   13. b   14. c   15. c   16. d   17. d   18. b   19. b   20. a   21. b   22. c   23. d   24. b   25. b   26. a   27. d   28. c   29. c   30. b   31. c   32. c   33. b   34. d   35. b   36. d   37. c   38. b   39. c   40. d   41. a   42. c   43. c   44. b   45. b   46. a   47. d   48. b   49. b   50. a

Free Response

1. (a) $104 \text{ g } O_2 = (104 \text{ g } O_2)\left(\dfrac{1 \text{ mol}}{32.00 \text{ g}}\right) = 3.25 \text{ mol } O_2$

   $$\begin{array}{ccccccc} X & + & O_2 & \longrightarrow & CO_2 & + & H_2O \\ & & 3.25 \text{ mol} & & 2 \text{ mol} & & 2.5 \text{ mol} \end{array}$$

   (multiply moles by 4)

   $$4X + 13 O_2 \longrightarrow 8 CO_2 + 10 H_2O$$

   Oxygen is balanced. By inspection, X must have 8/4 C atoms and 20/4 H atoms (2 C and 5 H).

   Empirical formula is C_2H_5.

   (b) Additional information needed is the molar mass of X.

2. (a)

   SO₂   O₂   SO₃

   (b) $25 \text{ g } SO_2\left(\dfrac{1 \text{ mol}}{64.07 \text{ g}}\right) = 0.39 \text{ mol } SO_2$

   $$5 \text{ g } O_2\left(\dfrac{1 \text{ mol}}{32.00 \text{ g}}\right) = 0.16 \text{ mol } O_2$$

   $$\text{mol ratio} = \frac{0.39}{0.16} = \frac{2.4 \text{ mol } SO_2}{1 \text{ mol } O_2}$$

   O_2 is the limiting reagent.

(c) False. The percentages given are not mass percentages. The percent composition of S in SO_2 is $(32/64) \times 100 = 50.\%$ S. The percent composition of S in SO_3 is $(32/80) \times 100 = 40.\%$ S.

3. (a) $\%O = 100 - (63.16 + 8.77) = 28.07\% $ O

Start with 100 g compound Z

C: $(63.16 \text{ g})\left(\dfrac{1 \text{ mol}}{12.01 \text{ g}}\right) = 5.259 \text{ mol}$

$\dfrac{5.259 \text{ mol}}{1.75 \text{ mol}} = 2.998$

H: $(8.77 \text{ g})\left(\dfrac{1 \text{ mol}}{1.008 \text{ g}}\right) = 8.70 \text{ mol}$

$\dfrac{8.70 \text{ mol}}{1.754 \text{ mol}} = 4.96$

O: $(28.07 \text{ g})\left(\dfrac{1 \text{ mol}}{16.00 \text{ g}}\right) = 1.754 \text{ mol}$

$\dfrac{1.754 \text{ mol}}{1.754 \text{ mol}} = 1.000$

The ratio of C:H:O is 3:5:1.
The empirical formula is C_3H_5O.
molar mass = 114; mass of empirical formula is 57
Therefore, the molecular formula is $C_6H_{10}O_2$.

(b) $2 C_6H_{10}O_2 + 15 O_2 \longrightarrow 12 CO_2 + 10 H_2O$

4. (a) $Ca_3(PO_4)_2$:

molar mass = 3(40.08) + 2(30.97) + 8(16.00)
= 310.3 %Ca in $Ca_3(PO_4)_2 = \dfrac{3(40.08)}{310.3}(100) = 38.7\%$

Let x = mg $Ca_3(PO_4)_2$
38.7% of x = 162 mg Ca
$x = \dfrac{(162 \text{ mg})(100)}{38.7} = 419 \text{ mg } Ca_3(PO_4)_2$

(b) $Ca_3(PO_4)_2$ is a compound.

(c) Convert 120 mL to cups

$(120 \text{ mL})\left(\dfrac{1 \text{ L}}{1000 \text{ mL}}\right)\left(\dfrac{1.059 \text{ qt}}{1 \text{ L}}\right)\left(\dfrac{4 \text{ cups}}{1 \text{ qt}}\right) = 0.51 \text{ cup}$

13% of x = 0.51 cup

$x = \dfrac{(0.51 \text{ cup})(100)}{13} = 3.9 \text{ cups}$

5. (a) Compound A must have a lower activation energy than compound B because B requires heat to overcome the activation energy for the reaction.

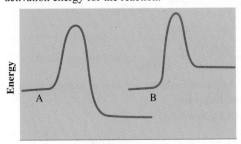

Reaction progress

(b) (i) $2 NaHCO_3 \longrightarrow Na_2CO_3 + H_2O + CO_2$
Decomposition of 0.500 mol $NaHCO_3$ requires 85.5 kJ of heat.

If 24.0 g CO_2 is produced, then

$(24.0 \text{ g } CO_2)\left(\dfrac{1 \text{ mol } CO_2}{44.01 \text{ g}}\right)\left(\dfrac{1 \text{ mol } H_2O}{1 \text{ mol } CO_2}\right)\left(\dfrac{18.02 \text{ g}}{1 \text{ mol } H_2O}\right)$
$= 9.83 \text{ g } H_2O \text{ produced}$

0.500 mol $NaHCO_3$ produces

$(0.500 \text{ mol } NaHCO_3)\left(\dfrac{1 \text{ mol } CO_2}{2 \text{ mol } NaHCO_3}\right)\left(\dfrac{44.01 \text{ g } CO_2}{1 \text{ mol } CO_2}\right)$
$= 11.0 \text{ g } CO_2$

Producing 11.0 g CO_2 required 85.5 kJ

Producing 24.0 g CO_2 requires $\left(\dfrac{24.0 \text{ g}}{11.0 \text{ g}}\right)(85.5 \text{ kJ})$
$= 187 \text{ kJ}$

(ii) $NaHCO_3$ could be compound B. Since heat was absorbed for the decomposition of $NaHCO_3$, the reaction was endothermic. Decomposition of A was exothermic.

6. (a) Double-displacement reaction

(b) $2 NH_4OH(aq) + CoSO_4(aq) \longrightarrow$
$(NH_4)_2SO_4(aq) + Co(OH)_2(s)$

(c) 8.09 g is 25% yield

Therefore, 100% yield = $(8.09 \text{ g } (NH_4)_2SO_4)\left(\dfrac{100\%}{25\%}\right)$
$= 32.4 \text{ g } (NH_4)_2SO_4$
(theoretical yield)

(d) molar mass of $(NH_4)_2SO_4 = 132.2 \text{ g/mol}$

theoretical moles $(NH_4)_2SO_4 = (32.4 \text{ g})\left(\dfrac{1}{132.2 \text{ g/mol}}\right)$
$= 0.245 \text{ mol } (NH_4)_2SO_4$

Calculate the moles of $(NH_4)_2SO_4$ produced from 38.0 g of each reactant.

$(38.0 \text{ g } NH_4OH)\left(\dfrac{1 \text{ mol}}{35.05 \text{ g}}\right)\left(\dfrac{1 \text{ mol } (NH_4)_2SO_4}{2 \text{ mol } NH_4OH}\right)$
$= 0.542 \text{ mol } (NH_4)_2SO_4$

$(38.0 \text{ g } CoSO_4)\left(\dfrac{1 \text{ mol}}{155.0 \text{ g}}\right)\left(\dfrac{1 \text{ mol } (NH_4)_2SO_4}{1 \text{ mol } CoSO_4}\right)$
$= 0.254 \text{ mol } (NH_4)_2SO_4$

Limiting reactant is $CoSO_4$; NH_4OH is in excess.

7. (a) $C_6H_{12}O_6 \longrightarrow 2 C_2H_5OH + 2 CO_2(g)$

Calculate the grams of $C_6H_{12}O_6$ that produced 11.2 g C_2H_5OH.

$(11.2 \text{ g } C_2H_5OH)\left(\dfrac{1 \text{ mol}}{46.07 \text{ g}}\right)\left(\dfrac{1 \text{ mol } C_6H_{12}O_6}{2 \text{ mol } C_2H_5OH}\right)\left(\dfrac{180.1 \text{ g}}{1 \text{ mol}}\right)$
$= 21.9 \text{ g } C_6H_{12}O_6$

25.0 g − 21.9 g = 3.1 g $C_6H_{12}O_6$ left unreacted

Volume of CO_2 produced:

$(11.2 \text{ g } C_2H_5OH)\left(\dfrac{1 \text{ mol}}{46.07 \text{ g}}\right)\left(\dfrac{2 \text{ mol } CO_2}{2 \text{ mol } C_2H_5OH}\right)\left(\dfrac{24.0 \text{ L}}{\text{mol}}\right)$
$= 5.83 \text{ L}$

The assumptions made are that the conditions before and after the reaction are the same and that all reactants went to the products.

(b) The theoretical yield is

$(25.0 \text{ g } C_6H_{12}O_6)\left(\dfrac{1 \text{ mol}}{180.1 \text{ g}}\right)\left(\dfrac{2 \text{ mol } C_2H_5OH}{1 \text{ mol } C_6H_{12}O_6}\right)\left(\dfrac{46.07 \text{ g}}{\text{mol}}\right)$
$= 12.8 \text{ g } C_2H_5OH$

$$\% \text{ yield} = \left(\frac{11.2 \text{ g}}{12.8 \text{ g}}\right)(100) = 87.5\%$$

(c) decomposition reaction

8. (a) double decomposition (precipitation)

(b) lead(II) iodide (PbI_2)

(c) $Pb(NO_3)_2(aq) + 2 KI(aq) \longrightarrow 2 KNO_3(aq) + PbI_2(s)$

If $Pb(NO_3)_2$ is limiting, the theoretical yield is

$$(25 \text{ g } Pb(NO_3)_2)\left(\frac{1 \text{ mol}}{331.2 \text{ g}}\right)\left(\frac{1 \text{ mol } PbI_2}{1 \text{ mol } Pb(NO_3)_2}\right)\left(\frac{461.0 \text{ g}}{\text{mol}}\right)$$
$$= 35 \text{ g } PbI_2$$

If KI is limiting, the theoretical yield is

$$(25 \text{ g KI})\left(\frac{1 \text{ mol}}{166.0 \text{ g}}\right)\left(\frac{1 \text{ mol } PbI_2}{2 \text{ mol KI}}\right)\left(\frac{461.0 \text{ g}}{\text{mol}}\right) = 35 \text{ g } PbI_2$$

$$\text{percent yield} = \left(\frac{7.66 \text{ g}}{35 \text{ g}}\right)(100) = 22\%$$

9. (a) Balance the equation

$$2 XNO_3 + CaCl_2 \longrightarrow 2 XCl + Ca(NO_3)_2$$

$$(30.8 \text{ g } CaCl_2)\left(\frac{1 \text{ mol}}{111.0 \text{ g}}\right)\left(\frac{2 \text{ mol } XCl}{1 \text{ mol } CaCl_2}\right) = 0.555 \text{ mol } XCl$$

Therefore, the molar mass of the XCl is

$$\frac{79.6 \text{ g}}{0.555 \text{ mol}} = 143 \text{ g/mol}$$

mass of (X + Cl) = mass of XCl

mass of X = 143 − 35.45 = 107.6

X = Ag (from periodic table)

(b) No. Ag is below H in the activity series.

10. (a) $2 H_2O_2 \longrightarrow 2 H_2O + O_2$

There must have been eight H_2O_2 molecules and four O_2 molecules in the flask at the start of the reaction.

(b) The reaction is exothermic.

(c) Decomposition reaction

(d) The empirical formula is OH.

CHAPTERS 10–11

Multiple Choice: 1. c **2.** a **3.** b **4.** a **5.** a **6.** b **7.** b
8. b **9.** c **10.** a **11.** d **12.** d **13.** a **14.** b **15.** c **16.** c
17. b **18.** c **19.** d **20.** d **21.** d **22.** a **23.** c **24.** c
25. b **26.** d **27.** b **28.** a **29.** b **30.** c **31.** c **32.** d
33. a **34.** a **35.** c **36.** d **37.** b **38.** c **39.** c **40.** c **41.** a

Free Response

1. The compound will be ionic because there is a very large difference in electronegativity between elements in Group 2A and those in Group 7A of the periodic table. The Lewis structure is

$$\left[M\right]^{2+} \begin{bmatrix} :\overset{\cdot\cdot}{\underset{\cdot\cdot}{X}}: \end{bmatrix}^{-} \\ \begin{bmatrix} :\overset{\cdot\cdot}{\underset{\cdot\cdot}{X}}: \end{bmatrix}^{-}$$

2. Having an even atomic number has no bearing on electrons being paired. An even atomic number means only that there is an even number of electrons. For example, carbon is atomic number six, and it has two unpaired p electrons: $1s^2 2s^2 2p_x^1 2p_y^1$.

3. False. The noble gases do not have any unpaired electrons. Their valence shell electron structure is $ns^2 np^6$ (except He).

4. The outermost electron in potassium is farther away from the nucleus than the outermost electrons in calcium, so the first ionization energy of potassium is lower than that of calcium.

However, once potassium loses one electron, it achieves a noble gas electron configuration and, therefore, taking a second electron away requires considerably more energy. For calcium, the second electron is still in the outermost shell and does not require as much energy to remove it.

5. The ionization energy is the energy required to *remove* an electron. A chlorine atom forms a chloride ion by *gaining* an electron to achieve a noble gas configuration.

6. The anion is Cl^-; therefore, the cation is K^+ and the noble gas is Ar. K^+ has the smallest radius, while Cl^- will have the largest. K loses an electron, and therefore, in K^+, the remaining electrons are pulled in even closer. Cl was originally larger than Ar, and gaining an electron means that, since the nuclear charge is exceeded by the number of electrons, the radius will increase relative to a Cl atom.

7. The structure shown in the question implies covalent bonds between Al and F, since the lines represent shared electrons. Solid AlF_3 is an ionic compound and, therefore, probably exists as an Al^{3+} ion and three F^- ions. Only valence electrons are shown in Lewis structures.

8. Carbon has four valence electrons; it needs four electrons to form a noble gas electron structure. By sharing four electrons, a carbon atom can form four covalent bonds.

9. NCl_3 is pyramidal. The presence of three pairs of electrons and a lone pair of electrons around the central atom (N) gives the molecule a tetrahedral structure and a pyramidal shape. BF_3 has three pairs of electrons and no lone pairs of electrons around the central atom (B), so both the structure and the shape of the molecule are trigonal planar.

10. The atom is Br ($35e^-$), which should form a slightly polar covalent bond with sulfur.

The Lewis structure of Br is $:\overset{\cdot\cdot}{Br}\cdot$

CHAPTERS 12–14

Multiple Choice: 1. b **2.** a **3.** b **4.** c **5.** a **6.** d **7.** d
8. b **9.** a **10.** b **11.** c **12.** a **13.** c **14.** c **15.** d **16.** a
17. c **18.** a **19.** a **20.** d **21.** b **22.** c **23.** a **24.** a **25.** d
26. d **27.** a **28.** b **29.** c **30.** a **31.** a **32.** c **33.** c **34.** c
35. a **36.** b **37.** b **38.** c **39.** d **40.** a **41.** c **42.** d
43. b **44.** c **45.** c **46.** c **47.** a **48.** c **49.** d **50.** b **51.** a
52. c **53.** d **54.** b **55.** b **56.** a **57.** d **58.** c **59.** b **60.** b

Free Response

1. 10.0% (m/v) has 10.0 g KCl per 100. mL of solution:

Therefore, KCl solution contains

$$\left(\frac{10.0 \text{ g KCl}}{100. \text{ mL}}\right)(215 \text{ mL})\left(\frac{1 \text{ mol}}{74.55 \text{ g}}\right) = 0.288 \text{ mol KCl}$$

NaCl solution contains

$$\left(\frac{1.10 \text{ mol NaCl}}{L}\right)\left(\frac{1 \text{ L}}{1000 \text{ mL}}\right)(224 \text{ mL}) = 0.246 \text{ mol NaCl}$$

The KCl solution has more particles in solution and will have the higher boiling point.

2. Mass of CO_2 in solution

$$= (\text{molar mass})(\text{moles}) = (\text{molar mass})\left(\frac{PV}{RT}\right)$$

$$= \left(\frac{44.01 \text{ g } CO_2}{\text{mol}}\right)\left(\frac{1 \text{ atm} \times 1.40 \text{ L}}{\dfrac{0.08206 \text{ L atm}}{\text{mol K}} \times 298 \text{ K}}\right) = 2.52 \text{ g } CO_2$$

mass of soft drink $= (345 \text{ mL})\left(\dfrac{0.965 \text{ g}}{\text{mL}}\right) = 333 \text{ g}$

ppm of $CO_2 = \left(\dfrac{2.52 \text{ g}}{333 \text{ g} + 2.52 \text{ g}}\right)(10^6)$

$\qquad\qquad = 7.51 \times 10^3 \text{ ppm}$

3. 10% KOH m/v solution contains 10 g KOH in 100 mL solution.

10% KOH by mass solution contains 10 g KOH + 90 g H_2O.

The 10% by mass solution is the more concentrated solution and therefore would require less volume to neutralize the HCl.

4. (a) $\dfrac{0.355 \text{ mol}}{0.755 \text{ L}} = 0.470 \text{ M}$

(b) The lower pathway represents the evaporation of water; only a phase change occurs; no new substances are formed. The upper path represents the decomposition of water. The middle path is the ionization of water.

5. Zack went to Ely, Gaye went to the Dead Sea, and Lamont was in Honolulu. Zack's bp was lowered, so he was in a region of lower atmospheric pressure (on a mountain). Lamont was basically at sea level, so his bp was about normal. Since Gaye's boiling point was raised, she was at a place of higher atmospheric pressure and therefore was possibly in a location below sea level. The Dead Sea is below sea level.

6. The particles in solids and liquids are close together (held together by intermolecular attractions), and an increase in pressure is unable to move them significantly closer to each other. In a gas, the space between molecules is significant, and an increase in pressure is often accompanied by a decrease in volume.

Liquid

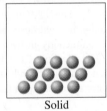
Solid

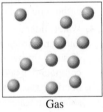
Gas

7. (a) The CO_2 balloon will be heaviest, followed by the Ar balloon. The H_2 balloon would be the lightest. Gases at the same temperature, pressure, and volume contain the same number of moles. All balloons will contain the same number of moles of gas molecules, so when moles are converted to mass, the order from heaviest to lightest is CO_2, Ar, H_2.

(b) Molar mass: O_2, 32.00; N_2, 28.02; Ne, 20.18

Using equal masses of gas, we find that the balloon containing O_2 will have the lowest number of moles of gas. Since pressure is directly proportional to moles, the balloon containing O_2 will have the lowest pressure.

8. Ray probably expected to get 0.050 mole $Cu(NO_3)_2$, which is

$(0.050 \text{ mol } Cu(NO_3)_2)\left(\dfrac{187.6 \text{ g}}{\text{mol}}\right) = 9.4 \text{ g } Cu(NO_3)_2$

The fact that he got 14.775 g meant that the solid blue crystals were very likely a hydrate containing water of crystallization.

9. For most reactions to occur, molecules or ions need to collide. In the solid phase, the particles are immobile and therefore do not collide. In solution or in the gas phase, particles are more mobile and can collide to facilitate a chemical reaction.

10. $\Delta t_b = 81.48°C - 80.1°C = 1.38°C$

$K_b = 2.53 \dfrac{°C \text{ kg solvent}}{\text{mol solute}}$

$\Delta t_b = mK_b$

$1.38°C = m\left(2.53 \dfrac{°C \text{ kg solvent}}{\text{mol solute}}\right)$

$0.545 \dfrac{\text{mol solute}}{\text{kg solvent}} = m$

Now we convert molality to molarity.

$\left(\dfrac{5.36 \text{ g solute}}{76.8 \text{ g benzene}}\right)\left(\dfrac{1000 \text{ g benzene}}{1 \text{ kg benzene}}\right)\left(\dfrac{1 \text{ kg benzene}}{0.545 \text{ mol solute}}\right)$

$\qquad\qquad\qquad\qquad\qquad\qquad = 128. \text{ g/mol}$

CHAPTERS 15–17

Multiple Choice: 1. c **2.** d **3.** d **4.** c **5.** c **6.** d **7.** a

8. d **9.** b **10.** c **11.** a **12.** d **13.** c **14.** b **15.** b **16.** d

17. c **18.** a **19.** a **20.** a **21.** b **22.** c **23.** a **24.** c **25.** c

26. d **27.** a **28.** b **29.** a **30.** b **31.** a **32.** c **33.** a **34.** b

35. c **36.** c **37.** b **38.** d **39.** a **40.** c **41.** a **42.** b

43. a **44.** b **45.** a **46.** a **47.** d **48.** a **49.** c **50.** d

51. b **52.** a **53.** a **54.** b

Balanced Equations

55. $3 P + 5 HNO_3 \longrightarrow 3 HPO_3 + 5 NO + H_2O$

56. $2 MnSO_4 + 5 PbO_2 + 3 H_2SO_4 \longrightarrow$
$\qquad\qquad\qquad\qquad 2 HMnO_4 + 5 PbSO_4 + 2 H_2O$

57. $Cr_2O_7^{2-} + 14 H^+ + 6 Cl^- \longrightarrow 2 Cr^{3+} + 7 H_2O + 3 Cl_2$

58. $2 MnO_4^- + 5 AsO_3^{3-} + 6 H^+ \longrightarrow 2 Mn^{2+} + 5 AsO_4^{3-} + 3 H_2O$

59. $S^{2-} + 4 Cl_2 + 8 OH^- \longrightarrow SO_4^{2-} + 8 Cl^- + 4 H_2O$

60. $4 Zn + NO_3^- + 6 H_2O + 7 OH^- \longrightarrow 4 Zn(OH)_4^{2-} + NH_3$

61. $2 KOH + Cl_2 \longrightarrow KCl + KClO + H_2O$

62. $4 As + 3 ClO_3^- + 6 H_2O + 3 H^+ \longrightarrow 4 H_3AsO_3 + 3 HClO$

63. $2 MnO_4^- + 10 Cl^- + 16 H^+ \longrightarrow 2 Mn^{2+} + 5 Cl_2 + 8 H_2O$

64. $Cl_2O_7 + 4 H_2O_2 + 2 OH^- \longrightarrow 2 ClO_2^- + 4 O_2 + 5 H_2O$

Free Response

1. $2 Bz + 3 Yz^{2+} \longrightarrow 2 Bz^{3+} + 3 Yz$

Bz is above Yz in the activity series.

2. (a) $2 Al(s) + 3 Fe(NO_3)_2(aq) \longrightarrow 2 Al(NO_3)_3(aq) + 3 Fe(s)$

(b) The initial solution of $Fe(NO_3)_2$ will have the lower freezing point. It has more particles in solution than the product.

3. (a) $pH = -\log[H^+] = -\log[0.10] = 1.00$

(b) mol HCl $= 0.050 \text{ L} \times \dfrac{0.10 \text{ mol}}{L} = 0.0050 \text{ mol HCl}$

Flask A

$Zn(s) + 2 HCl(aq) \longrightarrow ZnCl_2(aq) + H_2(g)$

HCl is the limiting reactant, so no HCl will remain in the product.

pH = 7.0

Flask B

No reaction occurs in flask B, so the pH does not change.

pH = 1.00

4. (a) $2 NaOH(aq) + H_2S(aq) \longrightarrow Na_2S(aq) + 2 H_2O(l)$

(b) $H_2S \rightleftharpoons 2H^+ + S^{2-}$ (aqueous solution)

$Na_2S \longrightarrow 2 Na^+ + S^{2-}$ (aqueous solution)

The addition of S^{2-} to a solution of H_2S will shift the equilibrium to the left, reducing the $[H^+]$ and thereby increasing the pH (more basic).

5. (a) Yes. The K_{eq} of AgCN indicates that it is slightly soluble in water, so a precipitate will form.

Net ionic equation:

$Ag^+(aq) + CN^-(aq) \rightleftharpoons AgCN(s)$

(b) NaCN is a salt of a weak acid and a strong base and will hydrolyze in water.

$CN^-(aq) + H_2O(l) \rightleftharpoons HCN(aq) + OH^-(aq)$

The solution will be basic due to increased OH^- concentration.

6. (a)

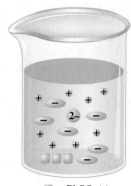

= $PbSO_4(s)$

(b)

No reaction—contents are merely mixed.

7. (a) $2 A_3X \rightleftharpoons 2 A_2X + A_2$

$$K_{eq} = \frac{(A_2X)^2(A_2)}{(A_3X)^2} = \frac{(3)^2(6)}{(4)^2} = 3.375$$

(b) The equilibrium lies to the right. $K_{eq} > 1$

(c) Yes, it is a redox reaction because the oxidation state of A has changed. The oxidation state of A in A_2 must be 0, but the oxidation state of A in A_3X is not 0.

8. (a) $X_2 + 2 G \rightleftharpoons X_2G_2$

$$K_{eq} = \frac{(X_2G_2)}{(X_2)(G)^2} = \frac{(1)}{(3)(2)^2} = 8.33 \times 10^{-2}$$

(b) Exothermic. An increase in the amount of reactants means that the equilibrium shifted to the left.

(c) An increase in pressure will cause an equilibrium to shift by reducing the number of moles of gas in the equilibrium. If the equilibrium shifts to the right, there must be fewer moles of gas in the product than in the reactants.

9. The pH of the solution is 4.5. (Acid medium)

$5 Fe^{2+} + MnO_4^- + 8 H^+ \longrightarrow 5 Fe^{3+} + Mn^{2+} + 4 H_2O$

CHAPTERS 18–20

Multiple Choice: 1. b 2. d 3. a 4. a 5. d 6. b 7. a
8. c 9. c 10. c 11. a 12. c 13. b 14. b 15. c 16. d
17. b 18. b 19. a 20. d 21. a 22. c 23. c 24. c
25. b 26. b 27. d 28. c 29. d 30. c 31. c 32. b
33. c 34. c 35. d 36. b 37. a 38. c 39. d 40. b
41. c 42. a 43. a 44. b 45. c 46. a 47. c 48. b
49. c 50. d 51. c 52. a 53. c 54. d 55. b

Free Response

1. The nuclide lost two alpha particles and one beta particle in any order. One series could be $^{223}_{87}Fr \xrightarrow{-\alpha} {}^{219}_{85}At \xrightarrow{-\beta} {}^{219}_{86}Rn \xrightarrow{-\alpha} {}^{215}_{84}Po$. $^{215}_{84}Po$ is the resulting nuclide.

2.

$$H_2N-CH-\overset{\overset{O}{\|}}{C}-NH-CH_2-\overset{}{C}-NH-CH_2-\overset{\overset{O}{\|}}{C}-NH-CH-C-$$

Tyr-Gly-Gly-Phe

3. Prostaglandin A_2 has two alkenes, an alcohol, a ketone, and a carboxylic acid (or carboxylate) group.

4. Water interacts mainly by hydrogen bonding, and molecules that can participate significantly in hydrogen bonding generally dissolve in water. The alcohol, $CH_3CH_2CH_2OH$, is the only compound listed that can significantly participate in hydrogen bonding as both a donor and an acceptor and should be significantly soluble in water.

5. Structural isomers

methyl propyl ether and 2-butanol;

butanoic acid and methyl propanoate

6. (a) An ester can be formed from a carboxylic acid and an alcohol.

Alcohol group: Ser, Tyr, or Thr.

Carboxylic acid group: Asp or Glu.

(b)

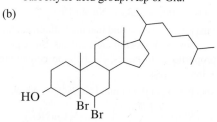

(c) Cholesterol is a lipid. Lipids are water-insoluble molecules. The alcohol group in cholesterol is only a small component of a large molecule.

7. (a) The zinc ion is called an activator.

(b) RNA contains ribose sugar; DNA contains deoxyribose sugar. RNA contains the base uracil; DNA contains the base thymine. RNA is generally single stranded; DNA is generally double stranded.

(c) Yes. All living things contain small amounts of carbon-14. Some people may have also been exposed to radioactivity such as from X-rays and have higher levels of radioactivity in their bodies.

8. (a)

```
        CHO
   H ——— OH ← secondary
  HO ——— H  ← secondary
   H ——— OH ← secondary
   H ——— OH ← secondary
       CH₂OH ← primary
```

(b) Tyrosine contains a para substituted aromatic ring.

(c) There are five amino acids in a pentapeptide. Four moles of water are lost in preparing a pentapeptide from five amino acids.

(d) Blood sugar is glucose. When hydrolyzed, maltose increases the glucose level more than lactose. Maltose hydrolyzes to two glucose units. Lactose hydrolyzes to one glucose unit and one galactose unit. Thus, it is better to cut down on maltose.

9. 96 g $\longrightarrow$ 48 g $\longrightarrow$ 24 g $\longrightarrow$
12 g $\longrightarrow$ 6 g $\longrightarrow$ 3 g $\longrightarrow$ 1.5 g

Each change involves a half-life. Six (6) half-lives are required to go from 96 g to 1.5 g. The half-life is about four days.

10. (a) The process is transmutation.

(b) Transmutation produces a different element by bombarding an element with a small particle such as hydrogen, deuterium, or a neutron. Radioactive decay is the formation of a different element by the loss of an alpha or beta particle from the nucleus of a radioactive element.

(c) No. Nuclear fission involves the bombardment of a heavy element with a neutron, causing the element to split (fission) into two or more lighter elements and produces two or more neutrons. A large amount of energy also is formed in a nuclear fission.

GLOSSARY

A

absolute zero $-273°C$, the zero point on the Kelvin (absolute) temperature scale. *See also* Kelvin temperature scale. [12.3]

acid ionization constant (K_a) The equilibrium constant for the ionization of a weak acid in water. [16.6]

activation energy The amount of energy needed to start a chemical reaction. [8.4, 16.3]

activity series of metals A listing of metallic elements in descending order of reactivity. [17.4]

actual yield The amount of product actually produced in a chemical reaction (compared with the theoretical yield). [9.5]

addition reaction In organic chemistry, a reaction in which two substances join together to produce one substance. [19.4]

alcohol An organic compound consisting of an —OH group bonded to a carbon atom in a nonaromatic hydrocarbon group; alcohols are classified as primary (1°), secondary (2°), or tertiary (3°), depending on whether the carbon atom to which the —OH group is attached is bonded to one, two, or three other carbon atoms, respectively. [19.7]

aldehyde An organic compound that contains the —CHO group. The general formula is RCHO. [19.9]

alkali metal An element (except H) from Group 1A of the periodic table. [3.2]

alkaline earth metal An element from Group 2A of the periodic table. [3.2]

alkane A compound composed of carbon and hydrogen, having only single bonds between the atoms; also known as saturated hydrocarbon. *See also* alkene and alkyne. [19.3]

alkene An unsaturated hydrocarbon whose molecules have at least one carbon–carbon double bond. [19.4]

alkyl group An organic group derived from an alkane by removal of one H atom. The general formula is C_nH_{2n+1} (e.g., CH_3, methyl). Alkyl groups are generally indicated by the letter R. [19.3]

alkyne An unsaturated hydrocarbon whose molecules have at least one carbon–carbon triple bond. [19.4]

alpha particle (α) A particle emitted from a nucleus of an atom during radioactive decay; it consists of two protons and two neutrons with a mass of about 4 amu and a charge of $+2$; it is considered to be a doubly charged helium atom. [18.2]

amino acid An organic compound containing two functional groups—an amino group (NH_2) and a carboxyl group (COOH). Amino acids are the building blocks for proteins. [20.4]

amorphous A solid without shape or form. [1.3]

amphoteric A substance having properties of both an acid and a base. [15.2]

anion A negatively charged ion. *See also* ion. [3.3]

anode The electrode where oxidation occurs in an electrochemical reaction. [17.5]

aromatic compound An organic compound whose molecules contain a benzene ring or that has properties resembling benzene. [19.5]

artificial radioactivity Radioactivity produced in nuclides during some types of transmutations. Artificial radioactive nuclides behave like natural radioactive elements in two ways: They disintegrate in a definite fashion, and they have a specific half-life. Sometimes called *induced radioactivity*. [18.3]

1 atmosphere The standard atmospheric pressure, that is, the pressure exerted by a column of mercury 760 mm high at a temperature of $0°C$. *See also* atmospheric pressure. [12.1]

atmospheric pressure The pressure experienced by objects on Earth as a result of the layer of air surrounding our planet. A pressure of 1 atmosphere (1 atm) is the pressure that will support a column of mercury 760 mm high at $0°C$. [12.1]

atom The smallest particle of an element that can enter into a chemical reaction. [3.1]

atomic mass The average relative mass of the isotopes of an element referred to the atomic mass of carbon-12. [5.6]

atomic mass unit (amu) A unit of mass equal to one-twelfth the mass of a carbon-12 atom. [5.6]

atomic number (Z) The number of protons in the nucleus of an atom of a given element. *See also* isotopic notation. [5.4]

Avogadro's law Equal volumes of different gases at the same temperature and pressure contain equal numbers of molecules. [12.4]

Avogadro's number 6.022×10^{23}; the number of formula units in 1 mole. [7.1]

B

balanced equation A chemical equation having the same number of each kind of atom and the same electrical charge on each side of the equation. [8.2]

barometer A device used to measure atmospheric pressure. [12.1]

bent structure Molecular structure in which the molecule looks like a V. The electron structure is tetrahedral but there are 2 pairs of nonbonding electrons. [11.10]

beta particle (β) A particle identical in charge (−1) and mass to an electron. [18.2]

binary compounds Compounds which are composed of two different elements. [6.4]

biochemistry The branch of chemistry concerned with chemical reactions occurring in living organisms. [20.1]

boiling point The temperature at which the vapor pressure of a liquid is equal to the pressure above the liquid. It is called the normal boiling when the pressure is 1 atmosphere. [13.3]

Boyle's law The relationship between pressure and volume of a gas. At constant temperature, the volume of a fixed mass of gas is inversely proportional to the pressure (PV = constant). [12.2]

buffer solution A solution that resists changes in pH when diluted or when small amounts of a strong acid or strong base are added. [16.8]

C

calorie (cal) A commonly used unit of heat energy; 1 calorie is a quantity of heat energy that will raise the temperature of 1 g of water 1°C (e.g., from 14.5 to 15.5°C). Also, 4.184 joules = 1 calorie exactly. *See also* joule. [4.5]

capillary action The spontaneous rising of a liquid in a narrow tube, which results from the cohesive forces within the liquid and the adhesive forces between the liquid and the walls of the container. [13.2]

carbohydrate A polyhydroxy aldehyde or polyhydroxy ketone, or a compound that upon hydrolysis yields a polyhydroxy aldehyde or ketone; sugars, starch, and cellulose are examples. [20.2]

carbonyl group The structure $\diagdown C = O$. [19.9]

carboxyl group The functional group of carboxylic acids:

$$\overset{\displaystyle O}{\overset{\displaystyle \|}{-C}}-OH.\ [19.10]$$

carboxylic acid An organic compound having a carboxyl group. [19.10]

catalyst A substance that influences the rate of a reaction and can be recovered essentially unchanged at the end of the reaction. [16.3]

cathode The electrode where reduction occurs in an electrochemical reaction. [17.5]

cation A positively charged ion. *See also* ion. [3.3]

chain reaction A self-sustaining nuclear or chemical reaction in which the products cause the reaction to continue or to increase in magnitude. [18.5]

Charles' law At constant pressure, the volume of a fixed mass of any gas is directly proportional to the absolute temperature (V/T = constant). [12.3]

chemical change A change producing products that differ in composition from the original substances. [4.2]

chemical equation A shorthand expression showing the reactants and the products of a chemical change (e.g., $2\ H_2O = 2\ H_2 + O_2$). [4.2, 8.1]

chemical equilibrium The state in which the rate of the forward reaction equals the rate of the reverse reaction in a chemical change. [16.2]

chemical formula A shorthand method for showing the composition of a compound, using symbols of the elements. [3.3]

chemical kinetics The study of reaction rates and reaction mechanisms. [16.1]

chemical properties The ability of a substance to form new substances either by reaction with other substances or by decomposition. [4.1]

chemistry The science of the composition, structure, properties, and reactions of matter, especially of atomic and molecular systems. [1.1]

colligative properties Properties of a solution that depend on the number of solute particles in solution and not on the nature of the solute (examples: vapor pressure lowering, freezing point depression, boiling point elevation). [14.5]

combination reaction A direct union or combination of two substances to produce one new substance. [8.3]

common ion effect The shift of a chemical equilibrium caused by the addition of an ion common to the ions in the equilibrium. [16.7]

compound A distinct substance composed of two or more elements combined in a definite proportion by mass. [3.3]

concentrated solution A solution containing a relatively large amount of dissolved solute. [14.4]

concentration of a solution A quantitative expression of the amount of dissolved solute in a certain quantity of solvent. [14.2]

condensation The process by which molecules in the gaseous state return to the liquid state. [13.2]

conversion factor A ratio of equivalent quantities. [2.5]

copolymer A polymer containing two different kinds of monomer units. [19.12]

covalent bond A chemical bond formed between two atoms by sharing a pair of electrons. [11.5]

critical mass The minimum quantity of mass required to support a self-sustaining chain reaction. [18.5]

curie (Ci) A unit of radioactivity indicating the rate of decay of a radioactive substance: $1\ Ci = 3.7 \times 10^{10}$ disintegrations per second. [18.4]

D

Dalton's atomic model The first modern atomic theory to state that elements are composed of minute individual particles called *atoms*. [5.1]

Dalton's law of partial pressures The total pressure of a mixture of gases is the sum of the partial pressures exerted by each of the gases in the mixture. [12.7]

decomposition reaction A breaking down, or decomposition, of one substance into two or more different substances. [8.3]

density The mass of an object divided by its volume. [2.8]

diatomic molecules The molecules of elements that always contain two atoms. Seven elements occur as diatomic molecules: H_2, N_2, O_2, F_2, Cl_2, Br_2, and I_2. [3.2]

dilute solution A solution containing a relatively small amount of dissolved solute. [14.4]

dipole A molecule that is electrically asymmetrical, causing it to be oppositely charged at two points. [11.6]

dipole–dipole attractions Forces of attraction between polar molecules as a result of the dipole moment within each molecule. [13.5]

disaccharide A carbohydrate that yields two monosaccharide units when hydrolyzed. [20.2]

dissociation The process by which a salt separates into individual ions when dissolved in water. [15.4]

DNA Deoxyribonucleic acid; a high-molar-mass polymer of nucleotides, present in all living matter, that contains the genetic code that transmits hereditary characteristics. [20.6]

double-displacement reaction A reaction of two compounds to produce two different compounds by exchanging the components of the reacting compounds. [8.3]

E

electrolysis The process whereby electrical energy is used to bring about a chemical change. [17.5]

electrolyte A substance whose aqueous solution conducts electricity. [15.4]

electrolytic cell An electrolysis apparatus in which electrical energy from an outside source is used to produce a chemical change. [17.5]

electron A subatomic particle that exists outside the nucleus and carries a negative electrical charge. [5.3]

electron configuration The orbital arrangement of electrons in an atom. [10.4]

electronegativity The relative attraction that an atom has for a pair of shared electrons in a covalent bond. [11.6]

element A basic building block of matter that cannot be broken down into simpler substances by ordinary chemical changes; in 1994, there were 111 known elements. [3.1]

empirical formula A chemical formula that gives the smallest whole-number ratio of atoms in a compound—that is, the relative number of atoms of each element in the compound; also known as the simplest formula. [7.4]

endothermic reactions A chemical reaction that absorbs heat. [8.4]

energy The capacity of matter to do work. [4.4]

enzyme A protein that catalyzes a biochemical reaction. [20.5]

equilibrium A dynamic state in which two or more opposing processes are taking place at the same time and at the same rate. [16.2]

equilibrium constant (K_{eq}) A constant representing the concentrations of the reactants and the products in a chemical reaction at equilibrium. [16.4]

ester An organic compound derived from a carboxylic acid and an alcohol. The general formula is $\text{R}-\overset{\displaystyle O}{\overset{\displaystyle \|}{\text{C}}}-\text{OR}'$. [19.11]

ether An organic compound having two hydrocarbon groups attached to an oxygen atom. The general formula is $\text{R}-\text{O}-\text{R}'$. [19.8]

evaporation The escape of molecules from the liquid state to the gas or vapor state. [13.2]

exothermic reactions A chemical reaction in which heat is released as a product. [8.4]

F

fats Esters of fatty acids and glycerol. *See also* triacylglycerol. [20.3]

formula equation A chemical equation in which all the reactants and products are written in their molecular, or normal, formula expression; also called a molecular equation. [15.7]

freezing point *See* melting point. [13.3]

frequency A measurement of the number of waves that pass a particular point per second. [10.1]

functional group An atom or group of atoms that characterizes a class of organic compounds. For example, $-\text{COOH}$ is the functional group of carboxylic acids. [19.2]

G

gamma ray (γ) High-energy photons emitted by radioactive nuclei; they have no electrical charge and no measurable mass. [18.2]

gas A state of matter that has no shape or definite volume so that the substance completely fills its container. [1.3]

Gay-Lussac's law of combining volumes When measured at the same temperature and pressure, the ratios of the volumes of reacting gases are small whole numbers. [12.4]

genes Basic units of heredity that consist primarily of DNA and proteins and occur in the chromosomes. [20.6]

ground state The lowest available energy level within an atom. [10.2]

groups or families (of elements) Vertical groups of elements in the periodic table (1A, 2A, and so on). Families of elements that have similar outer-orbital electron structures. [3.2, 10.5]

H

half-life ($t_{1/2}$) The time required for one-half of a specific amount of a radioactive nuclide to disintegrate; half-lives of the elements range from a fraction of a second to billions of years. [18.1]

halogens Group 7A of the periodic table; consists of the elements fluorine, chlorine, bromine, iodine, and astatine. [3.2]

heat Flow of energy due to a temperature difference. [2.7]

heat of fusion The energy required to change 1 gram of a solid into a liquid at its melting point. [13.4]

heat of reaction The quantity of heat produced by a chemical reaction. [8.4]

heat of vaporization The amount of heat required to change 1 gram of a liquid to a vapor at its normal boiling point. [13.4]

heterogeneous Matter without a uniform composition—having two or more components or phases. [1.4]

homogeneous Matter that has uniform properties throughout. [1.4]

homologous series A series of compounds in which the members differ from one another by a regular increment. For example, each successive member of the alkane series of hydrocarbons differs by a CH_2 group. [19.3]

hydrate A solid that contains water molecules as a part of its crystalline structure. [13.6]

hydrocarbons Compounds composed entirely of carbon and hydrogen. [8.4, 19.2]

hydrogen bond The intermolecular force acting between molecules that contain hydrogen covalently bonded to the highly electronegative elements F, O, and N. [13.5]

hydrogenation The process of adding hydrogen across a double or triple bond resulting in a more saturated compound. [19.4]

hydronium ion The result of a proton combining with a polar water molecule to form a hydrated hydrogen ion, H_3O^+. [15.1]

hypothesis A tentative explanation of certain facts to provide a basis for further experimentation. [1.2]

I

ideal gas A gas that behaves precisely according to the kinetic-molecular theory; also called a perfect gas. [12.6]

ideal gas law $PV = nRT$; that is, the volume of a gas varies *directly* with the number of gas molecules and the absolute temperature and *inversely* with the pressure. [12.6]

immiscible Incapable of mixing; immiscible liquids do not form a solution with one another. [14.2]

induced radioactivity *See* artificial radioactivity. [18.3]

intermolecular forces Forces of attraction between molecules. [13.5]

International System *See* SI, Systeme International. [2.5]

intramolecular forces Forces of attractions within a molecule, for example a covalent bond. [13.5]

ion A positively or negatively charged atom or group of atoms. *See also* cation, anion. [3.3]

ionic bond The chemical bond between a positively charged ion and a negatively charged ion. [11.3]

ionization The formation of ions, which occurs as the result of a chemical reaction of certain substances with water. [15.4]

ionization energy The energy required to remove an electron from an atom, an ion, or a molecule. [11.1]

ionizing radiation Radiation with enough energy to dislocate bonding electrons and create ions when passing through matter. [18.7]

ion product constant for water (K_w) An equilibrium constant defined as the product of the H^+ ion concentration and the OH^- ion concentration, each in moles per liter. $K_w = [H^+][OH^-] = 1 \times 10^{-14}$ at 25°C. [16.5]

isomerism The phenomenon of two or more compounds having the same molecular formula but different molecular structures. [19.3]

isomers Compounds having identical molecular formulas but different structural formulas. [19.3]

isotope An atom of an element that has the same atomic number but a different atomic mass. Since their atomic numbers are identical, isotopes vary only in the number of neutrons in the nucleus. [5.5]

J

joule (J) The SI unit of energy. *See also* calorie. [4.5]

K

ketone An organic compound that contains a carbonyl group between two other carbon atoms. The general formula is $R_2C = O$. [19.9]

kilogram (kg) The standard unit of mass in the metric system; 1 kilogram equals 2.205 pounds. [2.5]

kinetic energy (KE) The energy that matter possesses due to its motion; $KE = 1/2 \, mv^2$. [4.4]

kinetic-molecular theory (KMT) A group of assumptions used to explain the behavior and properties of gases. [12.6]

L

law of conservation of energy Energy can be neither created nor destroyed, though it can be transferred from one form to another. [4.4]

law of conservation of mass No change is observed in the total mass of the substances involved in a chemical reaction; that is, the mass of the products equals the mass of the reactants. [8.1]

law of definite composition A compound always contains two or more elements in a definite proportion by mass. [3.3]

law of multiple proportions Atoms of two or more elements may combine in different ratios to produce more than one compound. [3.3]

Le Châtelier's principle If a stress is applied to a system in equilibrium, the system will respond in such a way as to relieve that stress and restore equilibrium under a new set of conditions. [16.3]

Lewis structure A method of indicating the covalent bonds between atoms in a molecule or an ion such that a pair of electrons (:) represents the valence electrons forming the covalent bond. [11.2]

limiting reactant A reactant that limits the amount of product formed because it is present in insufficient amount compared with the other reactants. [9.5]

linear structure In the VSEPR model, an arrangement where the pairs of electrons are arranged 180° apart for maximum separation. [11.10]

line spectrum Colored lines generated when light emitted by a gas is passed through a spectroscope. Each element possesses a unique set of line spectra. [10.2]

lipids Organic compounds found in living organisms that are water insoluble but soluble in such fat solvents as diethyl ether, benzene, and carbon tetrachloride; examples are fats, oils, and steroids. [20.3]

liquid A state of matter in which the particles move about freely while the substance retains a definite volume; thus, liquids flow and take the shape of their containers. [1.3]

liter (L) A unit of volume commonly used in chemistry; 1 L = 1000 mL; the volume of a kilogram of water at 4°C. [2.5]

logarithm (log) The power to which 10 must be raised to give a certain number. The log of 100 is 2.0. [15.5]

London dispersion forces Attractive forces between nonpolar molecules which result from instantaneous formation of dipoles in the electron cloud. [13.5]

M

mass The quantity or amount of matter that an object possesses. [2.5]

mass defect The difference between the actual mass of an atom of an isotope and the calculated mass of the protons and neutrons in the nucleus of that atom. [18.6]

mass number (A) The sum of the protons and neutrons in the nucleus of a given isotope of an atom. [5.5]

matter Anything that has mass and occupies space. [1.3]

measurement A quantitative observation which requires both a number and a unit. [2.1]

meiosis The process of cell division to form a sperm cell and an egg cell in which each cell formed contains half of the chromosomes found in the normal single cell. [20.6]

melting point (or freezing point) The temperature at which the solid and liquid states of a substance are in equilibrium. [13.3]

meniscus The shape of the surface of a liquid when placed in a glass cylinder. It can be concave or convex. [13.2]

metal An element that is solid at room temperature and whose properties include luster, ductility, malleability, and good conductivity of heat and electricity; metals tend to lose their valence electrons and become positive ions. [3.2]

metalloid An element having properties that are intermediate between those of metals and nonmetals (e.g., silicon); these elements are useful in electronics. [3.2]

meter (m) The standard unit of length in the SI and metric systems; 1 meter equals 39.37 inches. [2.5]

metric system A decimal system of measurements. *See also* SI. [2.5]

miscible Capable of mixing and forming a solution. [14.2]

mitosis Ordinary cell division in which a DNA molecule is duplicated by uncoiling to single strands and then reassembling with complementary nucleotides. Each new cell contains the normal number of chromosomes. [20.6]

mixture Matter containing two or more substances, which can be present in variable amounts; mixtures can be homogeneous (sugar water) or heterogeneous (sand and water). [1.4]

molality (m) An expression of the number of moles of solute dissolved in 1000 g of solvent. [14.5]

molarity (M) The number of moles of solute per liter of solution. [14.4]

molar mass The mass of Avogadro's number of atoms or molecules. The sum of the atomic masses of all the atoms in an element, compound, or ion. The mass of a mole of any formula unit. It is also known as the molecular weight. [7.1, 9.1]

molar volume The volume of 1 mol of a gas at STP equals 22.4 L/mol. [12.5]

mole The amount of a substance containing the same number of formula units (6.022×10^{23}) as there are in exactly 12 g of ^{12}C. One mole is equal to the molar mass in grams of any substance. [7.1]

molecular formula The total number of atoms of each element present in one molecule of a compound; also known as the true formula. *See also* empirical formula. [7.5]

molecule The smallest uncharged individual unit of a compound formed by the union of two or more atoms. [3.3]

mole ratio A ratio between the number of moles of any two species involved in a chemical reaction; the mole ratio is used as a conversion factor in stoichiometric calculations. [9.1]

monomer The small unit or units that undergo polymerization to form a polymer. [19.12]

monosaccharide A carbohydrate that cannot be hydrolyzed to simpler carbohydrate units (e.g., simple sugars like glucose or fructose). [20.2]

N

natural law A statement summarizing general observations regarding nature. [3.3]

net ionic equation A chemical equation that includes only those molecules and ions that have changed in the chemical reaction. [15.7]

neutralization The reaction of an acid and a base to form a salt plus water. [15.6]

neutron A subatomic particle that is electrically neutral and is found in the nucleus of an atom. [5.3]

noble gases A family of elements in the periodic table—helium, neon, argon, krypton, xenon, and radon—that contains a particularly stable electron structure. [3.2]

nonelectrolyte A substance whose aqueous solutions do not conduct electricity. [15.4]

nonmetal An element that has properties the opposite of metals: lack of luster, relatively low melting point and density, and generally poor conduction of heat and electricity. Nonmetals may or may not be solid at room temperature (examples: carbon, bromine, nitrogen); many are gases. They are located mainly in the upper right-hand corner of the periodic table. [3.2]

nonpolar covalent bond A covalent bond between two atoms with the same electronegativity value; thus, the electrons are shared equally between the two atoms. [11.6]

normal boiling point The temperature at which the vapor pressure of a liquid equals 1 atm or 760 torr pressure. [13.3]

nuclear binding energy The energy equivalent to the mass defect; that is, the amount of energy required to break a nucleus into its individual protons and neutrons. [18.6]

nuclear fission A nuclear reaction in which a heavy nucleus splits into two smaller nuclei and two or more neutrons. [18.5]

nuclear fusion The uniting of two light elements to form one heavier nucleus, which is accompanied by the release of energy. [18.5]

nucleon A collective term for the neutrons and protons in the nucleus of an atom. [18.1]

nucleotide The building-block unit for nucleic acids. A phosphate group, a sugar residue, and a nitrogenous organic base are bonded together to form a nucleotide. [20.6]

nucleus The central part of an atom that contains all its protons and neutrons. The nucleus is very dense and has a positive electrical charge. [5.4]

nuclide A general term for any isotope of any atom. [18.1]

O

oils Liquid esters of fatty acids and glycerol. [20.3]

orbital A cloudlike region around the nucleus where electrons are located. Orbitals are considered to be energy sublevels (*s, p, d, f*) within the principal energy levels. *See also* principal energy levels. [10.2]

orbital diagram A way of showing the arrangement of electrons in an atom, where boxes with small arrows indicating the electrons represent orbitals. [10.4]

organic chemistry The branch of chemistry that deals with carbon compounds but does not imply that these compounds must originate from some form of life. *See also* vital-force theory. [19.1]

osmosis The diffusion of water, either pure or from a dilute solution, through a semipermeable membrane into a solution of higher concentration. [14.6]

osmotic pressure The difference between atmospheric pressure and an applied pressure that is required to prevent osmosis from occurring. [14.6]

oxidation An increase in the oxidation number of an atom as a result of losing electrons. [17.1]

oxidation number A small number representing the state of oxidation of an atom. For an ion, it is the positive or negative charge on the ion; for covalently bonded atoms, it is a posi-

tive or negative number assigned to the more electronegative atom; in free elements, it is zero. [17.1]

oxidation–reduction A chemical reaction wherein electrons are transferred from one element to another; also known as redox. [17.1]

oxidation state *See* oxidation number. [17.1]

oxidizing agent A substance that causes an increase in the oxidation state of another substance. The oxidizing agent is reduced during the course of the reaction. [17.1]

P

partial pressure The pressure exerted independently by each gas in a mixture of gases. [12.7]

parts per million (ppm) A measurement of the concentration of dilute solutions now commonly used by chemists in place of mass percent. [14.4]

Pauli exclusion principle An atomic orbital can hold a maximum of two electrons, which must have opposite spins. [10.3]

peptide linkage The amide bond in a protein molecule; bonds one amino acid to another. [20.4]

percent composition of a compound The mass percent represented by each element in a compound. [7.3]

percent yield The ratio of the actual yield to the theoretical yield multiplied by 100. [9.5]

period The horizontal groupings (rows) of elements in the periodic table. [10.5]

pH A method of expressing the H^+ concentration (acidity) of a solution; $pH = -\log[H^+]$, $pH = 7$ is a neutral solution, $pH < 7$ is acidic, and $pH > 7$ is basic. [15.5]

phase A homogeneous part of a system separated from other parts by a physical boundary. [1.4]

photons Theoretically, a tiny packet of energy that streams with others of its kind to produce a beam of light. [10.1]

physical change A change in form (such as size, shape, or physical state) without a change in composition. [4.2]

physical properties Inherent physical characteristics of a substance that can be determined without altering its composition: color, taste, odor, state of matter, density, melting point, boiling point. [4.1]

polar covalent bond A covalent bond between two atoms with differing electronegativity values, resulting in unequal sharing of bonding electrons. [11.5]

polyatomic ion An ion composed of more than one atom. [6.5]

polyhydroxyl alcohol An alcohol that has more than one — OH group. [19.7]

polymer A natural or synthetic giant molecule (macromolecule) formed from smaller molecules (monomers). [19.12]

polymerization The process of forming large, high-molar-mass molecules from smaller units. [19.12]

polypeptide A peptide chain containing up to 50 amino acid units. [20.4]

polysaccharide A carbohydrate that can be hydrolyzed to many monosaccharide units; cellulose, starch, and glycogen are examples. [20.2]

potential energy (PE) Stored energy, or the energy of an object due to its relative position. [4.4]

pressure Force per unit area; expressed in many units, such as mm Hg, atm, lb/in.^{2}, torr, and pascal. [12.1]

primary alcohol An alcohol in which the carbon atom bonded to the —OH group is bonded to only one other carbon atom. [19.7]

principal energy levels Existing within the atom, these energy levels contain orbitals within which electrons are found. *See also* electron, orbital. [10.3]

products A chemical substance produced from reactants by a chemical change. [4.2, 8.1]

properties The characteristics, or traits, of substances that give them their unique identities. Properties are classified as physical or chemical. [4.1]

protein A polymer consisting mainly of α-amino acids linked together; occurs in all animal and vegetable matter. [20.4]

proton A subatomic particle found in the nucleus of the atom that carries a positive electrical charge and a mass of about 1 amu. An H^+ ion is a proton. [5.3]

Q

quanta Small discrete increments of energy. From the theory proposed by physicist Max Planck that energy is emitted in energy *quanta* rather than a continuous stream. [10.2]

R

rad (radiation absorbed dose) A unit of absorbed radiation indicating the energy absorbed from any ionizing radiation; 1 rad = 0.01 J of energy absorbed per kilogram of matter. [18.4]

radioactive decay The process by which a radioactive element emits particles or rays and is transformed into another element. [18.1]

radioactivity The spontaneous emission of radiation from the nucleus of an atom. [18.1]

reactants A chemical substance entering into a reaction. [4.2, 8.1]

redox *See* oxidation–reduction. [17.1]

reducing agent A substance that causes a decrease in the oxidation state of another substance; the reducing agent is oxidized during the course of a reaction. [17.1]

reduction A decrease in the oxidation number of an element as a result of gaining electrons. [17.1]

rem (roentgen equivalent to man) A unit of radiation-dose equivalent taking into account that the energy absorbed from different sources does not produce the same degree of biological effect. [18.4]

representative element An element in one of the A groups in the periodic table. [3.2, 10.5]

resonance structure A molecule or ion that has multiple Lewis structures. *See also* Lewis structure. [11.8]

reversible chemical reaction A chemical reaction in which the products formed react to produce the original reactants. A double arrow is used to indicate that a reaction is reversible. [16.2]

RNA Ribonucleic acid; a high-molar-mass polymer of nucleotides present in all living matter. Its main function is to direct the synthesis of proteins. [20.6]

roentgen (R) A unit of exposure of gamma radiation based on the quantity of ionization produced in air. [18.4]

rounding off numbers The process by which the value of the last digit retained is determined after dropping nonsignificant digits. [2.3]

S

saturated hydrocarbon A hydrocarbon that has only single bonds between carbon atoms; classified as *alkanes*. [19.2]

saturated solution A solution containing dissolved solute in equilibrium with undissolved solute. [14.2]

scientific laws Simple statements of natural phenomena to which no exceptions are known under the given conditions. [1.2]

scientific method A method of solving problems by observation; recording and evaluating data of an experiment; formulating hypotheses and theories to explain the behavior of nature; and devising additional experiments to test the hypotheses and theories to see if they are correct. [1.2]

scientific notation A convenient way of expressing large and small numbers using powers of 10. To write a number as a power of 10 move the decimal point in the original number so that it is located after the first nonzero digit, and follow the new number by a multiplication sign and 10 with an exponent (called its *power*) that is the number of places the decimal point was moved. Example: $2468 = 2.468 \times 10^3$. [2.1]

secondary alcohol An alcohol in which the carbon atom bonded to the —OH group is bonded to two other carbon atoms. [19.7]

semipermeable membrane A membrane that allows the passage of water (solvent) molecules through it in either direction but prevents the passage of larger solute molecules or ions. [13.7, 14.6]

SI (*Systeme International*, or International System) An agreed-upon standard system of measurements used by scientists around the world. *See also* metric system. [2.5]

significant figures The number of digits that are known plus one estimated digit are considered significant in a measured quantity; also called significant digits. [2.2, Appendix I]

single-displacement reaction A reaction of an element and a compound that yields a different element and a different compound. [8.3]

solid A state of matter having a definite shape and a definite volume, whose particles cohere rigidly to one another, so that a solid can be independent of its container. [1.3]

solubility An amount of solute that will dissolve in a specific amount of solvent under stated conditions. [14.2]

solubility product constant (K_{sp}) The equilibrium constant for the solubility of a slightly soluble salt. [16.7]

solute The substance that is dissolved—or the least abundant component—in a solution. [14.1]

solution A system in which one or more substances are homogeneously mixed or dissolved in another substance. [14.1]

solution map An outline for the path of a unit conversion. [2.5]

solvent The dissolving agent or the most abundant component in a solution. [14.1]

specific gravity The ratio of the density of one substance to the density of another substance taken as a standard. Water is usually the standard for liquids and solids; air, for gases. [2.8]

specific heat The quantity of heat required to change the temperature of 1 g of any substance by 1°C. [4.5]

spectator ion An ion in solution that does not undergo chemical change during a chemical reaction. [15.6]

speed A measurement of how fast a wave travels through space. [10.1]

spin A property of an electron that describes its appearance of spinning on an axis like a globe; the electron can spin in only two directions, and, to occupy the same orbital, two electrons must spin in opposite directions. *See also* orbital. [10.3]

standard conditions *See* standard temperature and pressure. [12.5]

standard temperature and pressure (STP) 0°C (273 K) and 1 atm (760 torr); also known as standard conditions. [12.5]

Stock System A system that uses Roman numerals to name elements that form more than one type of cation. (For example: Fe^{2+}, iron(II); Fe^{3+}, iron(III).) [6.4]

stoichiometry The area of chemistry that deals with the quantitative relationships among reactants and products in a chemical reaction. [9.1]

strong electrolyte An electrolyte that is essentially 100% ionized in aqueous solution. [15.4]

subatomic particles Particles found within the atom, mainly protons, neutrons, and electrons. [5.3]

sublevels Each principal energy level is divided into sublevels. [10.3]

sublimation The process of going directly from the solid state to the vapor state without becoming a liquid. [13.2]

subscript Number that appears partially below the line and to the right of a symbol of an element (example: H_2SO_4). [3.3]

substance Matter that is homogeneous and has a definite, fixed composition; substances occur in two forms—as elements and as compounds. [1.4]

substrate In biochemical reactions, the substrate is the unit acted upon by an enzyme. [20.5]

supersaturated solution A solution containing more solute than needed for a saturated solution at a particular temperature. Supersaturated solutions tend to be unstable; jarring the container or dropping in a "seed" crystal will cause crystallization of the excess solute. [14.2]

surface tension The resistance of a liquid to an increase in its surface area. [13.2]

symbol In chemistry, an abbreviation for the name of an element. [3.1]

system A body of matter under consideration. [1.4]

T

temperature A measure of the intensity of heat, or of how hot or cold a system is; the SI unit is the kelvin (K). [2.7]

tertiary alcohol An alcohol in which the carbon atom bonded to the —OH group is bonded to three other carbon atoms. [19.7]

tetrahedral structure An arrangement of the VSEPR model where four pairs of electrons are placed 109.5° degrees apart to form a tetrahedron. [11.10]

theoretical yield The maximum amount of product that can be produced according to a balanced equation. [9.5]

theory An explanation of the general principles of certain phenomena with considerable evidence to support it; a well-established hypothesis. [1.2]

thermal energy A form of energy associated with the motion of small particles [2.7]

Thomson model of the atom Thomson asserted that atoms are not indivisible but are composed of smaller parts; they contain both positively and negatively charged particles—protons as well as electrons. [5.3]

titration The process of measuring the volume of one reagent required to react with a measured mass or volume of another reagent. [15.6]

total ionic equation An equation that shows compounds in the form in which they actually exist. Strong electrolytes are written as ions in solution, whereas nonelectrolytes, weak electrolytes, precipitates, and gases are written in the un-ionized form. [15.7]

transcription The process of forming RNA from DNA. [20.6]

transition elements The metallic elements characterized by increasing numbers of d and f electrons. These elements are located in Groups 1B–8B of the periodic table. [3.2, 10.5]

transmutation The conversion of one element into another element. [18.3]

transuranium elements An element that has an atomic number higher than that of uranium (>92). [18.3]

triacylglycerol (triglyceride) An ester of glycerol and three molecules of fatty acids. [20.3]

trigonal planar structure An arrangement of atoms in the VSEPR model where the three pairs of electrons are placed 120° apart on a flat plane. [11.10]

U

unsaturated hydrocarbon A hydrocarbon whose molecules contain one or more double or triple bonds between two carbon atoms; classified as *alkenes*, *alkynes*, and *aromatic compounds*. [19.2]

unsaturated solution A solution containing less solute per unit volume than its corresponding saturated solution. [14.2]

V

valence electrons An electron in the outermost energy level of an atom; these electrons are the ones involved in bonding atoms together to form compounds. [10.4]

vapor pressure The pressure exerted by a vapor in equilibrium with its liquid. [13.2]

vaporization *See* evaporation. [13.2]

vapor pressure curve A graph generated by plotting the temperature of a liquid on the *x*-axis and its vapor pressure on the *y*-axis. Any point on the curve represents an equilibrium between the vapor and liquid. [13.3]

vital-force theory A theory that held that organic substances could originate only from some form of living material. The theory was proved false early in the nineteenth century. [19.1]

volatile A substance that evaporates readily; a liquid with a high vapor pressure and a low boiling point. [13.2]

voltaic cell A cell that produces electric current from a spontaneous chemical reaction. [17.5]

volume The amount of space occupied by matter; measured in SI units by cubic meters (m^3) but also commonly in liters and milliliters. [2.5]

W

water of crystallization Water molecules that are part of a crystalline structure, as in a hydrate; also called water of hydration. [13.6]

water of hydration *See* water of crystallization. [13.6]

wavelength The distance between consecutive peaks and troughs in a wave; symbolized by the Greek letter lambda. [10.1]

weak electrolyte A substance that is ionized to a small extent in aqueous solution. [15.4]

weight A measure of Earth's gravitational attraction for a body (object). [2.5]

INDEX

A

Absolute zero, 31, 256
Acetic acid, 64, 136
Acetone, 467
Acetylene, 136, 453
Acids. *See also* Electrolytes
   anion names compared, 112
   definitions of, 338–341
   molarity of, 317
   naming of, 111–113
   neutralization of, 153, 352–354
   and pH, 349–351
   reactions of, 342–343
Acid ionization constant (K_a), 376–378
Acid rain, 356–357
Acid-type dry cell battery, 405
Actinide series, 204
Activation energy, 157, 372
Activity series of metals, 152, 401–403
Actual yield, 180
Acute radiation damage, 434, 435
Addition, significant figures in, 19–21
Addition reactions, 455
Agitation, solid dissolving rate and, 312
Air pollution, 267
Air pressure measurement, 250–251
Alanyltyrosylglycine, 495
Alcohols, 461–465
   ethanol, 463
   methanol, 462–463
   naming of, 463–465
   primary, 462
   secondary, 462
   tertiary, 462
Aldehydes, 467–468
Alkali metals:
   compound formulas formed by, 223
   and halide solubility, 309
   and periodic table, 50, 203
Alkaline earth metals, 50, 203
Alkaline-type dry cell battery, 405
Alkanes, 444–452
   isomerism of, 447–448
   naming of, 448–452
   structural formulas of, 446–448

Alkenes:
   addition reactions with, 455
   classification of, 452–453
   naming of, 453–454
   and unsaturated hydrocarbons, 444
Alkyl groups, 449–450
Alkyl halides, 460–461
Alkynes:
   classification of, 452–453
   naming of, 453–454
   and unsaturated hydrocarbons, 444
Allotropes, 267
Alpha particles (α), 418, 421–422
Alpha-particle scattering, 87–88
Aluminum:
   and atomic structure, 201
   chromate, 103
   fluoride formation, 221
   specific heat of, 71
Amatore, Christian, 4
Americium, 48
Amines, pharmaceutical uses of, 341
Amino acids, 492–497
Amino acid residues, 494
Ammonia, gaseous, 370
Ammonium phosphate, 103
Amorphous solids, 6
Amphoteric hydroxides, 343
Anhydrous crystals, 295–296
Anions:
   acid names vs., 113
   and compounds, 53
   formation of, 86, 105
   naming of, 84, 102
Anodes, 403
Argon, 48, 201
Aromatic compounds, 456–459
   disubstituted benzenes, 457–458
   monosubstituted benzenes, 457
   naming of, 457–458
   tri- and polysubstituted benzenes, 458
   and unsaturated hydrocarbons, 444
Aromatic hydrocarbons, 456–459
Arrhenius, Svante, 84, 338
Arrhenius acids and bases, 338

Art:
   oxygen in restoration of, 52
   synthesized molecules as, 235
Artificial radioactivity, 425–426
Artificial sweeteners, 293
Atmosphere (atm), 250–251
Atmospheric mass, 46
Atmospheric pressure, 250, 285–286
Atoms:
   definition of, 45
   Lewis structures of, 216–217
Atomic bomb, 431–432
Atomic clocks, 197
Atomic mass, 92–93
Atomic number (Z), 89–91
Atomic properties, periodic trends in, 213–216
Atomic radius, 214, 219–220
Atomic theory and structure, 82–94. *See also* Modern atomic theory
   and atomic mass, 92–93
   Dalton's atomic model, 83–84
   electric charge in, 84–85
   isotopes in, 89–91
   nuclear atomic model, 87–89
   structures of first eighteen elements, 198–201
   subatomic particles in, 85–87
Atomic weight, 92
Automobile storage battery, 405–406
Avogadro, Amadeo, 123
Avogadro's law, 259–260
Avogadro's number, 123

B

Baking soda, 342
Balanced chemical equations, 145–149, 168
Barium chloride dihydrate, 295
Barometers, 250
Bases. *See also* Electrolytes
   definitions of, 338–341
   neutralization of, 153, 352–354
   and pH, 349–351
   reactions of, 342, 343
Becquerel, Henri, 87, 418

Beef insulin, 495–496
Bent molecular structure, 237, 238
Benzenes:
    and addition reactions, 456
    disubstituted, 457–458
    empirical and molecular formulas, 136
    monosubstituted, 457
    tri- and polysubstituted, 458
Benzoic acid, 470
Berkelium, 48
Beryllium, 199
Beta particles (β), 418, 422
Binary acids, 111–112
Binary compounds:
    acids derived from, 111–112
    containing two nonmetals, 108
    flow chart for naming, 112
    ions from, 102
    with metals, 105–108
    naming of, 105–109
Biochemistry, 483–505
    amino acids and proteins, 492–497
    carbohydrates, 484–488
    DNA and genetics, 502–504
    enzymes, 497–499
    lipids, 488–491
    in living organisms, 484
    nucleic acids, 499–502
Biodiesel, 444
Black, Joseph, 71
Blood sugar, 485
Bohr, Niels, 101, 194–195
Bohr atom, 193–195
Boiling point, 286–287
Boiling point elevation constants, 320–324
Bonds, see Chemical bonds
Bond angle, 298
Bond length, 298
Boron, 199
Boron trifluoride, 235
Boyle's law, 252–256
Breeder reactor, 430–431
Bromides, 223
Bromine, 136
Brønsted, J. N., 338
Brønsted–Lowry proton transfer, 338–339
Büchner, Eduard, 497
Buffer solutions, 381–383
Butane:
    naming of, 449–450
    structural formula for, 447
Butyric acid, 470

C

Calcium chloride, 322
Calculations, significant figures and, 18–21
Calories, heat measurement and, 70
Capillary action, liquids and, 283
Carbohydrates, 484–488
    disaccharides, 486–487
    monosaccharides, 484–486
    polysaccharides, 488

Carbon:
    and atomic structure, 199
    footprints, 160
    isotope half-life, 419
    in organic chemistry, 442–445
    tetrahedral structure of, 442–443
Carbonates, acids and, 342
Carbon dioxide:
    in atmosphere, 159–161
    and covalently bonded molecules, 224
    for dry cleaning, 472
    empirical and molecular formula, 136
    geometric shape of, 235
    and oxygen in blood, 382
Carbonic acid, 342
Carbon monoxide poisoning, 151
Carbon tetrachloride, 228, 236
Carbonyl group, 459, 467
Carboxyl group, 469
Carboxylic acids, 469–470
Catalysts, 372–373
Cathodes, 403
Cathode rays, 85–86
Cations:
    from binary ionic compounds, 106–108
    and compounds, 53
    formation of, 86, 100
    and ion discovery, 84
Cavities, 367
Cellular genetic information process, 502
Cellulose, 488
Celsius scale, 31
Cerebroside, 491
CFCs (chlorinated fluorocarbons), 461
Chadwick, James, 86
Chain reaction, 429
Charles, J. A. C., 256
Charles' law, 256–258
Chemical bonds, 212–240
    in complex Lewis structures, 232–234
    in compounds with polyatomic ions,
        234–235
    covalent, 224–229
    and electronegativity, 226–229
    and formulas of ionic compounds,
        222–224
    hydrogen, 291–294
    intermolecular, 292
    ionic, 217–222
    and Lewis structures of atoms, 216–217
    and Lewis structures of compounds,
        229–232
    and molecular shape, 235–238
    and periodic trends in atomic properties,
        213–216
Chemical calculations, 167–182
    limiting reactant calculations, 176–181
    mass–mass calculations, 174–176
    mole–mass calculations, 173–174
    mole–mole calculations, 170–173
    stoichiometry, 168–170
    yield calculations, 180–181

Chemical changes:
    energy in, 69
    and properties of substances, 65–68
Chemical equations, 143–162
    calculations from, 168–181
    combination reactions, 150–151
    decomposition reactions, 151–152
    double-displacement reactions, 153–155
    general format, 144–145
    and global warming, 159–161
    and heat in chemical reactions, 156–158
    information in, 149–150
    molecular representations of, 68
    single-displacement reactions, 152–153
    and stepwise sequences, 146–148
    and stoichiometry, 168–170
    symbols commonly used in, 144
    types of, 150–156
    writing and balancing, 145–150
Chemical equilibrium, 363–384
    and buffer solutions, 381–383
    and catalysts, 372–373
    and concentration, 368–370
    definition of, 365–366
    equilibrium constants, 373–374
    ionization constants, 376–378
    ion product constant for water, 374–376
    Le Châtelier's principle, 366–373
    and pH control, 381–383
    rates of reaction, 364
    for reversible reactions, 365–366
    solubility product constant, 378–381
    and temperature, 371–372
    and volume, 370–371
Chemical formulas:
    of compounds, 54–55
    of ionic compounds, 103–104, 222–224
    percent composition from, 130–131
Chemical kinetics, 364
Chemical properties, 63–65
Chemists, 2–3
Chemisthesis, 344
Chemistry (term), 2
Chernobyl nuclear accident, 430
Chlorides, 223
Chlorinated fluorocarbons (CFCs), 461
Chlorination, 460
Chlorine:
    and atomic structure, 201
    and electron arrangements, 218
    empirical and molecular formulas, 136
    gas, 225
    name of, 48
    oxy-acids of, 113
    p electron pairing in, 225
    physical properties of, 64
    relative radii of, 219
    sodium atoms reacting with, 217–218
Chlorobenzene, 456
Chloroform, 461
Cholesterol, 491
Chromosomes, 503

Ci (curie), 427
Coal, 72–73
Cobalt, 48
Coins, modern technology and, 132
Colligative properties, of solutions, 320–324, 348
Combination reactions, 150–151
Combined gas laws, 260–262
Combustion:
  gasoline, 69
  hydrogen, 299
  magnesium, 157
Common ion effect, 380
Compounds. See also Quantitative composition of compounds
  and anions, 53–54
  and cations, 53
  chemical formulas of, 54–55
  composition of, 55–57
  empirical formulas of, 133–137
  and ions, 53–54
  Lewis structures of, 229–232
  molar mass of, 126–129
  molecular and ionic, 52–54
  molecular formulas of, 135–137
  and molecules, 53–54
  percent composition of, 129–132, 135
  with polyatomic ions, 234
  registered, 53
Concentrated solutions, 313
Concentration, 307, 312–320
  dilute vs. concentrated solutions, 313
  dilution problems for, 319–320
  and equilibrium, 368–370
  mass percent, 313–315
  mass/volume percent, 315
  molarity, 315–319
  volume percent, 315
Condensation, 285
Conservation of energy, 70
Conservation of mass, 145
Conversion factors, 23–24
Copernicus, 101
Copolymers, 473
Copper:
  atoms, 85
  isotopes of, 92
  mass of, 92
  and silver nitrate reaction, 155
  specific heat of, 71
Copper oxide formation, 66, 68
Cotton, 488
Covalent bonds, 224–229
Crick, Francis H. C., 499–500
Critical mass, 429
Crookes, William, 85
Crystals:
  surface area of, 311
  of table salt, 6
Crystalline solids, 6
C-terminal residue, 494–495
Curie (Ci), 427

Curie, Marie, 418
Curie, Pierre, 418
Currency, manufacturing of, 64

D

Dalton's atomic model, 83–84
  early thoughts on, 83
  and electrical charge, 84–85
  and subatomic parts of atom, 85–87
Dalton's law of partial pressures, 267–269
de Broglie, Louis, 195
Decomposition reactions, 151–152
Definite composition, law of, 56
Degrees in temperature measurement, 31
Density, 34–37, 270
Deoxyribonucleic acid, see DNA
Desalination, of seawater, 299
Diabetes, 252
Diatomic elements, 127–128
Diatomic molecules, 51, 100
Diborane, 136
Dichlorobenzenes, 457–458
Dilute solutions, 313
Dilution problems, 319–320
Dimensional analysis, 27–30
Dipeptides, 494
Dipoles, 227–228
Dipole–dipole attraction, 290–291
Disaccharides, 486–487
Dissociation, electrolytes, 345–346
Distillation, water, 299
Disubstituted benzenes, 457–458
Division, significant figures in, 18–21
DNA (deoxyribonucleic acid), 499–504
  and genetics, 502–504
  replication of, 502–504
  structure of, 500–501
  transcription of, 501–502
Double-displacement reaction, 153–156
Drug delivery, acid–base chemistry of, 341
Dry cell batteries, 405
Dry cleaning, 472

E

Earth, mass of elements in, 46–47
Egypt, ancient, 4
Electrical charge, 84–85, 87
Electrolysis, 66–68, 403
Electrolytes, 344–349. See also Acids; Bases; Salts
  colligative properties of, 348
  dissociation and ionization of, 345–346
  from ionization of water, 348–349
  strong and weak, 346–348
Electrolytic cells, 403–405
Electromagnetic field, 421
Electromagnetic radiation, 192–193
Electromagnetic spectrum, 192
Electrons:
  arrangements in noble gases, 217
  configurations of, 199, 201

  in covalent bonds, 224–226
  electrical charge of, 87
  energy levels of, 195–198
  Ion–electron method for balancing redox reactions, 398–399
  in ionic bonds, 217–222
  and pair arrangement, 238
  and periodic table, 201–206
  properties of, 86
  relative mass of, 87
Electronegativity, 226–229
Electroplating, of metals, 404–405
Elements, 44–58. See also Periodic table
  arranging by sublevel being filled, 204
  atomic numbers of, 89
  atomic structures of first eighteen, 198–201
  and atoms, 45
  broad categories of, 51
  collecting, 202
  in compounds, 52–57
  definition of, 45
  as diatomic molecules, 51
  distribution of, 46–47
  in inorganic compounds, 100, 102–103
  liquid and gaseous, 46
  mass in Earth's crust, seawater, and atmosphere, 46–47
  mass in human body, 47
  names of, 47, 48, 101
  natural states of, 45–46
  symbols of, 47–49
  transmutation of, 425
  transuranium, 426
Empirical formulas:
  calculating, 133–135
  molecular vs., 135–137
Endothermic reactions, 156–158
Energy. See also specific types
  in chemical changes, 69
  conservation of, 70
  definition of, 68–69
  of electrons, 195–198
  and mass in nuclear reactions, 433–434
  in real world, 72–74
Engine coolant, 322
Enriched uranium, 430
Enzymes, 497–499
Equilibrium, 365. See also Chemical equilibrium
Equilibrium constants (K_{eq}), 373–374
Esterification, 471
Esters, 471–472
Ethane, 445, 447
Ethanol, 461, 463
Ethers, 465–466
Ethyl alcohol:
  physical properties of, 287
  specific heat of, 71
  vapor pressure-temperature curves for, 286–287
Ethyl chloride, 287

Ethylene, 136, 452–453, 473, 474
Ethylene glycol, 463
Ethyl ether, 287
Evaporation, 284
Exothermic reactions, 156–158, 372
Experimental data, percent composition from, 135

F

Fahrenheit scale, 31
Falling water, conversion to electrical energy, 69
Fallout, 431
Families, periodic table and, 203
Faraday, Michael, 84, 456
Fats, 488–491
Fat cells, 489
Fatty acids, 489, 490
Fermentation, 497
Fission, nuclear, 427–429
Fission products, 427
Flerov, Georgy, 101
Fluorescence, 193
Fluorine, 198, 292
Food irradiation, 428
Formaldehyde, 136, 467
Formalin, 467
Formic acid, 470
Formula equations, 354–356
Fossil fuels, 159–161, 444
Fracking, 73
Francium, 48
Freezing point, 31, 287–288
Freezing point depression constants, 320–324
Frequency, of waves, 192
Fructose, 485–486
Functional groups, 444. *See also specific types, e.g.:* Alcohols
Fusion, nuclear, 432

G

Galactose, 485
Galactosemia, 485
Galvanic cells, 403–407
Gamma rays (γ), 418, 422–424
Gases, 248–276
    Avogadro's law for, 259–260
    Boyle's law for, 252–256
    Charles' law for, 256–258
    combined gas laws, 260–263
    common materials in, 7
    Dalton's law of partial pressures, 267–269
    density of, 35, 270
    formation of, 154
    gas stoichiometry, 270–274
    ideal, 266
    ideal gas law, 264–267
    and kinetic-molecular theory, 266
    mole–mass–volume relationships of, 262–263

pressure of, 249–252
properties of, 7, 249–252
real, 266–267
solubility in water, 309
and standard conditions, 260
Gaseous elements, 46
Gasoline combustion, 69
Gas stoichiometry, 270–274
    mass–volume calculations, 270–272
    mole–volume calculations, 270–272
    volume–volume calculations, 272–274
Gay-Lussac, J. L., 259
Gay-Lussac's law, 259–260
Geiger counter, 426
Geiger-Müller detecting tube, 426
Genes, 503
Genetic effects, of radiation, 434–435
Genetics, DNA and, 502–504
Glassware, volume measurements and, 26
Global warming, 159–161
Glucose, 136, 485, 488
Glutamate, 496
Glycerol, 463
Glycogen, 488
Glycolipids, 491
Glycosuria, 485
Glycylglycine, 494
Gold, 71
Goldstein, Eugen, 86
Gray, Theodore, 202
Greenhouse effect, 159–161
Ground state, energy levels and, 194
Groups, of elements, 203

H

Haber, Fritz, 370
Haber process, 370
Hahn, Otto, 427
Half-life ($t_{1/2}$), 419–420
Halogens, 50, 203
Halogenation, 460
Halothane, 461
Heat:
    in chemical reactions, 156–158
    and measurement of temperature, 30
    qualitative measurement of, 70–71
Heating curves, 289
Heat of fusion, 289
Heat of reaction, 156–157
Heat of vaporization, 289
Helium, 45–46, 48, 199
Heredity, 502–504
Heterogeneous matter, 7
Heterogeneous mixtures, 7–9
HFCs (hydrofluorocarbons), 461
Holoenzymes, 498
Homogenous matter, 7–8
Homogenous mixtures, 7–9
Homologous series, 445
Hot packs, 311
Human body, mass percent of elements in, 47

Hydrates, 295–296
Hydrazine, 136
Hydrocarbons:
    aromatic, 456–459
    energy from, 72–74
    and fossil fuels, 157
    names and formulas for, 73
    saturated, 444–445
    unsaturated, 444–445
Hydrocarbon derivatives, 459–461
Hydrochloric acid, 109, 111, 404
Hydrofluorocarbons (HFCs), 461
Hydrogen:
    combustion of, 299
    empirical and molecular formulas, 136
    and energy release, 69
    formation of hydrogen molecule, 224–225
    isotopes of, 89–90
    line spectrum of, 194
    properties of compounds containing, 291
Hydrogenation, 455, 490
Hydrogen atom:
    Bohr model of, 194
    modern concept of, 198
    orbitals for, 195
Hydrogen bonds, 291–294
Hydrogen carbonate–carbonic acid buffer, 383
Hydrogen chloride, 109, 111, 136
Hydrogen peroxide, 56, 152
Hydronium ion, 339
Hydroxides, amphoteric, 343
Hypothesis, scientific method and, 4

I

Ice:
    specific heat of, 71
    and water in equilibrium, 297
Ice cream, 324
Ideal gases, 266
Ideal gas law, 264–266
Immiscible liquids, 307
Induced radioactivity, 426
Inner transition elements, 204
Inorganic compounds, 98–115
    acids, 111–113
    binary, 105–109
    common and systematic names for, 99–100
    elements and ions, 100–103
    and ionic compound formulas, 103–104
    from polyatomic ions, 109–110
Insoluble precipitate formation, 154
Intermolecular bonds, 292
Intermolecular forces, 290–295
    dipole–dipole interactions, 290–291
    hydrogen bonds, 291–294
    London dispersion forces, 294–295
International System (SI), 21–23
International Union of Pure and Applied Chemistry (IUPAC), 100, 449
Intramolecular forces, 290

Iodine, 48, 136
Ions. *See also* Anions; Cations
  definition of, 53
  discovery of, 84–85
  and inorganic compounds, 100–103
  ionic bond formation, 227–228
  oxidation numbers for, 391
  polyatomic, 109–110, 234, 392
  spectator, 352
Ion–electron method (for redox reactions),
    398–399
Ionic bonds, 217–222
Ionic compounds. *See also* Salts
  characteristics of, 52–53
  predicting formulas of, 222–224
  writing formulas from, 103–104
Ionic equations, writing, 354–356
Ionic redox equations, balancing, 398–401
Ionization, of water, 348–349
Ionization constants (K_a), 376–378
Ionization energy, 214–216
Ionizing radiation, 426, 434
Ion product constant for water (K_w),
    374–376
Iron, 9, 71
Isomers, 447
Isomerism, 447
Isopropyl alcohol, 463
Isotopes of elements, 89–91
Isotopic notation, 90, 419
IUPAC (International Union of Pure and
    Applied Chemistry), 100, 449

J

Joliot-Curie, Frederic, 425–426
Joliot-Curie, Irene, 425–426
Joules, heat measurement and, 70

K

K_a *(acid ionization constant), 376–378*
Keiffer, Susan W., 33
Kekulé, August, 456
Kekulé structures, 456
Kelvin scale, 31
K_{eq} *(equilibrium constant), 373–374*
Ketones, 467–469
Kevlar, 232
Kilogram (kg), 24–25
Kinetics, chemical, 364
Kinetic energy, 69
Kinetic-molecular theory (KMT), 266
K_{sp} *(solubility product constant), 378–381*
K_w *(ion product constant for water),*
    374–376

L

L (liter), 26
Lactose, 487
Lanthanide series, 204

Laws:
  natural, 56–57
  scientific, 4–5
Law of conservation of energy, 70
Law of conservation of mass, 145
Law of definite composition, 56
Law of multiple proportions, 56
Lawrence, E. O., 425
Lead, 4, 71
Lead storage battery cells, 406
Le Châtelier, Henri, 366
Le Châtelier's principle, 366–373, 382
  and catalysts, 372–373
  concentration in, 368–370
  temperature in, 371–372
  volume in, 370–371
Length measurements, 22–23
Lewis, Gilbert, 216, 340
Lewis acids and bases, 340
Lewis structures:
  of atoms, 216–217
  complex, 232–235
  of compounds, 229–232
  formulas of, 223
  writing of, 229
Liberty Bell, 92
Light, electromagnetic radiation and,
    192–193
Light sticks, 372
Limiting reactant, 176–181
Linear molecular structure, 236, 238
Line spectrum, 194
Lipids, 488–491
Liquids, 282–301
  boiling and melting point, 286–288
  and capillary action, 283
  changes of state for, 288–290
  characteristics of, 283
  common materials in, 7
  densities of, 35
  evaporation of, 284
  hydrates, 295–296
  intermolecular forces in, 290–295
  measuring, 26
  and meniscus curves, 284
  properties of, 7, 283–286
  as state of matter, 283
  surface tension of, 283–284
  vapor pressure of, 285–286
  water as unique liquid, 297–299
Liquid crystals, 232
Liquid elements, 46
Liter (L), 26
Lithium, 199
Logarithm (log), 350
London dispersion forces, 294–295
Long-term radiation damage, 434
Lowry, T. M., 338
Lung cancer, 252

M

m (meter), 22

m *(molality), 322–324*
M *(molarity), 315–319, 322–323*
McMillan, Edwin M., 426
Macromolecules, 473–474
Magnesium:
  atomic structure of, 198, 201
  combustion of, 157
  mass of, 92
Magnesium chloride, 220
Magnesium oxide, 221–222
Maltase, 498
Maltose, 487
Mass. *See also* Molar mass
  atomic, 92–93
  conservation of, 145
  critical, 429
  and energy in nuclear reactions, 433
  measurements of, 24–25
  and moles of gases, 262–263
  in stoichiometric calculations, 173–176
  and volume, 35, 270–272
Mass defect, 433
Mass number, 90–91
Mass percent, 129, 313–315
Mass spectrometer readings, 92–93
Mass/volume percent, 315
Matter, 62–74. *See also specific states, e.g.:*
    Gases
  classification of, 7–9
  and energy, 68–70
  and heat, 70–72
  and hydrocarbons as energy sources,
    72–74
  particulate nature of, 5
  physical and chemical changes in, 65–68
  physical states of, 6–7
  properties of substances, 63–65
  and solving chemistry problems, 68
  states of, 249, 283, 288–290
Measurement(s):
  definition of, 14
  heat, 70–71
  length, 22–23
  mass, 24–25
  pressure, 249–251
  standard units, 22
  temperature, 15–16, 30–34
  uncertainty in, 15–16
  volume, 26
Medicinal chemistry, 4
Meiosis, 503
Meitner, Lise, 101
MEK (methyl ethyl ketone), 467
Melting point, 287, 288
Membranes, semipermeable, 298, 325
Mendeleev, Dimitri, 202
Meniscus, 284
Mercury, 46, 284, 286
Metals. *See also* Alkali metals
  activity series of, 152, 401–403
  alkaline earth, 50, 203
  atomic radii of, 219

Metals (cont.)
  in binary ionic compounds, 106–108
  electroplating of, 404–405
  and ionization energies, 215
  Noble, 45
  on periodic table, 50, 213
  and potassium hydroxide, 341
  reactions with, 342, 343
  and sodium hydroxide, 341
Metalloids, 50
Metal oxides, acids and, 153, 342
Meter (m), 22
Methanal, 467
Methane, 69
  in atmosphere, 160–161
  ball-and-stick model of, 236
  geometric shape of, 235
  structural formula for, 445, 447
  volume–temperature relationship for, 256
Methanol, 461–463
Methylbenzoic acids, 470
Methyl ethyl ketone (MEK), 467
Methylpropane, 447
Metric system, 21–23
  length measurements, 22–23
  mass measurements, 24–25
  prefixes for, 21–23
  standard units of measurement, 22
  unit conversions, 23–24
  volume measurements, 26
Meyer, Lothar, 202
Microbots, 177
Microchip technology, 177
Micromachinery, 177
Miscible liquids, 307
Mitosis, 503
Mixtures, 7–9
Models, natural laws vs., 57
Modern atomic theory, 191–207
  atomic structures of first eighteen
      elements, 198–201
  and Bohr atom, 193–195
  electromagnetic radiation in, 192–193
  electron energy levels in, 195–198
  electron structures and periodic table,
      201–206
  history of, 192
  Pauli exclusion principle in, 196
Molality (m), 322–324
Molarity (M), 315–319, 322–323
Molar mass:
  of compounds, 126–129
  definition of, 168
  determination of, 123–124
  and London dispersion forces, 294
Molar volume, 262–263
Mole:
  definition of, 122–124
  and mass of gases, 262–263
  relationship of molecule and, 168
  in stoichiometric calculations, 170–174
  and volume of gases, 262–263, 270–272

Molecular compounds, 52–53
Molecular formulas, 135–137
Molecular shape, 235
  bent structure, 237, 238
  determining using VSEPR, 237
  and electron pair arrangement, 238
  linear structure, 236, 238
  tetrahedral structure, 236–238
  trigonal planar structure, 236, 238
Molecules:
  and compounds, 53–54
  pressure and number of, 249–252
  relationship with mole, 168
Mole ratio, 168–169
Monomers, 473, 474
Monosaccharides, 484–486
Monosodium glutamate (MSG), 496
Monosubstituted benzenes, 457
Multiple proportions, law of, 56
Multiplication, significant figures in, 18–21

N

Nanorobots, 177
Natural gas, 72–73
Natural laws, 56–57
Natural radioactivity, 419–421
Natural states, elements in, 45–46
Neon, 200
Neptunium, 426
Net ionic equations, 354–356
Neutralization, 342, 352–354
Neutrons:
  characteristics of, 86
  determination of number, in atom, 90–91
  electrical charge of, 87
  relative mass of, 87
Nitrogen, 45, 136, 200
Nitrogen dioxide, 64
Noble gases, 50, 203, 217, 294
Noble metals, 45
Nonelectrolytes, 344–345
Nonmetals:
  atomic radii of, 219–220
  in binary ionic compounds, 108–109
  and ionization energies, 216
  on periodic table, 50
  periodic trends in, 213
Nonpolar covalent bonds, 227–229
Normal boiling point, 286–287
Normal butane, 447
N-terminal residue, 494–495
Nuclear atom, 87–89
Nuclear binding energy, 433–434
Nuclear chemistry, 417–437
  alpha particles, 421–422
  beta particles, 422
  biological effects of radiation, 434–435
  gamma rays, 422–424
  mass–energy relationship, 433–434
  nuclear energy, 427–431
  radioactive disintegration series,
      424–426

radioactivity discovery, 418–421
radioactivity measurement, 426–427
Nuclear energy, 427–431
Nuclear fission, 427–429
Nuclear fusion, 432
Nuclear power, 430–431
Nuclear power plants, 430–431
Nuclear radiation characteristics, 423–424
Nuclear scanning, 435
Nuclear waste, 430
Nucleic acids, 499–502
Nucleons, 419
Nucleoproteins, 499
Nucleotides, 499
Nucleus, atom, 88–89
Nuclides, 419

O

Observations, 5
Octane, 174
Oils, 488–491
Oil, crude, 446
Old Faithful, 33
Olive oil, 489
1 atmosphere (atm), 250–251
Orbitals, 195–197
  atomic orbitals, 195
  diagrams, 199, 201
  d orbitals, 196–197
  orbital filling, 200
  and Pauli exclusion principle, 196
  p orbitals, 196–197
  s orbitals, 196–198
  and spin, 195–197
Organic chemistry, 441–476
  alcohols, 461–465
  aldehydes and ketones, 467–469
  alkanes, 445–452
  alkenes and alkynes, 452–455
  aromatic hydrocarbons, 456–459
  beginnings of, 442
  and carbon atom, 442–445
  carboxylic acids, 469–470
  definition of, 442
  esters, 471–472
  ethers, 465–466
  hydrocarbon derivatives, 459–461
  polymers, 473–474
Organic halides, 460–461
Osmosis, 298, 325–326
Osmotic pressure, 325–326
Oxidation number, 391–395
Oxidation–reduction, 390–409
  and activity series of metals, 401–403
  balancing equations, 395–398
  balancing ionic redox equations, 398–401
  in electrolytic and voltaic cells, 403–407
  and oxidation number, 391–395
Oxidation state, 391
Oxides, 223
Oxidizing agent, 394
Oxy-acids, 112–113

Oxygen:
    and atomic structure, 200
    carbon dioxide exchange in blood, 382
    collected over water, 269
    corrosive effects of, 52
    electronegativity of, 298
    empirical and molecular formulas, 136
    gases, 45
    and hydrogen bonding, 292
    percent of atmosphere, 46
    physical properties of, 64
Oxytocin, 495–496
Ozone layer, 267

P

Paraffins, *see* Alkanes
Parrots, fluorescence and, 193
Partial pressures, 267–269
Particle size, dissolving rate and, 311
Particulate nature of matter, 5
Parts per million (ppm), 313
Pasteur, Louis, 497
Pauli exclusion principle, 196
Pauling, Linus, 226
Pentafoil, 235
Pentane, 174–175
Peptides, 494–495
Peptide linkage, 494
Percent composition, of compounds,
    129–132, 135
Percent yield, 180–181
Periods, of elements, 202–203
Periodic table, 49–52
    actinide series on, 204
    alkali metals on, 50
    alkaline earth metals on, 50
    and electron structures, 201–206
    groups/families on, 50, 203
    halogens on, 50
    inner transition elements on, 204
    lanthanide series on, 204
    metalloids on, 48
    metals on, 50
    noble gases on, 50
    nonmetals on, 50
    periods on, 202–203
    representative elements, 50, 203, 204
    transition elements on, 50, 203, 204
    and trends in atomic properties, 213–216
PET (positron emission tomography), 435
Petroleum, 72–73
pH:
    changes caused by HCl and NaOH, 382
    common applications of, 351
    of common solutions, 350
    control of, 381–383
    importance of, 351
    as logarithmic scale, 350
    scale of acidity and basicity, 349–350
    test paper, 351
Phases, matter and, 7
Phenyl group, 457

Phosphate system, as buffer in red blood
    cells, 383
Phospholipids, 491
Phosphorous, 201
Phosphorus pentachloride, 109
Photochromic glass, 400
Photons, 193
Physical changes, 65
Physical properties, 63–65
Physical states of matter, 6–7
Physiological saline solution, 325
Planck, Max, 194
Plastics, 473
Plutonium, 48, 426, 431
Polar covalent bonds, 225
Polyatomic compounds, 113
Polyatomic ions, 109–110, 234, 392
Polyatomic molecules, 100
Polyethylene, 473
Polyhydroxy alcohol, 462
Polymers, 473–474
Polymerization, 473
Polypeptides, 494–495
Polysaccharides, 488
Polysubstituted benzenes, 458
Polyunsaturated fats, 489
Polyvinyl chloride (PVC), 473, 474
Popcorn, 73
Positron emission tomography (PET), 435
Potable water, sources of, 299
Potassium hydroxide, metals and, 341
Potassium permanganate, 109–110, 307
Potassium sulfide, 103
Potential energy, 69
Power, nuclear, 430–431
ppm (parts per million), 313
Pressure:
    of gases, 249–252
    and molecule number/temperature of
      gases, 249–252
    and solubility, 310
    and volume of gases, 252–256
Primary alcohols, 462
Principle energy levels, 195–197
Problem solving, 68
Products, 67, 144, 145
Propane, 446–447
Properties (term), 63. *See also specific types*
Proteins, 492–497
Protons, 86
Pure substances, distinguishing mixtures
    from, 8–9
PVC (polyvinyl chloride), 473, 474

Q

Quadratic equation, 377–378
Quanta, 194
Quantitative composition of compounds:
    calculating empirical formulas, 133–135
    calculating molecular formula from
      empirical formula, 136–137

    empirical formula vs. molecular formula,
      133, 135–136
    molar mass of compounds, 126–129
    and the mole, 122–124
    percent composition of compounds,
      129–132
Quantum mechanics, 195

R

R (roentgen), 427
rad (radiation absorbed dose), 427
Radiation. *See also* Nuclear chemistry
    biological effects of, 434–435
    electromagnetic, 192–193
    ionizing, 426, 434
    solar, 160
    ultraviolet, 267
Radiation absorbed dose (rad), 427
Radiation units, 427
Radioactive decay, 419–421
Radioactive disintegration series, 424–426
Radioactivity:
    artificial, 425–426
    discovery of, 418–421
    induced, 426
    measurement of, 426–427
    natural, 419–421
    radioactive iodine, 435
Radium isotopes, 419
Rates of reaction, 364
Reactants, 67, 144, 145
Reaction rates, 364
Real gases, 266–267
Red blood cells, 325
Redox, *see* Oxidation–reduction
Reducing agent, 394
Reduction, 393. *See also* Oxidation–
    reduction
rem (roentgen equivalent to man), 427
Remineralizing therapies, 367
Representative elements, periodic table and,
    50, 203, 204
Resonance structures, 233
Reverse osmosis, 298, 325
Reversible chemical reactions, 365–366
Ribose, 486
RNA (ribonucleic acid), 501–502
Roentgen (R), 427
Roentgen, Wilhelm, 101, 418
Roentgen equivalent to man (rem), 427
Roentgenium, 425
Rounding off, 17–18
Rubbing alcohol, 464
Rubidium, 92
Rutherford, Ernest, 87–88, 101, 419, 425

S

Saccharides, *see* Carbohydrates
Salicylic acid, 470
Saline water, 297

Salts. *See also* Ionic compounds
  definition of, 343–344
  dissociation of, 345–346
Salt water, 84, 103
Sanger, Frederick, 495–496
Saponification, 490
Saturated fatty acids, 470
Saturated hydrocarbons, 444–445. *See also*
  Alkanes
Saturated solutions, 310–311
Scandium, 48
Schrödinger, Erwin, 195
Scientific laws, 4–5. *See also specific laws*
Scientific method, 4–5
  and hypothesis, 4
  and scientific laws, 4–5
  and theory, 4–5
Scientific notation, 14–15
Scintillation camera, 435
Scintillation counter, 426
Seaborg, Glenn, 101
Seawater, 46–47, 299
Secondary alcohols, 462
Semipermeable membranes, 298, 325
SI (International System), 21–23
Sickle-cell anemia, 504
Significant figures:
  and addition or subtraction, 19–21
  in calculations, 18–21
  defined, 16–17
  and multiplication or division, 18–21
  and rounding off, 17–18
  rules for counting, 16–17
Silicon, 5, 201
Silver nitrate and copper reaction, 155
Single-displacement reactions, 152–153
Smell, sense of, 252
Soap, 490
Sodium:
  and atomic structure, 200
  chlorine atoms reacting with, 217–218
  and electron arrangements, 218
  relative radii of, 219
Sodium atom, 198
Sodium bromide, 103
Sodium chloride, 322
  as compound, 53–54
  crystal, 219
  dissolution of, 308
  formation, 217–218
Sodium fluoride, 221
Sodium hydroxide, metals and, 341
Solar flares, 432
Solar radiation, 160
Solids:
  common materials in, 7
  definition of, 6
  densities of, 35
  dissolving rate of, 311–312
  physical properties of, 7
Solubility, 307–311
  of alkali metal halides, 309

of common ions, 308
  and pressure, 310
  and saturated solutions, 310
  and supersaturated solutions, 310
  and temperature, 309
  and unsaturated solutions, 310
Solubility product constant (K_{sp}), 378–381
Solutes, 306, 308–309
Solutions, 305–328. *See also* Solubility
  colligative properties of, 320–324
  common types of, 306
  concentration of, 307
  general properties of, 306–307
  osmosis and osmotic pressure in, 325–326
  saturated, 310–311
  solution maps, 25
  supersaturated, 311
  unsaturated, 310
Solution maps, for unit conversions, 23–24
Solvents, 306, 308–309
Space shuttle, 69
Specific gravity, 36
Specific heat, 71
Spectator ions, 352
Speed, of waves, 192
Sphingosine, 491
Spin, 195–196
Stable nuclides, 420
Standard conditions, 260
Standard temperature and pressure (STP),
  260
Stanford Linear Accelerator Center, 425
Starch, 488
Steroids, 491
Stirring, solid dissolving rate and, 312
Stock System, 106
Stoichiometry, 168–170. *See also* Gas
  stoichiometry
Stoney, G. J., 84–85
STP (standard temperature and pressure),
  260
Straight-chain alkanes, 445
Strassmann, Fritz, 427
Stratosphere, 267
Strong electrolytes, 346–348
Strontium-90 isotopes, 434
Structural formulas, of alkanes, 446–448
Subatomic particles, 85–87
  electrons, 86
  general arrangement of, 88–89
  neutrons, 86–87
  protons, 86
  subatomic particles, 87–88
Sublevels, energy, 195–198
Sublimation, 284
Subscripts, chemical formulas and, 54–55
Substances:
  definition of, 7–8
  properties of, 63–64
Substrates, enzyme, 498–499
Subtraction, significant figures in, 19–21
Sucrose, 487

Sugar, 64
Sulfates, 223
Sulfur, 9, 201
Sunglasses, 400
Super-ion battery, 407
Supersaturated solutions, 311
Surface area, dissolving rate and, 311, 312
Surface tension, of liquids, 283–284
Surgical implants, oxygen in cleaning of, 52
Suslick, Kenneth, 252
Symbols (of elements):
  common, 47–49
  from early names, 48
Systems, matter and, 7

T

$t_{1/2}$ *(half-life), 419–420*
Table salt, 127
Tantalum, 48
Temperature:
  and equilibrium, 371–372
  and gas pressure, 249–252
  and gas volume, 256–258
  measurement of, 15–16, 30–34
  and solid dissolving rate, 312
  and solubility, 309
Tertiary alcohols, 462
Tetrahedral structure, 236–238
Theoretical yield, 180
Theory, scientific, 4–5, 57
Thermal energy, 30, 445–446
Thomson, J. J., 85, 86
Thomson model of the atom, 86
Thorium, 48
Three Mile Island accident, 430
Titanium, 48
Titration, 352
Torricelli, E., 250
Total ionic equation, 354
Trans-*forming fats, 228*
Transition elements, 50, 203, 204
Transmutation, of elements, 425
Transuranium elements, 426
Triacylglycerols (triglycerides), 488–489
Trigonal planar molecular shape, 236, 238
Trigonal planar structure, 236
Tripeptides, 494

U

Ultraviolet radiation, 267
Umami, 496
Uncertainty, measurement and, 15–16
Unsaturated hydrocarbons, 444–445
Unsaturated solutions, 310
Unstable nuclides, 420
Uranium, 48
  disintegration series, 424
  enriched, 430
  isotopes, 418–419, 431
  oxide, 430

V

Valence electrons, 201, 203
Valence shell electron pair repulsion
    (VSEPR) model, 235–238
Vanadium, 48
Van de Graaf electrostatic generator, 425
Vaporization, 284
Vapor pressure, 285–286
Vapor pressure curves, 286–287, 321
Vasopressin, 495
Veggie Van, 444
Villard, Paul, 418
Vital-force theory, 442
Volatile substances, 286
Voltaic cells, 405–407
Volume:
    of equal masses, 35
    and equilibrium, 370–371
    and gas pressure, 252–256
    in gas stoichiometry, 270–274
    and gas temperature, 256–258
    and inverse PV relationship, 253
    measurement of, 26
    molar, 262–263
    volume percent, 315

Volumetric flasks, 316
VSEPR (valence shell electron pair
    repulsion) model, 235–238

W

Wastewater, reclamation of, 299
Water, 297–299
    composition of, 55–56
    as compound, 53
    dipole–dipole interactions in, 291
    electrolysis of, 66–68
    geometric shape of, 235
    H^+ and OH^- concentration relationship
        in, 375
    hydrogen bonding in, 291–292
    ice and water in equilibrium, 297
    ionization of, 348–349
    ion product constant for, 374–376
    meniscus of, 284
    molecular structure of, 298–299
    physical properties of, 287, 291, 297
    sources of, 299
    vapor pressure curves for, 321
    vapor pressure-temperature curves for,
        286–287

Water of crystallization, 295
Water of hydration, 295
Watson, James D., 499–500
Wavelength, 192, 193
Wave mechanics, 195
Weak electrolytes, 346–348
Weight, determination of, 24
Westphal, James A., 33
Wöhler, Friedrich, 442
Woody plants, energy and, 72

X

X-ray technology, 419, 434

Y

Yield calculations, 180–181

Z

Z (atomic number), 89–91
Zeolite, 160
Zinc, hydrochloric acid and, 152
Zinc-copper voltaic cell, 404–407
Zinc-mercury cells, 405
Zinc sulfate, 103